Steel — A Handbook for Materials Research and Engineering. Volume 2: Applications

A Handbook for Materials Research and Engineering

Volume 2: Applications

Edited by:
Verein Deutscher Eisenhüttenleute

With 447 Figures

Springer-Verlag
Berlin Heidelberg NewYork London Paris
Tokyo Hong Kong Barcelona Budapest
Verlag Stahleisen mbH Düsseldorf

Verein Deutscher Eisenhüttenleute
D-4000 Düsseldorf

Steel. A Handbook for Materials Research and Engineering
Edited by: Verein Deutscher Eisenhüttenleute (VDEh)
Berlin; Heidelberg; New York; Tokyo: Springer;
Düsseldorf: Verlag Stahleisen

Volume 2. Applications

ISBN 3-540-54075-x Springer-Verlag Berlin Heidelberg New York
ISBN 0-387-54075-x Springer-Verlag New York Heidelberg Berlin

ISBN 3-514-00378-5 Verlag Stahleisen m.b.H., Düsseldorf

This work is subject to copyright. All rights are reserved, whether the whole or part of the material is concerned, specifically the rights of translation, reprinting, re-use of illustrations, recitation, broadcasting, reproduction on microfilms or in other ways, and storage in data banks. Duplication of this publication or parts thereof is only permitted under the provisions of the German Copyright Law of September 9, 1965 in its current version, and a copyright fee must always be paid. Violations fall under the prosecution act of the German Copyright Law.

© Springer-Verlag Berlin Heidelberg and Verlag Stahleisen m.b.H., Düsseldorf 1993
Printed in Germany

The use of registered names, trademarks, etc. in this publication does not imply, even in the absence of a specific statement, that such names are exempt from the relevant protective laws and regulations and therefore free for general use.

Typesetting: Macmillan India Ltd., Bangalore
Printing: Mercedes-Druck GmbH, Berlin
Bookbinding: Lüderitz & Bauer, Berlin
61/3020-543210 – Printed on acid-free paper

List of Contributors

Responsible for Design and Execution

Dr.-Ing. Dr.-Ing. E. h. Walter Jäniche, Duisburg-Rheinhausen
Prof. Dr. rer. nat. Winfried Dahl, Aachen
Prof. Dr.-Ing. Heinz-Friedrich Klärner, Völklingen
Prof. Dr. rer. nat. habil. Wolfgang Pitsch, Düsseldorf
Dr.-Ing. Dieter Schauwinhold, Duisburg-Hamborn
Dr. rer. nat. Wilhelm Schlüter, Düsseldorf
Dr.-Ing. Hans Schmitz[†], Düsseldorf

Authors

Dr.-Ing. Hans-Egon Arntz, Dortmund-Aplerbeck
Dr. rer. nat. Fritz Aßmus, Hanau
Dr.-Ing. Klaus Barteld, Witten
Dipl.-Ing. Wilhelm Bartels, Peine
Dr.-Ing. Herbert Beck, Duisburg-Hochfeld
Ing. (grad.) Günter Becker, Duisburg-Ruhrort
Dipl.-Ing. Hans-Josef Becker, Witten
Prof. Dr.-Ing. Hans Berns, Düsseldorf-Oberkassel
Dr.-Ing. Harald de Boer, Duisburg-Hamborn
Dr.-Ing. Wolf-Dietrich Brand, Duisburg-Hamborn
Dipl.-Ing. Dieter Christianus, Gelsenkirchen
Prof. Dr. rer. nat. Winfried Dahl, Aachen
Dipl.-Ing. Richard Dawirs, Duisburg-Hamborn
Dr. rer. nat. Dipl.-Chem. Joachim Degenkolbe, Duisburg-Hamborn
Dr.-Ing. Hans-Heinrich Domalski, Witten
Prof. Dr. rer. nat. Hans-Jürgen Engell, Düsseldorf
Dr. phil. Dipl.-Phys. Heinz Fabritius, Duisburg-Huckingen
Dipl.-Ing. Helmut Fischer, Düsseldorf
Dr. rer. nat. Karl Forch, Hattingen
Dipl.-Ing. Heinz Fröber, Dortmund
Dr.-Ing. Ewald Gondolf, Völklingen
Prof. Dr. rer. nat. Hans Jürgen Grabke, Düsseldorf
Dr.-Ing. Dietmar Grzesik, Salzgitter
Dr. rer. nat. Klaus Günther, Duisburg-Hamborn
Dr.-Ing. Ives Guinomet, Dortmund
Dipl.-Ing. Hellmut Gulden, Siegen-Geisweid

List of Contributors

Dr.-Ing. Herbert Haas, Werdohl
Dr.-Ing. Hermann Peter Haastert, Duisburg-Huckingen
Dr.-Ing. Peter Hammerschmid, Duisburg-Rheinhausen
Dr.-Ing. Max Haneke, Dortmund-Hörde
Dr.-Ing. Winfried Heimann, Krefeld
Friedrich Helck, Duisburg-Hamborn
Dr.-Ing. Wilhelm Heller, Duisburg-Rheinhausen
Dr.-Ing. Bernd Henke, Dortmund
Dr. rer. nat. Dietrich Horstmann, Düsseldorf
Dr.-Ing. Hans Paul Hougardy, Düsseldorf
Dr.-Ing. Gerhard Kalwa, Duisburg-Huckingen
Dr.-Ing. Konrad Kaub, Dortmund
Dipl.-Ing. Heinz Klein, Völklingen
Dipl.-Ing. Walter Knorr, Bochum
Dipl.-Ing. Wilhelm Krämer, Duisburg-Rheinhausen
Dr. rer. nat. Dipl.-Phys. Max Krause, Bochum
Dr.-Ing. Karl-Josef Kremer, Siegen-Geisweid
Dr.-Ing. Rolf Krumpholz, Völklingen
Dr.-Ing. Werner Küppers, Krefeld
Dipl.-Ing. Jürgen Lippert, Bremen
Dipl.-Ing. Erich Märker, Düsseldorf
Dr.-Ing. Dipl.-Phys. Armin Mayer, Duisburg-Hamborn
Dr. rer. nat. Dipl.-Chem. Bernd Meuthen, Dortmund
Dr.-Ing. Lutz Meyer, Duisburg-Hamborn
Dr. rer. nat. Dipl.-Phys. Horst Müller, Saarbrücken
Dr.-Ing. Wolfgang Müschenborn, Duisburg-Hamborn
Dr.-Ing. Bruno Müsgen, Duisburg-Hamborn
Dr. rer. pol. Dipl.-Ing. Günter Oedinghoven, Osnabrück
Dr.-Ing. Rudolf Oppenheim, Krefeld
Dr. rer. nat. Werner Pappert, Dortmund
Prof. Dr. rer. nat. Werner Pepperhoff, Duisburg-Huckingen
Dr.-Ing. Jens Petersen, Salzgitter
Prof. Dr. rer. nat. habil. Wolfgang Pitsch, Düsseldorf
Dr. mont. Dipl.-Ing. Alfred Randak, Bochum
Dipl.-Phys. Hans Günter Ricken, Bochum
Dr.-Ing. Walter Rohde, Düsseldorf
Dipl.-Ing. Karl Sartorius, Völklingen
Dr. rer. nat. Dipl.-Phys. Gerhard Sauthoff, Düsseldorf
Dr. rer. nat. Wilhelm Schlüter, Düsseldorf
Dr-Ing. Herbert Schmedders, Duisburg-Hamborn
Dr. rer. nat. Karl-Heinz Schmidt, Bochum
Dr.-Ing. Dipl.-Phys. Werner Schmidt, Krefeld
Dr. rer. nat. Dietrich Schreiber, Hohenlimburg
Dr.-Ing. Johannes Siewert, Neuwied
Dr. rer. nat. Dipl.-Phys. Helmut Stäblein, Essen
Dipl.-Ing. Anita Stanz, Siegen-Geisweid

Dr.-Ing. Albert von den Steinen, Krefeld
Dr. rer. nat. Dipl.-Phys. Erdmann Stolte, Duisburg-Rheinhausen
Prof. Dr.-Ing. Christian Straßburger, Duisburg-Hamborn
Dipl.-Ing. Helmut Sutter, Neunkirchen (Saar)
Dr.-Ing. Gerhard Tacke, Bochum
Dr.-Ing. Klaus Täffner, Dortmund-Hörde
Dr.-Ing. Ulrich Tenhaven, Dortmund
Dr. rer. nat. Hans Thomas, Hanau
Dipl.-Ing. Friedrich Ulm, Düsseldorf-Oberkassel
Dipl.-Ing. Walter Verderber, Siegen-Geisweid
Dipl.-Ing. Klaus Vetter, Siegen-Geisweid
Dr.-Ing. Constantin M. Vlad, Salzgitter
Dr.-Ing. Hans Vöge, Siegen-Geisweid
Dr.-Ing. Klaus Vogt, Bochum
Dipl.-Ing. Helmut Weise, Dortmund
Dipl.-Phys. Karl Werber, Hattingen
Dipl.-Ing. Wilhelm Weßling, Siegen-Geisweid
Dipl.-Ing. Ingomar Wiesenecker-Krieg, Völklingen
Dipl.-Ing. Siegfried Wilmes, Düsseldorf-Oberkassel
Dr.-Ing. Peter-Jürgen Winkler, Duisburg-Huckingen

Preface

The two volumes of the STEEL Textbook have been designed as one unit. Nonetheless, either of the volumes can be used as independent works. It is therefore useful to repeat the preface to volume 1 here as an introduction to volume 2.

This STEEL Textbook is the outcome of reflections within the Materials Committee of the German Iron and Steel Institute on whether to republish the Manual of Special Steels, the highly commendable work by E. Houdremont. Discussions came to the conclusion, however, that for various reasons it was neither possible nor expedient simply to publish a follow-up edition of the famed *Houdremont*. To begin with and from today's vantage point, there no longer seemed to be any justification for restricting the work to special steels in the sense of the term as understood by E. Houdremont. The term "special steel" has never gained acceptance in official circles or standards. If we replace it by the term "high-grade steel", which is nowadays defined in standards, and this would appear permissible with certain qualifications, and if we bear in mind that the boundaries between high-grade steels and non-high-grade steels, the commercial and quality steels, although defined in standards (see part A), nonetheless in terms of engineering parameters are quite blurred, so it would seem only fitting for such a work to cover all the various grades of steel, also in view of the great significance of the non-high-grade steels. Because of the very many different grades of steel, this approach necessarily involves the collaboration of very many experts, in other words it entails a joint effort. Moreover, the vast, barely manageable quantity of literature in this field all of which can hardly be analysed by just one person, inevitably leads to the conclusion that there is a need to produce the new work as a joint effort.

With the definition extension to cover all forms of steel, it was also considered appropriate to change the overall layout of the work. The book is no longer subdivided according to alloying elements; instead, the guiding consideration is that the properties of steel primarily depend on the material's microstructure which, in turn, is influenced by the chemical composition. Consequently, part B of the new manual examines the possible types of microstructure and the conditions and mechanisms giving rise to them while taking account of chemical composition as well as thermal and mechanical treatment. Part C describes the interdependence between the previously described microstructures/chemical compositions and key service properties such as working and application properties (mechanical and chemical properties, suitability for heat treatment, formability, machinability, etc.) without, however, examining in any detail specific types or grades of steel. Following this treatment of fundamentals, part D then examines specific types of steel (general structural steels, heat-treatment steel, tool steel, etc.). It deals with the working- and application-related requirements in terms of properties, their designation and the specific engineering measures for obtaining the microstructure and properties serving as criteria for the respective steel grades. Parallel to this, the Textbook refers to the metallurgical interrelationships outlined in the previous parts, B and C. Preceding parts B, C, and D is part A examining the economic and engineering significance of the steels and part E detailing the influence of production conditions on the properties of steel. The required brevity prevented the authors from examining comprehensively all the details or grades of steel; instead, the aim has been to familiarize readers with metallurgical interrelationships between chemical composition, heat and mechanical treatment, microstructure and properties.

The new concept also meant that the previous title "Manual of Special Steels" had to be changed. Given the above-mentioned reflections on the basic format of the book, it has now taken the form of an instructional work, with instructional elements predominating, rather than a reference manual. Hence the new title "STEEL Textbook".

The vast field which had to be covered made it impossible to publish the work in one volume. The first volume contains parts A to C, and the second, parts D and E. Even though this divides the more theoretical explanations from the more practical presentations, this by no means indicates any interruption of the basic considerations behind the work. On the contrary, cross-references repeatedly underline its indivisibility.

The concept of the work was developed by W. Dahl, H.-F. Klärner, W. Pitsch, D. Schauwinhold, W. Schlüter and H. Schmitz[†]. In charge was W. Jäniche. This steering committee, headed by W. Schlüter, devoted considerable energy towards progressing the work and making sure that the individual articles matched the basic concept. Our warmest thanks to these gentlemen. Our special thanks also to the many members of the staff who, despite the load of routine work at the plants and research institutes, showed patient understanding for the needs of the steering committee, and who accordingly formulated their contributions. Our thanks also to Messrs. H. Schmitz[†], W. Schlüter and W. Liedtke[†] who with considerable effort and input managed to achieve a uniform editorial approach, despite the many contributors. Finally our thanks to the Springer-Verlag and Verlag Stahleisen for their cooperation and appreciation with respect to the magnitude of the project and the excellent layout and design of the "STEEL Textbook".

German Iron and
Steel Institute

Contents

Part C	**Steels with characterizing Properties for Specific Fields of Application**	1
C 1	**General Survey of Part C and its Aim**	3
	By W. Schlüter	
C 2	**Normal and High Strength Structural Steels**	6
	By B. Müsgen, H. de Boer, H. Fröber and J. Petersen	
C 2.1	General Remarks	6
C 2.2	Required Properties of Use	7
C 2.2.1	Required Application Properties	7
	Mechanical Properties under Static Loads. Mechanical Properties under Cyclic Loads. Resistance Against Brittle Fracture. Resistance to Weathering. Wear Resistance.	
C 2.2.2	Required Processing Properties	10
	Formability. Suitability for Shearing, Machining and Flame Cutting. Weldability. Suitability for Zinc Coating (Galvanizing).	
C 2.2.3	Requirements for Uniformity	12
C 2.3	Characterization of the Required Properties	12
C 2.3.1	Characterization of Application Properties	12
	Material Behavior in Tensile Testing. Material Behavior in Fatigue Testing. Resistance to Brittle Fracture. Resistance to Weathering. Resistance to Wear.	
C 2.3.2	Characterization of Processing Properties	22
	Formability. Machinability and Suitability for Flame Cutting. Weldability. Suitability for Zinc Coating (Galvanizing).	
C 2.4	Measures of Physical Metallurgy to Attain Required Properties	34
C 2.4.1	General Remarks	34
C 2.4.2	Influence of the Heat Treatment and the Resulting Microstructure on Properties	35
	Normalizing. Quenching and Tempering.	
C 2.4.3	Influence of the Chemical Composition on Properties and its Effect on Microstructure	39
C 2.5	Characteristic Steel Grades Proven in Service	51
C 3	**Reinforcing Steel for Reinforced Concrete and Prestressed Concrete Structures**	61
	By H. Weise, W. Krämer, W. Bartels and W.-D. Brand	
C 3.1	General Remarks	61
C 3.2	Reinforcing Bars and Wire-fabric	62
C 3.2.1	Required Properties	62
C 3.2.2	Characterization of the Required Properties	62
C 3.2.3	Measure Changes in Physical Metallurgy to Attain Required Properties	63
C 3.2.4	Characteristic Steel Grades Proven in Service [10]	66
C 3.3	Prestressing Steels	67
C 3.3.1	Required Properties of Use	67
C 3.3.2	Characterization of the Required Properties	70

C 3.3.3	Measures of Physical Metallurgy to Attain Required Properties	72
C 3.3.4	Characteristic Steel Grades Proven in Service	76
C 4	**Steels for Hot-Rolled, Cold-Rolled and Surface-Coated Flat-Rolled Products for Cold Forming**	77
	By Chr. Straßburger, B. Henke, B. Meuthen, L. Meyer, J. Siewert and U. Tenhaven	
C 4.1	Hot- and Cold-rolled Flat Products for Cold Forming	77
C 4.1.1	Hot- and Cold-rolled Sheet and Strip of Mild Unalloyed Steels	77
C 4.1.1.1	Hot-rolled Flat Products	77
	Required Properties. Characterization of the Required Properties. Measures for Attaining the Required Properties. Characteristic Steel Grades Proven in Service.	
C 4.1.1.2	Cold-rolled Flat Products	80
	Required Properties of Use. Characterization of the Required Properties. Measures for Attaining the Required Properties. Characteristic Steel Grades Proven in Service.	
C 4.1.2	Flat-Rolled Products of Structural Steels for General Purposes. and of High-strength Steels.	84
C 4.1.2.1	Hot-rolled Flat Products	84
	Required Properties. Characterization of Required Properties. Measures for Attaining Required Properties. Characteristic Steel Grades Proven in Service.	
C 4.1.2.2	Cold-Rolled Flat Products	90
	Required Properties. Characterization of Required Properties. Measures for Attaining Required Properties. Characteristic Steel Grades Proven in Service.	
C 4.1.3	Flat Products of Stainless Steels	96
C 4.2	Surface-coated Flat Rolled Products	97
	General Requirements for the Properties of Use. Characterization of Required Properties.	
C 4.2.1	Flat-Rolled Products with Metallic Coatings	99
C 4.2.1.1	Zinc-coated Flat Products	99
	Required Properties. Characterization of Required Properties. Measures for Attaining Required Properties. Characteristic Steel Grades Proven in Service.	
C 4.2.1.2	Aluminum-Coated Flat Products	105
	Required Properties. Characterization of Required Properties. Measures for Attaining Required Properties. Characteristic Steel Grades Proven in Service.	
C 4.2.1.3	Flat Products with Coatings Containing Tin and Chromium	109
	Required Properties. Characterization of Properties Required. Attaining Required Properties. Tinplate Grades Proven in Service.	
C 4.2.1.4	Flat Products with Lead Coating	111
	Required Properties. Characterization of the Required Properties. Measures for Attaining Required Properties. Characteristic Steel Grades Proven in Service.	
C 4.2.1.5	Flat Products with Other Metallic Coatings	112
	Flat Products with Electrolytically-deposited Metal Coatings. Flat Products with Metal Coatings Produced by Cladding. Flat Products with Coatings Produced by Vapor-Deposition.	
C 4.2.2	Flat-Rolled Products with Inorganic Coatings	114
	Phosphated and Chromated Flat Products. Enamelled Flat Products.	
C 4.2.3	Organic Coil-coated Steel Flat Products	116
C 5	**Heat-treatable and Surface-hardening Steels for Vehicle and Machine Construction**	118
	By G. Tacke, K. Forch, K. Sartorius, A. von den Steinen and K. Vetter	
C 5.1	General Remarks	118
C 5.2	Heat-treatable Steels (Steels for Quenching and Tempering), Steels for Flame, Induction and Immersion Surface Hardening and Nitriding Steels. ,	119
C 5.2.1	General Requirements for the Properties of Use of Heat-treatable Structural Steels, Their Characterization by Test Values and Possibilities to Attain Them	119

C 5.2.1.1	Properties which are Important for End Uses	119
	Strength Under Static Stresses. Toughness Properties. Fatigue Strength. Wear Resistance.	
C 5.2.1.2	Important Processing Properties	132
	Suitability for Heat Treatment. Machinability. Cold Formability. Weldability.	
C 5.2.2	Heat-treatable Steels (Steels for Quenching and Tempering)	134
C 5.2.2.1	General Selection Criteria	134
C 5.2.2.2	Steel Grades Proven in Service	134
C 5.2.3	Steels for Flame, Induction and Immersion Surface Layer Hardening	137
C 5.2.3.1	Special Aspects of Flame, Induction and Immersion Surface Layer Hardening	137
C 5.2.3.2	Steel Grades Proven in Service	140
C 5.2.4	Nitriding Steels	140
C 5.2.4.1	Special Features of Nitriding	140
C 5.2.4.2	Steel Grades Proven in Service	146
C 5.3	Steels for Heavy Forgings	147
C 5.3.1	Required Properties	147
C 5.3.2	Characterization of the Required Properties by Test Results	147
C 5.3.3	Measures for the Attainment of the Required Properties	148
C 5.3.3.1	Melting and Casting	148
C 5.3.3.2	Heat Treatment	150
C 5.3.3.3	Relationship Between Alloying Content, Microstructure and Mechanical Properties	151
C 5.3.4	Steel Grades Proven in Service	155
C 5.4	Case Hardening Steels	158
C 5.4.1	Requirements for the Base or Core Material to Suit Service and Processing Operations	158
C 5.4.1.1	Required Properties	158
C 5.4.1.2	Characterization of the Required Properties by Test Values	159
C 5.4.1.3	Measures for Achieving the Required Properties	159
C 5.4.2	Required Properties of the Case Hardened Layer and Their Production	161
C 5.4.2.1	Required Properties	161
C 5.4.2.2	Carburizing	163
C 5.4.2.3	Hardening	165
C 5.4.2.4	Stress Relieving	168
C 5.4.3	Behavior of the Composite Material of the Case and the Tough Core Under Service Conditions	169
C 5.4.4	Steel Grades Proven in Service	174
C 6	**Steels for Cold Forming (Forming e.g. by Cold Extrusion and Cold Heading)**	177
	By H. Gulden and I. Wiesenecker-Krieg	
C 6.1	General Remarks	177
C 6.2	Required Properties	177
C 6.3	Characterization of Required Properties	178
C 6.4	Measures of Physical Metallurgy to Attain the Required Properties	180
C 6.4.1	General Remarks	180
C 6.4.2	Chemical Composition and Microstructure	181
C 6.4.3	Cleanness	185
C 6.4.4	Surface Quality and Dimensional Accuracy of Steel Products	186
C 6.5	Characteristic Steel Grades Proven in Service	187
C 7	**Unalloyed Wire Rod for Cold Drawing**	195
	By H. Beck and C. M. Vlad	
C 7.1	Required Properties of Use	195
C 7.2	Characterization of the Required Properties	196
C 7.3	Measures to Attain Required Properties	199
C 7.3.1	General Remarks	199

C 7.3.2	Metallurgical Measures and Mill Practices to Produce Microstructures with Required Properties	200
C 7.4	Characteristic Steel Grades Proven in Service	204
C 8	**Ultra High-strength Steels**	208
	By K. Vetter, E. Gondolf and A. von den Steinen	
C 8.1	Definition of the Term "Ultra High-strength Steel" ("UHS") Steel)	208
C 8.2	Required Properties and Appropriate Test Methods	208
C 8.2.1	Application-derived Requirements	208
C 8.2.2	Requirements Imposed by Manufacturing (Processing) Technology	210
C 8.3	Meeting Requirements by Adjustment of Chemical Compositions and Appropriate Production Techniques	211
C 8.3.1	Basic Possibilities	211
C 8.3.2	Strength and Toughness Properties	212
C 8.3.2.1	Conventionally Heat-treatable Steels	212
C 8.3.2.2	Maraging Steels	216
C 8.3.2.3	Production of Conventionally Heat-treatable UHS Steels and Maraging Steels	218
C 8.3.3	Stress Corrosion Cracking, Hydrogen Embrittlement and Rust-resistance	219
C 8.3.4	Fatigue Strength	220
C 8.3.5	Processing Properties	221
C 8.4	Characteristic Grades of UHS Steel and Their Properties	223
C 9	**Elevated-temperature Steels** (Creep Resistant and High Creep Resistant Steels) and Alloys	226
	By H. Fabritius, D. Christianus, K. Forch, M. Krause, H. Müller and A. von den Steinen	
C 9.1	Required Properties	226
C 9.2	Characterization of the Required Properties by Results of Specified Tests	229
C 9.3	Physical Metallurgy Procedures to Attain the Required Properties	233
C 9.3.1	Ferritic Steels	233
C 9.3.1.1	Ferritic Steels for Slightly Elevated Temperatures	233
C 9.3.1.2	Ferritic Steels for Creep Resistance	237
C 9.3.2	Austenitic Steels	242
C 9.3.2.1	Significance of the Face Centered Cubic Lattice for the High-temperature Strength of Austenite	242
C 9.3.2.2	Effects of Chromium and Nickel	243
C 9.3.2.3	Effects of Carbon and Nitrogen	243
C 9.3.2.4	Effects of Niobium and Titanium	245
C 9.3.2.5	Effects of Molybdenum, Tungsten, Vanadium and Cobalt	246
C 9.3.2.6	Effect of Boron	249
C 9.3.3	High-temperature Nickel and Cobalt Alloys	249
C 9.4	Characteristic Material Grades Proven in Service	253
C 9.4.1	Ferritic Steels	253
C 9.4.1.1	Ferritic Steels for Slightly Elevated Temperatures	253
C 9.4.1.2	Ferritic Steels for Plates and Tubes	256
C 9.4.1.3	Ferritic Steels for Forgings and Bars	259
C 9.4.1.4	Ferritic Steels for Castings	261
C 9.4.2	Austenitic Steels	264
C 9.4.3	High-temperature Nickel and Cobalt Alloys	269
C 10	**Low Temperature Steels (Steels with Good Toughness at Low Temperatures)**	273
	By M. Hancke, J. Degenkolbe, J. Petersen and W. Weßling	
C 10.1	Required Properties of Use	273
C 10.2	Characterization of the Required Properties	274
C 10.3	Physical Metallurgy Procedures to Attain Required Properties	277

C 10.3.1	Ferritic Steels	277
	Chemical Composition. Measures to Achieve a Fine-grained Microstructrue. Absence of Non-metallic Inclusions (Cleanness), Influence of Tramp Elements. Heat Treatment. Rolling.	
C 10.3.2	Austenitic Steels	282
C 10.4	Characteristic Steel Grades Proven in Service	287
C 10.4.1	Ferritic Steels	287
C 10.4.2	Austenitic Steels	292
C 10.5	Processing of Low Temperature Steels	294
C 11	**Tool Steels**	**302**
	By S. Wilmes, H.-J. Becker, R. Krumpholz and W. Verderber	
C 11.1	Multiple Stresses in Tools	302
C 11.2	Properties Required of Tool Steels, their Characterization on the Basis of Test Values, and Measures of Physical Metallurgy to Attain them	303
C 11.2.1	Properties of Importance for Various Applications	305
C 11.2.1.1	Hardness at Low and High Operating Temperatures	305
C 11.2.1.2	Hardenability	311
C 11.2.1.3	Retention of Hardness	314
C 11.2.1.4	Compression Strength and Pressure Resistance	317
C 11.2.1.5	Fatigue Strength	320
C 11.2.1.6	Toughness at Operational Temperatures	321
C 11.2.1.7	Wear Resistance	329
C 11.2.1.8	Thermal Conductivity	334
C 11.2.1.9	Resistance to Thermal Fatigue	335
C 11.2.1.10	Corrosion Resistance	337
C 11.2.2	Steel Properties of Importance in the Manufacture of Tools	339
C 11.2.2.1	Dimensional Accuracy During Heat Treatment	339
C 11.2.2.2	Hot Formability	342
C 11.2.2.3	Cold Formability and Suitability for Hobbing	343
C 11.2.2.4	Machinability	346
C 11.2.2.5	Grindability	347
C 11.2.2.6	Polishability	349
C 11.3	Characteristic Steel Grades for the Different Fields of Tool Application	350
C 11.3.1	Steels for Plastics Molds	350
C 11.3.2	Steels for Pressure Casting Molds	354
C 11.3.3	Steels for Glass Processing	355
C 11.3.4	Steels for Cold-forming Tools	357
C 11.3.5	Steels for Forging and Pressing Dies	361
C 11.3.6	Steels for Hot Extrusion Press Tools	363
C 11.3.7	Steels for Machining Tools	365
C 11.3.8	Steels for Cutting Tools	368
C 11.3.9	Steels for Hard Tools	373
C 12	**Wear-resistant Steels**	**374**
	By H. Berns	
C 12.1	Key Properties and their Characteristics	374
C 12.2	Microstructural Considerations	374
C 12.3	Applications of Steels and Iron-base Alloys	379
C 13	**Stainless Steel**	**382**
	By W. Heinmann, R. Oppenheim and W. Weßling	
C 13.1	Required Design Properties of the Steels	382
C 13.1.1	Fields of Application for Stainless Steels	382
C 13.1.2	Resistance to Various Types of Corrosion	382
C 13.1.3	Mechanical and Technologically Significant Properties	383

C 13.2	Characterization of the Required Properties by Test Values.	384
C 13.2.1	Testing Corrosion Resistance.	384
C 13.2.2	Parameters Characterizing Mechanical and Technologically Significant Properties	390
C 13.3	The Importance of Microstructure and its Chemical Composition for Corrosion. Performance.	393
C 13.3.1	The Decisive Importance of Chromium Content	393
C 13.3.2	Influence of Other Alloying Elements on Corrosion Behavior	395
C 13.4	Attainment of the Desired Microstructure by Chemical Composition and Heat Treatment	396
C 13.4.1	Dependence of Structure Type on Content of Ferrite and Austenite Formers. Phase Diagram for Stainless Steels	396
C 13.4.2	Heat Treatment of the Characterizing Structure Groups.	399
C 13.5	Typical Steels Proven in Service	408
C 13.5.1	Ferritic Steels.	408
C 13.5.2	Martensitic Steels.	411
C 13.5.3	Austenitic Steels	416
C 13.5.4	Ferritic-austenitic Steels	421
C 14	**Steels for Use with Hydrogen at Elevated Temperatures and Pressures (High-pressure Hydrogen Resistant Steels)** . . . . By E. Märker	423
C 14.1	Hydrogen-induced Damage to Steel Used in High-pressure Applications	423
C 14.2	Basis of Physical Metallurgy to Attain High-pressure Hydrogen Resistant Steels.	424
C 14.3	Conclusions Regarding the Chemical Composition	426
C 14.3.1	Resistance to High-pressure Hydrogen	426
C 14.3.2	Consideration of the Processing Properties	427
C 14.4	Application-oriented Selection of the Steels	427
C 15	**Heat Resisting Steels**. . . . . By Weßling and R. Oppenheim	433
C 15.1	Required Properties of Use.	433
C 15.2	Characterization of the Required Properties by Test Values.	434
C 15.3	Chemical Composition and Microstructure of Applicable Steels	437
C 15.4	Characteristic Grades of Heat Resisting Materials	439
C 15.4.1	Ferritic Steels.	439
C 15.4.2	Austenitic Steels and Alloys.	440
C 15.4.3	Criteria for the Selection of Heat Resisting Material Groups	444
C 16	**Heating Conductor Alloys**. . . . . By H. Thomas	445
C 16.1	Required Properties and their Testing	445
C 16.2	Relationship Between Physical Metallurgy and Chemical Composition	446
C 16.2.1	Austenitic Heating Conductor Alloys.	446
C 16.2.2	Ferritic Heating Conductor Alloys	447
C 16.2.3	Influence of Special Alloying Additions.	448
C 16.3	Technically Established Heating Conductor Alloys.	449
C 17	**Steels for Values in Internal Combustion Engines** . . . . By W. Weßling and F. Ulm	452
C 17.1	Basic Properties of Valve Materials.	453
C 17.1.1	Requirements Resulting from Service Conditions	453
C 17.1.2	Requirements Concerning the Processing Properties	454

C 17.2	Characterization of the Requirements by Test Values	454
C 17.3	Measures for Satisfying the Requirements	456
C 17.3.1	Measures of Physical Metallurgy	456
C 17.3.2	Measures for Shaping Valve Parts and Protecting Their Surfaces	457
C 17.4	Presentation of Standard Valve Materials	458
C 17.4.1	Heat-Treatable Steels (Steels for Quenching and Tempering)	458
C 17.4.2	Austenitic Steels	460
C 17.4.3	Nickel Alloys	461
C 17.4.4	High-Temperature Behavior of the Three Material Groups in Comparison	462
C 18	**Spring Steels**	468
	By D. Schreiber and J. Wiesenecker-Krieg	
C 18.1	Required Properties	468
C 18.2	Measures of Physical Metallurgy to Attain Required Properties	469
C 18.3	Spring Steel Grades Proven in Service	472
C 18.3.1	Steels for Cold Formed Springs	473
C 18.3.2	Steels for Quenched and Tempered Springs	473
C 18.3.3	Elevated-temperature Strength Steels for Springs	475
C 18.3.4	Spring Steels with Good Toughness at Low Temperatures (Spring Steels for Low Temperature Service)	476
C 18.3.5	Stainless Steels for Springs	476
C 19	**Free-cutting Steels**	478
	By H. Sutter and G. Becker	
C 19.1	Characteristic Properties	478
C 19.2	Consequences Considering Physical Metallurgy	479
C 19.2.1	Nature of the Microstructure	479
C 19.2.2	Effects of Special Alloying Elements on Free-cutting Steels	479
C 19.3	Consequences for the Production of the Free-cutting Steels	482
C 19.4	The Free-cutting Steel Grades of Common Use and Their Properties	485
C 19.4.1	Steel Grades Characterizing the Application Range	485
C 19.4.2	Machinability	485
C 19.4.3	Mechanical Properties	487
C 19.4.4	Subsequent Processing and Heat Treatment	489
C 19.4.5	Other Properties	489
C 20	**Soft Magnetic Materials**	491
	By E. Gondolf, F. Assmus, K. Günther, A. Mayer, H. G. Ricken and K.-H. Schmidt	
C 20.1	Range of the Soft Magnetic Materials	491
C 20.2	Characteristics for the Evaluation of Soft Magnetic Materials	492
C 20.3	Soft Iron: Applications, Properties, Production, and Grades	494
C 20.3.1	Basic Requirements	494
C 20.3.2	Findings of Physics and Physical Metallurgy Regarding Influences on the Required Properties	494
	Saturation Polarization. Coercivity and Bloch-wall Movements. Anisotropy Energies. Implications for Manufacturing.	
C 20.3.3	Production Technique	497
	Melting and Deoxidation. Additional Processing and Heat Treatment.	
C 20.3.4	Magnetic Properties of Important Soft Iron Grades	
	Relay Materials. Heavy Forgings	499
C 20.4	Electrical Steel Sheet	501
C 20.4.1	Basic Requirements	501

C 20.4.2	Considerations of Physics and Physical Metallurgy.	502
	Magnetization Processes and Eddy Currents. Chemical Composition. Texture, Grain-oriented Electrical Steel Sheet. Consequences for the Fabrication.	
C 20.4.3	Production Processes of Electrical Steel Sheets	508
	Steel-making and Hot Rolling. Processing of the Hot-rolled Strip.	
C 20.4.4	Electrical Steel Sheets Currently Available on the Market and Their Properties. .	512
C 20.4.5	Comments on Other Materials.	514
	Recent Developments (Amorphous Metals, Silicon Steel with 6% Si). Silicon-alloyed Materials Besides Electrical Steel Sheets.	
C 20.5	Other Steels with Special Requirements for Their Magnetic Properties	515
C 20.5.1	Applications and Requirements for the Material Properties.	515
C 20.5.2	Relationships Between the Required Properties and the Microstructure.	516
C 20.5.3	Characteristic Examples of Other Steels with Special Magnetic Properties	516
	Steels for Generator Shafts. Steel Castings for Magnetic Applications. Stainless Chromium Steels for Magnetic Applications.	
C 20.6	Soft Magnetic Nonferrous Alloys	523
C 20.6.1	Applications and Requirements	523
C 20.6.2	Aspects of Physics and Physical Metallurgy	523
	Phases and Transformations. Magnetic Constants and Ordering Processes. Specific Shapes of the Hysteresis Loop.	
C 20.6.3	General Remarks on Processing and Products	526
C 20.6.4	Commercial Nonferrous Alloys for Magnetic Applications	526
	Iron-nickel Alloys with 70% to 80% Ni. Iron-nickel Alloys with 45% to 65% Ni. Iron-nickel Alloys with 35% to 40% Ni. Iron-nickel Alloys with Nickel Contents Near 30%. Iron-cobalt Alloys with 27% to 50% Co.	
C 20.7	Remarks on Further Crystalline Special Alloys and on Metallic Glasses	532
C 20.7.1	Special Iron Alloys.	532
C 20.7.2	Metallic Glasses	532
C 21	**Permanent Magnet Materials**.	534
	By H. Stäblein and H.-E. Arntz	
C 21.1	Required Properties	534
C 21.2	Permanent Magnet Characteristics	535
C 21.3	Metallurgical Fundamentals and Production	538
C 21.3.1	High-Iron Materials	538
C 21.3.1.1	Steels	538
C 21.3.1.2	Elongated Single Domain (ESD) Magnets	539
C 21.3.2	Medium-Iron Materials	539
C 21.3.2.1	Aluminum-Nickel and Aluminum-Nickel-Cobalt Alloys	539
C 21.3.2.2	Iron-Cobalt-Vanadium-Chromium Alloys	541
C 21.3.2.3	Chromium-Iron-Cobalt Alloys	541
C 21.3.2.4	Copper-Nickel-Iron Alloys	542
C 21.3.2.5	Oxide Materials (Hard Ferrites)	542
C 21.3.2.6	Neodymium-Iron-Boron Materials	543
C 21.3.3	Material Containing Little or No Iron	544
C 21.3.3.1	Rare Earth Metal-Cobalt Materials.	544
C 21.3.3.2	Platinum-Cobalt Materials	544
C 21.3.3.3	Manganese-Aluminum Materials.	545
C 21.4	Application Ranges.	545
C 22	**Nonmagnetizable Steels**	547
	By W. Weßling and W. Heimann	
C 22.1	Required Properties	547
C 22.2	Characterization of the Magnetic Properties by Test Results	548
C 22.3	Conclusions on the Chemical Composition and the Microstructure	549
C 22.4	Characteristic Steel Grades Proven in Service	552

C 23	**Steels with Defined Thermal Expansion and Special Elastic Properties**	556
	By H. Thomas and H. Haas	
C 23.1	Properties and Tests	556
C 23.1.1	Determining the Coefficient of Expansion	556
C 23.1.2	Determining the Modulus of Elasticity (Young's Modulus)	557
C 23.2	Considerations of Physical Metallurgy	558
C 23.3	Technically Proven Materials	562
C 23.3.1	Materials with Special Thermal Expansion	562
C 23.3.2	Materials for Thermobimetals	565
C 23.3.3	Constant-Modulus Alloys	565
C 24	**Steels with Good Electrical Conductivity**	566
	By K. Werber and H. Beck	
C 24.1	Ranges of Application of the Steels and Expected Properties of Use	566
C 24.2	Determination of Electrical Conductivity	568
C 24.3	Measures of Metallurgy and Physical Metallurgy to Attain a Good Conductivity	569
C 24.4	Characteristic Steel Grades	572
C 25	**Steels for Line Pipe**	574
	By G. Kalwa and K. Kaup	
C 25.1	Required Properties	574
C 25.2	Characterization of the Required Properties	574
C 25.3	Measures of Physical Metallurgy to Attain the Required Properties	578
C 25.4	Characteristic Steel Grades Proven in Service	581
C 26	**Ball and Roller Bearing Steels**	584
	By K. Barteld and A. Stanz	
C 26.1	Required Properties	584
C 26.2	Characterization of the Required Properties by Test Values	585
C 26.3	Measures of Physical Metallurgy to Attain the Required Properties	587
C 26.4	Characteristic Steel Grades Proven in Service	589
C 27	**Steels for Permanent-way (Track) Material**	593
	By W. Heller and H. Schmedders	
C 27.1	Required Properties of Use	593
C 27.2	Characterization of the Required Properties	594
C 27.3	Measures of Physical Metallurgy to Attain the Required Properties	595
C 27.4	Characteristic Steel Grades and Their Application	599
C 28	**Steels for Rolling Stock**	602
	By K. Vogt, K. Forch and G. Oedinghofen	
C 28.1	General Remarks	602
C 28.2	Required Properties of Use	602
C 28.3	Characterization of the Required Properties	603
C 28.4	Measures to Attain the Required Properties	603
C 28.5	Characteristic Steel Grades Proven in Service	606
C 29	**Steels for Screws, Bolts, Nuts, and Rivets**	608
	By K. Barteld and W.-D. Brand	
C 29.1	Required Properties of Use	608

C 29.2	Characterization of the Required Properties	610
C 29.3	Measures of Physical Metallurgy to Attain the Required Properties	612
C 29.4	Characteristic Grades Proven in Service	615
C 30	**Steels for Welded Round Link Chains**	621
	By H.-H. Domalski, H. Beck, and H. Weise	
C 30.1	Required Properties of Use	621
C 30.2	Characterization of the Required Properties	624
C 30.3	Measures to Attain the Required Properties	626
C 30.4	Characteristic Steel Grades Proven in Services	628

Part D The Influence of Production Processes on the Properties of Steel . . 634

D 1	**The Influence of Production Processes on the Properties of Steel and Steel Products – A Overview**	635
	By A. Randak	
D 2	**Crude Steel Production**	638
	By H. P. Haastert	
D 2.1	Raw Materials for Steelmaking	638
D 2.2	Influence of the Charged Materials in Steelmaking	641
D 2.3	Steelmaking	642
D 2.3.1	Melting and Refining	643
D 2.3.2	Deoxidizing and Alloying	645
	Deoxidizing and Alloying in the Steelmaking Vessel. Deoxidizing and Alloying in the Ladle.	
D 2.3.3	Ladle Metallurgy Processes for Steel Aftertreatment	646
	Gas Rinsing/Stirring Treatment. Injection Treatment. Heating. Vacuum Treatment.	
D 3	**Casting and Solidification**	650
	By P. Hammerschmid	
D 3.1	Characteristics of Common Methods of Casting	650
D 3.2	Casting and Solidification of Steel in Ingot Molds	651
D 3.2.1	Reoxidation, Flow, and Superheat	651
D 3.2.2	Crystallization Process	653
D 3.2.3	Heat Transfer and Solidification	654
D 3.2.4	Formation of Segregations	655
	Ingot Segregation. Microsegregation.	
D 3.2.5	Formation of Oxide Inclusions	657
D 3.2.6	Formation of Sulfide Inclusions	658
D 3.2.7	Formation of Blowholes	659
D 3.2.8	Formation of Pipe	660
D 3.2.9	Treatment of Cast Steel	660
D 3.3	Processes in Continuous Casting	661
D 3.3.1	Reoxidation, Flow, and Superheat	661
D 3.3.2	Heat Transfer During Solidification	662
D 3.3.3	Formation of Segregations	663
D 3.3.4	Influence of Electromagnetic Stirring on Segregations	664
D 3.3.5	Formation of Internal Cracks	665
D 3.3.6	Surface Defects	666
D 3.3.7	Significance of Casting Fluxes	667
D 3.3.8	Hot Charging and Hot Direct Rolling of Continuous Cast Steel	668
D 3.4	Comparison of Ingot Casting and Continuous Casting	668
D 3.5	Outlook: Continuous Casting Approaching Final Shape	668

D 4	**Special Methods for Melting and Casting**	670
	By H. Vöge	
D 4.1	Remelting Methods	670
D 4.2	Special Melting Methods for Heavy Forgings	675
D 4.3	Vacuum Melting Methods	676
D 5	**Hot Rolling**	678
	By K. Täffner	
D 5.1	Hot Rolling Process	678
D 5.2	Heating	679
D 5.2.1	Heating Conditions	679
D 5.2.2	Scaling	682
D 5.2.3	Surface Decarburization	682
D 5.2.4	Hot Shortness	684
D 5.2.5	Influences on the Microstructure	684
D 5.2.6	Hot Feeding and Direct Rolling	685
D 5.3	The Rolling Processes	685
D 5.3.1	Deformation Resistance	685
D 5.3.2	Formability	686
D 5.3.3	Control of Microstructure and Metal Properties	687
D 5.3.4	Improved Surface Characteristics	688
D 5.3.5	Yield	689
D 5.4	Finishing	690
D 6	**Hot Forming by Forging**	691
	By H. G. Ricken	
D 6.1	Aims of Forging	691
D 6.2	Forging Methods	691
D 6.3	Material for Forging	691
D 6.4	Working Conditions in Forging	692
D 6.4.1	Heating	692
D 6.4.2	Forging Conditions	693
	Effects on the General Quality Properties. Effects on the Mechanical Properties.	
D 6.4.3	Cooling from Forging Heat	697
D 6.5	Defects	697
D 7	**Cold Forming by Rolling**	699
	By J. Lippert	
D 7.1	Definition and Scope of Cold Rolling	699
D 7.2	Production Sequences in the Cold Rolling Mill	699
D 7.3	Effect of Process Variables in Cold Rolling on Material Properties	702
D 7.4	Heat Treatment of Cold Strip	703
D 7.4.1	General Information	703
D 7.4.2	Continuous Annealing	703
D 7.4.3	Box Annealing	704
D 7.5	Temper Rolling	704
D 7.6	Finishing for Dispatch	706
D 8	**Heat Treatment**	707
	By H. Vöge	
D 9	**Quality Management in the Production of Steel Works Products**	711
	By W. Rohde, R. Dawirs, F. Helck and K.-J. Kremer	
D 9.1	Concept of Quality Management	711

D 9.2	Measures for Quality Management	711
D 9.2.1	Quality Planning	711
D 9.2.2	Quality Inspection	712
D 9.2.3	Quality Control	713
D 9.3	Quality Management in Steel Making	718
D 9.4	Quality Management in the Deforming of Crude Steel	719
D 9.5	Quality Management in Steel Bar Production	721
D 9.6	Quality Management in Plate Production	724

Table 1. Comparison of the steel grades mentioned in Volume 1 and Volume 2 of "Werkstoffkunde Stahl" and identified here by their designations according to German standards (DIN Standard) with corresponding grades according to international, European and some other national standards 728

Table 2. Brief definition of the scope and contents of the DIN Standards mentioned in Volume 1 and Volume 2 of "Werkstoffkunde Stahl" and listed in column 1 of Table 1 744

Table 3. International and European standards as well as national standards of some non-German countries which are comparable with DIN. (The standards are identified just by their number; for abbreviated titles of the German standards refer to Table 2) 751

Glossary of Repeatedly Used Symbols and Abbreviations 759

References 763

Subject Index 813

Contents of Volume 1

Part A Microstructure of Steels
By W. Pitsch, G. Sauthoff (A1–A8), and H. P. Hougardy (A9)

A 1	Introduction
A 2	Thermodynamics of Iron and Iron-based Alloys
A 3	Nucleation
A 4	Diffusion
A 5	Typical; Microstructures in Steel
A 6	Kinetics and Morphology of Steel Constituents
A 7	Structure Development by Thermal and Mechanical Treatments
A 8	Comparative Summary of the Reactions in Steel
A 9	Description and Control of Transformations in Technical Applications

Part B The Properties of Steel in Dependence of Microstructure and Chemical Composition

B 1	Mechanical Properties By W. Dahl

B 2	**Physical Properties** By W. Pepperhoff	
B 3	**Chemical Properties** By H.-J. Engell and H. J. Grabke	
B 4	**Heat Treatability, Heat Treatment, and Properties**	
B 5	**Weldability** By H. P. Hougardy	
B 6	**Hot Formability** By P.-J. Winkler and W. Dahl	
B 7	**Cold Formability (Suitability for Cold Extension and Cold Heading)** By W. Schmidt	
B 8	**Cold Formability of Flat Rolled Products** By W. Müschenborn, D. Grzesik and W. Küppers	
B 9	**Machinability** By W. Knorr and H. Vöge	
B 10	**Wear Resistance** By E. Stolte	
B 11	**Cutting Tool Life** By H.-J. Becker	
B 12	**Surface Treatment Processes** By U. Tenhaven, Y. Guinomet, D. Horstmann, L. Meyer and W. Pappert	

How to use the Textbook

The work consists of two volumes. Volume 1 comprises parts A to C; volume 2, parts D and E, each with their consecutively numbered chapters starting from 1. Chapters, sections, equations, illustrations and tables always indicate the corresponding A, B, C, D or E. Equations, illustrations and tables additionally feature the chapter number preceding the consecutive number (except for part A where the illustrations and tables are numbered consecutively).

Cross-references within a volume or to another volume do not state the number of the volume but just the letter indicating the part and without any such reference as "part", "chapter" or "section". There are no cross-references to pages.

The literature references for all the parts and chapters are listed at the end of each volume – before the index. A footnote at the start of each chapter makes it easier to refer to the respective literature.

Concentrations are always referred to as mass percentages unless otherwise stated.

Measurements are always stated in SI units; measurements have been converted in the case of illustrations taken from older publications.

Chapters with several authors always list the chief author first followed by the others in alphabetical order.

Part C
Steels with Characterizing Properties for Specific Fields of Application

C 1 General Survey of Part C and its Aim

By Wilhelm Schlüter

Volume 1 with the first parts of "Material Science of Steel" dealt with the fundamentals of the formation of steel microstructure, and traced its dependence on the chemical composition and the temperature-time-sequences during the production and heat treatment of steel products. Relationships between the microstructure and the chemical composition on the one side and the most important properties of steel on the other were discussed and explained. In this part C of "Material Science of Steel" the basic information induced in parts A and B is applied to practical questions and problems, to indicate which steel types and steel grades have been developed generally, and been accepted by the market. The steels discussed have been developed mainly on the basis of practical experience. Knowledge of the scientific fundamentals has often been acquired later, but the application of these fundamentals has resulted in improvements and further developments which have brought steelmaking to the levels described in this book.

Within this book the different steel types are subdivided primarily on the basis of their major properties such as creep resistant, stainless, heat treatable and so on. This conception and idea is also expressed by the titles of the different chapters. Important applications of specific steel types stem from properties which may also be taken as further guidelines for subdivision. The connection between property and use of some steel types is so tight, that not the property but the application governs the characterization and denomination of the steel type. As an example, the decisive properties of ball and roller bearing steels are hardenability and wear resistance, but these steels are so much adapted to one single application and use, that this steel type is named accordingly.

The authors have tried to use a similar construction for each of the various chapters. Each chapter therefore starts with an outline of the requirements for the properties of application (e.g. strength properties) and on the processing properties (e.g. weldability) – taking into account the conditions of use of the steel type under consideration. Explanations of the test methods and resulting test values, that means test values by which a characterization of the required properties is possible and suitable, then follow. In explaining these matters and in describing the measures of physical metallurgy which must be taken in order to meet the requirements, frequent reference will be made to the fundamental information contained in the previous parts of this material science book. Finally the most characteristic grades of the respective steel types are dealt with and those steel grades for which the correlation of physical metallurgy can be made especially clear are discussed.

The above chapter construction is aimed at in principle but the reader will find a certain flexibility in the explanations. For instance, in dealing with one or another steel type it was sometimes necessary to add additional sections or to differentiate their importance in order to accentuate certain properties or relationships. In the final analysis the aim is to explain the facts and correlations in such a way that the reader is able to understand the approaches of physical metallurgy to solve any problem for a given task.

In dealing with the different steel types with this aim in mind no attempt was made to enumerate and specify within the respective sections all steel grades listed in DIN, SEW or similar data sheets. Inclusion of all of them would have resulted in the aim being obscured by a superabundance of data. In general, only the characteristic steel grades are dealt with but in dealing with some especially important steel types, some less-characteristic steel grades are named.

In discussing the required properties and their characterization of the different steel types in the different chapters it is not possible to avoid some overlapping because certain properties, for example, strength at ambient temperature, are very important for many steel types. These properties and the corresponding test methods, as well as the respective fundamentals of physical metallurgy, are listed in all chapters for the different steel types. However, to avoid excessive repetition, an attempt was made to deal with these properties in detail only in connection with the type of steel for which they are especially important, for instance the strength properties at ambient temperature were detailed mainly in connection with structural steels in chapter C 2. For other steel types, in addition to listing, a discussion of the properties may be included but only so far as the relationships are specific for the particular type of steel. Frequent reference will be made here to the fundamental data in volume 1, but this part will discuss mainly the relationships to be deduced from those fundamentals, which are specific for the steel type under consideration.

Several approaches could have been used to establish a sequence for the chapters devoted to the types of steel to be dealt with in this book. For part C of the "Material Science of Steel" the following pragmatic solution was chosen. First to be considered are those steel types in which important properties can usually be attained without heat treatment (chapters C 2 to C 4). Chapters on steel types for which a heat treatment is necessary follow (chapters C 5 to C 12). Subsequently, those steel types having properties which are more dependent on the chemical composition than are other steel types are discussed (chapters C 13 to C 19). In later chapters, steel types which are characterized mainly by their particular physical properties are covered (chapters C 20 to C 24). Finally, (in chapters C 25 to C 30) steel types of specific application are dealt with. There is some overlapping, e.g. ball and roller bearing steels which have been arranged in the last group, must be heat treated, therefore an integration within the second group would have been possible.

Although the application is very important, a special chapter on steels for the construction of pressure vessels is not included in part C. Such a chapter is omitted because – in contrast to most of the other fields of use – multitudes of steel types are used for pressure vessels (low-temperature steels, elevated-temperature steels, heat treatable steels (steels for quenching and tempering), normal- and high-strength structural steels, stainless steels, heat resisting steels, high-pressure hydrogen-resistant steels). If a special chapter on steels for pressure vessels had been included, much more overlappings would have resulted. In connection with pressure vessel steels reference is therefore made to the explanations in the corresponding other chapters.

C 2 Normal and High Strength Structural Steels

By Bruno Müsgen, Harald de Boer, Heinz Fröber and Jens Petersen

C 2.1 General Remarks

The term structural steel is very comprehensive. Therefore it is necessary to state that this chapter deals with structural steels used for house building, civil engineering, bridges, hydraulic structures and shipbuilding (including offshore technology). A series of these steels is known as "steels for general structural purposes". A group of unalloyed steels for mechanical engineering, that also belongs to the commercial structural steels, is covered and briefly discussed as well, but not the large number of steels designed specifically for mechanical engineering, such as heat treatable steels discussed in C 5. Concrete reinforcing steels also will not be dealt with here, but in C 3.

Repeated attempts have been made to *classify structural steels* according to different strength levels to simplify making a survey by groups, such as the group of normal strength steels or the group of high strength steels. Up to now no final agreements or definitions have been made in this field. However, it is conceivable to think of the following definition which takes shape from a joint proposal that only needs completion of a few details:

Normal strength and high strength structural steels are steels that fulfill certain requirements for strength, toughness and weldability. With regard to weldability these steels have a relatively low carbon content. If the minimum yield stress at ambient temperature is < 355 N/mm^2, the steels will be ranked with normal strength steels used in the as-rolled or normalized (annealed or normalizing rolled) condition. If the minimum yield stress at ambient temperature is ≥ 355 N/mm^2, the steels are to be included with high strength steels used in the normalized, thermomechanically-treated or quenched and tempered condition. Generally, the steels are fine-grained. A steel is fine-grained if its ferrite grain size, derived from metallurgical measurements and heat treatment, is 6 or finer (see Euronorm 103).

The heading of this chapter is to be understood in the sense of this characterization. Nevertheless, the internationally accepted classification of ship building steels (see C 2.5), which are also mentioned in this chapter, does not coincide with the above formulation. At the present stage there is no choice but to keep both patterns of classification side by side. Additionally it must be noted that also some of the steels mentioned in C 2.5, mainly those used in mechanical engineering

---

References for C 2 see page 763.

cannot be integrated into the above scheme. Suitability for electric arc and gas welding is not demanded from such steels, and the carbon content is not restricted.

As mentioned above, normal strength and high strength structural steels are suitable for a great number of applications. Therefore, it is not always easy to achieve a proper selection. In this context there are published papers [1–3] that consider other important aspects of processing. Generally it should be born in mind: less than ideal material is not just good enough, the satisfactory material is the best [4].

C 2.2 Required Properties of Use

If the properties of use are taken as a general term for the application properties (e.g. mechanical properties) and for the processing properties (e.g. weldability), special aspects to be considered with regard to the application are the strength, particularly the yield stress at ambient temperature, the toughness, and the weldability. These characteristics comprise the most important properties of use of the structural steels in question.

C 2.2.1 Required Application Properties

Mechanical Properties Under Static Loads

With regard to the application, mechanical properties are the first requirement from the structural steels considered in this chapter. First of all, values of the strength properties under tension load, that means values of the *yield stress* and the *ultimate tensile strength*, are needed, because they serve as basis for the calculation of construction.

The steels must resist the stresses calculated from the mechanical loads, without incurring fractures, cracks or inadmissible deformations. Service loads often are multi-axial loads. Behavior of a structure under such multi-axial stress conditions is seldom known, because usually only the strength values of uniaxial loads as determined in tensile testing are available. To be able to decide whether a steel can be applied under certain load conditions, a stress of comparison (reference stress) must be calculated according to the competent strength hypothesis from the main stresses of the most unfavorable multi-axial stress condition. The stress of comparison is related to the strength values, e.g. the yield stress [1]. After all it is necessary for the calculation (design) of statically loaded structures to provide for security compared to the yield stress or the tensile strength of the tensile test by a safety factor.

In *compression loading* of structures, the tensile strength and yield stress or the *compression strength* and *compression yield stress*, respectively, are of less importance than the *stability conditions* with regard to buckling, tipping and bulging. Within compressed areas of structures generally, the problems of exceeding the strength values fade into the background against the danger of achieving the limit of stability. Even at very low stresses, compression components can lose their stable equilibrium state. Under elastic service conditions the load limit depends exclusively on the geometry of the component and the *modulus of elasticity* of the material. However, outside of Hooke's range of deformation the special yield stress of the material must be considered with regard to the proof of stability. The boundary between the elastic and plastic area depends on the geometry and on the material. Therefore, a specified minimum yield stress is required with respect to the design of structures where stability may be endangered.

Next to its strength the ability to undergo plastic deformation due to external forces, without the material cohesion being disturbed is an important feature with regard to the bearing capacity of a material. The *deformability* of a material, which depends on the external load conditions (e.g. multi-axial load), makes it possible on the one hand, to bring the material into a desired shape during processing. On the other hand the deformability allows the material to attenuate local stress peaks resulting from the loads, by local yielding. Suitable characteristics of the deformability are the values of *elongation after fracture, reduction of area* or *deformation energy before fracture* of small scale specimens or structural members. Generally, high amounts of plastic deformation before a fracture occurs are desired.

Beside the main kinds of failure caused by the appearance of excessive plastic deformation, a crack, a fracture or onset of instability, a structure can also be made useless by chemical attacks (e.g. corrosion) or mechanical wear (e.g. abrasion). The two latter kinds of failure will not be discussed further here, because they are not caused directly by an external mechanical load, and not susceptible to an arithmetical treatment.

Mechanical Properties Under Cyclic Loads

Steels used for parts and structures such as in bridges, cranes and ships, that are subjected during service to loads of alternating amplitude and frequency, require adequate resistance to alternating mechanical loads which means a certain level of *fatigue resistance*.

When laying down requirements for admissible stresses in dynamically-loaded structures, the effects of shape, size, surface quality and treatment of the structure as well as environmental effects (e.g. corrosion) must be considered.

If, during use, a structural member has only a limited lifetime with relatively few cycles, then the design can be based on a higher load than the fatigue limit, namely the *fatigue strength for finite life*.

Unwelded steel with notchless surfaces shows a proportional relation between the tensile strength and the fatigue strength. According to the surface quality, the fatigue strength is far lower than the corresponding tensile strength during static loading.

If steels are exposed to a high number of cycles under severe notch conditions, then their fatigue strength is nearly independent of the tensile strength according to present knowledge, and the fatigue behavior of a structure cannot be improved by using a steel of high strength. Therefore the need exists to develop steels that show a higher fatigue strength when notches are present, as they usually are, particularly in welded joints. However, present experience shows this need cannot be filled by modifying materials. Improvements can be achieved by optimizing welding technology and by more favorable designs with respect to the insertion and distribution of forces.

The overwhelming portion (85 to 90%) of all failures are fatigue fractures. It may be pointed out that most of the fatigue fractures are not caused by material defects, but can be traced back to effects of geometry and surface, overloading etc. [5] (see also Table C 5.1).

Resistance Against Brittle Fracture

Besides the admissible design stress under static or dynamic loads, resistance against brittle fracture is an important property of the structural steels dealt with in this chapter. This statement is true above all, because the loads usually are multi-axial stresses and, additionally, often resemble an impact. Resistance against brittle fracture is decisive for the materials' behavior under multi-axial stresses. Safeguards against brittle fracture depend on the *toughness of the material* and on the *load conditions* of the structure.

The brittle fracture behavior of a material to be used for highly stressed structure is judged by whether a fast-running brittle crack is arrested by the material or results in a fracture. However, in most structures loads do not continue through running cracks. Nevertheless, it must be assumed that micro-cracks or sharp notches can be present. For the valuation of a structure's safety it is therefore important to know the limits of stress and temperature to which a material is allowed to be loaded, without a brittle fracture being possible even in the presence of a crack. The designer has to decide under which aspects the structure must be resistant to brittle fracture: safety against *crack initiation* or safety against *crack propagation* (see C 2.3.1).

Recommendations for the selection of material with regard to safety against brittle fracture exist in civil engineering. These recommendations are based on valuation of the stress condition of the structure under service loads, the importance of the structure, the lowest service temperature, the thickness of the material and the degree of cold working [6]. Approaches for the *calculation* of safety against brittle fracture are given in fracture mechanics (see C 2.3.1 and B 1.1.2.6).

Resistance to Weathering

For some applications *resistance against atmospheric corrosion* of structural steels is required to be sufficient to use them unprotected if possible, i.e. without painting or zinc coating. This requirement cannot be met by standard steels, and special steels have therefore been developed (see C 2.5). These steels have a chemical composition that produces a protective layer on the steel surface during weathering. The layer is formed and renewed continuously during weathering.

Wear Resistance

Only occasionally is a particular wear resistance required from normal-strength and high-strength structural steels. It is therefore sufficient to mention this property just briefly here.

C 2.2.2 Required Processing Properties

The structural steels dealt with in this chapter must be capable of being processed by the usual techniques of forming, dividing, joining and cutting. Moreover, zinc coatings for surface protection must be possible for certain steel types and applications. These processing properties are here dealt with in the above sequence, which corresponds more or less to the sequence of the possible processing steps. As noted in the introduction to C 2.2, weldability is the most important processing property of structural steels.

Formability

Generally, structural steels are *cold worked* at ambient temperature and the forming process (e.g. bending, bevelling, flanging, pressing, dishing) used is selected to produce the required quality and quantity of the desired work pieces at the lowest cost. Elevated temperatures may be used during cold working when the processing machine is not strong enough and therefore the yield stress of the material must be reduced or when particularly high degrees of deformation are necessary (see also B 7). With regard to cold formability (see for example [7]) the strain hardening and the simultaneous reduction in toughness of the material due to increasing deformation must be considered. Often these considerations decide, at what stage the steel must be annealed. The possibility of aging also must be observed for the assessment of cold formability and it must be kept in mind that, under certain conditions of deformation, cold-worked areas show a decreased yield stress (Bauschinger-effect) (see B 1.2.1).

After strong cold forming, heat treatment may be necessary to restore mechanical properties. Usually stress relieving is sufficient. However, such a heat treatment must not reduce mechanical properties of the material.

Hot forming (see B 6) should be feasible without failure of the material. As a rule, hot forming is carried out at the normalizing temperature of the steels. In hot forming, reheating, forming and cooling cause the steel to pass through several extended temperature ranges, each with different holding times. This treatment may change the microstructure and consequently the mechanical properties of the steel considerably. A subsequent heat treatment may be necessary to restore mechanical properties.

Suitability for Shearing, Machining and Flame Cutting

During processing structural steels can be shaped to the desired form by cutting in addition to cold and hot forming. Therefore cold shearing of the steels must be possible in the as-rolled and as-forged condition as well as in the normalized and in the quenched and tempered condition. The steels must also be machinable by use of suitable tools and cutting conditions. Assessment of machinability requires consideration of tool wear and life, surface quality produced on the work piece, the force required for machining the material and the shape of the chips [8] (see also B 9). Grinding may also be used for this work.

In many establishments, steel plates are prepared by flame cutting, which is used to separate work pieces or to bevel edges to make the groove for a welded joint. For economic reasons, flame cutting needs to be carried out without restrictions and above all without preheating. However, cutting conditions such as preheating temperature and cutting speeds, must be adjusted to suit the plate thickness and the steel grade. The carbon content must also be observed, to achieve a cut surface suited to requirements. If these precautions are taken, the surface quality of a flame cut edge will be in the range of quality 1 (smooth cut) according to e.g. DIN 2310 [8a].

Weldability

Weldability is required from steels to be used for the manifold applications in civil engineering. Steels with good weldability are required for economic reasons, meaning steels which can be welded with the least possible effort [9]. The manufacturer, who uses normal welding processes wants steels that can be welded without cracking or other limitations and above all without preheating even under unfavorable conditions.

Furthermore, the welded joints must have load capacities that are comparable to those of the base metal. Both the defects of the welded joints and their mechanical properties must be kept within certain limits to secure specified bearing capacities. These conditions can be achieved by suitable welding procedures.

To avoid defects in welded joints, certain production techniques must be chosen to secure sufficient melting off of the seam flanks, careful removal of the slag and adjustment of the welding parameters to the special crack susceptibility of the steel. The critical factors of the welding procedure (heat input, working temperature, hydrogen content and internal stresses) must be adjusted to suit the crack susceptibility of the steel [10].

The original microstructure and subsequently the properties of the steel are changed drastically in the heat-affected zone of welded joints. These changes must be considered with regard to the required mechanical properties of welded joints, because neither a loss of strength nor embrittlement can be accepted in the heat-affected zone.

To counter the risk of lamellar tearing [11] in welded structures, improved through-thickness properties are occasionally specified for structural steels. Lamellar tearing may be related to the anisotropy of the material, thought to be due mainly to long-stretched sulfide inclusions. These materials with improved through-thickness properties are needed above all in the form of heavy plates. Improved through-thickness properties can only be achieved by additional metallurgical procedures at increased cost. Therefore it should be kept in mind that the risk of lamellar tearing may be avoided not only by using materials with special properties perpendicular to the plate surface but also by good design and welding procedures.

Suitability for Zinc Coating (Galvanizing)

Structural parts are often zinc coated (galvanized) for protection against atmospheric corrosion. Therefore the manufacturer and the customer ask if the steel is suitable for zinc coating. This property depends not only on the chemical and physical nature of the base material, but to a large extent also on the pre-treatment of the steel surface, the composition and temperature of the zinc melt, the dipping time, the draw-out speed and the post-treatment (wiping, centrifuging, blowing out, post-heating). Therefore the parameters of the zinc-coating procedure must be adjusted properly to the steel grade to be used (see B 12).

C 2.2.3 Requirements for Uniformity

In modern industrial production the increasing mechanization and automatization require improved uniformity. Improved uniformity applies not only to the properties of materials discussed earlier, but also to the condition of the rolled products such as dimensional accuracy, straightness etc. Uniformity of properties and condition is required within individual rolled product as well as within a shipment.

The degree of *uniformity of the mechanical properties* is limited by metallurgical events during melting, casting and solidification. Narrow ranges of properties are not possible, especially with rimmed steels because differences of the chemical composition and consequently strong variations in the mechanical properties exist between top and bottom as well as between surface and core of a steel ingot.

Low dimensional variations (narrow tolerances) of products made from structural materials are important with regard to the functionality of steel constructions. Finishing of buildings for instance, with walls and ceilings is rendered more difficult, if the dimensions of the steel framework are not uniform. Uniformity of construction requires uniformity of the rolled product, particularly concerning dimensions and straightness.

The demand for increased uniformity of products, including their properties, can generally be fulfilled only to a limited extent because of metallurgical and general technical constraints. Technical efforts toward inprovement must be increased to raise uniformity.

C 2.3 Characterization of the Required Properties

The properties of a *structure* are determined not only by the material, but also by the design, the manufacturing and the environmental effects. Certain test characteristics of the *steel* are determined (see below), however, the behavior of a structure during service can only be estimated and not calculated from the results of material testing. Even estimation is only possible when all the details of product manufacture, the structural design and the service load are known.

In the following the characteristic values and their capacity to affect details of the required properties of the various steel groups of this chapter are explained.

C 2.3.1 Characterization of Application Properties

Material Behavior in Tensile Testing
The most important test for characterizing the steels treated in this chapter is the *tensile test* (see B 1.1.1.1). The strength values, the *yield stress* and the *tensile*

strength, as well as the *elongation* and the *reduction of area*, that are obtained from tensile testing, give information about the material's behavior under uniaxial tensile load. A high yield stress permits the production of light construction with good stiffness of the structure. High values of elongation after fracture correspond mostly to high values of elongation before necking (elongation at maximum load, uniform elongation) area and point to a good deformability of the steel. This characteristic is of great importance if the steel structure is loaded locally up to its yield stress or even above it. Stresses can then be released by yielding without a failure of the material.

In tensile testing the behavior of the steels discussed in this chapter is basically similar. Figure C 2.1 shows typical *stress-strain-curves* of normalized steels and of a quenched and tempered steel (StE 690) of different strength. Generally, also high strength quenched and tempered steels have a distinct yield point as do normal strength steels. The yield point elongation is about 2%, whereas the elongation before necking (uniform elongation) of quenched and tempered steels (StE 690, steel no. 4 in Table C 2.9) is between 8% and 10% and of normalized steels (e.g. St 52-3, steel no. 8 in Table C 2.5) is between 16% and 20%. The elongation during necking and the reduction of area are within the same order of magnitude for both steel types.

Along with rising yield stress the steel's *ratio of yield stress to tensile strength increases* (Fig. C 2.2). This ratio is often related to the steel's susceptibility to brittle fracture. However, the ratio of yield stress to tensile strength gives only a clue to what extent a uniaxial load on a structural steel at room temperature may be increased above the calculated admissible stress, without a fracture occuring. Under uniaxial tensile load, steels with a low ratio of yield stress to tensile strength can be stressed relatively higher than steels with a high ratio of yield stress to tensile strength [12].

The bearing capacity of a technical construction depends primarily on the yield stress provided that the structure is exclusively tensile loaded. For compression

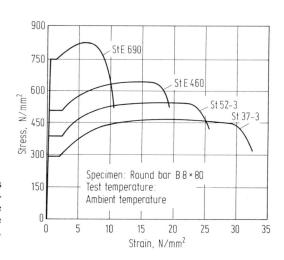

Fig. C 2.1. Stress-strain-curves of steels with different strength (StE 690 = quenched and tempered steel no. 4 in Table C 2.9. StE 460 = steel no. 7 in Table C 2.8. St 52-3 and St 37-3 = steels no. 8 and 5 in Table C 2.5).

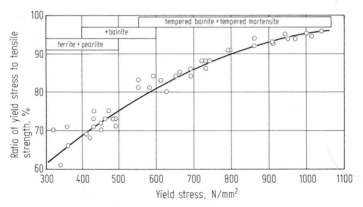

Fig. C 2.2. Relation between the ratio of yield stress to tensile strength ("yield stress ratio") and the yield stress of different structural steels in the as-rolled, normalized or quenched and tempered condition.

components on the other hand the stability conditions are also vital to determination of their optimum thickness. Thus Young's modulus is an important factor for such components. However, with regard to loads like bending, buckling, tipping and bulging, Young's modulus for all ferritic structural steels is about 206 000 N/mm² independent of their yield stress. As the elastic properties of steels are essentially unchanged by alloying elements, the often expressed requirement for a structural steel with a higher Young's modulus is unrealistic.

In structures made from high strength steels, higher elastic deflections may occur due to the reduced wall thickness. The higher deflection of high strength steel sections is often compensated for by a higher web which allows weight saving in spite of the unchanged modulus of elasticity. Occasionally, hybrid beams are used to reduce costs. The highly stressed flanges of hybrid beams are made of high strength steels and the webs, which have to resist bulging, are made of normal strength steels.

For structures that are loaded with buckling and bulging, the admissible compression stresses depend on the degree of slenderness. In the range of a low degree of slenderness substantially higher stresses are admissible for high strength steels than for normal strength steels, as Fig. C 2.3 shows, for buckling loads. Similar considerations apply when bulging may take place.

Material Behavior in Fatigue Testing

The basic procedure for the investigation of the behavior of a material during dynamic load is *Wöhler's method* (fatigue test; see B 1.2.2.1). The results gained from this test make clear that the resistance of steels to dynamic loads does not increase in the same manner as the strength values for static loads. Increases in the fatigue strength along with the yield stress depend on several factors, particularly on the stress ratio $\sigma_{min.}/\sigma_{max.}$ of the load and on the effect of notches of different forms.

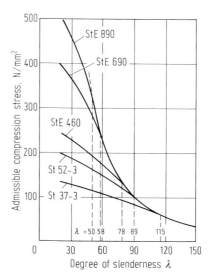

Fig. C 2.3. Stability behavior of some structural steels according to Tables C 2.5, C 2.8 and C 2.9. The admissible compression stresses with regard to buckling stresses (load case 1) were calculated on the basis of the bearing stresses. After [13]. Degree of slenderness λ = buckling length/gravity radius.

The *fatigue strength of the base metal* is often investigated by means of unnotched specimens. These specimens are free of design effects and, therefore, suitable in studies of the behavior of the base material, especially under dynamic loads. However, the fatigue strength of such specimens does not represent a definite material characteristic because it depends to a large extent on the surface quality of the specimens (Fig. C 2.4). Therefore, the surface quality must be described as exactly as possible with regard to the valuation of the fatigue strength measured in unnotched specimens. If the fatigue strength of steels with different strength is

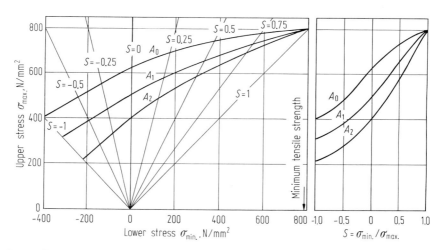

Fig. C 2.4. Effect of the surface condition on the fatigue strength (limiting number of the stress cycles = $2 \cdot 10^6$) of the steel StE 690 (steel no. 4 in Table C 2.9). A_0 = base metal with polished surface (R_t = about 2 μm), A_1 = base metal with ground surface (R_t = 12 μm to 19 μm) and A_2 = base metal with rolling scale (R_t = 30 μm to 100 μm). R_t = depth of roughness.

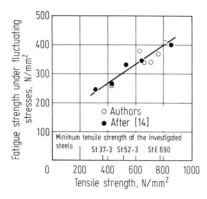

Fig. C 2.5. Fatigue strength under fluctuating (tension) stresses (limiting number of the stress cycles = $2 \cdot 10^6$) of solid bars with rolling scale of different structural steels according to Tables C 2.5 and C 2.9 in dependence on the tensile strength.

considered on the basis of unnotched specimens with rolling skin, a significant and linear dependence of both properties on each other is shown (Fig. C 2.5). The fatigue strength under fluctuating (tension) stresses rises with increasing tensile strength [15].

Local stress peaks occur in most technical structures due to the notch effect at places where the cross-section changes, e.g. drilled holes, offsets, grooves etc. The effect of such stress peaks is determined in tests under alternating tension – compression reversed (symmetrical) stresses and in tests under fluctuating (tension) stresses using notched specimens. In these conditions there is no longer a linear dependence between fatigue strength and tensile strength (Fig. C 2.6).

Welded joints are critical areas of dynamically loaded structures. Comprehensive parameter studies are necessary to determine the fatigue strength of welded joints and to obtain (assign) calculation characteristics for welded joints. As the situation is very complex, fatigue strength testing has been passed over to structural components. These tests clearly show the effects of the processing operations and of material variations [16].

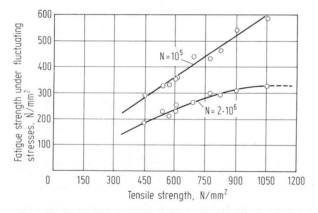

Fig. C 2.6. With punched bars of $\alpha_K = 2.45$ determined fatigue strength under fluctuating (tension) stresses (limiting number of the stress cycles $N = 10^5$ and $2 \cdot 10^6$) of different normalized and quenched and tempered fine-grained structural steels in dependence on the tensile strength.

The results of one load stage (step) tests (see B 1.2.2.1) of different stress conditions and steel grades can be presented as *standardized Wöhler curves* [17, 18]. The measured values shown are not represented by several individual curves that differ due to such factors as steel grade, construction and stress conditions, but are merged into a single curve. When the stress amplitude σ_a is related to the fatigue strength (limit) σ_D, then the standardized Wöhler curve is described solely by the two parameters limiting number of stress cycles $N_D = 2 \cdot 10^6$ and the slope k of the straight line that shows the fatigue strength up to this number of cycles. Results from fatigue strength tests of welded joints of StE 460 (steel no. 7 in Table C 2.8) and StE 690 (steel no. 4 in Table C 2.9) are well described by the standardized Wöhler curve (Fig. C 2.7).

Mechanical treatment, e.g. grinding of the welds, essentially improves the fatigue strength of welded structures. However, within the range of cycles where the curve of the fatigue strength is inclined no difference is observed. This fact is quite important where relatively few cycles occur during the life of the structure, e.g. for certain parts of a ship's hull. Welded joints that are levelled to the plate surface, achieve roughly the fatigue strength of the base material. The fatigue strength of welded joints can also be increased considerably through non-porous *remelting* of the critical zone along the weld seam with the transition from the weld to the base material by means of a TIG torch (TIG = tungsten inert gas) [19]. The reason is the removal of crack-like defects and slag inclusions that are probably responsible for the incipient fatigue crack (Fig. C 2.8).

Because of their stronger notch effect, fillet welds reduce fatigue strength to a larger extent than butt welds. In spite of this fact, fillet welds are used in nearly all steel structures because otherwise special profiles have to be rolled. The durability

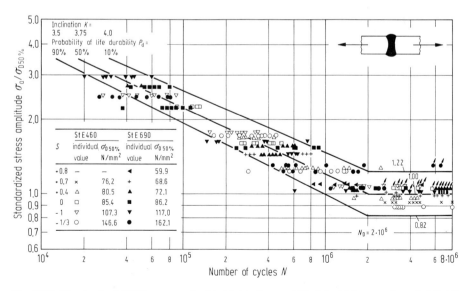

Fig. C 2.7. Standardized Wöhler-curve of welded joints (butt welds) of the high strength steels StE 460 (steel no. 7 in Table C 2.8) and StE 690 (steel no. 4 in Table C 2.9). $S = \sigma_{min.}/\sigma_{max.}$ After [17, 18].

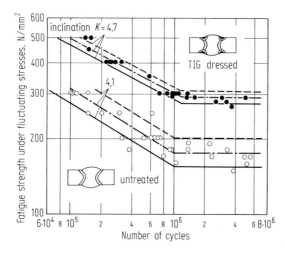

Fig. C 2.8. Effect of weld seam remelting by a TIG-dressing (electrical arc welding using a tungsten electrode in inert gas atmosphere) on the fatigue strength under fluctuating (tension) stresses of welded joints of the steel StE 690 (steel no. 4 in Table C 2.9).

of dynamically-loaded filler weldments depends to a large extent on the surface quality and the internal stresses in the area of the weld toe (transition zone). There values can be improved considerably if the critical surface areas can be plastically deformed after welding, e.g. through hammering or shot peening. The increase in the fatigue strength results mainly from strengthening of the material or by simultaneously setting up favorable systems of internal compression stresses. For high strength steels generally it seems that the induced internal compression stresses are vital.

Resistance to Brittle Fracture

Great numbers of test methods have been developed to investigate brittle fracture behavior (see B 1). In addition to quality control of the material the *notched-bar impact test* is particularly useful in evaluating the fracture behavior under the specified load conditions of test standards. The impact energy measured during testing is merely a qualitative characteristic for the evaluation of toughness. Figure C 2.9 shows the very different behavior of various steels, characterized by the dependence of the test impact energy on the temperature. From such curves it is possible by valuation of the microstructure of the fracture or by a certain specified value of the impact energy, e.g. 27 J, to define a temperature for the transition from tough to brittle fracture behavior. This so-called *transition temperature* T_{tr} (with the above definition $T_{tr\,27}$) is often used to characterize the brittle fracture behavior (see B 1.1.2.4).

The impact energy values of steels depend on their metallurgical, thermal and mechanical history. Therefore, the notched-bar impact test may also be used to assess the susceptibility of the toughness to changes caused by cold forming, aging or precipitation mechanisms. For instance, the preferred stretching of steel in one direction causes an anisotropy of the properties that manifests itself by large differences between the impact energy of longitudinal and transverse specimens (Fig. C 2.10) [20].

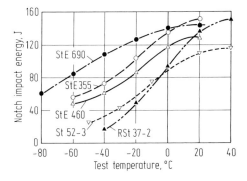
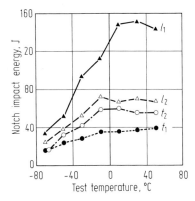

Fig. C 2.9. Notch impact energy-temperature-curves of different structural steels (see Tables C 2.5, C 2.8 and C 2.9), plate thicknesses from 15 mm to 25 mm (longitudinal ISO V-notch specimens).

Fig. C 2.10. Effect of the ratio of longitudinal rolling to transverse rolling on the notch impact energy of normalized heavy plates of St 52-3 (steel no. 8 in Table C 2.5) with 0.02% S. ISO V-notch specimens, l = longitudinal specimens, t = transverse specimens; index = ratio of longitudinal rolling to transverse rolling: 1 = 44:1, 2 = 1:1.

The material's resistance to the mechanism of *crack initiation* can be investigated by means of tensile or bend specimens of full plate thicknesses that are sharply notched or provided with natural cracks [21].

Figure C 2.11 shows the crack initiation behavior of heavy plates of StE 355 (steel no. 4 in Table C 2.8) and of StE 690 (steel no. 4 in Table C 2.9), determined by means of inside notched wide plate specimens. The maximum stress increases with decreasing test temperature, without a low stress fracture occuring. From the given stress-strain-curves and from the appearance of the fracture it can be concluded that, until a temperature T_a of $-30\,°C$ is reached, the phase of stable crack growth is present. The crack only grows to the same degree as the load is increased, and is stable until the fracture is completed (100% shear fracture).

It is interesting to note that this critical arrest temperature T_a according to the results of numerous comparison tests, corresponds well with the crack arrest temperature (CAT) (see B 1.1.2.4) resulting from the Robertson-test [22]. Below that critical temperature the crack propagates initially in a stable manner as a shear fracture, but after achieving a critical length it becomes unstable and runs through the further part of the specimen section as a brittle fracture. The crack opening displacement decreases considerably within this temperature range. The formation of the fracture as a shear fracture and its change to instability only after a phase of quasi-static crack growth, is preserved to about $-90\,°C$, the crack initiation temperature T_i. Below this temperature quasi-static crack growth does not take place. The crack starts as a brittle fracture and a crack opening displacement before fracture is scarcely observed.

The critical temperatures of normal strength and high strength structural steels demonstrate the relatively large range of this group of steels to resistance to brittle fracture (Table C 2.1). Bend tests using sharply notched specimens result in analogous statements [21].

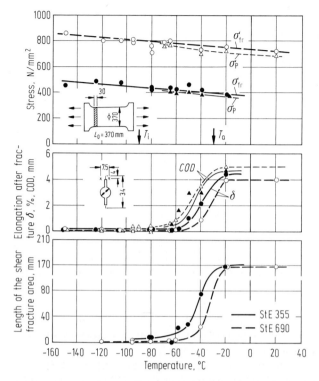

Fig. C 2.11. Results of wide plate tests at different temperatures of unwelded StE 355 specimens (steel no. 4 in Table C 2.8) and StE 690 (steel no. 4 in Table C 2.9). (σ_{fr} = stress at fracture, σ_p = stress at the limit of proportionality. Length of the shear fracture area = maximal extension of the shear fracture area, measured from the notch ground (saw cut) towards the small side of specimen). (T_i and T_a see text).

Table C 2.1 Crack initiation temperatures of normal strength and high strength structural steels determined in sharp notch bending tests [21]

Steel grade			Yield stress	Plate thickness	Microstructure	Crack initiation temperature
Designation	No.	in Table	N/mm²	mm		°C
RSt 37-2	4	C 2.5	255	30	ferrite + pearlite	−30 ... −40
St 52-3	8	C 2.5	370 ... 410	15 ... 30		−50 ... −70
StE 460	7	C 2.8	450 ... 470	30 ... 60	ferrite + pearlite + bainite	−20 ... −30
StE 690	4 a. 6	C 2.9	710 ... 790	25 ... 60	tempered martensite	−70 ... −120
StE 890	6	C 2.9	930 ... 955	15 ... 20	+ tempered bainite	−125 ... −135

Sometimes the knowledge of the *material's behavior against crack propagation* or the material's ability to arrest propagating cracks is requested. The Robertson test and the double tension test (see B 1.1.2.4) enable the material's behavior to be investigated under the load imposed by multi-axial stress conditions and the high deformation speed of a running crack. During the Robertson test, for instance, the full plate thickness material undergoes a tensile stress on the level of the admissible service stress and simultaneously is loaded by an externally-started brittle fracture. The lowest temperature at which a fast running crack is just arrested by the material, is determined.

A series of structural steel plates with a yield stress between 235 N/mm^2 and 900 N/mm^2 was subjected to the Robertson test. The 20- to 40-mm thick plates revealed crack arrest temperatures between $+10$ and $-57\,°C$ (Table C 2.2). A comparison between high strength fine-grained steels and commercial steels shows the superior safety against brittle fracture of high strength steels [23].

Whereas the expenses of Robertson brittle fracture tests is very high, the Pellini drop weight test (see B 1.1.2.4) has gained increasing importance during recent years because of the simpler manufacture of the specimens. Table C 2.3 shows NDT-temperatures of some high strength steels as scatter bands of the tested plate thicknesses.

Endeavours to provide a mathematical treatment of safety related to brittle fracture are made in *fracture mechanics*. In doing so, it is assumed that the material has a defect, and the conditions are determined under which such a defect (e.g. a small crack) propagates in a brittle manner or without deformation. The stress distribution around the crack tip is calculated by means of the theory of elasticity. Thus a relationship is established between defect size, stress and material characteristics. If two of these factors are known, the third can be calculated. An often raised question, for instance, is the determination of an admissible defect size for known load (stress) and a given material (see B 1.1.2.6).

Table C 2.2 Crack propagation behavior of normal strength and high strength structural steels during Robertson tests

Steel grade Designation	No.	in Table	Yield stress N/mm^2	Plate thickness mm	Test stress N/mm^2	Crack arrest temperature[a] °C
RSt 37-2	4	C 2.5	240	20	150	$+10 \ldots -5$
St 52-3	8	C 2.5	350	20	220	$-20 \ldots -25$
StE 420	6	C 2.8	420	40	300	$-11 \ldots -15$
StE 690	4 a. 5	C 2.9	780	30	470	$-53 \ldots -57$
StE 890	6	C 2.9	920	20	610	-53

[a] Crack arrest temperature = CAT of Robertson test

Table C 2.3 NDT-temperatures of high strength fine-grained structural steels determined by the drop weight tear test according to Pellini (see Fig. B 1.75 in B 1.1.2.4)

Steel grade Designation	No.	in Table	Yield stress N/mm^2	Plate thickness mm	Microstructure	NDT- temperature °C
StE 355	4	C 2.8	370 … 440	15 … 30	ferrite + pearlite	$-35 \ldots -60$
StE 460	7	C 2.8	440 … 490	30 … 40	ferrite + pearlite + bainite	$-40 \ldots -60$
StE 690	4 a. 5	C 2.9	720 … 800	25 … 75	tempered martensite + tempered bainite	$-50 \ldots -85$
StE 890	6	C 2.9	920 … 1020	15 … 35		$-40 \ldots -85$
[a]	7	C 2.9	960	25		$-65 \ldots -75$

[a] Known as HY 130

Resistance to Weathering

Resistance to weathering is affected by the chemical composition. More precise data can be obtained through weathering tests under real conditions, but these tests are costly and time-consuming so they are used only for basic investigations.

Development, formation time and protective effect of the covering layer on a weather resistant steel depend extensively on the atmospheric content. The latter has different effects which result mainly from the influences of large-scale climate (e.g. continental), small-scale climate (e.g. industrial area, city, country, coast) and orientation of the structural parts (e.g. exposed to the weather side or not, vertical or horizontal). The atmospheric load of pollutants must also be considered. The covering layer generally provides protection against atmospheric corrosion through weathering in industrial, urban and rural climate, as long as the limits of the emission values listed in §2.4 of the TA-Luft (GMB1 74/426) [23a] are not exceeded.

Resistance to Wear

Resistance to wear is only occasionally required from normal strength and high strength steels used for structural engineering so this material property is usually characterized by features which are well-known and easily tested. The chemical composition of the steel provides hints about resistance to wear. For good wear resistance, the microstructure should have a high proportion of pearlite so the amount of carbon present can serve as an indicator of this property. However, because of processing disadvantages and to safeguard against brittle fracture, the carbon content of this steel group is limited. Higher vanadium and niobium as well as chromium contents point to good wear resistance.

Eventually the wear resistance can be characterized roughly by the Brinell hardness number, plus consideration of certain restrictions (see B 10). This indication is possible because of the relation between hardness and resistance to wear. However, the chemical composition and the Brinell hardness of a material can characterize its wear behavior only qualitatively.

C 2.3.2 Characterization of Processing Properties

Formability

The possibilities for characterizing hot formability and cold formability of a steel are discussed in B 6 and B 7. Therefore it is not necessary to deal with the test procedures here. It seems to be more important to provide some general hints concerning the formability of normal strength and high strength steels and the effect of deformation on the properties of these steels.

Normal strength and high strength structural steels can be formed by similar methods. The form and dimensions of the products, and the available equipment for forming and heat treating will govern the choice of cold or hot forming. During cold forming of high strength steels, higher forces and enhanced spring back compared to steels with a lower yield stress must be considered. The original state of heat treatment of the steel remains with cold forming, but the properties of the steel will be changed, depending on the degree of deformation.

The deformability of a material depends on the possibility of creating dislocations and on their mobility. The dislocation density is increased by *cold forming*. As dislocations can hamper each others' movements, the yield stress increases while the deformability decreases. Consequently the yield stress of the material is increased and its toughness is decreased.

The amount of the property change is approximately proportional to the degree of deformation (Fig. C 2.12). The strengthening behavior of the different structural steels described here is approximately the same. Deformability characteristics change in the same measure as the material stress hardens. Particularly marked is the decrease in elongation before necking (elongation at maximum load, uniform elongation), whereas the reduction of area hardly changes.

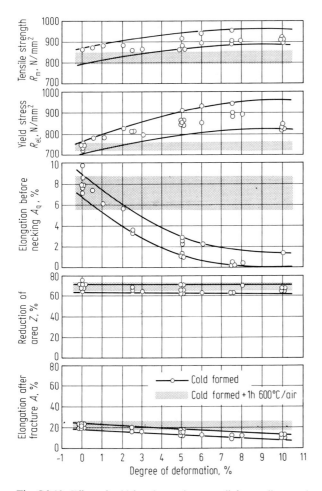

Fig. C 2.12. Effect of cold forming and stress relief annealing on the characteristics determined by the tensile test; 10 heats of the steel StE 690 (steel no. 4 in Table C 2.9).

A phenomenon that must be mentioned in this context is the Bauschinger effect [24] (see B 1.2.1), which is characterized by lowering of the yield stress when a cold deformation in a certain direction (e.g. by a tensile load) is followed by a deformation in the opposite direction (e.g. by a compressive load). The largest drop in the yield stress is observed after a predeformation of the order of 1 per cent (Fig. C 2.13). The effect occurs not only after cold forming at room temperature, but even after a minor deformation at an elevated temperature.

At any rate, the amount of the yield stress drop is reduced with increasing deformation temperature. The Bauschinger-effect may be compensated by a mechanical or thermal post-treatment. With a post heat-treatment the yield stress value of the non-deformed initial condition is approached with increasing annealing temperature and annealing time. Tests using large-scale wide plate specimens show that the bearing capacity of a slightly cold-formed structural part is not affected by the Bauschinger effect because it seems to occur only with small tensile specimens having limited material areas [24a].

Besides the admissible design stress that results usually from the yield stress, toughness is an important value affecting application of a steel. Increasing deformation moves the impact energy-temperature-curves to higher temperatures, which means that the transition temperature of the impact energy is raised [25]. Normalizing or quenching and tempering to remove the strengthening by cold working can seldom be carried out in practice because it may change the designed tolerances and affects the form-keeping unfavorably. Consequently mainly stress relief annealing is used. Figure C 2.14 shows that the usual temperatures used in stress relief annealing are too low to achieve a complete reduction of strain hardening through a recovery or recrystallization [26].

Residual stresses due to cold working or welding can only be released partly by *stress relief annealing*. The degree of relief that is attainable through annealing depends on the chosen heat treatment data and on the yield stress of the steel. The reduction of the residual stresses is most favorable with mild steels. But even with

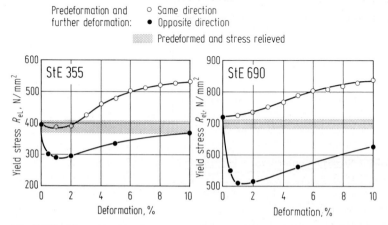

Fig. C 2.13. Effect of cold forming on the yield stress of the normalized steel StE 355 (steel no. 4 in Table C 2.8) and of the quenched and tempered steel StE 690 (steel no. 4 in Table C 2.9).

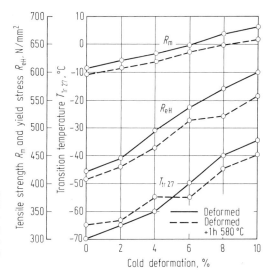

Fig. C 2.14. Effect of cold forming on the tensile strength R_m, yield stress R_{eH} and notch impact energy transition temperature $T_{tr\,27}$ (temperature at 27 J impact energy, longitudinal ISO V-notch specimens) of StE 355 (steel no. 4 in Table C 2.8).

high-strength steels a partial stress relief takes place after a short time annealing treatment at relative low temperatures (Fig. C 2.15) [26, 27].

Often a complete stress relief is not possible because of the disadvantageous effects of annealing on mechanical properties (Fig. C 2.16). The decreased values of yield stress and tensile strength as well as the increasing transition temperatures of the notch impact energy through stress relief annealing are due particularly to the diminution in tertiary cementite and some globulisation of the lamellar pearlite [27]. This behavior applies particularly with manganese steels. Microalloyed steels, especially quenched and tempered steels, generally behave differently because secondary hardening may occur with these steels during stress relief annealing, resulting in an increase of the yield stress. Only after annealing at higher temperatures is a decrease of strength and ductility observed.

Considering the *hot formability* of the steels and steel groups discussed here, it is important to distinguish between the behavior of normalized or normalizing rolled steels on the one hand and that of the quenched and tempered steels on the other.

The most important factor concerning the material's properties after hot forming in structural steels of the first cited group is the deformation temperature. To investigate its effect, annealing tests were carried out with a 10 mm thick heavy plate of St 52-3 (steel no. 8 in Table C 2.5) in the normalized condition. From Fig. C 2.17 it can be seen that the yield stress and the tensile strength are lowered by annealing temperatures up to about 1050 °C. Above 1100 °C the values increase again, but the yield stress remains below the initial value. The decrease of the notch impact energy due to increasing annealing temperature is unimportant only up to 1000 °C [28]. From these test results it can be concluded that the steel investigated may be superheated safely to about 1000 °C. Higher temperatures should be avoided during forming if normalizing is not to be carried out after forming [28, 29].

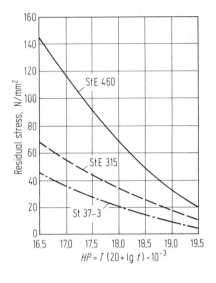 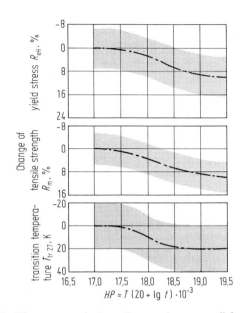

Fig. C 2.15. Stress relief of structural steels with different strength depending on the stress relief annealing conditions characterized by the Hollomon parameter HP (where T stands for temperature in K and t for time in h). StE 460 and StE 315 = steels no. 7 and no. 3 in Table C 2.8, St 37-3 = steel no. 5 in Table C 2.5.

Fig. C 2.16. Change of yield stress, tensile strength and notch impact energy transition T_{tr27} (see Fig. C 2.14) of unalloyed steels and manganese steels with minimum yield stress from 235 N/mm² to 355 N/mm² depending on the stress relief annealing conditions characterized by the Hollomon parameter. (The Hollomon parameter uses T for temperature in K and t for time in h).

Heat treatment effects in quenched and tempered steels disappear when hot forming temperatures above 650 °C are used and the material characteristics are changed (Fig. C 2.18). Cold forming is therefore generally preferred for operations on quenched and tempered steels. However, particularly with thick-walled parts, hot forming may be indispensable. Besides heat treatment after forming, the conditions during hot forming itself, namely the reheating temperature, holding time, deformation temperature and degree of deformation, have important effects on the microstructure.

Structural steels are generally reheated to temperatures as high as 900 °C to 1000 °C for hot forming. To establish the highest admissible temperature for this manufacturing step, the effects of reheating must be known. Investigations have shown that above about 1000 °C the austenite grain grows by one unit of DIN 50 601 or Euronorm 103 [29 a]. The reason for the grain growth is probably the initial dissolution of the nitrides in the temperature range mentioned. If the austenitizing reheating is followed by a complete quenching and tempering, the limited enlargement of the austenite grain will not be noticeable in the microstructure. Quenching and tempering restores the strength and toughness values of the initial state, and permanent deterioration of the material does not occur [30].

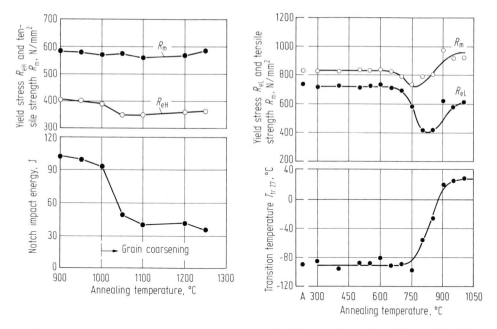

Fig. C 2.17. Effect of the annealing temperature (annealing time: 30 min) on the yield stress R_{eH}, the tensile strength R_m and the notch impact energy of longitudinal ISO-V-notch specimens at $-20\,°C$ of the steel St 52-3 (steel no. 8 in Table C 2.5).

Fig. C 2.18. Effect of the annealing temperature (annealing time: 30 min, air cooling) on the yield stress R_{eL}, tensile strength R_m and notch impact energy transition temperature $T_{tr\,27}$ (see Fig. C 2.14) of steel StE 690 (steel no. 4 in Table C 2.9). A = quenched and tempered starting condition (920 °C 1 h/water + 680 °C 2 h/air).

Machinability and Suitability for Flame Cutting

The property of *machinability* is of only subordinate importance for structural steels and usually it is not measured. To evaluate quantitatively the influence of a material on the life of the tool, a work operation such as turning or drilling is carried out at different cutting speeds but otherwise equal test conditions on work pieces of the same shape (see B 9).

Suitability for flame cutting may be judged by the chemical composition. However, discussion of details is superfluous because all the structural steels of this chapter are suitable for flame cutting, if the cutting conditions are adjusted to the specific chemical composition of the steel. Generally it is sufficient to preheat steels with higher carbon or alloy contents. The surface of the material at the flame cut edges can be critical because in a thin layer, flame cutting causes material changes that may result in hardening (Fig. C 2.19) [31]. The reasons for the hardening are a carbon pickup due to the process, the short time of reheating up to very high temperatures and the extremely rapid cooling to ambient temperature.

Hardening may also occur with steels which are not supposed to be hardenable based on their chemical composition. It has been shown that the carbon content of

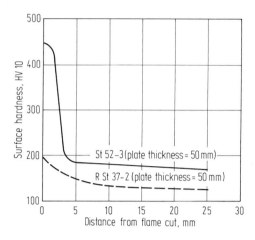

Fig. C 2.19. Hardness at edges that were flame cut without preheating, of steels with different chemical composition (R St 37-2 with 0.12% C and 0.67% Mn, see steel no. 4 in Table C 2.5; St 52-3 with 0.19% C and 1.57% Mn, see steel no. 8 in Table C 2.5).

an unalloyed structural steel with 0.13% C increases up to three times that value directly adjacent to the flame cut. A very steep concentration gradient is typical for an autogenous flame cut. The carbonisation results from diffusion of carbon out of the flame cutting slag [32].

Material changes within the area of the cut increase the crack susceptibility with rising carbon content and thickness of the plates. Edge cracks may occur during flame cutting, transportation or even at minor cold deformations. The risk of cracks may be diminished or neutralized through preheating or post-heating the flame cut parts. Subsequent welding removes the hardened zone of flame-cut material completely.

Weldability

The most important joining process for structural steels is the welding process. Weldability and its characterization is dealt with basically in B 5, which states the suitability of a material for welding generally means that it is possible to make economically welded joints which satisfy two conditions. The joints must not have defects (cracks) that impair the bearing capacity, and they must have mechanical properties similar to those of the base metal. To ensure fulfilment of these conditions, the chemical composition of the base metal is selected in such a manner that the heat input during welding will cause neither a loss of strength nor an embrittlement of the heat-affected zone or a crack. Consumables and welding conditions also must be adjusted to suit these requirements.

Hints for processing by welding include the relevant guide lines [33].

Many factors affect the weldability of a steel, (see e.g. DIN 8528) [33a], and there are many that bear on the possibilities and *methods to characterize the weldability*. Susceptibility to cracking must be regarded as a particularly important factor in addition to the chemical composition, the grain size and the cleanliness, to which this susceptibility is partly related, and which is dealt with by many test methods [34]. But to some extent these test methods are used mainly for examination of weldability. In contrast to these methods, it is easier to take the chemical

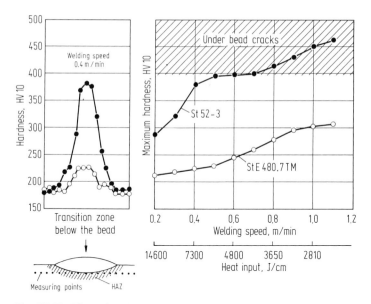

Fig. C 2.20. Effect of carbon content on hardness and cracking susceptibility of St 52-3 with 0.18% C (see steel no. 8 in Table C 2.5) and StE 480.7 TM with 0.08% C (thermomechanically treated steel for line pipes with 480 N/mm² minimum yield stress, see DIN 17 172 [35a]. (Hot strip thickness: 8.8 mm, welding position: down-hill, electrode: E Z e VII m.)

composition and above all the *carbon content* as an indication of the weldability. First of all it can be stated generally that the weldability of an unalloyed steel decreases with increasing carbon content (Fig. C 2.20) [35]. Unalloyed structural steels with carbon contents above 0.22% (product (check) analysis), also alloyed steels, must be preheated before welding to avoid hardness peaks (above 350 HB) by lowering the cooling rate and with that the hardening effect. In general, the preheating temperature must be higher with higher carbon contents.

The weldability of a steel depends further on the melting and deoxidizing methods used in its production, and above all on the cleanliness and grain size of the steel. Unkilled steel should not be used in thicker dimensions or for parts where there is a risk that segregated zones may be cut.

If the chemical composition is to be used to assess the weldability of a steel, it must be remembered that the dependence of the mechanical properties on the time-temperature-sequence of welding is different with low alloyed high strength steels than it is with unalloyed steels [36, 37]. With increasing cooling rates after welding, or with decreasing cooling time (see below), the toughness of the heat affected zone (HAZ) of low alloyed fine-grained steels with low carbon content is improved [37, 38] (Figs. C 2.21 and C 2.22). Simultaneously a short cooling time will produce the required high strength values. Consequently a specific range of cooling times must be chosen for welding these steels. The cooling time is determined by the steel, welding consumable and construction [39]. A minimum cooling time must be observed to avoid under-bead cracks. An upper limit of the cooling time must not be exceeded if the required strength and toughness properties are to be achieved within the HAZ.

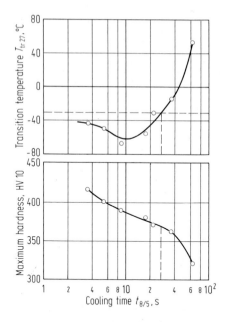

Fig. C 2.21. Effect of cooling time $t_{8/5}$ (cooling from 800 °C to 500 °C) on the hardness and notch impact energy transition temperature $T_{tr\,27}$ (see Fig. C 2.14) of heat treated (welding simulation) specimens of StE 690 (steel no. 4 in Table C 2.9) (peak temperature 1350 °C). After [37].

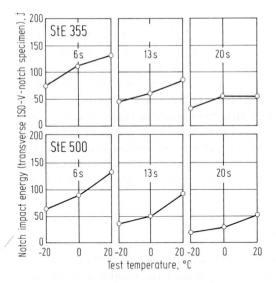

Fig. C 2.22. Effect of cooling time $t_{8/5}$ (see Fig. C 2.21) on the heat affected zone toughness of welded joints of normalized structural steels StE 355 and StE 500 (steel no. 4 and 8 in Table C 2.8).

The cooling time (generally indicated as the time required for a welded joint to cool from 800 °C to 500 °C: $t_{8/5}$) depends on the plate thickness, the working temperature and the heat input. The latter is a term that covers such welding data as voltage, current intensity and welding speed. The cooling time can be described with sufficient precision [40] (see also B 5) by mathematical equations.

To guard *against cracks*, above all against hydrogen induced under bead cracks, a certain working temperature T_0 must be chosen within the frame of welding conditions (heat input). In selecting the working temperature, the steel composition, the thickness, the hydrogen content of the weld metal and the level of residual stresses of the construction must be considered as the decisive influencing factors on cold cracking behavior. The tendency to cold cracking of welded joints is intensified with an increasing hydrogen content in the weld metal and an increasing content of alloying elements in the base metal and in the weld metal.

As mentioned briefly with regard to cold cracking, a lower limit must be specified for the cooling time. On the other hand an upper limit for heat input must be laid down to limit the maximum cooling time if the properties of the HAZ are to be controlled. The cooling time is determined by the working temperature and the plate thickness for a given steel type, a given welding process and a given welding consumable. Because of these relations, fundamental values for the effect of heat input, working temperature and thickness on the mechanical properties must be studied when welding conditions are being established. As the relationship between the effects of these parameters can be changed by the cooling time which characterizes the temperature-time-sequence, the mechanical and technological properties can be represented as dependent on the cooling time. Figure C 2.23 shows an example. From the experimentally established relation between the material properties in the HAZ and the cooling time, the range of the cooling time that must be observed during welding of a certain steel can be specified in accordance with the required properties to be met by the HAZ in any specific instance.

A high hardness in the HAZ of high strength steels is not a matter of concern as it is related to good toughness properties with these steels [41]. On the contrary, the portions of soft microstructure constituents caused by slow cooling, have poor toughness properties and must be avoided. Figure C 2.24 shows the relation between hardness and transition temperature of single pass welds of a high strength steel with 690 N/mm^2 minimum yield stress (steel no. 4 in Table C 2.9) [23]. For the same material, the specimens with high values of maximum hardness in the HAZ had favorable toughness values. Within a narrow zone along the weld seam the temperature during welding exceeds the tempering temperature of the quenching and tempering treatment, without a transformation of the microstructure taking place. A soft zone may develop in thin plates of very low alloyed steels during

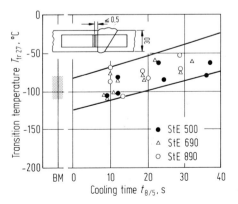

Fig. C 2.23. Effect of welding conditions on the notch impact energy transition temperature $T_{tr\,27}$ (see Fig. C 2.14) (transverse ISO V-notch specimen) in the heat affected zone of quenched and tempered structural steels (see steel no. 8 in Table C 2.8 and steels no. 4 and no. 6 in Table C 2.9). BM = base metal. Cooling time $t_{8/5}$ see Fig. C 2.21.

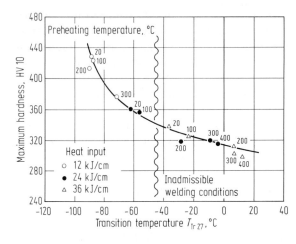

Fig. C 2.24. Relation between hardness and notch impact energy transition temperature T_{tr27} (see Fig. C 2.14, transverse ISO V-notch specimens) in the heat affected zone of one bead welds of 20 mm thick StE 690 plates (steel no. 4 in Table C 2.9).

welding with high heat input. However, even in unfavorable examples, the supporting effect of the adjacent zone is sufficient to provide for the full bearing capacity of the joint, because strengthening results within the weaker zone through the stress state [42].

In thick-walled welded constructions the steel may be stressed during welding in the through thickness direction to such an extent that so-called *lamellar tearing* occurs (Fig. C 2.25). Examples of such processing problems have occurred in mechanical engineering and offshore technology. This kind of failure refers to the fact that plates, strips and wide flats of normal production show a deformability in a direction perpendicular to the product surface that is not as good as in the other stress directions. Special measures can be used during production to achieve an improvement of the deformability. By these means the risk of lamellar tearing caused by loads perpendicular to the product surface (e.g. during cooling of fillet welds after welding) is diminished. This risk can also be restricted if the stress in the through-thickness direction is lowered by constructive measures and the welding procedure.

The safety of a construction depends ultimately on the bearing capacity of the welded joints. Therefore the mechanical behavior of welded joints, even when defects are present, has to be investigated thoroughly. Here, three aspects have to be observed: Behavior under overloading, behavior under dynamic loading and brittle fracture behavior.

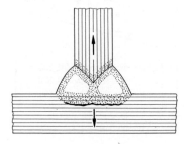

Fig. C 2.25. Schematic sketch of lamellar tearing.

The behavior of welded joints *under overload*, that is, when loaded above the admissible design stress up to the yield stress or even tensile strength, is covered by the tensile test and has been dealt with already in the context of the question about the effect of a hardness minimum. The bearing capacity of the joint is comparable to that of the base material. Wide plate specimens of welded joints also show that the yield stress of the welded joint is as high as that of the base material. The behavior of welded joints under *dynamic load* also has been dealt with already (C 2.3.2). The question about the *resistance of welded joints to brittle fracture* must be considered at full length. Therefore all the important brittle fracture tests must also be carried out on welded joints (Table C 2.4), to determine, on the basis of the deformation and fracture behavior, the lowest applicable temperature of a steel. In addition to these factors, it is necessary to consider the methods of fracture mechanics (see C 2.3.1 and C 1.1.2.6).

Suitability for Zinc Coating (Galvanizing)

For some steels and product forms of the steels in this chapter surface protection by zinc coating is often demanded. The suitability of a steel product for zinc coating depends on the steel composition. The thickness of the zinc layer increases primarily with rising silicon content, but also with increasing manganese content. The thinnest zinc layer is achieved on unkilled steels (no silicon addition) and steels with about 0.25% Si. Of course, a thinner zinc layer also means a reduced corrosion protection, but thick layers of an iron-zinc-alloy are brittle and tend to peel off during forming (see B 12).

Table C 2.4 Fracture behavior of welded joints of StE 690 (steel no. 4 in Table C 2.9), investigated according to various test methods with different specimens (notch always within the heat affected zone)

Test method	Specimen	Criterium	Transition temperature (typical values)	
			as welded °C	stress relieved °C
Notched-bar impact test	ISO V-notch longitudinal, unaged	$T_{tr\,27}$ [a]	− 60	− 45
Kohärazie test [b]	KO ($r \approx 0.01$ mm)	T_{LK0} [b]	− 45	− 30
Pellini test [c]	P-2 ($19 \times 51 \times 127$ mm^3)	NDT [c]	− 75	− 55
Robertson test [d]	($30 \times 350 \times 260$ mm^3) $\sigma_N = \sigma_S/1.5$	CAT [d]	− 55	− 35
Sharp notch bend test [e]	($65 \times 30 \times 300$ mm^3) ($r \approx 0.01$ mm)	T_i^e	− 80	− 70
Notch tensile test [f]	($20 \times 120 \times 450$ mm^2) ($r \approx 0.1$ mm)	T_i^e	− 75	− 75

[a] Temperature at 27 J impact energy
[b] Impact test with sharply notched specimens (see e.g. [43], also Fig. B 1.65 in B 1.1.2.4); T_{LK0} = temperature at 100% shear fracture using 5 m/s impact speed
[c] See Fig. B 1.75 in B 1.1.2.4
[d] See Fig. B 1.72 in B 1.1.2.4
[e] See e.g. [21]
[f] Rectangular specimen with inside notch

C 2.4 Measures of Physical Metallurgy to Attain Required Properties

C 2.4.1 General Remarks

The application properties of the steels dealt with in this chapter can be achieved by varying metallurgical procedures during melting, including alloying, hot forming and heat treating. The technological procedures for carrying out these measures during production, are permanently expanded and sophisticated.

During *steel melting* the alloying technique, also the deoxidation, desulfurization, sulfide shape control, degassing and process of solidification with regard to segregation are of particular importance. However, for the steels discussed here, these measures are used individually or in combination, only in accordance with the requirements.

An example of the effect of *hot forming* on properties is the change of anisotropy during heavy plate rolling on four-high-stands. The anisotropy of the toughness is substantially lowered if a plate is rolled with a small ratio of stretching to widening. Nevertheless, it is now possible to suppress the anisotropy of the mechanical properties much better through a desulfurization or sulfide shape control during steelmaking than by the subsequent hot forming process.

The recommendations in C 2.1 require all *measures* to be taken primarily with the aim of achieving the required strength, toughness and weldability.

There are several ways to increase, the *yield stress* of a steel. The theoretical limit, which lies at about 8400 N/mm^2 with steels on the basis of α-iron, has not yet been reached by a long way. The yield stress, regarded as the stress which shows clearly recognizable amounts of plastic deformation, is indicated by the movement of a larger number of dislocations within the crystals [44]. It is the aim of all strengthening measures to hinder the movement of these dislocations. The movement of the dislocations need to be hampered, but not completely blocked because blocking them would reduce the toughness and the deformability.

There are four basic mechanisms for the hindrance of dislocation movement, or strengthening of steels: solid solution strengthening, grain refinement, precipitation hardening and increase of the dislocation density. The *fundamentals* of physical metallurgy are dealt with in B 1. In *engineering* these mechanisms are utilized in steel production mainly through alloying and heat treating.

The *starting point for the application of these mechanisms* to the steels treated in this chapter, is a microstructure of ferrite with a certain small portion of pearlite. This microstructure results from the traditional processes of steelmaking without special measures, providing structural steels in the as-rolled state, with the basic properties: a certain strength combined with good toughness and weldability. The amount of pearlite (which is much stronger than ferrite), may be increased to raise the strength by increasing the carbon content (see Eq. (B 1.38) in B 1.1.1.3). However, his action will lower the toughness and weldability to unacceptable levels for application and processing of such steels. Preferred methods inlude first of all *solid solution hardening* (through appropriate alloying) and *grain refinement* (essentially through metallurgical measures). If a further increase of the strength (yield stress) is required, *precipitation hardening* (by microalloying elements) and an *increase in the dislocation density* (by quenching and subsequent tempering, that simultaneously comprises solid solution strengthening by carbon) may be used. The first three of these mechanisms for strengthening the matrix changes the proportions of ferrite and pearlite, and although the yield stress increases, the pearlite portion decreases more or less when high toughness and weldability are required. However, the application of the fourth mechanism (increasing the dislocation density by hardening) implies transformation to a microstructure of tempered martensite or bainite.

The following details must be considered under the aspects briefly discussed, and are dealt with according to the two technological measures of heat treatment and alloying technique. In doing so, an attempt is made to establish the individual relationships to the four strengthening mechanisms mentioned above.

C 2.4.2 Influence of the Heat Treatment and the Resulting Microstructure on Properties

Many rolled products in the as-rolled state already have the required properties. The material properties of rolled products can be improved further through thermomechanical treatment [44a] as well as through heat treatment. Applicable heat treatment, processes include normalizing, quenching and tempering and precipitation hardening.

Normalizing

For *normalizing* the work piece is austenitized and subsequently cooled in calm air (see B 4). The transformation results in equalization of the differences in the as-rolled microstructure, thus improving the uniformity of the ferritic-pearlitic microstructure substantially. Normalizing also results in grain refinement, particularly, if grain refining elements are added to the steel, e.g. during deoxidizing, that form nucleii and restrict the grain growth. Thus normalizing results in improved toughness and weldability along with high strength. The chemical composition of steels to be normalized, must be such that after normalizing the required ferritic-pearlitic microstructure is present. Only in exceptional cases is such a steel additionally tempered to reduce the hardness of undesired bainite portions that may be formed due to the higher cooling rate of thin plates in calm air.

Normalizing is superfluous if the *temperature during and after rolling is controlled* in such a manner that the rolling process is equivalent to normalizing [45, 46]. To achieve this aim, temperatures during normalizing rolling are controlled to such a degree that the plate or strip is finished at about 850 °C to 900 °C. The austenitic microstructure which becomes fine-grained at the end of the rolling procedure, recrystallizes completely between the individual rolling passes. Subsequent to rolling, a fine-grained ferritic-pearlitic microstructure whose mechanical properties equal those of the microstructure of the normalized state, is produced during air cooling. If the mechanical properties deteriorate during processing, the original state can be restored by a normalizing treatment.

This normalizing forming (during rolling) must not be confused with thermomechanical forming that should be classified as *thermomechanical treatment*. During thermomechanical forming the final deformation of the steel is carried out at temperatures at which the austenite is not, or not essentially, recrystallized. This sequence is aided by microalloying elements like niobium and titanium which retard the recrystallization [47, 47a]. Extremely fine ferrite grains are formed within the flattened austenite grains subsequent to rolling during cooling. Besides the grain refinement, the degree of precipitation of the microalloying elements and a high dislocation density contribute to a considerable increase in strength.

An essential advantage of this kind of treatment, (thermomechanical rolling), is the possibility of achieving a much better toughness when adjusting the steel to a certain strength level than when adjusting the same strength by means of the usual processes (see above), because a lower carbon content can be used. This work can be supported by combining the thermomechanical forming with accelerated cooling [47d]. Apart from steels that have been routinely thermomechanically formed (see C 4.1.2.1 and C 25), such treatment will be used increasingly for applications for which the steels of this chapter are used [47b, 47c]. When such thermomechanically rolled steels are processed, care must be taken to avoid hot forming or heat treatment (e.g. normalizing) at temperatures above about 600 °C if the enhanced strength properties are to be preserved without the toughnes being decisively improved.

Quenching and Tempering

A considerable increase of strength can be achieved along with good toughness by quenching and tempering. However, quenching and tempering is suited to only a few of the steels in this chapter. Nevertheless, some relations that are typical for the steels dealt with here may be pointed out to complete the explanations in B 4, since these steels, in spite of high strength, have a low carbon content with regard to toughness and weldability.

As described extensively in B 4, quenching and tempering comprises two separate steps of heat treatment. During the first step, quenching to produce fast cooling from austenitizing temperatures ensures that the austenite does not transform to ferrite and pearlite but to a transformation structure of the α-phase (martensite) that is supersaturated with carbon. From that structure, carbides are precipitated as finely-dispersed as possible at elevated temperature during the second step, the tempering.

Avoidance of the ferrite-pearlite formation is equivalent to the suppression of a diffusion controlled γ/α-transformation. The diffusion processes that result in the transformation to ferrite and pearlite take place usually in the temperature range from 800° to 600 °C. Through stabilizing the austenite, this temperature range can be crossed without a transformation occuring. Then at lower temperatures a lattice transformation, without diffusion, takes place from the cfc γ-lattice into a tetragonal distorted α-lattice, a phase of high hardness. Depending on the mechanism of forming the hardened microstructure the composition of the quenched and tempered steels is based on elements (nickel, manganese), that either stabilize the austenite in the sense of an expansion of the homogeneous field of the γ-solid solution towards lower temperatures or that hinder the diffusion. Of greatest importance is the diffusion of the carbon atoms, which is necessary for the ferrite-pearlite transformation to take place (chromium, molybdenum) [48].

The chemical composition of the quenched and tempered structural steels of this chapter must be of such a kind that the austenite transforms to martensite or bainite during quenching of the steel from the austenitizing temperature, which means that *hardening* takes place.

Figure C 2.26 describes the dependence of the transformation behavior of differently alloyed steels on the cooling rate. The effect of nickel on the transformation behavior is remarkable. The critical cooling rates can be extended by addition of nickel, which means nickel has a strong γ-stabilizing effect (s. also C 2.4.3). Because of the low carbon content of this type of steel the martensite transformation takes place at high temperatures (above 400 °C). The martensite therefore has only a little distortion and consequently the material is easily deformable [23]. Lower bainite contents have the same favorable properties. When heavy plates are to be quenched, the cooling rate and the alloying content are mutually adjusted in such a manner that the less tough upper bainite as well as pearlite and ferrite do not form during the quenching process.

Comparison of time-temperature-transformation- and time-temperature-property-diagrams (TTT- and TTP-diagrams) facilitates identification of the relationships between transformation structures and properties. The TTT-diagrams of the

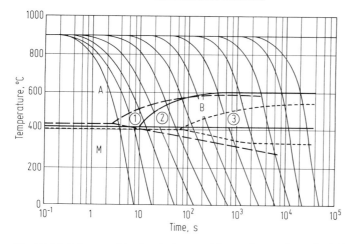

Fig. C 2.26. Influence of the chemical composition on the transformation behavior of StE 690 (steel with 690 N/mm² minimum yield stress, see steels no. 4 and 5 in Table C 2.9).

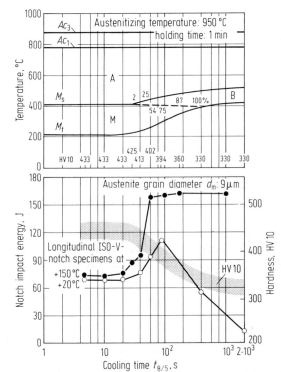

Fig. C 2.27. Transformation behavior and mechanical properties of StE 690 (see steel no. 4 in Table C 2.9). Cooling time $t_{8/5}$, see Fig. C 2.21.

steel StE 690 (for example steel no. 4 in Table C 2.9) according to Fig. C 2.27 can be subdivided into three stages. The steel transforms completely into martensite if the cooling time $t_{8/5}$ (cooling time from 800 °C to 500 °C) is below 30 s. If the cooling time $t_{8/5}$ lasts for more than 30 s, the austenite transformation begins at somewhat higher temperatures. It starts with a bainite formation which passes to martensite formation during further cooling. With longer cooling times the austenite transforms completely into bainite. A ferrite formation is not observed up to a cooling time $t_{8/5}$ of 2000 s [48a].

The TTP-diagram can also be subdivided into three stages. Martensite has the highest hardness. An increasing bainite content within the microstructure decreases the hardness. The notch impact energy of martensite is relatively poor, particularly as applied to the upper shelf energy of the weakly declined impact energy-temperature-curve. The transition area extends to low temperatures. This fracture behavior is known as "low energy tear" [49]. The occurrence of bainite causes an increase of the upper shelf energy. Due to increasing cooling time the hardened microstructures lose hardness and gain deformability, but become more susceptible to brittle fracture.

After quenching the steel is tempered below the temperature of the lower transformation point so that the *tempering* takes place within the temperature range of secondary hardening. Here the lattice defects that result from quenching are arranged in a very fine-grained substructure. Simultaneously, highly dispersed carbide and nitride precipitations are formed that cause a corresponding strength increase through precipitation hardening. The fine-grained substructure and the precipitations essentially determine the optimal microstructure of quenched and tempered steels, which have both high strength and good toughness.

The decisive factors for controlling the mechanical properties by means of the above heat treatment are the tempering temperature (Fig. C 2.28) and the tempering time. During tempering the yield stress and the deformability are changed.

The heat treatment of heavy plates is usually carried out after the hot rolled material has been cooled to room temperature. The plates are reheated for austenitizing and then quenched and tempered. However, there it is also possible to water quench directly from the rolling temperature [50]. Besides saving energy, the increase of the yield stress is remarkable. This higher yield stress may be caused by a higher dislocation density within the martensite when the steel is quenched from the rolling temperature. The application of this kind of treatment to high strength low alloyed steels containing chromium and molybdenum also results in higher strength values than the normal treatment, i.e. after reheating. However, the gain in strength carries with it a corresponding loss of deformability and toughness.

C 2.4.3 Influence of the Chemical Composition on Properties and its Effect on Microstructure

Chemical composition is the basis for the possibility of establishing different microstructures with their corresponding properties and utilizing the strengthening mechanisms in which heat treatment plays the decisive role. Under these considerations, the alloying and accompanying elements, that are associated with the steels

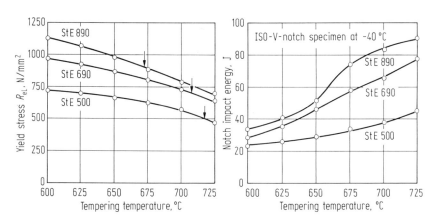

Fig. C 2.28. Tempering diagrams of quenched and tempered fine-grained structural steels (steels no. 1, 4 and 6 in Table C 2.9).

of this chapter, are dealt with in the following. In these discussions, references to an improvement in the hardenability or a decrease in the critical cooling rate (see B 4) by an element, comprise a short form of the statement that the element in question retards the γ/α-transformation during cooling. Consequently it is indicated that this element favors the formation of a hardened structure with the solid solution strengthening becoming effective, the dislocation density being increased and the grains being refined to a certain extent.

In dealing with the influence of the different elements it is necessary to distinguish (according to C 2.4.2) between steels that are used in the normalized state or in the quenched and tempered state. This characterization of the state is simultaneously a *short description of the resulting microstructure* that essentially determines the properties: the normalized state refers to a basically ferritic microstructure which includes pearlite islands, and the quenched and tempered state is based on an annealed martensite and/or bainite (see B 4).

Concerning as-rolled and normalized steels, the pearlite portion that is determined by the *carbon content* provides a substantial contribution to the strength. However, an increasing pearlite content gives rise not only to the strength but also to the notch impact transition temperature (Fig. C 2.29). Therefore the carbon content is restricted according to the application of the steel. These remarks apply particularly if weldability is required, because during welding the susceptibility of the HAZ to cold cracking increases with rising carbon content, due to the relation between carbon content and hardenability. In this connection a limiting value of about 0.25% C is often mentioned, and it must be observed that such a limiting value cannot be deduced from laws of physical metallurgy, but is based on experience which vary according to the particular welding conditions. Therefore such a value may change between certain limits. Besides, such a limiting value refers to the chemical composition shown by the product (check) analysis because it is more relevant to the welding behavior (e.g. within the HAZ) than the chemical composition obtained from the ladle analysis. But it is not only the carbon content that controls the susceptibility to cold cracking. Other elements (also welding parameters) also have an influence. In estimating the cold cracking behavior of structural steels attempts are made to combine them within a formula to produce a crack parameter that is comparable to the carbon-equivalent (see B 5). In doing so, it is clear that such formulae do not just express the laws of physical metallurgy, but also consider experience. Therefore such formulae will be completed and

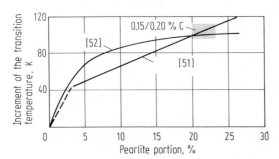

Fig. C 2.29. Dependence of the notch impact toughness transition temperature (at 85% brittle fracture of ISO V-notch specimens) on the pearlite content.

changed again and again. An example of an often used formula in this field is the crack parameter after Ito [53]

$$P_C(\%) = \%C + \frac{\%Si}{30} + \frac{\%Mn + \%Cu + \%Cr}{20} + \frac{\%Mo}{15} + \frac{\%V}{10} + \frac{d}{600} + \frac{H}{60}.$$

Here d is the plate thickness in mm and H the hydrogen content of the weld metal in $cm^3/100$ g. This crack parameter elucidates the strong effect of carbon on the cold cracking behavior as compared with other alloying elements. The strong effect of carbon on the notch impact transition temperature and on the cold cracking behavior makes it easy to understand attempts to develop pearlite-reduced high strength structural steels with a minimum yield stress of 355 N/mm². Figure C 2.30 shows results of IIW-tests [54], concerning the maximum hardness in the HAZ after bead on plate welding in two base metals with different carbon contents. It can be seen that the low carbon manganese-niobium steel, without being preheated, does not harden more than the properly weldable structural steel StE 355 (steel no. 4 in Table C 2.8) that has been preheated to 150 °C.

In normalized steels with a minimum yield stress between 420 N/mm² and 500 N/mm² it is not possible to reduce the pearlite portion as much as with steels having a minimum yield stress of 355 N/mm². To be able to restrict the carbon content of a normalized steel with a minimum yield stress of 500 N/mm² to about 0.2%, both nickel and manganese must be added for solid solution hardening, and precipitation hardening elements like vanadium and niobium also must be included [54a].

Water quenched and tempered steels gain their high strength first from the existence of carbon in a super-saturated solid solution after quenching. With rising tempering temperature an increasing precipitation hardening by carbides takes place. Both mechanisms overly the hardening caused by lattice defects due to the

Steel	Plate thickness mm	Yield stress N/mm²	%C	%Mn	%Nb	%Mn/%C	C_E%
○ StE 355	32	376	0.17	1.24	0.03	7	0.38
△ Mn-Nb	34	369	0.07	1.87	0.04	27	0.38

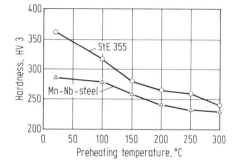

Fig. C 2.30. Effect of preheating temperature on maximum hardness in the heat affected zone of bead on plate welds according to IIW [54]. The steel StE 355 (see steel no. 4 in Table C 2.8) was preheated to 150 °C whereas the manganese-niobium-steel was not preheated before welding (C_E (%) = % C + % Mn/6).

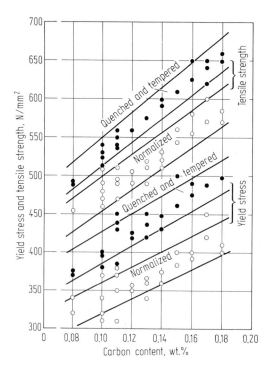

Fig. C2.31. Effect of the carbon content on the lower yield stress and the tensile strength of fine-grained structural steels with about 0.25% Si and about 1.3% Mn after normalizing and after quenching and tempering.

transformation, i.e. hardening through increasing dislocation density. The strength values depend on the carbon content (Fig. C 2.31), as happens with normalizing. In contrast to the normalized state, the yield stress of quenched and tempered manganese alloyed steels is about 70 N/mm² higher and the low temperature toughness is improved. The yield stress increase is somewhat stronger for higher carbon contents [55]. For the production of about 25 mm thick quenched and tempered plates with a minimum yield stress of 500 N/mm² a basically manganese (1.2%) alloyed steel needs 0.19% C. To be able to reduce the carbon content, more alloying elements must be added.

Solid solution hardening with *silicon contents* of about 0.3% is favorable to the strength and toughness of as-rolled and normalized steels [56]. Quenched and tempered steels generally contain 0.5% to 0.8% Si because this element decreases the critical cooling rate for hardening and increases the retention of hardness. In spite of this favorable effect the silicon content is likely to be restricted if weldability is a requirement. With regard to the hardenability of larger cross sections, higher silicon contents can only be avoided if, for example, a higher molybdenum content is provided. If the austenite transforms to martensite or lower bainite, then an increase in the silicon content does not affect the mechanical properties. Between steels with 0.2% Si and 0.7% Si there is virtually no difference in strength or toughness.

An important alloying element for improving the mechanical properties is *manganese*, which increases the strength and toughness of the ferritic solid solution. At constant strength the toughness and weldability of a normalized steel can be

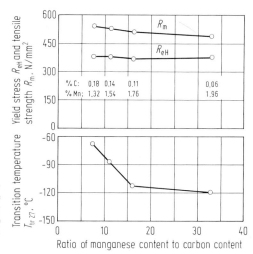

Fig. C 2.32. Influence of the manganese to carbon ratio on the yield stress R_{eH}, tensile stength R_m and notch impact energy transition temperature $T_{tr\,27}$ (see Fig. C 2.14) of normalized steels with 355 N/mm² minimum yield stress.

improved by the addition of manganese because, due to the solid solution strengthening effect of manganese, the carbon content and thereby the pearlite portion of the microstructure can be reduced. Examination of the properties of pearlite-reduced normalized structural steels reveals the advantage of a high manganese to carbon ratio [57]. Figure C 2.32 shows the effect of this ratio on the yield stress and tensile strength as well as on the notch impact transition temperature of normalized steels with a minimum yield stress of 355 N/mm². In quenched and tempered steels the effect of manganese is based essentially on a reduction of the critical cooling rate. Consequently manganese contributes to the suppression of the γ/α-transformation and thereby to the formation of hardened structures with their strengthening mechanisms. In quenched and tempered steels with about 0.16% C and 0.45% Si an increase of 0.1% manganese raises the yield stress by 30 N/mm² (Fig. C 2.33). Up to about 1.7% Mn both the strength and the toughness are increased. This effect is shown in Fig. C 2.33 by the declining impact transition temperature with increasing manganese content. The favorable effect of manganese is more pronounced in quenching and tempering than in normalizing. However, more than 1.7% Mn generally decreases the toughness (increase of the notch impact transition temperature in Fig. C 2.33) and further improves strength.

Molybdenum is not used generally in normal strength and high strength normalized steels, because it promotes the formation of bainite which is not desired in these steels. In contrast, molybdenum is a very important alloying element in quenched and tempered weldable steels. It improves the hardenability (see above), and, as a strong carbide former, gives secondary hardening (precipitation hardening). With an increase in the molybdenum content quenching and tempering produces a remarkable increase in the yield stress. The tensile strength behaves similarly, and the toughness also is slightly increased.

To be able to achieve uniform properties throughout the total cross section of a work piece during heat treatment, a homogeneous hardened structure (martensite and bainite) must be formed down to the core of the product. To achieve such

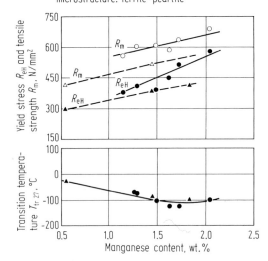

Fig. C 2.33. Effect of the manganese content on the yield stress R_{eH}, tensile strength R_m and the notch impact energy transition temperature $T_{tr\,27}$ (see Fig. C 2.14) of structural steels.

a structure, the cooling rate needed for the formation of the hardened structure, must be reached or exceeded at the core of the quenched piece. The possible cooling rate is determined by the available hardening process, so the steel composition must be adjusted to ensure that the steel's critical cooling rate is attainable by the equipment to be used. The thickness of the work piece also must be considered because its increase makes necessary the use of a steel with a lowered critical cooling rate. A lower rate may be achieved by use of higher alloying contents, and molybdenum is particularly suitable for this purpose. The relation between molybdenum content, cooling rate during quenching and steel properties is outlined in Fig. C 2.34 for steels with a basic composition of 0.15% C, 0.25% Si, 1.3% Mn and 0.5% Ni. The yield stress rises with increasing molybdenum content and increasing cooling rate (such as by means of decreasing plate thickness), while the impact transition temperature is pushed to lower levels. The higher the required mechanical properties of thick work pieces, the more molybdenum is needed. However, molybdenum contents above 0.5% cause temper brittleness due to the precipitation of the phase Mo_2C (see B 4) [48].

Nickel gives solid solution strengthening in normalized steels and increases their low temperature toughness. Because of cost, alloying of nickel generally is renounced for normal strength steels that are used at room temperature. In contrast, normalized fine-grained high strength steels with minimum yield stress values from 420 N/mm² to 500 N/mm² are made with about 0.6% Ni, to achieve, in combination with other alloying elements, strength and a good toughness. At higher nickel contents increased bainite formation is to be expected, and is

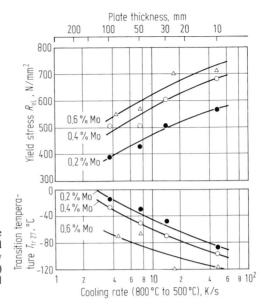

Fig. C 2.34. Effect of the cooling rate in the range from 800 °C to 500 °C on the yield stress R_{eL} and the notch impact energy transition temperature $T_{tr\,27}$ (see Fig. C 2.14) of a quenched and tempered structural steel with 0.2 to 0.6% Mo and 0.5% Ni.

undesirable in normalized steels. With quenched and tempered steels nickel contributes to improving the toughness, and reduces the critical cooling rate. Deeper hardening at equal cooling rate can thus be achieved with increased nickel content. For economic reasons generally, nickel additions of more than 1% are not used for steels with a yield stress below 700 N/mm² and a plate thickness below 50 mm [57a].

Chromium is not used in normalized steels of this chapter, since the attainable strength increase does not compensate for the loss of toughness and the reduced weldability. In quenched and tempered steels chromium also ranks with the elements that increase the strength by precipitation hardening due to carbide formation. Chromium is expected to improve hardenability in combination with an increased content of manganese or molybdenum. The effect of chromium was investigated taking, for example, a quenched and tempered steel with a high ratio of manganese to carbon. Figure C 2.35 shows the mechanical properties and Fig. C 2.36 illustrates the microstructure. Up to 3% Cr the increase of the yield stress and the tensile strength is nearly linear, that is about 7 N/mm² for every 0.1% Cr. The notch impact transition temperature remains more or less unaffected up to 1% Cr. In contrast, additions of more than 1% Cr affect the impact energy and its transition temperature adversely. Therefore quenched and tempered structural steels should not contain more than 1% Cr [55]. In addition to the deterioration of toughness a further disadvantage of higher chromium contents is the risk of cracking during welding. As Tekken tests have shown [58, 59], the susceptibility to cracking increases considerably above 1% Cr (Fig. C 2.37). Whereas heats of this test series with 0% and 1% Cr could be welded crack free at temperatures above +5 °C, the crack susceptibility at 2% and 3% Cr increased to such an extent that preheating was necessary at temperatures of 70 °C and 80 °C, respectively.

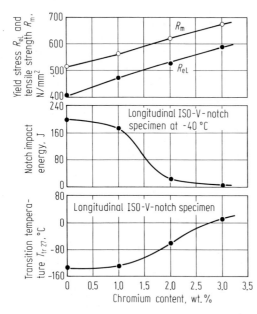

Fig. C 2.35. Effect of the chromium content on the mechanical properties of a quenched and tempered structural steel with 0.08% C and 1.65% Mn; plate thickness: 15 mm.

Precipitation hardening by additions of *copper* is occasionally utilized in normalized high strength fine-grained structural steels. The hardening due to copper (precipitation of ε-Cu) permits reduction of the carbon content, thereby improving the weldability. Steels with about 0.5% to 1% Cu need at least the same nickel content to avoid surface defects during rolling (risk of liquid-metal embrittlement) [60]. In normalized steels that are resistant to weathering, copper is an important alloying element in addition to chromium. Copper and chromium form complex phosphate and sulphate compounds that deposit as cover and protective layer between metal and rust. In quenched and tempered steels, copper can be utilized for strengthening through precipitation as well as in normalized steels.

In unkilled steels and in steels that are killed by silicon and manganese, the *nitrogen* is in solid solution and thus may cause aging (see A 6.2 and B 1.1.1.1). By the addition of aluminum e.g. during deoxidizing, both oxygen and nitrogen can be combined as aluminum oxide and nitride, respectively. Steels that have been deoxidized with a sufficient amount of aluminum are therefore more or less insensible to aging. An aluminum content of more than about 0.020% is considered sufficient for the steels discussed here. This value applies for the so-called portion of the acid soluble aluminum (due to the analytical procedure) available for fixing nitrogen, which is the portion that counts in connection with reducing the susceptibility to aging. The total content of aluminum that includes the portion used for fixing oxygen (deoxidation) is generally not much higher, due to modern techniques of deoxidation (see D 2). For this reason the total content of aluminum is often indicated whereas just the total contents are reported for the other accompanying and alloying elements. The reaction of nitrogen to aluminum nitride not only reduces the steel's aging susceptibility, but has a further effect beyond that: The

C 2.4 Measures of Physical Metallurgy to Attain Required Properties

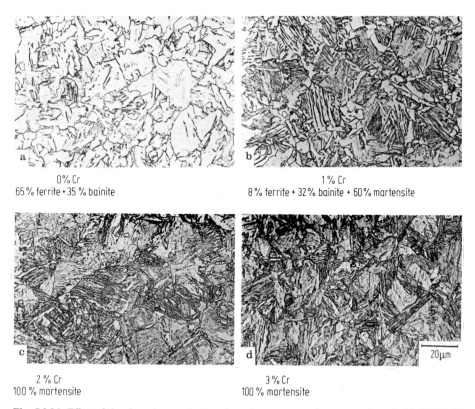

Fig. C 2.36. Effect of the chromium content on the microstructure of a structural steel with 0.08% C, 0.28% Si and 1.65% Mn after quenching; plate thickness: 15 mm. **a** 0% Cr: 65% ferrite + 35% bainite; **b** 1% Cr: 8% ferrite + 32% bainite + 60% martensite; **c** 2% Cr: 100% martensite; **d** 3% Cr: 100% martensite.

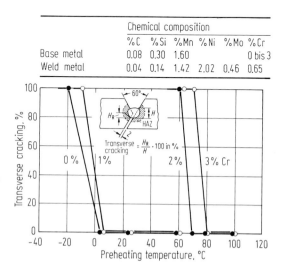

Fig. C 2.37. Effect of the chromium content on the cold cracking susceptibility of steels during the Tekken test [58, 59]. Heat treatment of the steels: 900 °C 1 h/water. Plate thickness: 15 mm. Welded with electrodes S H Ni 2 K 100 (3.25 mm dia.); voltage: 22 V, current: 115 A, heat input: 14.5 kJ/cm.

finely-dispersed aluminum nitrides provide grain refinement. During the temperature-time sequences in the course of processing (e.g. during welding in the HAZ) this dispersion restricts the growth of the steel's fine grains resulting from its heat treatment. Thus aluminum-killed steels are fine-grained steels with the characteristic properties of an enhanced yield stress, good toughness and good weldability.

Vanadium is an element whose compounds (e.g. nitrides and carbides) permit precipitation hardening. It is added to aluminum killed steels which obtain their characteristic properties through normalizing or normalizing rolling. The thermodynamics and kinetics of the precipitation of particles in steels containing aluminum, vanadium and nitrogen have been thoroughly investigated [61, 62]. Figure C 2.38 [63] shows the effect of vanadium on the yield stress and the notch impact transition temperature of normalized steels. To utilize the vanadium efficiently, these steels are produced with a nitrogen content of about 0.012%. No susceptibility to aging is given because the nitrogen is fixed by the available aluminum and vanadium.

Vanadium contents of more than 0.20% must be avoided with regard to weldability because, due to the temperature-time-sequences of welding, they result

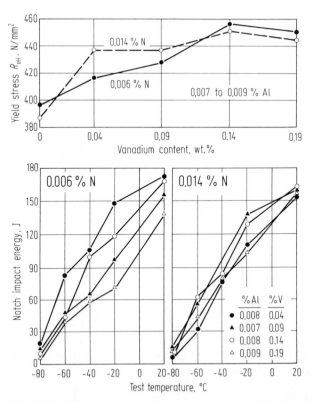

Fig. C 2.38. Effect of the vanadium content on the yield stress R_{eH} and the notch impact energy A_v (longitudinal ISO V-notch specimens) of normalized laboratory steels with 0.23% C, 0.43% Si and 1.2% Mn.

in unfavorable dispersions of the precipitations and may cause deterioration of the toughness in the HAZ.

Similar to its use in high strength normalized steels, vanadium is also employed as an alloying element in numerous high strength quenched and tempered steels. It works as vanadium carbide V_4C_3 through precipitation hardening, giving an effective strength increase. The carbon of these precipitations can be replaced partly or even completely by nitrogen because V_4C_3 and VN are isomorphous [64]. Vanadium contents of less than 0.15% exert the strongest effect. An addition of 0.1% V increases the yield stress by about 200 N/mm^2. As a result of this considerable strength increase the deformability decreases. Therefore in each individual application the question of whether a strength increase through a higher vanadium content is desirable [65], must be considered.

Niobium is a microalloying element that gives both grain refinement and precipitation hardening through the formation of niobium-carbo-nitride as well as vanadium. Compared with vanadium the important function of niobium is grain refinement. In normalized steels the best effect is achieved by a content of about 0.03% Nb.

In normalized steels the effects of a niobium content up to 0.03% and a vanadium content up to 0.15% are beneficial with regard to the mechanical properties, without negatively affecting the HAZ of welded joints. Negative effects might result from adjustment of the same strength level by higher contents of niobium or vanadium alone [66].

In high strength normalized fine-grained structural steels *titanium* reveals strong grain growth restricting and precipitation hardening effects that are based on the formation of TiN and TiC. However, with the carbon contents of about 0.15%, that are usual for these steels, a considerable deterioration of the HAZ toughness may be caused by the formation of coarse titanium carbides, particularly when welding plates of more than 20 mm thickness, and when welding with high heat input. The microalloying elements vanadium and niobium react less harmfully than titanium in this respect. In this connection titanium cannot be combined efficiently with other microalloying elements.

Under certain steelmaking conditions, small percentages of titanium, that do not produce precipitation hardening, form a dispersion of very fine titanium-nitrides in the steel matrix. This disperison prevents grain growth in the HAZ during welding of heavy plates to such an extent that high heat input welding for increasing performance is possible [67, 67a].

Titanium has another function in quenched and tempered structural steels, where expenditures for alloying elements such as molybdenum need to be limited. Equal properties at reduced molybdenum content may be achieved by use of titanium which can be added as a titanium-aluminum-silicon alloy. Titanium retards the transformation during hardening and produces precipitation hardening during tempering. Investigations have shown that titanium affects the yield stress of chromium-molybdenum-steels only in quantities above 0.02% Ti. The increase in the yield stress is considerable. In contrast, the strength values of steels that are additionally alloyed with 0.05% V, are scarcely changed. However, even a small titanium content decreases the toughness [55].

Boron is not alloyed in normalized steels because it influences the transformation behavior in such a manner as to hamper the formation of the ferrite-pearlite matrix.

Boron lowers the critical cooling rate in a manner similar to the actions of molybdenum and nickel. The element can be present in a steel both in solid solution and in fixed form. Because of its atomic radius of 0.097 nm (0.97 Å) the conditions for a solid solution within the steel matrix are unfavorable both on lattice places and on interstitial places. Therefore the dissolved boron is enriched within the temperature range of the austenite to a large extent on the grain boundaries [68, 69]. The effect is that the grain boundary energy is reduced, so that the formation of starting points for the transformation is retarded. The result is an improved hardenability of the steel. The boron content is usually below 0.01%.

Besides a poor solubility in iron, boron is characterized by a strong affinity with oxygen and nitrogen. Since boron can only effect an increase of hardenability if it is in solid solution, the nitrogen that is present in the molten steel must be fixed by another element. If the boron is present in the form of nitride, it no longer affects the formation of starting points and the consequent transformation behavior [70].

Sulfur ranks with the undesirable alloying elements found in the structural steels of this chapter. The manganese sulfides that prevent red shortness during hot forming, contribute considerably to the anisotropy of the toughness of manganese alloyed steels if they are stretched during hot forming. Figure C 2.39 shows the effect of sulfur on the impact energy of longitudinal and transverse specimens taken from heavy plates of grade St 52-3 (steel no. 8 in Table C 2.5) [28, 71]. Mechanical properties and weldability can be improved substantially through desulfurization and/or sulfide shape control [72]. The notch impact energy can be increased and the anisotropy of the toughness can be decreased by lowering the sulfur content (Fig. C 2.39). Figure C 2.40 gives the results of a survey of the attainable through thickness properties (characterized by the reduction of area) attained by different treatments. This chart reflects the material's resistance against lamellar tearing during welding.

Phosphorus embrittles the steel and therefore must be avoided as far as possible or must be removed from the steel by a suitable hot metal production and steelmaking procedure.

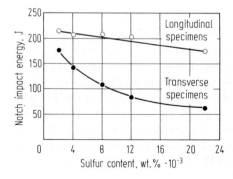

Fig. C 2.39. Effect of sulfur on the longitudinal and transverse notch impact energy (ISO V-notch specimen) of 50 mm thick plates of St 52-3 (steel no. 8 in Table C 2.5).

Fig. C 2.40. Effect of sulfur content and sulfide shape control on the reduction of area in the through thickness direction of StE 355 plates (steel no. 4 in Table C 2.8).

C 2.5 Characteristic Steel Grades Proven in Service

At the beginning of the normally-available series of normal strength and high strength structural steels stands a steel that has been the standard grade on the European continent since the introduction of processes for producing mild steel in a converter (air blowing, see D 2). According to today's prior art, this steel contains about 0.15% carbon and a corresponding portion of pearlite within the basically ferritic microstructure. The yield stress amounts to at least 235 N/mm² and the tensile strength to a minimum of 340 N/mm² (in former times at least 37 kp/mm², giving it the name St 37). This steel is such a steady reference point that in spite of the development of metallurgical processes, including resulting changes in chemical composition (e.g. lowering of the nitrogen content) and improvements in the degree of cleanliness giving a reduction of yield stress, again and again measures were taken to maintain the traditional yield stress.

Properties of this steel can be varied by changes in the carbon content, which means changes in the pearlite portion of the substantially ferritic microstructure to give somewhat lower or somewhat higher yield stress (whereby again and again as additional conditions good toughness and good weldability must be considered). However, a basically different steel results only through other strengthening mechanisms discussed in B 1 and in C 2.4. Primarily solid solution strengthening through additions of manganese and silicon (see C 2.4.3) and reduction of grain size through metallurgical measures are applied to structural steels. In this way, a steel with 355 N/mm² minimum yield stress and 490 N/mm² minimum tensile strength (formerly minimum 52 kp/mm², leading to the name St 52) has been developed. Both steels and their variations (see below) are to some degree the basis of the so-called *general structural steels* e.g. according to DIN 17 100, the most often used structural steels (see Table C 2.5) [45, 73]. These steels are used preferably for the applications mentioned in C 2.1, because they can be processed without difficulties.

As mentioned above, the first numbers after the two letters St (= steel) of the short name (designation) of the steels according to DIN 17 100 (Table C 2.5)

Table C2.5 Chemical composition and mechanical properties of steels for general purposes according to DIN 17 100 [45] (for details the Tables 1 and 2 of the standard have to be observed)

Steel grade		Deoxidation[a]	Treatment[b]	Chemical composition, ladle analysis					Mechanical properties of longitudinal specimens				
No.	Designation			%C max.	%P max.	%S max.	%N max.	Addition of nitrogen fixing elements	Tensile strength[c] N/mm²	Yield stress[c] N/mm² min.	Elongation[c] after fracture[c] A % min.	Notch impact energy[c,d] J min.	at °C
1	St 33	[e]	U, N	–	–	–	–	–	≥ 290	185	18	–	–
2	St 37-3	[e]	U, N	0.17	0.050	0.050	0.009	–	340 ... 470	235	26	27	+20
3	USt 37-2	U	U, N	0.17	0.050	0.050	0.007	–					+20
4	RSt 37-2	R	U, N	0.17	0.050	0.050	0.009	–					+20
5	St 37-3	RR	U N	0.17	0.040	0.040	–	yes					±0 −20
6	St 44-2	R	U, N	0.21	0.050	0.050	0.009	–	410 ... 540	275	22	27	+20 ±0
7	St 44-3	RR	U N	0.20	0.040	0.040	–	yes					−20
8	St 52-3	RR	U N	0.20	0.040	0.040	–	yes	490 ... 630	355	22	27	±0 −20
9	St 50-2	R	U, N	–	0.050	0.050	0.009	–	470 ... 610	295	20	–	–
10	St 60-2	R	U, N	–	0.050	0.050	0.009	–	570 ... 710	335	16	–	–
11	St 70-2	R	U, N	–	0.050	0.050	0.009	–	670 ... 830	365	11	–	–

[a] U = unkilled; R = killed; RR = killed + Al treated
[b] U = hot formed, untreated; N = normalized or normalizing formed
[c] The data are valid only for certain (different) thicknesses; for details see DIN 17 100, Tables 1 and 2
[d] Longitudinal ISO V-notch specimens
[e] Optional

indicate the customary minimum value of the tensile strength in kp/mm² which formerly was the characteristic property value. The numbers were not changed, although certain differentiations of the tensile strength took place. Furthermore the designations were not changed to today's valid unit N/mm², because the old short names exist in an immense number of documents and designs. Such changes would entail considerable expense. Since in general the yield stress in contrast to the tensile strength has become the more important value with regard to strength calculations (design), the yield stress value is used more and more within the short name (designation). As an example may be mentioned the high strength weldable steel StE 460 (see below). In its name the letter E ("*elastic-limit*") indicates that the following number represents the minimum value of the yield stress (in N/mm²).

The supplementary numerals 2 and 3 to the short names (designations) of the general structural steels corresponding to DIN 17 100, characterize the so-called "Gütegruppen" (quality groups), which, in addition to the minimum value of the yield stress are the most important distinctive marks. The two "Gütegruppen" (quality groups) of the steels that are assigned for welding (up to St 52-3 in Table C 2.5), differ with regard to their resistance to brittle fracture and weldability. This variation is due among other things, to the different maximum values of the phosphorus and sulfur contents, the nitrogen content and its fixation and thereby in a certain relation it is due to the kind of deoxidation (the steels of the "Gütegruppe" 3 are aluminum killed). The resistance to brittle fracture is also affected by the treatment (as-rolled compared with normalized or normalizing rolled). The differences between the "Gütegruppen" (quality groups) and between the states of treatment of a "Gütegruppe" (quality group) lie in the impact properties (see the different test temperatures in Table C 2.5). Measured values are shown in Fig. C 2.41. In this figure a steel of quality group the "Gütegruppe" 1

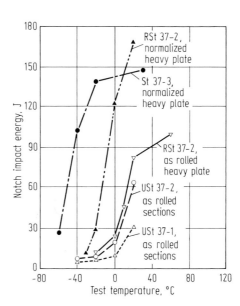

Fig. C 2.41. Notch impact energy-temperature-curves (longitudinal DVM-specimen) of St 37 of different "Gütegruppen" (quality groups) (see Table C 2.5) and heat treatment conditions.

(U St 37-1) still appears. The effective DIN 17 100 no longer contains steels of the "Gütegruppe" 1 [74]: The resistance to brittle fracture of steels of this "Gütegruppe" corresponds approximately to that of steels which were formerly produced by the Thomas process. This kind of steelmaking is no longer used in the Federal Republic of Germany and therefore corresponding steels are no longer listed in DIN 17 100.

For this reason the steels St 50, St 60 and St 70 according to DIN 17 100 are only available as "Gütegruppe" 2 (see Table C 2.5). Here the "Gütegruppe" (quality group) does not mark so much the resistance to brittle fracture, but the method of production (exclusion of the Thomas process) with its influence on the steel properties. These changes are connected with the fact that the steels are mainly applied in mechanical engineering and for purposes where weldability is not important, therefore no particular resistance against brittle fracture is required. For these reasons, carbon content and amount of pearlite within the basically ferritic microstructure to which these steels owe their higher strength, are not directly limited (see Table C 2.5), but are indirectly restricted by the specified elongation after fracture.

Steels for sheet piling [75] that are used in underground construction engineering and in hydraulic engineering, are closely related to the general structural steels (Table C 2.5). There are the steel grades StSp 37 (235 N/mm^2 minimum yield stress) StSp 45 (265 N/mm^2 minimum yield stress) and StSp S (355 N/mm^2 minimum yield stress). Resistance to brittle fracture is not required but the need for weldability can be assumed, though to a limited extent for StSp 45. In special cases sheet pilings of some steels corresponding to DIN 17 100 also can be supplied.

Another steel grade worthy of brief mention that fits into this chapter with respect to its application, but not its microstructure, is the *steel for colliery arches* specified in DIN 21 544 [75a]. This killed steel with about 0.3% C, 0.3% Si and 1% Mn (31 Mn 4), which is supplied in the form of special profiles, is a particular formulation in so far as it is supplied in three states of treatment: as-rolled (U); normalized (N) with a fairly high portion of pearlite within the ferritic microstructure corresponding to a minimum yield stress of 350 N/mm^2; and quenched and tempered (V) with the corresponding microstructure giving a minimum yield stress of 520 N/mm^2. With regard to processing (cold forming) it is of great importance that in each steel delivery a certain notch impact energy be present after aging (U: 18 J, N: 34 J, V: 48 J at 20 °C, when testing aged DVM-specimens). The major delivery of colliery arches steel is in the quenched and tempered state. Recently a thermomechanically treated steel with about 0.17% C, 1.7% Mn, 0.03% Nb and 0.15% V (17 MnV 7) has been developed [75b]. Its mechanical properties correspond to the quenched and tempered steel 31 Mn 4.

A side-line among general structural steels are the *steels resistant to weathering* WT St 37-2, WT St 37-3 and WT St 52-3, which have the same microstructure and mechanical properties as the corresponding steel grades corresponding to DIN 17 100. However, based on their chromium, copper and to some extent vanadium (with the WT St 52-3) contents these steels reveal a higher resistance to atmospheric corrosion [76]. When using these steels the remarks in C 2.3.2 must be observed with particular care [77].

The normal strength and higher strength *shipbuilding steels* (Tables C 2.6 and C 2.7 [78]) are also derived from the general structural steels St 37 and St 52-3, so they have a comparable microstructure. The minimum yield stress of the normal strength shipbuilding steels is uniformly equal to the value of the steel St 37. The grades A, B, D and E are produced in different ways which causes essentially differences in the resistance to brittle fracture and weldability. The higher strength shipbuilding steels comprise two strength levels, one of which corresponds to the steel St 52-3. Here the grades A, D and E of both strength levels are also characterized by different resistances to brittle fracture. Further details concern the specifications of the classification societies, e.g. the German Lloyd. The rules concerning general structural steels also apply to shipbuilding steels. In shipbuilding the most often used steel is the normal strength shipbuilding steel grade A, which comprises more than 80% of the total quantity of steel used for shipbuilding. Fine-grained normal strength and higher strength shipbuilding steels are applied only to particularly heavily-loaded parts of ships (e.g. grade E: E 32 and E 36).

Starting from the knowledge gained with the St 52-3, high strength normalized and high strength quenched and tempered fine-grained weldable steels have been developed by means of the procedures described in B 1.1.1.3 and C 2.4. These steels meet the highest requirements for strength and resistance to brittle fracture and have good weldability and they are therefore used for those constructions requiring these properties, for instance, in offshore oil-well drilling technology [78a].

Details of high strength *normalized weldable fine grain structural steels* are contained e.g. in DIN 17 102 (see Table C 2.8) [79]. This standard is not addressed exclusively to high strength steels (see C 2.1), for steels with a minimum yield stress below 355 N/mm^2 are also dealt with. Besides the standard series of steels, DIN 17 102 describes a series of corresponding steels for use at elevated temperatures and others with low temperature toughness as well as a special series with low temperature toughness. However, the steels of the last-mentioned series are mainly used for purposes other than those discussed here. In contrast to the general structural steels in DIN 17 100, the fine-grained structural steels reveal essentially better toughness properties. The most often used steels specified in DIN 17 102 are the steels StE 355 and StE 460.

Carbon and consequently the pearlite portion of the microstructure is no longer used to such an extent for strengthening normalized weldable fine-grained steels as it is with general structural steels. Even steels with a minimum yield stress of about 300 N/mm^2 contain about 1% Mn. Due to the resulting solid solution strengthening the pearlite portion can be reduced at equal strength levels. In this way a higher toughness is achieved in contrast to "Gütegruppe" (quality group) 3 of the general structural steels. Based upon St 52-3 which contains nitrogen fixing elements, e.g. aluminum, and, therefore has a certain amount of grain refinement, a further grain refinement can be achieved with the StE 355 through addition of niobium. Niobium also increases the yield stress through precipitation hardening. Consequently the carbon content and with it the pearlite portion of the microstructure can be reduced considerably while maintaining the same strength requirements. Toughness, resistance to brittle fracture and weldability are thus improved further by niobium additions (see Tables C 2.1 until C 2.3). Consistent extension of

Table C 2.6 Requirements for normal strength shipbuilding steels according to the "Unified Rules" (edition 1980) of the International Association of Classification Societies (I.A.C.S.). For some details, marked with *) the I.A.C.S. rules have to be observed

Steel grade[a]	Deoxidation	Chemical composition, ladle analysis[b]						Mechanical properties					Heat treatment
		% C max.	% Si	% Mn min.	% P max.	% S max.	% Al min.	Tensile strength N/mm²	Yield stress N/mm² min.	Elongation after fracture A*) % min.	Notch impact energy[c] l J min.	t at °C	
A	optional, but unkilled only for ≤ 12 mm thickness	0.23*)	–	2.5% C*)			–	400 to 490	235	22	–	–	as rolled
B	optional, but not unkilled ≤ 25 mm thickness semi-killed or killed, > 25 mm thickness fully killed	0.21	≤ 0.35	0.80*)	0.04	0.04	–				27	20	0
D		0.21		0.60			–				27	20	–10	normalized[d]*)
E	fully killed, fine-grained melting	0.18	0.10 to 0.35	0.70			0.019[e]				27	20	–40	normalized*)

[a] The steels have to be melted according to the basic open hearth, the electric arc or the oxygen blowing process or according to other approved processes
[b] Additionally it is applied to all grades: % C + 1/6% Mn ≤ 0.40%
[c] ISO V-notch specimens; l = longitudinal, t = transverse specimens
[d] If the required impact energy is proved (every 25 t), the normalizing may be replaced by normalizing rolling
[e] Content of acid-soluble aluminum. If the total content is determined, it must be at least 0.020% Al

Table C 2.7 Requirements for higher strength shipbuilding steels according to the "Unified Rules" (edition 1980) of the International Association of Classification Societies (I.A.C.S.). For some details, marked with *), the I.A.C.S. rules have to be observed

Steel grade[a]	Deoxidation	Chemical composition, ladle analysis[b]							Mechanical properties					Heat treatment
		% C max.	% Si	% Mn	% P max.	% S max.	% Al min.	% Nb	% V	Tensile strength N/mm²	Yield stress N/mm² min.	Elongation after fracture A*) % min.	Notch impact energy[c] J min. l t	at °C	
A 32	killed*)	0.18	0.10 to 0.50	0.90*) to 1.60	0.040	0.040	0.015[d*)	—	—	470 to 590	315	22	31 22	−20	
D 32														−40	
E 32														normalized[e*)
A 36								0.02 to 0.05	0.05 to 0.10	490 to 620	355	21	34 24	0	
D 36														−20	
E 36														−40	normalized[e*)

[a] The steels have to be melted according to the basic open hearth, the electric arc or the oxygen blowing process or according to other approved processes
[b] Additionally it is applied to all grades: ≤ 0.35% C, ≤ 0.20% Cr, ≤ 0.40% Ni and ≤ 0.08% Mo
[c] ISO V-notch specimens; l = longitudinal, t = transverse specimens
[d] Content of acid-soluble aluminum. If the total content is determined, it must be at least 0.020% Al
[e] If the classification society agrees normalizing of A 32, D 32, A 36 and D 36 may be replaced by normalizing rolling

Table C2.8 Chemical composition and mechanical properties of weldable fine grain structural steels, normalized (basic series[a]) according to DIN 17 102 [79]

Steel grade		Chemical composition, ladle analysis[b,c]					Tensile strength N/mm²	Mechanical properties[d]							
								Upper yield stress N/mm² min.	Elongation after fracture A % min.	Notch impact energy[e] at					
										−20 °C		0 °C		+20 °C	
No.	Designation	% C max.	% Si	% Mn	% P max.	% S max.				l J min.	t	l	t	l	t
1	StE 255	0.18	≤ 0.40	0.50 ... 1.30	0.035	0.030	360 ... 480	255	25	39	21	47	31	55	31
2	StE 285	0.18	≤ 0.40	0.60 ... 1.40	0.035	0.030	390 ... 510	285	24	39	21	47	31	55	31
3	StE 315	0.18	≤ 0.45	0.70 ... 1.50	0.035	0.030	440 ... 560	315	23	39	21	47	31	55	31
4	StE 355	0.20	0.10 ... 0.50	0.90 ... 1.65	0.035	0.030	490 ... 630	355	22	39	21	47	31	55	31
5	StE 380	0.20	0.10 ... 0.60	1.00 ... 1.70	0.035	0.030	500 ... 650	380	20	39	21	47	31	55	31
6	StE 420	0.20	0.10 ... 0.60	1.00 ... 1.70	0.035	0.030	530 ... 680	420	19	39	21	47	31	55	31
7	StE 460	0.20	0.10 ... 0.60	1.00 ... 1.70	0.035	0.030	560 ... 730	460	17	39	21	47	31	55	31
8	StE 500	0.21	0.10 ... 0.60	1.00 ... 1.70	0.035	0.030	610 ... 780	500	16	39	21	47	31	55	31

[a] For all grades there is not only the standard (basic) series but also the series for elevated temperatures, the low temperature toughness series and the low temperature special series
[b] All grades must be fine-grained through the steelmaking process, thus they contain usually min. 0.020% Al if not other nitrogen fixing elements (e.g. niobium, titanium or vanadium) are additionally alloyed
[c] The grades StE 355 up to StE 500 contain additionally alloying elements like copper, nickel, niobium, titanium, vanadium and others to achieve the required strength (for details see Table 1 of DIN 17 102)
[d] The data are valid only for certain (different) thicknesses (for details see Tables 3 and 5 of DIN 17 102)
[e] ISO V-notch specimens; l = longitudinal, t = transverse specimens

this development leads to structural steels with low carbon content that are used in offshore construction because of their very high requirements for processing behavior [35, 47, 80].

Normalized steels with a minimum yield stress of about 460 N/mm² and more, usually require a well-balanced chemical composition because they require both high toughness and good weldability to suit the applications in question. The main step to achieve sufficient strength is to increase the manganese content to 1.5% and about 0.15% V or other precipitation hardening elements also are added (Table C 2.8). Furthermore the precipitation hardening effect of adding about 0.5% Cu can be utilized, because the steel has a nickel content of about 0.6% designed to maintain the toughness and which is necessary in connection with copper [60]. By these means, the carbon content can be reduced, which is an advantage for toughness and above all, weldability.

Weldable fine-grained structural steels having high resistance to brittle fracture and with a ferritic-pearlitic microstructure can only be made with yield stress up to about 500 N/mm². Higher yield stress values can be reached only with a quenched and tempered microstructure resulting from a "soft" martensite based on a low carbon content and appropriate additions of alloying elements (see C 2.4.2). These limitations are due to the application requirements, which always require in addition to high strength, high toughness and good weldability. Some of these *quenched and tempered steels* with a minimum yield stress up to about 900 N/mm² are listed in Table C 2.9. The chemical composition of these quenched and tempered steels is essentially determined by the desired yield stress and toughness as well as the thickness of the product with respect to a full quenching and tempering treatment. The steel with a minimum yield stress of 500 N/mm² is an alternative to the corresponding normalized steel. Among the quenched and tempered steels, those with a minimum yield stress of 690 N/mm² are of the greatest importance. They present at a high yield stress the required high toughness and good weldability although the inclusion of alloying elements is low [81, 82].

As indicated in C 2.4.2, more and more thermomechanically treated steels with their specific properties (see above) are being made available for the fields of

Table C 2.9 Chemical composition and yield stress of quenched and tempered weldable structural steels

No.	Steel grade Designation	Characteristic alloying elements	% C max.	% Si	% Mn	% Cr	% Mo	% Ni	% B	% others	Yield stress[a] N/mm² min.
1	StE 500 V	Ni-Mo	0.15	0.4	1.2		0.3	0.4			510
2	StE 550	Cr-Mo-Zr	0.20	0.6	0.8	0.7	0.2			0.07 Zr	550
3	[b]	Ni-Cr-Mo	0.18	0.3	0.3	1.4	0.4	2.5			560
4	StE 690	Cr-Mo-Zr	0.20	0.6	0.8	0.8	0.3			0.07 Zr	690
5	StE 690	Ni-Cr-Mo-B	0.20	0.3	0.8	0.5	0.5	0.9	0.004	0.05 V	690
6	StE 890	Ni-Cr-Mo-V	0.18	0.3	0.7	0.6	0.3	1.7		0.07 V	885
7	[c]	Ni-Cr-Mo-V	0.20	0.3	0.5	1.0	0.4	4.0		0.10 V	910

[a] The values are valid usually up to a certain plate thickness
[b] Known as HY 80
[c] Known as HY 130

application of this chapter's steels. Recently the importance of these steels has increased to such an extent, that actual steel grades have been developed for certain applications. Besides a basic series, that must be mentioned in this chapter, these steels also include a series of low temperature steels and, for flat products only, a series of special low temperature steels [47b, 47c]. The steels of the basic series include minimum values for yield stress from 355 N/mm² to 550 N/mm² for flat products and 255 N/mm² to 500 N/mm² for profiles. The low carbon content, a basis of good weldability, characterizes these TM-steels. For example the steel with a minimum yield stress of 500 N/mm² for plates (BStE 500 TM) has the same mechanical properties as the steel StE 500 according to Table C 2.8, but the TM-steel contains ≤ 0.16% C in contrast to ≤ 0.21% C of the comparable normalized steel.

Because it is discussed in part C, product forms were almost omitted from this chapter, in which material properties are not extensively covered. Along with the explanations about steel grades in C 2.5 was included some information about plates and sections. The structural steels in question here are also processed into seamless and welded tubes. For this purpose the steel grades for making tubes are adjusted to the aforementioned steel grades. However, due to the product form, some values of the properties are specified slightly differently, and the kinds of properties that have to be proven are sometimes different. For details about tubes see [83–86].

C 3 Reinforcing Steel for Reinforced Concrete and Prestressed Concrete Structures

By Helmut Weise, Wilhelm Krämer, Wilhelm Bartels and Wolf-Dietrich Brand

C 3.1 General Remarks

Reinforced concrete steels are steels used as inserts in reinforced concrete and prestressed concrete [1]. There are several different varieties:

Reinforcing bars and *wire fabric* which are embedded in the ferroconcrete without stress and are stressed only by the dead weight and the live loads of the building; apart from the creeping and shrinking of the concrete.

Prestressing steels which are inserted in the concrete alone or in addition to the reinforcing bars and are later systematically set to exert compression stress on the concrete. These steel inserts also get additional stress from the dead weight and the live loads of the building. The creeping and shrinking of the concrete effect a reduction of stress.

Reinforcing steels are used practically exclusively for buildings which are subject to public building laws aimed at guarding the public safety and ensuring the application of commonly used, well tried (standardized) building material [2, 3]. New materials are subject to *certification by the public building authorities*. For reinforcing steel it is necessary to use *factory production control* to maintain regular standards of production. Controls include the quality control system of the manufacturer and inspection by a certification body. The survey contract must be approved by the public building authorities. The entire production of a manufacturer is subjected to quality control, but in exceptional cases single delivery lots also are subjected to it. The quality characteristics to be complied with are defined as characteristic values: that is, values below which some percentage of the batch test values is allowed to fall (as opposed to minimum values) (for example 5%); testing on a statistical basis and normal distribution of the test values is required. Because of the largely uniform shapes of the products (bars and wire rod) and the production process, the test values of reinforcing steel have statistically defined homogeneous groups of values, making possible the introduction of these characteristic values and the corresponding test methods, in contrast with, for example, structural steel with its different production forms.

References for C 3 see page 765.

C 3.2 Reinforcing Bars and Wire-fabric

C 3.2.1 Required Properties

The basis for design calculations in dimensioning of reinforcements for ferro concrete parts is the *yield stress* of the steel. Reinforcing steels are also used in buildings with not mainly static load; therefore sufficient *fatigue strength* in the range of fatigue under fluctuating tension stress is required.

Weldability is of particular importance as a processing property, as tack welding is often used to keep down costs. If necessitated by the design, load-bearing welded joints may be necessary too. Pressure welding methods such as flash butt welding, gas pressure welding and resistance spot welding – an essential for the production of wire fabric – are also used, in addition to fusion welding, especially metallic-arc welding and shielded-arc welding. Sufficient ductility must be present in the heat-affected zone for weldability to be acceptable.

Another important feature of reinforcing steels is *formability* (*bendability*). Reinforcing steel has to be suited to cold bending because these steels are often bent during construction. Formability by cold drawing or cold rolling is of special importance in reinforcing steel for wire fabric, if the optimal properties of the material are to be obtained at low cost.

Good bonding properties are essential for the interaction of the concrete and steel components if the loads (forces) from extension or stress in the reinforced concrete are to be transferred satisfactorily from one material to the other. Small parts of the loads (forces) at the surface of the steel are transferred by adhesive bond and the major part of the load is taken by the shearing bond which is provided by transverse ribs, and in addition, within twisted bars, by longitudinal ribs. The bond stresses for ribbed reinforcing bars specified in DIN 1045 [2] are much above the maximum bonding stresses and – as just mentioned – must be taken up by shearing bond between the ribs of the reinforcing steel and the concrete. The "related ribbed area" f_R has been defined as characteristic value for the bond effect of ribbed reinforcing steels: that is the projection of the ribs in a plane vertical to the axis of the bar, related to the surface of a cylindrical (smooth) bar of the nominal diameter. For standardized reinforcing steels the minimum values for f_R are between 0.039 and 0.056, depending on the bar diameter.

C 3.2.2 Characterization of the Required Properties

In the *tensile test* the *mechanical properties*, especially the *yield stress*, are tested. The mechanical tests of welded wire fabric must be made with samples having welds on at least one cross bar, to check the influence of the resistance spot welding operation on the base material. To investigate the *fatigue strength*, the bars are concreted in test beams which are subjected to alternating bending stresses, so that the steel bar is under fluctuating tension stress. The maximum stress in this test amounts to 70% of the actual yield stress and 2×10^6 cycles must be endured, the stress range being dependent on the steel product shape and the test beam. *Testing of weldability* is carried out by tensile tests on overlap and strap-joints; tensile, bending and shearing tests are also made on cross-welded joints. The test results may give some indication of the upper limits of the chemical composition. Bending tests on cross-welded joints have proven to be particularly effective tests of weld strength [4, 5]. Excessive hardening of the heat-affected zone can start premature cracking of the specimens.

Suitability for cold bending is tested by the *rebend test* with specimens in aged condition. In this test the steel is bent with a radius depending on the cross-section of the bar, subsequently aged (250 °C, 1/2 h, in cases of dispute 100 °C, 1/2 h) followed by rebending.

Immediate *proof of bonding properties* is made by the *beam test* performed in accordance with an agreement of the CEB (Comité Européen de Beton). In this test, slipping of the concrete reinforcing steel bar, concreted in the test-beam, is determined as a function of the effective bending stress. Further tests are made by measuring the profile of the ribs and calculating the related ribbed area (s.a.).

C 3.2.3 Measure Changes in Physical Metallurgy to Attain Required Properties

Focal points of properties requirements for reinforcing bars are *yield stress* and *weldability*. To begin with reinforcing bars, and neglecting heat treatment, the most economic means – according to B 1 – to attain a certain level of yield stress in the hot rolled condition is, to produce a *ferritic matrix* with *solid solution strengthening* by manganese and with a *proportion of pearlite* corresponding to the required yield stress; the proportion of pearlite being controlled by the carbon content. However, the application of this measure – increasing the proportion of pearlite – is limited by the requirements for toughness, formability and weldability. In flash butt welding or gas-butt welding, where the requirements for weldability are not very high, the steels specified may have relatively-high carbon content, which may be 0.4% C, so that the yield stress is obtained mainly from the proportion (fraction) of pearlite in the ferritic matrix. If there are higher requirements for weldability, that means general weldability representing the current state of the art, the carbon content must be limited; therefore the proportion (fraction) of pearlite in the microstructure is reduced. To reach the requirements for yield stress, the manganese content must be increased to raise the solid solution strengthening, and micro-alloying elements like niobium and/or vanadium must be added [6, 6a]. These elements, as compensation for the smaller proportion (fraction) of pearlite, produce *grain refinement* and *strength-increasing precipitations* (Fig. C 3.1). Another way to increase the yield stress with good weldability is to *increase the density of the dislocations* by cold straining of the ferritic-pearlitic initial microstructure. The hot rolled bars are brought to the required yield stress by stretching or twisting resulting in strain hardening (Fig. C 3.2).

Reinforcing bars with higher yield stress may also be produced by *temperature controlled cooling from rolling temperature* [8]. Using this procedure, the bar which, on the basis of its chemical composition would have a ferritic-pearlitic structure in

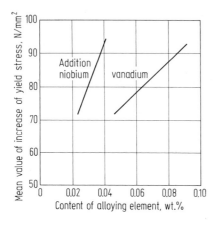

Fig. C 3.1. Influence of vanadium and niobium on the yield stress of reinforcing bars (dia. 28 mm) containing 0.18% C, 0.30% Si and 1.30% Mn. After [7].

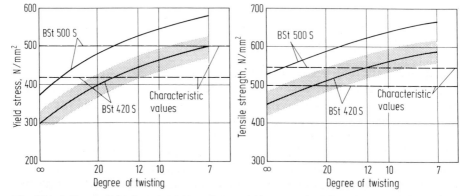

Fig. C 3.2. Influence of the degree of deformation on yield stress and tensile strength of reinforcing steel cold strained (strain hardened) by twisting. The degree of twisting is defined as the length of one thread of a longitudinal rib, related to the bar diameter.

the hot rolled condition, will be quenched from rolling temperature, that is from the austenitic condition, in a water cooling sequence designed so that a *martensitic surface zone* is formed. This zone subsequently is tempered in the range of 600 °C to 700 °C by the residual heat remaining in the core of the bar which converts into a ferritic-pearlitic structure. Apart from the chemical composition, this tempering temperature mainly influences the yield stress, so that the process parameters must be closely controlled. Figure C 3.3 shows the sequence of the cooling curves for different zones of the cross section. The tempering temperature of 660 °C in this example is reached after 8 s.

Special measures to fulfill the requirements for the other properties mentioned above are not necessary. The established steel production processes and the measures mentioned above for adjustment of yield stress and weldability, mean that most of the remaining property requirements are fulfilled. The surface conditions shown in Fig. C 3.4 provide some information about the bond effect.

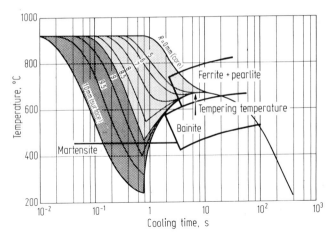

Fig. C 3.3. Time-temperature-transformation- (TTT-) diagram of a reinforcing bar containing 0.13% C and 1.21% Mn and time-temperature curves in different zones of the bar (defined by the distance R from the core) of a temperature controlled cooled bar of 20 mm. dia. After [8].

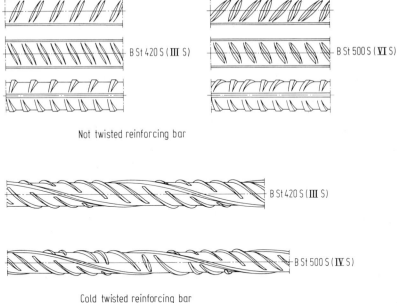

Fig. C 3.4. Examples for the surface condition (shape) of ribbed reinforcing bars. According to DIN 488 [10].

Adjustment of the required properties of *welded wire fabric* is carried out in roughly the same manner as for strain-hardened reinforcing bars. Unalloyed steels with ferritic-pearlitic microstructure and of coarse as well as fine grain are used. Because of the requirements for formability, gross segregations or differences in the

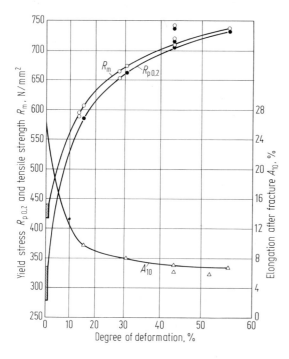

Fig. C 3.5. Change in the mechanical properties of hot rolled wire containing 0.15% C, caused by cold rolling with different degrees of deformation. After [9].

microstructure must be avoided. *Wire rod* for the production of welded wire fabric generally is shaped and straightened at high speeds on a cold rib mill in one production step. The influence of the degree of deformation on the mechanical properties is shown in Fig. C 3.5. With proper choice of chemical composition, degree of deformation and straightening parameters, the chemical and technological properties of the cold-rolled wire can be so adjusted that there is no need for tempering the wire fabric after welding which would otherwise be necessary to increase the elongation after fracture.

The high local thermal stress imparted to the cold strain-hardened material by resistance spot welding must not be neglected. To avoid on the one hand, embrittlement – increase of strength – and on the other hand, loss of strength of the cold-worked wire in the heat-affected zone (or to hold it down), welding must be carried out with as low a heat input as possible.

C 3.2.4 Characteristic Steel Grades Proven in Service [10]

A great variety of reinforcing steels with different microstructures and different levels of yield stress and weldability can be produced by the above-mentioned procedures. But the measures for improvement of the mentioned properties generally work against each other. The properties must therefore be optimized somehow, and the cost (for example, for alloying elements or after-treatment) plays an

Table C 3.1 Mechanical properties of reinforcing steels listed in DIN 488 Teil 1 [10]

Product form	Grade[a]		Characteristic values[b] for the mechanical properties		
	Designation[c]	Short sign[d]	Yield stress N/mm²	Tensile strength N/mm²	Elongation after fracture A_{10} %
Reinforcing bars	BSt 420 S	III S	420	500	10
	BSt 500 S	IV S	500	550	10
Wire fabric	BSt 500 M	IV M	500	550	8

[a] see C 3.2.4
[b] Values below which a certain fraction of the actual values are allowed to fall – here for normal distribution 5% of the values (production quantity) (see C 3.1)
[c] The number stands for the characteristic value of the yield stress in N/mm²
[d] For drawings and static calculations

important role. So, at present, of the variety of possible steels, those listed according to their mechanical properties in Table C 3.1, have prevailed. The basis for Table C 3.1 is DIN 488, Teil 1 [10]. The steel grades generally should also be weldable (see DIN 4099 [11]). In addition to Table C 3.1 Table C 3.2 shows data for the processing or the condition of treatment, the chemical composition and the weldability of the steels. Table C 3.2 also illustrates the variety of possibilities that can fulfill the requirements. The mechanical properties obtainable in reinforcing steels depend strongly on bar diameter and processing parameters. Therefore it is possible to produce, for example, with the same chemical composition, in the upper range of diameters, a hot-rolled microalloyed grade BSt 420 S, and in the lower range a grade BSt 500 S. The same variation is possible with heat treated (i.e. from rolling temperature controlled cooled) steel by varying the cooling parameters.

C 3.3 Prestressing Steels

C 3.3.1 Required Properties of Use

Prestressing steels, or tendons, are embodied in zones of prestressed concrete members which are subjected to tensile stresses resulting from dead weight and live loads [12].

As a result of the initial prestressing, these zones of the concrete are subjected to such high compressive prestress that even under a combination of the most unfavorable load stresses no tensile stresses (full prestressing) or only limited tensile stresses (limited prestressing) can occur in the concrete.

Tendons are incorporated and prestressed by two different methods [13]. Either they are tensioned between abutments and the concrete is placed immediately (pretensioning) or they are laid in special ducts around which concrete is placed and later prestressed against the hardened concrete; the ducts are finally filled with grout (post-tensioning).

Table C 3.2 Chemical composition and weldability of reinforcing steel according to DIN 488 Teil 1 [10]

Product form	Condition of treatment	Reference data for chemical composition[a]				Steel grade[b]	Weldability	Method of welding[c]
		% C	% Si max.	% Mn	Addition of microalloying elements			
Reinforcing bars	hot rolled microalloyed	0.22	0.60	1.30	Nb and/or V	III S and IV S	generally given	RA, GP, RP E, MAG
	cold formed			0.80	if necessary Nb and/or V			
	heat treated[d]			0.80				
Wire fabric	cold formed	0.15	0.60	0.50	–	IV M	generally given[e]	RP, E, MAG

[a] Ladle analysis
[b] Meaning of the short signs, see Table C 3.1
[c] RA = flash butt welding; GP = gas pressure welding; E = metallic-arc welding; MAG = active gas metal arc welding; RP = resistance spot welding
[d] Controlled cooling from rolling temperature, see C 3.2.3
[e] As concerns this product form dependent on processing (see methods of welding)

In service, prestressing steels are subjected to static and, in some applications, certain dynamic (alternating) loads. Most important, these steels must have a high elastic limit (0.01%-proof stress), so that the loss of prestress caused by creep and shrinkage of the concrete remains low relative to the initial prestress. The greater the amount of elastic strain attainable during prestressing, the lower this loss of prestress will be. Attainable strain, in turn, is dependent on the absolute elastic limit of the tendon.

Very high *yield stresses (0.2%-proof stresses)* and tensile strength values are necessary to ensure that, after shortening of the concrete as a result of creep and shrinkage, the strains in the steel and hence the precompression in the concrete remain sufficiently high. Characteristic stress-strain-curves of prestressing steels are shown in Fig. C 3.6.

In long-time loading in static tension, unelastic strains occur at lower stresses than during short-time loading. The extent of these strains depends on the level of applied stress, temperature and time, as well as on the type of steel. *This long-time behavior* results in a fall in the initial prestress of tendons in prestressed concrete structures, which is superimposed on the shortening of the concrete by shrinkage and creep. Related to the steel, therefore, the long-time behavior can be described either in terms of creep (change in length at constant stress) or relaxation (drop in stress at constant gauge length). Consequently, specific requirements are imposed on the creep behavior and relaxation of prestressing steel tendons.

Moving loads acting on prestressed concrete structures subject the tendons to fluctuating tension stresses which are superimposed on the applied prestress. *Fatigue strength* is therefore also important. In all methods of prestressing, the

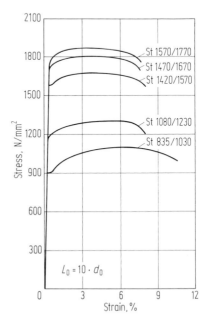

Fig. C 3.6. Stress-strain-curves of prestressing steels (steel-grades and designations see Table C 3.3).

measures used to anchor the tendons, e.g. bending, upsetting, wedging, result in a marked decrease in the fatigue strength of the tendons.

Behavior under *corrosive conditions* is also important. Like all carbon and low-alloy steels, prestressing steels are subject to atmospheric corrosion, i.e. during storage. A light *rust film* is not harmful but if scarring occurs, formability and fatigue strength will be impaired. In structural members prestressed by post-tensioning, unprotected tendons in not-yet grouted ducts may be subject to stress corrosion – sometimes without intensive surface corrosion – under the action of specific aggressive media (see B 3). This stress corrosion sometimes leads to cracking and, eventually, brittle fracture.

The mechanisms which produce this form of damage are still not fully clear [14]. In practice, however, all approved prestressing steels provide adequate resistance to stress corrosion cracking provided the relevant specifications are observed.

Regarding the *bond* [15], the requirements imposed on prestressing steels differ from those applied to reinforcing bars. A high degree of bond is necessary only in post-tensioning methods in which the tendons are anchored by adhesion and shearing bond, and in a few special applications (e.g. ground and rock anchors). Because of the magnitude of the forces to be introduced into the concrete, however, the bond stresses must be limited to prevent cracking of the concrete in the anchoring zone. To attain this "soft bond", the goemetry of the ribs differs from that of reinforcing bars and the rib surface area is smaller than that specified for reinforcing bars of the same size.

C 3.3.2 Characterization of the Required Properties

Tensile testing to determine *mechanical properties* is essentially the same as for reinforcing bars (see above), but in prestressing steels the 0.01%-proof stress is also determined.

Long time behavior under static stresses can be described either by the creep test (change in length under constant stress) or by the relaxation test (reduction in stress at constant gauge length). The relaxation test is increasingly favored over the creep test. It is carried out at initial stresses of between 60% and 80% of the actual tensile strength of the steel tested. Figure C 3.7 shows results of long-time relaxation tests performed on a heat-treated steel at different initial stress levels and temperatures. Figure C 3.8 shows the relaxation after 1000 hours for different steel grades and initial stresses.

Behavior under cyclic loading is determined both for prestressing steels alone and in conjunction with the anchoring and connecting elements. In testing prestressing steels the Wöhler method is used to check the range of fatigue under fluctuating tension stresses, to determine a Smith diagram for maximum stresses between the admissible stress and 90% of the yield stress or 0.2%-proof stress. Figure C 3.9 shows, as an example, the fatigue strength of a plain and a ribbed wire and of a wire with threaded ribs.

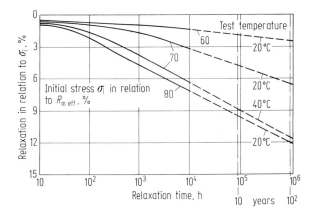

Fig. C 3.7. Relaxation values related to initial stress σ_i for prestressing wire St 1420/1570 according to Table C 3.3 (wire 7 mm dia.). $R_{m\,eff}$ is the actual tensile strength of the steel tested (dotted line = extrapolated).

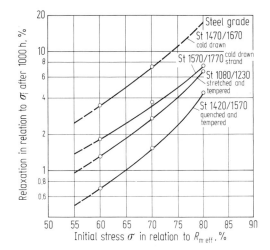

Fig. C 3.8. Relaxation values related to initial stress σ_i after 1000 h for different prestressing steels given depending on initial stress σ_i in % of actual tensile strength $R_{m\,eff}$ of the steel tested.

Different test methods are still being used to determine the behavior of prestressing steels under *stress corrosion cracking*, some leading to anodic, others to cathodic stress corrosion cracking. Following a recommendation by the "Prestressing Steel" committee of the Fédération Internationale du Béton Précontraint (FIP), standard international practice seems to be moving toward testing in ammonium thiocyanate at 45 °C at a stress of 80% of the tensile strength [16]. Even this test, however, is not capable of simulating actual conditions in the ducts accurately enough to permit conclusions to be drawn regarding the service behavior of the steels in the structure or for their further development.

In the course of a broad-based joint investigation [17] corrosion conditions in the ducts before grouting were examined in numerous examples. Based on these results, a test method reflecting actual conditions and using a neutral electrolyte

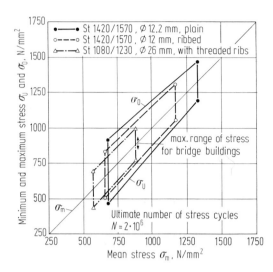

Fig. C 3.9. Fatigue strength diagram according to Smith (see Fig. B 1.136) for prestressing steels with different surface condition (shape) (see Fig. C 3.10 and Table C 3.3).

was developed and tried with the aim of replacing tests using concentrated solutions.

C 3.3.3 Measures of Physical Metallurgy to Attain Required Properties

Depending on the properties required, the main aim in the production of prestressing steel is to impart a yield stress or 0.2%-proof stress that is as high as possible. *Measures* taken to achieve these high levels *differ according to tendon form*. Up to about 16 mm thickness, prestressing steels are rolled on a wire rod mill, the rods being used either singly (prestressing wire) or in the form of strands; in the 16 mm to 36 mm diameter sizes the rods are rolled on a bar mill and used in the form of bars (Fig. C 3.10).

In manufacture of *wire*, two *different methods* are used to impart high yield stress: heat treatment of low-alloy steels, and drawing of plain carbon steels to strain-harden them.

In the first method, low-alloy steel is hot rolled in rolls having plain or ribbed grooves and subjected to quenching and tempering to impart the desired properties (Fig. C 3.11). The plain (smooth) or ribbed wires are welded together and are then continuously austenitized, hardened and tempered to give them a fine-grain microstructure containing finely dispersed precipitates.

In the second method, wire rod of ferritic-pearlitic microstructure that has been hot rolled in rolls having plain grooves is drawn at room temperature through a drawing die. The attendant reduction in cross sectional area increases the dislocation density and enhances the tensile strength and yield stress (Fig. C 3.12). Some wires are subjected to an additional cold forming operation. The drawn wires are finally heated to moderate temperatures (tempered wires), some of them under tensile stress (stabilized wires).

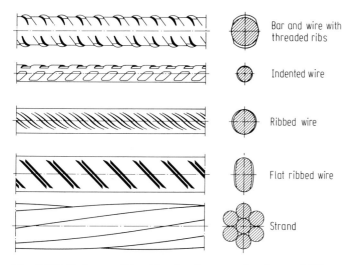

Fig. C 3.10. Surface condition (shape) and cross-sectional area of different prestressing steels.

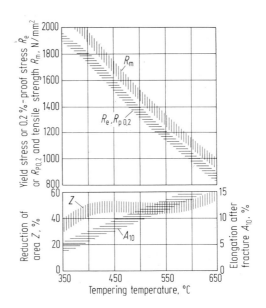

Fig. C 3.11. Tempering diagram for a quenched and tempered prestressing wire St 1420/1570 (see Table C 3.3).

In the manufacture of *strands*, drawn plain (smooth) wires are stranded with a specified length of lay on stranding machines. Tempering takes place after stranding, sometimes under tensile stress (stabilized strands). By staggering the welds of the individual wires it is possible to manufacture strands of almost unlimited length without impairing load-bearing ability.

A combination of two procedures is used to impart a high yield stress to *bars*. The chemical composition of the steels is such that after cooling from rolling temperature they display a purely pearlitic structure and thus already have high yield

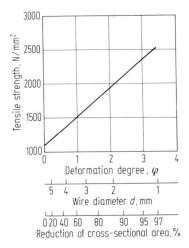

Fig. C 3.12. Tensile strength of a wire for prestressing steel with about 0.8% C and its dependence on reduction of cross-sectional area.

stress values, increased in some steels by the addition of vanadium to form strengthening precipitates. The bars are then strain hardened by stretching and finally heated to 250 to 350 °C to raise the elastic limit and yield stress still further.

In addition to these treatments to attain high yield stress values, all cold-worked prestressing steels are tempered to improve their *long-time behavior under permanent static loading*. For drawn wires and strands this tempering sometimes takes place under tensile stress to improve their less favorable relaxation behavior (see Fig. C 3.8).

To attain adequate *fatigue strength* in ribbed bars and indented wires, specific rib geometry requirements must be met. In particular, certain minimum fillet radii must be maintained to prevent high stress concentrations at points where the cross section of the bar changes.

To impart adequate *resistance to corrosion*, particularly to anodic stress corrosion cracking, the addition of up to about 2% Si to the steel has proved effective. As explained above, the factors influencing resistance to stress corrosion cracking are not yet fully understood and measures to counteract it cannot be covered here in detail. However, the resistance to stress corrosion cracking of the prestressing steels currently in use is sufficient, provided that specified storage and placing requirements are adhered to. After grouting the ducts or embedding the tendons in the concrete, reliable protection against stress corrosion cracking up to the present state of the art is ensured, provided the specifications for cement, aggregate and water are observed and the structural members remain free of critical-size cracks.

Numerous agents and methods also have been proposed and tested in recent years to provide effective corrosion protection for tendons which, for constructional reasons, have to be left under the action of prestress in ungrouted ducts for some considerable length of time [18].

As regards *bond*, reference may be made to rolling techniques designed to impart specific surfaces (see also Fig. C 3.10).

Table C 3.3 Prestressing steels and their mechanical properties

Form	Product Surface	Diameter mm	Steel grade Code[b]	Condition	Nominal values[a] of the mechanical properties		
					Elastic limit[c]	0.2%-proof stress N/mm²	Tensile strength	Elongation after fracture A_{10} %
Bars, round	plain or ribbed[d]	26 ... 36	St 835/1030	hot-rolled, stretched and tempered	735	835	1030	7
		15	St 885/1080		735	885	1080	7
		26 ... 36	St 1080/1230		950	1080	1230	6
Wire, round	plain or ribbed	6 ... 14	St 1420/1570	quenched and tempered	1220	1420	1570	6
	plain	8 ... 12.2	St 1375/1570		1130	1375	1570	6
	plain or indented	5.5 ... 7.5	St 1470/1670		1225	1470	1670	6
	plain or indented	5.0 ... 5.5	St 1570/1770	cold-drawn	1200 / 1325 / 1300	1570	1770	6
Strand	(7-wire)	9.3 ... 18.3	St 1570/1770	cold-drawn	1150	1570	1770	6

[a] Values below which a certain fraction of the actual values are allowed to fall – here with normal distribution 5% of the values (production quantity), also called characteristic values (see C 3.1)
[b] The two sets of figures indicate nominal 0.2%-proof stress- and tensile strength-values in N/mm²
[c] Determined as 0.01%-proof stress
[d] Threaded ribs

C 3.3.4 Characteristic Steel Grades Proven in Service

In line with the above-mentioned procedures aimed at imparting high yield stress values in prestressing bars, wires and strands, these products are supplied in various grades characterized mainly by yield stress and tensile strength (see also [19]). Details are given in Table C 3.3. Table C 3.4 contains approximate chemical compositions of prestressing steels.

Table C 3.4 Chemical composition of commonly-used prestressing steels

Product form	Approximate chemical composition[a]				
	% C	% Si	% Mn	% Cr	% V
Wire quenched and tempered	0.5	1.6	0.6	0.4	–
Wire and strands cold-drawn	0.8	0.2	0.7	–	–
Bars	0.7	0.7	1.5	–	0.3[b]

[a] Ladle analysis
[b] Only in St 1080/1230

C 4 Steels for Hot Rolled, Cold Rolled and Surface Coated Flat Rolled Products for Cold Forming

By Christian Straßburger, Bernd Henke, Bernd Meuthen, Lutz Meyer, Johannes Siewert and Ulrich Tenhaven

Flat rolled products defined in terms of their geometry [1] as having a roughly-rectangular cross section whose width is much greater than the thickness, account for a steadily increasing proportion of the many shapes taken by rolling mill products, today's figure being over 65%. This chapter discusses hot-rolled plate and strip with thicknesses of ≥ 3 to 16 mm and hot- and cold-rolled sheet and strip with thicknesses < 3 mm. Finished products are here regarded as cold rolled if they have undergone a reduction of cross section by at least 25% in cold rolling, without preheating. Regarding hot- and cold-rolled sheet and strip, the discussion includes products with and without metallic, nonmetallic or organic coating. Very light cold-rolled sheet ranging in thickness between 0.15 and 0.49 mm (blackplate) is also dealt with, especially product featuring a tin coating (tinplate) (see C 4.2.3.3). All these materials are called flat rolled products according to Euronorm 79 [1]. This term is also to be used in future in the relevant DIN standards; at present the term employed in these standards is mostly flat rolled material; therefore, this is the term that is also applied from time to time in the following text.

C 4.1 Hot- and Cold-rolled Flat Products for Cold Forming

For hot- and cold-rolled sheet and strip intended for production of drawn or formed components by deformation without heating, cold formability (see B 7.1 and B 8.1) is the most important property, whereas weldability is of only secondary importance. In addition to mild unalloyed formulations, steels with improved yield stress and tensile strength have been developed.

C 4.1.1 Hot- and Cold-rolled Sheet and Strip of Mild Unalloyed Steels

C 4.1.1.1 Hot-rolled Flat Products

Required Properties

As to hot-rolled sheet and strip made of mild unalloyed steels a distinction must be made between 1) sheet and strip for direct processing by cold forming and 2) strip

---

References for C 4 see page 766.

used as a semifinished material for the production of cold rolled strip. With both products the largest width to be rolled is currently limited to 2200 mm, the minimum thickness being about 1.5 mm.

The property requirements for hot-rolled flat products are predetermined by the kind of processing to be used and the intended applications, and are related primarily (see above) to their suitability for cold forming. While this statement is generally true for flat products that are processed directly (see B 8), as regards hot strip it concerns only cold rolling.

With both types of material, the products must have a suitable surface condition (to ensure good adherence of any subsequent surface coating). This, requirement means that oxide layers developed in hot rolling must be removed by pickling or by shot peening. For such products a well-defined surface structure normally found in cold rolled sheet is not suitable. For this reason, hot strip to be used for direct processing may be restricted to *applications* in which surface structure does not play a decisive role, e.g. for invisible parts in automobile construction. Sufficient flatness and the best possible dimensional accuracy are therefore of great importance. Skin pass rolling is included in many production sequences to improve flatness.

As indicated above, the main uses for hot-rolled flat material for direct processing are in automobile construction. The following examples represent the many applications: chassis parts, transverse control arms, side members, stub axle supports, wheel discs and blanking parts of every kind, also special shapes including door intrusion bars. Examples outside the automotive industry include gas cylinders, compressor housings and hydride storage units, as well as drawn and pressed bottoms for stationary tanks and boilers. Cold shaped sections are used for door frames and shelf construction, cold formed and expanded tubes for scaffolding. Crash barriers and sections for mobile crane jibs may also be mentioned.

Characterization of the Required Properties

As stated above, the most important property required for hot rolled *flat material for direct processing* is *cold formability*. To characterize this property, the test methods employed are essentially the same as for cold rolled sheet. These test methods are described in more detail in C 4.1.1.2 in relation to the importance of these products. The *tensile test* is the most important method for determining yield stress, tensile strength and elongation after fracture and the results of a hardness test can also be used in estimating strength properties. For a further description of cold formability also for the products dealt with here the (*modified*) *Erichsen cupping test*, and in special applications, the forming analysis tests (see B 8) are used.

The *microstructure* of steel is of utmost importance for cold formability. All the measures aimed at producing the desired structure are controlled very closely, so the measurement of oxidic cleanness is normally carried out only in random spot checks.

Assessment of properties based on tensile test data (see above) suffices when the production conditions specified for a certain steel grade are observed. In special instances, the *notch tensile test* supplies information on the amount and shape of sulphides produced by metallurgical procedures which become apparent through differences between the longitudinal and transverse properties of the material.

For the characterization of *hot rolled strip to be used in the production of cold rolled strip* generally only the *chemical composition* [2a] and – subject to the above-mentioned restrictions – the *microstructure* are employed.

Measures for Attaining the Required Properties

According to B 8 steel has good cold formability when the *microstructure* consists as far as possible of ferrite; in other words the carbon content is so low that only small proportions of pearlite are present in the microstructure and any cementite that develops does not precipitate in a coarse form. A uniform grain formation is also favorable. To produce such a structure the following steps are taken.

In rolling hot strip both for direct processing and for further processing by cold rolling the finishing temperature is maintained above the A_3-temperature to ensure that austenite transformation takes place completely after hot deformation.

Hot strip for direct processing is coiled at a temperature of about 650 °C, which is a compromise between the higher temperatures used for bonding nitrogen in aluminum-killed steels and the lower temperatures used to produce definite mechanical properties as well as the prevention of firmly adhering scale.

Hot strip for further processing by cold rolling is coiled at lower temperatures (below 600 °C) if the material is killed steel, to keep the aluminum in solution and to utilize the aluminum nitride by its controlled precipitation during the later annealing of the cold rolled strip in bell-type annealing furnaces or during treatment (at above 700 °C) in continuous annealing furnaces (see below) for production of a favorable surface texture (see B 8). Hot strip made of rimming steel, whose importance continues to decline, is coiled at temperatures of around 620 °C , to obtain the coarser grain favorable for cold forming.

Characteristic Steel Grades Proven in Service

In Table C 4.1 the steel grades used for hot-rolled flat stock for direct processing are listed together with their tensile strength and elongation after fracture values as indicative values. In addition, a scatter range for the *r*- and *n*-values of these steels is plotted in Fig. C 4.1. Further details will be found in DIN 1614 Teil 2 [2].

Table C 4.1 Steel grades and mechanical properties of hot-rolled flat material for direct cold forming (for further details see DIN 1614 Teil 2 [2])

Steel grade Designation	Deoxidation	Tensile strength N/mm² max.	Elongation after fracture	
			$A_{80\,mm}$ % min.	A
StW 22	Any	440	25	29
UStW 23	Rimmed	390	28	33
RRStW 23	Special deoxidation	420	27	31
StW 24[a]	Special deoxidation	410	30	34

[a] Yield stress max. 320 N/mm²

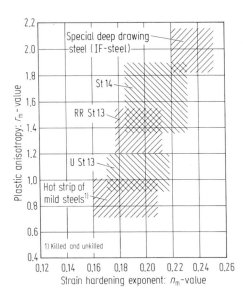

Fig. C 4.1. Comparison of r- and n-values of hot and cold rolled flat products of mild steels (indicative values).

C 4.1.1.2 Cold-rolled Flat Products

Required Properties of Use

Cold formability requirements for cold-rolled sheet and strip are even more stringent than those for hot-rolled flat products because more complicated blanks must usually be stamped from lower sheet thicknesses e.g. for passenger car bodies. For such work, an extremely good deep-drawing quality and/or stretch formability of the flat stock is required which sometimes can only be obtained only in special steel grades or by use of special measures in the processing. Formability requirements are lower, however, when other parts, such as office furniture, metal doors, exterior claddings and household appliances are to be produced. For these simpler products, a higher value is placed on good flatness and surface treatment suitability.

In connection with cold formability, requirements regarding freedom from Lüders' (stretcher) lines may be mentioned, such freedom depends in some degree on susceptibility of the steel to aging.

The *surface condition* takes second place after cold formability of the base material in considering the possibilities of using sheet or strip. A distinction must be made between type and finish of surfaces. The term type of surface means classes that display a varying degree of admissible surface defects. A basic differentiation is made between normal cold-rolled surfaces with small defects that are admissible and optimal surfaces which must have one side practically free from surface defects. The terms extra smooth, smooth, matt and rough are used to characterize surface finish.

Suitability for surface treatment with metallic or organic coatings, as described in B 8, is another important property of sheet products (see also B 12 and C 4.2).

Demands on the *cleanness* of sheet surfaces also must be considered. Cleanness can be characterized by the residual carbon content remaining at the surface after alkaline precleaning and by the abraded iron particles. There is also a link between the corrosion resistance of car body sheets and its influence on the capability for being phosphated and painted.

Roller seam welding is a sensitively-reacting joining method and requires especially-clean sheet surfaces. *Suitability for spot welding* of mild steels is assured by their chemical composition, with low contents of carbon and alloying elements.

The increasing automation of manufacturing methods in processing sheet material has resulted in a demand for favorable values and maximum *uniformity* of cold formability, surface quality, flatness and dimensional accuracy.

Characterization of the Required Properties

Cold formability of sheet and strip is provided essentially by data from *tensile testing*, i.e. by yield stress, tensile strength and elongation after fracture. The tensile test, therefore, is the most important test method carried out in practice. Moreover, *hardness tests*, such as Rockwell B or Rockwell F and T, are used to indicate the required strength properties; however, their results are not sufficient to be used as the only assessment criteria [3].

Behavior in aging may also be indicated from assessment of the yield point elongation in tensile tests (Lüders elongation) (see B 1).

Determination of *plastic anisotropy r* and the *exponent of strain hardening n* in tensile tests are finding increased application, although the procedures are not yet standardized. For a detailed explanation of these terms please refer to B 8.

The *surface condition*, i.e. kind of surface and, surface finish in particular, are assessed by visual judgment. Surface finish, however, may be measured directly by means of contact stylus instruments (according to Euronorm 49 [4]).

The above-mentioned possibilities of surface finish namely smooth, matt and rough must conform to specific values of centre line average height R_a (smooth: $R_a \leq 0.6$ µm; matt: R_a from 0.6 to 1.8 µm; rough: R_a at least 1.5 µm).

Application of surface treatments involving the application of decorative metal coatings demands an R_a-value of below 0.9 µm. Exposed parts of car bodies demand a limitation of the R_a-value to below about 1.5 µm for reasons associated with paint brilliance. For extremely deep-drawn products, such as oil pans, engine fuel containers or bath tubs, R_a-values of 1.5 to 2.8 µm are aimed at, so that the lubricant used becomes more effective during the forming operation.

Additional parameters for a more complete characterization of surface structure, peak or depth number and profile bearing portion may be used.

A simulating test method is the (modified) Erichsen *cupping test* the only test for assessment of cold formability which is standardized. Friction effects in this test conceal the material behavior to a large extent, as they do in other simulating test methods such as the cupping test. Deformation analysis (see B 8) tests are therefore sometimes conducted on component-like stamped parts in the research department or directly in the stamping plant.

Microstructure, including freedom from oxides, is basically the decisive factor for required sheet properties, in conjunction with surface properties. Metallographic tests to check production parts are carried out only as random spot checks because the chemical and physical procedures required to produce suitable structures are very carefully adhered to. Furthermore, control of sheet surface cleanness and application of anticorrosion oil to cold strip before shipment are not currently within the scope of continuous control.

Measures for Attaining the Required Properties

In B 8 and in C 4.1.1.1 above details are given of the microstructure to be aimed at to ensure fulfillment of requirements, especially as regards cold formability. Service property requirements are particularly high in cold rolled sheet and strip. To meet these demands, all specifications laid down for the production steps on the basis of the chemical and physical metallurgical correlations must be adhered to within the closest possible limits, so that the resulting products will have the most favorable structure and the most suitable surface condition.

The introduction of oxygen metallurgy and continuous casting as well as ladle metallurgy and vacuum treatment for steel grades with special quality requirements have created the best prerequisites for the production of improved quality, uniform steels. These manufacturing procedures result in steels with low carbon and manganese contents together with low inclusions of other or accompanying tramp elements. The result is a ferritic structure with very small proportions of pearlite or cementite and with favorable processing properties.

Further improvement of cold formability may be obtained by vacuum treatment which reduces the carbon content still more (to about 0.01% C). Such treatment reduces the pearlite or cementite content of the ferritic matrix. In addition an interstitial solid solution of carbon and nitrogen atoms in the ferrite is prevented by their bonding to titanium or niobium for example, which are added to the steel (IF-steel, see also B 8.3.2).

The already mentioned continuous casting process has contributed much to the uniformity of steel products. This process makes it possible – among other things (in addition to the fact that continuously-cast steel is always killed steel) – to improve the uniformity of the processing properties over the strip width and length even further. Special measures also ensure a far-reaching freedom from nonmetallic inclusions.

The undesirable effect of segregations in rimmed and therefore not continuously-cast steels may be limited by employing so-called bottle-top molds, to reduce the effervescent process.

Process conditions in *cold rolling*, above all the degree of cold reduction make it possible to attain a uniform grain structure and to influence the r- and n-characteristics that are important for cold formability. The structure of the hot-rolled strip from the range of about 1.5 to 6 mm thickness is work hardened in cold rolling to the specified final size. This work hardening may lead to tensile strength increases of 500 N/mm^2, depending on the steel grade. Thickness reduction normally is between 55% and 75%, and the degree of cold reduction also has an effect on property values resulting from recrystallization annealing (see B 8). Thus, for instance, at a thickness reduction of about 70% by cold rolling, the r-value is at a maximum after annealing in a box-type annealing furnace. For a continuous short-time annealing (of killed steel) a higher cold rolling reduction should be aimed at to achieve suitable *r*-values for subsequent processing.

The selection of suitable auxiliary materials and rolling conditions creates prerequisites for the production of clean sheet surfaces.

Strain hardening effects of cold rolling are removed by recrystallization *annealing*. Strip is annealed predominantly as *tight coils* in bell-type annealing furnaces at below the lower critical A_1-temperature. In killed steel the precipitation of aluminum nitride occurs during slow heating with sufficient exposure time in the range of elastic recovery and commencing recrystallization. Aluminum nitride here acts as a phase steering formation of the desired (111)-texture and a good *r*-value. The structure of elongated (stretched) crystallites formed in this process permits a check to be made on the described mechanism of precipitation that has taken place during the initial reformation of the structure.

For decarburizing or influencing the surface condition or for other reasons, steels are annealed as *open coils*. In such annealing the single wraps of the coil are kept apart by a wire during rewinding, so as to permit the gas employed to react more readily on the strip surface. Carbon contents of a few ppm can be achieved by this method. Such decarburized strips are used e.g. for single layer direct (white) enamelling.

Instead of annealing discontinuously in a bell-type furnace, *continuous annealing* can be performed in continuous *annealing furnaces* with high heating and cooling rates and a substantially-lower holding time at the annealing temperature. The carbon remaining in solution during the quick cooling may be precipitated by an in-line annealing treatment at about 400 °C with resulting improvements in aging behavior and mechanical properties of the continuously-annealed strip. Special precautions in the preceding steps of steel making (attainment of low contents of manganese, sulphur and oxygen), hot rolling (high coiling temperatures of more than 700 °C) and cold rolling are necessary for a favorable recrystallization behavior and the formation of a texture favoring deep drawability [5].

Skin pass rolling (*temper rolling*), which is re-rolling with minor degrees of thickness reduction (below 2%), eliminates the yield phenomenon which causes undesirable stretcher strains to appear in cold deformation. Skin pass rolling helps attain surface finish requirements (see above) and flatness is improved. The degree of reduction in skin pass rolling depends on sheet thickness, about 1% being the figure generally used. Sheet and strip made of steel having especially-low carbon and nitrogen contents or bonding of these elements by titanium or niobium (see IF-steel above), and therefore not showing a yield phenomenon, is skin pass rolled only to such a degree that the desired flatness and finish (roughness) are obtained.

Characteristic Steel Grades Proven in Service

Table C 4.2 lists the most important unalloyed low-carbon steels for cold rolled strip and sheet, together with their mechanical properties, as defined in DIN 1623 Blatt 1 [6].

By use of continuous casting (see above) the steel St 13 is also produced as a killed steel (like steel St 14), which is designated **RR St 13**. Confusion with the rimmed quality U St 13 can thus be prevented; the **latter** steel continues to be

Table C 4.2 Steel grades and mechanical properties of mild unalloyed steels for cold-rolled strip and sheet (for further details see DIN 1623 Teil 1 [6])

Steel grade Designation	Deoxidation	Tensile strength N/mm^2 max.	Upper yield stress N/mm^2 max.	Elongation after fracture $A_{80\,mm}$ % min.	Hardness HRB max.
St 12	Any	370	240	32	55
USt 13	Rimmed	350	225	34	52
RRSt 13	Special deoxidation	330	210	36	50
St 14	Special deoxidation	320	195	39	45

produced for those applications which require, for example, the special properties of the rimmed steel for surface treatment (rimmed surface zone of pure ferrite, see also D 3). Supplementary to Table C 4.2 in Fig. C 4.1 the ranges of normal anisotropy and strain hardening are listed for the various cold-rolled flat products made from these mild steels. The very wide range of values available in steel St 14 shows the far-reaching possibilities for application of this steel.

Parts needing extremely difficult forming operations also may be produced from the special deep drawing steel (IF-steel) included in Fig. C 4.1 and described above. This steel also possesses, in addition to the favorable r- and n-values, an extremely low yield stress (below 150 N/mm^2). For further details on this steel, see above and B 8.

C 4.1.2 Flat Rolled Products of Structural Steels for General Purposes, and of High-strength Steels

For some specific applications, formed parts produced from flat stock by cold forming often must have higher strength properties in the final state than can be attained in such parts made of unalloyed mild steels. For such parts, steels that possess higher yield stress and tensile strength values are required; these requirements apply to hot-rolled as well as to cold-rolled sheet and strip.

C 4.1.2.1 Hot Rolled Flat Products

Required Properties

The automotive industry in particular sets the standards for required properties in parts of the car body as well as in the chassis unit and the suspension unit. Reduced thicknesses are aimed at for these components to attain lower vehicle weights for fuel economy. Improvements in the payload-to-dead weight ratio are aimed at for utility vehicles such as mobile cranes, and dump trucks and for rolling stock. Compensation for wall thickness reduction in these components must be provided by an *increase in strength property values*.

The increased strength of the steel impairs formability and toughness (ductility) properties and affects processing behavior during forming and in welding. In principle, strength and toughness properties are linked inversely with one another so that the increase in strength can be taken only so far as minimum values of processing properties (such as *cold formability* in particular) permit. This inverse relationship exists quantitatively to very different degrees, depending on the strengthening mechanism employed with the various grades of high tensile steels (see B 1).

Characterization of Required Properties

Strength properties are tested by *tensile tests* employing the usual procedures. *Cold formability* as a processing property is characterized, as with the mild unalloyed-steels, by a series of different *test methods*. In addition to values from tensile tests (e.g. yield stress, tensile strength, elongation after fracture, r-value, n-value) and hardness tests, other test methods mainly simulating tests (modified Erichsen cupping test and cylindrical cupping tests) are employed with many variations. Shape and position of the forming limit curve also permit assessment of formability (see B 8).

Measures for Attaining Required Properties

Methods of increasing steel strength through modification of physical metallurgy have been described in connection with hot rolled products (predominantly hot strip and plate) (see B 1 and C 2). These methods use all qualified *strengthening mechanisms*: solid solution strengthening (including that effected by hardening), raising the density of dislocations (strain hardening), precipitation hardening and grain refinement.

Single effects or combinations of two or more of physical metallurgical mechanisms may be utilized since such combinations act in addition to each other [7]. Increased strength in *structural steels* [8] in the *hot rolled* condition is produced by an increased proportion of *pearlite* in the ferritic matrix in comparison to the amount in mild steels. As detailed in C 4.1.1 strengthening of the matrix is effected by attaining a *solid solution*, essentially including manganese, silicon and carbon and – in aluminum-deoxidized steels – by *grain refinement* due to the controlled precipitation of aluminum nitride, in combination with a rolling mill technology involving accelerated cooling to obtain fine-grained steel structure.

Similar remarks may be made about *weathering structural steels* which represent to some extent a sideline of the general structural steels. Despite a comparable microstructure, these steels differ in their chromium, copper and vanadium contents. These differences cause the formation of compact oxidic protective layers on the steel surface under the influence of the weather and result in an increased resistance to atmospheric corrosion (see C 2).

In comparison to the hot-rolled condition, improved processing properties in cold forming may be produced by a normalizing annealing or a normalizing rolling (formerly rolling at controlled temperatures) process [9–12]. In such processing, the *microstructure* is made *more uniform* and the *grain refining effect* is intensified.

In addition, the sulphur content may be reduced and/or the shape of the sulphides may be controlled by alloying and rolling operations. Stretching of sulphides to long bands in rolling is thus avoided or at least greatly reduced, together with their unfavorable effect on transverse and vertical properties.

Further improvements in formability and weldability requires use of steels other than general structural steels and comparable steel grades. Such improvements can be realized by a diminution of the pearlite content in the microstructure or a complete *elimination of pearlite* by reducing the carbon content to below 0.10% (low-pearlite or pearlite-free fine-grained steel) [13]. The increase in strength due to a harder second phase (pearlite) and of the reduction in solid-solution strengthening by the decreased carbon content are replaced by a combination of *grain refinement* and *precipitation hardening*, based on the effects of carbides or carbonitrides of the microalloying elements vanadium, niobium or titanium. Temperature controls employed in rolling and coiling in hot wide-strip mills and the controlled degree of reduction (pass sequence) are characteristics of thermomechanical processing (see C 2.4.2). Yield stress values of more than 500 N/mm^2 may be realized by these methods.

The microstructure of the microalloyed low pearlite steels results, despite a high strength, gives improved cold formability (Fig. C 4.2), welding and ductility (toughness) properties in service.

Much higher minimum values for the yield stress of from about 700 to 1000 N/mm^2 can be achieved by quenching (in a liquid) and tempering of the sheets before or after cold forming [14].

In addition, a high-tensile steel has been developed with a microstructure of nearly *carbon-free bainite* produced by thermo-mechanical treatment. This steel grade has already been tested in many applications. The yield stress lies between 700 and 850 N/mm^2 [15, 15a, 16].

Steels possessing a *purely ferritic matrix*, which are therefore, especially easy to form, may be strengthened be a suitable second phase in such a way that the cold formability is relatively little impaired. These so-called *dual-phase steels* attain their optimized microstructure [17–19] consisting of ferrite with from about 5 to 30% martensite dispersed in patches (Fig. C 4.3), by a special alloying and production technique during hot rolling or annealing. Martensite is obtained by accelerated cooling of a very low-carbon steel from the partly-austenitic state, i.e. from the

Fig. C 4.2. Double folding test sample (handkerchief test sample) **(a)** and practical operation simulating test of U-section **(b)** of 6-mm thick hot-rolled strip of microalloyed thermomechanically treated steel with yield stress of min. 380 N/mm^2.

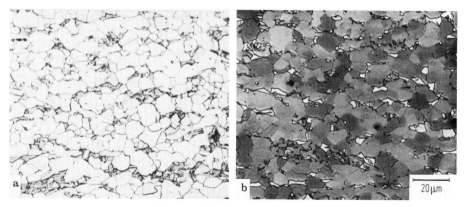

Fig. C 4.3. Microstructure of hot-rolled strip of a molybdenum-alloyed dual-phase steel. **a** Nital etching, **b** sodium thiosulphate etching.

$\alpha + \gamma$-two-phase region. In addition to martensite, small amounts of bainite also may be formed, and some residual austenite is also possible. Depending on the alloy composition (Table C 4.3) which determines the critical cooling rate, the dual-phase condition may be produced from rolling temperature immediately after hot-strip rolling [20] or by hot-strip annealing in a continuous-annealing furnace with accelerated cooling.

Dual-phase steels have distinctly better formability values compared to hot-rolled or normalizing-annealed or normalizing-formed hot strip or sheet of the same strength, in thicknesses up to about 5.0 mm. In the processed condition, tensile strength values of from 400 to 1000 N/mm^2 are achieved with simultaneous low yield stress values (yield stress to tensile strength ratio of 0.4 to 0.6). Fatigue strength and, in steel grades produced by hot strip annealing, weldability also are improved comparatively.

The favorable processing properties are based mainly on the ferritic matrix which is fine-grained, very low in carbon content and strengthened by substitution elements. The chemical composition, especially the amounts of the elements molybdenum, chromium, manganese and silicon, determine the transformation behavior of the steel and, in combination with the carbon content, define the hardness of the martensite formed from the residual austenite (see also B 8.3.2).

The part of the production process that covers cooling from rolling temperature requires a chemical composition that results in a transformation behavior corresponding to an extended incubation period in the pearlite formation. With lower alloying contents, i.e. using steels having less-sluggish transformation, a very low coiling temperature of below 400 °C is necessary after strip rolling.

Equipment for continuous annealing of hot strip permits the use of a steel with a chemical composition having reduced alloy content because of the very high cooling rate.

Steels with degenerated pearlite occupy an intermediate position between micro-alloyed and dual-phase steels. Up to 30% pearlite and upper bainite are dispersed in the ferritic matrix of these steels. Yield stress values of about

Table C 4.3 Chemical composition (indicative values) of some hot-rolled and cold-rolled (normal strength and) high strength steels for flat material intended for cold forming

Steel grade	Chemical composition				Remarks
	% C	% Si	% Mn	% other characteristic elements	
	Hot rolled (and heat treated if required)				
Structural steels for general purposes (normal strength steels) (St 37, St 44, St 52)	< 0.20	0.3 ... 1.5	< 1.5	Al: 0.020	
Low pearlite micro-alloyed steels	< 0.10	0.8 ... 1.4	< 0.5	Nb: < 0.1 Ti: < 0.2	thermomechanical treated
	< 0.22	1.2 ... 1.7	< 0.5	(V, Al)	normalized
Water quenched and tempered steels	< 0.20	< 1.3	< 0.8	Cr: < 1.5 Mo: 0.6 Ni: 1.5 Zr: 0.1 (B, Co, V)	
Low carbon bainitic steels	< 0.08	< 2.0 (3.0)	0.3 (0.6)	Mo: 0.3 B: 0.003 Ti: 0.02 (Nb, Cr)	
Dual-phase steels	0.06	0.3 (2.0)	0.7 (1.5)	Cr: 0.6 Mo: 0.4	produced from rolling heat
	0.12 ... 18	1.2 ... 1.8	0.3 ... 0.6	V: 0.15 (Nb)	continuously annealed
Steels with incomplete pearlite	0.14	0.10	1.1		
	Cold rolled (and heat treated if required)				
Structural steels for general purposes (normal strength steels) (St 37, St 44, St 52)	< 0.20	0.3 ... 1.5	< 1.5	Al: 0.020	
Steels for quenching and tempering	< 0.7	< 0.8 (2.0)	< 0.5	Cr: rd. 1 Mo: 0.3 (Ni, V)	
Micro-alloyed steels	< 0.1	0.8 ... 1.8	< 0.5	Nb: < 0.1 Ti: < 0.2 (V)	
Steels with increased phosphorus content	< 0.08	< 0.5	–	P: to 0.1	
Heavy temper rolled steels	< 0.08	< 0.5	–		
Partly recrystallized steels	< 0.08	< 0.5	–		
Dual-phase steels	0.08	1.3	0.5	Cr: 0.5	
	< 0.6 ... 0.1	0.3 ... 1.0	0 ... 1.0	P: to 0.15 N: to 0.02	alternative
	< 0.1	2.2	0 ... 0.3	Ti: 0.1 B: 0.001	compositions

420 N/mm² are obtained with no addition of micro-alloying elements by use of coiling temperatures between 400° and 500 °C. These steels are superior because of their very high ductility, which is due to the self-annealing effect in the slow cooling coil [20a].

Characteristic Steel Grades Proven in Service

General structural steels are made in grades defined by DIN 17 100 [8] and are dealt with extensively in C 2; indicative figures for chemical compositions of these steels are reproduced in Table C 4.3 for the sake of completeness. Specifications on weathering steels are also included in C 2.

High-tensile steels are detailed in Table C 4.3 which also lists their chemical composition. The mechanical and technological properties of the high-strength, fine-grained steels in the hot-rolled condition, intended for cold forming with a thickness of up to 16 mm, are listed in Table C 4.4 [10]. Differences between the properties of the normalizing annealed and the thermo-mechanically treated steels are discussed in the comments on the physical metallurgy principles included in C 2.4.2. These steels are used primarily to make components of vehicles [21] for example, for passenger car frames and chassis parts, truck side members, and wheel discs. Cold formability is a necessary but not vital property in some applications, such as in construction of large-diameter welded pipelines for transportation of inflammable gases and liquids (see C 25) [13, 22].

Table C 4.4 Grades and mechanical and technological properties of hot-rolled fine-grained steels for cold forming, thicknesses up to 16 mm (for further details see SEW 092 [10])

Steel grade Designation	Yield stress N/mm² min.	Tensile strength N/mm²	Elongation after fracture for thicknesses				Mandrel diam. D in folding (bending) test[a] (a = thickness of test piece)
			< 3 mm		≥ 3 mm ≤ 6 mm	≥ 3 mm	
			A_{50mm}	A_{80mm}	A_{50mm} % min.	A	
QStE 260 N	260	370 ... 490	26	24	28	30	$D = 0\,a$
QStE 340 TM	340	420 ... 540	21	19	23	25	$D = 0.5\,a$
QStE 340 N		460 ... 580	23	21	25	27	
QStE 380 TM	380	450 ... 590	20	18	21	23	$D = 0.5\,a$
QStE 380 N		500 ... 640	21	19	23	25	
QStE 420 TM	420	480 ... 620	18	16	20	21	$D = 0.5\,a$
QStE 420 N		530 ... 670	20	18	21	23	
QStE 460 TM	460	520 ... 670	15	14	18	19	$D = 1\,a$
QStE 460 N		550 ... 700	18	16	20	21	
QStE 500 TM	500	550 ... 700	13	12	16	17	$D = 1\,a$
QStE 500 N		580 ... 730	15	14	18	19	
QStE 550 TM	550	610 ... 760	11	10	14	15	$D = 1\,a$

[a] The test piece is to bent through 180°

C 4.1.2.2 Cold Rolled Flat Products

Required Properties

Cold-rolled sheet and strip of high tensile steels discussed here must have the same properties that have been discussed in connection with cold-rolled flat products of mild steels in C 4.1.1.2. In addition to the mechanical properties, cold formability is of prime importance. Requirements of high-tensile steels are no different from those for cold-rolled sheet and strip of mild unalloyed steels in regard to surface properties (kind, finish and cleanness of the surface) and suitability for surface treatment.

Characterization of Required Properties

High tensile steels are no different from mild steel cold-rolled sheet and strip as regards the test methods and the kinds of characteristic values to be determined for the mechanical properties, the simulating test methods, the application of the analysis of deformation or the recording of the microstructure (see C 4.1.1.2).

Measures for Attaining Required Properties

The production sequence for cold-rolled sheet and strip uses primarily the intermediate steps for hot-rolled strip. The first development step therefore used high-tensile hot strip methods for the subsequent cold rolling, annealing and temper rolling. The cold-rolling sequence however, was found to impair the strength properties of the hot-rolled condition as may be seen, for instance, in the micro-alloyed steel example shown in Fig. C 4.4. Despite the high yield stress values those of the elongation after fracture although reduced are nevertheless, still relatively good (Fig. C 4.5).

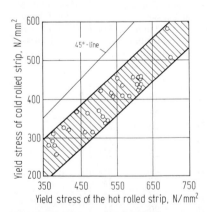

Fig. C 4.4. Decrease of the yield stress of micro-alloyed steels as hot rolled or thermomechanically treated strip by cold rolling, annealing and temper rolling. After [23].

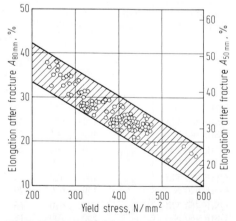

Fig. C 4.5. Decrease of the elongation after fracture of micro-alloyed steels in cold-rolled strip with increasing yield stress. After [23].

The second step in the development of high-tensile cold-rolled sheet and strip consisted of applying a carefully-directed influence in the cold rolling process.

Several groups of steel grades which now exist employ the strengthening mechanisms (see B1) and sometimes utilize them in combination, resulting in different properties. These groups include manganese-silicon-steels, steels with an increased content of phosphorus, micro-alloyed steels, heavily temper-rolled steels, partly-recrystallized steels, quenched and tempered steels, dual-phase steels and bake-hardening steels.

Attainable properties as well as existing differences in property combinations may be taken from Fig. C 4.6 in which test values for elongation after fracture and tensile strength of the various steel grades are plotted. Values for the mild, unalloyed deep-drawing steel St 14 (see Table C 4.2) are included for comparison. With increasing tensile strength, a decline in elongation after fracture, here representing cold formability, must be accepted.

The hatched areas characterizing the different steel groups make clear which broad range of properties can be covered. For the same tensile strength values with highly-different elongation values $A_{50\,mm}$ ranging from roughly 5 to 35% are attained. Practical utilization of these materials today centers on a yield stress range of about 260 to more than 350 N/mm², corresponding to a tensile strength of between about 400 and 700 N/mm².

Details of strengthening mechanisms are contained in B 1, as mentioned above. Their use in the field discussed here has led to the following developments:

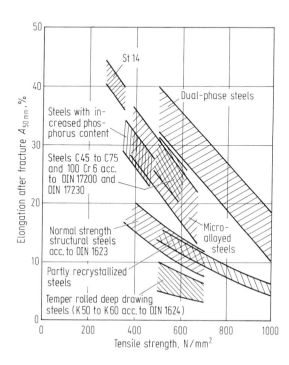

Fig. C 4.6. Correlation between tensile strength and elongation after fracture of cold-rolled sheet of different high-strength steels and mild steel St 14 according to DIN 1624 Blatt 1 [6]. (For the other standards named see [30, 32–34].)

Starting with mild deep-drawing steels (see C 4.1.1.2) tensile strength may be raised by *solid solution hardening*, above all with manganese and silicon contents. Steel grades thus obtained are comparable to the steels listed in C 4.1.2.1 (Table C 4.3) as regards their microstructures and chemical compositions. However, these cold-rolled and recrystallizing-annealed sheet and strip products show smaller strength increases than hot-rolled material.

Phosphorus may also be used for solid solution strengthening. Steels with increased phosphorus contents (up to roughly 0.1%) attain yield stress values up to about 430 N/mm². Moreover, phosphorus does not cause changes in the recrystallized texture of cold-rolled and annealed strip (see B 8), resulting in good deep drawability characterized by high *r*-values ($r_m = 1.6$).

Cold-rolled sheet and strip of steels subjected to *grain refinement* and *precipitation hardening* by *micro-alloying* with vanadium, niobium or titanium have yield stress values of 260 to 420 N/mm², and thus cover a wide range [23]. As a result, the ratio of yield stress to tensile strength increases with the raised tensile strength (Fig. C 4.7).

Increasing the degree of temper rolling to more than 6% raises the yield point beyond that of the cold-rolled and recrystallizing-annealed condition due to strain hardening of the structure by *increasing the density of dislocations*. Depending on the application, higher tensile strength values of up to 700 N/mm² may be obtained by still further increasing the degree of temper rolling.

Time-temperature-controlled recrystallizing annealing of cold-rolled strip mosty in the continuous annealing furnace, may be used to produce a microstructure which is the result of only *partial recrystallization* or recovery ("full-hard" condition). The cold-rolled and thereby strain-hardened structure is partially

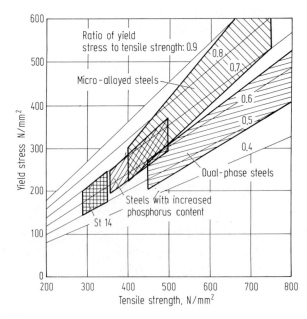

Fig. C 4.7. Correlation between tensile strength and yield stress of cold-rolled sheet of different steels.

retained and produces the higher strength properties, which are however, linked with strongly-reduced cold formability.

Steels having carbon contents of up to 0.7% and increased contents of alloying elements can be produced with a *hardened and tempered microstructure* accompanied by solid solution strengthening by carbon in supersaturation. Such steels have increased density of dislocations and refined grain resulting in high strength values. These steels however, can be utilized in sheet and strip for cold forming only if cold formability is improved by a carefully-directed rolling and annealing treatment. To this end, starting from a hot-rolled strip with preferably a bainitic microstructure, annealing is carried out after cold rolling in such a way that the cementite is spheroidized to 100% [24, 25], causing a highly-improved elongation after fracture.

For specific applications low alloy steels may be quenched and tempered with isothermal transformation resulting in a bainitic microstructure providing tensile strength values of 1500 N/mm^2. These steels possess an elongation after fracture $A_{50\,mm}$ that is increased by 5 to 10%, given equal tensile strength, compared to steels with a quenched and tempered microstructure [26].

Dual-phase steels possess favorable formability properties at relatively high tensile strength values as was described in the previous section under hot-rolled flat material. Cold rolled sheet made from these steels, especially in sizes below 1 mm thickness, is suitable for use in automobile construction. The sheet is produced after cold rolling by continuous annealing within the $\alpha + \gamma$-two-phase region in a continuous-annealing furnace at temperatures of 750 to 900 °C, followed by accelerated cooling. This process enables tensile strength values of 1000 N/mm^2 to be obtained.

The favorable cold formability of dual-phase steels in the form of cold-rolled sheet may be deduced from the following material behavior [27–29]:
- Low (initial) yield stress values (about 200 to 500 N/mm^2), measured against high tensile strength values (about 400 N/mm^2 to 1000 N/mm^2) and comparatively favorable values of elongation after fracture and consequently low spring-back after stamping.
- Absence of a pronounced yield point (no yield point elongation) and, therefore, freedom from stretcher strains.
- High capability for strain hardening (high strain hardening exponent), especially in the range of low deformations of below 5%.
- Possibility of an additional yield stress increase by about 100 N/mm^2 from paint baking (at roughly 200 °C) in the finished part.

Examination of the favorable strain hardening and formability properties of dual-phase steels suggests two reasons: Due to the martensite transformation, with resulting internal stresses of the structure, mobile dislocations are formed in the ferritic matrix. Mobile dislocations also are produced during a forming operation by a supplementary transformation of residual austenite into martensite.

Stress-strain-diagrams (Fig. C 4.8) of sheets of solid solution-hardened and precipitation-hardened steels show a behavior similar to that of mild deep-drawing steels with intensive strain hardening after recrystallizing annealing in bell type or

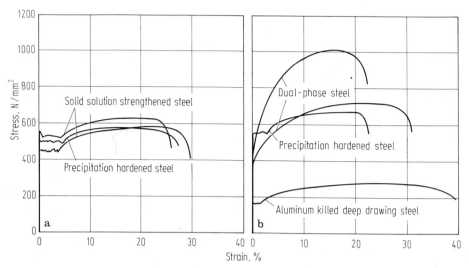

Fig. C 4.8. Stress-strain curves of cold-rolled sheet of different high-strength steels. **a** As annealed in bell-type furnace at 650 °C for 3 hrs slowly cooled and **b** as annealed in continuous-annealing furnace at 800 °C for 3 min. air cooled.

continuous annealing furnaces. Curves of dual-phase steels reflect the particular strain hardening and formability behavior described above.

After treatment in the continuous annealing furnace dual-phase steels have more or less marked aging effects caused by the carbon content dissolved in the ferrite. The result of this condition is that, for instance, in the course of the normal paint baking procedure during production of automobile body parts there is an additional increase of the yield stress of up to 100 N/mm^2 (a phenomenon known as "bake-hardening"). The effect can be controlled by varying the chemical composition, cooling rate and subsequent tempering treatments. This control aims at such a supersaturation of the carbon in the ferrite that natural aging does not take place during normal raw material storage periods and consequently values of the bake-hardening effect of 50 N/mm^2 are normal.

Steels with other than dual-phase microstructures may also be produced utilizing the bake-hardening effect. To this end physical metallurgy requires a certain content of solute carbon defined by the total carbon content of the steel, the cementite repartition and the cooling rate after recrystallizing annealing. The bake-hardening effect can be utilized in mild, unalloyed steels as well as in high-strength steels with an increased phosphorus content after annealing in both bell-type and continuous-annealing furnaces [29a–29o].

Characteristic Steel Grades Proven in Service

Steel grades with higher strengths resulting mainly from solid solution strengthening due to manganese and silicon are listed in DIN 17 100 [8] and can be supplied in sheet form with properties listed in DIN 1623 Teil 2 [30]. The material condition after cold rolling and recrystallizing annealing of sheet and strip, however, is marked by a smaller strength increase than in the hot-rolled condition (see C 4.1.2.1).

Steels with solid solution strengthening by phosphorus are alloyed at present with 0.03 to 0.1% P, giving yield stress values of at least 260 N/mm^2. Mechanical properties are given in Figs. C 4.6 and C 4.7.

Reference data on the chemical composition of micro-alloyed steels for cold-rolled sheet and strip are given in Table C 4.3. Mechanical and technological properties are listed in Table C 4.5 [31]. The capability of these steels both for cold rolling and cold forming is limited as the strength increases. However, cold working is possible basically with suitable changes to the tooling used. The steels are used primarily for load-bearing and functional parts with medium degrees of deformation. Resistance welding can be used without difficulty, but in shielded-arc welding conditions may need to be modified. With strips that are cold rolled after recrystallization reference is made to DIN 1624 [32]. For the processing discussed in this chapter, slightly temper-rolled (LG) and cold-rolled conditions (K 32 to K 60) are preferred. The elongation after fracture $A_{50\,mm}$ drops from the values of the annealed condition, about 30%, down to below 5% (Fig. C 4.6). Such high-tensile sheets are unsuitable for parts that demand high stretch-forming operations, due to the limited cold formability.

Partly recrystallized steels are also represented in Fig. C 4.6 with a range of values for tensile strength and elongation after fracture.

Cold-rolled sheet and strip with a microstructure in the quenched and tempered condition are usually produced from unalloyed steels with carbon contents of about 0.45 to 0.75% (Ck 45 to Ck 75 in DIN 17 200 [33]). Mechanical properties of these steels are also seen in Fig. C 4.6. Alloyed steels are also used, however, for example, steels used for ball and roller bearings containing about 1% and 1.5% Cr (100 Cr 6 according to DIN 17 230 [34]).

Dual-phase steels are available in several grades, depending on production methods, used to attain the characteristic microstructures, and indicative values for chemical composition are seen in Table C 4.3. A synopsis of mechanical properties is shown in Figs. C 4.6 and C 4.7. Cold-rolled dual-phase steels are also suited to

Table C 4.5 Grades and mechanical and technological properties of cold-rolled sheet and strip (thickness below 3 mm) with guaranteed minimum yield stress values for cold forming (for further details see SEW 093 [31])

Steel grade	Yield stress N/mm^2 min.	Tensile strength N/mm^2	Elongation after fracture $A_{80\,mm}$ % min.	Mandrel diam. D in folding (bending) test[a] (a = thickness of test piece)
ZStE 260	260	350 ... 480	26	$D = 0\,a$
ZStE 300	300	380 ... 510	24	$D = 0\,a$
ZStE 340	340	410 ... 540	22	$D = 0.5\,a$
ZStE 380	380	440 ... 580	20	$D = 0.5\,a$
ZStE 420	420	470 ... 620	18	$D = 0.5\,a$

[a] The test piece is to be folded (bent) by 180°

production of stretch-formed body parts, since zones with little deformation in such parts derive high strain hardening values and thus the desired stiffness.

The testing and introductory phase for cold-rolled sheet and strip made from high-tensile steels has not yet been completed so that a safe evaluation of the different steel grades can not yet be completed. The behavior of these steels in processing may be estimated from the comparative judgment in Table C 4.6.

The following high-tensile cold-rolled sheets are being tentatively used in the automotive industry today: micro-alloyed steels with increased phosphorus content, dual-phase steels and bake-hardening steels. The characteristic values of tensile properties for these steels can be derived from the comparative drawing in Fig. C 4.9 (reference point $R_m = 415$ N/mm^2); properties of bake-hardening steels are listed in [34a].

C 4.1.3 Flat Products of Stainless Steels

For applications that are subject to particular corrosive conditions, cold-rolled steel sheets for cold forming are available with rust and corrosion resistant (stainless) properties due to their chemical composition which includes high amounts of chromium, nickel and molybdenum. For these applications it is necessary to distinguish between ferritic and austenitic steels (see C 13).

Numerous applications in the car building and construction industries, chemical engineering and household appliance industries demonstrate the great importance of flat products for cold-formed parts especially those made from ferritic steels.

Details of alloy content, microstructure and properties of these steels are to be found in C 13. Results of tests applied to characterize cold formability cannot be distinguished from those for unalloyed steels.

Table C 4.6 Qualitative comparison of some processing properties of cold-rolled flat material of high strength steels with tensile strengths of 400 to 450 N/mm^2

Type of steel	Processing property[a] in		
	Cold forming		Resistance welding
	Deep drawing	Stretch forming	
Structural steels for general purposes (see DIN 1623 Teil 2 [30])	− to ○	− to ○	○
Steels with increased phosphorus content	+ to + +	○	○
Low-pearlite micro-alloyed steels	○ to +	○	+
Drawing steel, heavier temper-rolled	+	−	+ +
Dual-phase steels	○ to +	+ +	− to + +[b]
Steel St 14 for comparison (see DIN 1623 Teil 1 [6])	+ +	+ +	+ +

[a] + + = very good; + = good; ○ = fair; − = limited; − − = very limited
[b] Increased alloy contents in steels, yielding a dual-phase structure from rolling heat, possess higher hardenability and reduced weldability, therefore

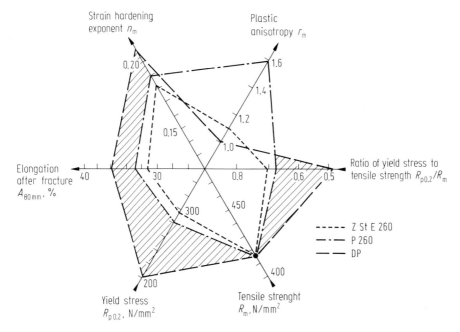

Fig. C 4.9. Comparison of the properties of cold-rolled sheet of micro-alloyed steel (ZStE 260, see Table C 4.5), steel with increased phosphorus content (P 260, see Table C 4.3) and dual-phase steels (DP, see Table C 4.3) (tensile strength R_m of all steels = 415 N/mm²).

Because of their peculiar strain-hardening behavior, austenitic steels possess good cold formability so that – as opposed to unalloyed steels – no reason exists for the development of steel grades suited particularly to cold forming. In this context therefore, the reference to steels dealt with in C 13 should be sufficient. About the same holds true for ferritic steels. Attention should be paid, however, to the difference in the strain hardening capacity of ferritic and austenitic steels, which becomes immediatey apparent in cold rolling, and more especially, in cold processing by deep drawing and stretch forming (see B 8).

In addition to flat products made completely of stainless steels such products are made from steels clad with stainless steel only on their surfaces (see B 12.4.2 and C 4.2.3.5).

C 4.2 Surface-coated Flat Rolled Products

Due to the chemically-base character of iron, unalloyed and low-alloy steels can oxidize and form rust on their surfaces. To preserve the utility value of objects and components made from such steels as long as possible they must be protected against the attack of aggressive media. Moreover, the material can be given a decorative look by surface treatment. Where it is technically possible and economically feasible, this surface treatment is carried out by the steel producer prior to delivery. *Numerous processes* are used for these surface treatments today, including metallic, non-metallic inorganic and organic coatings, depending on processing technology and final applications.

General Requirements for the Properties of Use

Cold formability is of prime importance, and for flat products without surface treatment, is largely determined by the basic material. The same is true for the *mechanical properties* linked to cold formability. Good cold formability of the basic material, however, can only be utilized effectively if the coating has not been overstrained by the forming operation, i.e. does not spall off in places. Thus *adhesion* is also an important requirement and is not important only for cold formability but also for the prime purpose of surface treatment, that is, corrosion protection as described above. Satisfactory *corrosion resistance*, therefore, is required from surface-coated flat products. Further demands are added to these four basic requirements, depending on product and final use, especially concerning such processing properties as weldability.

These requirements are not restated in the following text for each of the various surface-treated flat products. But they will be dealt with in sequence of measures for attaining required properties and supplemented where necessary.

The properties of flat products with surface treatment depend largely on the basic material and its preparation for the treatment, if one disregards corrosion resistance (see B 12). In principle all steels discussed in C 4.1 can be considered as basic material, that is primarily the mild steels for cold forming but also all the high-tensile steels mentioned above. Depending on the kind of surface treatment, it may be necessary to limit the properties of the basic materials in one way or another, for instance, the chemical composition, to meet requirements for properties of the surface-coated products. These limitations will be explained in discussing the various products.

Characterization of Required Properties

To characterize *cold formability* in principle the same methods are used as are applied to flat products without surface treatment. Frequently simple technological tests are also used (Fig. C 4.10 [35]). The cupping test is applied, as well as other methods such as measurement of the r- and n-values and the forming limit. In determining the forming limit, however, other failure criteria such as abrasion or flaking may be decisive. *Mechanical properties* are characterized by data from the normal tensile test (see B 1), including yield stress, tensile strength and elongation after fracture. Frequently the tensile test is replaced by a hardness test using different methods depending on product and thickness. *Adhesion* is tested by the folding test either in the simple way or in the double intensified test ("handkerchief-test"). Other test methods may be used to simulate stresses in plant processing.

There are no test methods that are for the measurement of *corrosion resistance* equally suitable for all the different possible products, due to the diversity of corrosive media and working conditions (e.g. temperature) to be considered. Test methods are therefore combined with corrosive stresses simulating more or less closely those encountered in service. Protection against corrosion in itself that

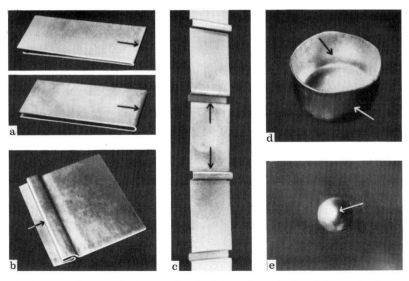

Fig. C 4.10. Different methods for testing the deformation behavior of zinc or aluminum coated sheet. **a** Folding; **b** Lock seam; **c** Profiling; **d** Cupping; **e** Ball impact indentation.

means disregarding e.g. adhesion or cold formability usually increases with increasing thickness of the coating. Therefore, to characterize corrosion resistance the thickness of the coating is often measured by dissolving the coating, weighing it in relation to the area and giving the data in g/m^2. Methods for dissolution vary and are devised in such a way that additional information becomes available. Recently non-destructive and metallographic methods have been utilized to measure layer thicknesses.

In the following text details of possible tests are dealt with only if they are important for, and specific to, the particular product and are beyond the scope of the above explanations.

C 4.2.1 Flat Rolled Products with Metallic Coatings

In practice, hot-dip processes and electrolytic deposition are of prime importance for applying metallic coatings on steel (see B 12.2 and B 12.3). Flat material thus produced shows generally similar behavior under corrosive stresses, although specific differences exist in the structures of the coating layers and consequently in processing and service behavior – due to differences in the properties of the coating metals.

C 4.2.1.1 Zinc Coated Flat Products

Zinc occupies the first place among the metals used for surface coating [36–40]. Presently, world wide roughly 20 Mill. t of steel are zinc coated annually.

A survey of *production processes* used for zinc-coated sheet and strip and characteristics of zinc coatings are listed in Table C 4.7. In hot dip galvanizing (zinc coating by immersion) it is necessary to distinguish between the discontinuous galvanizing of single pieces and the continuous galvanizing of strip. In electrolytic deposition of zinc, as in hot-dip galvanizing, the metal can be deposited on each side, on one side or with different thicknesses on both sides (differentially zinc coated) depending on the layout of electrodes and control of the electric field. The following text deals mainly with production of zinc-coated *strip* intended for processing by forming, welding and further surface treatments.

Basic material for zinc coating includes practically all the usual steels which are produced as cold rolled strip. However the silicon content of steel to be zinc coated must be limited since it can lead to undesirable reactions in hot-dip zinc coating (galvanizing) (see B 12.2.2). In Table C 4.7 the most important groups of the basic materials are listed in the order of cold formability and tensile strength qualities.

Table C 4.7 Survey of production and characteristic features of flat material with zinc and aluminum coatings

Process of zinc coating	coating thickness per side	*Aluminum coating process*	coating thickness per side
Electrolytical deposition of zinc two-sided different thickness[a] one-sided	1 … 15 μm	Hot-dip aluminum coating	15 … 50 μm
		Aluminum cladding by rolling	1 … 10% of sheet thickness
Hot dip zinc coating single pieces	50 … 200 μm	Vacuum vapor deposition	1 … 5 μm
continuous strip coating two-sided differentially coated[a] one-sided	10 … 50 μm		
Characteristics of the zinc coating		*Characteristics of the aluminum coating*	
Total thickness of coating Portion of zinc-iron alloy layer of coating thickness (in Galvannealing = 100%) Formation of zinc spangle spangle size surface relief crystallographic orientation		Total thickness of coating Portion of iron-aluminum alloy layer of coating thickness Composition of coating pure aluminum (type 2)[b] Al + 5 to 11% Si (type 1)[b] Surface condition	

Base materials		
Extra deep drawing steel Deep drawing steel Steel for drawing purposes Base steel High strength (elevated temperature strength) steels for drawing purposes Structural steels for general purposes	Increasing cold formability	Increasing strength

[a] One side has a higher nominal value for zinc coating than the other
[b] See p. 106

Required Properties

Regarding the required properties reference is made to page 98. In addition, suitability for resistance welding is required of zinc-coated sheet.

Characterization of Required Properties

The test methods discussed on page 98 for surface coated sheet are valid for zinc-coated flat products. Characterization of *corrosion resistance* by means of numbers however, remains problematic, due to the great number of corrosive media involved. Short-time (accelerated) tests or electrochemical examinations in the laboratory therefore, can yield only relative information. But results of field tests, in which corrosion losses of zinc occuring during outside storage under specific environmental conditions are recorded, are subject to relatively large scatter. Data on corrosion losses in different climates taken from the literature and deduced from the author's investigations, must therefore, be understood as being of informative value only (Table C 4.8 [41–43]). A characteristic of the corrosion behavior of zinc is the loss under certain conditions, which increases in proportion to the time of exposure. Consequently an initially-thicker zinc coating will last longer than a thinner coating as a corrosion protector.

Measures for Attaining Required Properties

Cold formability and, above all, *strength* of zinc-coated sheet and strip are determined substantially by the base material. However in contrast to uncoated annealed strip in the bell type furnace strip for zinc coating is run through a continuous-annealing furnace, resulting in marked differences in ductility and strength compared to uncoated steel. Smaller grain size, the non-development of a desired crystal orientation and the high cooling rate that may lead to quench aging impart to galvanized strip a generally increased tensile strength and decreased cold formability. To fulfill very high demands on cold formability therefore, measures must be taken in the steel plant or in the production of hot or cold strip and in galvanizing that match roughly the formability of the classical mild steel sheet grades [43a].

The same possibilities exist for the production of zinc-coated *sheet and strip with higher yield stress values* as for uncoated flat material (C 4.1.2.2). The limitations discussed below on the use of steel with silicon or phosphorus contents

Table C 4.8 Average corrosion losses of zinc (compared to unalloyed steel) in different climates

Climate	Zinc	Steel
	Loss in μm/year	
Country air	0.5 ... 2[a]	5 ... 25
Town air	2 ... 6	25 ... 50
Sea climate	2 ... 10[b]	25 ... 70
Industrial climate	3 ... 12[c]	25 ... 70

[a] In moist air, higher losses are possible
[b] Near the sea, higher losses are possible
[c] Under especially unfavorable conditions higher losses are possible

increased for strength reasons must, however, be given special attention here. Strips containing higher silicon percentages require corresponding changes in the operation of the continuous annealing furnace and in the composition of the zinc bath [44].

A special variant of high-strength zinc-coated strip is the so-called *"full-hard"* material which has been annealed in such a way that a recovery is required before zinc coating can proceed to ensure that most of the strength of the as-rolled strip is retained (see C 4.1.2.2). High demands for cold formability must not be made on this material, however.

Electrolytically zinc-coated sheet and strip have the same good cold formability as the uncoated flat material, especially since the same steel grades are employed because of the necessary production steps. If the cold rolled flat material has been cleaned carefully the electrolytically deposited zinc adheres so well to the base that the coated sheet and strip will withstand the same deformations as the initial product. Abrasion of zinc dust occurs only with higher deformations, because the zinc crystals are very small and intimately intergrown.

Cold formability of galvanized (zinc coated) sheet and strip is closely allied to the *adhesion* of the zinc coating during the forming operation. Good adhesion on one hand necessitates complete wetting of the steel surface by the molten zinc. On the other hand, such wetting leads to metallurgical reactions between steel and zinc even with the very short immersion times in the zinc bath. At about 5 s these times are 1 or 2 orders of magnitude shorter than those used in the galvanization of single pieces (see B 12.2.2). The very fast-acting diffusion between iron and zinc causes the formation of brittle intermetallic compounds. However, this formation can be impeded by addition of aluminum to the zinc bath to such an extent that only a 1 μm thick alloy layer is formed. Examples of a thicker and a desired thin alloy layer are shown in Fig. C 4.11. The ductile zinc coating is able to withstand heavy cold deformations without separation/flaking. In heavily-deformed regions, microscopic cracks may occur in the zinc layer (Fig. C 4.12), but the adhesion of the coating remains intact.

It is well known that concentrations of about 0.03% to 0.12% of silicon in zinc coatings gives rise to an anomalous growth of the alloy layer (Sandelin-effect, see B 12.2.2). Since the introduction of continuous casting the proportion of steel killed

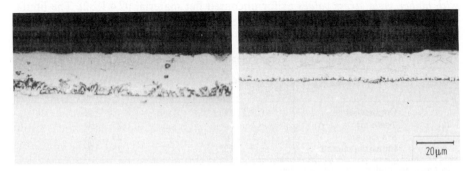

Fig. C 4.11. Formation of a thicker and a thinner zinc-iron-alloy layer in strip hot dip galvanizing (hot dip zinc coating).

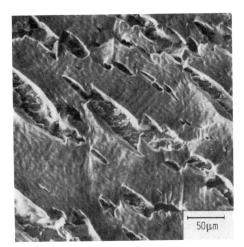

Fig. C 4.12. Crazing of a zinc coating caused by heavy forming operation of hot dip galvanized (zinc coated) sheet with excellent zinc adhesion on the steel (direction of deformation: ↕).

in the ladle by aluminum and inevitably containing a few hundredths of one percent of the element has increased steadily. In galvanizing sheets of continuously-cast steel, therefore, the adverse effect of silicon on the adhesion of the zinc coating can make itself felt under unfavorable conditions. Phosphorus exerts an influence similar to silicon [45], so it is necessary to take into account the joint influence of silicon and phosphorus on the alloy layer formation when flat products are to be produced from the recently-developed high-strength steels with increased phosphorus content (see C 4.1.2.2).

The good corrosion resistance of zinc-coated flat material is based on the fact that, under atmospheric conditions, zinc forms low-soluble basic zinc salts which adhere well to the metallic surface, and hence exert a good protective effect. Additionally, zinc offers cathodic protection to steel because, when defective points in the coating are attacked by a corrosive medium it corrodes more readily on account of its electrochemically more base character than the unprotected steel. The potential of this oxidation reaction is displaced electrochemically to such an extent that the corrosion of the steel ceases or is greatly retarded. This cathodic protection by zinc acts up to a distance of about 2 mm from the edge of the defect, and constitutes a special advantage of the corrosion mechanism of zinc at such defects; the effect is especially valid at cutting edges.

In through-alloyed (*galvannealed*) sheet (see B 12.2.2), corrosion may also lead to the formation of red rust owing to the presence of iron in the surface layer. However this effect does not impair the good corrosion resistance of the total protective layer.

The corrosion resistance of hot dipped zinc coatings is influenced to only a minor extent by small alloy additions to the zinc; higher additions, on the contrary, alter the corrosion chemical behavior of the coating. An example of this changed behavior is *Galfan*, which is coated with an alloy of zinc and 5% Al and imparts a greater corrosion resistance to the surface-treated sheet than a pure zinc coating [45a]. Another example is *Galvalume* whose coating consists of an alloy of about 45% Zn and 55% Al; this point is discussed in the section on aluminum coatings (see C 4.2.1.2).

Alloys containing about 10% Ni have been developed recently for electrolytically deposited coatings having improved corrosion resistance [45b]. To improve paint finishing electrolytically-deposited coatings of zinc-iron alloys have also been introduced recently [45c, 45d].

An additional surface treatment for structural members made of zinc-coated flat material becomes necessary whenever improved corrosion protection and/or an improved decorative appearance are required. The best-known example is a supplementary painting which is normally preceded by a phosphate and primer treatment.

Behavior in resistance welding of surface-coated sheet is essentially influenced by the interaction between the coating and the welding electrode. Optimization of welding conditions is aimed at securing a perfect weld and a satisfactory electrode life. These requirements can be obtained by the use of electrodes of copper-chromium-zirconium alloys, suitable design of electrodes, adequate electrode cooling, increase of electrode pressure and reduction of welding time.

Characteristic Steel Grades Proven in Service

In discussing the most important properties of zinc-coated flat material, more general remarks have already been made concerning certain steel types and steel grades. At this point, it is proposed to summarize once again the characteristic steel grades used in surface treatment with zinc. Literature is also referred to [46–48], in which the features important for fabrication of electrolytically- or hot-dipped zinc-coated sheet or strip are discussed. The same steel grades are basically available for electrolytic zinc coating as are processed in the non-surface coated flat form. Mild, readily cold-formable sheet grades are standardized in DIN 1623 Blatt 1 [6]. If continuous hot-dip galvanizing is employed, and assuming the same chemical composition of the steel strength values and somewhat reduced cold formability will be encountered. Four different sheet grades with increasing cold formability listed in Fig. C 4.13 for example, demonstrate the differences in the tensile test data resulting from production, either in a continuous hot-dip galvanizing line or after annealing in a bell-type furnace and electrolytic zinc coating. Data on mechanical properties, zinc coating and surface finishing of conventional steel grades are listed in DIN 17 162 Teil 1 [49].

For steels of higher yield stress basic materials can again be selected on the basis of the steel grades for uncoated flat material if electrolytic zinc coating is to be used. Hence possible choices among structural steels include those listed in DIN 17 100 [8] or sheet steel grades in SEW 093 [31] (see C 4.1.2.2). The higher strength level of galvanized strip (see above) is advantageous if a raised minimum yield stress is required.

The general structural steels (see DIN 17 162 Teil 2 [50]) are also available as well as other high-tensile steel grades. Internationally standardized steel grades suitable for electrolytically- or hot-dipped zinc-coated sheet are also described in several EURO- and ISO-standards [48].

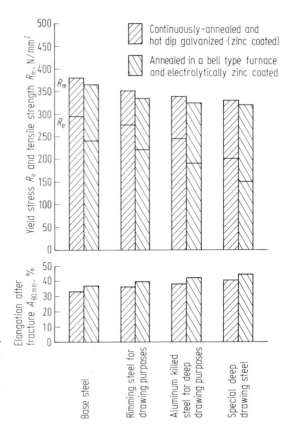

Fig. C 4.13. Characteristic results of tensile tests on electrolytically zinc coated or hot-dip galvanized mild sheet grades (Electrolytic zinc coating after annealing in a bell type furnace, hot-dip galvanizing after continuous annealing)

In recent years the use of zinc-coated sheet has increased, especially in the automotive industry. In accordance with the specific requirements and production possibilities for coating thicknesses, surface condition and cold formability, hot dip galvanized and electrolytically zinc-coated sheet grades are used. In addition to the soft and readily-formable steels high-strength steels also are being used increasingly in vehicle construction.

C 4.2.1.2 Aluminum-Coated Flat Products

Aluminum-coated flat material is used in structural components for which the strength of steel and the oxidiation resistance of aluminum are both required.

A survey of processes for coating with aluminum, characteristics of the aluminum coatings and basic materials used in surface coating with aluminum is presented in Table C 4.7 [51–53]. Hot-dip aluminum coating has gained the greatest importance among processes for such coating. For different areas of application two variants of the process are proven: 1) Hot dipping in a bath of pure molten aluminum (internationally known as Type 2) forming a Fe_2Al_5-alloy layer at the interface and 2) hot dipping in a bath of molten aluminum with an addition of silicon of 5% to 11% (internationally known as Type 1) forming an alloy layer of aluminum-iron-silicon-α-solid solution [54] (see B 12.2.3). The thickness and hardness of this layer are much smaller than those of the Fe_2Al_5-layer, so that formability of flat products with a coating of type 1 is better than that of type 2 (see Figs. B 12.9 and B 12.10).

Aluminum may also be coated on unalloyed steel using cladding by rolling (see also B 12.4.2). In this process, an aluminum foil is rolled continuously on steel strip in the cold state. This process is very flexible regarding sheet or foil thickness and chemical composition of the aluminum. No marked alloy formation takes place between substrate and coating in cladding and this condition has a favorable effect on forming behavior.

Vapor deposition of aluminum in vacuum (see also B 12.4.3) is suited only for limited applications due to the small thickness of the deposited coating.

Figure C 4.14 demonstrates the characteristic structure of aluminum coatings produced by hot-dip coating and cladding by rolling.

Similar principles apply to basic materials for aluminum coatings as to zinc-coated flat material. Both mild and high-strength steel grades may be used for hot-dip aluminum-coated sheet. After cladding by rolling a recrystallizing heat treatment is necessary, so only aluminum-free (rimming) steel can be used as basic material. Experience has shown that steels containing aluminum may encounter an undesired diffusion reaction.

Required Properties

Required properties for aluminum-coated steels are listed on page 98. In addition, it may be said that the demand for weldability of these products currently under development.

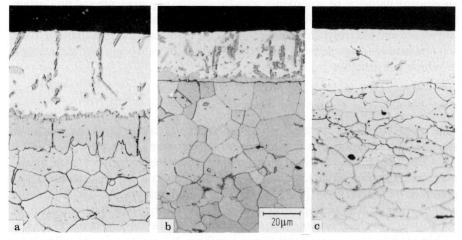

Fig. C 4.14. Formation of aluminum coating on steel in cross sections: **a** hot dip coated with pure aluminum (type 2); **b** hot dip coated with aluminum-silicon alloy (type 1); **c** clad with aluminum.

Characterization of Required Properties

The methods for testing surface-treated flat material discussed on page 98 are also used for aluminum-coated flat products. Some of the above-mentioned test methods using service-simulating stresses are aimed at the characterization of the corrosion resistance, such as a test that involves stress cycles similar to those in a vehicle exhaust system.

Measures for Attaining Required Properties

Cold formability of aluminum-coated flat material depends essentially on the basic material and on the behavior of the coating. In cladding flat products with aluminum by rolling good adhesion between coating and basic material is essential. In hot-dip aluminum coating, complete wetting and adequate reaction between aluminum and basic material must be secured by optimizing the process parameters. All simple methods of cold forming are then feasible without impairment of the coating. Heavier stretch-forming and deep-drawing procedures, however, lead to abrasion and shedding of the aluminum, which is increased by the well-known tendency of this metal to galling in the tool. Therefore, the coating should be kept as thin as possible for heavy cold forming. With thin aluminum coatings the good cold formability of the basic materials such as deep-drawing and special deep-drawing steels can be utilized. Production of even complicated drawn parts is made possible, as is proven by the application of hot dip aluminum-coated sheet in the production of passenger car exhaust systems. The design of the components, on the other hand, may be adapted to the limited formability of sheets with thicker aluminum coatings.

The *corrosion resistance* of aluminum-coated sheet is a result of the great stability of the less than 0.1-μm thick Al_2O_3-layer formed on the surface. This oxide layer is very dense and is renewed immediately when it is damaged. In atmospheric corrosion this oxide layer lends the aluminum surface great resistance to corrosive wear resulting in corrosion protection, even in different climates, that is superior by a factor of about 10 as compared to zinc-coated steel.

Corrosion resistance of aluminum-coated material is increased in similar proportions to zinc-coated sheet with increasing coating thicknesses. Hot-dip aluminum coated sheet of type 2 is more resistant to corrosion because it is generally produced with a heavier coating than type 1 and, moreover, the coating does not contain silicon. This advantage of a relatively thick coating consisting of pure aluminum may also be realized with sheet clad with aluminum by rolling.

The protection given by the thin aluminum-oxide layer on the aluminum coating is particularly effective at high temperatures, as is shown in Fig. C 4.15 [55].

A combination of hot-dip galvanized and hot-dip aluminum coated sheet that has good resistance to corrosion has become known, for instance, by the trade name *Galvalume*. This coating is made up of about 55% Al and 45% Zn (see also) B 12.2.4). Its resistance, however, does not match that of sheet with aluminum coatings under especially critical conditions, such as corrosion combined with oxidation [55a].

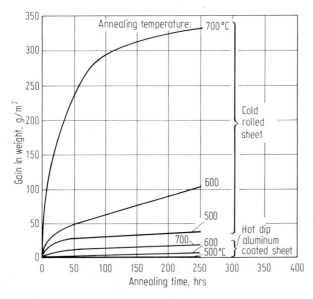

Fig. C 4.15. Gain in weight of cold-rolled uncoated sheet and of hot dip aluminum-coated sheet by annealing in oxidizing gas. After [55].

The processing of aluminum-coated flat material by *welding* is currently under development and is usually better than with zinc-coated sheet because of the behavior of the electrodes and the relatively high temperature of vaporization of the aluminum.

Although an aluminum coating represents a very effective and decorative surface refinement, an *additional surface treatment* may be considered. Temper rolling of the aluminum coated sheet, carried out generally for flattening of the aluminum surface, also has a favorable effect on the surface finishing. The aluminum coating also provides a highly suitable base for painting.

Aluminum-coated flat material can also be vitreous enamelled. Enamels for aluminum are distinguished from those used for steel by lower firing temperature and better formability [56].

Characteristic Steel Grades Proven in Service

Steels for aluminum-coated flat material may be selected from mild steel grades or from steel grades with increased yield stress. For sheet clad with aluminum by rolling, rimming low-carbon steel is used, as mentioned above. When mechanical properties are considered, account must be taken that in hot-dip aluminum-coated sheet, higher strength values and lower formability values are achieved due to the use of continuous annealing rather than the bell-type furnaces used for conventionally-annealed uncoated sheet (see Fig. C 4.13). This difference is reflected in Euronorm 154 [57] which contains values for hot-dip aluminum-coated sheet grades. Higher demands for cold formability can be satisfied by use of special deep-drawing steels, e.g. IF-steel (see C 4.1.2.1 and C 4.1.2.2).

For applications requiring increased yield stress values micro-alloyed steels may be used as an alternative to general structural steels. The presence of micro-alloying elements results in a high retention of hardness in tempering, so the improved yield stress of these steels at elevated temperatures in combination with the good oxidation resistance of the aluminum in the coating may be utilized for high temperature applications. A micro-alloyed special steel having satisfactory cold formability and good weldability possesses a roughly double the high yield stress in comparison to an unalloyed mild steel at temperatures up to 500 °C, as is demonstrated in Fig. C 4.16 [56].

C 4.2.1.3 Flat Products with Coatings Containing Tin and Chromium

Flat materials with a *tin coating* are produced as cold rolled strip with an electrolytically-deposited tin coating [58]. With a strip thickness in the range of 0.15 to 0.49 mm, the product, in the form of sheets cut from the strip, is called *tinplate* [59] and is by far the most important flat product with a tin coating.

Required Properties

As generally mentioned on page 98 the requirements for surface-treated flat products, i.e. for strength, cold formability and corrosion resistance, are also relevant for tinplate. In view of its final use, which is mainly for a wide range of packaging applications, tightness and impermeability to light, gases and liquids are also required, but these qualities are inherent properties of steel. For the manufacture of different packages, the *cold formability* of tinplate must be such that it can be seamed, beaded, flanged and rounded. In addition, it must be possible to produce punched and drawn parts such as can ends, crown corks, glass closures and drawn cans, even if the tinplate has been lacquered before fabrication. Adequate *corrosion resistance* is provided because tin is resistant to corrosion by most food products and is non-toxic. Tinplate should also be *solderable* and *weldable*.

Characterization of Properties Required

Among the methods generally stated on page 98 for testing *mechanical properties*, the determination of the Super Rockwell hardness HR 30 T [3] is most important

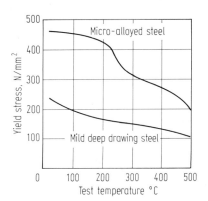

Fig. C 4.16. Yield stress at elevated temperatures of hot-dip aluminum coated sheet of mild and high strength (micro-alloyed) steels (Mild deep drawing steel: 0.063% C and 0.26% Mn; micro-alloyed steel: 0.115% C, 0.53% Mn and 0.03% Nb). After [56].

for tinplate. In addition, as specific tests for tinplate, spring-back tests [60] and the determination of earing on cups [61] are important. These tests may suffice if the material cannot be evaluated by qualification tests on the article to be produced, such as in making e.g. caps or drawn cans.

Measurement of the tin coating thickness is of major importance as far as *corrosion resistance* is concerned. Measurements are made by determining the weight of tin per unit of surface area [59], usually by chemical analysis of the tin dissolved from the surface [62, 63]. Several methods are available and need not be discussed here in detail [62–66]. Similar remarks apply to the determination of the amount of oil (in mg/m^2) applied to tinplate to improve its fabrication properties [66–68] without impairing its suitability for food packaging applications. The tin oxide layers formed during production and storage of tinplate can be measured by electrochemical reduction. The normal practice is to state the amount of current needed for reduction per unit of surface area [69, 70]. Tinplate is passivated in chromate solutions to avoid excessive oxidation. The thickness of the layers is measured by chemical or electro-chemical analysis of the chromium existing on the surface. It is also possible to determine the percentage of metallic chromium deposited during cathodic passivation by means of electrochemical methods [69, 71–73].

Attaining Required Properties

The base material, soft unalloyed steel, is of major importance for *strength* and *cold formability*. In can manufacture, after drawing of the cup, the wall thickness of the can body is reduced by ironing (stretching). Cold formability is improved by the tin coating which acts as a lubricant so that the material can be processed at high speeds. Such cans are used for beverage and food can applications.

Considering the *corrosion resistance* of tinplate, it should be pointed out that tin, in the presence of air (oxygen), is nobler than iron; the requirement for corrosion resistance of the outside of the can therefore, may be met only when it is stored in an almost dry environment [74]. If air is excluded, tin will remain less noble than iron and will, therefore, protect it against the usual slightly-acid products by slow dissolution [75–78]. This mechanism of reaction is also effective under a lacquer coating with very thin tin coatings.

For bonding, *solderability* (soft soldering) is a major requirement and is achieved by good wettability of the tinplate by the Pb-Sn solder and the quick alloy formation resulting from complete solubility of lead and tin. Tinplate can bodies may also be produced by resistance *roller seam welding*, using a non-reusable copper wire electrode to prevent welding problems due to alloying between copper and tin [79].

Tinplate Grades Proven in Service

Commercial tinplate grades are standardized e.g. in DIN 1616 [80] (see also Euronorm 145 [59]). The base material is mild, unalloyed steel; the grades are classified according to hardness and are produced from a steel of suitable chemical composition and annealed to obtain a microstructure with the desired hardness. Sheets of lower thicknesses are supplied also as "double-reduced" tinplate – the material is subjected to a second rolling pass instead of temper rolling [81]. Such material is classified by strength instead of hardness.

The tin coating is characterized by the mass of tin in grams per square meter of surface, normally between 2.8 and 11.2 g/m² [59]. Tin coatings with weights below 2.8 g/m² are also supplied for special applications. As with other metallic coatings, tinplate with differential coating can also be produced, i.e. tinplate with different coating thicknesses on opposite sides.

All tin coatings can be supplied with various (chemical) post-treatments. Cathodic passivation, with a chromium-chromium oxide layer, and dip passivation with the passivation layer produced by chemical reaction in a chromate solution have been generally accepted. Passivation treatments are used, for instance, if the lacquered sheet is intended to be subjected to considerable forming, as in the case of drawn cans.

In addition to tinplate, cold-rolled black plate is used with an electrolytic *chromium-chromium oxide coating* [81]. Coatings containing 50 mg Cr/m² to 100 mg Cr/m² as metallic chromium and 10 mg Cr/m² to 30 mg Cr/m² are deposited in the form of hydrated chromium oxide, and are, therefore, substantially thinner than those on tinplate. After lacquer coating on both sides, the material is processed into can parts taking advantage of the good adhesion provided by the base metal and the underfilm corrosion resistance [82, 83]. Can bodies may be produced by welding, after the coating in the seam area has been removed, or by cementing (sticking). Can components which are subject to heavy abrasion or severe forming should only be produced from lacquered chromium-chromium oxide-coated steel.

C 4.2.1.4 Flat Products with Lead Coating

Generally, lead coatings are deposited by hot dipping (see B 12.2.6). The base material consists of an aging-resistant steel. Due to its good resistance to gasoline, diesel oil and mineral oil, lead-coated sheet is used for the production of fuel tanks, oil filters and pipes in fuel systems. Applications in the constructional sector have also been reported [84].

Required Properties

As with other flat products with metallic coatings corrosion resistance is of major importance among the requirements of material properties of use. In addition to strength and cold formability, a certain level of weldability and solderability must also be provided by the material.

Characterization of the Required Properties

For the characterization of the required properties, reference should be made to page 98. Here again, the lead coating is tested to determine the corrosion resistance by measuring the weight per unit surface area. This value is obtained by determining the weight difference before and after removal of the lead coating [85, 86]. For testing the uniformity of the coating, salt spray testing according to

DIN 50 021 [87] is often used, preceded by a pre-treatment suitable for the material.

Measures for Attaining Required Properties

The base material is responsible for the strength and cold formability (see above). The cold formability is of specific importance in the manufacture of complicated drawn parts, such as fuel tanks. This requirement should be taken into consideration when selecting the base material.

The *adhesion* of the coating is important for both cold formability and corrosion resistance. Adhesion can be improved by adding 8 to 15% Sn and up to 3% Sb to the melted lead. These additions will cause chemical compounds to be generated on the steel surface, improving adhesion, and providing increased corrosion resistance in the coating, with higher tin coating weights (Terne plate) [85].

For *welding* lead-coated sheet, the wire welding process should be used [88]. Hard and soft soldering connections can also be used.

Characteristic Steel Grades Proven in Service

Only hot-dipped lead-coated sheet has been standardized [86, 88]. Apart from the mechanical properties, the standard grades differ in the lead coating, where the grade with a coating weight of 100 mg/m^2 is most important. Electrolytically (alloy) coated sheets are supplied with lead coatings between 2.5 and 7.5 μm corresponding to 27.5 and 82.5 g/m^2 of surface area. One-side coatings can also be produced [89].

C4.2.1.5 Flat Products with Other Metallic Coatings

Coatings on steel which differ from those described above not only in the metal used for the coating, but also sometimes in the coating technique employed are discussed in this section. A survey of metals, alloys, and production processes as well as characteristic properties of products is shown in Table C 4.9 [90]. Electrolytically-deposited coating and cladding by rolling have the widest range of applications.

Table C 4.9 Special properties and production processes for metal coatings on cold rolled strip [90, 94]

Coating metal	Special properties					Production process		
	Corrosion protection	Reflection	Decorative appearance	Electrical or thermal conductivity	Preparation for further surface treatment	Electrolysis	Cladding by rolling	Diffusion processes
Copper			×	×	×	×	×	
Brass			×		×	×	×	
Nickel	×	×	×		×	×	×	
Chromium	×	×	×			×		×
Stainless steel	×	×	×	×	×		×^{a}	

^{a} Also cladding by casting

Flat Products with Electrolytically-deposited Metal Coatings

Table C 4.10 lists some examples for application of electrolytically-coated cold strip which are of great economic importance.

As has been mentioned above, the corrosion resistance of metal coatings increases with layer thickness. In addition, paints (lacquers, clear varnishes) or coatings of several metal layers also are frequently applied. Roughness is likewise of great importance, i.e. the smoother and the more uniform the surface of the base metal and the surface-treated strip, the better the corrosion protection. Cold-rolled and recrystallizing-annealed sheet or strip intended for electrolytic surface treatment must conform to DIN 1623 or 1624 respectively [6, 32]. Copper, nickel and brass are the most widely-used metals or alloys for electrolytically deposited coatings [90, 91].

Outstanding properties are corrosion resistance, reflecting power for light or heat radiation, decorative appearance, and electrical and thermal conductivity [92].

Flat Products with Metal Coatings Produced by Cladding

The basic features of the various cladding processes have been treated in B 12.4.2. They will be discussed here only as concerns the production of flat material for cold forming. Nickel, copper, their alloys, and stainless and heat-resistant steels are used as cladding materials.

The cladding of mild steels with nickel, copper and their alloys is mostly carried out by rolling on one or each side [93]. Steels with carbon contents up to 0.12% are the usual base metal. The thickness of the coating amounts to 5 or 10% of the total thickness and may be different on both sides. Clad strip is supplied in the as-annealed condition or – to prevent stretcher strain markings – after skin-pass

Table C 4.10 Applications for electrolytically surface treated cold-rolled strip proven in service [90]

Coating metal	Layer thickness μm	Examples for application	
Copper	2.5	Seals, Table wear, Lamp sockets	sometimes with
	3.5	Trays, Smoking table utensils, Bundy-tubing	lacquer (clear varnish)
Brass	2.0	Hinges, Folder accessories, Table wear, Profiles for curtain rods	sometimes with
	10.0	Watch cases, Suitcase fittings, Toy cars, Flower stands, Purse bows	lacquer (clear varnish)
	15.0	Lamp shades, Handbag bows, Furniture fittings, Kitchen range fittings (further finishing of fabricated parts)	
	2.5	Letter-file accessories, Picture frames, Toilet case accessories, Mirror frames, Suitcase fittings	
Nickel	4.5	Cable tape, Radio tube parts, Sintered plate electrodes, Paint brush clamps, Flashlight (Pocket lamp) cases, Metal letter types, Brief case fittings	
	6.0	Chocolate molds, Grill housings, Metal mirrors, Kitchen range fittings, Cold rolled profiles, Prefinishing for consecutive part chromizing or silver-plating	

Table C 4.11 Mechanical and physical properties of steel clad by rolling [93]

Cladding metal	Two-sided cladding thickness %	Yield stress[c] N/mm²	Tensile strength N/mm²	Elongation after fracture A_{10} %	Brinell hardness[a] HB 2.5	Erichsen cupping index[b] mm	Density g/cm³	Modulus of elasticity N/mm²
Nickel	5	w 250 h 470	350 550	35 10	95 180	9 ... 10	7.95	195000
Copper- CuNi 15 Nickel- CuNi 20 Alloy CuNi 25	5	w 250 h 470	350 550	35 10	90 170	9 ... 10	7.95	195000
Copper	5	w 230 h 450	330 520	35 12	85 160	9 ... 10	7.95	190000
Brass Ms 90	5	w 230 h 450	330 520	35 12	85 160	9 ... 10	7.93	190000

[a] At about 1 mm total thickness and testing on clad layer
[b] At 0.5 mm product thickness
[c] w = soft, h = hard

rolling. Mechanical property values given in Table C 4.11 demonstrate the good cold formability of the material.

Cladding with stainless or heat-resistant steel is carried out in the production process by rolling or casting, cladding by rolling being the main process. Due to the high corrosion- and heat-resistance the composite material is well suited for applications such as silos, furnaces, pots, pans, automobile mufflers and tubing [94].

Flat Products with Coatings Produced by Vapor Deposition

Chromizing strip is produced with a surface diffusion layer containing about 25 to 30% Cr. The principles of the process have been described in B 12.4.3 (see also [95]). The strip is suitable for waste gas, heating and heat exchange systems due to its corrosion- and heat-resistance.

Tests normally applied to the metallic coatings covered in this chapter include the methods discussed on page 98 (see also [96, 97]).

C 4.2.2 Flat Rolled Products with Inorganic Coatings

Vitreous enamels as well as phosphate and chromate layers are included in the inorganic coatings (see B 12.5 and B 12.6). The two last-named act either as components in further surface treatment e.g. by painting (lacquering) and as adhesion-promoting agents or only as temporary corrosion protection.

Phosphated and Chromated Flat Products

Phosphating of cold-rolled surfaces is not carried out until the sheet material has been processed (e.g. to car body parts). After a pretreatment for the primer (anodic or cathodic electro-dip priming), the coating paint is applied.

Applying an inorganic coating on flat material with a metallic surface treatment is normally the final step in fabrication.

Electrolytically zinc-plated flat material (see C 4.2.1.1) is generally phosphate and chromate passivated as a pretreated for subsequent painting.

Chromating is used on hot-dip metallic-coated flat material and meets the demand for temporary corrosion protection. The danger of white rust formation on zinc-coated sheet in transport and storage is reduced by this method.

Enamelled Flat Products

Hot- or cold-rolled flat material is surface treated by enamelling for numerous applications. In B 12.5 it was demonstrated that the *suitability for enamelling* depends mainly on three properties of the part to be enamelled: good adhesion of the enamel, excellent surface quality and resistance to fish-scale flaking. Corresponding *requirements* cover the chemical composition of the steel, also the size, number and distribution of nonmetallic inclusions in the steel. *Adhesion* depends substantially on pickling behavior and tendency to oxidation in firing. In this connection, impurities such as phosphorus and copper and alloying elements such as titanium have an influence. Resistance to fish-scale-flaking requires, especially in double-sided enamelling, that nonmetallic inclusions forming phase boundaries and cavities in the structure are adequately available for the recombination of hydrogen.

An overview of production features and mechanical properties of characteristic steel grades for conventional two-sided enamelling (steels EK 2 and EK 4 according to DIN 1623 Teil 3 [98]) and for one-sided enamelling (steel ED 3) is given with Table C 4.12. Enamelled steels are used in many household appliances and components, in industry and in construction engineering.

New developments utilizing vacuum decarburization, deoxidation by aluminum and alloying with titanium have led to steels that are eminently suitable for one-sided enamelling [99].

Table C 4.12 Characteristic values for the mechanical properties of cold rolled sheet for vitreous enamelling

Steel grade Designation[a]	Data on production	0.2%-proof stress	Mechanical properties			
			Tensile strength	$A_{80\,mm}$ %	r_m	n_m
EK 2[b]	Rimmed	220	345	39	1.1	0.19
EK 4	Aluminum deoxidized in mold	180	335	41	1.5	0.21
ED 3[b]	Rimmed Cast in bottle top molds Decarburized[c]	170	300	44	1.6	0.22

[a] After DIN 1623 Teil 3 [98]
[b] Unaged condition
[c] By open coil annealing

C 4.2.3 Organic Coil-coated Steel Flat Products

Continuously organic-coated flat steel products (see B 12.7) may be used to advantage where corrosion resistance and decorative appearance are of primary importance. Thus these products have applications throughout the sheet processing industry.

Available forms include coiled strip in widths up to 1850 mm, sheets cut from strip, slit strip, and cut lengths (from sheet or slit strip) with a maximum product thickness for the substrate of 2 mm and sometimes greater thicknesses.

Further information on the types of products available, materials, characteristic properties, testing, storage and processing, also ordering requirements and marking is included in various publications [100, 101].

The more important steel substrates include cold-rolled sheet steel and strip and hot-dip zinc-coated sheet made to DIN, Euro and ISO standards (see Table C 4.13). Hot-dip zinc-coated sheet is used for exterior applications.

The decorative appearance, i.e. color, gloss, embossing, and multi-color printing, can be varied depending on the coating material selected. In addition to decorative appearance, coil-coated flat products provide functional properties characteristic of the suitability of the products for processing and end-use. These properties should be taken into account on selecting materials to be processed.

There are several kinds of one-sided zinc-rich pre-painted sheet steel differing in the coating system [105]. These products all have conductive zinc-rich epoxy coatings giving improved corrosion protection and good weldability.

Mechanical testing is similar to that used for the substrate steels. Special testing of characteristic properties, in accordance with ASTM and ISO testing procedures (see Table C 4.14), has been standardized by both the European Coil Coating Association (ECCA), Brussels, Belgium, and the National Coil Coaters Association (NCCA), Philadelphia, PA. Supplementary national standards for organic pre-coated sheet steel and the Euronorm 169-85 should also be mentioned. Work on a detailed ISO Technical Report has been started. Coil-coated flat products are usually supplied with finished surfaces. Therefore attention should be paid to appropriate tooling and processing parameters regarding cutting, forming, and joining of the finished surfaces [106, 107].

Table C 4.13 More common (important) substrates used for coil coating

Substrate	DIN[a]	EU	ISO
Cold-rolled unalloyed mild sheet			
for cold forming	1623 T.1 [6]	130 [102]	3574 [102a]
for structural purposes	1623 T.2 [30]	–	4997 [102b]
Hot-dip zinc-coated unalloyed mild sheet			
for cold forming	17162 T.1 [49]	142 [103]	3575 [103a]
for structural purposes	17162 T.2 [50]	147 [104]	4998 [104a]

[a] Only quality standards

Table C 4.14 Important properties of coil-coated flat products and ECCA-test-methods[a]

Test method	Property
T1 (1985)	Coating thickness[b]
T2 (1985)	Specular gloss[b]
T3 (1985)	Color difference[b]
T4 (1985)	Pencil hardness[b]
T5 (1985)	Resistance to cracking on rapid deformation[b]
T6 (1985)	Adhesion after indentation[b]
T7 (1985)	Resistance to cracking on bending (cylindrical mandrel)[b]
T8 (1985)	Resistance to salt spray fog[b]
T9 (1985)	Water immersion resistance
T10 (1985)	Resistance to accelerated weathering[b]
T11 (1985)	Metal marking resistance
T12 (1985)	Buchholz indentation[b]
T13 (1985)	Heat resistance
T14 (1985)	Measurement of chalking (Helmen method)[b]
T16 (1985)	Corrosion test for coil coated sheet in atmosphere containing SO_2
T17 (1985)	Determination of the adhesion of strippable films
T18 (1985)	Stain resistance of coil coated products

[a] ECCA = European Coil Coating Association
[b] See also EU 169–85 [100]

The products mentioned have been used successfully: in the building industry for both exterior and indoor applications such as roofing, siding, decking, and metal doors; in the general sheet processing industry, e.g. for furniture, shelving, and coverings of many kinds; in the automotive industry, e.g. for car-body panels and decorative parts; in the packaging industry, e.g. for drums, aerosol cans, and closures; and for jacketing.

Another special application includes "sandwich" materials with two outer coil-coated sheets and an organic core consisting of either a polyurethane rigid foam for building purposes or a thermoplastic material with sound-damping properties.

C 5 Heat-treatable and Surface-hardening Steels for Vehicle and Machine Construction

By Gerhard Tacke, Karl Forch, Karl Sartorius, Albert von den Steinen and Klaus Vetter

C 5.1 General Remarks

In the following, heat-treatable steels are understood to comprise in particular steels that can be heat treated, as well as those used for heavy forgings. In a more restricted sense, heat-treatment means hardening and tempering (quenching and tempering). Steels suitable for surface hardening include those suitable for flame-, induction- and immersion-hardening as well as nitriding and case-hardening steels. *These steel types are collected into one group because of their common use in machine and vehicle construction*, their suitability for high and dynamic or alternating stressing the recommended heat treatment procedures suited to the respective chemical compositions, cross-sections and applications.

Heat-treatable steels derive their essential *properties* from austenitizing, quenching and tempering. Their chemical composition is determined primarily by these processes, which are designed to produce martensite and/or bainite when quenching. There is no upper dimensional limit for the use of heat-treatable steels. In the authoritative German standards, properties are listed up to a diameter of 250 mm.

Steels intended for flame, induction and immersion surface hardening are usually heat treated (quenched and tempered) before being surface hardened which is often applied to only selected zones. The structure and applications at these steels therefore are closely similar to those of heat-treatable steels (steels for quenching and tempering). Nitriding steels are heat treated (quenched and tempered) before final surface treatment during which the surface layer is enriched with nitrogen. The chemical composition of these steels is, however, determined to some extent with a view to nitriding.

Steels for heavy forgings are used in the heat treated (quenched and tempered) condition on principle. (The few examples to which this rule does not apply will not be considered in this chapter.) As a matter of principle the chemical composition and the microstructure of these steels is always governed by the special problems relating to size and weight of the products. With lower loads, however, steels can be used for larger components for which heat-treatable steels in the narrower sense are normally used (see above).

Case-hardening steels are used in the hardened condition after carburising. Case hardening steels usually are given a lower carbon content than heat-treatable steels to ensure that the cores of the components are not too hard and have sufficient toughness in the heat-treated condition.

References for C 5 see page 769.

Heat-treatable steels, steels for surface hardening and nitriding steels generally, have tensile strengths between 500 and 1300 N/mm² in the service condition. Their high stress resistance is also emphasized by the toughness connected with the strength. Notch impact energy figures are generally used to characterize the toughness.

Most components made of case-hardening steels have tensile strengths of 800 to 1600 N/mm². As with case hardening, surface hardening by flame, induction or immersion is used to increase wear resistance and fatigue strength and/or to improve the running properties of components. The classical application of case-hardening steels is for highly-stressed gears.

Steels for hardening and tempering and for surface hardening are preferred for parts produced by forging, machining and/or cold forming from semi-finished and bar steel. The production of parts from flat steels is rare and steel sections from these steel types are hardly ever used.

The different processing methods often require different *processing characteristics*. These differences may result in a variety of requirements for the steel for the same application. Steels are also often classified according to their processing characteristics such as cold extrusion steels, etc. These steels can be either heat-treatable types or steels for surface hardening.

The *subdivisions of this chapter* follow the differences in the methods of heat treatment and their objectives. First the requirements concerning the properties for the end use (application) and those for processing, which are common to the various grades of steel, are dealt with. Data concerning the required properties which are generally applicable and their characterization by test results, which depend on their microstructure and chemical composition, need not then be repeated. Special data for heat-treatable steels, steels for flame, induction and immersion hardening and for nitriding steels are then dealt with. Steels for heavy forgings and case hardening steels have less in common with the above-mentioned group and somewhat more space is devoted to these two groups.

For an explanation of the heat treatment terms used in this chapter, such as harden and temper, see B 4 and e.g. DIN 17 014 [1].

C 5.2 Heat-treatable Steels (Steels for Quenching and Tempering), Steels for Flame, Induction and Immersion Surface Hardening and Nitriding Steels

C 5.2.1 General Requirements for the Properties of Use of Heat-treatable Structural Steels, Their Characterization by Test Values and Possibilities to Attain Them

C 5.2.1.1 Properties which are Important for End Uses

Steels listed in the title of this section are used primarily for components which are to be subjected to high mechanical stresses. These stresses can be caused by bending, tension, compression, torsion or shear forces. Depending on the respective operating conditions the capacity to withstand static stress is linked with certain strength characteristics such as *hardness, strength* (tensile, compression, bending, torsion and shear strength) and yield stress (proof stress). In addition, adequate *toughness properties* are required to ensure safety against failure due to brittleness. For power transmission components, both high wear resistance and good running characteristics also are often required.

In machine, and especially in vehicle, construction, *alternating stresses* occur which can lead to fractures even with stresses below the tensile strength. Alternating stresses, that is stresses which vary in amount and/or direction must therefore be considered during design of many components. Fatigue fractures are the main cause of breakdowns of machine and vehicle components and it is estimated that they account for 80 to 95% of all fractures [2, 3]. Only rarely can fatigue failures be traced back to the material and to faults in its production (see Table C 5.1 [4]), because fatigue strength of components depends to a large extent on the design, shape and surface finish of the parts (see B 1.2.2).

The properties stem from the microstructure which results from a combination of chemical composition and heat treatment (see B 1 and B 4).

Test results of tensile tests and notch impact tests are used to *characterize the quenched and tempered condition*. In contrast with tensile strength and yield stress, values of toughness cannot be used as a basis of calculation (design) although they indicate whether the expected properties of the selected steel grade, based on the tensile strength and size, have been obtained. In addition toughness values can be used to draw conclusions about the stressability of components where experiences with comparable applications exists.

Up till now, test methods of fracture mechanics have only been used to a limited extent for heat-treatable steels in general use. No standard test method has yet been generally accepted for the determination of fracture mechanics data for these relatively tough materials (see B 1).

Strength Under Static Stresses

The greatest strength occurs in heat-treatable steels having a fully-martensitic structure after hardening. The hardness and strength of the untempered martensite are determined by the carbon content and are largely independent of the alloy content (see Fig. C 5.1 [5]). Mixed structures consisting of martensite, bainite and possibly ferrite + pearlite have a reduced hardness which does not depend entirely on the carbon and martensite content [6]. The determining factor for the strength of mixed structures is the strength of the individual components of the structure and their volume proportions. Regression analyses show that it is possible to calculate the approximate strength characteristics of certain structures from the alloy compositions [7]. The uncombined nitrogen present in heat-treatable nitrogen-alloyed steels causes an additional increase in hardness of all types of structures (see Fig. C 5.2 [8]). The yield stress of a ferrite-pearlite structure can be effectively increased by the addition of elements such as vanadium and niobium which result in precipitation hardening even with additions of only about 0.1% [9, 10].

The final strength of heat-treatable steels (the strength in the condition of use) is determined by *tempering*. Tempering should produce a distinct *reduction of*

Table C 5.1 Classification of fatigue fractures examined by the material testing laboratory of the Allianz Versicherungsgesellschaft between 1957 and 1961 according to the causes of failures. After [4]

Type of damaged machine	Number of failures examined	Main causes of fatigue fractures (figures in brackets are contributory causes)									
		Material fault	Construction fault		Production fault during				Operation conditions		
			Unsuitable material	Design fault	pro-cessing	heat treatment	assembly	overhaul repair	Overload	Surface defects	Loosening	Wear etc.
Electrical machines	14	–	–	1	4 (4)	1 (2)	–	3 (1)	2	1	2 (2)	
Steam generators, Steam engines, Locomotives	12	1 (2)	–	1	3 (4)	–	– (1)	1 (–)	4	1	1	
Steam turbines	52	1 (2)	1	9 (5)	10 (6)	2 (1)	3 (1)	–	2 (1)	13 (3)	11 (5)	
Internal combustion engines	25	2 (1)	–	– (2)	7 (4)	2 (3)	1 (1)	–	3	7 (1)	2	1
Hydraulic power stations	8	1	–	1 (–)	1 (–)	– (1)	1 (–)	2 (–)	–	2 (1)	– (1)	
Lifting and conveying installations	80	4 (7)	– (2)	9 (3)	18 (19)	10 (10)	8 (5)	6 (2)	17 (7)	3 (3)	3 (2)	2
Pumps and compressors	35	1 (3)	–	6 (2)	10 (4)	1 (7)	3 (2)	2 (–)	4	3 (1)	3 (1)	2
Machine tools	32	1 (3)	–	2 (1)	10 (5)	1 (3)	3 (1)	1 (–)	7 (5)	2 (2)	5 (4)	
Machines for the production of consumer goods	34	–	–	4	8 (2)	6 (2)	6 (3)	3 (–)	3 (2)	3 (1)	1 (1)	1
Vehicles	11	– (1)	– (1)	– (1)	2 (2)	3 (–)	– (2)	2 (–)	2	–	1	
Total	303	11 (19) 3.63%	1 (3) 11.22%	33 (14)	73 (50)	26 (29) 47.52%	25 (16)	20 (3)	44 (15)	35 (12) 37.63%	29 (16)	6 (–)

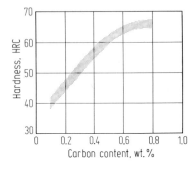

Fig. C 5.1. Maximum achievable hardness of martensite as a function of carbon contents. After [5].

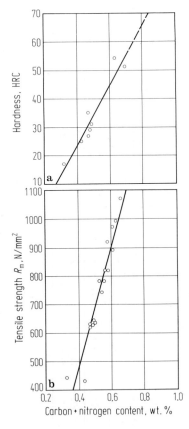

Fig. C 5.2. Effect of the sum of carbon and nitrogen content on hardness and tensile strength of nitrogen-alloyed steels. After [8]. Range of compositions of the steels tested: 0.19 to 0.53% C, 0.04 to 0.44% Si, 0.26 to 0.95% Mn, 0.04 to 0.24% Cr, 0.001 to 0.017% Al and 0.10 to 0.17% N. **a** Test-pieces 10-mm diameter quenched from 850°C in oil: Pearlite + bainite + martensite; **b** Test-pieces 20- to 50-mm diameter hot rolled: Ferrite + pearlite.

strength, i.e. the tempering temperatures must not be too low especially for a low *hardening degree* (the ratio of the attainable hardness to the highest possible hardness, see below). Finally the tempering temperature must be so chosen as to limit any temper brittleness.

Examples of the reduction of hardness by tempering shown for tempered end-quench specimens from a 42 CrMo 4 steel [8] in Fig. C 5.3, indicate that it depends both on the hardened microstructure and the chemical composition. Silicon and

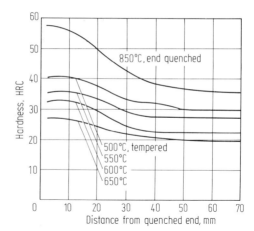

Fig. C 5.3. Effect of tempering after end-quenching on the change of hardness for the steel 42 CrMo 4. After [8].

manganese, as well as the special carbide formers molybdenum, vanadium and chromium, delay hardness reduction during tempering. Several authors deal with the mathematical formulation of the relationships between the tempering conditions, the composition of the steel and the tempering hardness or strength of steels [11, 12]. By tempering at a sufficiently high temperature, the possible variation of hardness across the cross-section can be made more uniform.

The *0.2%-proof stress* and the elastic limit of heat-treatable steels depend on the hardened microstructure and on the tempering temperature. Both these values are relatively low for untempered martensite and they reach their maximum values at tempering temperatures between 250 °C and 300 °C (see Fig. C 8.2).

Mixed structures containing upper bainite with ferrite + pearlite or proportions of residual austenite have only lower *ratios of yield stress to tensile strength* (or 0.2%-proof stress to tensile strength) even with higher tempering temperatures. Figure C 5.4 shows the relationship between the ratio of yield stress to tensile strength and the degree of hardening reached [13].

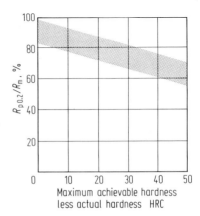

Fig. C 5.4. Relationship between the ratio of the proof stress $R_{p0.2}$ and the tensile strength R_m and the degree of hardening expressed as the difference between the highest achievable and the actual hardness. After [13].

Toughness Properties

The structure of heat-treatable steels also has a marked effect on their toughness, particularly their notch impact energy. For example, Fig. C 5.5 shows for a plain carbon and a low alloy heat-treatable steel the *effect of the microstructure* produced by different heat treatments [8]. The structure of the test materials used in Fig. C 5.5 can be described as follows.

The Ck 45 (Fig. C 5.5a) steel contains a pearlite-ferrite structure under all conditions. In the untreated (hot rolled) condition it consists of laminar pearlite within a matrix of free ferrite. Normalizing will refine and equalize the structure especially the ferrite component. The soft-annealed condition is similar to the untreated one except that only half the pearlite is spheroidized. Depending on the size, in the quenched and tempered condition a fine grain structure is present which consists of lamellar pearlite with small amounts of ferrite at the grain boundaries. The 42 CrMo 4 steel (Fig. C 5.5b) along with its higher hardenability, in the untreated condition, has a structure of bainite with portions of pearlite. When normalized the structure is finer and contains roughly similar portions of bainite, laminar pearlite and free ferrite. In the soft-annealed condition the carbides of the bainite and pearlite are largely spheroidized. Quenching and tempering produces a uniform fine structure of tempered martensite.

It becomes clear that the mixed structures containing ferrite, pearlite or spheroidized carbides result in a lower notched-bar impact energy, especially at lower test temperatures (see B 4). Despite a high yield stress and tensile strength, quenching and tempering produces by far the best toughness for both steels even though

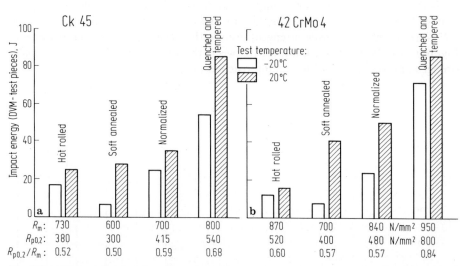

Fig. C 5.5. Effect of the microstructure (of the heat-treated condition) on the mechanical properties of heat-treatable steels (steels for quenching and tempering): **a** Steel Ck 45 with about 0.45% C; **b** Steel 42 CrMo 4 containing about 0.42% C, 1% Cr and 0.2% Mo. After [8]. Test-pieces taken from 80-mm diameter bars at a distance of 12.5 mm from the surface.

a martensitic transformation was not obtained in the Ck 45. Generally quenching and tempering produces the best combination of high strength with high toughness.

Experience shows that for components with lower requirements for toughness a ferrite-pearlite structure, which occurs in plain carbon heat-treatable steels in the normalized or hot formed (untreated) condition can be used. In recent years in particular, drop forgings have been produced and used successfully without the extra cost of additional heat treatment after controlled cooling in air from the hot-forming temperature. These forgings have processing and service properties suited corresponding to the application [14]. For higher toughness the better properties of the martensitic structure must be used, i.e. the components must be hardened or quenched and tempered.

Depending on the cross-section being treated and the cooling medium the hardenability of the steel determines the composition of the structure throughout the cross-section (see A 9 and B 4). The martensite content obtained during hardening or the ratio of the attained hardness to the highest achievable can be called *hardening degree* which must be the higher the greater is the stress on the component. The toughness values which are attainable in the quenched and tempered condition depend largely on the microstructure after hardening. The highest values are only obtained when the structure contains the highest obtainable amount of martensite before tempering, and when the hardness after quenching is as high as possible. Thus the hardening degree for a given chemical composition has a most important influence. Even small quantities of other constituents of the structure, e.g. pre-eutectoidic ferrite, pearlite or upper bainite considerably reduce the toughness (as example see Fig. C 5.6 [15–17]). In higher alloyed steels, in which structures of the lower pearlite or bainite stage are formed preferentially, the harmful effect of non-martensitic constituents of the structure is less and the effect is also reduced with increasing tempering temperature.

Even a fully bainitic structure can exhibit favorable toughness properties especially for very low and higher carbon contents above 0.5% (see B 4). For the range of steels of interest here, with medium carbon contents and tensile strengths below 1400 N/mm^2, bainite has no advantage over an optimally tempered martensitic structure.

Good notch impact energy values, especially at low temperatures, can *only* be achieved *with a fine-grain structure*. Additional data concerning investigations into the effect of the structure in heat-treatable steels can be found in [6, 18–21]. Knowledge of the relationship between the microstructures of steels and other metals and the fracture processes or toughness properties has been greatly increased (see B 1). This knowledge can be used to obtain higher values for the toughness-strength ratio for heat-treatable steels by use of different thermomechanical treatments, than by the usual quenching and tempering method [22] (see B 4).

Additional factors affect the toughness, *cleanliness* and *homogeneity* of critically stressed components requiring high strength after quenching and tempering, especially for large sizes. Elongated non-metallic inclusions, segregations and banding particularly affect the properties transverse to the direction of forming (see B 1).

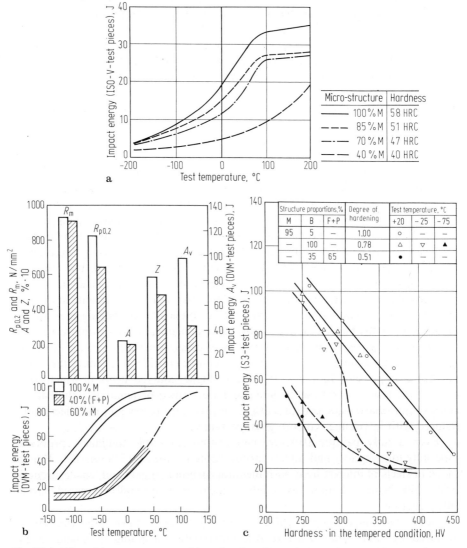

Fig. C 5.6. Effect of the structure composition of steel on the toughness properties in the quenched and tempered condition: **a** Steel SAE 1340, tempered to 35 HRC, after [15]; **b** Steel 50 CrV 4, after [16]; **c** Steel 42 CrMo 4, after [17]. SAE 1340: Steel containing 0.43% C, 1.79% Mn, 0.020% Al and 0.006% N. S3-test-piece similar to ISO-V test-piece with 3 mm deep notch.

When requirements for toughness properties and for uniformity of the properties throughout the component are high, specially clean and homogenous steels must be used. Several metallurgical processes have been developed for production of such steels in recent years (see D 4). The negative effect of non-metallic inclusions and segregations on the operating safety of components should not be exaggerated, however.

Toughness properties also depend on the *strength after quenching and tempering*, i.e. on the *tempering temperature*. However, while hardness and tensile strength of steels without marked secondary hardening fall steadily with increasing tempering temperature the relationship between toughness and tempering temperature is less simple. During tempering in certain temperature ranges brittleness phenomena may occur which disturb the direct relationship between toughness and strength (300 °C or 500 °C brittleness, tempering brittleness (see B 4)).

With the same microstructural condition and the same strength the toughness properties of heat-treatable steels depend on the alloy content. The effect is reduced after quenching with complete formation of martensite and tempering. The results from large-scale tests with through-hardened and tempered test-pieces of different heat-treatable steels show that for a certain strength after quenching and tempering the values for yield stress, reduction of area, elongation after fracture and notch impact energy are grouped in limited scatter ranges (see for example Fig. C 5.7 [23]). Generalizing these results has often led to the opinion that the properties of heat-treatable steels are determined exclusively by hardenability and tensile strength. However, numerous tests and practical experience prove that the chemical composition is by no means unimportant in achieving the required hardenability and tensile strength and that individual alloying elements or combinations have specific effects [15, 24]. These differences in the behavior of materials are not detected by all test methods but they may have an effect particularly under critical operating conditions such as high-speed stressing, multi-axis stress conditions, stress concentrations, low temperature and with higher strengths in the quenched and tempered condition.

Of *special importance* for toughness is the *carbon content*. With equal proportions of microstructural constituents and equal tensile strength, heat-treatable steels with a low carbon content usually have a higher ductility and toughness (for example see Fig. C 5.8 [16, 25]). The carbon content must be adjusted so that the required strength in the quenched and tempered condition can be obtained with favorable tempering temperatures, i.e. those which do not cause brittleness. For the strength range below 1400 N/m^2 the most favorable carbon content is between 0.25% and 0.5%.

A *manganese content* considerably above 1% brings down the reduction in area and the notch impact energy especially when manganese is used as a substitute for chromium [26, 27]. However, manganese increases the tendency to temper brittleness, and production of melts of high cleanliness and a low primary segregation with a high manganese or silicon content has some difficulties. Steels which are alloyed primarily with chromium are therefore greatly preferred to heat-treatable steels alloyed primarily with manganese [26].

By facilitating the cross slip processes *nickel* improves the low temperature toughness and reduces the impact transition temperature [28]. Martensitic heat-treatable steels which are tough at sub-zero temperatures therefore may contain up to 10% Ni and can then be used at temperatures down to -200 °C (see C 10). Owing to the tendency of nickel-chromium steels containing more than 3% Ni to form residual austenite and to produce temper brittleness as well as their cost compared with differently-alloyed steels, their use has been largely discontinued [29, 30].

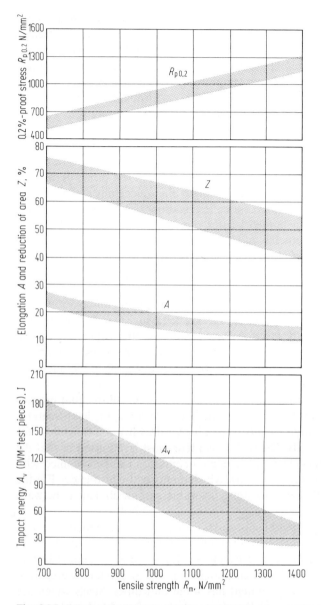

Fig. C 5.7. Relationship between tensile strength and other mechanical properties of quenched and tempered martensitic steels (longitudinal test-pieces). After [23].

Even a small addition of *molybdenum* considerably increases the hardenability and reduces the tendency to temper brittleness. Heat-treatable steels containing *nickel, chromium and molybdenum* in different alloying combinations now fulfill the highest requirements in many countries. The advantages of multiple alloy steels are made clear in Figs. C 5.9 to C 5.11 [15, 31]. The differences become most apparent for a hardening degree < 1 (see Fig. C 5.10).

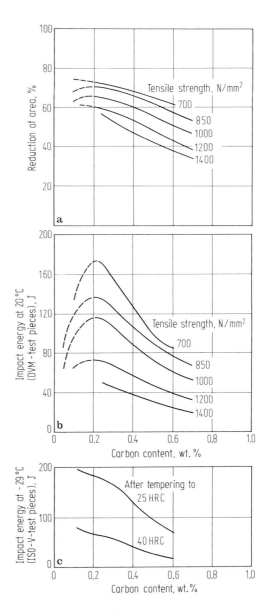

Fig. C 5.8. Effect of the carbon content on the toughness properties of heat-treatable steels having the same strength after quenching and tempering. **a** and **b** Result of the tests at +20 °C on 82 steel grades in the range of 0.12% to 0.7% C, up to 2.4% Si, up to 2.1% Mn, up to 3.4% Cr, up to 5.2% Ni, up to 0.6% Mo, up to 0.4% V and up to 3.5% W after quenching and tempering to 450 °C to 750 °C for 2 hr, after [16]; **c** Steels of the series SAE 86xx at −29 °C. Range of composition: 0.15% to 0.30% Si, 0.70% to 1.0% Mn, 0.4% to 0.6% Cr, 0.15% to 0.25% Mo and 0.4% to 0.7% Ni. The test-pieces were tempered to 25 or 40 HRC from the fully martensitic condition. After [25].

Fatigue Strength

The *fatigue strength* (fatigue strength under reversed stress) of heat-treatable steels can be calculated approximately from the tensile strength. Deviations from these calculated values for small unnotched longitudinal test pieces are due to additional influences such as size, notch effect, surface, shape of cross section and anisotropy.

Apart from tensile strength the *microstructure* has the most important material effect also on this property. A martensite tempered as high as possible has the best

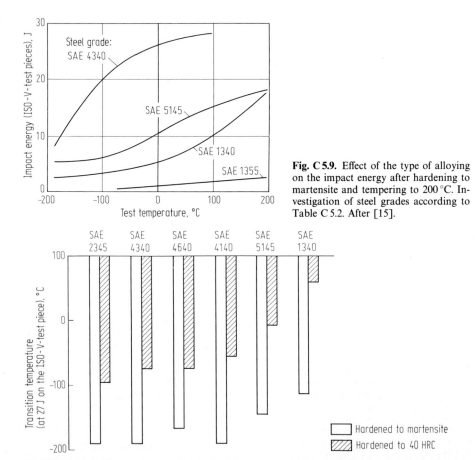

Fig. C 5.9. Effect of the type of alloying on the impact energy after hardening to martensite and tempering to 200 °C. Investigation of steel grades according to Table C 5.2. After [15].

Fig. C 5.10. Transition temperature of the impact energy (for 27 J on ISO-V-test-pieces) of differently alloyed steels according to Table C 5.3 under the additional influence of the degree of hardening. After [15].

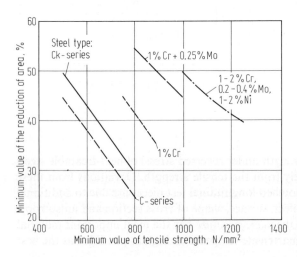

Fig. C 5.11. Relationships between the minimum values of tensile strength and reduction of area for the quenched and tempered condition for the steel series of DIN 17 200. After [31].

Table C 5.2 Chemical composition of steels tested in Fig. C 5.9

Steel	% C	% Si	% Mn	% Cr	% Mo	% Ni	% Al	% N
SAE 1340	0.43	0.24	1.79	0.07	0	0.11	0.020	0.006
SAE 1355	0.56	0.29	1.82		(Coarse grained steel)			
SAE 4340	0.38	1.57	0.88	0.84	0.32	1.82	0.027	0.07
SAE 5145	0.45	0.17	0.68	0.95	0.02	0.04	0.024	0.001

Table C 5.3 Chemical composition of steels tested in Fig. C 5.10

Steel	% C	% Si	% Mn	% Cr	% Mo	% Ni	% Al	% N
SAE 1340	0.43	0.24	1.79	0.07	0	0.11	0.020	0.006
SAE 2345	0.44	0.26	0.83	0.06	0.03	3.55	0.055	0.010
SAE 4140	0.41	0.23	0.82	0.05	0.16	–	0.018	0.005
SAE 4340	0.40	0.29	0.82	0.85	0.25	1.72	0.017	0.010
SAE 4640	0.42	0.24	0.76	0.18	0.26	1.78	0.032	0.008
SAE 5145	0.45	0.17	0.68	0.95	0.02	0.04	0.024	0.001

fatigue strength for a high degree of hardening (see Fig. C 5.12 [32]). *Non-metallic inclusions* can impair the fatigue strength by facilitating the formation of cracks, especially with higher tensile strength. However, a harmful effect need only be feared if large inclusions are present or if they accumulate in critically-stressed regions, e.g. near the surface. For highly-stressed components specially-melted steels may be preferred (see D 4) but only if the much more important factors such as design and component manufacture are taken into account.

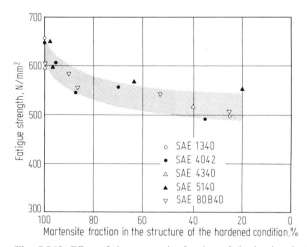

Fig. C 5.12. Effect of the martensite fraction of the hardened structure on the fatigue strength of differently alloyed steels after quenching and tempering to a hardness of 36 HRC. After [32]. The steel SAE 80B40 (= AISI-TS 80B40) is a fine-grained steel containing 0.42% C, 0.98% Mn, 0.31% Cr, 0.13% Mo and 0.35% Ni which was melted with the addition of a compound containing aluminum, boron, titanium and vanadium. The composition of the other SAE steels is similar to the data in the Tables C 5.2 and C 5.3.

Owing to the large number of factors involved it is not possible to give reliable strength values for specified cyclic stresses with individual materials. For the same hardening degree, the same degree of cleanliness, the same homogeneity and the same strength in the quenched and tempered condition, chemical composition of heat-treatable steels has no direct effect on the fatigue strength. In addition to the above-mentioned relationships for calculating the fatigue strengths under reversed stress of small accurately-machined test pieces from the tensile strength, *fatigue strength diagrams* have been established for the standardized heat-treatable steels which give an approximation of the acceptable stress as a function of the mean stress [3, 33–35].

For most applications, the maximum stresses occur at the surface of the parts and the surface is particularly susceptible to damage causing notch sensitivity. Thus, the fatigue strength can be increased above the value for the quenched and tempered condition by *special surface treatments*. In addition to surface hardening methods (see C 5.2.3, C 5.2.4 and C 5.4) processes such as shot peening and surface strengthening by rolling can be used to produce mechanical strain hardening and favorable residual compression stresses in the surface layer [35a, 35b, 35c].

Wear Resistance

The wear resistance of quenched and tempered steels depends primarily on the hardness, i.e. on the tensile strength. For higher wear resistance requirements, one of the surface-hardening processes must be applied (see C 5.2.3, C 5.2.4 and C 5.4). Wear resistance is a property which depends less on the material than the system, so it will not be discussed here further (see [36–39] and B 10).

C 5.2.1.2 Important Processing Properties

Suitability for Heat Treatment

It is obvious that the steels to be used must be suitable for the intended heat treatments (see B 4). For quenching and tempering, which is common to all the groups of steels, *hardenability* is an important property. The results obtained from end-quench tests [40] allow important conclusions to be drawn for individual steels regarding the possibility of obtaining the desired quenching and tempering result [41]. Behavior during tempering is characterized by retention of hardness and resistance to temper embrittlement.

Machinability

As the steels most suited to quenching, tempering and surface hardening are mostly used for production of solid components of many shapes by extensive machining, their machinability is of great importance with regard to production costs. In the quenched and tempered condition machinability is determined mainly by the *tensile strength* but also by toughness, *sulphur content* and the presence of other elements which have a favorable effect on machinability such as lead, selenium, bismuth and tellurium (see B 9). In mass production, for economic and uniform machinability it is usual to restrict the sulphur content to between about 0.020 and 0.035%. For more highly-stressed components, however, for which steels with a considerable proportion of alloying elements are used, a lower sulphur content is preferred with a view to improved toughness and fatigue strength. Lower sulphur is also to be recommended for the avoidance of inclusions stringers in flame, induction or immersion hardened or nitrided surfaces.

Cold Formability

Cold forming of these steels is usually done in the *soft-annealed condition*. However, when quenching and tempering to a low yield stress, forming can also be carried out in that condition.

In addition to some structural parameters the strength of a ferrite-pearlite structure depends, mainly on the *chemical composition* of the steel (see B 4). The tensile strength and chemical composition also determine the stress-strain diagram or the "flow stress" (the true stress for a given true plastic strain) obtained for different degrees of deformation. It is not possible, however, to adjust the chemical composition only for cold formability, the suitability of the material for subsequent heat treatment, i.e. quenching and tempering as well as surface hardening, must also be considered. With the more highly-alloyed heat-treatable steels and nitriding steels there is only little leeway for such consideration. With less highly-alloyed steels the chemical composition can be optimized by achieving the necessary hardenability with those alloying elements which increase the strength in the soft-annealed condition and the strain hardening only slightly. Therefore the carbon content and elements which particularly strengthen the ferrite such as silicon should be reduced as far as possible. A favorable element is boron which, when dissolved and with a content of only 0.002%, makes possible a noticeable increase in hardenability without affecting the strength in the as-rolled or the soft-annealed condition.

Chapters B 7, C 6 and C 29 are devoted to cold deformability in general and to the specially-developed steel grades for cold deformation in particular.

With heat-treatable and surface hardening steels machining and cold deformation processes often are used in combination.

Weldability

For the group of steels treated here, welding, also has a certain importance, especially for the joining of component elements that are produced separately. Although the required carbon content for quenching and tempering as well as for flame, induction and immersion hardening limits the welding possibilities, the use of processes with a low energy input or high energy density such as electron beam and friction welding appear to be favorable. For applications in which suitability for welding is of great importance, special steels of lower carbon content such as steels for pressure vessels or chains have been developed, and are considered in C 9 and C 30.

C 5.2.2 Heat-treatable Steels (Steels for Quenching and Tempering)

C 5.2.2.1 General Selection Criteria

Owing to the often opposing factors affecting the properties of heat-treatable steels, compromises are often necessary in their selection. For example good processing properties must sometimes be obtained by limiting the application properties and vice versa. Economic factors require that the sum of the material and processing costs be minimized and that expenditures be confined to that needed for a particular application.

The fundamental opposing tendencies of strength and toughness force a compromise in the adjustment of the strength in the quenched and tempered condition and/or different properties in the surface and core zones of the components. Often the stress varies across the section and is greatest at the surface so that there is no need for an optimum microstructure in the core. The following types of steel are selected according to the differences in the intensity of stress likely to be encountered in the finished components:

- Unalloyed steels of medium carbon content in the forged or as-rolled condition with a ferrite-pearlite structure.
- Unalloyed steels in the normalized condition with a ferrite-pearlite structure (for properties see e.g. DIN 17 200 [31]).
- Unalloyed or alloyed steels of medium carbon content in the quenched and tempered condition with
  a) a varying degree of hardening (amount of martensite) over the cross section where applicable and
  b) a strength adjustable within certain limits,
- Unalloyed or alloyed steels of which the surface layers can be improved, compared with the quenched and tempered or normalized condition, by flame, induction or immersion hardening or nitriding, and where the composite parts of the base material and the surface layer reach specific properties.

C 5.2.2.2 Steel Grades Proven in Service

Owing to the many factors at work, it is hardly possible to draw clear conclusions for a few optimum steels. On the one hand there is no uniform on the evaluation of the steels and the criteria for their selection, and on the other the availability and cost of raw materials differs from region to region. This situation has resulted in the development of steels in different countries of partly similar and partly completely different types. International standardization introduced after the Second World War has so far not resulted in a comprehensive adaptation of or agreement on specific new types of steel [42, 43]. For mass production it is sometimes possible to utilize less-effective steels and still to obtain acceptable service properties by a strong effort in control and testing. On the other hand, when individual parts and small batches are produced, it is generally more economic to choose safer steels and to save the cost of trials and tests [44].

The most important conclusion to be drawn from C 5.2.1 is that the properties of heat-treatable steels for quenching and tempering depend on the microstructure and, especially the toughness, additionally on the alloy composition. By "microstructure" is meant here primarily the structure after hardening which, in addition to the size of the quenched and tempered cross section and the quenching medium depends on the important material property of "hardenability". The *classification of heat-treatable steels* according to hardenability thus corresponds for a given size to a grading of the structures. Depending on the strength after quenching and tempering and on special requirements as to toughness and fatigue strength a certain degree of hardening (see C 5.2.1.1) must be reached, extending over a specified distance from the surface of the component or throughout the cross section. Account must be taken of whether the components are stressed across the whole section or mainly at the surface. For highly-stressed parts the degree of hardening must be > 0.9 as shown for example in Fig. C 5.13 where tensile strength is taken as the stress criterion [45]. For typical highly-stressed automobile parts in a cross-sectional position at 3/4 of the radius a martensite content in the microstructure of 90% is required [46]. Table C 5.4 lists characteristic grades of heat-treatable steels with their designation, chemical composition and appropriate standard designations. DIN 17 200 gives the ranges of hardenability for the end-quench test of the steel grades considered there. The standard also includes ranges for scatter bands of restricted height and for unalloyed steels restricted ranges for certain distances from the quenched end [31].

For less-critical components there is no need to obtain martensitic hardening across the whole cross section of the part. Unalloyed steels for quenching and tempering can be used in various degrees of hardenability [47]. For components that are susceptible to cracking the possibility of using oil for quenching may be an advantage compared with steels of lower hardenability which can only be quenched in water.

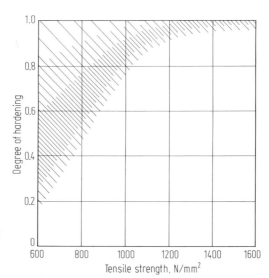

Fig. C 5.13. Required degree of hardening as a function of the necessary strength in the quenched and tempered condition. After [45].

Table C 5.4 Typical heat-treatable steels (steels for quenching and tempering)

Steel grade Designation	Comparable steel grades[a]			Chemical composition[c]				
	DIN 17 200 [31]	Euronorm 83 [42]	ISO/R 683[b] [43]	% C	% Mn	% Cr	% Mo	% Ni	% V
Ck 45[d]	+	2 C 45	I C 45 e	0.42 … 0.50	0.50 … 0.80	–	–	–	–
28 Mn 6	+	28 Mn 6	V 1	0.25 … 0.32	1.30 … 1.65	–	–	–	–
30 MnCrB 5[e]	–	–	–	0.28 … 0.33	1.10 … 1.40	0.20 … 0.45	–	–	–
41 Cr 4[d]	+	41 Cr 4	VII 3	0.38 … 0.45	0.60 … 0.90	0.90 … 1.20	–	–	–
42 CrMo 4[d]	+	42 CrMo 4	II 3	0.38 … 0.45	0.60 … 0.90	0.90 … 1.20	0.15 … 0.30	–	–
30 CrMoV 9	+	–	–	0.26 … 0.34	0.40 … 0.70	2.30 … 2.70	0.15 … 0.25	–	0.10 … 0.20
34 CrNiMo 6	+	35 CrNiMo 6	VIII 3	0.30 … 0.38	0.40 … 0.70	1.40 … 1.70	0.15 … 0.30	1.40 … 1.70	–
34 NiCrMo 16	–	34 NiCrMo 16	VIII 6	0.30 … 0.37	0.30 … 0.60	1.60 … 2.00	0.25 … 0.45	3.70 … 4.20	–

[a] + means: Steel is listed in the standard; – means: Type not listed in the standard
[b] Roman numerals give part of respective ISO standard, arabic numerals the designation of the steel grade in this part
[c] In addition ≤ 0.40% Si, ≤ 0.035% P ≤ 0.030% S
[d] Corresponding steel grades containing 0.020% to 0.035% S are Cm 45, 41 CrS 4 and 42 CrMoS 4
[e] In addition about 0.001% to 0.004% B

Steels that are positioned at the transition between the unalloyed and the heat-treatable steels of medium hardenability are those which are alloyed with *manganese* only, e.g. 28 Mn 6, or the steels containing manganese and boron [48] which may also have small addition of chromium, e.g. 30 MnCrB 5.

For medium-hardenability steels, grades containing 1% Cr, e.g. 41 Cr 4, and especially the *chromium-molybdenum series* containing 1% Cr and 0.25% Mo, e.g. 42 CrMo 4 have proved their worth. Owing to the favorable properties and the economic production methods developed for these chromium-molybdenum steels, multiple alloyed heat-treatable steels of medium hardenability with nickel added have not established themselves to the same extent in Germany as in other western industrial nations. A nickel-free steel of higher hardenability which will provide high strength in the quenched and tempered condition and for large sizes is the grade 30 CrMoV 9. The heat-treatable steels alloyed with *chromium, nickel and molybdenum*, containing about 0.33% carbon are regarded in Germany as having the highest capability, e.g. 34 CrNiMo 6. Heat-treatable alloy steels of this types are used abroad with a comparatively lower chromium content but containing up to 4% Ni, e.g. 34 NiCrMo 16.

Table C 5.5 lists the mechanical properties in the heat treated condition for different ranges of sizes; additional properties can be taken from the standards [31, 42, 43].

C 5.2.3 Steels for Flame, Induction and Immersion Surface Layer Hardening

C 5.2.3.1 Special Aspects of Flame, Induction and Immersion Surface Layer Hardening

To produce improved component properties by flame, induction or immersion surface hardening compared with quenching and tempering *some special requirements* must be taken into account. The aim is to obtain a *complete formation of martensite in the surface layer*. The surface hardness then depends on the carbon content of the steel or on the carbon fraction which is in solution during austenitizing (comp. Fig. C 5.1). As heating is faster and the austenitizing time shorter than for quenching and tempering in most instances higher hardening temperatures are required for sufficient austenitizing. The initial structure also has an influence. Particularly rapid heating is used for the surface-hardening processes in which the critical cooling rate is often achieved without additional quenching media only through thermal conduction into the cold core of the component [49, 50].

Surface hardness often is reduced slightly by tempering at 140 to 200 °C. The component may be heated right through or only at the hardened surface layer. The aim is mainly to relieve stress peaks and to reduce cracking susceptibility during straightening and grinding.

The thickness of the hardened surface layer (the hardness penetration) obtained by flame, induction and immersion surface hardening which can be characterized, for example, by a specified limit hardness (e.g. Rht 550 HV [51]) depends on the depth of the austenitized layer, the cooling rate and the hardenability of the steel.

Table C 5.5 Mechanical properties of the steel grades listed in Table C 5.4 in the quenched and tempered condition for two ranges of diameters

Steel grade Designation	Mechanical properties[a]									
	Diameters over 16 up to 40 mm					Diameters over 40 up to 100 mm				
	$R_{p\,0.2}$ N/mm² min.	R_m N/mm²	A % min.	Z % min.	A_v[b] J min.	$R_{p\,0.2}$ N/mm² min.	R_m N/mm²	A % min.	Z % min.	A_v[b] J min.
Ck 45	430	650...800	16	40	30	370	630...780	17	45	30
28 Mn 6	490	690...840	15	45	45	440	640...790	16	50	45
30 MnCrB 5	590	800...950	13	40	40	480	700...850	15	45	40
41 Cr 4	660	900...1100	12	35	40	560	800...950	14	40	40
42 CrMo 4	750	1000...1200	11	45	40	650	900...1100	12	50	40
30 CrMoV 9	1020	1200...1450	9	35	30	900	1100...1300	10	40	35
34 CrNiMo 6	900	1100...1300	10	45	50	800	1000...1200	11	50	50
34 NiCrMo 16	1050	1250...1450	9	40	40	950	1150...1350	10	45	45

[a] Applies to longitudinal test-pieces whose center line is 12.5 mm from the surface of the part
[b] Applies to DVM test-pieces

C 5.2 Heat-treatable Steels

Compared with hardening after heating the component right through, this kind of surface hardening even with plain carbon steels, allows greater depths of hardness to be produced. Heat removal from the core is not required which contributes to the cooling process, increasing the cooling rate in the surface zones. However it is scarcely possible to exceed a hardness penetration of 3 mm for plain carbon steels. The dependence of the hardness produced by induction hardening on the initial microstructure is shown by an example in Fig. C 5.14 [52].

Flame, induction and immersion surface hardening do not affect the microstructure of the core material which is generally in the quenched and tempered condition although in unalloyed steels it may be in the normalized condition. To be suitable for this kind of surface hardening a steel must also resist crack formation and distortion as well as grain coarsening. In addition to being largely unaffected by overheating and suitable hardenability, the hardening process (austenitizing and quenching method) must be matched to the shapes of components to be produced and to the desired hardness change over the cross section.

The toughness of hardened surface layers is lower than that of the core material and for the whole cross section it is also affected by the strength and toughness of the core. Because of testing difficulties, few indications of toughness values for flame, induction or immersion hardened test pieces hardly are available [36, 53].

These surface hardening methods are used to *improve wear resistance and fatigue strength*. Important parts are played in these improvements by the increase in surface hardness, matching the depth of hardness to the overall cross section of the component and the introducton of compressive internal stress into the surface layer. Flame, induction and immersion surface hardening is better suited for hardening components only partly at the most highly-stressed areas than are case hardening and nitriding. Extensive literature exists on values of the fatigue strength of flame, induction and immersion surface hardened specimens and components (e.g. [3, 53–56]).

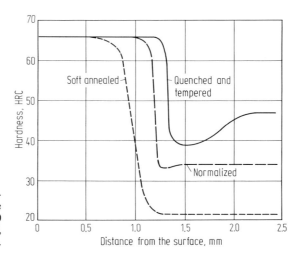

Fig. C 5.14. Effect of the initial structure for induction hardening on the variation of hardness for a SAE 1070 steel containing about 0.70% C, 0.20% Si and 0.80% Mn. After [52].

C 5.2.3.2 Steel Grades Proven in Service

In principle all heat-treatable steels (steels for quenching and tempering) can be used for flame, induction and immersion surface layer hardening provided their *carbon content* can guarantee the *required surface hardness* and their *hardenability* the *required core properties and hardness penetration* depending on the size. With increasing carbon content the susceptibility to quench cracking increases and the toughness of the core material is reduced, so the carbon content should be no higher than is required to reach the required surface hardness. For most applications the proportion of alloying elements is lower than for case-hardening and nitriding steels.

Table C 5.6 gives details of some characteristic *representatives of steel grades* used in Germany. The carbon content of the unalloyed steels used varies between 0.35 and 0.75%. For higher requirements fine-grain steels are used and are supplied in several hardenability grades. Details of hardenability values can be seen in the material standards [31, 57–59]. If higher core strengths are needed for larger component sizes or if hardness penetrations exceeding about 3 mm are required, alloy steels must be used. Suitable grades available contain 0.5% or 1% Cr, e.g. 45 Cr 2 and 38 Cr 4, or 1% Cr with 0.25% Mo, 49 CrMo 4 for example. The steels containing molybdenum have the considerable advantage of a higher retention of hardness so that residual stresses can be reduced during tempering. Where there is a high susceptibility to cracking of the components to be hardened, the alloy steels with sufficient hardenability can be cooled gently in an oil emulsion instead of being quenched in water.

The *mechanical properties* in the quenched and tempered condition for two size ranges and the achievable surface hardness are shown in Table C 5.7 for some selected steel grades.

C 5.2.4 Nitriding Steels

C 5.2.4.1 Special Features of Nitriding

In general usage as well as in standards it is usual to refer to steels as nitriding steels when the hardness of their nitrided surface is particularly high and they are to be treated by gas nitriding. Because of the different requirements and the improvement of many properties however, nitriding is used with many other steels such as tool steels, high-speed steels, ultra-high-strength steels and stainless steels. For components in the vehicle and machine construction industries, in addition to nitriding steels numerous low-alloy and carbon case hardening and heat-treatable steels are nitrided. This confers an increased capacity to withstand stress combined with a satisfactory toughness compared with the quenched and tempered condition although highest surface hardness values may not be obtained. For these types of steel, salt bath nitriding may be used although gas nitriding, too, with its numerous variations of short duration is being used more and more.

As with flame, induction and immersion surface hardening the condition of the core material does not change with nitriding. *A quenched and tempered initial*

Table C 5.6 Typical steel grades for induction and flame surface hardening

Steel grade Designation	Comparable steel grades[a] in			Chemical composition[c]				
	DIN 17212 [57]	Euronorm 86 [58]	ISO/R 683[b] [59]	% C	% Si	% Mn	% Cr	% Mo
Cf 53	+	C 53	XII 5	0.50...0.57	0.15...0.35	0.40...0.70	—	—
45 Cr 2	+	45 Cr 2	XII 6	0.42...0.48	0.15...0.40	0.50...0.80	0.40...0.60	—
38 Cr 4	+	38 Cr 4	XII 7	0.34...0.40	0.15...0.40	0.60...0.90	0.90...1.20	—
49 CrMo 4	+	—	—	0.46...0.52	0.15...0.40	0.50...0.80	0.90...1.20	0.15...0.30

[a] + means: Steel is listed in the standard; — means: Type not listed in the standard
[b] Roman numerals give part of respective ISO standard, arabic numerals the designation of the steel grade in this part
[c] In addition ≤ 0.025% P and ≤ 0.35% S, by choice also 0.020% to 0.035% S

Table C 5.7 Mechanical properties of the steel grades listed in Table C 5.6 in the quenched and tempered condition and surface hardness after flame or induction hardening

Steel grade Designation	Mechanical properties[a]									Surface hardness[b] applies to diameter up to mm
	Diameters over 16 up to 40 mm				Diameters over 40 up to 100 mm					
	$R_{p\,0.2}$ N/mm² min.	R_m N/mm²	A % min.	Z % min.	A_v^c J min.	$R_{p\,0.2}$ N/mm² min.	R_m N/mm²	A % min.	Z % min.	A_v^c J min.	HRC min.	
Cf 53	430	690...830	14	35	—	400	640...780	15	40	—	57	100
45 Cr 2	540	780...930	14	45	42	440	690...830	15	50	42	55	100
38 Cr 4	630	830...980	13	45	42	510	740...880	14	50	42	53	100
49 CrMo 4	—	—	—	—	—	690	880...1080	12	50	35	56	250

[a] Applies to longitudinal test-pieces whose center line is 12.5 mm from the surface of the part
[b] Applies to the condition after quenching and tempering (for Cf 53 after normalizing) and surface hardening followed by stress relieving at 150 to 180 °C for about 1 hr
[c] Applies to DVM test-pieces

condition is advantageous both for good nitriding and for the optimum interaction of the surface and core material in the nitrided component. The same principles apply to quenching and tempering nitriding steels as have been described for heat-treatable steels in general. Unalloyed heat-treatable steels are also nitrided in the normalized condition. The strength required in the core material is related to its task of supporting the outer surface layer and preventing it from being indented under high surface pressures. The retention of hardness of nitriding steels must be sufficient to ensure that the hardness of the quenched and tempered condition is maintained after the nitriding process. This retention of hardness can be obtained with higher strength steels by alloying with chromium, molybdenum and vanadium. In order to lessen a tendency towards brittleness when nitriding for long periods at temperatures around 500 °C, steels intended for gas nitriding usually contain about 0.2% Mo.

Nitrided layers consist of the compound and the diffusion layer. In the compound layer which usually has a thickness of up to 50 µm, iron is converted completely into iron nitride or iron carbonitride by taking up nitrogen. Towards the core the adjoining diffusion layer contains the nitrogen dissolved in the iron or precipitated in the form of the finest nitrides [63]. The structure and nitride phases of the layers are influenced intentionally by the nitriding process and medium with the intention of optimizing certain properties [64] by addition of other elements to the nitriding medium, particularly carbon.

The *hardness of the compound and the diffusion layers* depend largely on the chemical composition of the steel as well as on the nitriding temperature. While unalloyed, heat-treatable steels can be hardened to about 400 HV, addition of alloying elements which form hard nitrides, such as chromium, molybdenum, vanadium and aluminum, can produce hardnesses up to 1100 HV as shown in Figs. C 5.15a and C 5.15b [65, 66]. A moderate increase in the hardness of the nitrided case, as well as its thickness, is also obtained by decreasing the carbon content.

The differences between *different nitriding processes* have little effect on the surface hardness [67]. The depth of penetration of the nitrogen, and hence the depth of hardness due to nitriding, depend mainly on the nitriding temperature and duration of treatment, but are also affected by the type and content of the nitride formers dissolved in the steel (see Fig. C 5.16 [67]). Strong nitride formers reduce the depth of penetration of the nitrogen. High-alloy types of steel such as hot working and high-speed steels permit only thin diffusion layers. Surface hardness and depth of nitriding hardness also depend on the condition of the quenched and tempered steel. With higher tempering temperatures, chromium, molybdenum and vanadium are largely combined as carbide and are not fully available for forming nitrides so that the nitriding hardness is reduced [68, 69].

When testing surface hardness and the depth of hardness due to nitriding [51], account must be taken of the dependence of the hardness values on the force applied in testing, especially where the nitrided layers are thin [67].

Hard nitrided layers have a *lower toughness* than the base material. Increasing the content of nitride formers and the resulting greater hardness, increases the brittleness of the nitrided layers. In addition, the toughness depends on the kind of

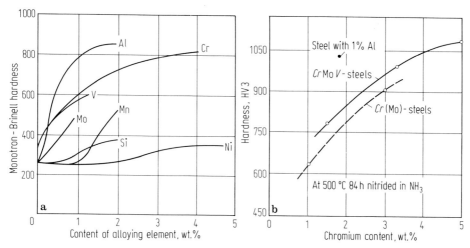

Fig. C 5.15. Effect of some alloying elements on the surface hardness after gas nitriding. **a** after [65], **b** after [66].

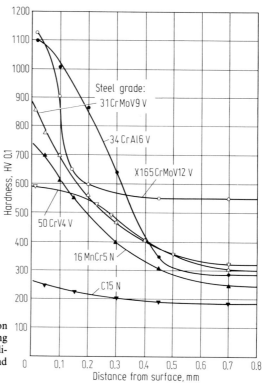

Fig. C 5.16. Effect of steel composition on the variation of hardness after gas nitriding at 500 °C for 60 h. After [67]. Initial condition N = normalized, V = quenched and tempered.

the nitride phases. Just as with case-hardening steels the toughness of surface-nitrided specimens depends largely on the strength and toughness of the core material and the depth of hardness of the surface layers (see Fig. C 5.17 [66, 70]).

Like other surface hardening processes, nitriding results in a high *increase in fatigue strength* especially for notched samples, that is for a steep variation in the stress from the surface to the core. This behavior corresponds to a reduction in notch sensitivity. With nitriding it is particularly the diffusion layer which produces an increase in the fatigue strength. For this reason unalloyed steels for quenching and tempering are often nitrided in a salt bath and then must be cooled as fast as possible to hold the nitrogen in the supersaturated solution. Fatigue strength and toughness of the resulting unstable material condition can often be affected unfavorably by precipitation-related aging phenomena caused by plastic deformation or by a moderate increase in temperature above ambient. Higher fatigue strength values can be obtained by nitriding of steels having a high core strength (see C 5.8). For impact-stressed components that must withstand stresses above the fatigue limit but applied less frequently, the fatigue strength for finite life is not increased to the same extent by surface-hardening processes as is the fatigue limit. This rule applies especially to nitriding.

Optimization of the fatigue strength of surface-hardened components requires that the hardness change, the strength and toughness of the core material and a favorable structure of the case are matched to the particular load application. The

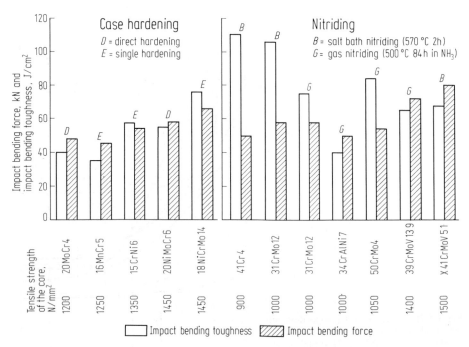

Fig. C 5.17. Comparison of the results of impact bending tests on different steels after case hardening or nitriding. After [66, 70].

Table C 5.8 Typical steel grades for nitriding

Steel grade Designation	Comparable steel grades[a]			Chemical composition[c]					
	DIN 17 211 [60][d]	Euronorm 85 [61]	ISO/R 683[b] [62]	% C	% Mn	% Al	% Cr	% Mo	% Ni	% V
Ck 45	[e]	[e]	[e]	0.42...0.50	0.50...0.80	–	–	–	–	–
42 CrMo 4	[e]	[e]	[e]	0.38...0.45	0.50...0.80	–	0.90...1.20	0.15...0.30	–	–
15 CrMoV 59	+	–	–	0.13...0.18	0.80...1.10	–	1.20...1.50	0.80...1.10	–	0.20...0.30
31 CrMo 12	+	31 CrMo 12	X1	0.28...0.35	0.40...0.70	–	2.80...3.30	0.30...0.50	–	–
39 CrMoV 13 9	–	39 CrMoV 13 9	X2	0.35...0.42	0.40...0.70	–	3.00...3.50	0.80...1.10	–	0.15...0.25
34 CrAlNi 7	+	–	–	0.30...0.37	0.40...0.70	0.80...1.20	1.50...1.80	0.15...0.25	0.85...1.15	–

[a] + means: Steel is listed in the standard; – means: Type not listed in the standard
[b] Roman numerals give part of respective ISO standard, arabic numerals the designation of the steel type in this part
[c] In addition $\leq$ 0.40% Si, $\leq$ 0.030% P and $\leq$ 0.030% S
[d] According to draft of March 1985 (In the meantime: Ed. April 1987)
[e] Contained in the respective standards for heat-treatable steels = steels for hardening and tempering (see Tables C 5.4 and C 5.5)

built-up layer e.g. the compound layer zone after nitriding, is of special importance for the rolling contact fatigue strength [38, 64, 71]. Reference is made to the literature regarding the values of the fatigue strength of nitrided specimens and components [37, 38, 54, 67, 70, 72, 73].

In addition to hardness, phase structure and porosity are also important for the *wear resistance of nitrided cases* which therefore depends also on the nitriding process and its execution. The compound layers of nitrided steels offer special advantages for reducing adhesion wear (galling).

C 5.2.4.2 Steel Grades Proven in Service

Table C 5.8 shows data for examples of steels of different alloys used for nitriding. The steels Ck 45 and 42 CrMo 4 are listed to represent the large group of the usual steels suitable for quenching and tempering. Table C 5.9 gives data for the *mechanical properties* of the initial quenched and tempered condition and the achievable surface hardness after nitriding (see Fig. C 5.15b).

The steel 31 CrMo 12 and particularly the *chromium-molybdenum-vanadium steels* have high hardenability and retention of hardness so that even in large sizes they permit through-hardening and high core strength values. Depending on the requirements arising from the application and from processing, the values for the core strength can be varied within certain limits.

The steel 34 CrAlNi 7 represents the steel grades alloyed with 1% Al which were developed specially for gas nitriding. The nitrided layers of these steels can reach the highest surface hardness but on the other hand have the lowest ductility. The steels containing aluminum in addition have the disadvantage of requiring special melting procedures to ensure a high degree of cleanliness.

Table C 5.9 Data (for the steels listed in Table C 5.8) of mechanical properties in the quenched and tempered condition and the surface hardness after nitriding

Steel grade Designation	Diameter mm	Mechanical properties[a]				Hardness of the nitrided surface about
		$R_{p0.2}$ N/mm² min.	R_m N/mm²	A % min.	A_v[b] J min.	
CK 45	≤ 40 > 40 ≤ 100	430 370	650...800 630...780	19 20	30 30	350
42 CrMo 4	≤ 100 > 100 ≤ 160	650 550	900...1100 800...950	12 13	40 40	600
15 CrMoV 5 9	≤ 100 > 100 ≤ 250	750 700	900...1100 850...1050	10 12	35 40	800
31 CrMo 12	≤ 100 > 100 ≤ 250	800 700	1000...1200 900...1100	11 12	45 45	800
39 CrMoV 13 9	≤ 70	1100	1300...1500	8	25	900
34 CrAlNi 7	≤ 100 > 100 ≤ 250	650 600	850...1050 800...1000	12 13	35 40	950

[a] Applies to longitudinal test-pieces whose center line is 12.5 mm from the surface of the part
[b] Applies to DVM test-pieces

C 5.3 Steels for Heavy Forgings

C 5.3.1 Required Properties

Heavy forgings are used primarily in the manufacture of turbine and generator components, in the construction of pressure vessels, in the building of engines and ships and in other heavy machinery. They generally transmit high forces and are often formed symmetrically around the axis of rotation. In high speed rotating machinery, these components are frequently subjected to large centrifugal forces. Among the largest and most heavily-stressed components are turbine and generator shafts used in the construction of power stations. The dimensions of these components approach the limits of today's manufacturing capabilities. The arising problems have been and are the starting points for progress and innovation in the development of materials and processing techniques.

On the whole the required properties for heavy forgings depend on the application. High yield-stress values, coupled with high demands for toughness, in relation to the large dimensions, are of primary importance. Special demands may be made on the variation of material characteristics across the section. Property requirements such as elevated or high-temperature strength, fatigue strength, surface hardenability, magnetic induction or non-magnetizability, corrosion resistance, resistance to high pressure hydrogen, weldability and certain residual stress conditions may also be added.

The *additional requirements* which are made on the *integrity* of the whole or certain parts of the steel workpiece are characteristic for heavy forgings, and are related to the respective stress conditions. For turbine rotors and generator shafts the requirements particularly concern toughness and freedom from flaws. These requirements are based on the safety risk and the extraordinary damage which can occur in the event of a failure [74].

C 5.3.2 Characterization of the Required Properties by Test Results

Large forgings used in ship, engine and machine construction exhibit marked differences in the mechanical properties between the surface regions and core as a function of the tempered condition; raw ingots may weigh up to 500 t. Even the more through hardened, most highly-stressed forgings for power station construction have different mechanical properties depending on the position of the test pieces taken from the part. The *positions of the test-pieces* are therefore of special importance when specifying the mechanical properties and their tests.

For many components the stresses in the outer parts of the section are greater than those in the inner parts. It is therefore appropriate, also from the point of view of the application, to determine the material properties in the outer zone, generally within one sixth of the diameter measured from the outside. When describing the direction of the specimen as longitudinal, transverse, tangential or radial it is not meaningful to refer these terms to the external dimensions of the component. It is preferable to specify them in relation to the direction of the maximum grain flow in the component. For components that will rotate at high speed, and whose inner portions also are highly stressed, the determination of the mechanical properties in the center of the cross-section and in the transition zone is specified for specimens from trepanned cores [74–76].

Mechanical *tests in the longitudinal direction* with specimens taken from the longitudinal axis, establish the properties of the position generally less affected by segregation and with the lowest cooling rate when hardening. *Radially-trepanned cores* may be taken partly vertically as blind holes to a specified depth and partly through the whole thickness of the forging through the center or at a certain angle deviating from the radial direction into the main segregation zone of the forging. Test-pieces of this type are partly tensile-tested as a whole (over the whole length) and partly tensile-tested in several sections along their length to identify the worst positions of the section and the differing mechanical properties in the different zones of the section. In the past these tests have had considerable importance for establishing the relationships between the mechanical properties and segregation and for finding flakes in forgings. Nowadays these test-pieces must also be regarded as particularly suitable for identifying segregations in the transition zone of the cross section as well as the mechanical properties in this critical area.

Characteristics such as nil ductility transition (NDT) temperature and fracture toughness (K_{Ic} value of fracture mechanics) are used to describe the applicability of forgings for power station construction in addition to the usual criteria of 0.2%-proof stress and the transition temperature of impact energy values. The results of creep tests at elevated temperatures and magnetic induction tests of generator shafts may also be used.

The most important *non-destructive test* for heavy forgings is *ultrasonic testing*, which is used to detect any cavities, larger inclusions and cracks as well as indicating the position and approximate size of such defects. Generally the pulse-echo technique is used at an appropriate frequency for ultrasonic transmission. With forgings that require lengthy manufacture operations, ultrasonic tests are carried out as early as possible and may be repeated several times up to the acceptance test. Magnetic powder tests and dye penetrant tests are also important to ensure that the machined surface of heavy forgings is free from flaws.

For the *measurement of residual stresses* in the surface zones of forgings the annular core method has proved itself. A small annular slot is milled into the surface and the stress relief strain is measured by means of a previously-applied strain gage [77].

Thermal stability tests may be [78] carried out on medium and high-pressure turbine shafts to ensure that they do not show unbalance even at varying operating temperatures.

C 5.3.3 Measures for the Attainment of the Required Properties

Processing steps for the manufacture and attainment of the required properties are basically the same for large forgings as for rolled and forged products of lower

C 5.3.3.1 Melting and Casting

Melts for casting forging ingots up to 500 t in weight are generally produced in electric arc furnace, or by the basic oxygen steelmaking process or by combinations of these.

From the point of view of favorable ingot solidification, and taking the subsequent forging operations into account, *ingot geometry* and especially the length-diameter ratio and the hot top volume play a decisive part. Ratios of length to diameter substantially exceeding 2 are generally avoided and the larger the ingot diameter becomes the nearer the ratio should be to 1.

Special metallurgical measures are need to keep the extent of segregation to a minimum (Fig. C 5.18), so as to obtain the densest and, as far as possible, defect-free ingots, which will ensure further defect-free processing.

Since the first large-scale experiments at the beginning of the fifties [79] *vacuum treatment* of the steel or the casting in a vacuum of ingots for heavy forgings has been adopted worldwide and almost without exception. Among the numerous technical methods available, pouring stream degasification processes for high quality forging ingots of alloy steels are most widely used. Heat treatment after

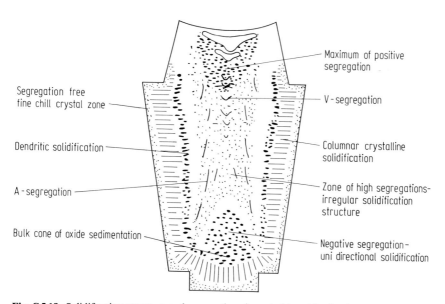

Fig. C 5.18. Solidification structure and segregations in a steel ingot for forgings.

forging can be considerably simplified and flaking prevented by producing a low hydrogen content in the steel. Use of vacuum carbon-deoxidation in place of the classic precipitation deoxidation can markedly reduce the number and diameter of the A segregation strings (see Fig. C 5.18). In addition the concentration deviations from the mean chemical composition, which occur in the strings, can be held down. The shape of the manganese sulphides also is converted more to the spheroidal form of type 1 thus reducing the impairment of the mechanical properties. Preconditions for reducing the content of oxide inclusions in the steel are also improved [80–83].

Other possibilities for improving ingot quality are *extreme desulphurization* and *modification of the sulphide shape*, e.g. by combining sulphur with calcium. This also reduces the directionality of the mechanical properties of the forgings [84].

The production of ingots that are as flawless as possible for good forgeability is also improved by the use of different methods for *hot topping feeder insulation* and *heating*. In particular a closed solidification of the ingot is obtained at least to the hot top line [85]. When casting very large ingots the use of controlled *after pouring* of low-carbon and low-alloy steel during solidification of the ingot retards core segregation [86]. More expensive methods include remelting of the core zones [87] and complete *remelting* of heavy ingots by electro-slag remelting [88] whereby largely segregation-free and dense ingots can be produced. Experiments to improve the primary structure by *forced cooling* of the mold from the outside however, have not so far led to a decisive success. *Hot forming* of heavy forgings is considered in D 6 so that further details need not be given here.

C 5.3.3.2 Heat Treatment

Heat treatment to produce a microstructure with the required properties usually consists of *quenching and tempering* with preceding, and sometimes multiple, *normalizing*, and it is carried out before or after machining operations. Grain refining in higher alloy steels is restricted, so that several transformation annealing operations (Fig. C 5.19) may become necessary [89] partly to make adequate

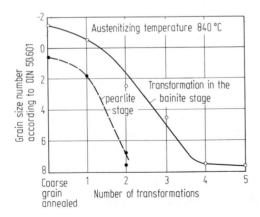

Fig. C 5.19. Change of grain size of a quenched and tempered 26 NiCrMoV 14 5 steel with the number of austenite transformations (laboratory tests). After [89].

ultrasonic testing possible. In particular A-segregation strings tend to remain coarse-grained. Cooling for hardening is carried out in air, oil or water depending on chemical composition, part dimensions and required properties. Water quenching either by immersion or by spraying (partly under rotation) is widely used nowadays on large parts with toughness requirements.

In unalloyed heat-treatable steels in the range of sizes of large forgings no martensitic or bainitic *hardness* is obtained by quenching in water even in the surface zone. In many of the heat-treatable alloy steels used for heavy forgings, with water quenching the formation of martensite can be only expected in the surface zone. Mostly bainite structures with varying proportions of ferrite are produced in the core, especially with oil hardening. Through-hardening free from ferrites can be obtained in these large sections only with higher alloy steels, meaning steels containing about 2 to 4% Ni or those with a higher chromium and molybdenum content (see Table C 5.10).

It is not always possible to use the highest cooling rate technically possible because high internal stresses build up during hardening *owing to the high temperature differences between the surface and the core* which subject the surface and later the core – after the stress reversal – to tensile stresses. If there is insufficient control over the heat treatment processes, these stresses can lead to the fracture of the part. To reduce the risk, one approach is to make calculations of the stress variation in the part during cooling.

Some of the steels used for heavy forgings can tend towards *temper embrittlement*. This problem is of special importance for the steels which are through-hardened most completely and which contain 3.5% Ni, 1.5% Cr, 0.3% Mo and 0.1% V. Precautions can be taken by keeping the proportion of the elements that increase temper embrittlement namely, phosphorus, antimony, tin and arsenic, as low as possible [90, 91] and by avoiding addition of silicon by using the vacuum carbon-deoxidation process [89].

C 5.3.3.3 Relationship Between Alloying Content, Microstructure and Mechanical Properties

The relationship between alloying content, microstructure and mechanical properties is dealt with below using the example of nickel-chromium-molybdenum-vanadium steels.

Figures C 5.20 and C 5.21 show the *effect* of different proportions of *chromium and nickel on the 0.2%-proof stress and on the transition temperature of the impact energy*. These figures are based on the two types of structure, bainite and martensite, produced by simulated heat treatment. The martensite is characteristic of the surface zone and the bainite of the core zone of large shafts. As increase in the chromium or nickel content causes the transition temperature of impact energy to be lowered for both the martensite and bainite structures. The effect on the bainite structure in the core is particularly great. Furthermore the 0.2%-proof stress values are reduced somewhat where the effect of the martensite structure is greater so that within the range of the usual proportions the 0.2%-proof stress values are roughly the same for both types of microstructure.

Table C 5.10 Common steel grades for large forgings, their chemical composition and order of their applicability by the largest diameter, 0.2% proof stress and impact energy

Steel grade Designation	SEW[a]	Chemical composition							Mechanical properties			
		% C	% Si	% Mn	% P, % S[b]	% Cr	% Mo	% Ni	% V	Diameter[c] mm	$R_{p\,0.2}$ N/mm² min.	Test piece	A_v min.
Ck 45	550[d]	0.42...0.50	≤ 0.40	0.50...0.80	2					> 250 ≤ 500	325	DVM	38
28 Mn 6	550[d]	0.25...0.32	≤ 0.40	1.30...1.65	2					> 250 ≤ 500	345	DVM	41
20 MnMoNi 4 5	550	0.17...0.23	≤ 0.40	1.00...1.50	1					> 250 ≤ 500	390	DVM	41
42 CrMo 4	550[d]	0.38...0.42	≤ 0.40	0.60...0.90	2	0.90...1.20	0.45...0.60			> 500 ≤ 750	390	DVM	38
23 CrMo 5	555	0.20...0.28	≤ 0.30	0.30...0.80	2	0.90...1.20	0.15...0.30			≤ 750	400	ISO-V	47
22 NiMoCr 3 7	640	0.17...0.25	≤ 0.35	0.50...1.00	4	0.30...0.50	0.50...0.80	≤ 0.60		≤ 700	420	ISO-V	41
28 NiCrMo 5 5	555	0.26...0.32	≤ 0.30	0.15...0.40	5	1.0...1.3	0.25...0.45	0.60...1.20		≤ 750	500	ISO-V	55
20 CrMoNiV 4 7	555	0.17...0.25	≤ 0.30	0.30...0.80	5	1.1...1.4	0.80...1.0	1.0...1.3	≤ 0.03	≤ 750	550	ISO-V	31
34 CrNiMo 6	550[d]	0.30...0.38	≤ 0.40	0.40...0.70	2	1.40...1.70	0.15...0.30	1.40...1.70	≤ 0.15	> 500 ≤ 1000	490	DVM	41
30 CrNiMo 8	550[d]	0.26...0.34	≤ 0.40	0.30...0.60	2	1.80...2.20	0.30...0.50	1.80...2.20		> 500 ≤ 1000	590	DVM	41
23 CrNiMo 7 4 7	555	0.20...0.26	≤ 0.30	0.50...0.80	5	1.7...2.0	0.60...0.80	0.90...1.2	0.25...0.35	≤ 1000	600	ISO-V	47
26 NiCrMoV 8 5	555	0.22...0.32	≤ 0.30	0.15...0.40	5	1.0...1.5	0.25...0.45	1.8...2.1	0.05...0.15	≤ 1000	600	ISO-V	63
32 CrMo 12	550	0.28...0.35	≤ 0.40	0.40...0.70	1	2.80...3.30	0.30...0.50			> 750 ≤ 1250	490	DVM	34
30 CrMoNiV 5 11	555	0.28...0.34	≤ 0.30	0.30...0.80	5	1.1...1.4	1.0...1.2	0.50...0.75	0.25...0.35	≤ 1500	550	ISO-V	31
X 21 CrMoV 12 1	555	0.20...0.26	≤ 0.50	0.30...0.80	3	11.0...12.5	0.80...1.2	0.30...0.80	0.25...0.35	≤ 1500	600	ISO-V	24
26 NiCrMoV 11 5	555	0.22...0.32	≤ 0.30	0.15...0.40	5	1.2...1.8	0.25...0.45	2.4...3.1	0.05...0.15	≤ 1800	600	ISO-V	71
26 NiCrMoV 14 5	555	0.22...0.32	≤ 0.30	0.15...0.40	5	1.2...1.8	0.25...0.45	3.4...4.0	0.05...0.15	≤ 1800	700	ISO-V	71
33 NiCrMoV 14 5	550	0.28...0.36	≤ 0.40	0.20...0.50	1	1.00...1.70	0.30...0.60	3.20...4.00	≤ 0.15	> 1500 ≤ 2000	685	DVM	34

[a] The list of references contains the exact source for SEW 550 under [92], SEW 555 under [93] and SEW 640 under [97]
[b] The figures give the following maximum permissible content:
1: ≤ 0.035% P and S each, 2: ≤ 0.035% P, ≤ 0.030% S, 3: ≤ 0.25% P, ≤ 0.020% S, 4: ≤ 0.020% P and S each, 5: ≤ 0.015% P, ≤ 0.018% S
[c] The largest diameter listed in the standard for the respective steel grade with the values for 0.2%-proof stress and notch impact energy at + 20 °C valid for this diameter
[d] For dimensions up to 250 mm diameter DIN 17 200 applies [31]

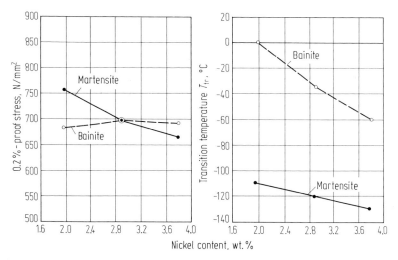

Fig. C 5.20. Effect of nickel content on the 0.2%-proof stress and the transition temperature of the impact energy of nickel-chromium-molybdenum-vanadium steels. After [89]. Test melts contained 0.23% C, 0.29% Si, 0.39% Mn, 0.013% P, 0.012% S, 1.80% Cr, 0.47% Mo, 0.007% Sn and 0.14% V. Heat treatment
– of the test-pieces from the surface zones: 900 °C/oil + 610 °C 30 h/air: martensite, tempered;
– of the test-pieces from the core: 900 °C/with 50 °C/h to 610 °C 30 h/air: bainite and 10% martensite, tempered.
The transformation temperature T_{tr} was chosen as the temperature at which the cross-section area of the fractured ISO-V test-piece exhibits a 50% crystalline appearance.

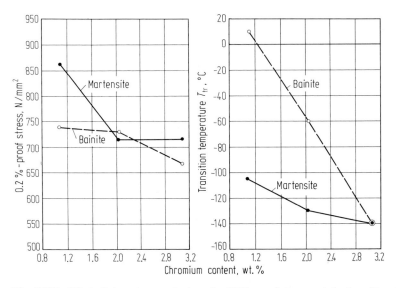

Fig. C 5.21. Effect of chromium content on the 0.2%-proof stress and the transition temperature of nickel-chromium-molybdenum-vanadium steels. After [89]. Basic composition of the test melts as in Fig. C 5.20 containing, however, 3.4% Ni. For the heat treatment of the test-pieces and the transformation temperature the same applies as for Fig. C 5.20.

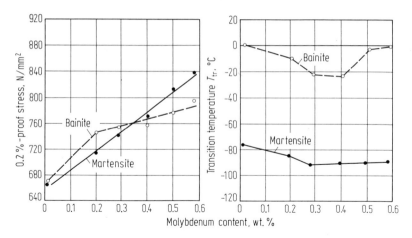

Fig. C 5.22. Effect of molybdenum content on the 0.2%-proof stress and the transition temperature of nickel-chromium-molybdenum-vanadium steels. After [89]. Basic composition of the test melts as in Figs. C 5.20 and C 5.2 but containing 1.85% Cr and 3.46% Ni.
Heat treatment
- of the test-pieces from the surface zones: 910 °C/oil + 640 °C 15 h/oil: martensite, tempered;
- of the test-pieces from the core: 910 °C/h to 640 °C 15 h/oil: bainite with ≤ 10% martensite, tempered. For the transition temperature see Fig. C 5.20.

Figure C 5.22 shows the *effect of the molybdenum content* on the 0.2%-proof stress and on the impact energy as well as on the transition temperature. The yield stress can be increased considerably by increasing the molybdenum content. Moreover by raising the molybdenum contents to about 0.3 to 0.4% the transition temperature is improved slightly. The reduction in the tendency to temper embrittlement caused by molybdenum, as measured by the displacement ΔT_{tr} of the transition temperature of the impact energy to the higher values shown in Fig. C 5.23 is of special importance. Here, too, an optimum occurs in the range of about 0.3 to 0.5% Mo, and for contents above 0.5% a reduction of the effect, i.e. a stronger embrittlement, is indicated.

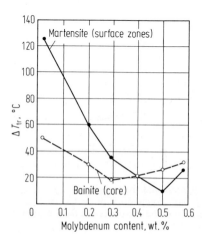

Fig. C 5.23. Effect of the molybdenum content on the temper embrittlement of the test melts from Fig. C 5.22. After [89]. The quenching and tempering shown in Fig. C 5.22 is followed by step cooling: 600 °C 4 h → 540 °C 15 h → 525 °C 24 h → 495 °C 48 h → 465 °C 72 h → 315 °C/air. ΔT_{tr}, the difference between the transition temperature T_{tr} for the brittle condition obtained by the step cooling and the transition temperature T_{tr} when avoiding the temper embrittlement (here 600 °C 4 h/water) was chosen as a characteristic for the embrittlement.

The relationships illustrated here apply similarly to the other heat-treatable alloy steels used for large forgings.

C 5.3.4 Steel Grades Proven in Service

Steels are selected for different applications first on the basis of the required values of yield stress and tensile strength, the toughness requirements and component cross-sections.

For minimum values of yield stress up to about 300 N/mm^2, plain carbon steels may be considered for diameters up to 500 mm (see Ck 45 and 28 Mn 6 in Table C 5.10). Higher requirements for toughness within the same range of forging dimensions and strengths are met by low alloy grades of which the steels 20 MnMoNi 4 5 and 42 CrMo 4 listed in the table are examples.

To the extent to which the heat-treatable cross section and the requirements for strength and toughness increase, higher degrees of alloying must be selected until for a 2000-mm diameter, steels containing 3.5% Ni, 1.5% Cr, 0.5% Mo and 0.1% V are reached. With these grades, yield stresses of about 700 N/mm^2 can be obtained, with favorable toughness properties.

In Germany steels for heavy forgings are mostly standardized in the SEW 550-Steels for larger forgings – [92], 555-Steels for larger forgings for turbine and generator plant components – [93] and 640-Steels for components in the primary circuit of nuclear power plants – [97]. Some of the steels listed in these specifications are also listed as heat-treatable steels (steels for quenching and tempering) in DIN 17 200 [31] for diameters up to 250 mm.

Table C 5.10 lists a selection of the steels now preferred for large forgings. The steels are arranged in order of the largest diameters for which they have been proved suitable and which are mentioned in the SEW's (see above). To characterize their recommended ranges of application, only the 0.2%-proof stress values and the impact energy values at ambient temperature of these steels are shown for the largest cross section suitable for quenching and tempering according to the SEW's. Reference must be made to the data shuts (SEW's) for detailed information about the steps in the mechanical properties, depending on the diameter of the quenched and tempered cross section and the dependence of elongation, reduction of area and impact energy on the direction of the maximum grain flow.

Depending on the alloy combination for heavy forgings, plain steels are used as well as steels alloyed with manganese, chromium and molybdenum as well as with nickel, chromium, molybdenum and vanadium. Some of these steels can be assigned to systematically-designed series of alloys. For plain carbon steels there is thus a virtually complete series of carbon contents between about 0.2 and 0.65% and for the 1% Cr and 0.25% Mo steels there is a series extending from 24 CrMo 4 to 50 CrMo 4. For the nickel-chromium-molybdenum-vanadium steels from the SEW 555, the nickel content extends from about 2 to 3.5%.

Because of their international importance reference may also be made to the largely chromium-free steels for heavy forgings containing about 3.5% Ni and 0.5% Mo as standardized in ASTM A 469 [94]. These steels do not attain the favorable combinations of the 0.2%-proof stress and toughness values of the nickel-chromium-molybdenum steels spelled out in SEW 555.

In the following some examples are given for *the application of the steels* listed in Table C 5.10 and the mechanical properties important for their selection. In particular in SEW 550 [92] 18 different steels are available for the larger forgings used in *machine and ship construction* which vary greatly in shape, weight and functions. Selection from these steels can be determined by requirements such as weldability which narrow the choice to steels with a low carbon content. If parts of the forgings are to be surface hardened, the minimum carbon content is governed by the required surface hardness.

For *crankshafts* the selection of the steel depends mainly on the through-hardenability derived from the strength requirement. For this application, plain carbon steels such as Ck 45, chromium-molybdenum steels such as 42 CrMo 4 on chromium-nickel-molybdenum steels such as 34 CrNiMo 6, are used.

For the production of *low-pressure turbine and generator shafts* the nickel-chromium-molybdenum-vanadium steels are mainly used, with three standard grades containing 2, 3 and 3.5% Ni. Up to the largest diameters of about 2000 mm the 26 NiCrMoV 14 5 with 3.5% Ni is used which fulfills the highest requirements for strength and toughness properties. The values listed in Table C 5.11 give the properties that have been obtained in the cores of large shafts where the required yield stress was a minimum of 700 N/mm^2. A comparison between the data of the first and the fourth shaft shows the difficulty encountered in obtaining good toughness values in the cores of large-diameter forgings which also have a high yield stress.

A summary of *fracture mechanics* characteristics obtained by tests over many years is shown in Fig. C 5.24 [95]. These K_{Ic}-values from the core zones of large generator shafts clearly place the 26 NiCrMoV 14 5 steel above both the 26 NiCrMoV 8 5 and the chromium-free variant 24 NiMoV 14 5.

Table C 5.11 Test results from test-pieces taken from the cores of turbine and generator shafts made from 26 NiCrMoV 14 5 steel. After [89]

Type of part	Diameter	0.2%-proof stress	Transition temperature of the notch impact energy[a]	NDT-temperature from the drop weight test
	mm	N/mm^2	°C	°C
46-t-Generator shaft	1065	726	−45	−65
54-t-Turbine shaft	1235	726	−25	−40
52-t-Turbine shaft	1520	700	−10	−30
74-t-Turbine shaft	1615	774	−5	−30
200-t-Generator shaft	1809	657	−25	−55
Part for investigations	890	1015[b]	+20[b]	–
		900[c]	−10[c]	–

[a] With ISO-V-test-pieces for 27 J
[b] Tempering temperature 580 °C
[c] Tempering temperature 590 °C

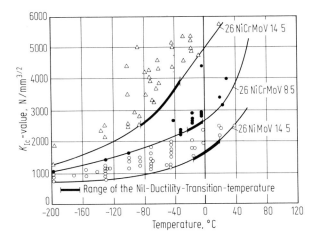

Fig. C 5.24. Fracture toughness (K_{Ic} value) of steels for turbine and generator shafts as a function of temperature. After [95].

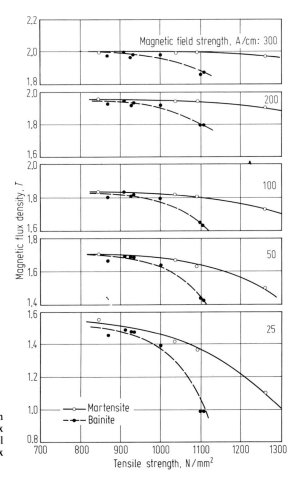

Fig. C 5.25. Relationship between tensile strength and magnetic flux density of the 26 NiCrMoV 14 5 steel for different microstructures and flux densities. After [96].

For *generator shafts* in addition to the mechanical properties account must be taken of the magnetic characteristics. The objective is to obtain the highest possible values for magnetic flux density, high permeability and high saturation magnetization. Figure C 5.25 shows the variations in magnetic flux density as a function of tensile strength for bainitic and martensitic microstructures for the 26 NiCrMoV 14 5 steel [96]. The flux density is superior to the 26 NiCrMoV 8 5 for all kinds of microstructures for strength values up to about 1050 N/mm^2.

High-pressure turbine shafts operate at temperatures up to about 550 °C. The *creep resisting steels* required have a medium carbon content of 0.25% and are alloyed on the basis of chromium-molybdenum-(vanadium) (see also C 9). For these steels the aim is not to obtain a martensitic structure because the bainitic structure obtained by quenching and tempering gives better creep strength properties. On the other hand, the toughness properties at ambient temperatures are noticeably lower.

If there are special requirements for *weldability* the carbon content must be reduced appreciably. For example, the 23 CrNiMo 7 4 7 steel is used for welded turbine shafts. The steels 20 MnMoNi 5 5 and 22 NiMoCr 3 7 are preferred for pressure vessels for nuclear reactors, on which extensive joint and surface layer (built up) welds must be carried out. Both these steels are listed in the SEW 640 [97] (but recently the 20 MnMoNi 5 5 is the only steel being discussed and used in this connection, see [97]).

C 5.4 Case Hardening Steels

Case hardening steels are used for components whose surface is to be carburized completely or in defined areas which are then quenched and tempered (stress-relieved) usually at low temperatures. Components thus treated have a hard surface layer and a softer and tougher core. The extra hardness of the surface layer gives a high resistance to mechanical wear and the tougher core can withstand steady and impact-type stresses. The high surface hardness, in combination with the internal stress conditions formed by heat treatment of the composite part, increase the fatigue strength. The carbon content of case-hardening steels is therefore 0.3% maximum and, depending on the strength and size requirements, the steels range from plain carbon to medium alloyed.

C 5.4.1 Requirements for the Base or Core Material to Suit Service and Processing Operations

C 5.4.1.1 Required Properties

The uncarburized heat-treated core material must contribute to the capacity of the case-hardened component to withstand steady or impact-type stresses without serious permanent deformation and, of course, without breaking. The material must therefore have a sufficiently high yield stress, tensile strength and toughness. For these properties the microstructure and the chemical composition of the steel are vitally important.

The *requirements for hardenability* vary, depending on the application of the component. For parts which are subjected primarily to sliding or friction wear the requirements are low but they are high for components that will be subjected to high steady, impact-type or alternating stresses.

The *processing* of material in the form of semi-finished or bar steel (but less often including wire or sheet) is carried out by hot forming, cold forming and/or machining. Most components made from case-hardening steels are produced in large batches, in the automobile industry for example, so that a large degree of uniformity of the properties is required to facilitate subsequent processing [98].

Among the most important requirements for a case-hardening steel are *capability for carburization and surface hardenability* [99]. The retention of a fine grain when the steel is carburized is necessary primarily to permit direct hardening, i.e. hardening direct from the carburizing temperature.

C 5.4.1.2 Characterization of the Required Properties by Test Values

The hardenability of alloy steels is checked with the *end-quench test* (DIN 50 191 [40]) by using the hardening temperatures listed e.g. in DIN 17 210 [100]. This standard contains reference values for the properties determined by tensile tests on blank-hardened round specimens of different diameters, also for plain carbon case-hardening steels.

With reference to subsequent processing, *hardness tests* and *metallographic investigations* can give information depending on the condition of the material as delivered.

The carburization capability and surface hardenability depend largely on the process and the steel grade. The *austenitic grain size* which can be expected following the carburizing process is checked for each melt as a matter of routine, especially where direct hardening is intended [101].

During the development of the materials and manufacturing processes, provision can also be made for the measurement of mechanical data of *case-hardened specimens*. For mass production it is normal for the user to test each melt to determine its behavior during carburization, and hardening.

C 5.4.1.3 Measures for Achieving the Required Properties

With case-hardening steels containing between about 0.07 and 0.30% carbon the *maximum hardness* obtained with fully martensitic hardening [102, 103] is from about 37 to 55 HRC. In contrast with the previously considered heat-treatable steels in which the required strength and toughness is adjusted by quenching followed by tempering at higher temperatures, *tempering* resp. stress relieving of case hardened components can take place *at only low temperatures* (around 200 °C) when the object is to obtain as high a hardness as possible of the surface layer. With this treatment, the hardness and, as shown for the example of the 20 MnCr 5 steel,

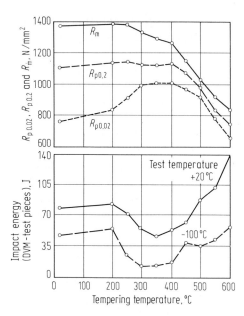

Fig. C 5.26. Tensile strength, 0.2%- and 0.02%-proof stresses and impact energy of the steel 20 MnCr 5 after quenching and tempering. After [104]. Heat treatment of the 25-mm diameter test-pieces: 850 °C 0.5 h/water; tempering time 2 h.

in Fig. C 5.26 [104] the 0.2%-proof stress, tensile strength and notch impact energy of the base material are little affected. The microstructure obtained during hardening thus determines the properties of the core material. Thus the hardenability characteristics of the case-hardening steels have a direct effect on the service properties.

The level of hardenability is important here, as is the *choice of alloying elements*. The composition of the solid solution governs the behavior of the material in the hardened condition which is summarized by the term toughness. In case-hardening steels, manganese and chromium provide about an equal increase in hardenability. For a larger required alloy content with a view to increased toughness, chromium is to be preferred or a combination of the two is appropriate. Molybdenum in place of manganese improves ductility and a nickel content of 1.5% and over contributes to a further increase. Adequate toughness with maximum core strength is obtained with steels alloyed with chromium, nickel and molybdenum.

For case-hardening steels a knowledge of the *dependence of the grain size of austenite on the temperature* is necessary in connection with the carburization treatment which must be carried out at high temperatures. Coarse austenite grain on the one hand impedes the diffusion of carbon. On the other hand, the coarse grain increases distortion during hardening and reduces the impact toughness of case-hardened components [105]. Direct hardening may be performed after carburization. When no heat treatment is included for regenerative annealing before hardening (to refine the grain) attaining the required mechanical properties in the core and the case makes it necessary to use steels which will retain their fine grain.

The most usual method for the production of steels having a retarded *grain growth* is the addition of a greater proportion of *aluminum* to the melt. The addition of *titanium* or *niobium* to aluminum- and nitrogen-containing steels can further raise the temperature at which coarse graininess starts [106, 107], and will slightly reduce hardenability. These steels enable the carburization temperature to be raised which is desirable from an economic point of view.

For the subsequent processing of material supplied by the steel producer, forming methods (drop forging, cold forming) can be used to make blanks having dimensions close to the finished component and with the direction of the fibers suited to the loads that will be imposed. Low-carbon soft case-hardening steels are very suitable for *cold forming* (e.g. cold extrusion), and will provide blanks having very smooth surfaces. These methods also make the best use of the material.

The final form of the component before case hardening is usually produced by machining which is of considerable economic importance in mass production. Great demands are therefore made on the quality and uniformity of the machinability of case-hardening steels. With the carbon content used in case hardening steels the best *machinability* generally is obtained with a structure of uniformly-distributed ferrite and pearlite which is best obtained by an *isothermal transformation in the pearlite stage* at the temperature of the shortest transformation time.

For case-hardening steels with high hardenability the time required for complete transformation in the pearlite stage is naturally greatly extended. Before machining, these steels can be subjected to a *soft annealing process*, when there is no danger of "smearing" owing to the higher ferrite strength due to the alloy content, compared with plain carbon and low-alloy steels. On the other hand by *matching the alloying elements* [108, 109] silicon, manganese, chromium, molybdenum and nickel, steels have been developed which combine a high hardenability with a short transformation time in the pearlite stage so that the isothermal transformation at this stage becomes more affordable.

Machinability can be improved to some extent by producing a coarse grain, by use of higher austenitizing temperatures [110], possibly direct from the transformation heat [111] (*coarse grain annealing*). Experience of the effect of coarse grain annealing on the machinability vary. Improvements have been found [105] but in an evaluation of a large series of trials in practice [112] there were no advantages worth mentioning.

C 5.4.2 Required Properties of the Case Hardened Layer and Their Production

C 5.4.2.1 Required Properties

The surface layer must have the highest *wear resistance*. In general terms a very hard surface is advantageous for wear resistance requiring a martensitic structure which can be obtained from an appropriate *carbon content* of the case and a matched hardening. With plain carbon and alloy steels the maximum hardness of the martensite is reached with a carbon content of 0.6% [102] so that regarding the hardness of the martensitic solid solution, there appears to be no need for carburization to a higher carbon content. However, it is well known that wear resistance can be further increased by the dispersion of hard particles, such as carbides, in the matrix. For this reason, a higher carbon content is provided in the surface layer in parts which are mainly subject to sliding wear.

Higher carbon contents, of more than 0.6 to 0.9% C depending on the alloying elements, lead to the formation of retained *austenite* if, after the carburizing treatment or austenitizing before hardening the carbon is in solution in the austenite. The importance of the retained austenite varies depending on the size of the component, the thickness of the carburizing case and especially the type of stress to which the component is to be subjected. For a purely sliding load retained austenite contents up to 20% can be tolerated safely and for certain steel compositions even higher proportions may be advantageous [113]. With a rolling movement including high and especially varying contact loads the resistance to pitting rises with an increase in residual austenite (Fig. C 5.27 [114]). The fatigue strength can be reduced (Fig. C 5.28 [115]) or increased [113] as a function of the chemical composition of the residual austenite in the case and the type of stress or pre-stress on the component.

The *fatigue strength* is reduced (Fig. C 5.29 [116]) by internal oxidation during carburizing [117–119] and in the same way the incipient crack and fracture stresses in the static [118], and the maximum load in the impact bending test [116], are reduced.

The *thickness of the carburized case* which can vary between a few tenths and several millimeters must be matched to the function and the size of the component. In principle, for primarily reversed stresses it can be kept small and for sliding or rolling stresses and for taking up high and possibly alternating surface contact loads it must be larger. The thickness must also be considered in conjunction with the strength of the core material not so much because of the danger of pressing through the case-hardened layer, a rarely-observed phenomenon, but for reasons described below.

The surface hardness of the case and its overall thickness are sufficient for the *characterization of the case* on simple components mainly subjected to wear. For less simple components with complex stressing (which applies to the great majority of components made from case-hardening steel) measured values for the hardness at the outer surface (measured with low testing forces) and the variations in

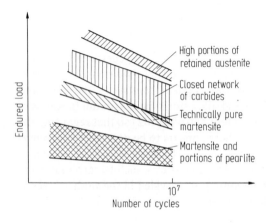

Fig. C 5.27. Effect of the surface microstructure of case-hardened gears on the pitting tendency (expressed by the sustained load). After [114].

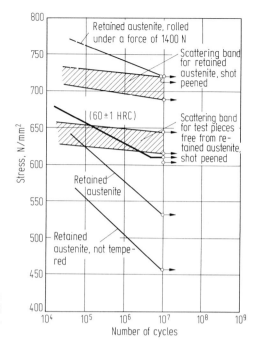

Fig. C 5.28. Effect of retained austenite on the fatigue strength under reversed bending stresses of case-hardened notched test-pieces. After [115].

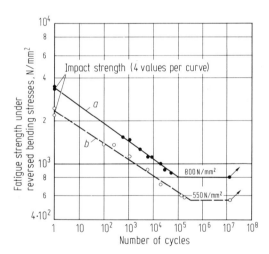

Fig. C 5.29. Effect of internal oxidation on impact strength and fatigue strength under reversed bending stresses of test-pieces of 15 CrNi 6 after case-hardening with 0.08 mm of surface removed (curve a) and without removal of surface (curve b). After [116].

hardness in the case as well as the depth of the case [51] must be considered together under the term case hardenability.

C 5.4.2.2 Carburizing

Different *processes* are used for carburizing the surface layers [120]. The depth and carbon content of the case are controlled by the carburizing treatment parameters although they are also influenced by the base material. Reference has already been made to the effect of the austenite grain size which exists during the carburizing process on the carbon diffusion. This effect is greater than that of the steel's chemical composition [121, 122].

The carbon content in the case is fundamental to the structure of the outside of the case of case-hardened components. The content changes with the *carburizing temperature* and *duration* as well as with the composition of the steel. Longer carburizing periods can lead to supercarburizing (formation of carbides), depending on the carburizing material used [123]. Supercarburizing occurs particularly in case-hardening steels, that are rich in chromium such as manganese-chromium and the chromium-nickel steels. Figure C 5.30 [124] shows the effect of the amount of chromium on the carbon content in the case after carburization for different grades of alloy steels.

During the carburizing treatment oxidation may appear within a thin surface layer and is called internal or surface *oxidation* [125, 126]. The preconditions for surface oxidation are a minimum oxygen potential in the carburizing furnace and alloying elements in the steel which have a higher affinity for oxygen than iron. These preconditions generally exist in common industrial carburizing processes and in the steel. In addition to chromium a strong influence is also exerted by silicon [127]. Plain carbon and nickel or molybdenum alloy steels should show the least tendency towards internal oxidation. On the other hand some carburizing processes will provide largely oxygen-free atmospheres in the carburizing chamber by means of inert gases or a vacuum [128], and are suitable even for large-scale production.

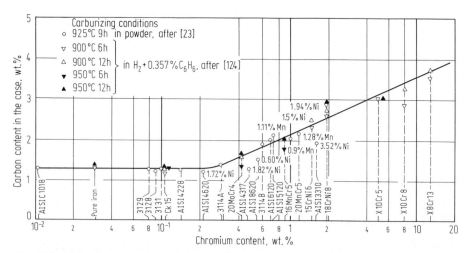

Fig. C 5.30. Effect of the chromium content of steels on the carbon content in the surface zone for different carburizing conditions. After [124].

C 5.4.2.3 Hardening

When hardening the carburized part it must be remembered that the *case* consists of materials with a continuously-varying carbon content so that the *transformation behavior* within the case is different from that of the *base material*. Figure C 5.31 [129] shows time-temperature-transformation (TTT) diagrams for continuous cooling for melts of steel X MnCr 5 (containing roughly 1.1% Mn and 1.0% Cr) with carbon contents X between 0.13 and 1.08%. The steel in Fig. C 5.31a corresponds with 16 MnCr 5, i.e. a standard base material. The steel in Fig. C 5.31b with 0.45% C is almost eutectoid, and that in Fig. C 5.31c is hypereutectoid in which, as the same hardening temperature of 860 °C was used for all the samples, undissolved carbides are present. In addition to changes in temperature and the incubation and transformation times of pearlite and bainite stages the Ac_3 or the Ac_{1e} and the M_s temperatures are reduced with increasing carbon content.

The *transformation stresses* formed during the decomposition of the austenite in any zone of the case affect the subsequent transformation reactions in zones of different carbon contents [130]. Tensile stresses accelerate and compressive stresses retard the transformation process. Nevertheless the characteristic differences of Figs. C 5.31a to c show the behavior of carburized components with subsequent heat treatment.

For *single and double hardening* (see B 4 and e.g. DIN 17 210) the temperatures Ac_3 and Ac_{1e} (which decrease with increasing carbon content), characterize the appropriate hardening temperatures for base and case-hardened materials.

Figure C 5.30 showed that the carbon content of the steel cases increases with their chromium content. Similarly, as shown in Fig. C 5.32 where the dependence of case hardening steels is plotted against their mean chromium content, the *retained austenite content* increases (in spite of their different alloy composition) especially at temperatures above 900 °C. These temperatures are used for carburizing (followed by direct hardening) for economic reasons. The result of the increased residual austenite content is a reduction in hardness towards the surface as can be seen in Fig. C 5.33a for the steel 16 MnCr 5 [121] after direct hardening. Compare Fig. C 5.33a with Fig. C 5.33b which shows the change of hardness after single hardening from 825 °C. With this lower hardening temperature, less carbon is dissolved in the austenite so that less or no retained austenite remains. Comparing the maximum hardness or the start of the reduction of hardness wth the respective carbon content shows that the hardness is shifted towards the lower carbon content with increasing chromium content of the steel, Fig. C 5.34 [131]. At the same time the range of carbon contents can be seen in which a hardness of 700 HV is reached ($C^D_{700\,HV}$, see Fig. C5.34). This hardness value is owing to an excessive residual austenite content (in the diagram to the right of the maximum hardness C^D_{HVmax}). The hardness is also minimized because of the decreasing hardenability (to the left of the carbon content for maximum hardness). The spread of the range of carbon content for good case hardenability can be regarded as a *measure of the suitability of a steel for direct hardening* [132].

The hardening method to be used also depends on the permissible *dimensional changes during hardening* [133–135]. In principle the dimensional changes increase

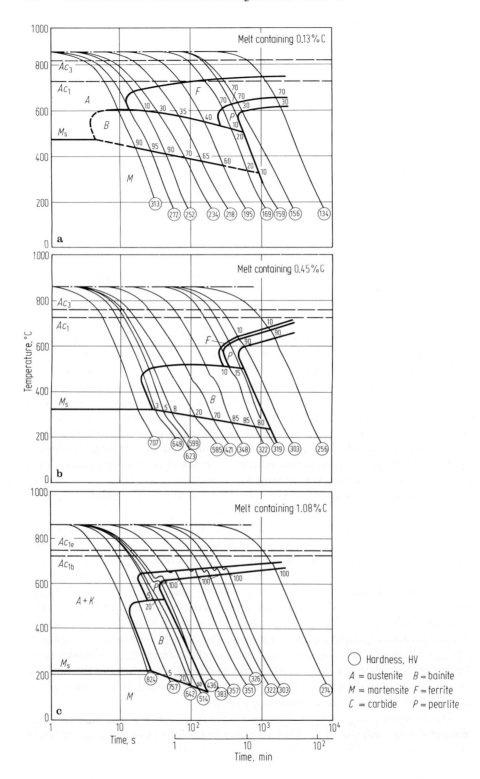

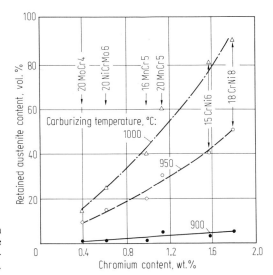

Fig. C 5.32. Effect of chromium content on the fraction of retained austenite in the case of case-hardening steels after carburizing. (Carburizing time 3 h). After [122].

with increasing hardenability and strength of the base material. Similarly an increasing hardening temperature and frequency of heating and cooling of the carburized component have the same effect. What is important for trouble-free production is not so much the absolute amount of the dimensional change, which can be controlled by suitable dimensional allowances on the blank, but the uniformity of the distortion. This uniformity is improved by the use of material from melts having a reduced spread of hardenability bands (see e.g. DIN 17 210) and a defined grain size. Material from melts with greater variations in the austenite grain size show a greater spread of distortion in the case-hardened condition [105]. Noise measurements carried out on power dividers on trucks lead to a similar conclusion and show the superiority of fine-grain melts compared with those of mixed grain sizes [112].

In addition to the influences on the surface of case hardened parts discussed above, an important characteristic is the *carburizing depth* and hence the change of hardness below the surface which is designed to suit the intended use of the component. Apart from the carburizing treatment, the steel alloys and the hardening or further heat treatment determine the change in hardness for which the so-called *depth of hardened case* [51] represents a characteristic value. Figure C 5.35 [136] shows the hardness curves for different steels for which a reference hardness of 615 HV was selected as a measure of the hardness depth. For manufacturing inspection, a non-destructive method of measurement of hardness depth is also used [137].

Fig. C 5.31. Time-temperature-transformation diagrams for continuous cooling of melts containing about 0.3% Si, 1.1% Mn and 1.0% Cr – as in 16 MnCr 5 – for different carbon contents (hardening temperature 860 °C): **a** 0.13% C; **b** 0.45% C; **c** 1.08% C. After [129].

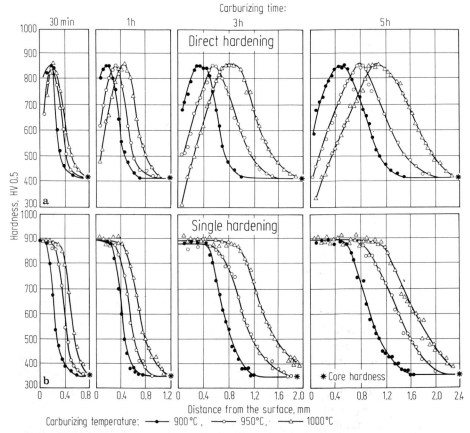

Fig. C 5.33. Hardness variations in the case of a 16 MnCr 5 steel carburized in a salt bath **a** after direct hardening from the carburizing temperature, **b** after single hardening from 825 °C. After [121]. Hardening in oil, tempering 2 h, in each example.

C 5.4.2.4 Stress Relieving

After hardening case-hardened components are usually subjected to *stress relieving at temperatures between 150 and 200 °C*. If the required tolerances are very narrow, this treatment prevents the components from changing size by aging [138]. Stress relieving is also necessary for parts which require grinding if grinding cracks are to be avoided, [136] as the tetragonal martensite, present after hardening, has a greater thermal coefficient of expansion than the martensite with a body-centered cubic lattice resulting from stress relieving [139]. Grinding cracks can be formed by purely thermal stresses or by transformation stresses so that attention must also be paid to the grinding conditions.

In conjunction with Fig. C 5.26 it was also found that the *hardness and tensile strength* of the base material remain virtually unchanged by tempering at temperatures up to 200 °C. As Fig. C 5.35 [136], and in a general form Fig. C 5.36 [121] show, this effect does not apply for the cases of higher carbon contents for which in

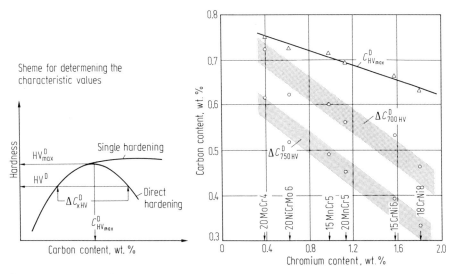

Fig. C 5.34. Diagram for assessment of the direct hardenability of case-hardening steels containing different amounts of chromium. After [131].

fact the maximum hardness and depth of hardness are decreased. On the other hand, in the same way as for the base material (Fig. C 5.26), the stress relieving treatment used for case-hardened specimens markedly increases the proof stress values for small amounts of plastic deformation. These changes may also contribute to the explanation of the varying susceptibility to grinding cracks in specimens which are only hardened and those which are also stress relieved [136].

Other effects of stress relieving will be considered in conjunction with *fatigue strength*. This property is closely associated with the internal (residual) stress condition in the case-hardened component. The behavior of the combination of the properties of both the core and the case will now be considered.

C 5.4.3 Behavior of the Composite Material of the Case and the Tough Core Under Service Conditions

A typical example will be used to explain the *required properties needed for a case-hardened component*, namely a *gear wheel* for an automobile transmission, subjected to complex stresses [116, 140].

The bending moment acting on a gear tooth with every revolution requires at the stressed root sufficient strength, i.e. an adequate fatigue limit, overload capacity in the range of finite fatigue life and impact resistance (i.e. toughness) for single impact loads. For the flank load capacity, adequate surface fatigue resistance, which means resistance to alternating contact loads (pitting), and sufficient general wear resistance are required, properties that have already been discussed.

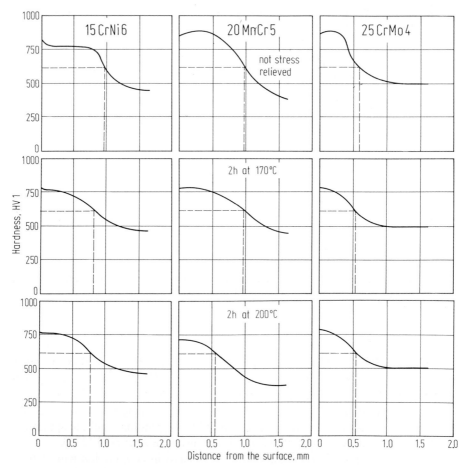

Fig. C 5.35. Effect of stress relieving on the variation of hardness in the surface zone of case-hardened steels. After [136].

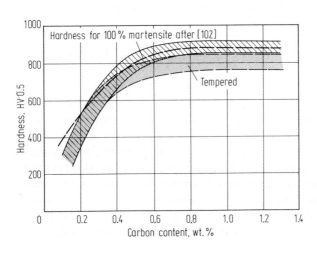

Fig. C 5.36. Change of surface hardness of carburized case-hardening steels by tempering for 2 h at 200 °C compared with the hardened condition as a function of the carbon content. After [121]. Results with 16 MnCr 5, 20 MnCr 5, 20 MoCr 4, 15 CrNi 6, 18 CrNi 8 and 20 NiCrMo 6; Carburization: 0.5 to 5 h at 900 °C to 1000 °C; hardening: quenched from 825 °C in oil.

The fracture ductility values obtained from tensile testing or the energy absorbed during a notch impact test characterizing the toughness of a metallic material are characteristic for a case-hardening steel, in the blank-hardened and stress-relieved condition but not for the behavior of a case-hardened component. Figure C 5.37 [141] shows the extent to which the *energy required to knock off the tooth* of a spur gear changes after case hardening compared with the blank-hardened condition. The required energy decreases with increasing depth of case. Important in this illustration also is the proportion of the energy absorbed by elastic deformation (the curve marked "elastic limit") which increases for this size of tooth for a case depth up to 0.5 mm and decreases thereafter. From this results an effect of the ratio of case hardening depth to the thickness or diameter of the component or the uncarburized base material [142]. A gear tooth is required to withstand increased impact loads within the elastic limit as permanent deformation prevents proper meshing of the gear teeth and can lead to cracking of the hard case in the root of the teeth. In addition, a small reserve of plastic deformation energy helps to prevent brittle fracture of the tooth [140].

Measurement of the *"toughness" of thin, hard cases* presents difficulties [143]. The results from impact testing of specimens which have been carburized right through [142] or by means of fracture mechanics [144] are not immediately applicable to components. Measurements of the fracture toughness of specimens which are only case hardened at the surface still require interpretation. Static and dynamic bending tests which measure the stress needed to initiate a crack [142, 145] or the impact force to fracture [116] are not specified in the standards. These properties are more characteristic than the previously-preferred impact energy values. Values for the cracking and fracture forces strongly depend on the depth of

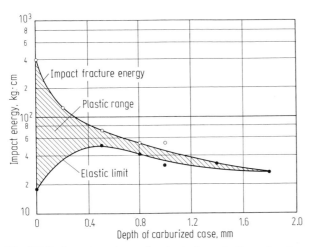

Fig. C 5.37. Impact toughness of carburized gear teeth as a function of the case hardening depth. After [141].

the case or its ratio to the entire cross-section and on the core strength [142, 145]. If the case carbon content and hence the retained austenite content, internal oxidation, change of hardness near the surface, case hardening depth, notch geometry and core strength are kept within the usual limits for highly-stressed gears as for the tooth impact test [116], conclusions can be drawn concerning their fatigue behavior in the range of finite life (Fig. C 5.29) and on the effect of the base material. However, the stated limits show that although the test is useful for a specific application it is less suitable for a general assessment of case-hardened components or of the steel being used.

The result of the composite material test largely depends on the alloy-dependent toughness of the low-carbon solid solution of the base material and the microstructure of the case-hardened layer, which is also affected by the alloying elements. In this connection it may be noted that *boron* can become *effective in case-hardening steels in two ways*. It is well known that small additions of boron dissolved in steel considerably increase hardenability and can thus contribute to the saving of alloying elements [146]. It has also been found [147, 148] that when boron is combined in steel as nitride, the impact strength and toughness of case-hardened parts are improved. Up till now no definite explanation has been found for this phenomenon which occurs only with case hardening. The effect of boron on the properties is used in steels of different compositions.

From the changes in the characteristic value, designated elastic limit in Fig. C 5.37, it can be concluded [140] that the start of *permanent deformation is influenced by the internal stress condition* of the component. During the hardening of carburized parts, high internal compressive stresses form in the case (see B 4). The highest tensile stresses occur in the outer zone when the tooth is subjected to bending. The high internal compressive stresses in the case are decisive for the start of cracking with higher external loads, a reduction of notch sensitivity and, together with the high hardness or strength of the case, for the high fatigue resistance (under alternating bending stresses) of case-hardened gears or similarly-stressed components, as is shown by a comparison between blank and case-hardened specimens of the 16 MnCr 5 steel in Fig. C 5.38 [149]). If the outer case does not have maximum hardness and internal compressive stresses owing, for instance, to a higher content of retained austenite (Fig. C 5.28), because of oxidation of the case, (Fig. C 5.29) or if the hardness and internal compressive stresses are reduced by the usual stress relieving (Fig. C 5.35 [136]), the effect on the fatigue limit and fatigue strength for finite life is unfavorable. On the other hand introduction of internal compressive stresses by mechanical means (e.g. shot peening) can improve the fatigue limit considerably [115, 136, 149].

Although the highest internal compressive stresses in the outer case and thus the highest *fatigue strength* are to be expected with a relatively-thin case and a low core strength [140], the high contact load on the tooth flank, which results in the danger of pitting, requires that the case be of sufficient thickness and that the core strength under the case also be sufficient. The results shown in Fig. C 5.39 [150] and other collected experiences from gear manufacture and testing show [119, 140] that, under the total stresses which occur, the highest fatigue resistance of the tooth is achieved with a core strength of between about 1000 N/mm^2 and 1400 N/mm^2.

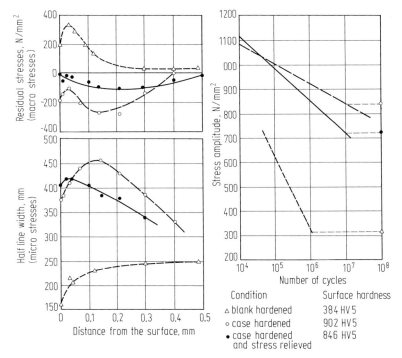

Fig. C 5.38. Effect of the case hardening conditions and of stress relieving (1 h at 170 °C) on internal stresses and fatigue strength under reversed bending stresses of test-pieces made from 16 MnCr 5. After [149].

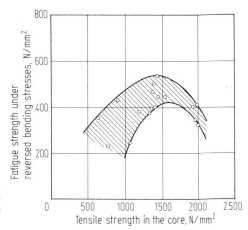

Fig. C 5.39. Fatigue strength under reversed bending stresses of gears (module $m = 3$) with different core tensile strengths. After [150].

However, newer investigations show that, for lightly-notched specimens in contrast with smooth case-hardened ones, there is no relationship between fatigue strength and core strength (within the apparently-usual limits for high performance gears) [151].

In conclusion it should be mentioned that *the above discussion applies particularly to highly-stressed gears* on which most investigations have been concentrated. Only a few of the above comments apply to most of the components made from case-hardening steels and their operation in service. In the same way as for heat-treatable steels, (steels for hardening and tempering) the durability (service life) of case-hardened parts and their base materials depends largely on the design and manufacture of the component. This dependence applies even more to the interaction between the base material and the case whose properties can be largely independent of the base material.

C 5.4.4 Steel Grades Proven in Service

Table C 5.12 [100, 153, 154] shows the *chemical composition* of case-hardening steels irrespective of the time of their introduction; it is intended to give a survey of the range of alloys available. The most important alloying elements are manganese, chromium, molybdenum and nickel. Steels containing vanadium and titanium as well as tungsten in countries in which molybdenum is in short supply are also produced. Increasing knowledge of the behavior of case-hardened components; the saving of expensive or occasionally-scarce alloying elements; the avoidance of case supercarburization and excessively-high residual austenite content; the introduction of economic processes suitable for mass production for carburizing and hardening as well as the required case hardenability and retention of fine grain even with high carburizing temperatures; have all influenced the development of case-hardening steels from both the alloying and the metallurgical points of view.

The strength of a component for a given size and heat treatment is determined in the first place by the *hardenability* of the steel which depends on the *amounts of the alloying elements*. The heat treatment as well as the selection of the alloying elements manganese, chromium, molybdenum and nickel, singly or in combination, are decisive for the ductility of the base material (compare the fracture ductility characteristics of the steel grades e.g. in DIN 17 210 taking the strength ranges into account), of the case and of the complete component. The scatter bands of the hardenability contained e.g. in DIN 17 210 show that steels are available to suit different requirements for hardenability. In addition the user and steel producer can now agree upon hardenability bands reduced to two-thirds of the original range.

For plain carbon steels there are no scatter bands because the test using the usual end-quench specimens for these steels of low hardenability, is not suitable. Chromium steel 17 Cr 3 has a slightly higher hardenability, the lowest of the alloy steels.

The steels 16 MnCr 5 and 20 MoCr 4 have roughly the same hardenability. Introduction of the 16 MnCr 5 which is common today, arose from a shortage of molybdenum and nickel but the 20 MoCr 4 containing less chromium was developed for direct hardening [140, 152]; its special suitability for this hardening process can be seen in Fig. C 5.34. The higher chromium content of the

Table C 5.12 Examples of case hardening steels of different alloy-type

Characteristic alloying elements	Comparable steel grades in			Chemical composition[a]				
	DIN 17210 [100]	Euronorm 84 [153]	ISO/R 683 [154]	% C	% Mn	% Cr	% Mo	% Ni	% B
Unalloyed	Ck 15	2 C 15	C 15 E 4	0.12 ... 0.18	0.30 ... 0.60				
Chromium	17 Cr 3	(15 Cr 2)	–	0.14 ... 0.20	0.40 ... 0.70	0.60 ... 0.90			
Manganese-chromium	16 MnCr 5	16 MnCr 5	16 MnCr 5	0.14 ... 0.19	1.00 ... 1.30	0.80 ... 1.10			
Molybdenum-chromium	20 MoCr 4	20 MoCr 4	–	0.17 ... 0.22	0.70 ... 1.00	0.30 ... 0.60	0.40 ... 0.50		
Chromium-molybdenum	22 CrMoS 3 5	–	–	0.19 ... 0.24	0.70 ... 1.00	0.70 ... 1.00	0.40 ... 0.50		
Chromium-molybdenum with boron additive	(23 CrMoB 3 3)	–	–	0.20 ... 0.25	0.70 ... 0.90	0.70 ... 0.90	0.30 ... 0.40		×
Nickel-chromium-molybdenum	21 NiCrMo 2	20 NiCrMo 2	20 NiCrMo 2	0.17 ... 0.23	0.65 ... 0.95	0.40 ... 0.70	0.15 ... 0.25	0.40 ... 0.70	
Chromium-nickel	15 CrNi 6	14 CrNi 6	–	0.14 ... 0.19	0.40 ... 0.60	1.40 ... 1.70		1.40 ... 1.70	
Chromium-nickel-molybdenum	17 CrNiMo 6	17 CrNiMo 7	18 CrNiMo 7	0.15 ... 0.20	0.40 ... 0.60	1.50 ... 1.80	0.25 ... 0.35	1.40 ... 1.70	
Nickel-chromium	(14 NiCr 14)	14 NiCrMo 13	–	0.10 ... 0.17	0.40 ... 0.70	0.55 ... 0.95		3.25 ... 3.75	

[a] In addition ≤ 0.40% Si, ≤ 0.35% P and ≤ 0.035% S; but for several steel grades a sulphur content of between 0.020% and 0.035% can be used, as for example, the steel 22 CrMoS 3 5 is listed

22 CrMoS 3 5 which also contains about 0.5% Mo produces its higher hardenability which is comparable with that of 15 CrNi 6. The nickel-chromium steel 14 NiCr 14 was standardized as early as 1928 under the designation ECN 35 [155, 156]. To save nickel, the steel 15 CrNi 6 was introduced with the same hardenability at a later date [157]. Among the steels with three alloying elements, 21 NiCrMo 2 was created due to the availability of scrap after World War II. In DIN 17 210 the steel 17 CrNiMo 6 has the highest hardenability and is used in place of the previously more common 18 CrNi 8. Finally 23 CrMoB 3 3 can be mentioned as a representative of the boron-treated steels. Although it has a slightly lower alloying content, this steel has a hardenability similar to that of 22 CrMoS 3 5.

C 6 Steels for Cold Forming (Forming e.g. by Cold Extrusion and Cold Heading)

By Hellmut Gulden and Ingomar Wiesenecker-Krieg

C 6.1 General Remarks

General information on cold forming is to be found in B 7 and in B 8. The principal methods of cold forming are cold extrusion and cold heading or upsetting. The essential advantages of cold forming are low material wastage, high dimensional accuracy of parts, good surface finish, favorable grain flow, increased strength in the cold formed condition and the possibility of automated processing for mass production. A further advantage compared with hot forming is that there is no need for heating to higher working temperatures. These advantages have resulted in the use of cold forming to produce many components for the automotive and machine construction industries. Typical examples of components produced are gears, shafts, piston pins, mechanical tubing and cylinders. Nuts, bolts and other parts are produced by cold heading.

C 6.2 Required Properties

The most important property of steels for cold forming is *cold formability* (for details see C 7), which is the result of low *flow stress* k_f* and good *deformability*. The flow stress characterizes the material's specific resistance against the deformation, i.e. it determines the forces which must be applied by the forming tools. Good ductility is necessary for sufficient die filling and for preventing discontinuities. Figure C 6.1 shows a longitudinal section through a cold formed part showing such internal discontinuities (chevrons), which may be due to insufficient ductility or to an excessive reduction ratio (true plastic strain).

Fig. C 6.1. Longitudinal section of a bar with center bursts (chevrons) after cold extruding (1:1).

References for C 6 see page 773.
* The term "flow stress" means the true stress necessary to produce a given true plastic strain.

In addition to ductility the steel must have all the properties required for good performance in service, i.e. *strength* (also strength at elevated temperatures and fatigue strength), *toughness, corrosion and wear resistance* and heat treatability, particularly *hardenability*, as applicable. The essential properties offered by the manufacture of parts from cut-to-length bar or wire rod are good machinability and good internal soundness, good *surface quality*, low decarburization depth and high dimensional accuracy of the finished product. Mass production requires maximum material *uniformity*.

The requirements and properties listed above are somewhat conflicting. In general, however, the choice of steels is governed by the intended use of the component rather than by the manufacturing process. In this section only those properties ae discussed which are of major importance with regard to cold forming.

C 6.3 Characterization of Required Properties

The decisive parameter for cold forming, i.e. the *flow stress* k_f, required for a particular rate of plastic deformation true plastic strain, is determined by the initial strength and by the effect of strain hardening. The flow stress k_f is usually measured in a cylindrical compression (upsetting) test [1–3] and is represented in dependence on the true plastic strain f (see Eq. (B 7.1)) as the *true stress-true strain-curve**. Such curves for various steels are shown in [3–7]. Figure C 6.2 shows the scatter bands of true stress-strain-curves for some alloyed and unalloyed steels [5] (see also B 1 and B 7). The k_f value for a given heat-treated condition can also be computed with sufficient accuracy on the basis of chemical composition alone or in combination with tensile strength [3, 5, 8, 9] (see also Fig. C 7.5).

In practice, determination of cold formability the tensile strength of a material instead of the flow stress (see above) is normally used as the basis, as the tensile strength is easier to measure.

Deformability is assessed on the basis of the *reduction of area* in the tensile test. In connection with cold formability, reduction of area values from 60 to 70% mean very good and 50 to 60% mean sufficient deformability [10, 11]. Occasionally the elongation after fracture is also used for assessing the deformability.

The flow stress and deformability of a steel are determined by its microstructure, which is the result of chemical composition and heat treatment. The decisive properties of the microstructure in this context are the percentage and shape of ferrite and pearlite, the *degree of spheroidization* and size of carbides also being assessed. The degree of spheroidization is the proportion of globular carbides (produced by annealing) to the total amount of carbides. Evaluation can be done in accordance with SEP 1520 [12, 13].

The cold formability of small-gauge wire rod and bar can also be assessed on the basis of the *cold-upsetting test* (compression test). Here a cylindrical sample is cold headed to 1/3 or 1/4 of its initial height. Details on this test procedure are to be

---

* This term is used in this chapter also in the abreviated version "true stress-strain-curve".

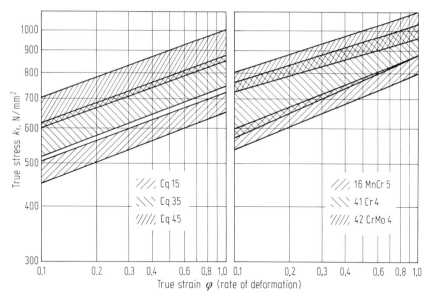

Fig. C 6.2. Scatter bands of true stress-true strain-curves of unalloyed and low alloyed steels after spheroidize annealing to globular cementite (GKZ). After [5]. (Cq 15 and 16 MnCr 5, see Table C 6.2, Cq 35 and 41 Cr 4 see Table C 6.3, Cq 45: unalloyed steel with about 0.45% C and 42 CrMo 4: steel with about 0.42% C, 1% Cr and 0.2% Mo.

found, for example, in DIN 1654, Teil 1 [14]. Shear fissures at 45° to the axis of the sample indicate insufficient ductility of the stock. Usually such samples are used to detect internal and external flaws which may open up during cold forming (Fig. C 6.3).

The most stringent requirements are made concerning *surface quality* of the steel products to be cold formed. The usual metallographic and nondestructive testing methods as well as the cold-upset test mentioned above are employed to assess this quality. Permissible crack or flaw depths are determined according to classes (see, e.g., the classes listed in the draft of SEP 055 [16]) and a decarburization depth limit is specified (see, e.g., DIN 1654 Teil 1 [14]) when working with steel bars. Additional requirements for surface roughness and strain-hardening of the rim zone are usually specified for machined steel bars.

Fig. C 6.3. Specimen with surface defects after the compression (upsetting) test. After [15]. **a** Without defects; **b** with marks; **c** with small cracks; **d** with medium cracks; **e** with heavy cracks.

Many components are heat treated by being quenched and tempered or case-hardened after being cold formed. The *hardenability* of steels for quenching-and-tempering and of case-hardening steels is generally indicated by their hardness in the end quench test [16a] or by their strength in a blank-hardened or quenched-and-tempered condition. In addition, the hardenability (i.e. full hardening) of steels for cold forming is assessed by the hardness present in the core of a round bar of a certain diameter after oil hardening. (Depending on steel grade and size, minimum hardnesses of 40 to 48 HRC may be required. See also the minimum hardness values required to comply with DIN 1654 Teil 4 [17] in steels for quenching and tempering as a function of the diameter.)

C 6.4 Measures of Physical Metallurgy to Attain the Required Properties

C 6.4.1 General Remarks

The properties of steels for cold forming are determined primarily by the intended use of the components to be made from them, usually by specifying certain values for mechanical properties. The respective measures are discussed in the chapters dealing with the steels in question here, e.g. the steels for general structural purposes or the steels for quenching and tempering (see C 2 and C 5). Service properties are the main criteria in selecting steels for cold forming. However, several special measures are taken which have a favorable influence on cold formability without affecting other requirements. These measures may consist, for example, in attaining particular microstructures in steels and in fine tuning the chemical composition, e.g. with regard to the strength of the ferrite or the spheroidization ability of the carbides. Thus the *chemical composition* may be mentioned as a distinct variable in addition to the *microstructure*. The surface condition concerning stocks and final products is also important, so that, in the final analysis, distinct steel grades are created (see C 6.5).

In considering all the measures undertaken to influence cold formability, the sequence of operations must be distinguished. If the product is in its final condition after being cold formed so, the steel must have a certain cold formability from the very beginning, i.e. in its initial condition. If the component is to be subjected to heat treatment after cold forming, to attain the final condition so, the steel can be made with an initial condition especially suitable for cold forming, with the final properties the component is to possess being disregarded for the time being.

Another distinction must be made between steels with a ferritic matrix and those with an austenitic matrix. With ferritic, unalloyed or low-alloy steels, it is essential that, whatever the final condition is to be (e.g. quenched and tempered), these steels usually are cold formed in a condition marked by a ferritic-pearlitic (carbidic) structure. The following comments refer to this ferritic state, which may represent an intermediate state.

C 6.4.2 Chemical Composition and Microstructure

The *chemical composition* of ferritic steels, determines their microstructure and thus has a bearing on their properties. Carbon and alloy additions to such steels increase the flow stress (see above), which is a decisive factor in connection with cold forming (Fig. C 6.2). Carbon has a major impact as it determines the pearlite content which is decisive for the strength and reduction of area (Fig. C 6.4). Carbide-forming elements such as chromium, molybdenum, vanadium and titanium also contribute to increased strength and thus to flow stress via the carbides, although to a lesser extent. Other alloying elements such as manganese are also effective through solid solution hardening. However, this effect applies only to steels of lower carbon and higher ferrite content, e.g. case-hardening steels [19–21]. Figure B 1.37 shows how alloying elements increase the yield stress of ferrite.

The influence of chemical composition on the flow stress of steels with up to 0.6% C in a soft-annealed condition is shown in Fig. C 6.5. The strengthening effect of carbon is outstanding, but the effect of the other elements studied, i.e. molybdenum, silicon, nickel, manganese and chromium, on the flow stress is much less, and decreases in this order. The carbide-forming elements molybdenum and chromium contribute less to the flow stress at higher true plastic strains (deformation).

The following *measures for improving cold formability* have been determined on the basis of research results: with steels for case-hardening and quenching and tempering the maximum silicon content (normally 0.40%) is limited to 0.15%. The resulting decreased ferrite strength [20, 21] makes for a lower flow stress and longer tool life in cold forming. In addition, maximum amounts of accompanying elements, such as copper, nickel and molybdenum, are occasionally specified for unalloyed steels. In view of the fact that the content of these elements in steel is usually low today, their effect on cold formability can be considered to be negligible

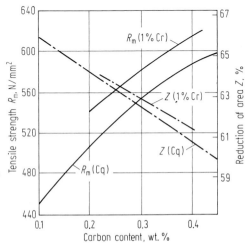

Fig. C 6.4. Tensile strength and reduction of area of unalloyed Cq-steels (Cq) and alloyed steels with 1% Cr for quenching and tempering (1% Cr), hot rolled and annealed to spheroidized cementite (GKZ) in dependence on the carbon content. After [18].

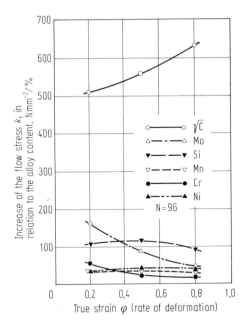

Fig. C 6.5. Effect of alloy-elements on the true flow stress of steels in the soft annealed condition. After [8].

[19]. Chromium is very often added to steels which are to be heat treated after being cold formed. Chromium offers the advantage that it only slightly increases the flow stress of steels in an annealed condition, while greatly improving hardenability [8, 18]. In addition, chromium steels can be more easily spheroidized, so that higher reduction of area values are obtained (Fig. C 6.4).

Boron has also proven favorable for cold forming steels. This element is added to certain steels for increasing hardenability. Solute boron delays the formation of ferrite and thus increases hardenability, but has no effect on strength in the annealed condition [22, 23]. Thus, despite increased hardenability, the flow stress of boron steel is the same as or even lower than that of boron-free steel (Fig. C 6.6). By using boron the content of other hardenability-increasing elements can be reduced, and cold formability can be improved or larger-sized material can be quenched and tempered because of the increased hardenability [15, 24–26]. The hardenability-increasing effect of boron decreases with increasing carbon content, so that boron steels with a carbon content exceeding 0.50% are seldom used.

Apart from the chemical composition, the *microstructure* is the most important variable with regard to cold formability. Steels with low contents of carbon and alloying elements are often cold formed in a *ferritic-pearlitic condition*, which exists after hot-rolling or normalizing. Characterized by lamellar pearlite, this condition is desirable when cold forming is to be followed by extensive machining. If more highly alloyed steels are also to be cold formed in this microstructural condition then a pronounced ferritic-pearlite structure is desirable. Pearlite patches are evenly distributed in such a microstructure which is obtained by annealing at about 900 °C to 1000 °C followed by controlled cooling and transformation to the pearlite phase ("BG-annealing"). A true stress-strain-curve for this state is shown in Fig. C 6.7 for a steel containing about 0.16% C, 1.2% Mn and 1% Cr (16 MnCr 5).

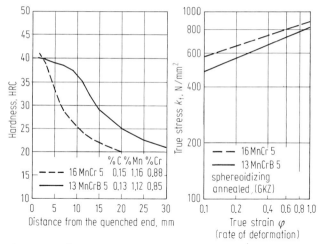

Fig. C 6.6. Influence of boron on the hardenability (signified by the hardness curve obtained in the Jominy end-quench test) and on the true stress-true strain-curve (flow curve) of case-hardening steel with 1.3% Mn and 0.9% Cr (16 MnCr 5 and the 13 MnCrB 5).

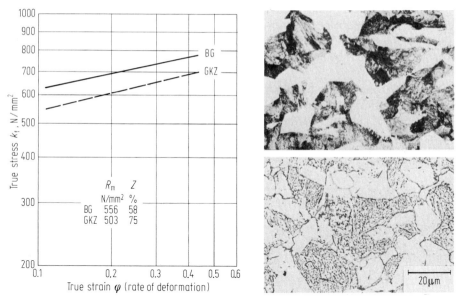

Fig. C 6.7. Influence of the heat-treatment and the microstructure produced by it on the true stress-true strain-curve of a steel with 0.16% C, 1.2% Mn and 0.9% Cr (16 MnCr 5). **a)** BG = heat treated to a microstructure of lamellar pearlitic patches with regular distribution in the ferritic matrix. **b)** GKZ = spheroidizing annealed to globular cementite. R_m = Tensile strength; Z = Reduction of area.

Considerable improvement in cold formability can be achieved when the pearlite carbides are spherically shaped by annealing. By this treatment, the tensile strength R_m and the flow stress k_f are lowered, and the reduction of area Z is increased (Fig. C 6.7). Most steels therefore are cold formed in this state, i.e. spheroidizing-annealed ("GKZ-annealing").

The importance of the carbide form for the cold formability increases as carbon content increases. Carbide form is the most important variable [27] for steels with a carbon content exceeding about 0.35% C. Steels for quenching and tempering are thus almost always cold formed in a spheroidized condition. Spheroidize-annealing consists of holding the material for a sufficient time adequate to the dimension at temperatures below the A_1-temperature followed by slow cooling. To accelerate spheroidization, annealing is occasionally initiated in the two-phase region between the A_1- and A_3-temperatures or is accomplished at about the A_1-temperature. Carbide spheroidization is determined by the chemical composition, annealing temperature and time, the thickness of the carbide layers and the number of grain boundaries and discontinuities [28–30].

Higher alloyed steels are easier to spheroidize because they have a finer lamellar pearlite or possibly even a martensite-bainite initial microstructure. Spheroidization of unalloyed steels is facilitated, with increasing carbon contents, when cold forming has been done beforehand [30]. Generally speaking, spheroidization is greatly increased in its dependence on the true plastic strain (deformation) when cold forming is done beforehand, so wire rod and bars are often annealed after a preliminary drawing.

For a given steel grade, annealing time and annealing temperature govern the degree of spheroidization, as shown in Fig. C 6.8 for an unalloyed steel containing about 0.45% C (Ck 45). Complete spheroidization is often not necessary for cold forming, 70% often being sufficient, and this factor should be taken into account for cost savings.

An initial martensite-bainite microstructure is quickly annealed into a well-spheroidized microstructure with evenly distributed carbides, due to the even distribution of carbon and the resulting short diffusion paths. Therefore, when a high degree of spheroidization is desired, hardening is occasionally done before spheroidize-annealing.

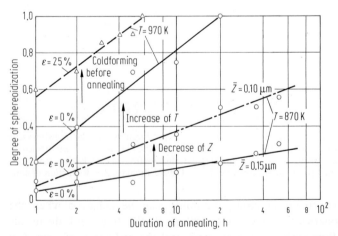

Fig. C 6.8. Factors of influence for the relation between annealing-time and degree of spheroidizing below A_1 of an unalloyed steel with 0.45% C (Ck 45-AISI C 1045). After [28]. T = Annealing time, Z = Thickness of cementite lamella, ε = Cold reduction ratio before annealing.

The initial microstructure before spheroidize-annealing has an influence not only on the annealing time required, but also on the properties in the annealed condition [29]. Ferrite-pearlite microstructures containing free ferrite result in a minimum of flow stress. A martensite microstructure is transformed into ferrite with evenly-distributed carbides. Improved deformability is the result, despite high strength, as is indicated by the higher values of reduction of area which are related to this microstructure (Fig. C 6.9).

Steels for quenching and tempering today are also treated in the form of wire rod coils and cold formed in this condition [31, 32]. Here strain hardening is used to achieve the component strength required.

Steels intended for cold forming having an austenitic matrix mainly are alloyed with chromium-nickel and chromium-nickel-molybdenum. With its face-centered cubic lattice structure austenite shows a very low flow stress at low true plastic strain values (deformations), which, however, increases rapidly with increasing true plastic strain (deformation) as the result of strain hardening. During cold forming the austenite is partially transformed into martensite, thus increasing the strength additionally [18, 33, 34], resulting in non-linear true stress-strain-curves. Higher nickel contents stabilize the austenite and reduce strain hardening (Fig. C 6.10), which is the reason why steels with a stable austenitic structure are preferred for cold forming (e.g. steels with nickel contents over 10%) (see also B 7.3.2 and B 8.3.2).

C 6.4.3 Cleanness

Nonmetallic inclusions must be borne in mind when subjecting steels to difficult cold forming operations, as they may lead to stress concentrations and cracking due to their notch effect. Under the usual deformation conditions the standard

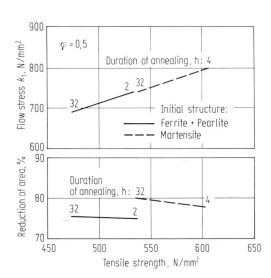

Fig. C 6.9. True flow stress k_f (true plastic strain $\varphi = 0.5$ rate of deformation) and reduction of area in dependence on the tensile strength for different annealing time to spheroidic cementite from different initial microstructures (ferrite-pearlite or martensite). After [29].

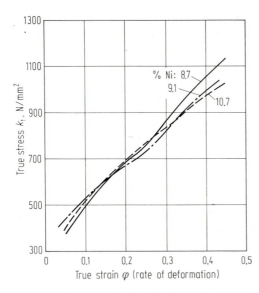

Fig. C 6.10. True stress-true strain-curves of austenitic steels with 0.05% C, 18% Cr and different Ni-content 8.7%–10.7%. After [34].

maximum sulphur levels (e.g. ≤ 0.035%) specified for the steels in question generally do not have any adverse effect. If machining is to be done after deformation, then specified sulphur contents may even range from 0.020% to 0.035%. In special instances the number and size of sulphide inclusions are limited by specifying low sulphur contents; maximum levels of 0.010% S can be achieved. Globular sulphides can be more favorable for cold formability than elongated manganese sulphides [21]; they are produced by adding calcium, selenium or tellurium.

The same holds true for oxide inclusions: globular oxides impede deformation less than linear ones [21, 35]. In the metallurgical processes employed today, however, the degree of oxide cleanliness achieved is so high that cold formability is not affected.

C 6.4.4 Surface Quality and Dimensional Accuracy of Steel Products

The *surface quality* of the hot rolled bar employed is of paramount importance for the costs and the reject rates involved in cold extrusion. However, surface flaws resulting from the production process cannot be completely prevented. Therefore permissible crack or flaw depths according to classes and a decarburization depth limit are specified (see C 6.3). Machined bar is used when extremely exacting requirements must be met. Surface roughness and strain hardening of the rim zone are also subject to specification. A rough surface is favorable for good phosphate coat bonding, therefore, the roughness of bright bars is sometimes increased by blasting.

Extremely close dimensional tolerances almost always must be maintained in connection with bars. Permissible dimensional variations of as-rolled bars often do not suffice but must be further limited. Closer dimensional tolerances are achieved by new rolling processes [36]. Tighter tolerances can be held by cold drawing (bright steel).

C 6.5 Characteristic Steel Grades Proven in Service

As discussed in C 6.4.1, steels for cold forming are characterized essentially by their application properties. The steels most used are those which have proven themselves in other fields and which may be subjected to additional, more exacting requirements in connection with cold formability. This designation applies not only to steel as stock but to steel products as well. Corresponding information is to be found in C 6.2 and C 6.4. Steels which meet the entire set of requirements, i.e. which are particularly suitable for cold forming, are treated e.g. in DIN 17 111 [37], DIN 1654 [14, 17, 38–40] and Euronorm 119 [114] as well as in ISO-Standard 4954 [42].

These steels can be divided into three groups according to the various requirements to be met by the components, i.e. the properties of use of steels:
1. Unalloyed steels with a ferritic-pearlitic microstructure which derive the required component properties from cold forming without heat treatment.
2. Unalloyed and low-alloy steels which are generally first transformed into a condition of good cold formability, then cold formed followed by heat treatment to achieve the desired properties. Examples are case-hardening steels, steels for quenching and tempering, nitriding steels and ball and roller bearing steels.
3. High-alloy steels with a ferritic or austenitic microstructure which, depending on their intended use, possess properties such as strength at elevated temperatures, high-temperature strength, heat resistance and corrosion resistance. Ferritic steels are cold formed in an intermediate condition, as are the steels listed in 2. Austenitic steels are cold formed in the condition as received.

Steels not intended for heat treatment are *unalloyed* for cold forming and are used for the mass production of cold formed parts such as screws, nuts, bolts and rivets (see e.g. DIN 17 111 [37]), as shown in Table C 6.1. Due to their low carbon content these steels have a low amount of pearlite in their ferritic matrix resulting in low initial strength and high deformability even in the untreated condition. Selection of the corresponding steel grade is governed by carbon and manganese content, due to the fact that the initial strength and the local true plastic strain (deformation) (strain hardening) are decisive for the final strength in certain areas of the finished part.

Steels are produced killed or unkilled. The microstructure of unkilled steels (e.g. UQSt 36 and UQSt 38 in Table C 6.1) is not uniform due to segregation, thus the components may show considerable scattering in strength. Greater uniformity of cold-formed component properties can be achieved by using specially killed steels (e.g. QSt 32-3 to QSt 38-3 in Table C 6.1). Steels of this type are largely used in the cold extrusion of light-duty components in the automotive and other manufacturing industries, e.g. for spark-plug bodies and shock absorber tubes.

Table C 6.1 Chemical composition and mechanical properties of unalloyed steels for cold forming, not intended for heat treatment [37, 38]

Steel grade (Designation)	Chemical composition[a]			Mechanical properties					
				for the untreated condition		for the normalized condition			
	% C	% Si	% Mn	Tensile strength N/mm² max.	Reduction of area % min.	Yield stress N/mm² min.	Tensile strength N/mm²	Elongation after fracture A % min.	Notch impact energy[b] J min.
QSt 32-3[c]	≤ 0.06	≤ 0.10	0.20 … 0.40	400	60	170	290 … 400	30	27
QSt 34-3[c]	0.05 … 0.10	≤ 0.10	0.20 … 0.40	400	60	180	310 … 420	30	27
UQSt 36	≤ 0.14	Spuren	0.25 … 0.50	430	60	180	320 … 430[d]	30	27
QSt 36-3[c]	0.06 … 0.13	≤ 0.10	0.25 … 0.45	430	60	200	320 … 430	30	27
UQSt 38	≤ 0.19	Spuren	0.25 … 0.50	460	55	220	360 … 460[d]	25	27
QSt 38-3[c]	0.10 … 0.18	≤ 0.10	0.25 … 0.45	460	55	220	360 … 460	25	27

[a] Content of phosphorus and sulphur each max. 0.040%
[b] ISO-V notch test specimens (longitudinal) at 20 °C
[c] Killed with Al (≥ 0.020% Al$_{total}$) or with other elements, having a similar effect for deoxidizing and fixing nitrogen
[d] Normalized condition is unusual

Table C 6.2 Chemical composition and mechanical properties of case-hardening steels for cold forming [39]

Steel grade (Designation)	Chemical composition[a]					for the condition GKZ[b]		Mechanical properties for the blank-hardened condition (Diameter of testpiece 30 mm)			
	% C	% Si	% Mn	% Cr	% Mo	Tensile strength N/mm² max.	Reduction of area % min.	Yield stress N/mm² min.	Tensile strength N/mm²	Elongation after fracture A % min.	Reduction of area % min.
Cq 15	0.12 ... 0.18	0.15 ... 0.35	0.25 ... 0.50	–	–	470	65	355	590 ... 790	14	45
17 Cr 3	0.14 ... 0.20	0.15 ... 0.40	0.40 ... 0.70	0.60 ... 0.90	–	530	60	450	700 ... 900	11	40
16 MnCr 5	0.14 ... 0.19	0.15 ... 0.40	1.00 ... 1.30	0.80 ... 1.10	–	550	60	590	780 ... 1080	10	40
20 MoCr 4	0.17 ... 0.22	0.15 ... 0.40	0.60 ... 0.90	0.30 ... 0.50	0.40 ... 0.50	550	60	590	780 ... 1080	10	40
21 NiCrMo 2[c]	0.17 ... 0.23	0.15 ... 0.40	0.60 ... 0.90	0.35 ... 0.65	0.15 ... 0.25	590	59	590	830 ... 1130	9	40
13 MnCrB 4[d]	0.12 ... 0.16	≤ 0.20	1.00 ... 1.20	0.80 ... 1.10	–	540	60	600	800 ... 1100	10	40

[a] P- and S-content each max. 0.035%, except steel 13 MnCrB 4 with 0.020%–0.035% S
[b] GKZ = spheroidizing annealed to globular cementite (carbide)
[c] Besides 0.40% to 0.70% Ni
[d] Besides 0.0010% to 0.0030% B

Steels intended for heat treatment after cold forming comprise selected *unalloyed* and *low-alloy* steels suited to case hardening, quenching and tempering, induction hardening or nitriding, depending on their intended use.

A selection of the *case-hardened steels* most commonly used is shown in Table C 6.2, which is adapted from DIN 1654 Teil 3 [39]. Case-hardening steels are particularly suitable for cold forming. The low carbon content of these steels and the resulting low amounts of pearlite in the ferritic matrix (and thus the low initial strength in an untreated condition or after spheroidize-annealing) make it possible to produce cold formed components with high true plastic strains (deformations). Optimizing chemical composition, e.g. achieving low silicon and sulphur contents (see C 6.4), can be useful in connection with complicated components. Improvements to ductility by use of various chemical compositions is limited to the extent allowed by the requirements for core hardenability and machinability. As a rule, cold forming of unalloyed case-hardening steels is done in the untreated condition, i.e. with the steel having a normal ferritic-pearlitic microstructure. Alloyed case-hardening steels are preferably cold-formed after being spheroidize-annealed. Depending on component shape and the sequence of processing operations, cold forming can also be done in a condition of microstructure better suited for machining with individual workpieces. As noted in C 6.4.2, this condition is achieved by controlled cooling of a steel from an annealing temperature of between 900 and 1000 °C, making for an even distribution of the pearlite patches in the ferritic matrix ("BG-annealing"). A microstructure obtained in this manner is also referred to as a black/white microstructure (mottled structure). This heat treatment is used in producing the steels 16 MnCr 5, 13 MnCrB 4 and 20 MoCr 4 according to Table C 6.2, the maximum strength value for this microstructure being about 600 N/mm^2. Using boron-alloyed case-hardening steels (e.g. 13 MnCrB 4) for cold forming (as detailed in C 6.4.2) offers the special advantage that boron increases hardenability without cold formability being adversely affected, as is it may be with conventional alloying additions.

The steels Cq 15 and 17 Cr 3 are preferred for production of cold-extruded piston pins, valve spring retainers, bearing sleeves and self-tapping screws. The alloyed case-hardening steels 13 MnCrB 4, 16 MnCr 5 (also with boron addition) and 20 MoCr 4 are used in automobile transmissions for pinion gears, medium-duty sliding gears and bevel gears; steel 21 NiCrMo 2 (among others, see DIN 1654 Teil 3 [39]) is used for higher-duty components of the same type.

Steels for quenching-and-tempering commonly used for cold forming are shown in Table C 6.3 as adapted from DIN 1654 Teil 4 [17]. The choice of a suitable material is governed by the tensile strength which is assessed on the basis of the hardness indicated by the end quench test. Alternatively, with unalloyed and low-alloy steels of low hardenability, the core hardness of a round sample having a diameter corresponding to the intended may be used. The majority of steels for quenching and tempering are used for producing cold formed bolts and similarly-shaped components, i.e. in the form of round steel bars, so this hardness test sample matches the demands of many practical applications.

No details need be given of the microstructure of steels in a quenched and tempered condition, as they are detailed in C 5. Here only the use of steels for bolts,

Table C 6.3 Chemical composition, mechanical properties and hardenability of steels for quenching and tempering, intended for cold forming [17]

Steel grade (Designation)	Chemical composition[a]					for the condition GKZ[c]		Mechanical properties[b] for the quenched and tempered condition (> 16 ≤ 40 mm dia.)					Hardenability	
						Tensile strength N/mm² max.	Reduction of area % min.	Yield stress N/mm² min.	Tensile strength N/mm²	Elongation after fracture A % min.	Reduction of area % min.	Notch impact energy[e] J min.	Hardness of the core HRC min.	Dia.[d] mm max
	% C	% Si	% Mn	% Cr	% B									
Cq 22	0.18 ... 0.24	0.15 ... 0.35	0.30 ... 0.60	—	—	500	60	295	490 ... 640	22	50	39		
Cq 35	0.32 ... 0.39	0.15 ... 0.35	0.50 ... 0.80	—	—	570	60	365	580 ... 730	19	45	29	40	8
38 Cr 2	0.34 ... 0.41	0.15 ... 0.40	0.50 ... 0.80	0.40 ... 0.60	—	600	58	440	690 ... 840	15	45	29	40	16
41 Cr 4	0.38 ... 0.45	0.15 ... 0.40	0.50 ... 0.80	0.90 ... 1.20	—	620	57	665	880 ... 1080	12	45	29	40	26
34 CrMo 4[f]	0.30 ... 0.37	0.15 ... 0.40	0.50 ... 0.80	0.90 ... 1.20	—	610	58	665	880 ... 1080	12	50	34	45	20
22 B 2	0.19 ... 0.25	0.15 ... 0.40	0.50 ... 0.80	—	0.0008 ... 0.0050	500	60	370	590 ... 740	18	50	60	40	9
35 B 2	0.32 ... 0.40	0.15 ... 0.40	0.50 ... 0.80	—		570	60	510	690 ... 830	16	45	40	40	18
25 MnB 4	0.21 ... 0.28	0.15 ... 0.40	0.90 ... 1.20	—	min. 0.0008	520	60	(580)	(750 ... 900)	(14)	(55)	(40)	40	14
37 CrB 1	0.35 ... 0.40	0.15 ... 0.35	0.50 ... 0.80	0.25 ... 0.40		580	59	590	740 ... 880	15	45	35	40	24

[a] P- and S-content each max. 0.035%
[b] Numbers in brackets are approximated values only
[c] GKZ = spheroidizing annealed to globular cementite (carbide)
[d] Diameter up to which after oil-quenching a hardness of 40 HRC or 45 HRC min. can be achieved for the core
[e] ISO-U-notch test specimens (longitudinal) at 20 °C
[f] Besides 0.15% to 0.30% Mo

the most important type of components in connection with these steels, is to be discussed in more detail, as a supplement to the discussion in C 29.

The properties of bolts made of the steels discussed here are specified according to strength classes contained in DIN 267 Teil 3 [43] and in DIN ISO 898 Teil 1 [44] (see also C29). The specified mechanical properties can be achieved via strain hardening alone in steels belonging to strength classes up to 6.8 (tensile strength of at least 600 N/mm^2) (see the steels in Table C 6.1). Quenching and tempering is necessary to achieve the values of the strength classes 8.8 to 14.9 (tensile strength of min. 800 N/mm^2 to min. 1400 N/mm^2). Steels possessing sufficient hardenability are required to achieve the specified mechanical properties after cold forming for various dimensions. Tensile strength increases with increasing carbon and alloy contents in untreated as well as in annealed steels. Cold formability, however, decreases, and hardenability-increasing additives must therefore be used in accordance with component dimensions. Unalloyed steels and alloyed boron, chromium- and chromium-molybdenum-steels are used for cold heading of components for the fastener (screw) industry. Classified by alloy series, these steels have various carbon contents and thus their hardenability can be suited to the requirements of the various strength classes and component cross-sections. Unalloyed steels (e.g. Cq 35 in Table C 6.3) can be employed in components of strength class 8.8 (tensile strength of min. 800 N/mm^2), however, only for parts having thin dimensions (dia. $\leq$ 12 mm). By adding 0.5% chromium (e.g. steel 38 Cr 2 in Table C 6.3), the size range can be extended to include medium cross-sections (dia. $\leq$ 18 mm), with cold formability somewhat reduced due to the alloy content. Hardenability is increased considerably by micro-additions of boron (e.g. in the steels 35 B 2 and 37 CrB 1 in Table C 6.3), without cold formability being adversely affected. These boron steels are used for bolts of strength class 8.8 in the medium-size range (dia. $\leq$ 18 mm) and for those of strength class 10.9 (tensile strength of min. 1000 N/mm^2) in the thin cross-section range (dia. $\leq$ 8 mm) instead of unalloyed steels and low-alloy chromium steels (with about 0.35% or 0.50% Cr).

Cold forming of unalloyed and boron-alloyed steels as well as of the manganese boron steel 25 MnB 4 (see Table C 6.3) can also be done in an untreated condition if necessary, if the true plastic strain φ (deformation) is not too high with regard to toughness properties. Alloyed steels for quenching and tempering usually are cold formed in a spheroidize-annealed condition with a degree of spheroidization of at least 70%.

Steels for quenching and tempering mentioned above are used for cold formed components which operate at temperatures below 300 °C. Alloyed *steels for elevated temperature service* are used for temperatures of up to about 540 °C. These steels contain about 0.24% C, 1.1% Cr and 0.3% Mo or about 0.21% C, 1.3% Cr, 0.7% Mo and 0.3% V (steels 24 CrMo 5 and 21 CrMoV 7 5 in accordance with DIN 17 240 [45], see also C 9), and are used for the manufacture of cold formed nuts and bolts. *Nitriding steels and ball and roller bearing steels* are also cold formed. Nitriding steel containing about 0.3% C, 2.5% Cr, 0.2% Mo and 0.15% Mo (31 CrMoV 9, see DIN 17 211 [46], see C 5) is used for cold-formed piston pins. Ball and roller bearing steels, e.g. containing 1.05% C and 1% Cr (105 Cr 4 in

Table C 6.4 Chemical composition and mechanical properties of austenitic stainless steels, intended for cold forming [40]

Steel grade (Designation)	Chemical composition[a]					Mechanical properties for the quenched condition			
	% C max.	% Cr	% Mo	% Ni	% other	Yield stress N/mm² min.	Tensile strength N/mm²	Elongation after fracture A % min.	Notch impact energy[b] J min.
X 5 CrNi 19 11	0.07	17.0 ... 20.0	–	10.5 ... 12.0		185	500 ... 700	50	60
X 5 CrNiMo 18 10	0.07	16.5 ... 18.5	2.00 ... 2.50	10.5 ... 13.5		205	500 ... 700	45	60
X 2 CrNiMoN 18 13	0.03	16.5 ... 18.5	2.50 ... 3.00	12.0 ... 14.5	N: 0.14 ... 0.22	300	600 ... 800	40	60
X 10 CrNiMoTi 18 10	0.10	16.5 ... 18.5	2.00 ... 2.50	10.5 ... 13.5	Ti: ≥ 5 × % C	225	500 ... 750	40	60

[a] Besides in all steel grades max. 1% Si, max. 2% Mn, max. 0.045% P and max. 0.030% S
[b] ISO-U-notch test specimens (longitudinal) at 20 °C

accordance with DIN 17230 [47]), are used for cold formed rollers and balls. For this group of steels cold formability requires a spheroidize-annealed condition, with spheroidization being as complete as possible.

Suitable *high-alloy steels* must be used for cold heading and cold extrusion of components for special applications requiring high temperatures and aggressive media. Of the large number of these corrosion-resistant, high-temperature, high-strength or heat-resistant steels only a small group is suitable for cold forming. A selection of steels most commonly used is listed in DIN 1654 Teil 5 [40]. Steels are classified here according to their microstructure into ferritic, martensitic and austenitic steels. The microstructure also indicates the degree of deformability. *Ferritic and martensitic steels* are cold formed in a spheroidize-annealed condition and their behavior is similar to that of unalloyed steels. In comparison, *austenitic steels* possess higher ductility which, when subjected to plastic deformation, has a larger usable range between yield stress and tensile strength in a quenched and tempered condition. This group of steels is thus preferred for cold forming and for difficult forming operations. Table C 6.4 shows some steel grades belonging to this group.

Austenitic chrome-nickel steels with and without molybdenum, e.g. containing about 0.05% C, 19% Cr and 11% Ni or about 0.05% C, 18% Cr, 2% Mo and 12% Ni (steels X 5 CrNi 19 11 and X 5 CrNiMo 18 10 in Table C 6.4), are used for nuts, bolts and other parts likely to be exposed to corrosion by water, the atmosphere and acids. Higher contents of nitrogen of from 0.12% to 0.22% are also used in these steels to improve austenite stability and increase strength. Thus, a higher yield stress can be achieved in the quenched condition for a steel, for example, containing about 0.02% C, 18% Cr, 2.8% Mo, 13% Ni and 0.18% N (steel X 2 CrNiMoN 18 13 in Table C 6.4), although the carbon content is limited to 0.03% for increased resistance to intercrystalline corrosion: ULC (ultra-low carbon) steels. These steels are thus particularly suitable for complicated forming operations.

In comparison, austenitic steels stabilized with titanium or niobium, e.g. containing about 0.08% C, 18% Cr, 2.8% Mo, 12% Ni and titanium (steel X 10 CrNiMoTi 18 10 in Table C 6.4) have less-favorable cold formability due to their titanium or niobium content needed for fixing carbon and for improving their resistance to intercrystalline corrosion.

C 7 Unalloyed Wire Rod for Cold Drawing

By Herbert Beck and Constantin M. Vlad

C 7.1 Required Properties of Use

Steels may be made into steel wire by a series of deformations and heat treatments [1–5]. The unalloyed wire rod considered in this chapter has carbon contents of up to about 1%. It is hot rolled as round rod with diameters ranging from 5.5 mm to 30 mm or with square, flat or other cross sections. Normally, the rod is supplied in the form of irregularly-reeled coils.

Good deformability of the rod (see B 7), i.e. *cold drawability* is the most important property required [5]. The *surface quality* of the wire rod should be suited to the drawing operations and to the finished product, and is closely related to the drawability.

Other requirements imposed on the wire rod are essentially defined by the *mechanical* and *technological properties* required in the drawn wire. The final properties depend on the degree of cold deformation and the heat treatments used in processing.

As far as the *technological properties* are concerned the *formability* of the drawn wire is of prime importance in any subsequent processing. Widely different stresses involved in bending, stretching, twisting, stranding, spring coiling, braiding, heading, cold rolling, and weaving dictate minimum requirements for yield stress and tensile strength, on elongation after fracture and reduction of area, behavior in reverse bending tests, in torsion, wrapping and upset heading tests. To comply with these requirements the wire rod must have a good deformability. If formability is insufficient the wire must be heat treated before or at some stage in the process, e.g. by annealing before weaving or braiding or, in wire for highly stressed springs, by patenting before the final drawing operations. The *transformation behavior* of steels, therefore, must be considered for these heat treatments as well as during cooling from the rolling temperature. Transformation behavior is important also in the quenching and tempering of spring wire, coiled springs or other components.

Springs subjected to high dynamic loads must have good mechanical properties especially good *fatigue properties*. These properties may be obtained with *surfaces* free from injurious defects and a high degree of cleanness, e.g. a low level of harmful *inclusions*.

Weldability of steels is also important. The ends of wire rod coils are generally butt welded and annealed before drawing. For some finished products, e.g. concrete

---

References for C 7 see page 774.

reinforcing wire mesh, the wire net is resistance spot welded. High carbon wire grades are not well suited to welding, so allowances must be made for occasional breaks at the weld during drawing. The inhomogeneous structure of the weld itself must also be considered and no welds are permitted in some products after the last patenting operation.

C7.2 Characterization of the Required Properties

Drawability of wire rod is understood as the limit in reduction of cross section at which the wire begins to break excessively often on further drawing, or at which the ductility of the wire has decreased to such an extent that its further formability is reduced. One limit, therefore, is characterized by the minimum diameter which may be reached under operating conditions (Fig. C7.1a). Drawability depends on the chemical composition, microstructure and surface condition of the steel. Deformation in drawing takes place predominantly in the ferrite and to a much lesser extent in the pearlite. On the other hand, the ferrite/pearlite ratio is determined by chemical composition, size and cooling rate from the hot rolling temperature or from the austenitization temperature used during the last heat treatment. This ratio and the grain size, along with distance and thickness of cementite lamellae determine the drawability [6]: the microstructure should be as fine as possible. A high *tensile strength may serve as an indicator* for the assessment of the drawability.

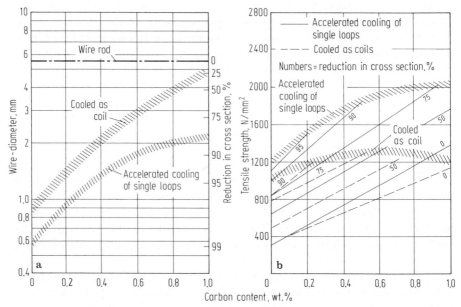

Fig. C7.1. Limits of drawability of wire rod for various carbon content levels. **a** Limit for predrawing of wire rod of 5.5 mm dia. to smaller sizes without intermediate patenting. **b** Maximum obtainable tensile strength of wire drawn from rod within the limits of drawability according to Fig. **a**.

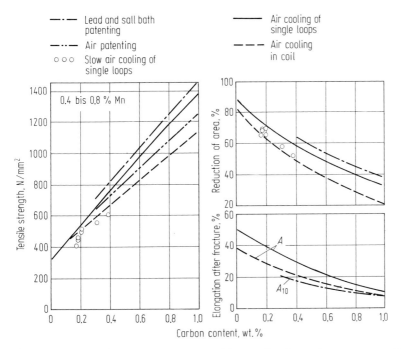

Fig. C 7.2. Mechanical properties of wire rod of 5.5 mm dia. after various treatments. After [2, 5, 27].

Typical (mean) values of the mechanical properties which may be expected from a 5.5 mm diameter wire rod after various treatments are shown in Fig. C 7.2. In addition to an unsatisfactory microstructure, coarse *inclusions* of hard undeformable oxides, sulphides or carbides (like tertiary cementite in low carbon steels) are prone to cause failures in drawing or processing.

Depending on the desired final size, tensile strength and other mechanical properties, the wire rod may be drawn direct to final size, provided that it has been subjected to suitable cooling after leaving the finishing stands of the hot rolling mill, or annealed or patented before or at some stage during drawing. Of course, the desired residual deformability is of great concern. Two extremes are usually encountered. One is characterized by the fact that the drawing is carried out without consideration of tensile strength, ductility or residual formability to the smallest possible wire diameter (Fig. C 7.1a). The other extreme is to be seen in the production of high tensile strength wire with satisfactory ductility and residual formability from the rod without any intermediate heat treatment. The most frequent instance of optimal ductility, as shown in Fig. C 7.1b is at about 75%.

Depending on size and carbon content the *surface quality* of the rod is mostly tested by upsetting, reverse torsion, acid pickling, or magnaflux inspection. In the upsetting (compression) test a straight sample of the rod with a length equal to 1.5 times the diameter is hot deformed by upsetting at a temperature of around 900 °C to about 50% of its length. The reverse torsion test is carried out to a minimum of specified numbers of reversing torsions depending on size and carbon content.

Examples of test requirements are included in DIN 17 140 [7]. Measures to control the surface quality of rolling mill products are discussed in D 5.3.4.

Eddy current surface inspection with on-line monitoring of the moving rod [7a] is used in several plants. The elevated temperature of the rod at this point, rules out magnetic disturbances. However, at this stage of the development eddy-current inspection is not sensitive enough to detect very fine surface defects below a certain size even though such small defects can still be detrimental to the product [7b].

At the same position as the eddy-current unit, behind the rolls, a *dimension-measuring* instrument may also be installed.

Mechanical properties of wire rod are assessed by conventional methods, preferably by tensile testing (see B 1). Attempts sometimes are made to relate properties to chemical composition by an equation. If this is done with adequate precautions and taking account of limiting conditions, useful information may be obtained about the influence of the elements concerned. A long-term statistical evaluation of tensile strength data obtained from wire rods within the most frequent size range of 5.5 to 10.0 mm dia. with 0.1 to 1.0% C and 0.3 to 0.8% Mn showed the following relationship between tensile strength and chemical composition:

$$R_m (N/mm^2) = 267 + 1015 (\% C) + 111 (\% Si) + 199 (\% Mn)$$
$$+ 1220 (\% P) - 804 (\% S) - 264 (\% Al) + 104 (\% Cu)$$
$$+ 231 (\% Cr) - 11.3 (mm\ dia.)$$

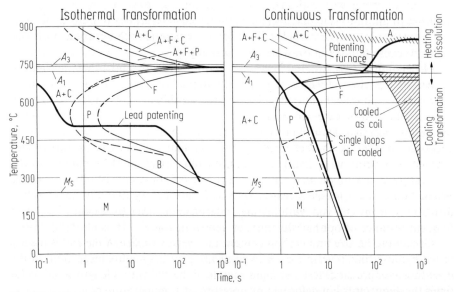

Fig. C 7.3. Time-temperature-transformation diagram and time-temperature-dissolution diagram of unalloyed steel with 0.75% C (D 75-2 after DIN 17 140 [7]) for the characterization of the structural transformations in wire rod on heating and cooling to the patenting process and after rolling. After [5].

This relationship was calculated by a single producer from values of wire rod control cooled in the fanned-out state as single loops by forced air. The relationship may be adapted by other producers to fit their particular conditions.

The influence of other elements on tensile properties may be neglected as they are present at consistent low levels in modern basic oxygen steels.

The *transformation behavior* and *hardenability* may be characterized by the usual methods (see B 4). Effects of heating may be displayed in time-temperature-austenitization diagrams (Fig. C 7.3). The subsequent cooling processes may be shown in time-temperature-transformation (TTT)-diagrams for isothermal transformation or continuous cooling depending on cooling process and size of rod (Fig. C 7.3 a and b). These diagrams should be regarded only as an indication because they do not take into account the different conditions in the austenitization and dynamic recrystallization (see A 7.2) of test samples as compared to actual wire rod after rolling.

The *weldability* may be characterized by methods which have been dealt with in a general manner in B 5.

C 7.3 Measures to Attain Required Properties

C 7.3.1 General Remarks

The *drawability* of steel wire rod is improved basically by a finer *microstructure*, disregarding minor details. This need for a finer microstructure is true regardless of the carbon content of the unalloyed wire rod grades or the heat treatment, i.e. from the almost pure ferrite in low carbon steels up to the pearlite of the high carbon steels. To control the drawability, therefore, it is necessary to use manufacturing procedures which will yield small ferritic grain size in low carbon, and fine lamellar pearlite in high carbon, steels: ferritic-pearlitic structures follow the same rule.

Although they are derived from basic *physical metallurgy* principles, in practice, the necessary procedures are strongly affected by the *process technology*. It therefore seems logical not to separate these two viewpoints when basic features are considered in detail. On the contrary, the technology will often be emphasized because within the last 25 years the heat treatment associated with the technology used to make much unalloyed rod has been moved back to the rolling mill from the wire drawing plant [8]. As a result of this technology transfer, many wire drawing plants were able to reduce the percentage of wire to be patented substantially [8a]. The wire industry of one country for instance had reduced the patenting factor from 1.42 in 1970 to 0.5 in 1980 [8b]. This development resulted in the imposition of more stringent requirements on the drawability of the wire rod. The technological processes connected with this step are also used for alloy steels [9], which are not discussed in this chapter.

Because of its decisive influence on the pearlite content in the microstructure, the highly different requirements for *mechanical properties* of wires can be met in the simplest and most economic way by changing the carbon content. Due regard must also be paid to the cooling rate imposed on the rod after hot rolling. In general it can be stated that the fineness of the microstructure which favors drawability also improves the mechanical properties (see B 1). Measures to attain good drawability and mechanical properties, therefore, frequently merge into each other and cannot be discussed separately.

C 7.3.2 Metallurgical Measures and Mill Practices to Produce Microstructures with Required Properties

The *basic metallurgical* principles for the production of microstructures with specific required properties are discussed in D 2 and D 3. The most important aim of *metallurgical procedures* is control of the required chemistry to maintain sufficient consistency within and between different heats of a steel grade.

The *chemical composition* must be chosen in such a way that a microstructure supporting the required mechanical properties and desired drawability is obtained under the production parameters in use, esp. the cooling conditions after hot rolling. Effects of the elements on mechanical properties are discussed in B 1 and B 4. The level of the tensile strength in the unalloyed steels discussed in this chapter is essentially determined by the grain size and by the ratio of cementite to the ferritic matrix, which in turn depends on the carbon content. The amount of cementite required is largely determined by tensile strength needs.

The use of higher manganese contents to increase tensile strength by solid solution strengthening is limited because high manganese contents markedly retard the transformation of the austenite in the pearlite range. Levels of the order of 1.5% or higher may cause a partial transformation of the austenite to bainite or even to martensite on accelerated cooling. Although the tensile strength is increased by the bainite and martensite, the presence of these constituents has an impairing effect on the drawability.

Elements that increase hardenability such as chromium and molybdenum also must be watched closely even at levels as low as 0.1%, particularly, when the wire rod is subjected to an accelerated cooling, because of danger of martensite formation.

Tensile strength can be improved by raising the phosphorus content (see the formula in C 7.2) in low carbon steels which are to be subjected to comparatively little deformation such as those used for concrete reinforcing bars.

Metallurgical means are also being used to obtain low *segregation* in *clean steels*. Only a little possibility exists to influence the structure of the cast steel in ingot casting by changing the size of the ingot. Structure is improved in continuous casting, particularly when the strand is stirred electromagnetically in the solidification zone [10]. This stirring method leads to reduced primary segregations, improved cleanness due to elimination of most of the impurities of exogenous nonmetallic inclusions and other deoxidation products as well as an improved surface quality [10a]. Zones with primary segregations formed in the core of the strand depending on the rate of solidification reduce the deformability of medium and high carbon steels. Because of their near-martensitic hardness these steels are prone to breaking in drawing. Segregations of carbon act in the same manner when present as hypereutectoid cementite precipitates in eutectoid steels. Heavily-segregated manganese and chromium also may cause breakage of such steels because of the martensite formation after accelerated cooling [10b].

The effects of phosphorus and sulphur as impurities are increased particularly by their enrichment in segregation zones. The sulphur content must be limited for applications involving the highest deformations because ability to withstand fine wire drawing operations and lateral deformation, as in sharp bends or cold rolling to flat shapes, may increasingly be restrained by rising amounts of sulphides [11, 12].

Oxygen is present bound as an oxide depending on deoxidation and precipitation (see D 2). Influence of the content of nonmetallic oxide inclusions on drawability is of growing importance when the wire diameter reaches the size of the almost undeformable oxide particles. Rimmed ingot casting processes – which are of decreasing importance – are reserved for very soft low-carbon wires which can be drawn to extraordinarily fine sizes because of their thick and clean rim zone.

These very low carbon steels are an exception to the above-mentioned rule on grain size since a coarser grain size of the ferrite (pearlite is almost not present) gives a high formability even after severe reductions with concomitant low tensile strength.

Aluminum additions lead to binding of oxygen, improving cleanness and surface quality. Aluminum also combines with nitrogen [13], counteracting grain coarsening in patenting thick wire sizes by the formation of aluminum nitride with its grain-refining effect.

The *rolling* of rod diameters at least up to about 12.5 mm is now carried out in continuous mills [14]. Compared to older Garrett-type machines, these continuous mills maintain a uniform rod temperature over the entire length of wire in the coil. Furthermore twisting of the rod is avoided by the HV-arrangement of the rolls in the finishing block. This arrangement makes possible a scratch free surface provided a good pass design is employed [15]. Hard metal alloy sintered tungsten carbide, cermet rolls wear less readily than those of cast metal, yield a smoother surface and allow permissible tolerances to be maintained over a far larger number of coils. These features in turn produce greater consistency in the amount of drawing required and thus in the wire properties.

Cooling of the rod *after rolling* is of great importance in obtaining the microstructure most suitable for drawability, strength and deformability. In older mills the wire rod was generally cooled in coil after reeling. This procedure did not produce the desired fine-grained structure of fine-lameller pearlite for drawability especially in large rod sizes and high carbon contents (more than about 0.35%).

In the past, therefore, medium high and high carbon wire rods were given a patenting treatment to increase the drawability [16]. In this process individual strands of rod or wire are cooled from austenitization temperature relatively rapidly in molten lead, molten salt or air. The transformation of the austenite to pearlite occurred isothermically in a temperature range in which very fine pearlite lamellae are formed. This sequence is shown schematically in the following table:

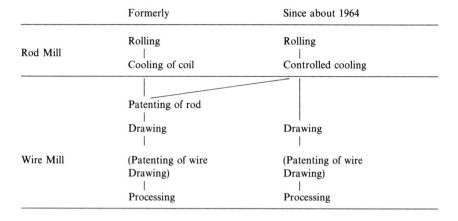

Cooling after hot rolling in the rod mill is presently carried out in such a way that a microstructure with a good drawability comparable to that of patented wire can be produced directly from the rolling heat. The accelerated cooling of the rod increasingly suppresses the precipitation of the proeutectoid ferrite during the austenite transformation and a fine-lamellar pearlite is formed, favoring drawability. In Fig. C 7.2 a comparison is made between the mechanical properties of the typical wire size of 5.5 mm dia. of various steel grades after different treatments exemplifying the modern possibilities as compared to lead or air patented wire.

A stepwise cooling of the rod, such as in a cascade, has been introduced into mill operations after the final pass and before the laying reel or head. Several sections are used to equalize the temperature across the diameter intermittently between different cooling sections despite the high delivery speeds of the order of 80 m/s or greater [17]. By this method the scatter in the temperature over the whole coil length is reduced to a very narrow range and thus consistent temperatures can be maintained. Thus, the microstructure and the associated mechanical properties are maintained over long periods of time within a very narrow range. Of course the final cooling of the wire must be conducted with equal consistency, which was hardly possible when the reeled coil was cooled, even with small sizes, because the outer rings cooled more quickly than the inner ones. The wire rod is now cooled with forced air under close control after laying in single non-concentric rings on a conveyor [8, 8a, 18].

Below a diameter of about 8 mm it is possible not only to reduce the scale layer by rapid cooling but also to obtain a microstructure capable of being drawn consistently, independent of the coil weight. Heat transfer from the core to the surface slows down with increasing rod size. As a consequence, the transformation of the austenite takes place at higher temperatures and after longer time periods. The volume fraction of proeutectoid ferrite becomes larger, pearlite lamellae are thicker and drawability is impaired. Some improvements have been introduced by blowing air or water into the laying reel or onto the coil behind the reel [19]. To improve the rod properties it was necessary to use fanned-out non-concentric loops or rings on a moving chain- or roller-conveyor and subsequent cooling by an air blast [8] despite the use of a lower conveyor speed than the rolling speed. Microstructure and mechanical properties are at least equivalent if not superior, to those of air patented stock [20]. This process is called "Stelmor" and has found the widest acceptance due to its operational simplicity and its good overall performance.

A few other processes aiming in the same direction may be mentioned. In the "Demag-Yawata"-process air is blown on individual rings in a vertical shaft [21]. In the "Schloemann"-process [22] and the "Krupp"-process [23] the rod cools in still air behind a horizontal laying head reel as a standing or lying single ring on a chain conveyor. In the "Fluidized-bed"-process the rod is fanned out on a conveyor which is run through a temperature-controlled fluidized bed of zirconia sand [24]. The "ED"- and "EDC"-process (= Easy Drawing Conveyor) utilize the relatively-mild quenching effect of the vapor blanket formed when hot rods are laid in near-boiling water containing surface-active agents [25].

The "Salt-Bath-Patenting"-process differs somewhat from that used in the wire industry. The rods are fed from the laying tube directly into the salt bath or are fanned out as in the "Stelmor"-process and, after a short equalization period, are run through one or two salt baths [24, 25a–25c]. The transformation, therefore, takes place in a controlled, nearly isothermal manner at temperatures in the neighborhood of the pearlite nose of the TTT diagram (see Fig. C 7.3) analogous to lead patenting. After sufficient cooling, the salt is washed off and the rod is dried by its residual heat content.

In contrast to the processes so far mentioned, the "Temprimar"-process deliberately aims at a fine lamellar pearlitic core zone with a martensitic surface layer [26]. The freshly-formed martensite is tempered during the cooling period of the rod in the reeled coil. The tempered martensite is not detrimental to fabrication of the wire if the martensitic rim does not take up more than one third of the cross section and the volume fraction of simultaneously-formed proeutectoid ferrite is not higher than 1%. For a more detailed description of these and other cooling processes refer to the literature [18].

Mention should also be made of the production of reinforcing rod and bar with a tempered martensite rim layer by controlled cooling after rolling [26] (see C 3). The production of dual-phase steels (see A 7.4, A 9.4 and B 4.2.1) is applied to low alloy steels, e.g. for high-strength fasteners [26a].

The only recognizable development in quenching and tempering of wire rod has existed as a separate process step for several years. In this process, complete coils of any rod size are subjected to vibrations during quenching in oil [26b]. In principle the process seems suited to quenching directly from the rolling heat of heavy coils [9].

In contrast to processes for the medium- and higher-carbon steel grades there exist for lower-carbon steel grades the necessity of cooling at a somewhat lower cooling rate in order to improve formability by overaging. This slower cooling has been achieved in the "Stelmor"-process by reduced water cooling before reeling and use of air cooling on the conveyor to lower strength values. In addition to this retarded cooling a slow cooling is applied by covering the conveyors with insulated, heat-resistant lids [27]. The extensive precipitation of proeutectoid ferrite which results produces a satisfactory drawability even with low alloy steels.

C 7.4 Characteristic Steel Grades Proven in Service

Wire rod grades may be divided into three classes based on their carbon content [4a].

1. *Low carbon mild steel rods* are normally used after drawing in the soft-annealed condition or thermally treated (annealed) and redrawn to specified hardness (half- or full-hard condition) and with specific surface treatment (by galvanizing, or by coating with copper, nickel, chromium, or plastics). Wire for general usage may also have an (extra smooth) clean bright finish for electroplating of finished parts after further processing. Wire for general purposes includes such applications as pins, chain link fences, coat hangers or straightened and cut pieces for bending purposes, to name just a few. A wire grade intended for an unusual kind of machining is the so-called wire wool which must have a suitable chemical composition and closely controlled strength values as well as a very high degree of freedom from non-metallic inclusions.

All these steel grades demand accelerated cooling after hot rolling, and the extra-soft wires need a retarded or even a slow cooling to minimize carbon and nitrogen embrittlement caused by quench aging. This very low tensile wire is drawn (with in-process annealing if necessary) into fine round wire e.g. for the manufacture of netting or woven products. The steel contains less than about 0.06% C and is produced as rimmed steel with little segregation and impurities, and must have a clean surface to ensure good drawability. If this steel grade is to be continuously cast with aluminum deoxidation, the carbon content must be limited to an absolute maximum of 0.03%. This steel will provide mechanical properties comparable to ingot casting and an even better drawability to more than 99% reduction of the cross section from the rod without annealing [27a]. The wire to be rolled into glass also belongs to this group of very soft wires. It should be drawable to a fine size, readily weldable and possess a very clean and smooth surface.

Preformed staple wire of larger sizes, to be used in automatic stapling machines employed in furniture fabrication or housebuilding is characterized by a high and very uniform temper. High strength values are obtained in these wires by restricting the chemical composition of the killed steel within a narrow range, an appropriate accelerated cooling of the rod and cold deformation by drawing and flat rolling. Other mild steel grades rolled on wire rod mills are those made for concrete reinforcement (see C 3), cold heading and cold extrusion (see C 6), free machining (see C 19), electrical conductors in communication lines for telephone and telegraph, fuses and relays, (see C 24), chains (see C 30) and welding wires and electrodes (see B 5).

2. *Wire rod* steel grades *containing a medium-high carbon content* (roughly from 0.30 to 0.60%C) are often supplied in spheroidized condition for applications requiring good hardenability or in a hard-drawn condition e.g. for screws or other cold deformed parts. Wires with special profile cross sections such as carding wire are cold rolled after drawing. For applications that require quenching and tempering of the wire parts the chemical composition must be selected carefully, especially for elements influencing through-hardening such as manganese, chromium and molybdenum. This is even more critical if cold deformation is to be performed after quenching and tempering. Hard drawn wire for use as spokes or springs that are subjected to low static loads and relatively infrequent stress repetitions also belongs to this group.

3. *Wire rod of high carbon steel* grades (usually more than 0.5% C) is drawn predominantly to more than 70% reduction of cross section because of its high strain hardening capacity. In the as-drawn condition this material is used as high tensile wire. Deformability properties during drawing as they are influenced by the various processing steps in the wire plant are described quite amply in the literature [2–5]. Figure C 7.4 gives such an example. Several indices may be calculated for the assessment of the residual deformability [3, 28, 29]. Wire for strands and ropes, for

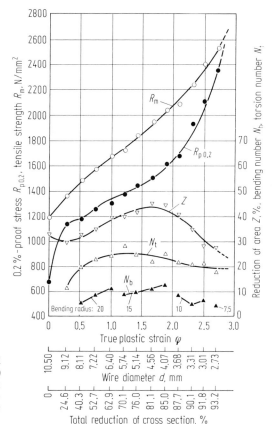

Fig. C 7.4. Mechanical and technological properties of patented wire (10.5 mm dia.) containing 0.86% C and 0.75% Mn in dependence on the "deformation degree"- true plastic strain $\varphi = \ln(d_0^2/d^2)$. After [6].

electrical conductors (see C 24), for helical springs (see C 18) and rubber reinforcement will be discussed here as examples of some of the most important applications.

Wires for the fabrication of strands and ropes [30] are drawn from wire rods with a tensile strength of about 800 to 1200 N/mm^2 to values of 1370 to 1960 N/mm^2 (see DIN 2078 [31]). To provide locking means for cables intended for load carrying in bridges, cableways or dredges the wires are drawn and cold rolled to wedge, Z- or S-shape and supplied in about the same classes of tensile strength values (DIN 779 [32]). Cold forming, particularly of the Z-shape, imposes high demands on the steel, which must therefore be produced with an exceptionally high uniformity of microstructure across the diameter as well as along the entire length and with a very high freedom from impurities. The load-carrying capacity of the rope is determined primarily by the breaking strength of the rope wire, by the structural design and by the method of manufacture of the rope itself (DIN 3051) [33]. Special specifications with a narrow range for permissible loads exist for ropes for cableways and shaft hoists as well as for ropes in lifting appliances (see DIN 21 254 [34] and DIN 15 020 [35]).

Electrical conductors consist of a core of galvanized (zinc coated) steel sheathed with copper or aluminum. Aluminum conductors steel reinforced (ACSR) consist of galvanized steel wire cable sheathed with aluminum wire. The steel wire is galvanized prior to cabling to prevent both the atmospheric corrosion and the local contact corrosion between steel and copper or aluminum sheathing. The steel core can be drawn direct from Stelmor wire rod [35a]. For further details on steel for electrical conductors see C 24.

The production of spring wire grades A and B according to DIN 17 223 Teil 1 [36] starts from an unalloyed wire rod as specified in DIN 17 140 Teil 1 [7]; for grades C and D preferably unalloyed wire rods are used which when compared to the former grades possess a higher cleanness. In addition for the spring wire of grade D an improved surface is also required [37]. Most of the spring wire is patented in process and drawn; only a small percentage is oil quenched and tempered. Alloyed steels are usually applied to the manufacture of those quenched and tempered wire and springs (see C 18). To these alloyed steels which are not within the scope of this chapter only the comment is given that they cannot be processed in the patented condition because they possess predominantly a hardened microstructure. Therefore, they are spheroidized prior to drawing or other forming operations. Criteria for the selection of the spring wire grade suitable to the particular application are named in DIN 2088 [38] and 2089 [39] in greater detail [40].

Spring wire is drawn to increasing tensile strength values in sizes ranging from 20 to 0.07 mm dia. The tensile strength must be kept within narrow limits between and within the different coils. Minimum values are specified for the reduction of area at fracture, for the number of torsions and for the wrapping test. After fabrication of the springs by bending, wrapping, twisting or other forming procedures, the finished parts are stressed repeatedly up to their design stresses, in compression springs up to blocking. As a consequence, the elastic limit is surpassed and the spring undergoes plastic deformation (setting). The tempering of the spring (frequently carried out before setting) implies in fact an accelerated aging with its concomitant reduction of internal stresses induced by the drawing. Furthermore by heating to temperatures of about 200 to 275 °C with its aging effect the elastic limit is raised to about 60% of the tensile strength, providing an additional safeguard against plastic deformation during working of the spring.

Rubber reinforcing wires such as bead-wire and steelcord in tires or in high pressure hoses and conveyor belts, have carbon levels of about 0.7%. Tire-bead wire is usually drawn from rod with in-process patenting to 0.95 mm dia. The wire must possess high mechanical and technological properties with restricted narrow scatter ranges which are tested in the torsion test and the reversed bend test. Beyond these requirements, a suitable bronze- or copper-plated finish must be produced to satisfy specifications for weights of coating and rubber adhesion. For the production of the tire bead reinforcement the wire usually is coiled with rubber to rings, which lie next to each other in the manner of a parallel wire cable, clad by tissue and vulcanized.

Steel cord wires make up the reinforcement under the tread and in the sidewalls of heavier steel cord tires. The small size of 0.15 to 0.35 mm dia. requires the steel to have an exceptionally high freedom from undeformable inclusions (oxides) [41, 41a]. Due to this required consistency a very narrow scatter range must be maintained for the contents of all alloying and trace elements. After the intermediate in-process patenting the wire is drawn to tensile strength values of 2600 to 2900 N/mm^2 and stranded. The strands are laid or bunched to cord construction and sometimes even wrapped with a wrapping wire. Efforts have been made recently to increase the tensile strength of the wires to levels of two newly-created higher classes from 3000 to 3300 N/mm^2 and even from 3300 to 3500 N/mm^2 by using cross section reductions during drawing after the intermediate patenting in excess of 96% [41b]. It is of interest to note here that a surface treatment by a heavy brass plating after patenting is necessary for a good rubber adhesion. Finally in these processes the highest value is set for consistency of mechanical properties and freedom from defects because of the very high speeds and stresses employed in drawing, stranding, bunching and wrapping. A rather important property of the drawn wire is a high bending fatigue limit which in turn is linked to the above-mentioned properties such as the tensile strength and the freedom from surface defects and nonmetallic inclusions.

C 8 Ultra High-strength Steels

By Klaus Vetter, Ewald Gondolf and Albert von den Steinen

C 8.1 Definition of the Term "Ultra High-strength Steel" ("UHS" Steel)

Ultra high-strength ("UHS") steels combine very high tensile strength and yield stress values (reaching or exceeding 1200 N/mm^2) with an assortment of properties that may be classified under the general heading "toughness", in that they are normally required of structural materials [1]. The special properties of UHS steels derive from various physical metallurgical processes, and they can be developed in steels which belong to either of two distinct groups. One of these groups is steels that harden as a result of the formation of carbon martensite, and need to be tempered (in this chapter these materials will be called "conventionally heat-treatable UHS steels"). The other group is steels that harden as a result of the formation of nickel martensite, which is virtually carbon-free. These steels are subjected to a precipitation-hardening or "aging" treatment (for brevity, these materials will be called "maraging steels").

C 8.2 Required Properties and Appropriate Test Methods

C 8.2.1 Application-derived Requirements

UHS steels are utilized in the production of components that must have high load-carrying capacities in combination with the smallest possible cross-sections and with lightweight. These steels must therefore exhibit a high *strength/density* *ratio* – equivalent to a large "break length". If their tensile strengths reach about 1600 N/mm^2, the strength/density ratios of these steels actually become comparable to those of the strongest low-density non-ferrous alloys available at the present time, namely alloys based on magnesium, beryllium, aluminum and titanium. UHS steels also exhibit toughness properties that are at least as good as those of the above mentioned non-ferrous alloys, and they also occupy less space. UHS steels are therefore utilized for particularly highly-stressed components in

---

References for C 8 see page 776.

automotive and aerospace engineering, as well as in nuclear engineering and defense technology. In addition to these component-oriented applications, UHS steels are also used as tool materials.

High strength is accompanied by reduced deformability so that it is difficult to effect the release of stress concentrations at abrupt changes in cross-section, or at any notches or cracks that may be present in the component. The *toughness* of UHS steels thus has to be assessed by reference to criteria other than those that apply to structural steels with conventional strength properties. Different test methods therefore have to be used to determine toughness in UHS steels. These methods produce ductility values measured in tensile tests (elongation and reduction in area), amounts of energy absorbed in notched-bar impact tests and, in particular, associated ductile/brittle transition temperatures, which represent criteria that are not wholly valid because their relevance diminishes with increasing yield stress [2]. However, these values are still used for comparative assessments, and for checking product uniformity, partly because they involve relatively low-cost test procedures. The *notched-specimen tensile strength* is an adequate criterion for the qualitative assessment of toughness in UHS steels, provided that the notches are made sharp enough. An even better measure of toughness is given by the ratio of the tensile strengths of notched specimens to those of plain ones.

UHS steels can exhibit a phenomenon known as "delayed brittle fracture". This effect is attributable to their sensitivity both to hydrogen embrittlement and to stress-corrosion cracking [3–5]. Essential information is generated by tests in which notched specimens are subjected to sustained tensile loads. The *notched-specimen constant load rupture strength*, determined in this way, decreases with increasing sensitivity of the material, increasing notch sharpness and increasing hydrogen content. The latter represents hydrogen that was initially present in the steel, or that migrates into it during the actual test, the outcome being that the notched-specimen static fatigue limit depends very much on the nature of the surrounding medium. Delayed brittle fractures of this kind can occur at stress levels far below the 0.2%-proof stress values of the steels in question.

Since UHS steels are frequently used in automotive and aeronautical engineering, it is necessary to possess data describing their *behavior under cyclic stresses*. The need for such data is all the greater because these steels no longer exhibit the approximately-constant ratio of fatigue strength to tensile strength that is typical of softer materials. Unlike conventional structural steels, the very hard steels do not exhibit a sharply-defined fatigue limit, and cyclic stressing within the fatigue range can lead to fracture after widely differing numbers of stress cycles [6]. In addition to the fatigue strength for infinite life (fatigue limit) interest also attaches to the behavior in the finite-life stress range, depending on the prevailing stress levels, and on the design lives of the components in question. The finite-life fatigue behavior should preferably be determined on notched specimens, at both high and low stress-cycling frequencies, and under test conditions that reproduce any relevant corrosion effects and simulate the expected service conditions as closely as possible (see B 1).

Qualitative or comparative assessment of UHS materials is possible on the basis of the parameters and types of test mentioned. *Parameters defined in linear-elastic fracture mechanics* can be used for the numerical design of components, or for estimating component lives, given the availability of information describing the prevailing stress fields and the sizes and geometries of any cracks that may be present (see B 1). The fracture toughness (K_{Ic}) or – if corrosive conditions also prevail – the critical stress level for stress corrosion cracking (K_{ISCc}) must be known for a given material if components are to be designed properly [7, 8]. Furthermore, if cyclic stressing is expected, the design calculations must also take account of the crack propagation rate (da/dN) under whatever ambient conditions may be present.

Fracture toughness values and parameters measured in conventional tests (tensile, or notched-bar impact bending) do not respond to certain material-dependent factors in the same sense, and are not mutually convertible. It therefore becomes necessary to use more than one test method to assess the toughness of a given material [9].

Reflecting the range of applications for which UHS steels are used, the demands that are made on their properties are limited to *temperatures* that normally prevail on the Earth's surface and in the lower atmosphere, and that are essentially centered on room temperature. Only in exceptional circumstances will UHS steel components be subjected to higher temperatures.

The availability of both *corrosion resistance* and non-rusting properties and an enhanced resistance to stress corrosion cracking would be particularly advantageous in several applications of UHS steels. However, there is no ideal way to achieve ultra high strength in combination with good anticorrosion properties. Current availability of corrosion-resistant UHS steels is hence limited to a number of materials that exhibit a certain slowness to rust allied to a degree of insensitivity to humid air or (polluted) air of industrial regions, as well as to the salt-laden atmospheres of coastal regions, or just above the sea.

C 8.2.2 Requirements Imposed by Manufacturing (Processing) Technology

UHS steels are shaped to produce components that fulfill many different functions. The processes that are used for manufacturing components from *low-strength, conventionally heat-treatable steels* (see C 5) can also be applied to UHS steels. Suitable manufacturing processes include machining, cold working, welding, heat treatment and various surface treatments. The two families of materials thus have similar requirements in terms of workability and machinability. Use of material-compatible design and manufacturing techniques is particularly important for UHS steels if satisfactory service performance is to be ensured, especially as manufacturing errors can result in exceptionally undesirable effects in these materials. UHS steels are generally used for critical, highly-stressed components, so their service properties often have primary importance and comparatively high manufacturing costs are usually accepted.

C 8.3 Meeting Requirements by Adjustment of Chemical Compositions and Appropriate Production Techniques

C 8.3.1 Basic Possibilities

Based on practical experience, two groups of UHS steels are used to meet the various requirements: "conventionally heat-treatable UHS steels" and "maraging steels".

The strength that is developed in the first-mentioned group is due mainly to the presence of carbon martensite, produced by a quench hardening treatment, as in all *conventionally heat-treatable steels*. UHS steel components normally experience high service stresses so it is essential that martensite hardening should occur throughout the entire cross-section. These steels must therefore be alloyed to levels that are high enough to produce suitable hardenability characteristics. The strength of the finished component is determined by the tempering treatment, so the tempering and the chemical composition must be matched to ensure that the resulting toughness properties will be adequate for the desired strength range. Where the tensile strength values are required to exceed about 1400 N/mm^2, the tempering temperature must be kept low, or the composition of the steel must be adjusted to ensure that it exhibits good hardness-retention when it is tempered. To avoid embrittlement of the martensite that occurs in the region of 300 °C, tempering should be performed either within a lower temperature range, namely from 150 to approximately 200 °C, or above 500 °C. Groups of steels have been developed with chemical compositions that render them particularly suitable for one or other of these two tempering ranges.

The addition of about 18% Ni to the low-carbon *maraging steels* ensures formation of a soft martensite when cooling takes place from about 800 °C. This soft martensite starts to form from the austenite at temperatures in the region of 200 °C (see Fig. C 8.1). On being reheated, this martensite does not transform back to austenite until the temperature exceeds 500 °C. This hysteresis in the transformation temperatures on cooling and heating is the basis for the precipitation-hardenability of these steels. Although austenite is capable of dissolving large proportions of alloying elements such as cobalt, molybdenum, titanium, etc., their solubilities in martensite are very low. However, the alloying elements can no longer precipitate from the martensite as it forms by transformation at temperatures in the region of 200 °C, because their diffusion rates are then too low. If the resulting supersaturated martensite is heated, ordering and pre-precipitation processes occur, especially as the temperature nears the limit at which retransformation will take place, usually between 450 and 500 °C. These processes have a significant precipitation-hardening effect (see A 6.4 and B 1). Numerous publications are available regarding the nature of the precipitates in this hardening process, but their authors do not all reach the same conclusions [11–16]. However, it is generally accepted that the phases FeTi, Ni$_3$Mo and Fe$_2$Mo are involved in the precipitation-hardening process.

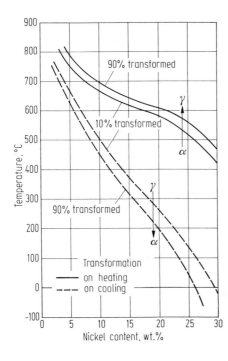

Fig. C 8.1. Iron-nickel phase diagram for nickel contents ranging from 0% to 30%. After [10].

The alloy contents of maraging steels must be adjusted so as to ensure that
- not even small proportions of delta ferrite occur,
- no undesirable intermetallic phases occur (i.e. phases which cause embrittlement),
- an increase in strength can be brought about as a result of precipitation hardening, and
- if the intention is to avoid an additional, sub-zero cooling treatment, the transformation to martensite should be completed on cooling to a temperature higher than room temperature.

If chromium is added to improve rust-resistance, appropriate adjustments will have to be made to the other alloying-element contents.

C 8.3.2 Strength and Toughness Properties

C 8.3.2.1 Conventionally Heat-treatable Steels

Factors that influence structures and properties of conventionally heat-treatable UHS steels are in principle the same as those which apply to their low-strength counterparts, as discussed in C 5. These factors include transformation behavior, hardenability, and temper brittleness.

High strength is easy to develop by an appropriate hardening treatment of cooling or quenching, followed by low-temperature tempering. For example, high strength in spring steels is achieved by using high carbon contents. However, if

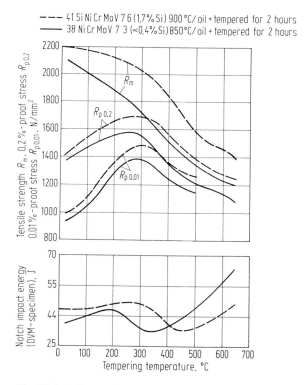

Fig. C 8.2. Influence of silicon content on variations in the mechanical properties of conventionally heat-treatable UHS steels with tempering temperature. After [17].

enhanced toughness properties are required, the *carbon and alloying-element contents have to be specially matched*. For two steels which have essentially similar compositions, save for different silicon contents, Fig. C 8.2 shows how the tensile strength, the 0.2%- and 0.01%-proof stress values, and the notch impact energy values vary with the *tempering temperature* [17]. For the 38 NiCrMoV 7 3 steel, which contains about 0.25% Si, these data show that an advantageous combination of properties is obtained when tempering is performed at 200 °C. The higher 1.7%, silicon content of the 41 SiNiCrMoV 7 6 steel, retards the processes that take place during tempering [18], and causes the martensite embrittlement to be displaced towards higher temperatures [19, 20]. The result is that steels of this type can be tempered at temperatures as high as approximately 330 °C, without any loss of toughness. The strength of the martensite, both after the hardening treatment and after tempering at about 200 °C, depends almost entirely on the carbon content, so the tensile strength and 0.2%-proof stress for these steels, can be adjusted only by altering the carbon content (see Fig. C 8.3a). Similar relationships are found with silicon contents that exceed 1.5% when tempering is performed at about 300 °C. For a given carbon content, the two strength parameters of the higher-silicon materials lie about 100 N/mm² to 150 N/mm² above the values in the figure.

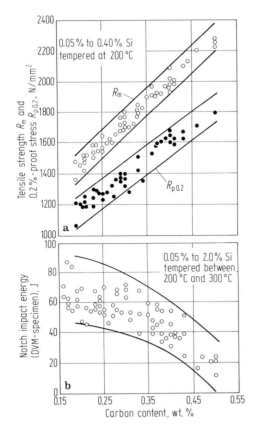

Fig. C 8.3. Influence of carbon content on mechanical properties of conventionally heat-treatable steels tempered at low temperatures subsequent to martensitic hardening. **a** Tensile strength and 0.2%-proof stress. After [17, 21–23]. **b** Notch impact energy. After [17, 21].

Taking notch impact energy as an example, Fig. C 8.3b shows, for comparable types of steel, that – as on tensile strength and yield stress – the *toughness* is as strongly dependent on carbon content. However, toughness is also affected by alloying elements, more than is strength. Particularly good results have been obtained from the combination of nickel with chromium and molybdenum – sometimes with further additions of silicon, cobalt and vanadium. In such alloys the nickel contents lie within the range from 1.5 to 8%. The toughness exhibited by this group of alloys is liable to be reduced by additions of manganese and silicon, which increase the solid solution hardening. Toughness is also reduced by additions of the carbide-forming elements molybdenum and chromium, but such additions cannot be omitted altogether because of the need to preserve an adequate degree of hardenability [24]. At the highest nickel contents, cobalt additions of up to 5% are beneficial because they prevent excessive depression of the M_s temperature and the undesirable retention of austenite during the hardening treatment [24].

Figure C 8.2 shows that if the steel 38 NiCrMoV 7 3 is *tempered at temperatures above 500°*, it will have good toughness properties, but only a comparatively-low tensile strength. If tensile strength values exceeding 1400 N/mm² are required after tempering at such temperatures, it becomes necessary to make *further alloying additions*, introducing elements that *increase the as-tempered hardness*. A precipitation-hardening process can prove useful in increasing the as-tempered hardness, either by forming finely-dispersed special carbides, or by forming intermetallic phases from certain alloying elements. In particular, where tensile strength values exceeding about 1700 N/mm² are required, consideration must be restricted to steels containing comparatively-large proportions of elements that form special carbides, such as are used in hot-work tool steels or nitriding steels. By way of an example, Fig. C 8.4 shows how the mechanical properties of two grades of steel, namely X 41 CrMoV 5 1 and X 32 NiCoCrMo 8 4, vary with tempering temperature [25, 26]. In the latter steel the decrease in strength is merely displaced towards higher tempering temperatures, but a secondary hardening effect occurs in the chromium-molybdenum-vanadium steel when tempering is performed at temperatures in the region of 500 °C. The toughness values of this steel pass through a minimum at the tempering temperatures associated with the greatest *secondary hardening*. For this reason, conventionally heat-treatable steels that exhibit secondary hardening are generally tempered at temperatures higher than the one at which maximum hardness is developed.

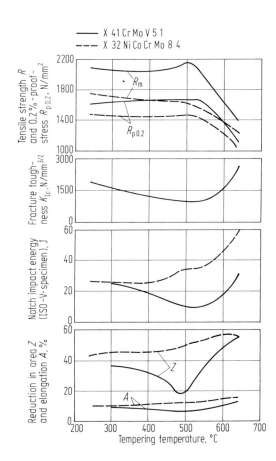

Fig. C 8.4. Influence of tempering temperature on mechanical properties of two UHS steels that exhibit good hardness retention (specimens tempered 2×2h). Steel X 41 CrMoV 5 1, hardened 1020 °C/air cooling. After [25]. Steel X 32 NiCoCrMo 84, hardened 840 °C/oil quenching followed by subzero cooling for 2 h at −75 °C. After [26].

The ability of steels that exhibit secondary hardening to retain their hardness during tempering is due to the precipitation of special carbides as tempering proceeds. The alloying elements that form such carbides are vanadium, molybdenum, chromium, tungsten, titanium and niobium. Copper can also assist the precipitation-hardening processes, if present in excess of about 1%. The achievable strength properties thus depend on both the carbon content *and* the alloying-element contents [27–30]. Either martensite or bainite can be considered as a hardened microstructure. The effective alloying-element content is reduced by any carbides or nitrides that fail to dissolve during the preceding austenitizing treatment. The degree of secondary hardening thus depends on the austenitizing temperature. The condition of the special carbides at precipitation largely determines the achievable toughness properties. However, these properties are not so clearly dependent on the carbon content or on the tensile strength, as they are with conventionally heat-treatable steels that have been tempered at low temperatures [31, 32].

C 8.3.2.2 Maraging Steels

The strength of maraging steels is derived from the *martensite* in the solution-annealed condition (which has a tensile strength in the region of 1000 N/mm^2) plus the strength increment added by *precipitation hardening*. In steels containing 18% Ni, the contribution from precipitation hardening is increased considerably by titanium (see Fig. C 8.5 [33]) and molybdenum (see Fig. C 8.6), although aluminum and niobium are also effective [34–37]. The part played by cobalt in the precipitation-hardening process is less easy to discern. Although cobalt alone does not have any significant precipitation hardening effect in nickel steels, it nevertheless intensifies the hardness-enhancement produced by molybdenum. It is suspected that, via

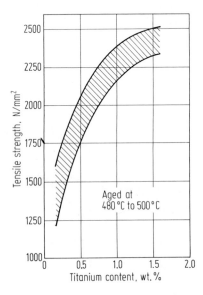

Fig. C 8.5. Influence of titanium content on tensile strength of maraging steels containing 18% Ni, 8% to 12% Co and 4% to 5% Mo, determined in the precipitation-hardened condition. After [33].

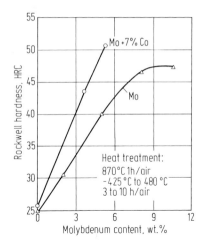

Fig. C 8.6. Influence of molybdenum, on the one hand, and of molybdenum + 7% Co, on the other, on the maximum hardness of steels containing 18.5% to 20.1% Ni. After [11].

states of ordering, cobalt promotes the formation of localized nickel-rich zones, reducing the solubility of the molybdenum. As a result of the transformation to martensite, the structure contains many defects which tend to accclerate the diffusion-controlled precipitation-hardening processes. Thus, aging the suitably pretreated X 2 NiCoMo 18 8 5 steel at 525 °C, increases the Vickers hardness by about 200 HV within the first minute [1]. Figure C 8.7 shows how the hardness of suitably pretreated steel X 2 NiCoMoTi 18 12 4 varies with aging time and aging temperature [1]. The decrease in hardness at the higher temperatures, and after comparatively-long aging times, is attributable to over-aging and austenite reversion.

The *toughness* of maraging steels depends on the type of alloy, in that it is governed by both the composition of the matrix and the kind and content of the particular elements that promote precipitation hardening. Steels containing 18% Ni, plus various additions of cobalt, molybdenum and titanium, have been found to exhibit very attractive toughness properties.

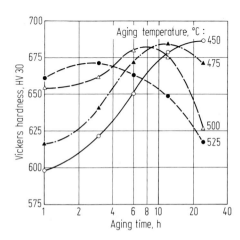

Fig. C 8.7. Hardness of X 2 NiCoMoTi 18 12 4 steel versus aging time, for various aging temperatures. Specimens air-cooled from 820 °C, 1 h, before being aged. After [1].

Maraging steels are generally *aged* at temperatures within the range over which the greatest precipitation hardening occurs. Should insufficient toughness be developed on aging within this temperature range, it is better to use a grade of steel that has a lower maximum strength, than to vary the aging temperature. For reasons that have yet to be explained, some steels with high cobalt and titanium contents suffer a pronounced reduction in toughness on being aged at temperatures between about 350 and 470 °C, i.e. below the range of maximum hardness [1, 33, 38].

C 8.3.2.3 Production of Conventionally Heat-treatable UHS Steels and Maraging Steels

UHS steels must exhibit both strength and toughness and this dual requirement can be met only by microstructures that are both uniform and as nearly free from defects as possible. These steels must therefore be *melted* and *cast* under conditions that ensure low contents of phosphorus, sulphur and gas contents. Furthermore, production techniques must be selected with a view to minimizing the content of non-metallic inclusions, and to obtaining cast structures that exhibit as little segregation as possible (see D 4).

In general, the *optimum structure* is one which is *uniformly martensitic*, because such a structure allows homogeneous slip to occur. Comparatively small amounts of stable residual austenite can improve the toughness properties, especially if they take the form of fine halos surrounding islands of martensite. Excessive amounts of residual austenite can sometimes be suppressed by means of sub-zero cooling (see the results for the steel X 32 NiCoCrMo 8 4 in Fig. C 8.4).

As in all steels, the toughness properties of UHS steels are enhanced by a *fine-grained structure*. So-called "superfine grain" can be obtained by repeatedly subjecting the material to a treatment sequence comprising rapid heating, short-time austenitization and hardening. This treatment has the further effect of increasing the strength, particularly the yield stress [39, 40].

The toughness and/or the strength can be improved by various *thermomechanical treatments*, which can thus be used for improving the toughness/strength ratio. Ausforming techniques [1, 41–44], as well as strain-aging treatments [45] applied to workpieces that have first been brought to a martensitic or bainitic condition enable extremely high strength values to be achieved. However, these ausforming and strain-aging treatments are not widely used, partly because they are applicable only to components with simple shapes, and partly because they are both difficult to perform and liable to create further problems when the treated components are subsequently processed. Better chances of achieving enhanced properties are offered by a high-temperature thermomechanical treatment in which the steel is hot-worked in the region of the Ac_3 temperature and then is hardened. This treatment produces dynamic polygonization after which the material is hardened before a recrystallization treatment [46]. The improvement of the properties by this treatment is due to the formation of a fine, uniform microstructure in association with certain dislocation structures inherited by the martensite from the deformed austenite, which reduces local strain concentrations over distances of the order of several microns (see B 1 and B 4).

On comparing the *toughness properties of various groups* of UHS steels, it is found that, for a given strength, the values for elongation and notch impact energy scarcely differ although the maraging steels are superior to the other types of alloy in terms of reduction in area and fracture toughness (see Fig. C 8.8).

C 8.3.3 Stress Corrosion Cracking, Hydrogen Embrittlement and Rust-resistance

Neither the conventionally heat-treatable UHS steels nor the chromium-free maraging steels are corrosion-resistant. They become increasingly susceptible to stress corrosion cracking or hydrogen embrittlement as their tensile strengths increase, and as notch effects and externally-imposed stresses become more severe. With UHS steels, serious problems can be caused by even very *small hydrogen contents*, such as may result from exposure to media that are only weakly corrosive, such as distilled water, mains water or certain organic liquids, or even moist air [5, 25, 47–51]. Hydrogen penetration during electroplating aimed at providing corrosion protection has proved to be the primary cause of many delayed brittle fractures in UHS steel components.

Within each individual group of materials, the *influence of chemical composition* is secondary to the influence of strength level. Different investigators have often obtained conflicting results when ranking the susceptibilities of UHS steels to attack by particular media. Resistance to stress corrosion cracking in various steels generally improves with increasing fracture toughness, so maraging steels behave better than conventionally heat-treatable UHS steels of the same tensile strength. Various precautions such as surface protection, internal compressive stresses, reduction of notch effects can be taken to avoid damage by stress corrosion

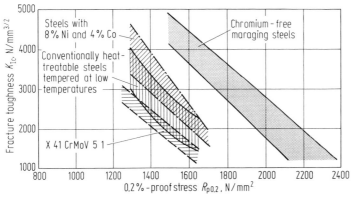

Fig. C 8.8. Relationship between fracture toughness K_{1c}, and 0.2%-proof stress $R_{p\,0.2}$, for four types of UHS steels of different alloying systems. Data from the literature and the authors' own research.

cracking. However, the precautions to be taken in any given instance require careful allignment with whatever experience is available.

Enhanced *resistance to stress corrosion* and reduced susceptibility to rusting can be obtained by using *maraging steels containing chromium*. It has been found that a considerable degree of rust-resistance can be obtained by adding as little as 9% Cr, in combination with the other necessary alloying elements [52]. Up to the present, investigations to develop grades of steel that exhibit similar corrosion resistance in association with comparatively high strength levels, e.g. exceeding 1700 N/mm^2 have not been successful. This lack of success is due to the fact that such materials are beset by other disadvantages.

C 8.3.4 Fatigue Strength

Basically the processes of material fatigue in UHS steels and softer steels are the same. There is no straightforward way to rank different UHS steels, or groups of steels in order of their fatigue strengths. Various metallurgical and technological factors cause the relevant measurements to be subject to scatter that is many times greater than the differences between individual materials.

Tempering at the highest possible temperatures is advantageous because it more or less *eliminates residual stresses*, which are harmful. The highest fatigue strength is often exhibited by the temper resistant chromium-molybdenum-vanadium steels properties. The achievable fatigue strength in tests exceeding 10^6 stress cycles, rises with increasing tensile strength for specimens without notches, or components characterized by high quality of surface finish and only slight cross-sectional irregularities.

Fatigue strength can decrease quite considerably as *notch effects* increase, irrespective of whether they are due to external factors, reflecting the shape or design of the component in question, or to inhomogeneities within the material itself. For example, the fatigue strengths of notch-free specimens, measured by the rotating bar method, can vary randomly between 600 and 1000 N/mm^2, while the corresponding tensile strength values vary between 1600 and 2100 N/mm^2. When components are to be manufactured from UHS steels it is advisable to design them so that low notch factors are achieved as a result of appropriate shaping and surface conditioning. There is no point in specifying exceptionally-high requirements in terms of material purity and freedom from inclusions unless notch effects can be avoided.

For a given tensile strength, *fine-grained, homogeneous martensitic or bainitic* microstructures yield the *best fatigue strength*, because they are highly resistant to overaging and have high capacities for work-hardening.

Compressive stresses at the surface of the component also have a beneficial effect and can be produced by shot peening or cold rolling, also by nitriding. Figure C 8.9 shows that salt-bath nitriding of UHS steels having good hardness-retention properties, heat-treated to various strength levels, offers a means of achieving significant improvements in the fatigue strength of notched specimens, as measured by the rotating-bar method [17]. In comparison with steels having

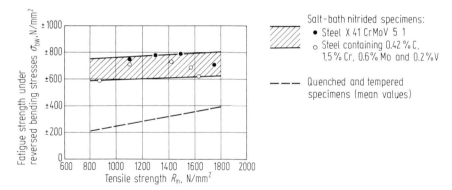

Fig. C 8.9. Influence of salt-bath nitriding on the rotating bar-bending fatigue strength (10^7 cycles) of conventionally heat-treatable steels with good hardness retention properties, heat-treated to various tensile strength values (specimen shape: double-shoulder bar with form factor $\alpha_k = 2$). After [17].

conventional strength values in the fully heat-treated condition, the high-tensile properties of UHS steels can be utilized to the best advantage in alternating stress conditions, since their fatigue strength declines relatively sharply under higher tension mean stresses (sensitivity to mean stress) [53].

UHS steel components, even if designed for infinite fatigue lives do not always prove superior to components made from conventional steels. When notch effects intervene, significant *advantages* can be obtained in the range of lower numbers of stress-cycles with components that are designed on the basis of *fatigue strength for finite life*.

In contrast to soft steels, in which the crack propagation stage represents by far the greater part of the total life, UHS steels exhibit enhanced *resistance to crack initiation*. However, crack propagation is particularly dependent on the nature of the environment. Even in X 1 NiCrMo 10 9 3 steel, a maraging steel containing 9% Cr, the fatigue strength declines very markedly when cyclic stressing occurs in a corrosive environment. Tested in air, under rotary bending conditions, smooth specimens of X 1 NiCrCoMo 10 9 3 steel are capable of withstanding as many as 10^7 stress cycles, with an amplitude of 600 N/mm². Notched specimens will exhibit the same endurance if the alternating stress amplitude is reduced to 280 N/mm². However, if these tests are conducted in a 3%-NaCl-solution, the corresponding stress-amplitude values fall to only 380 and 140 N/mm² [52].

If additional information is required regarding fatigue properties, reference may be made to the comprehensive body of literature available on this subject, which also includes fatigue strength data for many other grades of steel [1, 25, 30, 34, 54–57].

C 8.3.5 Processing Properties

As a rule, UHS steels are not worked in the high-strength condition that will normally be established before the components in question are finished, ready for

Table C 8.1 Comparison of the processing operations to which UHS steels are normally subjected

Maraging steels	Conventionally heat-treatable steels
Rolling or forging	Rolling or forging
Air-cooling	Slow cooling necessary, in a suitable facility
	Soft-annealing
Rough machining	Rough machining
Solution-annealing (about 830 °C/air)	Stress-relief annealing, normalizing
Fine finishing to final dimensions	Fine finishing (precision machining), but leaving workpiece slightly oversize
Aging to increase hardness (about 500 °C/air) (slight, uniform distortion (contraction))	(Hardening: 840 °C to 1040 °C/oil or air) (distortion, decarburization, risk of cracking) Tempering (repeated several times if appropriate) Machining to final dimensions
Surface treatment (if appropriate)	Surface treatment (if appropriate)

installation. Instead, the steel producers usually deliver UHS materials in a condition that facilitates cold-working or machining. As can be seen from Table C 8.1, the sequences of processing operations that users must perform on these materials differ according to whether the material in question is of the "conventionally heat-treatable" or "maraging" type. Here the maraging steels have advantages because they require fewer and easier treatment operations involving less risk.

In contrast to the treatment procedure employed for conventionally heat-treatable UHS steels, maraging steels only need a simple aging treatment at a temperature not exceeding about 500 °C, to establish the high-strength condition in which the finished components will be used. Any surface oxidation or dimensional changes occurring during this aging treatment will be so slight that there will generally be no need for any further finishing operations.

Particular significance should also be attached to the good *cold formability* of the maraging steels, and to the fact that they are only slightly susceptible to strain hardening. These steels will thus tolerate very high reductions in cross-section by either cold rolling or drawing, with no need for intermediate annealing. Moreover, they can be aged after cold-forming, without having to be solution-annealed for a second time. If this procedure is followed, the final strength will exceed that of material that has not been cold-worked prior to being aged, but the toughness will be reduced.

The maraging steels are also found to exhibit superior *welding behavior* because they do not harden during the cooling phase that follows the welding operation, except in narrow zones within the transition to the parent metal, where precipitation-hardening is very prone to occur. The greater part of the weld remains soft, so there is virtually no risk of cracking [58–61]. On the other hand, conventionally heat-treatable UHS steels have carbon and alloying-element contents that are too high to permit straightforward welding. Expensive special measures such as preheating, stress relieving, annealing and repetition of the hardening and tempering operations must therefore be used with these steels [1] (see B 5).

At strength levels above 1400 N/mm^2, becoming increasingly difficult, although different machining processes are affected to varying degrees. In addition to requiring carefully selected tools, which must be both hard and sharp, machining of UHS steels requires special cutting conditions and rigid machine tools, as well as correspondingly-rigid cutting tools and tool-clamping devices [58, 62, 63]. New machining processes are becoming increasingly important and economically attractive. These processes include electroerosive and electrochemical methods, electron beams and lasers.

C 8.4 Characteristic Grades of UHS Steel and their Properties

Despite the fact that they are produced only in relatively small quantities, *large numbers of UHS steels* have gained international recognition. The development of these materials has exploited the fundamental knowledge discussed in the preceding Sections. Tables C 8.2 to C 8.5 list the key properties of a selection of UHS steels that are representative of the two main groups identified at the beginning of this review. In the Federal Republic of Germany, DIN Standards have so far been issued only for those UHS steels intended for aerospace applications.

The steel grades 38 NiCrMoV 7 3 and 41 SiNiCrMoV 7 6 belong to the group known as *conventionally heat-treatable UHS steels*. These steels are hardened by quenching in oil, after which they are tempered at only 200 °C, or 300 °C for the higher-silicon grades (see Fig. C 8.2). Many other variants have similar compositions, and are thus heat-treatable to various strength levels that are determined by their carbon contents. Steels with reduced carbon contents down to 0.25%, are covered by Standards which are mostly non-German. These steels have minimum tensile strength values ranging between 1400 N/mm^2 and 1800 N/mm^2. Steel X 32 NiCoCrMo 8 4 is typical of several steels developed in the USA, containing 8% Ni and 4% Co. These steels are known to include a series with various carbon contents ranging between 0.20 and 0.45% [24, 64, 65]. Grades with carbon contents at the upper end of the range are intended to be tempered at low

Table C 8.2 Chemical compositions of selected UHS steels (guide values)

Steel grade Designation	% C	% Si	% Mn	% Co	% Cr	% Mo	% Ni	% Ti	% V
	Conventionally heat-treatable steels								
38 NiCrMoV 7 3	0.38	0.25	0.65	–	0.80	0.35	1.85	–	0.12
41 SiNiCrMoV 7 6	0.41	1.65	0.55	–	0.80	0.40	1.50	–	0.10
X 32 NiCoCrMo 8 4	0.32	0.10	0.25	4.5	1.0	1.0	7.5	–	0.09
X 41 CrMoV 5 1[a]	0.41	0.90	0.30	–	5.0	1.3	–	–	0.50
	Maraging steels								
X 2 NiCoMo 18 8 5[b]	0.02	0.05	0.05	7.5	–	4.8	18.0	0.45	–
X 2 NiCoMo 18 9 5[b]	0.02	0.05	0.05	9.0	–	5.0	18.0	0.70	–
X 2 NiCoMoTi 18 12 4[b]	0.01	0.05	0.05	12.0	–	3.8	18.0	1.60	–
X 1 NiCrCoMo 10 9 3[c]	0.01	0.05	0.05	3.3	9.0	2.0	10.0	0.80	–

[a] Undergoes secondary hardening
[b] Also contains about 0.1% Al
[c] Stainless

Table C 8.3 Characteristic mechanical properties of the UHS steels listed in Table C 8.2 (data from measurements at 20 °C, on specimens[a] in the hardened and tempered or precipitation-hardened condition, as appropriate)

Steel grade Designation	Tensile strength R_m N/mm²	0.2%-proof stress $R_{p0.2}$ N/mm²	Elongation after fracture A %	Reduction of area Z %	Notch impact energy (DVM-specimen) A_v J	Fracture toughness K_{Ic}[b] N/mm$^{3/2}$
38 NiCrMoV 7 3	1850	1550	8	40	35	1600…2600
41 SiNiCrMoV 7 6	1950	1650	8	40	30	1600…2400
X 32 NiCoCrMo 8 4	1550	1350	10	50	40	2400…3800
X 41 CrMoV 5 1	1900	1600	8	40	30	1200…1800
X 41 CrMoV 5 1	1700	1400	9	45	35	2000…2500
X 41 CrMoV 5 1	1500	1300	10	50	40	2500…3000
X 2 NiCoMo 18 8 5	1850	1750	8	50	35	2800…3500
X 2 NiCoMo 18 9 5	2100	2000	7	45	25	1800…3000
X 2 NiCoMoTi 18 12 4	2400	2300	6	30	15	1000…1500
X 1 NiCrCoMo 10 9 3	1550	1500	10	55	45	3800

[a] Longitudinal specimens, machined from bar about 50 mm. dia.
[b] Values at the upper ends of the ranges are typical of specially-melted materials

Table C 8.4 Characteristic values for the elevated-temperature 0.2%-proof stress of a selection of UHS steels

Steel grade Designation	0.2%-proof stress ($R_{p0.2}$) at N/mm²					
	20 °C	200 °C	250 °C	300 °C	350 °C	400 °C
X 32 NiCoCrMo 8 4	1400	1350	1330	1300	1250	1170
X 41 CrMoV 5 1	1600	1500	1430	1390	1350	1300
X 2 NiCoMo 18 8 5[a]	1750	1600	1500	1450	1400	1330
X 1 NiCrCoMo 10 9 3	1500	1320	1270	1220	1180	1140

[a] For any of the higher-strength maraging steels with 18% Ni, the ratio of the elevated-temperature 0.2%-proof stress to the room-temperature 0.2%-proof stress will be the same as for this material.

Table C 8.5 Heat-treatment data for a selection of UHS steels

Steel grade Designation	Soft-annealing °C	Normalizing °C	Hardening[a] °C	Solution-annealing[a] °C	Tempering or aging °C
38 NiCrMoV 7 3	670…720	880…920	850…880/oil	–	180…220
41 SiNiCrMoV 7 6	700…740	900…940	880…910/oil	–	280…320
X 32 NiCoCrMo 8 4	600…650	880…920	840…870/oil, W[b]	–	530…570
X 41 CrMoV 5 1	820…860	(900…1000)	1000…1040/oil, A	–	550…640
X 2 NiCoMo 18 8 5	–	–	–	800…840/A	470…490
X 2 NiCoMo 18 9 5	–	–	–	800…840/A	480…500
X 2 NiCoMoTi 18 12 4	–	–	–	800…840/A	490…510
X 1 NiCrCoMo 10 9 3	–	–	–	820…850/A	470…490

[a] Coolant: A = air, W = water
[b] Sub-zero cooling is recommended, at about −75 °C

temperatures. To attain the highest strength levels, steel X 32 NiCoCrMo 8 4 and certain lower-carbon grades are tempered at temperatures above 500 °C, because they have good hardness-retention and structural stability (see Fig. C 8.4).

The X 41 CrMoV 5 1 steel permits *secondary hardening* in association with good temper resistance, and can be heat-treated to achieve various strength and toughness properties by tempering at appropriate temperatures above 540 °C [25, 30, 50] (see Fig. C 8.4 and Table C 8.3). This chromium-molybdenum-vanadium alloying system has also served as a basis for the development of several grades that are particularly suitable for nitriding.

Since the introduction of the *maraging steels* in the early 1960's, various compositions have assumed some temporary measures of importance. However, in the long term, consistent success has been obtained with the three chromium-free grades listed in Table C 8.2. These steels contain 18% Ni, 8% to 12% Co, 4% to 5% Mo and 0.4% to 1.6% Ti, and they cover the tensile strength range from about 1750 N/mm^2 to 2450 N/mm^2 (see Table C 8.3). Relatively recent developments have been directed towards variants with higher cobalt and molybdenum contents, and these steels have usable tensile strengths in excess of 2800 N/mm^2 [37, 66–68].

The last of the steels listed in Table C 8.2, grade X 1NiCrCoMo 10 9 3, is a newly-developed *stainless steel* [52]. Unlike other rust-resistant steels, this new steel has a chromium content of only 9%, which benefits its toughness properties, even at quite low temperatures. Extended-duration, room-temperature immersion tests, some in condensate water and others in a 3%-NaCl-solution, failed to provoke rusting within upwards of 2000 hours. In tests designed to induce stress corrosion cracking, on both smooth and notched specimens in artificial seawater, no specimen fractured under tensile stresses up to 90% of the yield stress, even with test durations ranging from 1600 h to 2000 h. The fact that a K_{ISCc} value of 3200 N/mm$^{3/2}$ was determined in the NaCl-solution is indicative of high resistance to stress corrosion cracking [52, 69].

Table C 8.3 contains reference values for the tensile strength, 0.2%-proof stress, elongation and reduction in area, notch impact energy and fracture toughness of the steels selected for discussion. These values are characteristic of either the conventionally heat-treated condition or the precipitation-hardened condition, depending on the type of steel.

In common with those conventionally heat-treatable UHS steels that are tempered at comparatively high temperatures, maraging steels are characterized by good *elevated-temperature strength properties*, up to at least 400 °C (see Table C 8.4). Conventionally heat-treatable UHS steels that are tempered at low temperatures are unsuitable for elevated-temperature service.

Although all the commonly used UHS steels exhibit satisfactory toughness properties at temperatures down to about −40 °C, the maraging steels can be used even at considerably lower temperatures, at which they continue to exhibit good *toughness*. This insensitivity to temperature is particularly marked in the steel X 2 NiCoMo 18 8 5.

Table C 8.5 presents *heat-treatment data* for the eight materials selected as representative of the two groups of UHS steels discussed in this review. These steels enjoy the widest acceptance at the present time.

C 9 Elevated-temperature Steels (Creep Resistant and High Creep Resistant Steels) and Alloys

By Heinz Fabritius, Dieter Christianus, Karl Forch, Max Krause, Horst Müller and Albert von den Steinen

C 9.1 Required Properties

Elevated temperature steels, creep resistant steels and high creep resistant materials are applied wherever components or installations are *exposed to mechanical loads at high temperatures*. Examples are thermal power plants for production of electrical energy, gas turbines and aviation power units, chemical plants, e.g. of the crude oil chemistry and crude oil manufacturing industries, and industrial furnaces. Plates, tubes, forgings and castings are utilized in great amounts in the construction of such installations.

Thermal power plants today are operated with steam in the *temperature range* of about 300 °C to 650 °C. Fresh steam temperatures in conventional power plants are mainly in the range of 500 °C to 560 °C, but a series of plants with fresh vapor temperatures up to 650 °C was built, in the Fifties. Stationary gas turbines e.g. in air heaters can have temperatures of 700 °C and more. Fast breeders or high-temperature reactors will also operate at these or higher temperatures. In one test reactor of the last-mentioned type, helium is used as a coolant with an operating temperature of about 950 °C [1]. Aviation power units and chemical plants with operating temperatures of up to about 1100 °C are at the top of the temperature scale.

Many of the hot-working components in these units carry a substantially-constant mechanical load at uniform constant operation temperatures. Examples are the steam generators and steam pipes used in the base-load-operation of thermal power plants. Here the decisive mechanical load is caused by the steam pressure. Components for constant load are designed on the basis of *characteristic data*, determined by static tests. The kind of test used depends on the component type and its operating temperature. For components with high operating temperatures, creep limits and creep rupture strength are of primary importance. For components with low operating temperatures the yield stress at elevated temperature is included in characteristic data for design calculations. The limiting temperature, at which the design data changes from the yield stress at elevated temperatures to long-time (creep) values, is taken from the intersection of the yield stress-temperature-curve and the 10^5 h-creep stress rupture strength-temperature-curve.

The *demand for elevated-temperature steels for screws* is for a *high-relaxation resistance*. Moreover it is an important requirement in the application of screw steels that they remain free from long-time embrittlement after quenching and tempering to suitable strength ranges.

The operating conditions of hot working plants are characterized by *alternating mechanical stresses and temperatures*. If components are exposed to changing thermal conditions temperature differences will develop between their internal and external regions. These differences will increase with the thickness of the components and will depend on the time-related changes to the surface temperature. The forces generated lead to mechanical stresses which can increase the total stress beyond the elastic range. These components must be so designed that they can be produced economically and with optimum use of the material. Specific resistance to development of cracks by alternating strains is required in such material. Based on results of tests under alternating loads conducted in conditions similar to those encountered in service [2] the components are designed so that a minimum number of starting procedures without crack is possible during the calculated (projected) life-time. In addition to the low frequency stress cycles, higher-frequency cyclic stresses may be caused by oscillations and pulsations of the operating medium. These variations must also be taken into account in the design of the relevant components. Knowledge of the material behavior under constant and alternating load is completed by measurements of the crack propagation rate.

The *ductility (toughness) of creep-resistant steels is of special interest*, for instance for stress release in the heat-affected zone (HAZ) of weldments, in zones of manufacturing-related stress concentrations or in the area of inclusions and microcracks. In these examples the local relaxation resistance must be low, to reduce stresses. There must also be sufficient ductility (toughness) in the entire operating temperature range to ensure that an overstressed component will first deform in a ductile manner in a long stress phase before a crack occurs. This sequence can be recognized during inspection and allows timely protective measures to be taken. Increased ductility (toughness) behavior in high-temperature steels usually is accompanied by reduced creep resistance, so an adjusted material condition can be only a compromise. Both the toughness and the creep resistance may change during the load time, so knowledge of the extent of these changes during the expected life-time is of great importance.

Products made from creep-resistant steels often experience high *stresses* at temperatures that are *very far below the operating temperature*. These stresses may occur during centrifugal tests performed at room temperature by manufacturers of turbine shafts with 25% over-speed, and they may also appear during cold starting of turbines and during routine control by the plant operator. Similar stress events occur in hydraulic pressure tests. Characteristic values of fracture-mechanics like crack-initiation and crack-propagation behavior of the materials at room-temperature are of importance for proof of technical safety in these tests.

Depending on the kind of the component, the behavior of materials under impact stress – either at increased or at ambient temperatures – is also important. Sufficient notch impact strength is demanded not only in the delivery condition but also after long thermal stress has been experienced during operation.

The high requirements for safety of nuclear power plants have led to rising demands on the toughness of material to be used for reactor components, especially for reactor safety vessels (containment). In selecting material for such applications the aim is to decrease the importance of the inevitable small cracks by providing a tough matrix. Thus the German recommendation in 1975 was to apply the ASME Boiler Code to nuclear containment vessels. This code specifies the nil ductility transition temperature (NDTT) of embrittlement determined by a room-temperature toughness test.

The demands on the *conditions* of welded steels, especially those used in the construction of pressure vessels, also have been raised in recent years. Until recently, control of the notch impact energy in the heat-affected zones of welds was not required but since 1976 data sheets for inspection of such vessels published by the "Arbeitsgemeinschaft Druckbehälter" (Working party for pressure vessels) have stipulated that the notch impact energy in the heat-affected zone of a weld must have a value of at least 50% of the value prescribed for the base material. The long-time behavior of welded joints with regard to strength and toughness in the temperature range in which creep occurs is of great importance.

A further demand is for the adjustment of microstructure and strength properties so that they are as uniform as possible over the cross section for components with great wall thicknesses. This objective is aimed at by choosing a favorable chemical composition, by procedures during the production of forging ingots (deoxidation-, pouring, remelting processes) and by other measures during heat treatment.

In all plants that operate at high temperatures, materials are in contact with gaseous, liquid or solid agents. These materials must be resistant to reactions with these agents, that means resistant to corrosion absorption of carbon and nitrogen, decarburization, the effect of high pressure hydrogen and to erosion, because these reactions all lead to general or local wall thickness reduction, crack formation or other impermissible changes to the material properties. In particular examples the corrosion behavior may be of such great importance that it almost governs the material selection and the yield stress values to be aimed at. Then instead of the strength at elevated temperatures for instance, the heat resistance or the resistance to high pressure hydrogen will be placed in the forefront. Materials developed to suit these requirements are treated in C 14 and C 15.

Requirements for yield stress values at elevated temperatures are limited by the fact that the magnetite protective layer growing on the internal vessel wall during operating conditions bursts during the cooling period at compression stresses above about 600 N/mm^2, resulting in strong local corrosion [4, 5]. Therefore, the nominal stress is limited to 1700 N/mm^2 [6] based on a notch factor $a_k = 3.5$. This limitation means that the yield stress at elevated temperature which can be utilized for the particular steel is located at about 340 N/mm^2.

Knowledge of physical properties, e.g. for the calculation of temperature stresses, is also part of the list of requirements [7]. Decisive criteria for materials to be used in the reactor core or in the primary cycle of nuclear plants, are certain nuclear-physical properties and the behavior in the presence of neutron radiation [8].

C 9.2 Characterization of the Required Properties by Results of Specified Tests

The *important tests* for characterization of the properties of creep-resistant materials at elevated temperatures are the *creep-stress rupture test* under constant mechanical stress and *relaxation test* with constant strain. Fundamental explanations of these tests can be found in chapter B 1.3.3. Schmidt [9] in particular, reports in detail on the test procedures, devices and evaluation methods. On the basis of long experience with creep and creep rupture tests DIN standards also are established for the design of equipment to be used for these tests [10].

Creep tests determine the duration under load before attainment of certain permanent strains (elongations) and sample fracture. The results are summarized in different diagrams according to the object of the evaluations, as shown for instance in Figs. B 1.159 to B 1.161.

Determination of significant and reliable material data for design periods of 200 000 to 300 000 h, which today are the basis for creep-resistant components, requires extraordinary long testing times. Many programs therefore are undertaken to gain *characteristic data for long loading times* by extrapolating the results of short-time tests [9, 11]. Up to now no extrapolation procedure is known which draws conclusions only from the numerical compilation and evaluation of physical-metallurgical changes in the stressed material. All pure mathematical procedures (methods) have the disadvantage that they rely on the curve function, that is determined in the range covered by test results beyond this range. Thus these procedures cannot be applied, for instance, if there is a turning point in the creep rupture curve [12]. The most reliable procedure still is manual graphical evaluation with competent weighting of the points to be used. Experience has shown however that with this procedure the interpolation time proportion must not exceed the value of 3 [10, 13].

Under the above restrictions, test times of about 70 000 to 100 000 h (i.e. from 8 to 11 years) are necessary for determination of material characteristics covering the above-mentioned design (layout) times. For these periods the material behavior must be determined by measurements at many points and at different temperatures; steels in common use and promising further-developed materials, together with their welded joints must be tested. Because of the extent and the costs such tests are undertaken in Germany mainly by joint work in the frame of two working parties of steel producers, plant engineers and plant operators [14]. Several reports on the results have already been published [15–17].

To derive usable long-time values for a material or a material condition, specimens of different origin must be tested so that the *scattering range of the products applied in practice* can be roughly assessed. With rising numbers of measuring points a limited scatter range of the values can be distinguished. Experience shows, for instance, that the creep rupture strength in stress direction (in the stress-time-coordinate system) is scattered by $\pm 20\%$ around the mean value. Because of this scatter range, great differences in life-time may result even from a flat creep rupture strength curve (Fig. C 9.1). Reasons for the scattering are

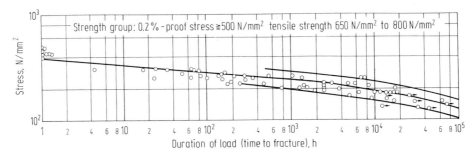

Fig. C 9.1. Scatter band of creep rupture strength values of steel X 20 CrMo(W)V 12 1 at 550 °C. After [16]. Range of composition: 0.17% to 0.26% C, 0.28% to 0.46% Si, 0.43% to 0.60% Mn, 11.5% to 12.1% Cr; 0.98% to 1.12% Mo, 0.28% to 0.37% V and 0.12% to 0.59% W.

concealed by unimportant differences in the sample properties and test conditions. Differences in properties of the specimens result mainly from permissible variations in chemical composition and heat treatment. Differences in test conditions weigh heavily especially with deviations of the test temperature from the nominal value.

To reduce the time and costs required for protracted creep and creep stress rupture tests, many extrapolation methods and *short-time test procedures* have been proposed. For instance, it is proposed that 45 h-tests be used for determination of the DVM-creep limit [18]. This procedure had to be abandoned because of faulty developments and false assessment of the material. Evaluation of short-term procedure proposed by Rajacovicz is not yet completed [19]. In this procedure the yield stresses given by slow hot tensile tests with constant straining rates are measured and extrapolated linearly to short straining rates. Figure C 9.2 shows for example the determination of a 1%-creep limit for a duration of load of 100 000 h for the steel X 22 CrMoV 12 1 [19].

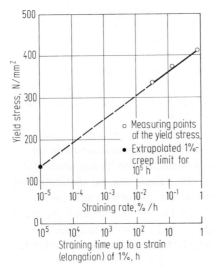

Fig. C 9.2. Determination of the 1%-creep limit for 100 000 h at 550 °C for steel X 22 CrMoV 12 1. After [19].

Short time tests are inappropriate to predict the properties of materials after long loading periods, but they might be *adequate in special applications*. It frequently happens that it is necessary to compare the creep behavior under specified conditions of materials of the same type after the same kind of heat treatment, where the behavior of the material used as a standard is known from long-time tests. In such instances, creep-rupture tests with a duration of up to 1000 h are taken as a basis for acceptance tests.

Great importance is attached to *creep and creep rupture tests of welded joints* and – for completion – tests with samples consisting only of weld material. Circumferential seams are the main part of constructional weldments so the internal pressure tests that simulate operational conditions of tubes is important. The results agree very well with the values for the base material [20]. With longitudinal seams in cylindrical hollow bodies, interior pressure causes very strong stresses transverse to the seam. Creep test specimens with a transverse seam within the gauge length can be used for evaluation of these stresses. A comprehensive review of test results [20] from such specimens shows creep stress rupture strength values in the lower range of the scatter band of the base material. With long-lasting loads especially, fractures are very often found in the heat-affected zones.

For the design (layout) of plants to use operational media other than air it is essential to know, whether values of the long-term creep behavior measured in air can be applied. Therefore, *comparative tests must be conducted, for instance in reactor-helium* [21] *or in turbine-hot gas* [22], and these tests require considerable and additional costs. The tests are performed either with constant strain rate [23] or as creep rupture tests with constant stress.

To control the behavior of creep-resistant screw materials, *isothermal stress relieving tests* (*relaxation tests*) are performed. For this purpose specimens are extended (strained) at testing temperature up to a certain level. The extension (strain) is then kept constant and the gradually decreasing stress necessary to preserve the extension (strain) is measured. In addition, a process that is nearer to the practical operation is used, in which a threaded bolt with 2 nuts is tightened within a tube-like shell [24]. After the specified load time the screw is released and the elastic recovery of the bolt is measured. In addition to the determination of material properties isothermal stress-relieving tests have importance, for instance, for the optimization of stress-release annealing [25].

The values of elongation after fracture and reduction of area determined by creep-stress rupture tests serve to characterize *the toughness of creep-resistant steels*. Other indicators are values of the reduction of area of slow-speed tensile tests at elevated temperatures, comparisons of times to fracture of notched and plain creep-rupture test specimens at similar loads and ratios of creep rupture strength values measured with notched and plain test specimens. To an increasing extent also the results of fracture-mechanical investigations are included [26]. Determinations of the notch impact energy and of the transition temperature after long-term annealing also are often necessary.

Comparison of *times to fracture of plain and notched creep-rupture specimens*, that is, the ratio of the creep rupture strength values measured with notched and plain test specimens give indications of the influence of a multi-axial stress distribution. With a tough material condition, notched specimens show a longer time to fracture than plain ones. On the contrary, in an embrittled condition, a range is developed that is, dependent on material strength, testing stress and testing temperature, where notched specimens fracture earlier than plain ones. In this range, elongation after fracture and reduction of area pass a minimum value. Figure C 9.3 shows results of tests on plain and notched specimens of a chromium-molybdenum-vanadium steel in two heat treatment conditions, one of which shows a tendency to embrittlement [27]. Results of the described tests, which were performed mostly on notched specimens complying with DIN 50 118 [10], cannot be used to draw conclusions on creep and creep-rupture behavior at high stress concentrations and multiaxiality, for instance, in the presence of a deep crack. Thus a material may still be characterized as tough on the basis of results obtained from the usual notched creep-rupture specimens, whereas specimens with artificial flaws (cracks) show embrittlement [27].

To clarify the creep-stress rupture behavior of specimens with a growing flaw (crack) requires the *determination of fracture mechanical characteristic values* [9]. Knowledge of such characteristics for instance, is important in heavy open die forgings if it is presumed that crack nuclei, or inhomogeneities that may develop

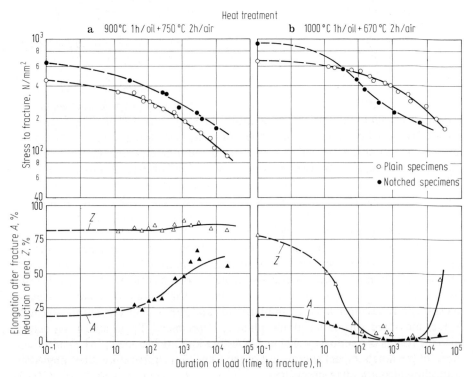

Fig. C 9.3. Results of creep-stress rupture tests with plain and notched samples at 550 °C of 21 CrMoV 5 7 in two heat treatment conditions. **a** with 702 N/mm^2; **b** with 1062 N/mm^2 tensile strength at room temperature. After [27].

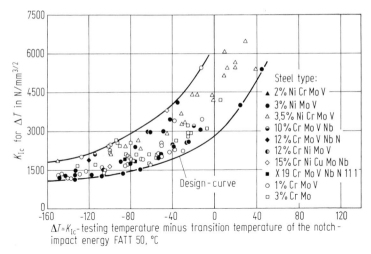

Fig. C 9.4. Fracture toughness values K_{Ic} dependent on a reference temperature for important forging steels. After [28].

into such nuclei, exist in large steel masses. The most widely-applied stress intensity factor K_I is, for instance, determined by measurements of the crack opening displacement. As shown in Fig. C 9.4 a useful compilation of results is achieved by plotting the fracture toughness K_{Ic} over the difference ΔT between the K_{Ic}-testing temperature and the transition temperature FATT (see B 4.6.2) for 50% crystalline fracture [28].

In the design of high-temperature components, data on the *long-term behavior* of creep-resistant materials *under alternating load conditions* are of increasing importance. Hence e.g. stress and strain cycles in the surface areas of turbine shafts during starting and slowing down procedures as well as during power changes require determination of the numbers of stress cycles that can be tolerated before the first flaw appears, at given strain amplitudes. The required testing technique has been adjusted as far as possible to suit the cyclic stress sequences occurring in practice. A summary description of this field of problems appears in [29].

C 9.3 Physical Metallurgy Procedures to Attain the Required Properties

C 9.3.1 Ferritic Steels

C 9.3.1.1 Ferritic Steels for Slightly Elevated Temperatures

Fine grain steels are well known for applications in temperature ranges up to about 400 °C. The most important alloying element in these steels is manganese. In addition, depending on the strength desired, smaller amounts of nickel, copper and molybdenum may be added. Elements such as aluminum, nitrogen, niobium and vanadium also may be added, either individually or in combination, to attain the fine grain character. The steels are normalized or air or fluid quenched and tempered.

Much information is available with regard to the *influencing mechanisms of the alloying elements* mentioned [30–33], so that only some short remarks need to be made here. Manganese and nickel shift the austenite transformation to lower temperatures and longer times, and produce a certain grain refinement. Formation of bainite is favored, and therefore nickel especially is added to steels for quenching and tempering to improve the through hardenability. Both manganese and nickel increase the strength and toughness of the solid solution of ferrite. Copper influences austenite transformation in a similar way but produces a precipitation hardening effect. Aluminum, nitrogen, niobium and vanadium have a grain-refining effect typical of this steel group. This effect results from formation of nitrides and carbides which are only dissolved at high temperatures. The fine-grained structure increases the yield stress and improves the toughness properties. Niobium and vanadium act to increase the strength values especially at elevated temperatures, by precipitation of carbides and carbonitrides. Molybdenum has similar effects in the same direction by aiding solid-solution hardening and formation of bainite.

To attain good weldability of fine grain steels, *the carbon content* must be limited. The fine grain effect also has a favorable influence on the brittle fracture behavior in the heat-affected zone (HAZ) of the weld.

In recent years the aim of material development in this field was directed to a lesser extent toward an increase in the strength at elevated temperatures but, in steels for pressure vessels operating at elevated temperatures, was directed more toward the improvement of the weldability. In the special example of steel for nuclear power applications the bias was additionally toward increasing the toughness.

Initiatives to *improve material toughness* were clearly seen in the manufacture of steels for reactor pressure vessels. For this application, no high yield stress values at elevated temperatures are demanded (about 300 N/mm^2 at 145 °C), so comparatively low-alloy, fine-grain structural steels can be used. On the other hand the steel selected must meet the high requirements for toughness. Steels based on manganese-nickel- and manganese-nickel-vanadium have been applied successfully to these purposes. Improvements to the toughness properties of high strength steels alloyed with vanadium, requires the correct adjustment of nitrogen and aluminum contents as well as reductions in the sulphur content. Ladle metallurgy has been used to adjust sulphur contents to low values and for fixation of residual sulphur in the form of fine spheroidal sulphides, to increase the toughness values even further [34]. Furthermore, by remelting, the advantage of a low sulphur content can be combined with low segregation and high cleanliness, and a fine-grain primary structure can be effected by quick solidification. The values of the notch-impact energy are also improved by these procedures.

At the end of 1976 it was decided in Germany to reduce the nominal stress requirements in the design of reactor safety vessels (containments). This change requires the application of especially tough materials with low yield stress and strength values. A normalized manganese-nickel steel, developed with regard to the new standard, has a mean content of 0.16% C, 1.5% Mn and 0.70% Ni. Because of its relatively low carbon content in combination with a favorably adjusted combination of manganese- and nickel content this steel has good toughness properties which can be further improved by electroslag remelting or desulphurization in the ladle.

In connection with toughness it is necessary to consider long-term and temper brittleness.

The reason for the *long-term embrittlement in the temperature range of 350 °C to 400 °C* is not yet fully understood. Beside the influence of tramp elements such as arsenic, antimony, tin and phosphorus, which seem to have a qualitative relationship with temper embrittlement, the literature hints at [35] an embrittlement by precipitation of the Ni_3Al-phase. This intermetallic phase has been noted after aging at 370 °C, in steels with 1% to 3% Ni and 0.03% to 0.06% Al.

The phenomenon of temper embrittlement has been described in B 4.2.1 and B 4.6.2. The reason for these embrittlement phenomena is seen in an equilibrium segregation at the grain boundaries of the tramp elements in steel, mainly phosphorus and tin. The effects of these elements are enhanced by added solid solution formers, mainly nickel, but also manganese and silicon by the carbide-forming element chromium. Molybdenum contents of about 0.4% act against the tendency to embrittlement, whereas higher molybdenum contents may increase the tendency [36, 37]. Molybdenum additions to fine-grain structural steels dealt with here to produce the strength values required at elevated temperatures are also favored because they suppress temper embrittlement in the base material and in the heat-affected zones of welded joints. Only with steel 22 NiMoCr 3 7 is the intended molybdenum range of 0.50% to 0.80% Mo slightly too high. Temper embrittlement is to be expected in heat-affected zones of welded joints in low-alloyed molybdenum-free grades, after stress-relief annealing. This condition must be counteracted by an appropriate selection of temperature and time for the treatment.

Limits in the technique used for welding of fine-grain steels for elevated temperatures became apparent at the end of the Sixties as technical manufacturing requirements demanded the use of thicker and thicker walls. With *walls thicker than about 70 mm*, difficulties arose due to development of *intercrystalline cracks* in the HAZ *during stress-relief annealing* [38]. This crack phenomenon is called "stress-relief cracking" or "reheat-cracking"; with regard to the basic process, in the German language the expression "Ausscheidungsriß" (precipitation. crack) has been chosen for this phenomenon (s. B 5.4.4, page 525). Formation of these intercrystalline cracks has its origin in internal residual stresses in the weld zone combined with reduced formability (ductility) of the HAZ-microstructure. Coarse grain zones transformed to martensite are very much exposed to dangers of this kind. Special carbides or carbonitrides, brought to solution by the effect of the welding heat, are again precipitated in the HAZ-structure during stress-relief annealing. These carbides reduce the possibility of lattice slipping (sliding) in the HAZ so that relaxation must proceed essentially by grain boundary (sliding) in an intercrystalline crack [39].

The phenomenon of under-cladding cracks, which have been found in the HAZ after cladding by welding with austenitic materials of components for reactor pressure vessels of steel 22 NiMoCr 3 7 is similar to these intercrystalline HAZ-cracks [38a].

Quantitative data regarding the most important factors of influence in the formation of these cracks have been obtained mostly from relaxation tests according to Murray [40] or from short-term creep-rupture tests [41]. In these tests, samples with simulated HAZ-structures are exposed to stresses of up to 80% of the yield stress at room temperature and then annealed at between 500 °C and 700 °C to relieve the stress.

Fundamental investigations [42, 43] of the influence of different *elements forming special carbides*, resulted in the discovery that there are *critical limiting concentrations*. When these limits are exceeded the relaxation behavior falls off steeply. With a base composition of about 0.20% C, 1.4% Mn and 0.70% Ni the limiting values according to Fig. C 9.5 amount to 0.5% for chromium and molybdenum, and the comparative values for vanadium and niobium are essentially lower [44]. The explanation of physical metallurgy for these distinctly-marked limiting concentrations is the start of precipitation of carbide phases of type MC, M_2C and M_7C_3 and of the corresponding carbonitride phases. The retarding effect of these phases on the relaxation decreases as soon as they are sufficiently spheroidized. Fe_3C is quickly precipitated and spheroidized and takes no part in retarding the

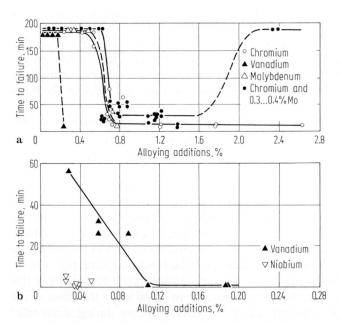

Fig. C 9.5. Influence of different alloying additions on the relaxation behavior of two steel grades at 640 °C in a test according to Murray [40]. After [44]. **a** Base material with ·0.14% to 0.22% C, 0.20% to 0.50% Si, 1.20% to 1.50% Mn, < 0.02% Cr and 0.50% to 0.90% Ni; **b** Base material with 0.11% to 0.19% C, 0.20% to 0.50% Si, 1.40% to 1.50% Mn, 0.10% to 0.50% Cr and 0.10% to 0.50% Ni.

relaxation. Carbides and carbonitrides of vanadium and niobium and the Mo_2C-phase, on the contrary, are spheroidized only slightly or not at all, even after long annealing times.

The slowdown in deformation is due to precipitation of special carbides in the grains, so the degree of lattice disorder in the HAZ-microstructure is a factor determining the relaxation behavior. With a given steel composition the *martensitic structure* does not favor reduction of residual (internal) stresses during stress-relief annealing. This result in stress is caused by the high number of lattice defects and by the dependence on good nucleation behavior during carbide precipitation. With rising amounts of bainite and further transition to a bainite-ferrite-structure, the relaxation behavior is improved as shown by the example in Fig. C 9.6 [44].

C 9.3.1.2 Ferritic Steels for Creep Resistance

Alloying techniques are combined with appropriate heat treatment to increase strength at elevated temperatures above that of the unalloyed steel. Alloying additions have several effects. For instance, they can modify the properties of the matrix, and of the transformation and precipitation behavior and thus of the microstructure including the grain size and dislocation structure. Details of the superposing effects of alloying additions on the microstructure and thus on the mechanical properties may be obtained from A 2, A 6 and B 1.

The easiest way to improve the strength at elevated temperatures to a modest extent consists of increasing the manganese content. Manganese is always present in steels for metallurgical reasons. The main effect of this element is in solid-solution hardening and in its interaction with the interstitially-included nitrogen in the iron-manganese-solid solution [45]. The influence of *nitrogen* is eliminated if it is removed from the solid solution by nitride formation, as described in Fig. C 9.7 for 450 °C [46]. The 10^4 h-creep-stress rupture strength of steels killed by silicon is

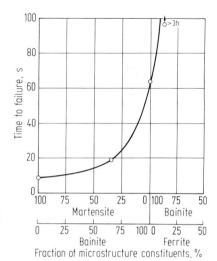

Fig. C 9.6. Influence of the microstructure on the time to failure of 16 MnNiMoV 5 3 in the relaxation test according to Murray [40] at 640 °C. (Steel with 0.17% C, 0.30% Si, 1.90% Mn, 0.26% Mo, 0.74% Ni and 0.15% V). After [44].

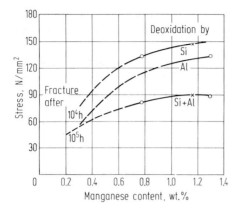

Fig. C 9.7. The influence of manganese and the deoxidation by silicon or aluminum on the creep-stress rupture strength at 450 °C of unalloyed steels. After [46].

higher than that of steels treated with aluminum, by which nitrogen is fixed. After 10^5 h the creep-stress rupture strength values of both kinds of steel are similar because the nitrogen precipitates as silicon nitride during the stress procedure [47]. During tempering or stress-relief annealing this nitride may speroidize rapidly and impair the creep behavior [48]. An increase of the manganese content beyond about 1.5% does not lead to any real advantages [46].

Molybdenum has a much greater effect on creep behavior according to Fig. C 9.8: traces as small as 0.05% increase the creep-stress rupture strength of the otherwise unalloyed steel at 400 °C and 450 °C by about 40% [49]. An increase of

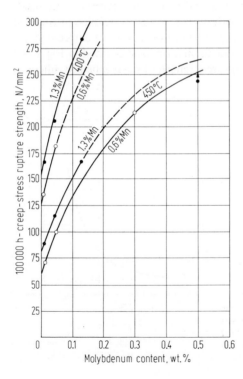

Fig. C 9.8. The influence of molybdenum on the 100 000 h-creep-stress rupture strength of aluminum-killed steels with 0.16% C and 0.6% or 1.3% Mn at 400 °C and 450 °C. After [49].

the content beyond 1% does not further improve the strength at elevated temperatures [50]. The influence of molybdenum is partly explained by reactions in the solid solution [45, 50]. Longer times and elevated temperatures, however, even with as little as 0.3% Mo effect the precipitation of a molybdenum carbide of the type M_2C, so that the long-term behavior of molybdenum steels is thus influenced. Nitrogen has a favorable influence on the creep rupture strength values [45, 51].

The toughness properties of molybdenum steels deteriorate during creep stress due to precipitation of the M_2C-carbide; a steel with 1% Mo becomes totally brittle [52], so steels alloyed only with molybdenum have contents not exceeding 0.5% Mo. Another disadvantage is a tendency to form graphite at temperatures above about 440 °C. Such formation occurs with both unalloyed and manganese steels. The process is advanced by aluminum and may lead to damage [53].

Both these disadvantages can be prevented by addition of *chromium*. Even 1% Cr is sufficient, in the presence of 1% Mo to produce good toughness values in the creep rupture test [52]. Elimination of the tendency toward graphite formation is based on the generation of more stable, chromium-rich carbides, for instance by enrichment of chromium in the cementite [53]. The formation of molybdenum carbide alone on the other hand does not prevent graphite formation [51].

Compared with molybdenum the influence of chromium on the creep and creep-rupture behavior is low as can be seen from the results of investigations on steels with 1.1% and 2.4% Cr at 550 °C [52]. With higher additions the long-term values are only slightly increased. Thus the same creep rupture values are reported for the predominantly transformation-free ferritic and the heat-resistant steels with 6% to 26% Cr, independent of the chromium content [54]. Although the creep rupture strength can be raised in steels containing for instance, about 13% Cr, which are capable of transformation, by quenching and tempering, steels alloyed only with chromium are not used for elevated temperature duty.

Very important and well-known are the *chromium-molybdenum steels*, which may be alloyed with 0.5% to 2% Mo and 0.5% to 12% Cr. This large range of composition results in considerable differences in the transformation behavior of the particular materials [55]. The unalloyed steels show a ferritic-pearlitic microstructure after cooling from the normalizing temperature, and the amount of pearlite is reduced in favor of bainite with rising alloying content. Steels with 5% and 9% Cr are partly- or fully-martensitic after fairly-rapid cooling. During tempering, precipitations form, mostly carbides of the types M_2X and $M_{23}C_6$ in a single sequence, the distribution and amount depending on the chemical composition.

The base microstructure present after cooling from the austenitizing temperature consists of different amounts of ferrite, pearlite, bainite and martensite and, together with solid solution hardening and the phases predominantly precipitated during tempering, determines the creep-stress rupture strength. The contributions of these influencing factors differ from steel to steel, but it often was not possible to determine them exactly or, until recently, to distinguish them from one another. The influence of the grain size is low in ferritic steels [56].

Heat treatment must lead to a microstructure with mechanical properties as favorable as possible within justifiable technical and economical expenditure, plus sufficient toughness and suitable creep rupture strength values. Precipitations have the greatest creep retarding effect when they exist as very fine, eventually coherent parts in dense dispersions. Such a microstructure is far from the thermodynamic equilibrium. Thus, precipitations tend to change at higher temperatures by growing and coagulation, and by transformation or re-dissolution and re-precipitation, to more-stable structures with coarser particles of the original or other carbide types. These changes can impair the elevated-temperature strength. It is therefore tried develop *suitable alloying and heat treatment procedures* to produce structures that hardly change up to the highest application temperatures for the steel grade in question.

As to the creep-stress rupture strength the *top position* among *chromium-molybdenum steels* is occupied by the higher-alloyed materials with 8 to 12% Cr, which also have good resistance to scaling. For this group an increase of the molybdenum content above 1% still produces a considerable rise in the elevated-temperature strength [57], which, even with 2% Mo perhaps does not reach its highest value. The ever-present martensitic structure of fully-transformable steels after cooling from the solution temperature is no guarantee of a high strength at elevated temperatures. Some of these steels are only partly transformable and contain more or less large portions of delta ferrite [57].

The elevated temperature strength of *chromium-molybdenum steels* can be further improved by addition of *vanadium*, alone or together with *niobium* [50]. Both elements have a considerably-higher affinity to carbon and nitrogen than the elements treated up to now, and lead to a precipitation of stable carbides and nitrides of the type MX. Their effect on creep resistance depends on their size and distribution, i.e. on the heat treatment.

The special importance of precipitation hardening is obvious from reports of investigations into the precipitation process [58] and the influence of the particle distance [59] for a steel with 0.5% Cr, 0.5% Mo and 0.3% V. The vanadium carbide, and even more, the niobium carbide [50] have a low tendency toward coarsening, a property which is essential in the preservation of high elevated-temperature strength over long periods at higher temperature.

Chromium-molybdenum-vanadium steels with about 1% Cr have a wide field of application as materials for screws and turbines as well as for steel castings. The bases of physical metallurgy for their properties, especially for their creep resistance, have been fully investigated. The microstructure after quenching and tempering has a great influence on their room temperature toughness and creep-stress rupture strength (Fig. C 9.9 [60]). The transformation diagram contains 5 cooling curves from 950 °C. After cooling the specimens were tempered to nearly the same tensile strength. The figure shows the notch impact energy at room temperature and the notch impact energy transition temperature as well as the relationship between the creep resistance K and the creep resistance of the martensite K_M (structure 1) for the different structures. By varying the cooling rate a structure can be produced with high room temperature toughness but low creep-stress rupture strength (martensitic structure) or with low room-temperature toughness but high creep-stress rupture strength (upper bainite 4). The highest creep resistance is developed in the upper bainite with homogeneous and fine distribution of flaky

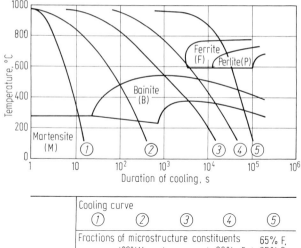

Fig. C 9.9. The influence of different cooling rates from the austenitizing temperature on the structure and properties at room temperature of a chromium-molybdenum-vanadium steel with about 1% Cr. After [60].

vanadium carbides having mean particle distances of at most 0.1 µm and particle sizes between about 0.005 µm and 0.03 µm [61].

The *fracture ductility of chromium-molybdenum-vanadium steels* during long stress periods can be endangered by a too high austenitizing temperature, as seen from the example in Fig. C 9.3. In general 960 °C to 970 °C is regarded as the upper limit. Extremely-high contents of tramp elements like antimony, tin, arsenic, copper and others, also can influence the fracture ductility unfavorably [62].

To improve through hardening and toughness, up to about 4% Ni is added to the chromium-molybdenum (vanadium)-heat treatable steels (containing about 1% Cr). In contrast to the grades poor in nickel of the same tensile strength above about 340 °C, these steels then show quickly-decreasing creep-stress rupture strength values. These changes depend on the lower molybdenum and vanadium contents of such alloyed steels. On the other hand, the nickel content will probably effect a quicker coarsening of the strengthening M_2X-precipitations as it does with the chromium-vanadium steels containing 12% Cr [63]. Furthermore with these steels attention must be paid to temper brittleness of which more details may be found in B 4.6.2.

In the chromium-molybdenum-vanadium-steels with about 12% Cr a former martensitic structure with $M_{23}C_6$-carbides exists after quenching at temperatures between 1020 °C and 1070 °C and tempering at temperatures between 700 °C and 750 °C. The carbides take up the dissolved vanadium and therefore no vanadium carbide is observed in these steels [64]. Addition of the strong carbide-forming element niobium leads to the formation of stable carbides, that is carbon nitrides of the type MX and thus to high long-term creep-stress rupture strength values. The same is true for steels alloyed with vanadium and niobium with 8% to 10% Cr [65].

Other alloying elements like silicon, titanium and tungsten are seldom regarded as essential additions in ferritic elevated-temperature steels. The effect of *tungsten* is very similar to that of molybdenum. Titanium has a favorable influence on the elevated high-temperature strength [66], but titanium-containing steels were mostly utilized as substitute materials in times of crisis when molybdenum was in short supply, and they were later displaced again by chromium-molybdenum steels. Chromium-molybdenum-titanium steels with 7% to 8% Cr and with very considerable creep and creep rupture values also have not succeeded [67]. *Silicon additions* to improve the creep-stress rupture strength of steels are unusual but they are used sometimes to increase resistance to scaling.

All elevated-temperature steels that will be in long-term service at temperatures above about 550 °C must show sufficient scaling *resistance*, usually attained by alloying with chromium. See also B 3.3.3 and C 15.

C 9.3.2 Austenitic Steels

C 9.3.2.1 Significance of the Face Centered Cubic Lattice for the High-temperature Strength of Austenite

The face centered cubic lattice structure of the solid solution of austenitic steels differs from the body centered cubic lattice of ferritic steels by the denser packing of the metal atoms. Combined with the lower diffusion coefficient, which is nearly two tenth powers lower than with α-iron, the low stacking fault energy *results in the higher high-temperature strength*.

Compared with ferritic steels lattice type and alloying of austenitic steels result in modified physical, chemical and mechanical properties. There is also the higher thermal expansion coefficient, the lower thermal conductivity, the increased oxidation and corrosion resistance and the comparatively low 0.2%-proof stress of the normal alloyed austenitic steels.

The origin of high-temperature steels goes back more than 60 years, starting from stainless steels with about 18% Cr and 8% Ni. With more knowledge of the behavior of metallic materials at higher temperatures, further development of creep-resistant austenitic steels started nearly 40 years ago and led to a large number of new materials [68]. The effect of alloying, especially with regard to the creep and creep-rupture behavior is discussed in the following [69].

C 9.3.2.2 Effects of Chromium and Nickel

Compared with stainless steels containing about 18% Cr and 8% Ni, the *chromium content* of most lower-alloyed high-temperature steels *is reduced* and the *nickel content is increased* to 13% or 16%. These changes serve to prevent ferrite formation and to provide greater stability of austenite at higher temperatures.

Higher *ferrite amounts* aggravate the hot formability. Elements like chromium, molydenum and niobium are enriched in the ferrite and it decomposes at temperatures above 500 °C into the carbide $M_{23}C_6$, austenite of lower chromium content and the intermetallic sigma-phase FeCr. With local chromium enrichment, which may result from dissolution of the chromium-rich carbide $M_{23}C_6$, the sigma-phase may also form from the austenite. Figure C 9.10 shows that the formation of sigma-phase begins in the ferrite and later in the austenite; with steels of about 16% Cr and 13% Ni there is considerable delay compared with steels of 18% Cr and 10% Ni [70].

Together with the chromium content the amount of the *sigma-phase* increases and its formation is accelerated by mechanical stress [71]. The precipitating phase mostly at the grain boundaries has its greatest influence on the notch impact energy at room temperature [72], and its influence on the creep and creep-rupture behavior seems to be low. Greater amounts of sigma-phase form only after relatively-long lasting stresses, so that the question of embrittlement of steels for components of long-lasting equipment is important.

C 9.3.2.3 Effects of Carbon and Nitrogen

Because of their small atomic diameter carbon and nitrogen occupy interstitial sites and they contribute fine precipitations to *increasing the creep resistance*. Figure

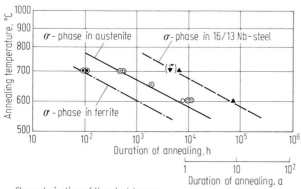

Characterization of the steel types:
- ○ 18/10 Ti after [70 a]
- ⊘ 18/10 Ti and 18/10 Nb after [70 b]
- ◎ ⊙ 18/10 Ti after [70 c]
- ⊖ ● 18/10 Nb after [70 c]
- ▲ 16/13 Nb after [70 a]
- (▼) 16/13 Nb without σ-phase after [70 c]

Fig. C 9.10. Commencement of the formation of sigma-phase in stabilized austenitic chromium-nickel steels. After [70–70c].

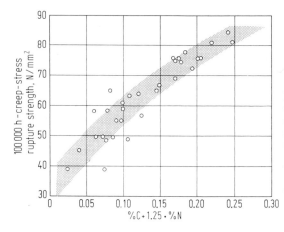

Fig. C 9.11. Influence of the carbon and nitrogen content on the 100 000 h-creep-stress rupture strength at 650 °C of non-stabilized austenitic chromium-nickel-steels. After [73]. Creep-stress rupture strength values have been extrapolated from 1000 h-tests. Chemical composition of the tested steels: 0.018% to 0.09% C, 0.32% to 0.79% Si, 0.28% to 2.05% Mn, 18.38% to 19.57% Cr, 9.48% to 11.3% Ni and 0.028% to 0.14% N.

C 9.11 shows this effect in a steel with about 19% Cr and 11% Ni [73]. With regard to sufficient creep-stress rupture strength a carbon content of, for instance, 0.04% is provided under consideration of the nitrogen usually existing in these steels. The creep-stress rupture strength values for 10^5 h plotted in C 9.11 are mostly obtained by extrapolation of creep rupture strength values determined by tests of only 1000 h. Therefore the dependence shown is valid only with reservations. To the extent that the $M_{23}C_6$-carbides coagulate with longer test periods, their influence on the creep resistance is reduced.

In unstabilized austenitic steels the *solubility of nitrogen as opposed to carbon* is higher and its tendency to precipitation is lower so that its influence on creep resistance lasts longer. This effect is indicated in Fig. C 9.11 by the factor 1.25 for nitrogen. Austenitic steels with elevated nitrogen contents are therefore preferred. The example of a molybdenum-alloyed steel in *Fig. C 9.12* [74] shows the increase in the creep-stress rupture strength by a nitrogen addition of about 0.20%; in molybdenum-free steels the effect is similar. The greatest effect on the creep-stress rupture strength of molybdenum-alloyed steels is attained with 0.1% nitrogen [65].

Some limits to the nitrogen and carbon contents are imposed by *an embrittlement that starts at excessively-strong nitride or carbide formations*. Figure C 9.12a shows that the elongation after fracture in creep rupture tests is reduced with too-high nitrogen contents. More distinctly Fig. C 9.12b shows the decrease in deformability (ductility) after thermal strain, by the notch impact strength measured at room temperature. The reason for the decrease lies in the sequence of the precipitation process in comparison with nitrogen-poorer steels. The chromium-nitride Cr_2N or – in the presence of niobium – the mixed nitride (Nb, Cr) N are formed and they affect the deformability of the solid solution. Because of these effects, the nitrogen content of high-temperature steels is limited to about 0.15%.

Nitrogen addition also increases the yield (proof) stresses at room and at elevated temperatures [75], eliminating disadvantage of the normal alloyed austenitic steels.

Precipitation of chromium carbides at the grain boundaries, may cause susceptibility to intercrystalline corrosion. To prevent this susceptibility, e.g. with

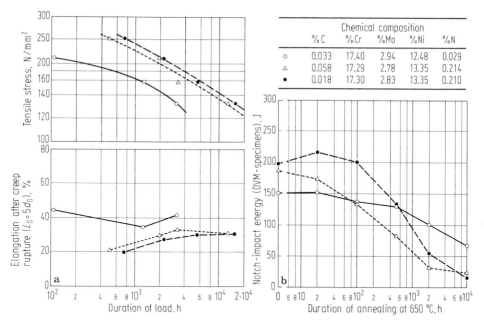

Fig. C 9.12. Influence of the nitrogen content of non-stabilized chromium-nickel-molybdenum steels. **a** on the creep-stress rupture strength at 650 °C, **b** on the notch-impact energy at +20 °C after annealing at 650 °C. After [74].

components of high wall thickness which cool very slowly after solution annealing or after welding, the carbon content can be reduced and the nitrogen content raised. These measures will improve the mechanical properties [76].

C 9.3.2.4 Effects of Niobium and Titanium

It is well known that, intercrystalline corrosion, originating from the formation and *precipitation of chromium carbide*, can be prevented in austenitic stainless by addition of the *elements titanium or niobium*, which have more affinity to carbon than chromium. This method is also used for high-temperature austenitic steels. These additions also increase the creep-stress rupture strength. In high-temperature steels this "carbon stabilization" is effected predominantly by niobium, as it has been proved that niobium normally contributes more than titanium to an increase of the creep-stress rupture strength [77, 78].

In the original state of these steels, i.e. after quenching from 1050 °C, undissolved niobium carbide exists in coarse form [79]. Although the formation of niobium carbide corresponds to the equilibrium state at operating temperature with the carbon in solution at first the chromium-rich carbide $M_{23}C_6$ is formed because of the considerably more numerous chromium atoms present and their greater mobility. Hence a temporary susceptibility to intercrystalline corrosion may result [80], but in general has no operational disadvantages [81]. This carbide is re-dissolved in the further course of stress and the carbide NbC is formed which corresponds to the equilibrium condition.

At higher temperatures the niobium carbide precipitating in the interior of the grains in a fine-dispersion can strengthen the solid solution in such a manner that during deformation cracks may occur at the weaker grain boundaries. Such phenomena have been observed in welded components [82], where niobium carbide has been dissolved by the welding heat in the adjacent base material. When the material returns to operational temperature, more carbon and niobium is available in a finer form for re-precipitation. The structure may be changed by a post weld heat treatment at 900 °C to 950 °C, which converts the niobium carbides to a coagulated shape. In case components with thick walls and with steep cross sectional transitions, which are disposed to greater temperature fluctuations causing deformations, similar defect phenomena may occur by stress-induced precipitations of niobium carbide [83]. Long experience with installations (components) made of steels stabilized by niobium [81, 84] show, however, that with suitable design and appropriate operation damage may be avoided.

In the *high-temperature steels, rich in nickel, titanium fulfills a further purpose.* With rising nickel content, up to 25% and more, the solubility of carbon in austenitic steel decreases. In a steel with about 30% Ni the solubility amounts only to one third of the amount in steels with 18% Cr and 10% Ni. Then below 800 °C, $M_{23}C_6$ forms preferentially, whereas the precipitation susceptibility of the cubic carbides, also of titanium carbide, is reduced. Therefore, the titanium content of such steels cannot be regarded from the viewpoint of stabilization. With a sufficiently-high content of titanium and aluminum (% Ti + % Al > 0.5%) the coherent precipitating intermetallic γ'-phase $Ni_3(Al, Ti)$ occurs. The lattice parameter differs only slightly from that of the solid solution. The coherence tension is therefore low and hardness, yield (proof) stress and creep resistance are only slightly increased. With titanium contents above 1% as are usual in high-temperature nickel alloys (see C 9.4.3) and with a higher ratio of titanium content to aluminum content, the misfit and coherence tension become greater, and thus the yield (proof) stress and creep resistance especially at low and medium temperatures are increased. On the other hand the coarsening of precipitations is accelerated so that with longer loading at higher temperatures the creep resistance, at first very high, decreases quicker as with materials not subject to precipitation hardening via the γ'-phase.

As the solid solution gets depleted in nickel by precipitation of the γ'-phase, the sigma-phase may occur in these steels. If the silicon content is not reduced the G-phase $Ti_6Ni_{16}Si_7$ [85], precipitates at the grain boundaries and may lead to a lower fracture deformation [86].

C 9.3.2.5 Effects of Molybdenum, Tungsten, Vanadium and Cobalt

Based on the size of their atoms and their lower affinity to carbon compared with titanium and niobium, the elements molybdenum, tungsten, vanadium and cobalt are among those that occupy the lattice sites in the austenitic solid solution and slightly increase the yield (proof) stress. The increase in the recrystallization temperature which is additionally effected and a change to the course of the precipitation procedure lead to an increase in the creep rupture strength.

In simple non-stabilized chromium-nickel steels the formation of intermetallic phases at higher temperatures is prevented mainly by adjustment of the chromium and nickel contents. Addition of *molybdenum* may cause precipitation of Laves (Fe_2Mo)- and chi-phase ($Fe_{36}Cr_{12}Mo_{10}$). In addition, sigma-phase and besides $M_{23}C_6$, carbides of type M_6C may occur [87, 88]. If niobium is absent the formation of sigma- and Laves-phase is more likely than chi-phase [89].

Formation of intermetallic phases, undesirable especially with regard to toughness behavior, can be decreased or prevented by reduction of the molybdenum content. New investigations show that the molybdenum content of 2.0% to 2.5% in non-stabilized chromium-nickel-molybdenum steel can be reduced to nearly 0.75% [90] without lowering the creep rupture strength.

Addition of *tungsten* instead of molybdenum leads to the formation of Laves-phase Fe_2W [91]. This hexagonal phase precipitates especially in the interior of the grain [92] and can thus contribute to an increase in creep resistance. With the same mass contents the mole contents of tungsten, because of its higher atomic weight, is smaller than that of molybdenum. Therefore the tendency to formation of intermetallic phases in a tungsten alloyed steel is lower.

Vanadium, known among ferritic steels as a stronger carbide forming element does not form a stable carbide in austenitic steels. For a short time, however, vanadium carbide, that is vanadium carbon nitride forms, but it dissolves in favor of $M_{23}C_6$ after annealing. Therefore vanadium is not appropriate for use in preventing intercrystalline corrosion. On the other hand vanadium is added to niobium stabilized steels that have higher nitrogen contents. Some of the vanadium is precipitated as niobium vanadium carbon nitride and thus contributes to an increase in the creep resistance. Vanadium reduces the resistance to scaling, therefore vanadium-containing steel should not be utilized for long-term applications at temperatures above 650 °C.

Cobalt increases the recrystallization temperature and the solubility of the carbides during solution heat treatment so that a stronger carbide precipitation is observed during aging and in service. Hence, steels alloyed with 10 or 20% Co often have higher carbon contents and higher additions of carbide-forming elements.

The creep and creep rupture as well as the physical-metallurgical behavior of steels containing 20% Co because of their additional alloying contents conform rather to the cobalt alloys (see C 9.4.3). Therefore, only a steel with 10% Co (see *Fig. C 9.13*)) will be discussed here. In this steel, precipitations occur in a manner similar to that of the steels reviewed above. Because of the high carbon content of 0.40% in this steel, the amount of precipitated carbides is essentially higher. Carbides and the Laves-phase $Fe_2(Mo, W, Nb)$ are the basis of the high-temperature strength. Despite the presence of molybdenum and tungsten, neither the sigma- nor the chi-phase is found in this steel, as the chromium, molybdenum and tungsten contents remaining in solid solution after fixing in the carbide and Laves-phase obviously do not suffice to form these phases (sigma and chi).

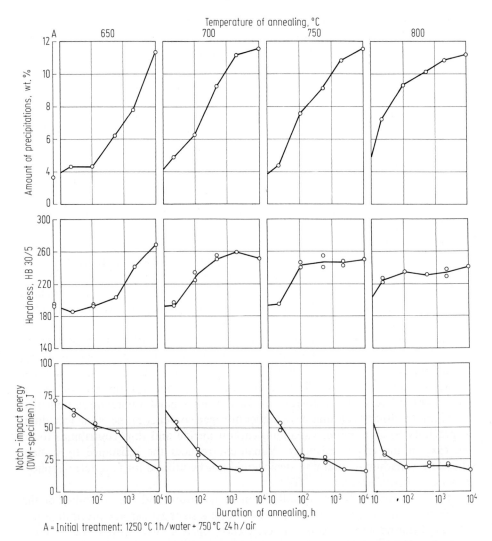

Fig. C 9.13. Influence of the annealing temperature and time on the amount of precipitations, hardness and notch-impact energy of a steel with 0.40% C, 10% Co, 17% Cr and 13% Ni. After [88].

Figure C 9.13 shows that the change of hardness and of the notch impact energy at room temperature have close relationships with the precipitation processes [93]. The hardness depends not only on the amount of precipitations but also on their dispersion rate, as is shown by the hardness decrease after longer annealing times at higher temperatures. The amount of precipitations governs the notch-impact energy, as shown by the fact that the notch-impact energy does not increase again, even after long annealing times nor with pronounced coagulation of the precipitations. Despite the low notch impact energy at room temperature, the cobalt-containing steels are characterized by good deformability characteristics after long-term stress at higher temperatures.

C 9.3.2.6 Effect of Boron

From a comparison of the atomic radii of carbon (0.082 nm), nitrogen (0.086 nm) and boron (0.095 nm) it can be expected that the insertion of boron atoms at interstitial lattice sites is difficult. Therefore these atoms are found preferably at disturbed areas in grain boundary zones which hinders grain boundary slip and diffusion processes in these zones. Grain boundary precipitations, e.g. of $M_{23}C_6$ or sigma-phase therefore, are strongly retarded by low boron contents. These conditions prevent formation of grain boundary zones, which are poorer in alloying elements, and foster a higher creep resistance of the solid solution. This increased creep resistance is effected for instance by a coherent precipitation or by strain hardening, and may occur without premature fracture by tearing up the grain boundaries.

A boron content as low as 0.0005% has an influence on the diffusion behavior at the grain boundaries of austenitic steels [94], so an increase in the creep-stress rupture strength may be expected even with low boron contents. This increase is seen in stabilized steels as well as in non-stabilized steels but is lower with chromium-nickel-steels without further additions than with molybdenum- or tungsten-containing steel [95]. It is further stated [95a], that with non-stabilized molybdenum-containing steels under long-term stress the effect is decreased because of precipitations of the boron in the carboboride $M_{23}(C, B)_6$.

Boron also offers the possibility *of improving the mechanical properties* of austenitic steels *by mechanical pre-strain hardening*. These additions also increase the 0.2-proof stress and creep rupture strength up to temperatures at which recovery begins and on recrystallization. Pre-strain hardening is effected by cold-deformation or by deformation just below the recrystallization temperature (hot-cold-deformation). The number of dislocations thus caused in the crystal lattice, which are nucleation points, is increased and aids the precipitation of $M_{23}C_6$ or NbC inside the grain during annealing or under creep stress. The creep resistance of the solid solution thus increased can only be utilized so far as the grain boundaries show a suitable creep resistance and do not tear too soon. But boron contributes strongly to this condition. Figure C 9.14 demonstrates this effect with the example of niobium-stabilized pre-strain hardened steels [69]. The steel without boron plotted in Fig. C 9.14a shows lower values for elongation after fracture in the creep-stress rupture test, and the creep-stress rupture strength curve for notched specimens after an intersection is considerably lower than that for plain specimens. Such a steel cannot be considered for long-term applications. On the other hand, steel with about 0.1% B (Fig. C 9.14b) shows a much more favorable behavior confirmed by long-lasting experience in service [82]: Beside high fracture deformation values of the plain specimens, for notched specimens longer times to fracture are found.

C 9.3.3 High-temperature Nickel and Cobalt Alloys

Since the beginning of the development of steels for use at high operating temperatures, chromium, nickel and cobalt were always of great importance. Steel

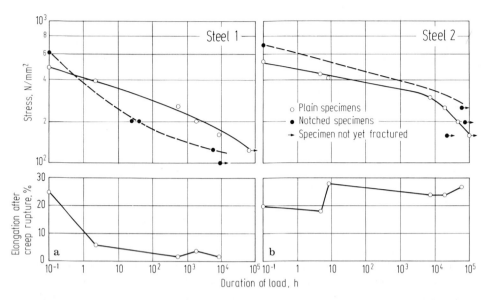

Fig. C 9.14. Creep-stress rupture behavior at 650 °C of austenitic steels after strain hardening. **a** Steel X 8 CrNiNb 16 13 with 0.065% C, 16.74 Cr, 0.61% Mo, 12.31% Ni and 0.76% Nb, $R_{p\,0.2}$ at RT: 540 N/mm^2; **b** Steel X 8 CrNiMoBNb 16 13 with 0.065% C, 17.45% Cr, 1.50% Mo, 13.58% Ni, 0.55% Nb and 0.116% B, $R_{p\,0.2}$ at RT: 600 N/mm^2. After [69].

research was therefore pushed forward into the field of non-ferrous metal alloys. Subsequently, for technical, organizational and economical reasons it became necessary to agree upon *a differentiation between steels and non-ferrous metal alloys*. In Euronorm 20–74 this differentiation is made by the definition for ferrous materials: "Ferrous materials are metal alloys where the mean weight portion of iron is higher than that of any other element". However, the production and application of high-temperature materials covers both steel and non-ferrous metal alloys so it seems reasonable to include the high-temperature non-ferrous metal alloys in this book on steel.

Nickel and cobalt together with chromium form alloys that have a face-centered cubic lattice, which is known to be a requirement for good mechanical properties at higher temperatures. Several books and reports give information on the basic principles of physical metallurgy and on production, processing and applications of metals [96–105].

The high temperature cobalt- and nickel-alloys mostly contain iron only as an accompanying element, although some alloys of this type have iron contents up to 20%. Chromium contents between 10% and 30% give the necessary resistance to high-temperature corrosion which can be improved still further by addition of rare earth elements. Both material groups also contain molybdenum and tungsten in solid solution or as carbides that increase the high-temperature strength properties. The cobalt alloys contain nickel to stabilize the face-centered cubic lattice. One part of the nickel alloys contains cobalt which reduces the stacking fault energy of the solid solution and improves the precipitation hardenability by the γ'-phase

Ni$_3$(Al, Ti). This improvement is effected by changing the temperature dependence of the solubility of aluminum and titanium. Both material groups differ in the amount of the carbon content, which is low in the nickel alloys and higher in most of the cobalt alloys.

Strengthening by carbides is in the foreground in the cobalt alloys. In cobalt alloys of high carbon content used for castings the primary carbides M$_{23}$C$_6$ and MC have the shape (form) shown by Fig. C 9.15 [106]. At these sizes it can not be expected that strengthening is effected blocking the dislocations. Instead, it is assumed that this net of primary carbides is effective as a kind of local fiber reinforcement, which also explains the higher creep rupture strength of cast versus hot-deformed cobalt alloys of similar composition. With forging alloys based on cobalt, structures with fine distribution of the secondary carbides are attained by solution annealing at temperatures around 1200 °C and subsequent aging at 750 °C to 800 °C.

Nickel alloys are composed so that they are strengthened by the precipitation of the intermetallic γ'-phase which is formed during aging, after solution annealing, as can be seen from Fig. C 9.16 [103]. Like the solid solution, this γ'-phase is also face-centered-cubic and it precipitates coherently.

In many technical nickel alloys, more elements may be inserted into the γ'-phase; titanium, niobium or tantalum may be inserted in place of aluminum; on the other hand, iron or cobalt may be inserted instead of nickel. Furthermore tungsten, molybdenum and chromium are soluble in the the γ'-phase. The creep resistance of the nickel alloys is influenced by the different compositions, sizes, shapes and distributions of the precipitated particles, the volume fraction and the distortions (braving) present at the phase boundary areas between γ'-phase and γ-solid solution. Other factors include the coherence tension, the height of which depends on the difference between the lattice parameters in the solid solution and

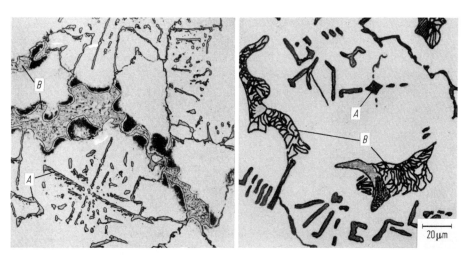

Fig. C 9.15. Structure of a cobalt-precision cast alloy with 0.6% C, 21.5% Cr, 10% Ni, 3.5% Ta, 0.2% Ti, 7.0% W and 0.5% Zr. The arrows A mark MC-carbides, B mark M$_{23}$C$_6$-carbide islands (eutectic). After [105, 106].

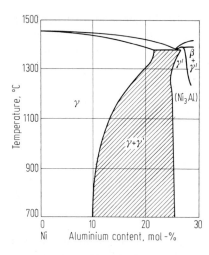

Fig. C 9.16. The phase diagram nickel-aluminum. After [105].

in the γ'-phase [96–98, 101–105]. Special attention must be paid to the essential influence of the volume fraction of the γ'-phase on the creep-stress rupture strength (see Fig. C 9.17 [97]). The fraction of the precipitations is essentially higher, for instance in the elevated-temperature steels. Nickel alloys with more than about 50 vol.-% γ'-phase cannot be hot formed by the usual processes, but must be formed by precision casting.

The structure stability of nickel and cobalt alloys governs their creep and *creep rupture behavior*. Modifications to the structure result from aging processes, i.e. by

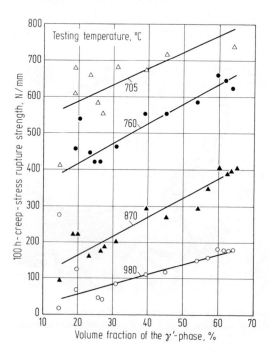

Fig. C 9.17. Influence of the volume fraction of γ'-phase on the 100 h-creep-stress rupture strength of different high-temperature nickel alloys. After [97].

agglomeration of the carbides and the γ'-precipitation as well as from the transformation or reformation of intermetallic phases. These types of modifications may proceed relatively quickly due to the increased diffusion possibilities at high operating temperatures. In nickel alloys with higher chromium contents there is a disintegration of the solid solution into a chromium-rich body centered cubid (α-) solid solution. A nickel-rich face centered cubic (γ-) solid solution may also occur as a result of the extraction (removal) of nickel from the matrix resulting from the γ'-formation. It is also possible that the chromium-rich sigma-phase (compare C 9.3.2) may form. The titanium- or niobium-rich γ'-phase is metastable. With a high ratio of titanium content to aluminum content, or with niobium-rich alloys, the equilibrium phases Ni_3Ti (eta-phase, hexagonal) or Ni_3Nb (delta-phase, orthorhombic) form incoherently with acicular shapes or on the grain boundaries. These phases do not improve high-temperature strength and may lead to brittle fractures. Because of the higher defect frequency at the grain boundaries, diffusion processes can run off quicker with the consequence that precipitations form here preferably and the zone near the grain boundary is depleted of alloying elements. The precipitations, especially if they are distributed like a film (thin layer) at the grain boundaries, and the alloy-depleted seams with their reduced creep resistance can cause premature fractures of the material.

As with austenitic steels the addition of small amounts of *boron*, and of *zirconium*, can retard diffusion at the grain boundaries and thus contribute to a stability of the zone. In this context reference may be made to *directional solidification* [107] of turbine blades produced *by precision casting*. The solidification proceeds mainly in the longitudinal direction of the blades so that grain boundaries transverse to the ultimate main stress direction are largely avoided.

C 9.4 Characteristic Material Grades Proven in Service

C 9.4.1 Ferritic Steels

C 9.4.1.1 Ferritic Steels for Slightly Elevated Temperatures

The series of normalized weldable fine-grain structural steels specified by DIN 17 102 [108] includes a group with measured yield stress values up to 400 °C. The series in total is treated in chapter C 2, because of its general importance, so data can be limited here to the essential features. The standard covers flat and profile products in thicknesses up to 150 mm. The minimum values given for the yield stress up to 16 mm thickness, which are utilized for the designation of these steel grades, lie between 255 N/mm² and 500 N/mm². This strength range is covered by 8 grades, from WStE 255 in steps of 35 N/mm² to 40 N/mm² up to WStE 500. For temperatures between 100 °C and 400 °C, minimum values are listed for the 0.2%-proof stress, depending on the product thickness. At 400 °C values for WStE 255 are 108 N/mm², and for WStE 500 255 N/mm² is listed for the smallest thickness range below 70 mm. The chemical composition of this steel series is in accordance with the explanations in C 9.3.1.1. Depending on the strength range, the highest admissible carbon content is 0.18% to 0.21%. The upper limit of the manganese content is 1.3% to 1.7% and additions of chromium – ≤ 0.30% –, copper – ≤ 0.20% –, molybdenum – ≤ 0.10% –, nickel – between 0.2% and 1.0% – as well as elements like aluminum, niobium, titanium and vanadium as fine-grain producing elements are indicated.

In addition to the steels of DIN 17 102 there is another series of elevated-temperature *fine-grain structural steels designed for higher-service*, the most common of which are listed in Table C 9.1. This series also includes normalized and tempered steels. Water-quenched and tempered materials 16 MnMoNi 5 4, 22 NiMoCr 3 7 and 20 MnMoNi 5 5 are also listed. With the exception of 15 MnNi 6 3 they are all designed for temperatures up to 400 °C and cover at 300 °C a range of yield stress values of 345 N/mm^2 to 430 N/mm^2. For these steels a notch-impact energy value (ISO-V-specimen, transverse) at 0 °C of min. 31 J is guaranteed. Only in the steel 20 MnMoNi 5 5 is a higher value of 41 J stated. Reference is made again to the possibility of improving the toughness of steels on the basis of manganese-nickel-vanadium by several metallurgical measures (see C 9.3.1.1). Processing difficulties caused by reheat-cracking in the heat-affected zone have not occurred with the vanadium or niobium containing steels in the characteristic plate thicknesses used since 1974.

Results of creep and creep-stress rupture tests for temperatures between 400 °C and 500 °C are available for some elevated-temperature fine-grain structural steels like 13 MnNiMo 5 4, 16 MnMoNi 5 4, 15 NiCuMoNb 5 and 22 NiMoCr 3 7.

After it was found that the occurrence of cracks in the HAZ during stress-relief annealing could be connected with critical limiting contents of some special carbide-forming elements, some lower alloyed steels were introduced. In the elevated-temperature fine grain structural steels intended for higher requirements at temperatures up to 400 °C, early attempts to improve weldability of heavy wall thicknesses on the simultaneous alloying of molybdenum and vanadium has been abandoned. Instead, researches have adjusted the required strength values of *manganese-molybdenum-nickel based*, water-quenched and tempered steels. The molybdenum content was adjusted to the critical limiting concentration (see above). This water-quenched and tempered steel 16 MnMoNi 5 4 has been used sucessfully for fabrication of vessels with high wall thicknesses operating at elevated temperatures.

Development of advantageously-weldable air-quenched and tempered steels has also progressed independently. A useful alloy which contains manganese, chromium, molybdenum, nickel and niobium with chromium- and molybdenum contents below the critical concentration and a very low niobium addition of nearly 0.01% as well as a reduced carbon content of a maximum of 0.16%, is the steel 13 MnNiMo 5 4. In this steel, the yield stress at elevated temperature has been measured at 345 N/mm^2 at 350 °C so that it still covers the range which is utilizable [109] according to TRD 301 [110]. With wall-thicknesses up to 100 mm the steel is further characterized by a yield stress at RT of min. 400 N/mm^2, a tensile strength between 580 N/mm^2 and 750 N/mm^2 and a notch-impact energy (ISO-V-specimen) at 0 °C of min. 32 J. The steel also is resistant to formation of underclad

Table C9.1 Survey of the most common ferritic steels for slightly elevated temperatures and higher requirements

Steel grade Designation	Chemical composition						
	% C	% Mn	% Cr	% Mo	% Nb	% Ni	% V
13 MnNiMo 54	≤ 0.16	1.00...1.60	0.20...0.40	0.20...0.40		0.60...1.00	
17 MnMoV 64	≤ 0.19	1.40...1.70		0.20...0.40	0.01	0.50...1.00	0.10...0.19
15 NiCuMoNb 5 5[b]	≤ 0.17	0.80...1.20	≤ 0.30	0.25...0.50	0.015...0.040	1.00...1.30	
12 MnNiMo 55	≤ 0.15	1.10...1.50	≤ 0.30	0.20...0.50		0.80...1.60	≤ 0.05
11 NiMoV 53	≤ 0.15	1.20...1.50		0.20...0.50		1.20...1.80	0.06...0.13
16 MnMoNi 54	≤ 0.18	1.10...1.65		0.20...0.50		0.50...1.20	
22 NiMoCr 37	0.17...0.25	0.50...1.00	0.30...0.50	0.50...0.80		0.60...1.20	≤ 0.03
20 MnMoNi 55	0.17...0.23	1.00...1.50	≤ 0.30	0.45...0.60		0.50...0.80	≤ 0.03
15 MnNi 63	0.12...0.18	1.20...1.65	≤ 0.15	= 0.08		0.50...0.85	≤ 0.02

[a] For all steel grades: 0.10% to 0.60% Si, ≥ 0.015% Al and ≤ 0.020% N
[b] 0.50% to 0.80% Cu

cracks. Creep-stress rupture tests in the temperature range of 450 °C to 550 °C resulted in values which up to 500 °C can be compared with those of steel 13 CrMo 4 4 (see Table C 9.2). There is no long-term embrittlement at temperatures up to 400 °C.

At the same time as the work was proceeding on adjustment of the contents of special carbide-forming elements to the limiting concentrations that have been determined as critical for the development of precipitation cracks, experiments were conducted with regard to the *underclad cracks* in steel 22 NiMoCr 3 7. This work was to study utilization of the molybdenum range of 0.50% to 0.80% only in the lower permissible range and the effects of reducing the carbon content as far as possible. Recommended values for the carbon content of this steel for components in the primary circuit of nuclear power plants were 0.20%, for the molybdenum content 0.55% and – in general – a very low content of tramp elements (temper brittleness). In the nuclear power industry, the steel 22 NiMoCr 3 7 was mainly replaced by the steel 20 MnNoNi 5 5, which is also liquid-quenched and tempered. Welding of this steel is simplified because of the molybdenum content being adjusted to 0.5% and the reduced carbon content, offers clear advantages with regard to the problem of precipitation cracks during stress-relief annealing. The insignificantly higher value for the notch-impact energy should not be interpreted to mean that the steel 20 MnMoNi 5 5 has better properties of through hardening and tempering than the steel 22 NiMoCr 3 7; rather, it offers certain advantages. Concerning toughness, in addition to the minimum value of 41 J for steel 20 MnMoNi 5 5 mentioned – a NDT-temperature of $\leq 0\,°C$ is required.

In the steel 20 MnMoNi 5 5 restrictions of the chemical composition are specified for special applications in the nuclear power technology [111]. These restrictions relate, as with the 22 NiMoCr 3 7 steel, mainly to the molybdenum content and to the amount of tramp elements.

Steel 15 MnNi 6 3 is the result of a newer development in the manufacturing of reactor safety shells (vessels). Normalized plates with thicknesses up to 50 mm and most suitable composition have been produced, for instance, with reduced sulphur content, a yield stress at RT of 364 N/mm^2 and a yield stress at 145 °C of 314 N/mm^2. The average NDT-temperature is $-40\,°C$, and the mean value of the notch-impact energy (ISO-V-specimens) at $-20\,°C$ is above 100 J and at $+5\,°C$ above 200 J. Values for lateral expansion (spread) fulfill easily the RT-NDT-conception [111]. Stress-relief annealing at 560 °C to 580 °C yields very favorable results. In this condition tensile tests at 145 °C according to the upper temperature limit for reactor safety shells (vessels) showed minimum yield stress values of 385 N/mm^2.

C 9.4.1.2 Ferritic Steels for Plates and Tubes

Quality specifications and technical delivery conditions of ferritic steels for plates and tubes are contained in the respective standards [112]. The most common of these are compiled in Table C 9.2. The unregistered unalloyed steels and the *manganese steels* have the lowest long-term (creep) values and are seldom used in the creep rupture range. The favorable influence of nitrogen in the dissolved

Table C9.2 Common ferritic steel for plates and tubes

Steel grade Designation	Chemical composition						
	% C	% Si	% Mn	% Cr	% Mo	% Ni	% V
15 Mo 3	0.12...0.20	0.10...0.35	0.40...0.80		0.25...0.35		
13 CrMo 4 4[a]	0.10...0.18	0.10...0.35	0.40...0.70	0.70...1.10	0.45...0.65		
10 CrMo 9 10[a]	0.08...0.15	≤ 0.50	0.40...0.70	2.00...2.50	0.90...1.20		
14 MoV 6 3	0.10...0.18	0.10...0.35	0.40...0.70	0.30...0.60	0.50...0.70		0.22...0.32
12 CrMo 19 5	≤ 0.15	≤ 0.50	0.30...0.60	4.0...6.0	0.45...0.65		
X 12 CrMo 9 1	≤ 0.15	0.25...1.00	0.30...0.60	8.0...10.0	0.90...1.10		
X 20 CrMoV 12 1	0.17...0.23	≤ 0.50	≤ 1.00	10.0...12.5	0.80...1.20	0.30...0.80	0.25...0.35

[a] The steel is listed in DIN 17 155 (Ausg. Okt. 1983) partly with some other values

condition is not considered in establishing (fixing) the long-term values of German Standards, as nitrogen can be removed by nitride formation resulting from addition of aluminum or during stress-relief annealing [113]. Aluminum could be added, for example, to attain a fine-grain structure with high toughness.

With the transition to the *steel 15 Mo 3* a disproportionately great rise of elevated temperature strength is achieved regarding the cost of low alloying additions (see Fig. C 9.8). This grade of steel, therefore, has found wide application in the temperature range between 400 °C and 500 °C. This steel is the oldest elevated-temperature steel in Germany and has been proved since 1929, despite the fact that its toughness decreases under long-term creep stress. The similar steel 16 Mo 5 with higher molybdenum content must be evaluated less favorably in this respect, especially as it tends to notch embrittlement [113].

Steel 13 CrMo 4 4 known since 1930, has been applied especially in the construction of steam boilers in a temperature range of 500 °C to 530 °C and, in comparison to 15 Mo 3, has increased creep and creep-stress rupture values and improved toughness behavior. This steel has a microstructure of ferrite and bainite, more seldom pearlite, the quantity of which depends on the cooling rate after normalizing. The creep and creep-stress rupture behavior are rather independent of the heat treatment, as well as of a tempering temperature within the usual limits [113, 114]. Also cold deformation does not influence the creep stress rupture strength. On the contrary, the fracture deformability values in the lower range of normal application temperatures are lower. This influence can be eliminated by tempering [113].

Steel 10 CrMo 9 10 shows a better resistance to scaling because of its higher chromium content. It can therefore be used at temperatures of up to about 590 °C. The creep and creep-stress rupture values of this steel are also higher so that it can be used economically at temperatures above about 530 °C where steel 13 CrMo 4 4 is out of the question. The steel is characterized by a very good toughness behavior under long-term creep stress and has a more varied microstructure than steel 13 CrMo 4 4 resulting in influence on the long-term values [113, 114]. But these modifications are only of importance at temperatures below 550 °C.

Because of its scaling properties, the *molybdenum-vanadium steel 14 MoV 6 3* is only usable for continuous services at temperatures of up to about 560 °C. Because of its high creep rupture strength, this steel is particularly useful for steam piping in the temperature range of 540 °C. Precipitation hardening by vanadium carbide provides the required high strength at elevated temperatures [113], but is deleterious for the toughness behavior. So the same phenomenon is visible with these steels as with molybdenum steels that elongation after fracture and reduction of area in creep-stress rupture tests decrease with time [115]. But this decrease can be kept within allowable limits by suitable heat treatment [115] and by addition of about 0.5% Cr [116]. The toughness values are further reduced by cold deformation [113]. With suitable melting arrangements and appropriate heat treatment the notch-impact energy is maintained at normal demand levels, i.e. in the delivery condition of smooth (plain) tubes and after their further processing by hot bending and welding [116]. In the heat-affected zone of the welded joints there is a range of reduced creep-stress rupture strength that has led occasionally to damage under unfavorable stress conditions with this steel and with chromium-molybdenum steels [113, 117]. Only circumferential welds stressed by internal pressure do not have the tendency to early (premature) failure [117].

For applications at *temperatures above 550 °C steels with increased chromium content* are preferred because of the higher demand for resistance to scaling. The grades 12 CrMo 19 5 with 5% Cr and 0.5% Mo and X 12 CrMo 9 1 with 9% Cr and 1% Mo in the soft annealed state, are known especially as high-pressure hydrogen resistant steels for the petroleum industry (see C 14). Steel grade X 12 CrMo 9 1 in the quenched and tempered condition is also regarded for the construction of steam boilers but because of its relatively low creep-stress rupture strength replaced by the better steel X 20 CrMoV 12 1 with 12% Cr.

This martensitic *steel X 20 CrMoV 12 1* has been known and approved in Germany for 30 years. This steel has the highest elevated-temperature strength of all ferritic steels. An unsufficient solution of carbides during solution annealing reduces the long-term values, a temperature of at least 1020 °C must be used [113]. Transformation of the austenite performs in general totally in the martensitic stage even with slow cooling of thick cross-sections with cooling times between 800 °C and 500 °C of 1 h and more. Rising carbon contents and perhaps also nitrogen contents result in a reduction of the critical cooling times [113]. The comparatively high martensitic hardness, together with the low martensitic temperature of about 280 °C, may lead to hardening flaws during welding particularly if thicker cross sections are welded. Preheating is therefore recommended before welding and afterwards the workpiece should be cooled only to 150 °C to 100 °C to ensure a rather complete transformation of the austenite into martensite and preventing hardening flaws; subsequent tempering must be performed immediately. Greater amounts of retained austenite are not allowed because austenite does not dissociate under certain conditions even during tempering and transforms to martensite. During cooling, martensite as mentioned, tends to crack formation [113]. Workpieces that are in the solution-annealed martensitic state must be handled carefully and should not be stored in damp (humid) surrounding if stress corrosion cracking is to be avoided.

C 9.4.1.3 Ferritic Steels for Forgings and Bars

Steels commonly used for forgings are listed in Table C 9.3. The steels 26 NiCrMo 8 5, 26 NiCrMo 11 5 and 26 NiMoV 14 5 are used for *components with great wall thicknesses*, with increased strength values and especially large-diameter shafts for low-pressure turbines. The upper limiting temperature for long-term applications – taking into account a possible decrease in toughness – is at about 340 °C to 350 °C. The properties of these materials are compiled in SEW 555 [118].

For *bolts and nuts* for elevated temperatures steels 24 CrMoV 5 5 and 21 CrMoV 5 11 are utilized. These steels have been in use for some years and are being gradually replaced by steels 21 CrMoV 5 7 and 40 CrMoV 5 7 specified in DIN 17 240 [119].

In the production of *heavy forgings for steam power plants* the steels 20 CrMoNiV 4 7, 28 CrMoNiV 4 9 and 30 CrMoNiV 5 11 according to SEW 555

Table C 9.3 Common steels for forgings and bars

Steel grade Designation	Chemical composition						
	% C	% Si	% Mn	% Cr	% Mo	% Ni	% V
26 NiCrMo 8 5	0.22...0.32	≦ 0.30	0.15...0.40	1.0...1.5	0.25...0.45	1.8...2.1	≦ 0.15
26 NiCrMoV 11 5	0.22...0.32	≦ 0.30	0.15...0.40	1.2...1.8	0.25...0.45	2.4...3.1	0.05...0.15
26 NiCrMoV 14 5	0.22...0.32	≦ 0.30	0.15...0.40	1.2...1.8	0.25...0.45	3.4...4.0	0.05...0.15
24 CrMoV 5 5	0.20...0.28	0.15...0.35	0.30...0.60	1.2...1.5	0.50...0.60	≦ 0.60	0.15...0.25
21 CrMoV 5 11	0.17...0.25	0.30...0.60	0.30...0.50	1.2...1.5	1.0...1.2	≦ 0.60	0.25...0.35
21 CrMoV 5 7	0.17...0.25	0.15...0.35	0.35...0.85	1.2...1.5	0.65...0.80		0.25...0.35
40 CrMoV 5 7	0.36...0.44	0.15...0.35	0.35...0.85	0.90...1.2	0.60...0.75		0.25...0.35
20 CrMoNiV 4 7	0.17...0.25	≦ 0.30	0.30...0.80	1.1...1.4	0.80...1.0	0.50...0.75	0.25...0.35
28 CrMoNiV 4 9	0.25...0.30	≦ 0.30	0.30...0.80	1.1...1.4	0.80...1.0	0.50...0.75	0.25...0.35
30 CrMoNiV 12 1	0.28...0.34	≦ 0.50	0.30...0.80	1.1...1.4	1.0...1.2	0.50...0.75	0.25...0.35
X 21 CrMoV 12 1	0.20...0.26	≦ 0.50	0.30...0.80	11.0...12.5	0.80...1.2	0.30...0.80	0.25...0.35
X 11 CrNiMo 12	0.08...0.15	0.10...0.50	0.50...0.90	11.0...12.5	1.5...2.0	2.0...3.0	0.25...0.40[a]
X 19 CrMoVNbN 11 1	0.16...0.22	0.10...0.50	0.30...0.80	11.0...11.5	0.50...1.0	0.30...0.80	0.10...0.30[b]

[a] Further addition: 0.02% to 0.05% N
[b] Further addition: 0.05% to 0.10% N and 0.15% to 0.50% Nb

[118] are available. These steels may be regarded as the preliminary final stage of a long optimization period with research work performed, as a result of close contact between steel producers and steel users, appliers. These steels are the most economic elevated-temperature steels for this application when considering the low alloying content and the strength values obtained from them.

The highest alloyed non-austenitic elevated-temperature materials are the *martensitic steels with about 12% Cr* and further alloying additions. These steels were first used in the construction of chemical plants and later developed further by additional alloying of molybdenum, vanadium or tungsten [120]. Gradual optimization for different application in turbine engineering led finally to the steels for forgings X 21 CrMoV 12 1, X 11 CrNiMo 12 and X 19 CrMoVNbN 11 1 [121]. The first-mentioned of these steels has been used since 1950 according to SEW 555 [118] for high-pressure turbine shafts in power plant engineering. According to SEW 670 [122] the steel has been used for heavier forgings in different application ranges and, with the designation X 20 CrMoV 12 1 according to DIN 17 243 [123], as a weldable steel for making shaped pieces. The second steel is a material for the aircraft engine industry, i.e. for compressor discs and -blades; the third steel – according to DIN 17 240 [119] – serves for the production of bolts and nuts with increased elevated-temperature strength.

C 9.4.1.4 Ferritic Steels for Castings

The quality specifications and technical delivery conditions for ferritic elevated-temperature steel castings are contained in DIN 17 245 [124]. The most-used steels are listed in Table C 9.4.

Th *choices of alloying elements* for these steels generally correspond to those for rolled and forged steels. Because of the lack of deformation the carbide distribution is more coarse than it is with rolled steel. In austenitizing, a smaller proportion of carbides is brought into solution with the effect that less carbon is available for the formation of secondary carbides, which determines the creep behavior. For this reason the carbon content of elevated-temperature steels for castings is in general a little higher than it is in rolled steel. The amount of carbon is limited by the requirement for weldability and is generally below that of the comparable steels for forgings.

The series of standardized steels for castings is led by the *unalloyed steel GS-C 25*. The creep-stress rupture strength of this steel is low, so that it is preferably applied in the range of yield stress values up to 400 °C. The creep-stress rupture strength values do not differ from those of the unalloyed rolled and forged steels.

In steels for castings it also is desirable to increase the creep-stress rupture strength by *addition of molybdenum*. Slightly higher molybdenum content of 0.4% compared with the rolled steel 15 Mo 3 provides the improved creep rupture strength values through adequate formation of secondary carbides without the danger of embrittlement [125, 126]. Therefore the creep-stress rupture strength of GS-22 Mo 4 lies in the scatter band of steel 15 Mo 3.

The embrittlement of steels caused by a additional *alloying with chromium* molybdenum content above 0.4% can be avoided. Steels like GS-17 CrMo 5 5 and

Table C 9.4 Chemical composition of common elevated-temperature steels for castings (ladle analysis)

Steel grade Designation	Chemical composition								
	% C	% Si[a]	% Mn[b]	% P maximum	% S maximum	% Cr	% Mo	% Ni[b]	% V	Others
GS-C 25	0.18 to 0.23	0.30 to 0.60	0.50 to 0.80	0.020	0.015	≤ 0.30	–	–	–	–
GS-C 22 Mo 4	0.18 to 0.23	0.30 to 0.60	0.50 to 0.80	0.020	0.015	≤ 0.30	0.35 to 0.45	–	–	–
GS-17 CrMo 5 5	0.15 to 0.20	0.30 to 0.60	0.50 to 0.80	0.020	0.015	1.00 to 1.50	0.45 to 0.55	–	–	–
GS-18 CrMo 9 10	0.15 to 0.20	0.30 to 0.60	0.50 to 0.80	0.020	0.015	2.00 to 2.50	0.90 to 1.10	–	–	–
GS-17 CrMoV 5 11	0.15 to 0.20	0.30 to 0.60	0.50 to 0.80	0.020	0.015	1.20 to 1.50	0.90 to 1.10	–	0.20 to 0.30	–
G-X 8 CrNi 12	0.06 to 0.10	0.10 to 0.40	0.50 to 0.80	0.030	0.020	11.50 to 12.50	max. 0.50	0.80 to 1.50	–	[c]
G-X 22 CrMoV 12 1	0.20 to 0.26	0.10 to 0.40	0.50 to 0.80	0.030	0.020	11.30 to 12.20	1.00 to 1.20	0.70 to 1.00	0.25 to 0.35	[d]

[a] In ordering a vacuum-treated steel the silicon content may remain under the lower limit
[b] If it seems reasonable considering through hardening (in quenching and tempering) or the microstructure, an increase of the manganese content up to 1.1% or of the nickel content up to 0.70% may be agreed upon on ordering. This variation is not valid for the grades G-X 8 CrNi 12 and G-X 22 CrMoV 12 1
[c] To limit the δ-ferrite content, nitrogen up to max. 0.050% may be added
[d] A tungsten content of max. 0.50% is permissible

GS-18 CrMo 9 10 have been developed for this purpose. Their creep and creep-rupture strength properties correspond with those of comparable rolled and forged steels. The combined effect of alloying elements of the material GS-18 CrMo 9 10 leads to a bainitic structure, so the demands for high toughness and non susceptibility to brittle fracture at comparatively high creep-rupture strength are well met.

Of the *steel grades for castings alloyed by vanadium* the most popular is the steel with about 1% Cr – known and standardized in Germany as GS-17 CrMoV 5 11. This grade has a creep-rupture strength that is far above that of the chromium-molybdenum steels and it is combined with good toughness properties. The structure to be adjusted by quenching in oil or by accelerated air cooling consists of upper bainite with a maximum of 20% ferrite. With wall thicknesses over 300 mm the manganese content can be raised to about 1% to improve a through hardening in quenching and tempering. Sometimes nickel also is added [127]. A temporary creep-stress embrittlement occasionally observed after testing times of 1 to 2×10^4 h, depends mostly on the presence of martensite after too-rapid cooling during hardening [128, 128a].

Similar or slightly over-alloyed filler metals are used for welding. The creep-stress rupture strength of welded joints made with welding (filler) metal of the same kind as the base material is partially at the lower limit [129] or even below the scatter band of the base material. This condition can be improved by the use of over-alloyed weld material. Over-alloying is not possible for GS-17 CrMoV 5 11 and for GS-18 CrMo 9 10 is not desired because the weld material must be free of vanadium. With these steels it is apparent that an increase of the carbon content in the weld material improves the creep values of the assembly (joint) [130, 130a].

Heat treatment after welding generally consists of tempering during which the temperature must be maintained above the precipitation temperature range of the special carbides that cause brittleness, at least in the range of 660 °C to 680 °C, depending on the alloy content. Bigger weldments sometimes are double-quenched and -tempered and filler metals that are over-alloyed or contain higher amounts of carbon are suitable. The fracture sites of creep-rupture tested specimens especially of GS-17 CrMoV 5 11, after long testing times are shifted from the weld material to the heat affected zone (HAZ), so the double-quenching and -tempering also improves the creep rupture properties [130a].

In Great Britain the chromium-molybdenum-vanadium-cast steel grade with 0.12% C, 0.4% Cr, 0.5% Mo and 0.3% V is preferred. By reduction of the carbon content and addition of alloying elements that increase the hardenability, weldability also is improved. After cooling in still air and tempering a ferrite-pearlite-structure is observed [131, 132] with a creep behavior which corresponds to that of the above-mentioned 1% Cr steel. The strength values at room temperature, however, are essentially lower and the toughness properties are reduced. The steel with 2.25% Cr and 1% Mo is chosen as filler metal. In post weld heat treatment the tendency to form stress relief cracks must be taken into account [133].

Investigations of the influence of deoxidation and of tramp-elements on the sensitivity of this steel to stress relief cracks and creep embrittlement have shown that deoxidation with titanium has favorable effects in each respect. A low content of tramp elements improves the creep ductility and resistance to stress relief cracks, but does not compensate for the raised costs in manufacturing [132].

The *martensitic steel for castings G-X 22 CrMoV 12 1* corresponds in composition and properties to the equally-alloyed rolled and forged steel. With its creep properties this steel is at the top of the ferritic steel grades for castings. The creep rupture strength of the cast steel is stated to be somewhat lower than for rolled and forged steel. From current creep tests it can be expected that the differences will vanish with longer working times [134]. Adequate toughness values can be obtained with ferrite-free structures [135]. Castings made from this steel are used at temperatures up to 650 °C for gas- and steam-turbines. Further alloying with nitride formers has been tried, to improve the creep properties especially in the upper temperature range [135a]. For turbines for mean peak loads which must be switched on and off frequently, this grade has advantages because the wall thicknesses and thus the stress at temperature cycles are reduced so that the velocities during switching on and off can be increased [135a].

For turbines operating in the wet steam range of power stations with light water reactors *a steel for castings* with 12% Cr was developed. This steel is *resistant to corrosion and erosion* and has a very high insensitivity to brittle fracture because of its low carbon content and its special heat treatment. This G-X 8 CrNi 12 is not applied in the creep range [136].

C 9.4.2 Austenitic Steels

The steels mentioned as examples in *Table C 9.5* are to a great extent contained in specifications [119, 137], which also contain the characteristic values that govern the capacity to withstand stresses. Here using *Fig. C 9.18* reference shall be made only to the creep *rupture strength for 100 000 h*. At these levels the materials differ essentially only in their absolute height and in temperature dependence.

The non-stabilized steels X 6 CrNi 18 11 and X 6 CrNiMo 17 13 listed in Table C 9.5 have only recently been used in Germany as high-temperature materials. These steels correspond well to the steels 304 and 316 of the American Iron and Steel Institute, much-used in North America, but to increase the austenite stability the chromium content is limited to lower values and the nickel content is raised. Applications of these steels have been increased by a rising demand for weldability. Possible disadvantages in welding of heavy cross-sections of niobium-stabilized steels were mentioned above.

Although fully-austenitic weld material tends to the formation of micro-cracks [138], because of local segregations at the grain boundaries, no damages have so far resulted from this cause so far as is known [139]. These micro-cracks can be prevented by application of *filler metals*, which result in a weld material containing about 3% to 8% ferrite. In a molybdenum-containing weld material this condition may lead to difficulties because of embrittlement by formation of sigma-phase [140]. It was therefore proposed [141] that the composition of the filler metal be modified by reduction of the contents of chromium, molybdenum and nickel to the extent that no sigma-phase can be formed.

C 9.4 Characteristic Material Grades Proven in Service   265

Table C9.5 High-temperature austenitic steels[a]

Steel grade	Chemical composition											
Designation	% C	% Al	% B	% Co	% Cr	% Mo	% N	% Nb[c]	% Ni	% Ti	% V	% W
X6CrNi18 11	0.06				18.0	≤ 0.50			11.0			
X6CrNiMo17 13	0.06				17.0	2.25			13.0			
X8CrNiNb16 13	0.07				16.0			≥ 10·%C[b]	13.0			
X8CrNiMoNb16 16	0.07				16.5	1.8		≥ 10·%C[b]	16.5			
X6CrNiWNb16 16	0.07				16.5			≥ 10·%C[b]	16.5			3.0
X8CrNiMoVNb16 13	0.07				16.5	1.3	0.1	≥ 10·%C[b]	13.5		0.70	
X10NiCrMoTiB15 15	0.10		0.005		15.0	1.15	0.1		15.5	2.1		
X8CrNiMoBNb16 16	0.07		0.08		16.5	1.8		≥ 10·%C[b]	16.5	0.45		
X40CrNiCoNb17 13[d]	0.40			10.0	16.5	2.0		3.0	13.0			2.5
X12CrCoNi21 20	0.12			20.0	21.0	3.0	0.15	1.0	20.0			2.5
X10NiCrAlTi32 20	≤ 0.12	0.35			21.0	1.25			32.0	0.35	0.30	
X5NiCrTi26 15	≤ 0.08	≤ 0.35	0.007		14.5				26.0	2.1		

[a] If not otherwise marked, the chemical composition figures are mean values.
[b] Additional Nb ≥ 10 · % C + 0.4% ≤ 1.2%
[c] The amounts relate to the sum % Nb + % Ta
[d] The steel is also produced with 13% or 19 % Cr

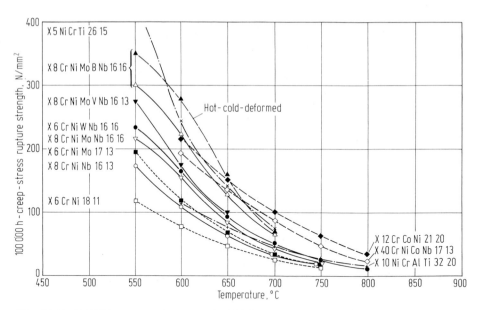

Fig. C 9.18. 100 000 h-creep rupture strength of high-temperature austenitic steels as mentioned in Table C 9.5 (mean values).

To prevent the susceptibility to intercrystalline corrosion of welded components from non-stabilized steels, an exchange of carbon for nitrogen led to the steels *X 3 CrNiN 18 11 and X 3 CrNiMoN 17 13* with about 0.03% C and 0.1% N. Long-term tests have shown [142] that the nitrogen-containing material shows an improved creep and creep-stress rupture and toughness behavior in comparison with X 6 CrNiMo 17 13.

The higher creep-stress rupture strength of the niobium-alloyed steels X 8 CrNiNb 16 13 and X 8 CrNiMoNb 16 16 is more clearly seen in from Fig. C 9.19 in contrast to Fig. C 9.18. Both steels and X 8 CrNiMoVNb 16 13 have been proven in Germany since 1951 in the production of steam boilers. However, according to Fig. C 9.20 [139]. The notch-impact energy of these steels at room temperature and after service stresses is decreased considerably by precipitation processes.

The chemical composition of series of steel grades for castings corresponds to the above-described rolled and forged steels, for instance the steels G-X 6 CrNi 18 11, G-X 6 CrNiMo 17 13, G-X 7 CrNiNb 16 13 and G-X 8 CrNiMoVNb 16 13. In castings, the structure, and especially the final grain size, will be adjusted by the solidification conditions in the mold since grain regeneration is not possible because of lack of deformation and recrystallization. The low-melting point constituents therefore remain on the grain boundaries, so that very careful melting is necessary to ensure high cleanness [139]. The creep-stress rupture strength of cast steel conforms in general to that of the forged or rolled steels of the same type [143, 144]. By increasing the carbon content and thus also the amount of carbides, the creep-stress rupture strength can be improved

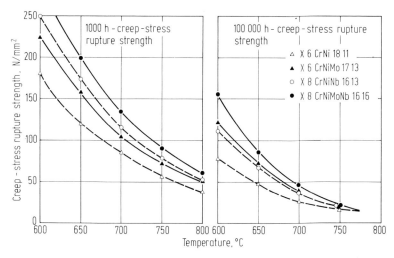

Fig. C 9.19. Mean values of the creep-stress rupture strength of four steels according to Table C 9.5 with different content of nickel, molybdenum and niobium.

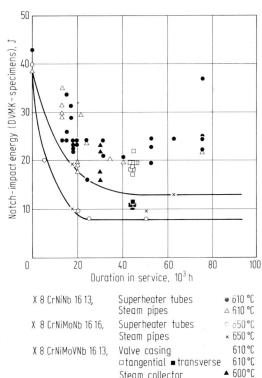

Fig. C 9.20. Notch-impact energy at room temperature of steels according to Table C 9.5 after long-time operational stress in service. After [139].

although with very much reduced toughness. This feature is utilized preferably with the heat resisting steel grades for castings.

The *steels X 6 CrNiWNb 16 16 and X 12 CrNiWTi 16 13* are alloyed with tungsten instead of molybdenum. The last-mentioned steel is often melted with a small boron addition and is used for certain applications, e.g. for turbine blades, in a pre-strain hardened condition. The steel X 6 CrNiWNb 16 16 with a *higher nitrogen content* has almost the same creep rupture strength [145] as the similar steel alloyed with nitrogen *X 8 CrNiMoVNb 16 13*. The vanadium-containing steel may tend to increased scale formation so that at temperatures above 650 °C with the same requirement for creep rupture strength, the tungsten-alloyed steel should be used.

The titanium-stabilized boron-containing *steel X 10 NiCrMoTiB 15 15* [146] is vacuum melted and therefore has a low nitrogen content. Consequently titanium precipitates essentially as carbide, i.e. preferably at the dislocations. The dislocation and precipitation density and therefore also the creep resistance are increased by means of a preliminary cold deformation [147]. The precipitations finely-distributed in the grains form "traps" for helium, which is generated by radiation in the nuclear reactor at temperatures above about 500 °C. These conditions result in (n, α)-reactions and prevent the helium from travelling to the grain boundaries, thus causing high-temperature embrittlement [148] by helium bubble formation. Simultaneously the numerous dislocations in the strain-hardened solid solution may retard the growing of pores, resulting in the so-called swelling, caused by radiation. Therefore this steel in the cold pre-strain hardened condition is considered as a material for the jacket tubes of fuel elements with high temperatures, e.g. in rapid-breeder reactors.

In non-stabilized steels with boron contents of 0.007% and higher, grain boundary cracks may occur in the base material close to the weld seam under unfavorable conditions (heavy ingots and strong segregations, high final temperatures during deformation). These cracks can be explained by the formation of a boron-rich eutectic phase which solidifies later as the solid solution. Niobium-free steels have a lower sensitivity to welding cracks because the niobium is engaged in the formation of the boron-rich eutectic.

Hence the steel X 8 CrNiMoBNb 16 16 from Table C 9.5 with high boron content cannot be welded by fusion processes. This steel has the highest creep-stress rupture strength of all steels so far discussed and, in the pre-strain hardened condition, has high creep-stress rupture strength at simultaneously high creep fracture deformation values and notch creep-stress rupture strength. In this condition the steel is used for turbine blades and for high relaxation resistance requiring parting line (joint) screws [119] in turbine manufacturing. The improvement in creep-stress rupture strength and especially of the 1%-creep limit remains at up to nearly 700 °C. After a cold-hot-deformation this steel is superior to the cobalt-alloyed steels up to about 650 °C (see Fig. C 9.18).

Cobalt-containing materials of which X 40 CrNiCoNb 17 13 and X 12 CrNiCo 21 20 are mentioned in Table C 9.5, have the highest creep-stress rupture strength of all austenitic steels at temperatures above 650 °C and under long-term stress. X 40 CrNiCoNb 17 13 is used for gas-turbine discs and rotors as well as for valve casings. Steel X 12 CrNiCo 21 20 is also applied as material for highly-stressed outlet valves in combustion engines [149], as an aviation material 1.4974 in plate shape in power units and with material number 1.4957 for a precision casting steel G-X 15 CrNiCo 21 20 20.

The nickel-rich *steel X 10 NiCrAlTi 32 20* is a material of multiple applications. With a carbon content below 0.03% this steel is regarded as stainless, and with higher carbon content as a heat-resistant and high-temperature steel [150]. The steel has relatively-high creep-rupture strength at temperatures around 900 °C, so that it can be applied in coal gasification and in the utilization of nuclear process heat.

The high creep-stress strength of *steel X 5 NiCrTi 26 15* (the last-mentioned in Table C 9.5) under long-term stress at lower (see Fig. C 9.18) and at short-term stress at higher temperatures govern its application to elevated-temperature springs [151] and use for construction of discs, screws and plates in the shorter-living aviation gas turbine. This material – in its base composition nearly 40 years old [152] – is the only one of the cobalt-free austenitic steels in the precipitation-hardened condition shows high creep-stress rupture strength and high strength values measured by tensile tests. The usual austenitic steels have a 0.2%-proof stress at room temperature of about 200 N/mm^2 or less but in case of this material this value is above 600 N/mm^2.

C 9.4.3 High-temperature Nickel and Cobalt Alloys

High-temperature alloys are used in manufacture of steam- and gas-turbines, of aviation power units (e.g. for compressors and turbine discs, turbine blades and shafts) and units for the chemical industry with working temperatures up to 1100 °C.

A useful compilation of such alloys with data on the chemical composition and the creep rupture strength is published annually by the American Society for Metals [153]. According to this publication, the nickel alloys are more numerous than other formulations.

	Nickel-alloys	Cobalt-alloys
Forging alloys	44	10
Precision cast alloys	35	12

Alloys proven in service are listed in Table C 9.6. The alloys specified for forgings are hot and cold deformable, but not the precision casting alloys.

Characteristic for the *forgeable alloys based on nickel* is that the carbon content is below 0.1%. Chromium contents between 15% and 20% guarantee resistance to oxidation and hot gas corrosion. The contents of elements such as aluminum and titanium that contribute to the formation of precipitation hardening γ'-phase $Ni_3(Al, Ti)$ amount to 0.2% up to 4.3% Al and 1% to 3.5% Ti depending on the alloying. With rising alloying contents the creep-stress rupture strength is increased and the steels can be used at higher temperatures. These properties can be further increased in the precision cast alloys based on nickel by still higher aluminum and titanium contents (in total about 7% to 10%) and by addition of molybdenum combined with niobium, tantalum and tungsten depending on the alloying.

The forgeable alloys based on cobalt as shown for example in Table C 9.6 as well as the precision casting alloy G-Co Cr22 Ni10 W Ta contain at least 0.1% C, 20% Cr but only 10% to 20% Ni. Instead, grade Co Cr20 Ni20 W contains 4% each of molybdenum, niobium and tungsten. Steel Co Cr20 W15 Ni otherwise contains 15% W, whereas the precision cast alloy additionally contains 3.5% Ta, 0.2% Ti, 7% W and 0.5% Zr. The elements boron and zirconium, added in small amounts to the most forgeable and precision castable alloys, contribute to the improvement of strength- and creep-strength behavior of the alloys because of their effect at the grain boundaries.

To meet the quality requirements for high-temperature materials, special *measures are necessary during production*. Above all the chemical composition must be very carefully adjusted within narrow limits. The highest degree of freedom from oxides and nitrides possible and reduced contents of deleterious tramp elements as for instance lead, bismuth and tellurium, should be aimed at. Contents of sulphur and phosphorus should be reduced as far as possible. The alloys are mostly melted in a vacuum induction furnace using raw materials as pure as possible. If materials are to be hot deformed they are again remelted, usually in a vacuum arc furnace.

The best possible properties are obtained in hot forged or cast components by a heat treatment consisting of solution annealing at temperatures between 900 °C and 1230 °C and hot aging of one or more stages between 620 °C and 850 °C. For some precision-cast alloys no heat treatment is necessary.

The strength of the forgeable nickel-alloys at room temperature extends from about 1180 N/mm^2 to 1400 N/mm^2, that of the cobalt-alloys from about 960 N/mm^2 to 1100 N/mm^2, and that of the precision casting alloys from about 800 N/mm^2 to 1100 N/mm^2.

One of the important characteristic values of high-temperature alloys is the creep-stress rupture strength. The *temperature dependence of the 1000 h-creep-stress rupture strength* of forging alloys according to Table C 9.6 is described in Fig. C 9.21 and Fig. C 9.22 gives the same data for the precision casting alloys. Obviously forging and precision-cast alloys based on cobalt in contrast to those based on nickel show an inferior dependence of temperature on the creep-stress rupture strength.

Further characteristic values important for the engineer are the creep limits. For instance, longer times under operational stress strains, (elongations) of, for example, max. 0.1% to 0.5% must not be exceeded.

The alloys should show *additional properties* depending on the operational stresses, such as resistance to hot gas corrosion, low susceptibility to thermal fatigue by quick or slow temperature changes and high fatigue strength at low and high cycle frequencies [154, 155].

The *future development* of high-temperature alloys is aimed at reaching still more production (processing) safety considering the cost and availability of such raw materials as cobalt, molybdenum and niobium [156]. New processes are being developed to produce more-efficient alloys e.g. by *application of powder-metallurgy*

Table C 9.6 Chemical composition of high-temperature nickel- and cobalt-alloys (examples)

Alloying designation AMCMA[c],[b]	Chemical composition[a]											
	% C	% Al	% Co	% Cr	% Fe	% Mo	% Nb	% Ni	% Ta	% Ti	% W	% Zr
Forging alloys												
NiCr19NbMo Ni-P 100-HT	0.05	0.5		18	18	3	5	54		1		
NiCr20TiAl Ni-P 95-HT	0.05	1.4		20				75		2.4		
NiCr20Mo	0.06	1.4		20		4.5		70		2.4		
NiCo19Cr18MoAlTi	0.03	1.4	14	20		4.5		56		3		
NiCr18CoMo Ni-P 94-HT	0.08	2.9	18.5	18		4		53		2.9		
NiCo18Cr15MoAlTi	0.08	4.3	18	15		5.2		53		3.5		
CoCr20Ni20W	0.40		42	20	<4	4	4	20			4	
CoCr20W15Ni Co-P 92-HT	0.10		52	20	1			10			15	
Precision cast alloys												
G-CoCr22Ni10WTa	0.60		54	22				10	3.5	0.2	7	0.5
G-NiCr13Al6MoNb Ni-C 98-HT	0.05	6		12	1	4.5	2	72		1		0.1
G-NiCr16Co8AlTiNb	0.07	3.4	8.5	16		1.8	1	62		3.4	2.5	0.1
G-NiCo15Cr10MoAlTi[d]	0.15	5.5	15	10		3		60		4.7		
Ni-C 104-HT G-NiCo10W10CrAlTa	0.15	5.5	10	9		2.5		60	1.5	1.5	10	0.1 / 0.05

[a] The figures for the chemical-composition are mean values
[b] On the base of the designation system of non-metal-alloys in DIN-standards not yet mentioned therein
[c] Association Européenne des Constructeurs de Material Aerospatial
[d] Besides 1 % V

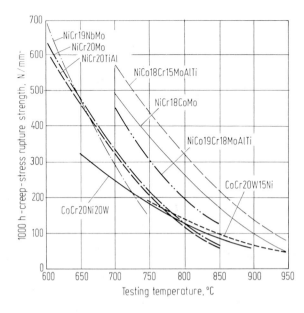

Fig. C 9.21. 1000 h-creep-stress rupture strength of hot deformed nickel- and cobalt alloys according to Table C 9.6. After [153].

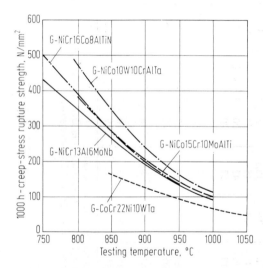

Fig. C 9.22. 1000 h-creep-stress rupture strength of precision cast alloys according to Table C 9.6. After [153].

[154–157]. As an example may be mentioned use of powder of Ni Cr20 Ti Al and yttrium oxide Y_2O_3. After intensive mixing in a mill process – the so-called mechanical alloying – processing by pressing and sintering leads to an alloy which can be utilized at mean temperatures in the precipitation hardened condition and at high temperatures in the dispersion hardened condition.

C 10 Low Temperature Steels (Steels with Good Toughness at Low Temperatures)

By Max Haneke, Joachim Degenkolbe, Jens Petersen and Wilhelm Weßling

C 10.1 Required Properties of Use

Steels used for low-temperature service are constructional steels with satisfactory toughness properties at those temperatures. This toughness permits their use for applications at low temperatures (below $-10\,°C$).

Such applications are common in cryogenic technology, which has grown in importance following the introduction of new processes in energy, chemical and food industrial sectors. The energy sector employs hydro-carbon liquifying processes on a large scale [1]. The boiling temperature of light hydro-carbons ranges between $-42\,°C$ and $-161\,°C$. Metallurgical and chemical processes consume large quantities of oxygen obtained by the fractional distillation of liquid air at temperatures of about $-200\,°C$. This same process also yields inert gases. Petrochemical industries use liquifying plants for separating mixtures of hydrocarbons.

Liquified gases are also used in rocket and space technology, nuclear research and electro-technical fields. In these applications, temperatures may be as low as 1.8 K. All these applications require installations, vessels for transport and storage and pipe lines that are constructed from steels with good cold-fracture toughness properties [2–4].

These steel grades must not only exhibit satisfactory *toughness at low temperatures*. To meet vessel pressure requirements or mechanical stresses (as for example for screws) high demands are also put on *strength properties*. Manufacture of cryogenic equipment also includes welding processes so it follows that the majority of these steel grades must also have good *welding properties* (*weldability*).

The selection of the proper steel grade for reliable low-temperature service requires a profound knowledge of the mechanism by which mechanical properties, in particular toughness, is affected by low service temperatures. Brittleness in steel and welds must occur only at levels that are well below the lowest service temperature.

Steels for low temperature service that means steels with good toughness at such temperatures comprise – depending on required toughness properties – unalloyed and alloyed ferritic and austenitic grades. Figure C 10.1 details fields of applications of some of the more important steels for low-temperature service.

References for C 10 see page 782.

274   C 10 Low Temperature Steels

Steel grade[1]	Yield stress at RT N/mm² min.	Notch impact energy[2]		Applications												
		Test temperature °C	J min.	Butane	Propane	Propene	Carbon dioxide	Ethane	Ethene	Methane	Oxygen	Argon	Nitrogen	Hydrogen	Helium
				±0°C	-42°C	-47°C	-78°C	-89°C	-104°C	-164°C	-183°C	-186°C	-196°C	-253°C	-269°C
				Boiling temperature												
T StE 255 – T StE 500	255 – 500	-50	27												
11 Mn Ni 5 3	285	-60	41												
13 Mn Ni 6 3	355	-60	41												
10 Ni 14	345	-100	27												
10 Ni 14 V	390	-120	27												
12 Ni 19	420	-140	35												
X 7 Ni Mo 6	490	-170	39	Range of applications												
X 8 Ni 9	490	-196	39												
Austenitic steels	240 – 340	-196	55												

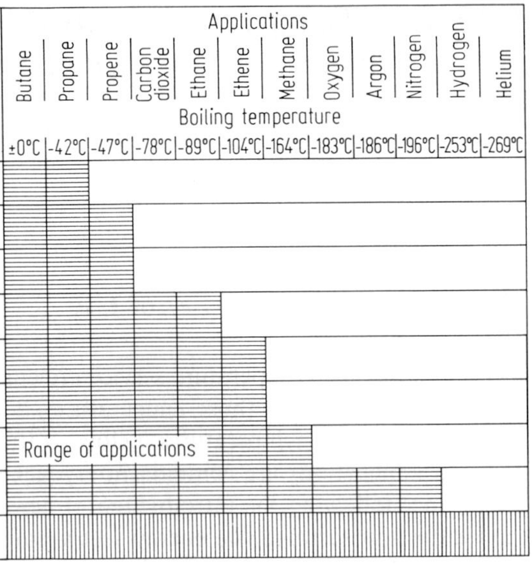

[1] Chemical composition see Table D 10.3
[2] ISO-V-notch longitudinal specimens, average value from 3 individual tests.

Fig. C 10.1. Ranges of application in liquid gas technology of steels with good toughness at low temperatures. After [2].

In using ferritic steels attention must be paid to the fact that with decreasing temperatures, the body-centered cubic lattice of the ferrite experiences sharply-reduced toughness properties when passing through a certain narrow temperature range (transition temperature). The critical temperature range of ferritic steels depends not only on the chemical composition of the steel but also to a large degree on multi-directional stresses and stress velocities (see B 1).

Austenitic steels with a face-centered cubic lattice also exhibit a decrease in toughness with decreasing temperatures but to a lesser degree than steels with a body-centered cubic lattice. What is more important is that the above-mentioned sharp decrease (steep drop) of toughness does not occur.

The installations in the field of low temperatures making use of the special steel grades in question are frequently controlled by the relevant authorities. Steels used for these application therefore must be approved by national and international certification organizations (see Table C 10.1).

Guidelines on the use of such steels for the construction of pressure vessels are contained e.g. in the German AD-Merkblatt W 10 (5).

C 10.2 Characterization of the Required Properties

Good *toughness*, the main suitability criteria for service at low temperatures, is preferably evaluated by the *notched bar impact test* to which these steels are subjected at the lowest possible temperatures.

Table C 10.1 Steels for low-temperature service with good toughness at low temperatures according to different standards and rules [10–12, 24]

FR Germany DIN 17 280	Euro Standard EU 129	England BS 4360 BS 1501	France NFA 36-208 Bureau Veritas	Norway Det Norske Veritas	USA American Bureau of Shipping ASTM
TSt E 355[b]	Fe E 355 KT	4360 Grade 50 E	–	NV E 36 S	–
11 MnNi 5 3	Fe E 245 Ni 2	–	0.5 Ni A	NV 2-4	V-051/060
13 MnNi 6 3	Fe E 355 Ni 2	–	0.5 Ni B	NV 4-4	–
14 Ni 6	Fe E 335 Ni 6	–	1.5 Ni	–	–
10 Ni 14	Fe E 355 Ni 14	1501-503	3.5 Ni	NV 20-0	A 203 Grade D
12 Ni 19	Fe E 390 Ni 20	–	5 Ni	NV 20-1	A 645
X 7 NiMo 6	–	–	–	–	–
X 8 Ni 9	Fe E 490 Ni 36	1501–509/510	9 Ni	NV 20-2	A 353/553

[a] For chemical compositions see Table C 10.3
[b] As an example for the series of steels with good toughness at low temperatures contained in DIN 17 102 [10] and with minimum yield stress values of 255 N/mm² to 500 N/mm²

Since results of notched bar impact testing do not suffice to predict the *service behavior of full-size components*, additional tests have been developed to simulate the toughness of components under service conditions i.e. so-called type tests. These type tests demonstrate the behavior of the steel separately according to the crack initiation and to the arresting behavior of the crack. The most important tests are the sharp notch bending test, the notch tensile test, and the "Kohärazie" test (sharp notch impact test) used to pinpoint the critical crack initiating temperature, and the Pellini drop weight test, the Robertson test and again the "Kohärazie" test (sharp notch impact test here with modified conditions) to characterize the crack arresting behavior. For details of these testing procedures see B 1. Table C 10.2 contains results of tests on some of the more important steels for low-temperature service.

Material *strength properties* are commonly determined by *tensile tests* (see also B 1). Decreasing temperatures increase the yield stress and the tensile strength. Design of vessels for low-temperature service is based on test results determined at room temperature so it follows that the strength properties at service temperature can not be utilized fully.

Figure C 10.2 shows for some of the more important low-temperature service steels, the relationship between yield stress and temperature. As can be seen the yield stress at actual service temperatures may be considerably higher than at room temperature.

Testing of austenitic steels includes determining the tensile strength and yield stress, generally determined as 0.2%-proof stress, and the 1% proof stress at 20 °C. The 1%-proof stress value is commonly used for design purposes to allow a better utilization in view of the considerable plastic reserves of these steels.

Commonly applicable testing procedures to characterize the *weldability* of steels for low temperature service are not available (for basic principles see B 5). Testing weldability is generally performed under simulated service conditions. Because of the importance of the welding behavior and the different character of these steels, general remarks are not dealt with here. Some of the more important steel grades are covered in a special chapter (see C 10.5).

Table C 10.2 Fracture behavior of steels for low temperature service, plate thickness 25 mm [6]

Test method	Characteristic temperature (assessing criteria)	Test results, assessing temperature (°C) of steel[a]						
		TTStE 36 N	13 MnNi 63 N	10 Ni 14 N	10 Ni 14 V	12 Ni 19 NN	12 Ni 19 V	X 8 Ni 9 V
Crack propagation								
Impact test[b]	$T_{tr\,27}$[c]	− 55	− 100	− 125	− 160	− 140	− 170	− 196
"Kohärazie" test[d]	T_{LKO}[e]	− 10	− 20	− 65	− 105	—	− 90	− 196
Pellini test[f]	NDT[f]	− 38	− 60	− 100	− 120	− 130	− 130	− 196
Robertson test[g]	CAT[g]	− 40	—	—	—	—	—	—
Crack initiation								
"Kohärazie" test[d]	T_{LBO}[h]	− 30	− 60	− 95	− 125	—	− 150	− 196
Sharp notched impact test	T_{ij}[i]	− 70	− 80	− 90	− 120	—	− 150	− 196
Notched tensile test	T_j[j]	− 70	− 85	—	—	—	—	− 196

[a] N = normalized, V = quenched and tempered
[b] ISO-V-notch specimen
[c] $T_{tr\,27}$ Transition temperature for 27 J notch impact energy
[d] Impact test on sharp notched specimen
[e] T_{LK0} = Temperature for 100% shear-fracture (impact speed: 5 m/s)
[f] See Fig. B 1.75 in B 1.1.2.4
[g] See Fig. B 1.72 in B 1.1.2.4
[h] T_{LB0} = Temperature for 100% shear-fracture (impact speed 0.1 m/s)
[i] See for example [8]
[j] T_i = crack initiation temperature (specimen of medium dimensions) T_i = limiting temperature, below which the fracture occurs as in stable cleavage fracture at low stresses, without a preceding stage of quasistatic crack propagation as shear fracture

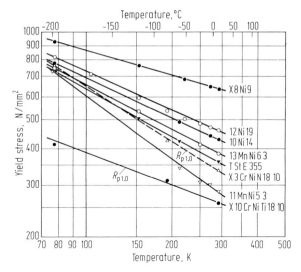

Fig. C 10.2. Yield stress of different steels for low temperature service (see Table C 10.3) depending on the temperature. After [9].

Additional service properties are the coefficient of thermal expansion, thermal conductivity, specific heat, electrical resistance and modulus of elasticity. All these properties are determined by common testing procedures that are partly detailed in B 2.

The physical properties of individual ferritic steels i.e. unalloyed and manganese- and/or nickel-alloyed steels differ only to a small extent with the exception of the steel grade with 9% Ni (see also X 8 Ni 9 in Table C 10.3). This grade has a higher electrical resistance and a lower modulus of elasticity compared with other steel grades. Thermal conductivity decreases almost linearly with increasing nickel additions.

All ferritic steels with good toughness at low temperatures exhibit an elastic modulus of 207 kN/mm² at room temperature with the exception of the grade containing 9% Ni and having a Young's modulus of 186 kN/mm². The corresponding value for austenitic steel grades is 200 kN/mm². The modulus of elasticity is only slightly influenced by varying temperatures. A value of 207 kN/mm² has been recorded at $-196\,°C$ with a steel containing 9% Ni.

C 10.3 Physical Metallurgy Procedures to Attain Required Properties

C 10.3.1 Ferritic Steels

Chemical Composition

Good toughness at low temperatures is the most important property governing applications. The most effective method to achieve this low-temperature toughness in ferritic steels is to produce a fine-grained microstructure and on the whole a general refining of the microstructure for example by the generation of pearlite with decreased spacing. A fine-grained microstructure can also be achieved by adding alloying or tramp elements that influence indirectly the transformation or the recrystallization process. The effect of different alloying elements on the fine-grained microstructure, which is important for the low temperature toughness, is discussed, together with observations on side-effects, in the following chapters.

The first element to be discussed is *carbon*, which decreases the toughness due to the rising quantities of pearlite with increasing carbon additions (Fig. B 1.80) [13]. Steels for low-temperature service therefore require low carbon contents below about 0.2% to limit the formation of pearlite which should be present with a low spacing (fine development of the cementite lamella). Low carbon values improve welding properties. High carbon contents may lead to formation of martensite during cooling of welds in the heat-affected welding zone. This undesired effect would cause an increase in hardness.

Silicon increases the yield stress and tensile strength but additions above about 0.6% have an adverse effect on toughness. Low silicon additions improve the transition temperature of the toughness only slightly. In contrast to carbon, silicon has almost no effect on hardenability and therefore scarcely influences properties in the heat-affected welding zone.

An important alloying element that improves toughness of steel is *Manganese* which forms a substitution solid solution with iron. Manganese additions of 1% raise the critical temperature, at which transition from ductile to brittle fracture occurs, by about 50 °C (Fig. C 10.3). This rise is due to a reduction of the transformation temperature which leads to smaller secondary grain sizes and more finely-structured pearlite. With low carbon levels, i.e. high ratios of carbon and manganese contents, steel grades with up to 2% Mn exhibit excellent toughness properties. Manganese additions above 2% generally decrease toughness properties because the transformation undergoes a change that leads to the formation of bainite. Increasing additions of other alloying elements, for example nickel, lower this transformation [1]. Yield stress and tensile strength increase with rising manganese additions due to solid solution hardening. However, the influence of manganese in solid solution on toughness is negligible.

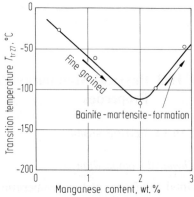

Fig. C 10.3. Influence of the manganese content on the transition temperature of the notch impact energy $T_{tr\,27}$ which is the temperature at which the impact energy (ISO-V-notch specimens (here longitudinal specimens) is 27 J. Chemical composition of the steel: 0.05% C, 0.25% Si, 0.01% P and 0.04% Al. After [1].

To guarantee satisfactory toughness at very low temperatures further additions of alloying elements to the already-present manganese becomes necessary. *Nickel* is particularly suited to lower the transition temperature from ductile fracture to brittle fracture. The influence of nickel on toughness of steels with low carbon contents and small nickel additions (below 2% Ni) is due to the lowering of the transition temperature which gives rise to a decrease of the ferrite grain size. In this connection it may be noted that nickel favors cross slipping (see B 1.11 and Fig. B 1.14) and this improves the toughness. Higher additions of nickel produce bainite and martensite even at normalizing temperatures. Fast cooling rates tend to accelerate this effect, which by the subsequent tempering operations produce a fine grained microstructure with a very good toughness (see B 4). With nickel additions of about 9% toughness is further improved by the presence of small quantities of austenite that is newly formed during the tempering treatment and remains stable. If these steels are subjected to heat-treatment operations it must be remembered that nickel lowers the Ac_1- and Ac_3-transformation temperatures considerably [14].

Up to very high additions nickel improves the toughness and forms a stable austenitic microstructure particularly in combination with higher chromium and maganese additions. This structure does not exhibit a more or less sharp drop in toughness values in a certain temperature range (transition temperature) which means there is no temperature-induced embrittlement.

In ferritic steels, nickel lowers the transition temperature of the notch impact energy in the normalized condition by about 7 °C for each percent of nickel in the range from 1% to 9%, nickel, and water quenching and tempering lowers it by about 11 °C (Fig. C 10.4). The beneficial effect of nickel on toughness is already used in manganese-alloyed steels. Additions of 0.6% Ni in combination with a favorable ratio between carbon and manganese contents produce a steel suitable for low-temperature service (13 MnNi 6 3, see Table C 10.3). This steel is superior in

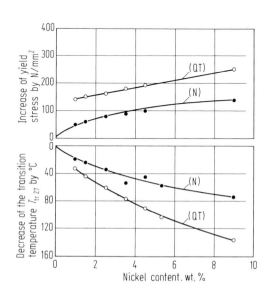

Fig. C 10.4. Influence of nickel on the mechanical properties of a steel with 0.14% C, 0.3% Si, 0.9% Mn, 0.018% P, 0.017% S and 0.02% Al. Plate thickness 30 mm, tested in the normalized (N) and water-quenched and tempered condition (QT) ($T_{tr\,27}$ see C 10.3, here transverse specimen).

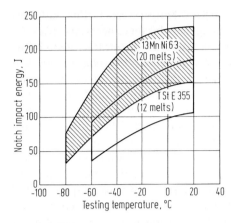

Fig. C 10.5. Comparison of the notch impact energy (ISO-V-notch longitudinal specimens) of steel grade 13 MnNi 6 3 with steel grade TSt E 355 (see Table C 10.3).

toughness to a steel containing manganese additions only (TSt E 355, see Table C 10.3) as can be seen from Fig. C 10.5, where notch impact energy is plotted against temperature. Normalized manganese-nickel-steels (13 MnNi 6 3) have transition temperatures of −80 °C in the notch impact test (ISO-V-notch specimens).

Steels with nickel additions of 3.5%, 5% or 9% have such excellent toughness values at low temperatures that they are suited for service at temperatures between −100 °C and −200 °C (see C 10.4). Below this temperature range, austenitic steel grades must be used.

The influence of carbon on toughness is also evident in steels with high nickel contents (Fig. C 10.6).

A further increase of toughness can be achieved in nickel-alloyed steels by small additions of *molybdenum* which promotes an even finer microstructure.

Measures to Achieve a Fine-grained Microstructure

Of equal technical importance to the above-described indirect method of grain refining by adding alloying elements and heat treatment is the direct generation of

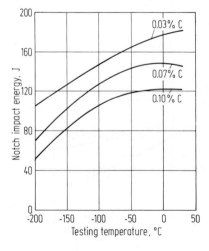

Fig. C 10.6. Influence of the carbon content on the notch impact energy (ISO-V-notch longitudinal specimens) of steels with 9% Ni.

a fine-grained ferrite structure. This grain refinement can be achieved by adding certain alloying elements and by controlling both rolling and heat-treatment temperatures. Suitable alloying elements for grain refinement are nitride- or carbonitride-forming elements such as aluminum or niobium. In combination with nitrogen and/or carbon these elements will initiate precipitations that delay grain growth during annealing cycles in the austenite stage. These precipitates also act as nuclei in the transformation from austenite to ferrite during cooling, thus reducing the ferrite grain size (secondary grain size). The refinement of the ferritic grain size causes the yield stress to increase (see B 1.1.1.3).

Absence of Non-metallic Inclusions (cleanness), Influence of Tramp Elements

In addition to the influence of the intentionally-added alloying elements on the toughness the metallurgical and chemical cleanness also influences the toughnesss. Any lowering of the sulphur content leads to a lower transition temperature (see C 2).

Like sulphur, phosphorus has a detrimental effect on the toughness properties of ferritic steels. The influence of phosphorus on the transition temperature depends on the chemical composition of the steel. Figure C 10.7 shows the influence of phosphorus in combination with varying manganese contents; 0.01% P shifts the transition temperature to an average of about 10 °C.

Heat Treatment

Together with the chemical composition the heat treatment has a distinctive bearing on the microstructure and thus on material properties. Normalizing causes grain refinement and an enhanced degree of uniformity, and both of these results lead in better toughness properties. A further improvement of properties is achieved by fast cooling after annealing in the austenite stage. Fast cooling is

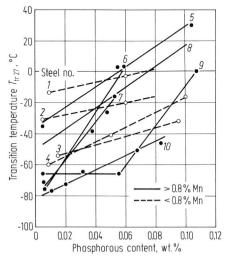

Steel no.	%C	%Si	%Mn	% Other
1	0.20	0.05	0.45	
2	0.18	0.05	0.75	
3	0.18	0.49	0.43	0.023 Al
4	<0.14	0.25	0.45	
5	0.18	0.34	1.24	<0.010 Al
6	0.40	0.20	1.50	0.20 Cr; 0.15 Mo; 0.30 Ni
7	0.08	0.20	0.90	
8	0.30	0.25	1.00	0.96 Cr; 0.49 Mo; 1.0 Ni
9	0.18	0.34	1.24	~0.050 Al
10	0.15	0.30	1.35	0.004 Al

Fig. C 10.7. Influence of the phosphorus content on the transition temperature of the notch impact energy $T_{tr\,27}$ (ISO-U-notch specimen) of different steel grades.

particularly effective if the steel contains sufficient alloying elements to allow the formation of martensite or lower bainite during cooling. If cooling is followed by tempering a very fine microstructure of the quenched and tempered condition will result, with excellent toughness properties.

Rolling

Good toughness properties depend very much on the grain size so it is only natural that rolling procedures are taken to achieve grain refinement and thus a lowering of the transition temperature of the notch impact energy.

Lowering the final rolling temperatures will produce smaller austenitic grains during recrystallization. Subsequently small ferritic grains are generated at transformation during rolling. Plate rolling at low temperatures will produce excellent toughness properties in the as-rolled condition (Fig. C 10.8). Only rolling temperatures well below the upper transformation point can reverse this effect in consequence of strain hardening.

The fine-grained microstructure of the as-rolled material also is a good starting condition for heat treatment operations and improves the toughness of the normalized steels. Both the final rolling temperatures and the degree of deformation in the lower temperature range also have a distinct effect on the toughness.

C 10.3.2 Austenitic Steels

The face-centered cubic austenitic solid solution exhibits good formability at low temperatures, equivalent to excellent toughness properties. This formability is due

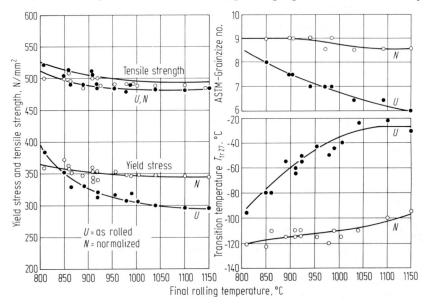

Fig. C 10.8. Influence of the final rolling temperatures on grain size and mechanical properties of steel grade 11 MnNi 5 3. Plate thickness 15 mm ($T_{tr\,27}$ see C 10.3).

to the number of possible crystal slip systems which is higher in the face-centered than in the body-centered cubic lattice.

This effect explains why austenitic steels in contrast to ferritic steels do not show a sharp toughness drop over a certain temperature range. However, the austenite must remain stable during cooling and not transform into martensite i.e. into a body-centered cubic lattice. The stability of austenite can be rated by the temperature at which martensite begins to form (M_s-temperature, M_s-point). The lower the M_s-temperature, the more stable is the austenite. All the more important elements which are added to austenitic steels lower the M_s-point i.e. improve its stability, as can be seen from the following equation which was established by means of dilatometer – tests at very low temperatures on austenitic steels containing 18% Cr and 8% Ni [15]:

$$M_s \,(°C) = 1305 - 1665 \,(\% \text{ C} + \% \text{ N}) - 28 \,(\% \text{ Si})$$
$$- 35.5 \,(\% \text{ Mn}) - 41.5 \,(\% \text{ Cr}) - 61 \,(\% \text{ Ni}) \,.$$

Higher molybdenum additions also tend to stabilize the austenite at very low temperatures. Molybdenum has no detrimental effect on the toughness at low temperatures if the austenitic structure is free of ferrite as a result of an appropriate combination of alloying elements. If the austenitic structure is not completely stable, referred to as metastable, martensite may be formed above the M_s-point by cold deformation. The degree of martensite formation depends on the rate of deformation and the temperature in addition to the chemical analysis. The limiting temperature M_d is the temperature above which no more transformation takes place. It has been suggested that the temperature $M_d 30$ i.e. the temperature at which after a 30% cold deformation 50% of the austenite has transformed into martensite be used [16]:

$$M_d 30 \,(°C) = 413 - 462 \,(\% \text{ C} + \% \text{ N}) - 9.2 \,(\% \text{ Si}) - 8.1 \,(\% \text{ Mn})$$
$$- 13.7 \,(\% \text{ Cr}) - 9.5 \,(\% \text{ Ni}) - 18.5 \,(\% \text{ Mo}) \,.$$

Investigations have been carried out into the undesired transformation of austenite into martensite in low-carbon steels with high alloying additions in various combination of manganese, chromium and nickel. These investigations showed that a twofold reaction was taking place. At a temperature of $-10\,°C$, steel with additions above 15.5% Cr and 9% Ni and with a low manganese content does not transform directly from austenite to martensite but in the sequence austenite–ε-martensite–α-martensite [17]. Thus the transformation proceeds more completely the lower the contents of carbon and nitrogen in the solid solution, the lower the temperature below the respective M_s point, the higher the grade of plastic deformation and the lower the deformation temperature may be. The hexagonal ε-phase is isomorphous in manganese- and in chromium-nickel-steels [18]. Figure C 10.9 presents the mechanical properties and the microstructure characteristics of the steels under investigation at temperatures of $+20\,°C$ and $-196\,°C$ [18]. In metastable austenitic steels microstructure undergoes a transformation during mechanical testing by "shear martensite". The high 0.2%-proof stress values and notch impact energies of steels with high manganese additions are due to the

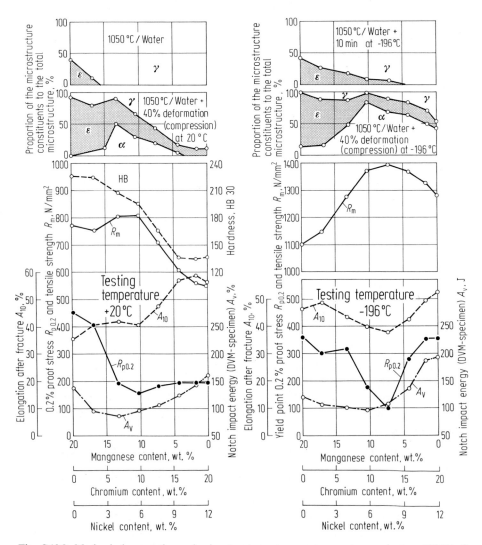

Fig. C 10.9. Mechanical properties and microstructure at room temperature and at $-196\,°C$ of manganese- and chromium-nickel-steels with about 0.03% C, 0.02% Si and 0.01% N after water quenching from 1050 °C. After [18].

considerable resistance of the ε-phase against deformation which partly exists in the structure after quenching. The high notch impact energy of chromium-nickel-steels however is possibly a specific nickel alloying effect in combination with a hardening by the transformation process. Very low 0.2%-proof stress values can be explained as the result of a transformation ductility which occurs during deformation between the M_s- and M_d-temperature (see above) [18, 19]. The highest values of fracture elongation are also recorded in this temperature range.

Increasing nickel additions improve the low-temperature toughness of chromium-nickel-steels. Part of nickel may be replaced by nitrogen because nitrogen improves the low-temperature stability of the austenite (Fig. C 10.10) [20]. These steels also exhibit higher yield stresses at room temperature than the common austenitic steels thus permitting increased strength values to be used for design calculations.

The effect of higher manganese additions to austenitic steels alloyed with nitrogen and containing 18% Cr and 10% Ni is not uniform (Fig. C 10.11) [21]. Increasing manganese additions will retard the formation of martensite if the austenitic phase is not yet completely stable and the notch impact energy will increase (Fig. C 10.11b). Manganese additions to a stable austenitic structure result, however, in a slight decrease of the notch impact energy values (Fig. C 10.11b); this behavior is initiated by small additions of manganese while nitrogen content of the steel is high (Fig. C 10.11c).

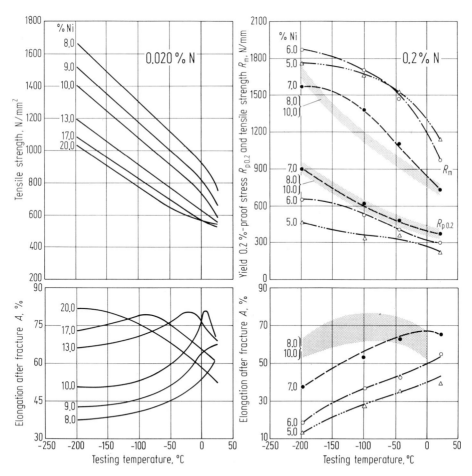

Fig. C 10.10. Influence of the nickel and nitrogen content on mechanical properties at low temperatures of nonstabilized austenitic steels with 0.04% C, 0.40% Si, 1.60% Mn, 18% Cr and X % Ni. After [20].

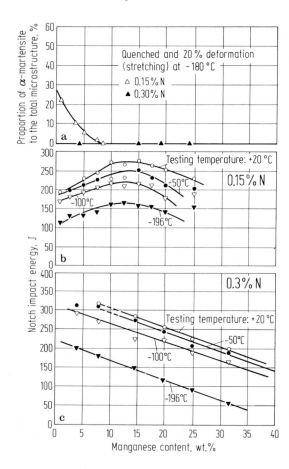

Fig. C 10.11. Influence of the manganese content (**b** and **c**) on the notch impact energy (ISO-V-notch longitudinal specimens) and **a**) on the stress-induced martensite fraction in the microstructure of austenitic chromium-nickel nitrogen-steels with 18% Cr and 10% Ni at low temperatures. Specimens from round bars, with 20 mm dia. After [21].

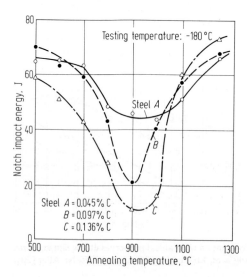

Fig. C 10.12. Influence of the annealing temperature (1 h) on the notch impact energy of austenitic chromium-nickel steels with 18% Cr and 10% Ni and varying carbon content (14 mm square bars). Treatment: 1250 °C/water to simulate conditions in the heat-affected weld zone. V-notch specimens notch angle: 60°, depth 5 mm, notch radius 0.2 mm. After [22].

Mention has already been made about the strong influence of carbon on austenite stability. The precipitation of chromium-carbide $Cr_{23}C_6$ may induce the austenite to become metastable if the nickel content is not sufficiently high to counteract this process. The carbon content of austenitic steels for low temperature service should therefore be limited to values that make the precipitation of chromium-carbide unlikely during slow cooling or in heat-affected weld zones (Fig. C 10.12) [22].

C 10.4 Characteristic Steel Grades Proven in Service

Many different steel grades are used for applications that require good toughness at low temperatures [23]. The more important of these grades will be discussed in the following. Table C 10.3 [10, 11, 24] shows the chemical compositions of these steels.

C 10.4.1 Ferritic Steels

Low-carbon manganese steels subjected to grain refinement and normalizing treatment exhibit excellent toughness properties at low temperatures. Typical of this category is a steel with < 0.18% C and about 1.3% Mn (TSt E 355 in Table C 10.3). In the normalized condition of these steels, transition temperatures of about $-50\,°C$ (notch impact energy of 27 J with ISO-V-notch longitudinal specimens) can be achieved. Thermo-mechanical rolling procedures, improved metallurgical cleanliness and a high ratio of manganese to carbon content permit this limit to be shifted further.

Manganese-nickel-fine-grained steels with nickel additions of 0.5% to 0.8% were developed for the construction of liquid-pertroleum-gas- (LPG)-tankers with service temperatures of $-50\,°C$. Lowering the carbon content also lowers the pearlite fraction and results in an improved toughness and better processing behavior i.e. better welding properties. Minimum values of the yield stress of 285, 315 and 355 N/mm^2 allow safe and weight-saving parts to be designed. Sufficient notch impact energy is achieved at test temperatures down to $-60\,°C$. Therefore a steel belonging to this series with < 0.16% C, about 1.3% Mn and 0.65% Ni (13 MnNi 6 3 in Table C 10.3) can be employed at service temperatures down to $-55\,°C$ fully using the safety factors (according to the rules contained e.g. in AD W10 [5]). For less-severe service stresses even lower temperatures are permitted.

Transformation characteristics of this type of steel are shown in the TTT-diagram for continuous cooling (Fig. C 10.13) where the most important microstructures also are indicated. Also belonging to this series are steels with $\leq 0.13\%$ C, about 0.9% Mn and 0.65% Ni (11 MnNi 5 3) and with $\leq 0.12\%$ C, about 1.2% Mn and 0.65% Ni (12 MnNi 6 3).

Technical and economical importance is attached to a nickel-steel with $\leq 0.12\%$ C and about 3.5% Ni (10 Ni 14 in Table C 10.3) that is used for

C 10 Low Temperature Steels

Table C10.3 Chemical composition of steels for low-temperature service (ladle analysis)

Steel grade	% C max.	% Si	% Mn	% P max.	% S max.	% Cr	% Mo	% Ni	% V max.
TStE 355[a,b]	0.18	0.10 to 0.50	0.90 ... 1.65	0.030	0.025				0.10
11 MnN 5 3[c]	0.14	≤ 0.50	0.70 ... 1.50	0.030	0.025				
13 MnNi 6 3[c]	0.16	≤ 0.50	0.85 ... 1.65	0.030	0.025	—		0.30 ... 0.80[d]	
14 NiMn 6	0.18		0.80 ... 1.50			—		0.30 ... 0.85[d]	
10 Ni 14	0.15	[e]	0.30 ... 0.80			—		1.30 ... 1.70	0.05
12 Ni 19	0.15		0.30 ... 0.80	0.025	0.020	—		3.25 ... 3.75	
X 7 NiMo 6	0.08	≤ 0.35	0.60 ... 1.40			—	0.20 ... 0.35	4.50 ... 5.30	
X 8 Ni 9	0.08		0.30 ... 0.80			—	≤ 0.10	5.0 ... 10.0	
								8.0 ... 10.0	
X 5 CrNi 18 10[f]	0.07						≤ 0.50	9.0 ... 11.5	
X 3 CrNiN 18 10[g]	0.04					17.0 ... 19.0	≤ 0.50	9.0 ... 11.5	
X 6 CrNiNb 18 10[f,h]	0.08	≤ 1.0	≤ 2.0	0.045	0.030		≤ 0.50	9.0 ... 12.0	
X 6 CrNiTi 18 10[f,i]	0.08						≤ 0.50	9.0 ... 12.0	
X 3 CrNiMoN 18 14[g]	0.04					16.5 ... 18.5	2.4 ... 3.0	12.5 ... 15.0	
Ni 36	0.10	≤ 0.50	≤ 0.50	0.030	0.030	—	—	35.0 ... 37.0	

[a] As an example for the series of steels with good toughness at low-temperatures contained in DIN 17 102 [10] and with minimum yield stress values of 255 N/mm² to 500 N/mm².
[b] Further details see DIN 17 102
[c] Max. 0.05 % Nb
[d] In case of low product thicknesses the lower limit may go to 0.15% Ni
[e] Further details see DIN 17 280 [11]
[f] Further details see DIN 17 440 and 17 441 [12]
[g] Additionally 0.10%–0.18% N
[h] Max. 1.0% Nb
[i] Max. 0.8% Ti

manufacture, transport and storage e.g. of liquid carbon-dioxide ($-78.5\,°C$) ethane ($-88.6\,°C$) or, depending on the stress conditions, ethene ($-103.6\,°C$). Provided the steel has a high degree of metallurgical cleanliness and low percentages of phosphorus, sulphur and oxygen, good notch impact energies (≥ 27 J on ISO-V-notch longitudinal specimens) can be obtained at $-120\,°C$. This steel is commonly supplied in the normalized condition. The transformation behavior is illustrated in Fig. C 10.14. For the storage of ethen in ships a steel with $\leq 0.15\%$ C and about 5% Ni (12 Ni 19 in Table C 10.3) is supplied in either the normalized or water-quenched and tempered condition is particularly suitable. The influence of different heat-treatment operations on the level of the notch impact energy-temperature-curves for this steel is clearly indicated in Fig. C 10.15. The transformation characteristics with continuous cooling are shown in the TTT-diagram (Fig. C 10.16). If special and additional steps are taken, the field of application of this steel with 5% Ni can be widened up to the boiling temperature of methane, that means to $-161\,°C$.

The aforementioned steel 12 Ni 19 can be modified by increasing the nickel content to 5.5% and the manganese content to about 1.2% and by adding 0.2% molybdenum. A 3-stage quenching and tempering with two quenching operations (860 °C/water +700 °C/water) and tempering (620 °C/air) produces a microstructure with excellent toughness. At $-160\,°C$, this steel grade X 7 NiMo 7 (see Table C 10.3) achieves a notch impact energy of at least 43 J on ISO-V-notch longitudinal specimens, and 27 J on ISO-V-notch transverse specimens. At

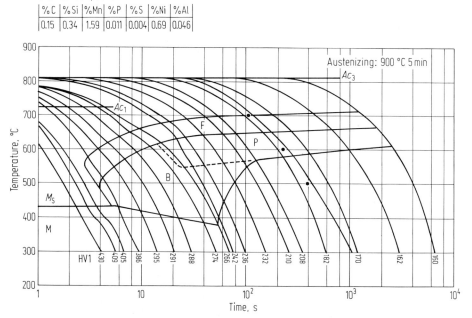

Fig. C 10.13. Time-temperature-transformation (TTT)-diagram, of steel 13 MnNi 6 3 (see Table C 10.3).

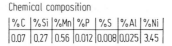

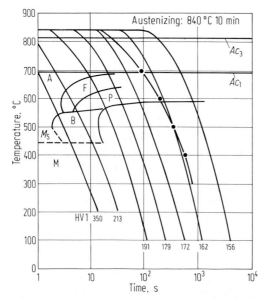

Fig. C 10.14. Time-temperature-transformation (TTT)-diagram of steel 10 Ni 14 (see Table C 10.3).

−170 °C the steels show a value of at least 29 J on ISO-V-notch longitudinal specimens. These values are in the vicinity of the values obtained with a steel containing 9% Ni.

This type of steel with a carbon content of ≤ 0.10% and a nickel content of 9% (X 8 Ni 9 in Table C 10.3) offers the highest toughness of all ferritic steels. This toughness is achieved through air-cooling or water-quenching and tempering. The steel satisfies all requirements set for the transport and storage of natural gas and for the liquefaction of air. Compared with austenitic steels the 9% Ni steel offers a considerably higher tensile strength which allows weight-saving designs. The toughness is excellent down to very low temperatures.

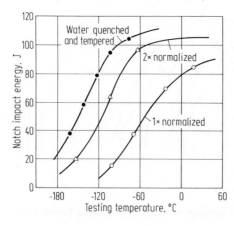

Fig. C 10.15. Influence of the heat treatment on the notch impact energy of steel 12 Ni 19 (see Table C 10.3) (ISO-V-notch transverse specimens).

Chemical composition

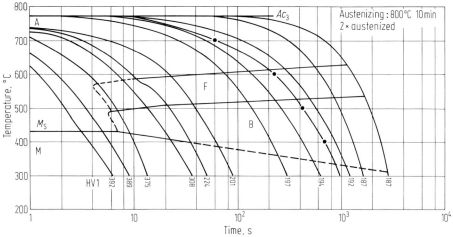

Fig. C 10.16. Time-temperature-transformation (TTT)-diagram, of steel 12 Ni 19 (see Table C 10.3).

Thus this steel type achieves with notch impact energy values at $-196\,°C$ ISO-V-notch longitudinal specimens), which characterize this steel by a minimum value of 39 J. From Fig. C 10.17 it can be seen that this steel exhibits good toughness even below $-200\,°C$. Tests at $-240\,°C$ (ISO-V-notch longitudianl specimens) showed impact energies higher than 40 J.

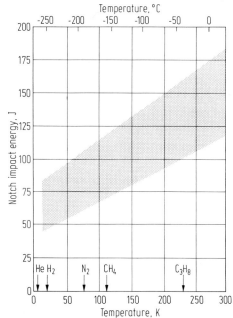

Fig. C 10.17. Scatter band of the notch impact energy values (ISO-V-notch longitudinal specimens) of steel X 8 Ni 9 (see Table C 10.3) as plate of 15 to 20 mm thickness.

The toughness of the steel X 8 Ni 9 is decisively influenced by small amounts of austenite which forms during tempering. The fraction of the stable austenite increases from about 7% to 43% with rising tempering temperatures from 5 00 °C to 625 °C at a holding time of 4 h. Increasing quantities of austenite also increase the toughness. This increase must be taken into consideration when heat treatment procedures are calculated. Transformation characteristics of the steel grade X 8 Ni 9 are shown in Fig. C 10.18. The amount of the nickel content and the heat treatment used also influence the crack-arresting behavior of the steel [23].

C 10.4.2 Austenitic Steels

Austenitic steels are appropriate for service at very low temperatures with high demands on toughness properties [12, 15]. These requirements are corresponded by good toughness, strength properties and processing behavior for the construction of vessels and accessories. Toughness properties remain sufficient even after very long service at low temperatures [26]. A commonly-employed austenitic steel is titanium stabilized with $\leq 0.08\%$ C, about 18% Cr and 10% Ni (X 6 CrNiTi 18 10 as shown in Table C 10.3). Another steel with similar toughness properties contains additions of nitrogen and $\leq 0.04\%$ C, about 18% Cr and 10% Ni (X 3 CrNiN 18 10 as in Table C 10.3). This steel type has high strength properties which allow weight-saving designs. This and another steel with $\leq 0.04\%$ C, about 17.5% Cr, 13% Ni, 2.7% Mo and 0.14% N (X 3 CrNiMoN 18 14 shown in

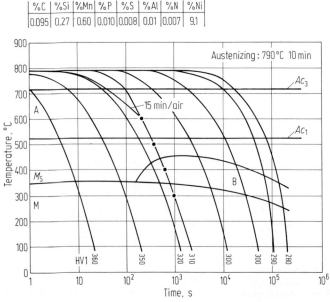

Fig. C 10.18. Time-temperature-transformation (TTT)-diagram of steel X 8 Ni 9 (see Table C 10.3).

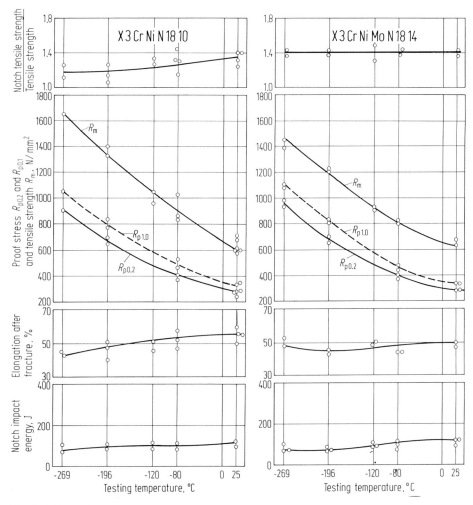

Fig. C 10.19. Mechanical properties of plates and bars in 12 to 80 mm thickness at low temperatures of steel X 3 CrNiN 18 10 and X 3 CrNiMoN 18 14 (see Table C 10.3) (Transverse tensile specimens from plates; longitudinal specimens from 80 mm square bars. Impact specimens: V-notch-transverse specimens, notched tensile specimens: $\alpha_k = 4$). After [27].

Table C 10.3), has been approved after testing at temperatures down to $-271.5\,°C$ Fig. C 10.19 [27].

The impact energy values of both steels are equivalent. However, tensile testing reveals some differences. The metastable microstructure of the molybdenum-free steel tends to higher tensile test values and a lower ratio between notch tensile strength and tensile strength. The use of a molybdenum-alloyed austenitic steel for low temperature service may be necessary if a non-magnetic behavior at low temperatures is required. If fully-austenitic filler electrodes are used such a steel will show no ferrite or hot cracking even in the heat-affected zone, since molybdenum and manganese increase the resistance of fully-austenitic steels to hot cracking.

An austenitic steel with 36% Ni (Ni 36 in Table C 10.3) exhibits a very low coefficient of expansion i.e. $1.5 \pm 0.5 \cdot 10^{-6}\,K^{-1}$ in addition to good toughness down to the boiling temperature of liquid helium. For this reason, the steel is particularly suited to absorb shocks at cryogenic temperatures, for example for vacuum-isolated liquid gas pipe lines employed for high energy fuels in space technology. Cold-rolled strip in steel grade Ni 36 is suited to manufacture of inner panels of the diaphragm tanks for fluid gas tankers by automatic welding of flat shapes. Austenitic chromium-nickel-steels tend to form buckles as a result of thermal stresses which must be equalized by a corrugated shape of the panels [28].

C 10.5 Processing of Low Temperature Steels

Steels for low-temperature service can be processed by the same methods as other ferritic and austenitic contructional steels by hot- and cold-forming processes and thermal cutting and welding. However, possible influences on the microstructure and mechanical properties must be taken into account.

Due to the low carbon contents and depending on the steel grade also the low manganese contents, steels with high nickel contents show little tendency to harden. The applicable cutting speeds decrease with increasing nickel content since the formation of easy-running oxide slags is adversely affected by increasing nickel contents. Austenitic steels for low-temperature service do not require preheating prior to mechanical- or thermal cutting operations. Hardening does not occur.

Welding is of particular importance. All the above-mentioned *ferritic steels* are weldable by all conventional welding methods, which means electric arc-manual welding with rod electrodes (E), submerged arc-welding (UP) and gas shielded arc-welding (MAG or MIG). Apart from observing common technical welding rules it is advisable, when welding steels with high nickel additions, to keep working temperatures below 80 °C to avoid hot cracking, because the load bearing capacity at higher temperatures of the high-alloyed filler metal is limited and hot cracking may occur.

The chemical composition and material characteristics of the *filler metals* must correspond with those of the base metal. Only rod electrodes with basic covering or equivalent wire-powder-combinations should be used. Hydrogen has a good solubility in nickel steels, so care must be taken to limit the hydrogen contents to avoid cold cracking (Fig. C 10.20).

Achieving good toughness also requires the weld metal to be free from oxidic and sulphidic nonmetallic inclusions which means that a good cleanness must be aimed at. The filler metals should correspond with the base metal so far as possible. Filler metals contain between 1.0% and 2.5% of nickel to improve the toughness (Table C 10.4). Welding of steel 10 Ni 14 with 3.5% Ni (Table C 10.3) can be performed with ferritic filler metals having nickel contents of about 2.5% or with austenitic filler metals. Limiting factors governing the use of ferritic filler metals are the service temperature of the component, the kinds of stresses in the weld (predominant static or predominant non-static) and the welding position.

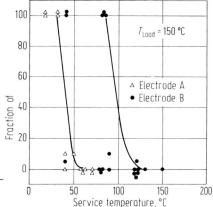

Fig. C 10.20. Cold cracking behavior of pure weld metals in the implant-test. After [29].

Electrode:	Chemical composition					
	% C	% Si	% Mn	% Cr	% Mo	% Ni
A	0.06	0.18	1.31	–	–	1.05
B	0.04	0.15	1.65	0.51	0.39	1.73

Table C 10.4 Filler metals for steels for low-temperature service (steels with good toughness at low temperatures) with about 0.6% Ni

Welding process	Average chemical composition					Type of shielding, powder or gas	Max. dia. of welding rod mm	Welding current
	% C	% Si	% Mn	% Mo	% Ni			
Electric arc-manual welding (E)	0.05	0.3	1.0	–	1.0	Basic	5.0	Direct current
Submerged arc-welding (UP)	0.10	0.15	1.0	–	1.0 2.5	Basic	4.0	Direct current or Alternating current
Metal-active gas welding (MAG)	0.10 0.10	0.5 0.6	1.2 1.7	– 0.4	2.5 1.0	CO_2 Mixed gases	1.6	Direct current

Steels with nickel contents between 5% and 10% and austenitic steels require exclusively austenitic filler metals. Two predominantly-used filler metal alloys are a) an alloy with a high nickel content (about 65% Ni) and b) an alloy with a low nickel content (about 13% Ni). Direct welding current is normally used; for special work, alternating current can be used. Table C 10.5 contains information on the average chemical composition of austenitic filler metals and recommended largest electrode rod diameters. In addition to about 12% Ni the electrodes contain additions of chromium, manganese and tungsten. Weld metals with high nickel content usually also contain molybdenum. Gas-shielded arc welding is performed with gases based on argon which may contain additions of helium, oxygen, carbon dioxide and nitrogen.

Weld metals with a high nickel content have almost the same thermal expansion coefficient as the base metal (X 8 Ni 9 according to Table C 10.3): 9.4×10^{-6} mm/°C. The thermal coefficient of expansion of chromium-nickel alloyed weld metals differs from that of the steel X 8 Ni 9.

Table C 10.5 Austenitic filler metals for steels for low temperature service (steels with good toughness at low temperatures) with nickel contents of 5% to 10%

Welding process	Electrode	Average chemical composition						Type of shielding, powder or gas	Max. dia. of welding rod mm	Welding current
		% Cr	% Mn	% Mo	% Nb	% Ni	% W			
Electric arc-manual welding (E)	A	17	8	–	–	12	3.5	Basic	4.0	Direct current Alternating current
Submerged arc-welding (UP)								Basic	3.0	Direct current Alternating current
Metal-inert gas welding (MIG)								Mixed gases	1.6	Direct current
Electric arc-manual welding (E)	B	15	–	6.0	2.5	68	1	Lime	4.0	Direct current Alternating current
Submerged arc-welding (UP)								Lime	2.0	Direct current
Metal-inert gas-welding (MIG)								Mixed gases	1.2	Direct current

Information on the thermal expansion coefficients of different grades of filler metals is contained in Fig. C 10.21. Cooling from room temperature to service temperature may cause additional longitudinal tension stresses in addition to existing residual stresses resulting from the welding process in the weld material. During cooling to service temperatures the yield stress will rise more than the load stresses. The residual stresses are unlikely to have harmful effects on the component. Investigations carried out on the fatigue behavior of welds at room temperature and at $-165\,°C$ have shown that the test results from austenitic weld metals with high nickel contents and those from weld metals with low nickel contents are in the same scatter band. It follows that components welded with filler metals of the grades E 18 14 Mn 9 W 3 B 20 and UP-X 2 CrNiMnMoN 20 16 can safely be subjected to fatigue stresses.

Attention must be paid to the fact that increasing nickel contents may cause increasing magnetism during welding operations. Magnetic fields in the weld groove may divert the arc, in particular when employing direct current, thus causing welding problems, especially in the roots of joints. This phenomeon can be avoided by using rod electrodes, that can be welded with alternating current and permit a good bridging of the weld groove. When welding with direct current it is possible to neutralize existing magnetic fields by employing counter-magnetic fields (by so called Oerstites).

The influence of the cooling time on the notch impact energy of heat-affected welding zones and the additional effect of stress-relieving heat treatment is illustrated in Figs. C 10.22 and C 10.23 for the steel grades 13 MnNi 6 3 and X 8 Ni 9

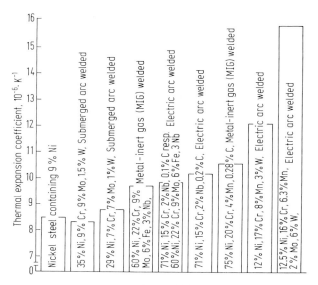

Fig. C 10.21. Thermal expansion coefficients (+ 20 °C down to − 196 °C) of base metal (steel with 9% Ni, see Table C 10.3) and weld metal of different filler-metals.

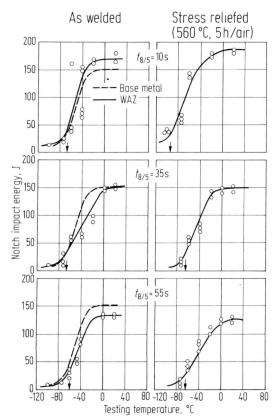

Fig. C 10.22. Influence of cooling time $t_{8/5}$ (see B 5) on the notch impact energy (ISO-V-notch transverse specimens) in the heat-afffected zone (HAZ) of steel 13 MnNi 6 3 (see Table C 10.3). The arrows indicate the transition temperature $T_{tr\,27}$ (see Fig. C 10.3).

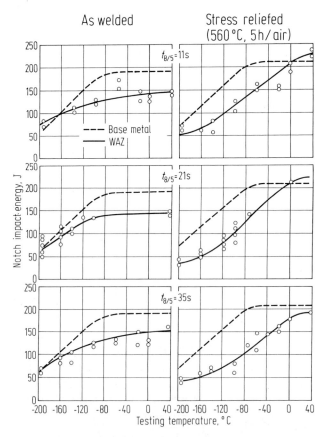

Fig. C 10.23. Influence of cooling time $t_{8/5}$ (see B 5) on the notch impact energy (ISO-V-notch transverse specimens) in the heat-affected zone (HAZ) of steel X 8 Ni 9 (see Table C 10.3).

(Table C 10.3). Shortened cooling-times lead to improved toughness values in the heat affected zone of ferritic manganese steels, even if the steels contain low amounts of nickel. Stress-relieving heat-treatment operations cause a further decrease of the notch impact energy transition temperature. Cooling time however, has only a small influence on the notch impact energy in the heat-affected zone of steel grade X 8 Ni 9. Stress-relieving heat-treatment operations have almost no harmful effect on the toughness of the weld and therefore are not necessary to modify the toughness properties of the weld. When using austenitic filler materials with low nickel contents stress-relieving heat-treatment may even have adverse effects because carbon tends to diffuse from the base metal into the austenitic weld metal, resulting in formation of chromium-carbides that may cause embrittlement. It is common practice to assess the toughness properties of welds by measuring the impact energy of specimens with varying notch positions in the weld transition zone. Common notch positions are the center of the weld metal, at the fusion line and 1 to 5 mm from the fusion line in the heat-affected zone. Testing temperatures are generally about 5 °C to 7 °C lower than the service temperature.

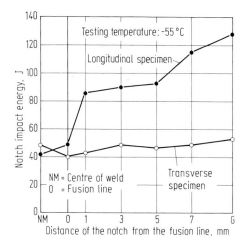

Fig. C 10.24. Influence of the notch position on impact energy (ISO-V-notch specimens) in the heat-affected zone of steel 13 MnNi 6 3 (see Table C 10.3). Plate thickness 40 mm, submerged-arc welding, wire filler material: S 2 Ni 2, powder: 8 B 536, X-weld, flat position, $t_{8/5}$: 14 s.

Figure C 10.24 indicates the notch impact energies in the welded joint of a 40-mm thick plate of the steel grade 13 MnNi 6 3 (see Table C 10.3) taking into consideration the position of the specimens, which were taken in the main rolling direction and transverse to it. The influence of elongated sulphide inclusions is here still clearly evident. Low sulphur contents have a beneficial effect in that steel of a high metallurgical cleanliness exhibits a nearly isotropic behavior. The decrease of notch impact energy in the heat-affected zone is due to a coarse grain structure developed during the welding in the transition zone. The width depends on the cooling time $t_{8/5}$ (see B 5) and decreases with decreasing $t_{8/5}$. Welding of steels with higher nickel contents (for example X 8 Ni 9) the notch impact energy in the heat-affected zone (Fig. C 10.25) of all weld positions is improved.

When designing welded components the properties of both the base metal and the weld metal are of equal importance. This aspect is to be considered in particular in welding the steel grade X 8 Ni 9 since the yield stress of currently-available filler metals is generally lower than that of the base metal. Additions of tungsten or

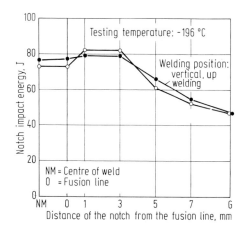

Fig. C 10.25. Influence of the notch-position on impact energy (ISO-V-specimens) in the heat-affected zone of steel X 8 Ni 9 (see Table C 10.3) as plate with 14.5 mm thickness (Manual arc welding with a high nickel electrode, V-type grove flat position or vertical up welding $t_{8/5}$: 5 s).

nitrogen to weld materials have the effect of increasing the yield stress to the level of the base metal. Examples for the properties of the weld metals and those of the welded joints are shown in Table C 10.6. Hardness values measured in the heat-affected zone of nickel steels are generally below 360 HV10. Higher hardness values have no influence in view of the considerable toughness of the developing low-carbon nickel-martensite.

When evaluating bending-test results of specimens of welds from steel X 8 Ni 9 it must be considered that the tensile strength of the weld metal is less than that of the base metal. For this reason, elongation (strain) takes place exclusively in the weld metal, so only small bending angles are obtained prior to the appearance of cracks. The acceptance criteria here is not the bending angle but the elongation (strain) in that part of the specimen that is subjected to tension stresses.

Austenitic chromium-nickel steels can be welded without preheating. The aforementioned general principles are applicable to all austenitic steel grades including weld groove preparation. Care has to be taken to select filler metals of a similar chemical composition to the base metal. Filler metals for shielded-arc welding and submerged-arc welding are usually rod electrodes with

Table C 10.6 Mechanical properties of weld metal and welded joint of steel X 8 Ni 9 (see Table C 10.3) (submerged arc-welding, 11 kJ/cm heat input)

Filler with metal	Mechanical properties of					
	weld metal[a]			welded joint[b]		
	0.2%-proof stress N/mm^2	Tensile strength N/mm^2	Elongation after fracture A %	0.2%-proof stress N/mm^2	Tensile strength N/mm^2	Elongation after fracture A %
17% Cr, 12% Ni, 3.5% W	420	590	40	495	650	16
15% Cr, 68% Ni, 2.5% Nb	390	575	44	410	600	18

[a] Longitudinal specimen from the weld metal of the joint
[b] Specimen transverse to the joint, weld metal in the center of the specimen

Table C 10.7 Mechanical properties at room temperature of the weld metal of consumable electrodes with lime-basic covering for welding austenitic steels for low temperature service (Test procedure see DIN 32 525 [32])

Designation of the welding rod electrode according to DIN 8556 [30]	0.2%-proof stress N/mm^2 min.	1.0%-proof stress N/mm^2 min.	Tensile strength N/mm^2	Elongation after fracture A % min	Notch impact energy (ISO-V-notch specimen)		
					20 °C	−196 °C	−269 °C
19 9	330	350	500 ... 600	35	70	32	−
19 9 L	375	390	490 ... 640	35	95	30	−
19 12 3 L	395	410	540 ... 640	33	95	30	−
20 16 3 MnL	450	470	600 ... 700	35	80	50	40
S-NiCr 16 FeMn[a]							

[a] See DIN 1736 T.1 [31]

lime-basic covering. These filler metals are approved for service temperatures down to −196 °C. Mechanical properties of the corresponding weld metals are shown in Table C 10.7. Austenitic steels alloyed with nitrogen must be welded with the specially-developed filler metals E 20 16 3 Mn 7 N nC B 20 (see also X 2 CrNiMnMoN 20 16 in German standard 8556 [30] which is approved for service temperatures down to −269 °C (liquid helium).

Welded joints in austenitic steels do not generally require any heat treatment since no transformations are to the expected. If stress-relieving operations are required, particularly of heavy cross-sections, it must be remembered that the precipitation of chromium carbides may produce metastable austenite. To avoid this problem it is recommended that the base metal and the filler metal selected have low carbon contents of max. 0.04%, and that the nickel content is so harmonized that it ensures a stable austenitic structure even at cryogenic temperatures.

The δ-ferrite content of austenitic weld metals which is considered to be the basic resistance to hot cracking must be limited for low temperature service components. For this reason, chromium-nickel-steel grades are usually employed as filler materials. An exception however, is the filler material E 20 16 3 Mn7 N nC B 20 (see above) which is alloyed with manganese, molybdenum and nitrogen because molybdenum and manganese increase the resistance of welds to hot cracking. Thus components which require welding of large cross-sections should be fabricated from a molybdenum-alloyed steel having a stable austenitic structure and containing ≤ 0.04% C, about 17.5% Cr, 13.5% Ni, 2.7% Mo and 0.14% N (X 3 CrNiMoN 18 14, see Table C 10.3). To safeguard sufficient toughness in welds and in heat-affected zones the filler material mentioned above should be employed. This material combination is necessary when the service temperatures of liquid hydrogen and liquid helium are to be encountered. In nuclear installations requiring non-magnetic materials it is necessary to keep hysteresis losses at a minimum. The strength values of this filler material correspond with the properties of the base metal even after post-heat treatment.

C 11 Tool Steels

By Siegfried Wilmes, Hans-Josef Becker, Rolf Krumpholz and Walter Verderber

C 11.1 Multiple Stresses in Tools

The *importance of tool steels* far exceeds their commonly-identified use for hand tools. Almost all the objects of daily life are manufactured with the help of tool steels. German tool steel production amounts to about 1% of total steel production in quantity and about 4% of value.

The range of the possible stresses in tool steels is extraordinarily wide. This range is illustrated by the variety of fields of application of the tools, which extend from machining and cutting and forming by stamping, pressing or forging to the forming of shapes from the molten state in glass, plastics or metals and by die casting.

The stresses vary considerably from one type of tool to the other, and require the use of many different types of steel. For this reason, examination of the chemical composition or characteristic alloying elements do not indicate *whether a steel belongs to the tool steel group* as they do for example with stainless steels. The wide variety of applications means that tool steels are used for many other fields. Thus, unalloyed tool steels are also used for pistons in compressed-air machines, antifriction bearing steels are used for cutting tools, hot work tool steels are used as high-tensile steels in aircraft construction, many steels for plastics molds have the same alloy structure as steels for quenching and tempering, and glass casting molds often have the same composition as scale resistant alloys.

Even when the *properties* are considered, there are *none* which are *typical for all tool steels*. Although tool steels often have a high carbon content allowing them to be made hard and wear-resistant, large numbers of tool steels with very low carbon contents and much softer are employed for glass casting molds, steels having medium carbon contents and hardnesses are used for hot-forming tools and plastics molding tools.

A characteristic common to all tool steels cannot be found either in the chemical composition or the properties, so a tool steel can *only* be *identified on the basis of its use for particular tools*. Consequently, according to DIN 17 350 [1] all steels that are suitable for machining or processing materials, and for handling and measuring workpieces are tool steels and, on the basis of historical developments and the resulting agreements, belong to the special steel category [2].

The different fields of application and the different stresses to which tool steels are subjected make it necessary to classify the steels more precisely so that the

---

References for C 11 see page 783.

requirements and properties can be described. In the past, this classification has been made on the basis of such aspects as the alloy content – unalloyed tool steels –, the type of heat treatment – water, oil or air hardening –, according to the stressing – high speed tool steels – and, in particular, according to the application temperature – hot- and cold-work tool steels. Modern production techniques have now led to a great deal of overlapping with respect to stressing, making the fields of application for these traditional classifications indistinct.

The following description of the grades of tool steel, therefore, uses the definition according to tool types, as outlined in DIN 8580 [3], with narrow, clear requirements. This procedure results in a *classification of the tool steels* according to their application:

for *basic forming*, such as in plastics molds, glass casting molds or die casting dies;

for *deforming*, examples of which are forging tools, die sets and stamping tools;

for *cutting*, as used for machining and cutting tools, and

for *composite stressing*, particularly for hand tools.

The chart in Table C11.1 shows the classifications chosen here and provides *qualitative information on the properties required of tool steels in the individual groups of application.*

Attention should first be drawn to a special factor in the assessment of tool steels. In the application of most steel groups, the behavior under operational stress can be predicted on the basis of the material parameters such as, for example, the capacity of a structure to withstand stresses from strength tests, or the corrosion behavior from potential measurements. However, it is usually not possible to predict the efficiency (performance) or to classify or group tool steels according to capability. Up to now, no methods of testing tool steels under the different and combined stresses to which they are subjected in operation have been developed with sufficient accuracy that they can be used to estimate the capability of the tool. Even if practical applications efficiency (performance) are used to assess tool performances, it rarely leads to universally valid results. The reason is, on the one hand, that the influence of the material or of the condition of the material is largely concealed by the influences of the operating conditions. On the other hand, these influences cannot be filtered out by means of statistical examinations because tools are often individual pieces of equipment. In small tool series, the scope of an examination usually is not large enough to produce safe statistical results. Therefore, data on tool life or tool efficiency (performance) are not free from subjective influences.

C 11.2 Properties Required of Tool Steels, their Characterization on the Basis of Test Values, and Measures of Physical Metallurgy to Attain them

For all tools general requirements can be specified that should be fulfilled by the material used. The quality of the material employed should be such that:

C 11 Tool Steels

Table C 11.1 Requirements on the properties of use of tool steels

Properties to be taken into consideration[a]	basic working[b] cold		basic working[b] hot		Application of tools for working cold			Application of tools for working hot				for separating cold-hot		for other purposes
	Molds for plastics	Pressure casting	Glass molds	Stamping punching	Extrusion	Rolling	Forging Hammer	Forging Press	Extrusion[c]	Extrusion[d]	Machining	Cutting	Hand tools
Hardness	●	◐ ●	○ ◐	●	◐ ●	●	◐ ●	◐ ●	◐ ●	◐ ●	● ●	● ○	◐
Hardness at elevated temperatures	○		◐	-	-	-	-	-	-	-	●	-	-
Hardenability	◐	● ● ●	◐ ● ●	○ ○ ●	● ◐ ●	○ ○ ●	● ● ●	● ● ●	● ● ○	● ● ◐	● ● ●	● ◐ ●	◐ ○ ○
Retention of hardness	○		○		○						◐	◐	
Compression strength	◐	● ◐	◐	● ● ●	● ● ●	● ● ●	● ● ●	● ● ●	◐ ● ◐	● ◐ ●	● ◐ ●	● ●	●
Fatigue strength	◐	●	○ ○		○ ●		●	●	○		●		
Toughness			◐										
Toughness at elevated temperatures													
Wear resistance	●	◐ ◐	○ ◐	●	◐ ●	●	○ ●	○ ●	○ ●	○ ●	● ●	● ◐	○
Wear resistance at elevated temperatures	○										●		
Maintenance of the cutting edge												◐	◐
Thermal conductivity	●	● ●	◐ ●				◐ ●	● ●	◐ ●	◐ ●	● ●	● ○	
Resistance to thermal fatigue	○	●									●		
Corrosion resistance	◐	◐ ●	-	-	○ ●								
Dimensional stability	●	● ○	◐ ●	●	○	○	○	○	○	○	◐ ○	●	● ◐
Hot formability	○			◐	○						●		
Cold formability	◐		○ ◐	◐ ●	○ ●	○	●	●	◐	◐	●	●	
Machinability	●	● ○	◐	●	●	●	● ○	● ○	● ○	● ○	◐ ●	● ●	○ ○
Grindability	◐			●	●	●	○	○	○	○	● ◐ ○	● ● ○	
Polishability	●		◐										

[a] —: not required, ○: of little importance, ◐: of medium importance, ●: of great importance
[b] That means a forming proceeding from, e.g. the liquid state (casting). [c] Same direction of force and flow of the material. [d] Opposite direction of force and flow of the material

- the tool will not break under the stresses to which it is subjected,
- the tool will not suffer permanent deformation during operation,
- the tool surface will remain unchanged for as long as possible during use and will not be harmed or destroyed as a result of wear or corrosion.

To fulfil these requirements, steel properties of differing degrees of importance are necessary, depending on the type of tool, as can be seen from Table C 11.1. For each of these properties, the following sections will deal with
1. the reasons for requiring the property,
2. the methods for testing the property, and
3. the measures of physical metallurgy for influencing or producing the properties.

Descriptions of the steel grades that have proven to be particularly reliable in the various applications of tools are shown in C 11.3 according to tool types, on the basis of the diversity of stresses and the resulting differences of requirements in regard to the steel properties.

C 11.2.1 Properties of Importance for Various Applications

C 11.2.1.1 Hardness at Low and High Operating Temperatures

1. As a rule, tool steels are associated with high hardness. However, the examination of the different types of tool shows that the *hardness of a tool* must only be high *in relation to the hardness of the material* to be machined or processed; in fact, the hardness is often very low or of an order of magnitude related to quenched and tempered structural steels. Figure C 11.1 [4] illustrates these relationships. No general rules can be stated in respect to the differences in hardness required between material and tool. Normal hardness values vary between about 200 HV

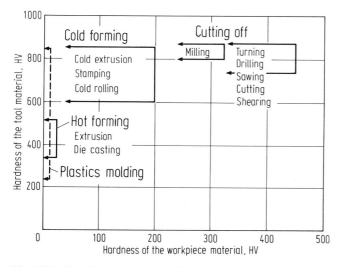

Fig. C 11.1. Tool hardness and material hardness in different processing procedures. After [4].

for glass mold steels at the lower level, and 900 HV for forming and machining tools at the upper level.

Even if the *hardness* value of a tool has no general and assessment significance, it is the most important *characteristic* of a tool from which its potential application can be recognized. The hardness allows inferences to be drawn on the working stress limit and, thus, on the *shape stability* of a tool. Because of the shape stability, the hardness must at least be so high that the yield stress (flow stress) or 0.2%-proof stress, which are to a certain degree connected with the hardness, is above the highest load stress on the tool. Figure C 11.2 shows the relationship between hardness and *flow stress* generally applicable to all tool steels [5]. Thus, the requirement of shape stability could be fulfilled by always making the tools as hard as possible. This method is not practicable in reality because the hardness is not an independent material property and, apart from the yield stress (flow stress) also influences other characteristics. For instance, the *toughness* is usually reduced which means that the susceptibility to fracture becomes greater with increasing hardness. Due to their limited toughness, low-tempered tool steels with high hardness can fracture without deformation below the theoretical yield stress (flow stress). Figure C 11.2 seems to show that, below a tempering temperature characteristic for every steel, the yield (flow) stress decreases with increasing hardness. Independent of this relationship, however, high hardness is basically of advantage for all tools since, in general, the *wear resistance* increases with increasing hardness.

The *hardness of a tool steel* decreases to a varying degree with *increasing temperature*. For tools used at ambient temperature or low temperatures, or cooled during operation, the change in the hardness with increasing temperature has no influence on the service properties. Tools which operate above about 200 °C, must have as high a hardness as possible at elevated temperatures to ensure that the shape stability and the wear resistance remain adequately high when the tools are in use. Figure C 11.3 illustrates how the hardness changes [6, 7] in some tool steels, although to a varying degree, with the service temperature. The strength requirement of the tools also drops with increasing operating temperature because the

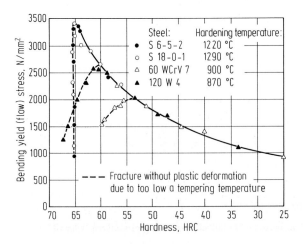

Fig. C 11.2. Relationship between hardness and bending yield (flow) stress of hardened tool steels. After [5]. The broken lines indicate insufficient tempering. (Concerning the chemical composition of the grades of steel examined, attention is drawn to the tables in C 11.3 and in DIN 17 350 [1].)

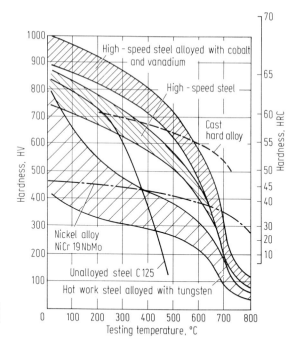

Fig. C11.3. Hardness at elevated temperatures of different types of tool steels. After [6, 7].

materials being processed become softer. Nevertheless, the hardness values of the martensitic steels above about 600 °C operating temperature are no longer sufficient to cope with the stress. At these high temperatures, reliable hardnesses are to be found only in some austenitic steels and in nickel and cobalt alloys which, on the other hand, are not suitable for tools for low operating temperatures due to their low hardness.

2. Conventional *methods* are used *to test the hardness* of tool steels: The Brinell method [8] for low hardness values such as are found in the soft-annealed or quenched and tempered condition, up to a strength of about 1500 N/mm² in the quenched and tempered condition, the Rockwell method [9] for the measurement of higher hardness values, and the Vickers method [10] for a wide range of hardnesses. If no visible traces of hardness measurements must remain, as when measuring the hardness of cold rolls with high mirror polish, hardness testing by the Shore rebound method can be employed [11, 12].

Measurements of hardness at temperatures above ambient temperature are not standardized. The testing of hardness at high temperatures is very complicated because the test temperature must be set and held constant and the surface must be protected from oxidation. In most cases, hardness tests at elevated temperatures are, therefore, carried out only for research purposes [13].

3. From the conversion possibilities spelled out in DIN 50150 [14], it is apparent that hardness values, within certain limits, depend on each other and behave in the same direction. Thus it can be seen that all *measures for influencing the hardness* can be traced back to the possibility of changing the strength.

Hardness values between 180 and 300 HB are appropriate for the *working of tool steels* by machining. If the tools are manufactured by hobbing, the hardness should be below 220 HB and, for larger deformation work, down to 120 HB. Reductions in hardness are achieved by soft annealing which produces a microstructure of a ferritic matrix with interstitial carbides. Figure C 11.4 shows examples of soft-annealed structures of various grades of tool steel. The hardness of the tool material is particularly influenced by the shape, size and distribution of the carbides in the ferrite. The greater the distances between the carbide particles and the larger the diameter of the spheroidized carbides, the lower is the hardness. If particularly soft types of microstructure are to be produced, the hardness of the ferritic matrix must also be taken into consideration. This value depends on the amount of substitutively and interstitially dissolved alloying elements [15, 16]. Of the alloying elements which influence the hardenability, chromium has the least influence on the solid-solution hardness, as can be seen from Fig. C 11.5 [17], so very soft tool steels are alloyed only with chromium. The carbon and tramp element contents are also purposely kept very low. To obtain sufficiently high hardnesses in tools made from very soft materials, such tools must be subjected to thermochemical processes, particularly to carburization.

In principle, all the *mechanisms described in B 1 for improving strength* can be used for tool steels: grain refinement, increasing the dislocation density, solid-solution formation and particle hardening. However, grain refinement and increasing of the dislocation density by cold working play no appreciable role as far as tool steels are concerned. Strength improvement by "simple" solid-solution formation (as opposed to solid-solution strengthening in martensite formation) does not suffice for adjusting very high hardness values. Of greater importance than the substitutional elements is carbon which is soluble in the ferrite in only very small quantities. Carbon is generally present in the form of cementite or special carbides and influences the material hardness by their different shape and distribution.

The *most important possibility for improving hardness* in tool steels is through the mechanisms that are effective during quench hardening through *martensite formation*: a strong solid-solution strengthening of the bcc tetragonally deformed iron lattice supersaturated with carbon, and an increase in the dislocation density during martensitic transformation. The attainable hardness of the iron matrix increases with the content of dissolved carbon; it reaches the highest value at about 0.6% C with about 65 HRC (Fig. C 11.6) [18–21]. Higher carbon contents promote carbide formation; particularly in alloyed tool steels, they lead to increasing residual austenite contents during hardening that reduces the hardness of the hardened structure (Fig. C 11.7).

Particle hardening of tool steels is effected by heterogeneous *precipitation of finely dispersed special carbides* of molybdenum, chromium and vanadium. These special carbides can easily form in the plentiful imperfections of the martensite, precipitate from the supersaturated solution during annealing in the temperature range around 500 °C. This precipitation causes the hardness to increase again after an initial drop with increasing tempering temperature. Details of the retention of hardness are shown in C 11.2.1.3.

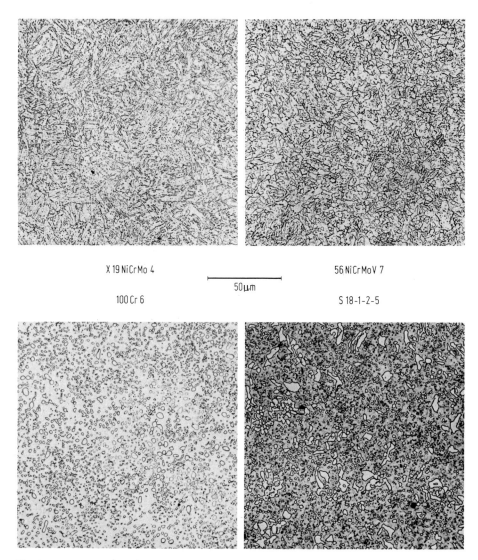

Fig. C 11.4. Microstructure of soft-annealed tools steels. (Concerning the chemical composition of the grades of steel examined, attention is drawn to the tables in C 11.3 and in DIN 17 350 [1].)

The *hardness at elevated temperatures*, which is important for tools operated at higher temperatures, is always lower than the hardness at ambient temperature. The decline in hardness with increasing temperature is directly connected with the improvement of diffusion in the metallic lattice, in particular, with the improved mobility of dislocations. The decline is less in austenitic steel grades because their more densely packed fcc lattice hinders diffusion more than the bcc lattice of the martensitically hardenable steels. The hardness at elevated temperatures is increased by alloying elements that contribute to solid-solution strengthening or

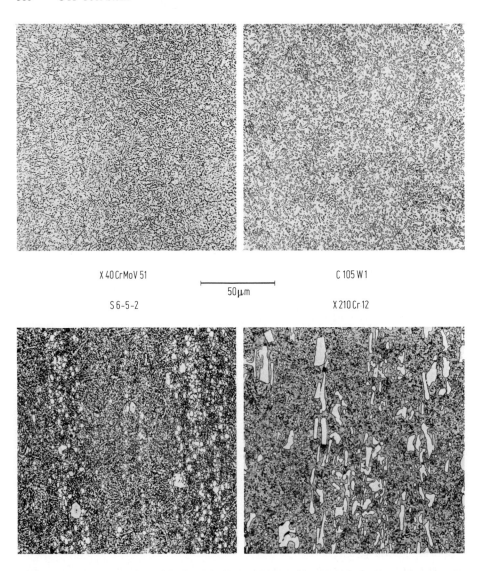

which become effective as a result of particle hardening through the precipitation of special carbides or intermetallic phases. Both influences impede the mobility of dislocations. As long as the operational temperature remains lower than the tempering temperature, the hardness at elevated temperatures does not change in the relatively short period of use and life of the tools. Only if tools are used for long periods at temperatures of 500 °C or higher, does the influence of particle hardening decrease as a result of particle growth [22], known as Ostwald ripening (see A 6.2.6), and the hardness at elevated temperatures decreases.

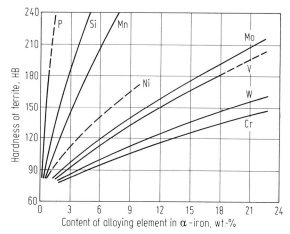

Fig. C 11.5. Influence of alloying elements on the solid-solution strengthening of ferrite. After [17].

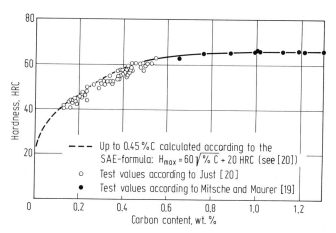

Fig. C 11.6. Influence of the carbon content on the hardness attainable by quenching. Summarized in accordance with the literature [18–21].

C 11.2.1.2 Hardenability

1. The importance of hardness as the most essential property of tool steels is equivalent to that of the hardenability, which includes both the *maximum achievable hardness* ("Aufhärtbarkeit") and the depth of hardening (hardness penetration; ("Einhärtbarkeit") (see DIN 17 014 T.1 [23]). The *requirements for hardenability* vary depending on the stress on the tool, its cross-section and its shape. Good hardenability of the steels is frequently required and can quite possibly be lower with tools having smaller cross-sections than with tools having large cross-sections, if the stressing is the same. Tools with high compression stress always require good hardenability. A good depth of hardening is also required to decrease distortion during hardening. Steels having a good depth of hardening (hardness penetration) do not require abrupt cooling from the austenitizing temperature, which means

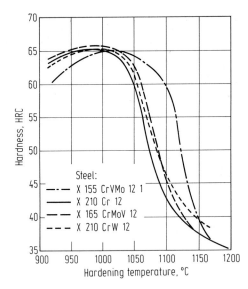

Fig. C 11.7. Decrease of hardness at high hardening temperatures through retained austenite in the martensitic microstructure of tool steels with 12% Cr. (Concerning the chemical composition of the grades of steel examined, attention is drawn to the tables in C 11.3 and in DIN 17 350 [1].)

that the thermal stresses caused by temperature differences between surface and core, that transform themselves into deformations, are low.

It should be mentioned that the hardenability or the hardness penetration indicates only the variation in hardness over the cross-section of a piece, and allows no inferences to be drawn on the microstructure. Accordingly, it cannot be presumed that, with a through-hardening steel, the microstructure is also uniform. Evaluation of the uniformity of properties, therefore, depends not only on the hardenability but also on the microstructure. For instance, with the same hardness, a tempered martensite has better toughness properties than a bainitic or pearlitic microstructure. This relationship also applies to the toughness at elevated temperatures at which hot work tool steels with high bainite contents have a tendency to hot embrittlement, whereas steels of the same hardness that transform completely to martensite have increasing values of reduction of area and of elongation after fracture with increasing temperature, that is to say they are hot ductile (tough) [24].

For *fractureresistance* some tools require a thin surface zone with very high surface hardness but a soft core, so a steel with good maximum hardness at low depth of hardening (hardness penetration) is required. The steels 145 V 33 in Table C 11.6 and C 105 W 1 in Table C 11.11 are examples of these *surface (shell) hardening* steels, which, unlike the through-hardening steels, have compressive stresses on the surface as shown in Fig. C 11.8. These steels are used for tools subject to bending or impact, particularly because of their lack of susceptibility to cracking [25, 26]. Due to their soft core the compressive loadability of these tools is low. For details of hardenability, refer to B 4.

2. *Hardenability testing* is rarely carried out for tool steels. The testing is most commonly used with shell hardening steels, where the depth of hardening (hardness penetration) is determined on the basis of hardened and fractured specimens [27]. The thickness of the fine martensitic shell provides a measure of the depth of

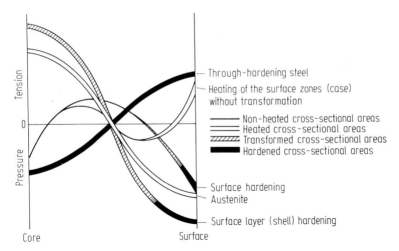

Fig. C 11.8. Residual (internal) stresses at ambient temperature of different types of steel after heat treatment. After [25].

hardening (hardness penetration). Testing of structural steel hardness by end quench tests [28] is not generally suitable for tool steels because the good depth of hardness (hardness penetration) of this steel group does not allow differences to be distinguished with end quench test methods [29]. For this reason, the depth of hardening (hardness penetration) of tool steels is usually assessed on the basis of time-temperature-transformation diagrams, taking the piece size and the possible cooling rate into consideration (see A 9 and B 4). Methods to calculate the hardenability [20] are generally not employed.

3. All alloying elements which increase the austenite stability and so reduce the transformation tendency, are suitable for *increasing hardenability*. In this respect, the improvement of the hardenability is comparable to the hardenability increases in structural steels. However, with tool steels, the choice of alloying elements depends to a great extent on desirable or undesirable side effects such as carbide formation, carbide hardness, decarburization tendency, nitridability, deformability, or influencing of the transformation temperature. For instance, if good hardenability without carbide formation is required, it can be obtained by adding nickel, with the disadvantage that the transformation temperature drops and soft annealability is adversely affected. If silicon is used instead, the steel is more likely to decarburize. When chromium, molybdenum, tungsten and vanadium are used, carbide formation results; at the same time, steels alloyed in this way are easy to nitride.

Carbide quantities in tool steels constitute in hot work tool steels up to 5 vol.-%, in high-speed steels up to 20 vol.-% and in ledeburitic steels with 12% Cr up to 25 vol.-%. The hardenability of these steels can also be improved by increasing the hardening temperature or by prolonging austenitizing, as a result of which further carbide quantities are dissolved in order to reduce the transformation tendency of the austenite. With carbon contents above about 0.7%, the result of increasing the hardening temperature is that the austenite can become so stable that at ambient temperature there are large quantities of residual austenite and the hardness is decreased.

C 11.2.1.3 Retention of Hardness

1. As a result of tempering and the associated martensite decomposition hardness decreases. The lower the decrease in hardness, the more the steel retains its hardness. With a number of steels, the hardness continuously decreases during tempering, as is shown in Fig. C 11.9. With other steels the hardness increases again due to the precipitation of special carbides above about 400 °C (Fig. C 11.10). No *demands are placed on the retention of hardness* in respect of cold work tools such as, for example, cutting tools. If, however, the tool can heat up during use or the tool operates at higher temperatures, as does, for instance, a die-casting die, the retention of hardness must be such that no decrease in hardness occurs. In assessing the retention of hardness, not only the temperature but also the duration of temperature affecting the tool, which is comparable to a tempering effect, must be taken into consideration. The tempered condition determined by the tempering temperature and time should be the same or better than that to be expected as a result of operating temperature and stressing time. With tools which operate above about 500 °C, the tempering effect on the hardness of a long operational time can be avoided if the operating temperature remains 80 to 100 degrees K below the tempering temperature applied over a two-hour tempering time.

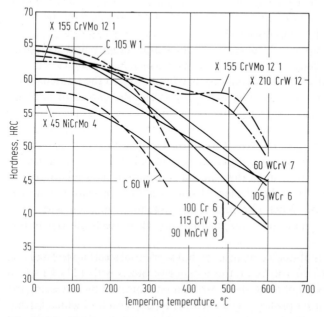

Fig. C 11.9. Change of hardness in current cold work steels (according to DIN 17 350) [1] with the tempering temperature after quenching from medium hardening temperatures. Tempering time 2 h.

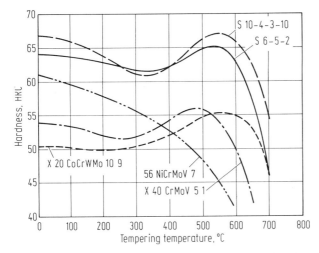

Fig. C 11.10. Tempering curves of current hot work tool steels and high-speed steels. (Concerning the chemical composition of the steel types tested, attention is drawn to the tables in C 11.3 and DIN 17 350 [1].)

The *retention of hardness* is also *important* in some special heat treatment procedures. To give sufficient support to the hard surface layers if the hardness of a tool is increased by *nitriding* or *coating processes* to improve wear resistance of the skin, the retention of hardness of the steel should be so high that the hardness does not decrease, or only as little as possible when the steel is heated to nitriding or coating temperature, which as a rule is 450 to 600 °C.

2. The *retention of hardness is tested* by producing a tempering curve that depicts the dependence of the hardness on the tempering temperature with about 2 h tempering time (see Figs. C 11.9 and C 11:10). Therefore, the influence of time on the hardness in the tempered condition, which can make itself noticeable above about 450 °C through diffusion-controlled changes in type and shape of the precipitated special carbides during the life of the tools, has not been taken into account in the tempering curves of Figs. C 11.9 and C 11.10. With a very long tempering time, the tempering curves can be displaced to a considerable degree. In the range of 450 °C to 750 °C, the reciprocal influence of tempering temperature (T) and tempering time (t) on the hardness can be calculated from the equation $P = T_{(K)} \cdot (20 + \lg t_{(h)})$ [30].

This *relationship between tempering temperature and tempering time* is chiefly taken into account with hot work tool steels, with the aid of the tempering parameter P and by "primary" tempering curves ("Anlaßhauptkurven") [31]. Only for this steel group can operating conditions be such that they are equivalent to a long tempering period. The influence of the tempering time is more noticeable at high temperatures due to the increasing diffusion rate than at low operating temperatures. Where tool operating temperatures are below 450 °C, influences of time on the retention of hardness need not be taken into account during the life of the tool.

3. The *retention of hardness can be influenced* by alloying elements which delay martensite decomposition or lead to precipitation of special carbides. These carbides include molybdenum, tungsten, vanadium, and chromium, which delay the decline in hardness during tempering and, at high contents, lead to a secondary increase in the hardness with a maximum between 500 and 600 °C [32] (see Fig. C 11.11 [17]). The effect of the alloying elements on the retention of hardness increases in the order tungsten, molybdenum, vanadium, as shown by Fig. C 11.12 which uses as an example a steel with 0.3% C and 2.5% Cr [33]. The influence of chromium is considerably less, as can be seen by comparison with Fig. C 11.11. The change of hardness, strength and toughness properties at the secondary hardening

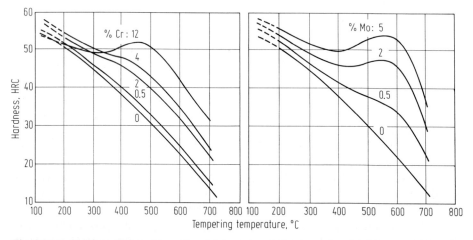

Fig. C 11.11. Influence of chromium and molybdenum on the hardness of steels with about 0.35% C in the tempered condition. After [17].

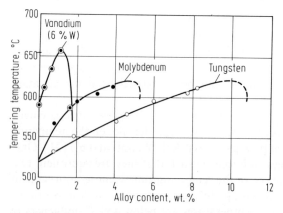

Fig. C 11.12. Influence of molybdenum, vanadium and tungsten on the retention of hardness of steel with 0.3% C and 2.5% Cr (for vanadium: additionally 6% W). After [33]. The tempering temperature was selected for each steel in such a way that the tensile strength was 1500 N/mm^2.

maximum is caused by carbide precipitations and their transformation to other carbide types [34–37]. Using as an example a steel with 0.32% C, 1.3% Mn, 1.0% Cr, 1.7% Mo, 1.5% Ni, 0.7% V, and 2.6% W, Fig. C 11.13 shows that the increase in hardness (tensile strength) initially results from the carbide MC rich in molybdenum and tungsten, precipitated as a coherent phase. At temperatures above the secondary hardening maximum, this carbide is transformed to acicular M_2C-carbide, from which in turn the coagulated, very stable M_6C-carbide develops. Upon development of the M_6C-carbide, the hardness (tensile strength) rapidly decreases. If the tempering temperature is increased even more and reaches the soft annealing temperature range, particularly with high chromium steels, the $M_{23}C_6$-carbide forms, which is linked with very low hardnesses.

Nickel, silicon and, above all, cobalt, also have an influence on the retention of hardness; however, their effect is not due to the formation of special carbides, but to a change in the solubility of carbon in the austenite and a reduction in the possibility of diffusion during martensite decomposition.

With all steels having higher carbide contents, the retention of hardness is improved by increasing the hardening temperature, which results in a higher alloy martensite [38, 39].

C 11.2.1.4 Compression Strength and Pressure Resistance

1. The *compression yield (flow) stress*, i.e. the crushing yield stress or, if applicable, the 0.2%-compression proof stress of a tool steel provides information on the *shape stability* under compression stress; in connection with tools, one often refers to compression strength.

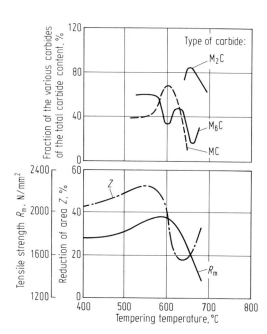

Fig. C 11.13. Influence of the tempering temperature on the carbide content and the mechanical properties of a hot work tool steel with 0.32% C, 1.3% Mn, 1.0% Cr, 1.7% Mo, 1.5% Ni, 0.7% V and 2.6% W. After [34–36].

The *requirements in respect of the compression yield stress* of tool steels vary within wide limits. Very low compression yield strengths are necessary if the tool is to be formed by hobbing (see C 11.2.2.3). Apart from this exception, it is desirable that tool steels have as high a compression loadability as can be tolerated without plastic deformation and without fracture. This requirement of a high compression loadability is placed on all cutting, machining and forming tools. The requirements for compression yield stress are between 1500 N/mm² for ordinary cutting and stamping work and 3500 N/mm² for cold extrusion tools used for difficult forming work.

2. The *loadability* of hardened tools under compression stress can be determined by the *compression test* described in DIN 50106 [40]. However, the technical elastic limit here established is not used to determine the compressive stressability of tools. Usually, the compressive stressability is compared with, and derived from, the yield (flow) stress under bending load which is easily measurable by the static bend test; a simple way to determine this value in steels having low deformability is described in SEP 1320 [41].

A comparison of various measured values for the yield (flow) stress under bending loads for tool steels shows that after tempering at sufficiently high temperatures, all steels manifest the *same relationship between hardness and yield (flow) stress* (Figs. C 11.2, C 11.14 and C 11.15). Independent of the steel grade the yield (flow) stress of tool steels, therefore, at a certain hardness value, is nearly the same [42].

This dependency does not apply to low-tempered hardened conditions; in secondary hardening steels, e.g. high-speed steels, it applies only if tempering has taken place above the hardness maximum. With low tempering temperatures, the toughness of the steels, characterized by the plastic bending energy, can be so low that they fracture before reaching the theoretical yield stress, without visible

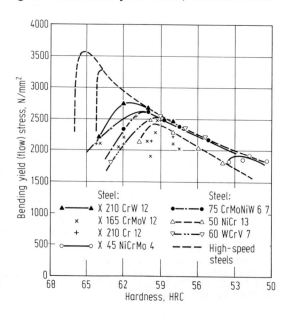

Fig. C 11.14. Relationship between bending yield (flow) stress and hardness in respect of hardened and tempered tool steels. After [5] and own investigations. (Concerning the chemical composition of the grades of steel examined, attention is drawn to the tables in C 11.3 and in DIN 17 350 [1].)

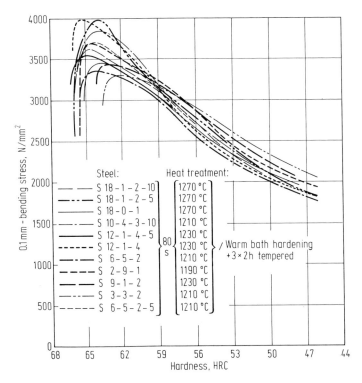

Fig. C 11.15. Dependency of the 0.1 mm-bending stress on the hardness of different high-speed steels. After [42]. (Concerning the chemical composition of the grades of steel examined, attention is drawn to the tables in C 11.3 and in DIN 17 350 [1].)

deformation. The yield (flow) stress under bending load which, for all steels, rises initially with increasing hardness, then appears to drop again above a hardness value characteristic for every type of steel. The scattering of the measured values reflects the relationship between hardness and yield stress, and is chiefly attributable to the fact that the yield stress, as primary deviation from the elastic straight line, cannot be determined accurately enough. In the ledeburitic chromium-alloyed steels additional scattering can be found. Due to their high percentage of very coarse and angular carbides, these steels apparently have a toughness so low that, even under ideal tempering conditions, they can fracture before reaching the yield (flow) stress (Fig. C 11.14).

3. Tools requiring ability to withstand high compressive stress in operations such as stamping, cutting, extrusion and other machining require a high hardness or strength. The steels used, therefore, have *carbon contents* of more than 0.5%. Heavy-duty extrusion tools must have an appreciable carbide content and show a *secondary hardness maximum* during tempering, because only in this way, through particle hardening, can the required high yield (flow) stress values be attained.

Carbon contents and alloying additions that exceed the composition of the high-speed steels or the ledeburitic chromium steels, do not increase the compressive stressability further.

C 11.2.1.5 Fatigue Strength

1. Tools are seldom subject to static stress. As a rule, *the stress* resulting from tension, pressure, bending or torsion, *changes* its degree and direction when impact and shock loads are imposed in rapid sequence. The degree of stress to which a tool is subjected is often higher than its *fatigue strength*; fractures can also be caused by stresses which are far below the yield stress. If tools do not fail as a result of wear or overstressing, the fracture surfaces often have the character of fatigue fractures. A good fatigue strength is particularly required for tools having a high hardness and that are subject to high stress and a higher number of stress cycles. Stamping tools, drop forging dies and rolls are examples of such tools. It must, however, be mentioned that no pronounced fatigue limits are reported for tool steels of very high hardness because the influence of notch stresses, which increase with the hardness, leads to very scattered values.

2. *Testing of the fatigue strength* is carried out in the same way as for structural steel; for details, refer to B 1.2.2.1.

3. The *possibilities for influencing the fatigue strength* for tool steels are similar to those for structural steels.

Since the *notch sensitivity* becomes greater with increasing tensile strength or hardness, non-metallic inclusions are of considerably more importance for tool steels than for structural steels of lower tensile strength. To improve the fatigue strength, tool steels therefore, are often manufactured by electroslag remelting. The beneficial influence of this treatment on the fatigue strength under reversed (symmetrical) bending stresses of the high-speed steel S 6-5-2 can be seen in Fig. C 11.16 [43]. Carbides have a similar effect on the fatigue strength to that of non-metallic inclusions. Spheroidal carbides are more beneficial than angular and elongated carbide shapes.

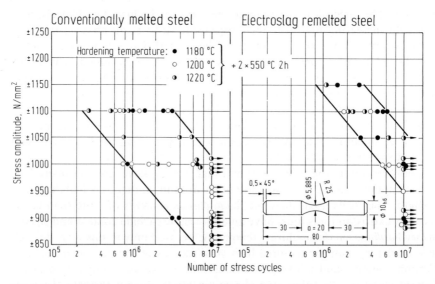

Fig. C 11.16. Fatigue strength under reversed (symmetrical) bending stresses of high-speed steel S 6-5-2 after conventional melting and after electroslag remelting. After [43].

Decarburization must be avoided without fail because it reduces the hardness and, thus, the fatigue strength under reversed stresses in the surface region. Decarburization also reduces the fatigue strength through tensile stresses that are present in decarburized surface zones and superimpose themselves on the cyclic alternating stresses.

The microstructures must be as homogeneous as possible. A pure martensitic microstructure has a higher fatigue strength in the tempered condition than a mixed microstructure of the same hardness.

In considering fatigue behavior it must be remembered that the influences of the material properties are masked by those of the shape of the tool and its surface treatment (finish). For this reason, fatigue fractures in tools are generally easier to prevent by changing the tool shape and by reducing surface roughness than by metallurgical means.

C 11.2.1.6 Toughness at Operational Temperatures

1. With static loads and uniform stress distribution, the service stress of a tool can be close to the yield stress of the tool material. However, these conditions practically never exist. On the contrary, dynamic stresses occur, with differing rates of stressing up to and including impact stress, irregular stress distribution with local stress peaks and conditions of multi-axial stress conditions; even with nominal stresses which are far below the yield strength these stresses can lead to brittle cleavage fractures. With tools subject to these types of stress, the *term "toughness"* is understood as meaning the ability of a tool to release stress peaks by small local plastic deformations and, thus, to prevent crack formation. Toughness is, however, not a specific material characteristic but a generic term for all influences which concern the resistance of a tool to fracture.

With increasing hardness, the yielding (flowing) capability of the steels *decreases* at a superproportional rate. With hardnesses above 55 HRC, this capability is already so low that, under the stresses mentioned, minor cracks can form and lead to fracture. To reduce this danger as much as possible, a compromise must often be made between requirements for shape stability and, thus, high hardness, on the one hand, and toughness or resistance to fracture on the other. In so doing, the efforts to produce a long life of the tool through high hardness and, thus, high wear resistance, often result in very low toughness reserves being accepted. The toughness of very hard tool steels, consequently, must be given particular attention to.

2. Toughness is often difficult to define on the basis of results of specific test methods, so the characterization of toughness is not uniform. Of the numerous methods developed [44], those based on the determination of the plastic formability have proven to be the most reliable comparative methods. Figure C 11.17 schematically illustrates how the toughness of tool steels can be assessed on the basis of its deformation behavior [4]. Brittle behavior is demonstrated by materials whose yielding (flowing) capability is so low that stress peaks cannot be reduced and the steel fractures even at nominal stresses below the yield (flow) strength.

The *toughness* of tool steels that are used with service hardnesses below about 55 HRC is usually *assessed* on the basis of the elongation after fracture and the

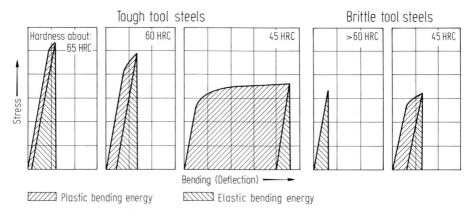

Fig. C 11.17. Stress-bending (deflection) curves of tough and brittle tool steels. After [4].

reduction of area from tensile tests at ambient temperature or at elevated temperatures. However, these data provide only limited information on the material behavior under multi-axial stress or at high rates of stress; they are, therefore, more appropriate for the evaluation of the general material quality, the uniformity of the material or the heat treatment carried out. For the assessment of the toughness behavior of the tools with these hardness values under service conditions, impact energy on notched and un-notched specimens has proven more significant.

The *test methods* just mentioned are unsuitable for testing steels *with hardnesses above 55 HRC*, because the measured values are so low, due to the low formability, that adequate differentiation between the various materials and material conditions is not possible. Even the test methods recommended and used earlier for the definition of the toughness of hard steels, such as the impact bending test with un-notched specimens and the impact torsion test, have not been adopted due to the high scattering of the measured values obtained. On the other hand, methods with which, due to a reduced strain rate, plastic deformation is promoted, have proven reliable. In this connection, particular mention should be made of the static bending test [44, 45] and the static torsion test [42, 46]. With the yield (flow) strength and plastic bending energy or torsional energy determined by these methods, the toughness of hard tool steels can be compared (Fig. C 11.18).

It has been shown that, independent of the steel alloy, the yield (flow) strength values which characterize the maximum stressability, are almost the same in steels of the same hardness, as shown in Figs. C 11.2, C 11.14 and C 11.15. For this reason, when comparing different materials having the same hardness, only the plastic bending energy or torsional energy need be considered in assessing toughness (Fig. C 11.19 [6, 42]). However, even with these values, a general characterization of the material behavior under service conditions is not possible because not enough is known or can be determined about the stress on the tool. This stress, as a rule, does not correspond to that of the static bending and torsion tests. Nevertheless, the bending and torsion tests are very good aids in establishing the material or the heat-treatment condition.

C 11.2 Properties Required of Tool Steels

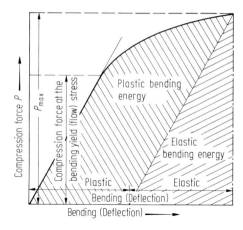

Fig. C 11.18. Force-bending (deflection) diagram for the determination of the plastic deformability according to SEP 1320 [4].

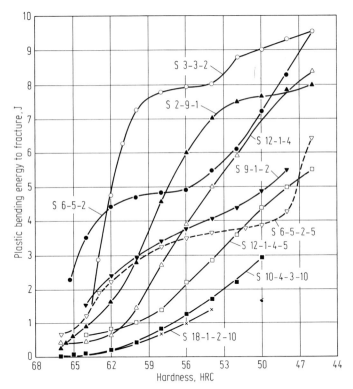

Fig. C 11.19. Comparison of the toughness of different high-speed steels on the basis of the plastic bending energy to fracture. After [42]. (Concerning the chemical composition of the grades of steel examined, attention is drawn to the tables in C 11.3 and in DIN 17 350 [1].)

3. The toughness properties of tool steels are only of interest in respect to the service conditions of the tools, i.e. to the hardened or hardened and tempered condition. Consequently, in addition to the variations from the initial microstructure that influence the toughness, those of the heat treatment must also be

taken into account. A microstructure which is largely homogeneous generally has the best toughness properties. Inhomogeneities in the microstructure, no matter whether they were present in the initial microstructure or were produced by the subsequent heat treatment, reduce toughness.

The *influence of the chemical composition* on the toughness behavior, evaluated on the basis of the plastic fracture bending energy for hard material conditions between 55 and 65 HRC, is shown in Fig. C 11.20 [47]. It can be seen that, on the basis of this toughness assessment, tool steels can be divided into two groups, namely, into carbide-rich ledeburitic steels with low plastic bending energy to fracture and steels having only low carbide quantities in the hardened condition. As far as the ledeburitic steels are concerned, with equivalent hardness, high-speed steels with smaller spheroidized carbides are tougher than the ledeburitic chromium steels whose carbides have a rough, angular shape. The higher the rise in the hardness values, the more similar the toughness properties of all steels become at a very low level.

The *degree of homogeneity of the initial microstructure*, i.e. the extent of the ingot and crystal segregation, and, in a wider sense, the type, quantity and distribution of the inclusions such as oxides, sulphides and carbides, is influenced to a very high degree by the steel production process. There are, however, limitations to the improvement of the homogeneity with conventional processes. A considerable reduction of the ingot segregations is possible by remelting with consumable electrodes. In ledeburitic tool steels, this remelting process results in a more uniform distribution of the carbides throughout the cross-section.

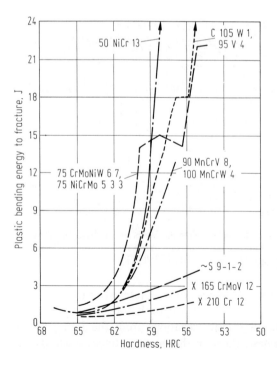

Fig. C 11.20. Comparison of the plastic bending energy to fracture of different tool steels as a function of the hardness in the tempered condition. Specimens of 5 mm dia., 75 mm support distance and single-point loading. After [47]. (Concerning the chemical composition of the grades of steel examined, attention is drawn to the tables in C 11.3 and in DIN 17 350 [1].)

The size of the carbides depends predominantly on the cooling rate during the solidification of the ingot. With a given ingot size, therefore, the carbide size cannot be appreciably influenced [48]. Insofar as ledeburitic types of steel are not concerned, crystal segregation can also be eliminated to a large degree by homogenizing (diffusion annealing) the ingot. The influence of time and temperature on the decrease of crystal segregations during homogenizing is shown in Figs. C 11.21 and C 11.22 [49]. The improvement of the degree of homogeneity by remelting and homogenizing is particularly effective in tools subject to high stresses transverse to the direction of deformation. Due to the crystal segregations a pronounced longitudinal fiber is found in bars after hot forming, which leads to considerable toughness differences in the longitudinal and transverse direction. By reducing segregations, the transverse properties which, particularly in tools from conventionally produced steels, are a critical weak point, are remarkably improved. Using the hot work tool steel X 40 CrMoV 5 1 (see Table C 11.3) as an example, Fig. C 11.23 shows how – statistically seen – elongation after fracture, reduction of area and notch impact energy for the tensile strength range of 1400 to 1600 N/mm² can be improved by remelting processes [43]. Remelting processes alone, however, improve the toughness properties in the transverse direction only in the core zone. A desirable approximation of the toughness in the transverse direction to that in the longitudinal direction is only possible by additional homogenizing (diffusion annealing), which reduces the degree of segregation (see Fig. C 11.24) [50].

An even more far-reaching possibility of reducing inhomogeneities which occur in conventional production, particularly of ledeburitic tool steels, is the *powder-metallurgical process*. In the past, this process acquired a certain importance in the

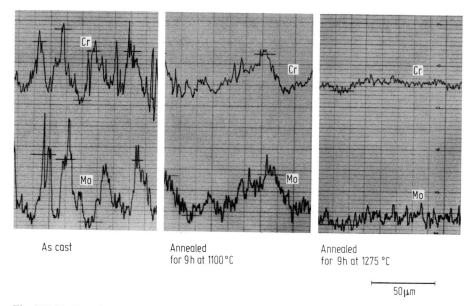

Fig. C 11.21. Crystal segregations (measured with the microprobe) in the core of a 600 mm dia. ingot of X 40 CrMoV 5 1 (see Table C 11.3) remelted by the electroslag refining process and their removal by homogenizing (diffusion annealing). After [49].

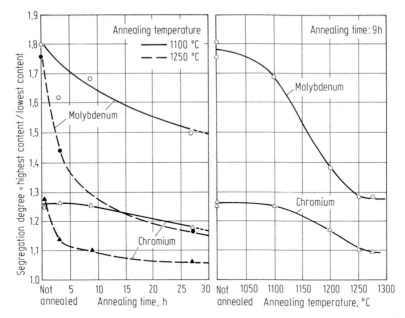

Fig. C 11.22. Influence of temperature and time on the removal of crystal segregations of chromium and molybdenum by homogenizing (diffusion annealing) in the hot work tool steel X 40 CrMoV 5 1 as in Fig. C 11.21. After [49].

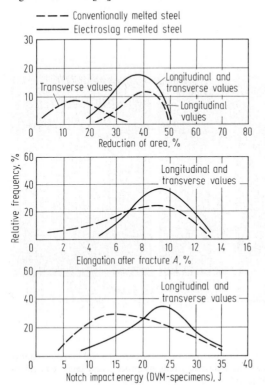

Fig. C 11.23. Toughness measurements on the hot work tool steel X 40 CrMoV 5 1 as in Table C 11.3 produced according to different melting processes. After [43].

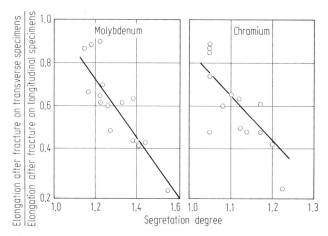

Fig. C 11.24. Influence of the degree of segregation changed by homogenizing (diffusion annealing) on the ratio of elongation after fracture on transverse test specimens to the elongation after fracture on longitudinal test specimens. Results on steel X 40 CrMoV 5 1 as in Table C 11.3, quenched and tempered to a tensile strength of 1600 N/mm². After [50].

production of high-speed steels, but it is also used today for high-carbide, wear-resistant cold work tool steels [51]. Tool steels produced in this way, compared to those conventionally produced, have a microstructure with smaller and completely uniformly distributed carbides. Due to this homogeneous microstructure (Fig. C 11.25), which cannot be further improved, in addition to a uniform deformation behavior during heat treatment and the better grindability mentioned in C 11.2.2.5, also have a better and more uniform toughness in all directions.

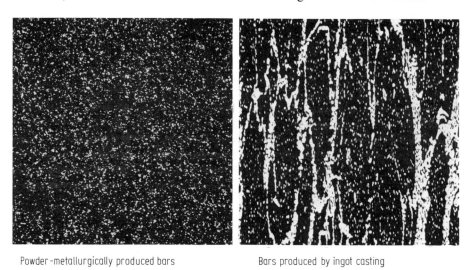

Fig. C 11.25. Microstructures in 100 mm dia. bars of high-speed steel S 6-5-3 (see Table C 11.8) produced by different processes (evaluation point: 25 mm below the surface).

The *toughness properties* can also be unfavorably influenced by the proeutectoidic carbide precipitations [52] occurring during heat treatment, by premartensitic transformation products, particularly by the upper bainite microstructure [24], and tendencies towards embrittlement during tempering [34].

The *carbide precipitations on the grain boundaries* that occur during cooling from the austenitizing temperature, can be suppressed or restricted when cooling from the hardening temperature is carried out as quickly as possible in the proeutectoidic carbide precipitation temperature range. Proeutectoidic carbide precipitations, however, do not always have a detrimental effect on the toughness, as has been shown by an investigation on the influence of the cooling rate on the toughness of high-speed steels [53].

Higher amounts of bainite in the hardened microstructure have a detrimental effect on the toughness of the widely-used secondary hardening hot work tool steels. However, hot work tool steels differ from each other considerably in their transformation behavior in the bainite stage, as can be seen from Fig. C 11.26 [54].

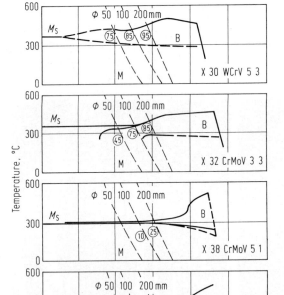

Fig. C 11.26. Transformation behavior of some hot work tool steels in the bainite range during continuous cooling. After [54]. (Concerning the chemical composition of the grades of steel examined, attention is drawn to the tables in C 11.3 and in DIN 17 350 [1].)

In the steels X 30 WCrV 5 3 and X 32 CrMoV 3 3 (see Table C 11.3), even at diameters under 50 mm considerable fractions of bainite are encountered, but with the steels X 38 CrMoV 5 1 and X 40 CrMoV 5 1 a pure martensitic microstructure with better toughness is achieved up to appreciably larger diameters. What is important here is the influence of chromium which postpones the commencement of the transformation phase in the bainite stage to longer times. But the carbon not fixed in the carbide also improves the possibility of producing a pure martensitic microstructure during hardening. For this reason, the steel X 38 CrMoV 5 1 has, with the same dimensions, a higher amount of martensitic structure during hardening than the steel X 40 CrMoV 5 1, where a larger amount of carbon is fixed to vanadium.

In respect of *hot work tool steels*, the toughness properties at ambient temperature are of less interest than those at elevated operational temperature. It should be noted that with some steels – for instance, with X 30 WCrV 5 3 and X 32 CrMoV 3 3 – hot embrittlement occurs at about 600 °C. This effect is not very marked at low tempering stages of about 1300 N/mm^2 but becomes painfully apparent at very high tempering stages of about 1700 N/mm^2 through a very intense drop in the toughness values. Higher hardening temperatures increase hot embrittlement. Hot work tool steels with about 5% Cr, such as X 38 CrMoV 5 1, manifest no hot embrittlement [24] at 600 °C. The 300 °C-embrittlement common to other types of steel, particularly structural steels, and the temper brittleness occurring between 500 °C and 650 °C has little significance in tool steels due to their alloy composition. In secondary-hardening steels attention must be paid to the decline in toughness that occurs in the area of the secondary hardness maximum during tempering. For this reason, high-speed steels are tempered at about 20 °K above the secondary hardness maximum [42, 44]. This decline in toughness is not only the result of the well-known interaction between toughness and hardness but depends to a high degree on the carbide reactions that occur. Thus, as can be seen from Fig. C 11.13, the toughness minimum, e.g. with a hot work tool steel having 0.32% C, 1.3% Mn, 1.0% Cr, 1.7% Mo, 1.5% Ni, 0.7% V and 2.6% W, does not occur in the stage of highest strength but in the range of the MC-carbide dissolution and the increasing formation of M$_2$C-carbides [34].

C 11.2.1.7 Wear Resistance

1. The reason for the limited life of tools is usually *undesirable wear and tear of the working surfaces and cutting edges*. Tool materials, therefore, should have a wear resistance that is suited as much as possible to the stressing to which they are to be subjected.

Tool wear is not simply of a mechanical, physical or chemical nature, but a *combination of various types of wear* (see B 10). In working solid materials by cutting, machining, drawing, punching, and similar processes, tool wear is primarily caused by abrasive (micro-cutting) and adhesive (cold welding) wear. Tools subject to high cyclic compression stress, such as flanging tools, usually fail as a result of wear by disruption (Zerrüttung) [55, 56]. Heat checking which occurs in dies used in pressure die casting and glass processing as a result of the continuous changes in temperature, can be considered as a combination of disruption wear and oxidation wear. According to a more recent study, the formation of heat checking is not attributable to material disruption but to the creep produced by the thermal stress in the surface [57]. With die casting dies local blasting wear and cavitation wear are also encountered.

Heavy abrasive wear can occur in plastics molds for plastics as a result of hard ingredients in the plastics such as glass fibers, and, in working with some types of plastics, the surface can also be attacked by chemicals as a result of decompositions.

The influence of temperature on the wear resistance must also be taken into consideration with hot work tools. As a rule, the wear resistance decreases with increasing temperature. During hot forming, the oxide on the material to be formed greatly increases wear on the tool.

There is a *close connection between the wear resistance* and the *cutting tool life* (see B 11). In forming by machining, e.g. turning, drilling or milling, and cutting off, the cutting tool life is reduced chiefly by wear. Apart from mechanical abrasion, abrasion through pressure welding, through softening of the cutting material, as well as chemical and electrochemical influences can also accelerate wear of the cutting edges of machining tools.

2. As indicated in B 10, *testing of wear resistance* in the laboratory is problematical. Service testing therefore is preferred for evaluating wear resistance. An example of service testing is the determination of the wear resistance of die steels under operating conditions by the so-called pin (stud) test procedure [58]. With this procedure, pins from the steels to be examined are inserted in a carrier tool, so that several die materials can be tested simultaneously. Thus, the influence of the chemical composition, the heat treatment condition, the hardness and the die temperature on wear behaviour can be examined under the same operating conditions.

Testing of wear resistance of machining and cutting tools and, in a wider sense, testing of cutting tool life is performed in the same way as testing of machinability. Either the measured volumetric amount of wear or the linear wear characteristics, such as width of wear band, depth of crater and similar parameters, are used to measure the tool wear (see B 9 and B 11). For cutting tools it is also customary to determine the amount of wear by geometric or volumetric measurements on the cutting edge [59].

3. With reference to the qualifying remarks on the relationship between hardness and wear resistance in B 10, it can be said that the most important *possibilities of improving the wear resistance* of tool steels lie in increasing the matrix hardness and in the inclusion of hard particles in the steel matrix. It should be noted, however, that both an increase in the hardness and inclusions of hard particles reduce the toughness and increase the danger of fracture (see C 11.2.1.6). Depending on the importance of toughness for the tool concerned, there are corresponding limits to increases in the wear resistance.

As a result of *alloying with carbide-forming elements*, such as chromium, molybdenum, tungsten, vanadium, niobium or titanium, other, harder types of carbide form, in addition to the cementite whose highest attainable hardness does not essentially exceed that of the martensite at about 900 HV. The chromium carbide M_7C_3, attains maximum hardness values of about 1500 HV. The molybdenum-tungsten carbide M_6C reaches a hardness of about 1650 HV. The highest hardness values are found with the vanadium, niobium and titanium carbides, which attain more than 2000 HV [13, 60]. The carbides in tool steels usually contain larger amounts of iron, so that lower hardness values than those stated above are often found.

Wear resistance increases with both the hardness and the *quantity of carbides*, as indicated by Fig. C 11.27 [61]. The carbide quantity is in the order of 5 vol.-% in the soft-annealed eutectoidic steels and can amount to more than 20 vol.-% in ledeburitic steels [61, 62]. Steels with substantially higher carbide contents cannot be produced in a conventional way because the formability limit is reached with a carbide content of about 25 vol.-% (see C 11.2.2.2). Apart from this aspect, the susceptibility to fracture becomes so high with carbide contents above 20 vol.-% that these steels can only be used for simple tool shapes which do not require a high degree of toughness.

In conventionally-produced ledeburitic tool steels, the carbides are not uniformly distributed. The carbides of chromium M_7C_3, of tungsten and molybdenum M_6C, and of vanadium MC, arrange themselves during solidification into a net-shaped eutectic whose mesh size increases with slower solidification, i.e. with increasing ingot size. The carbide network is, of course, destroyed by deformation and stretched to long, ever-thinner lines. The irregular distribution of the carbides

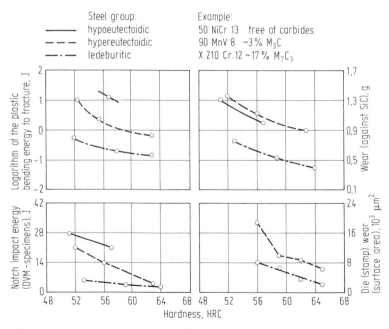

Fig. C 11.27. Toughness and wear of differently tempered cold work steels. After [61]. (Concerning the chemical composition of the grades of steel examined, attention is drawn to the tables in C 11.3 and in DIN 17350 [1].)

cannot, however, be completely eliminated through this mechanism. The disadvantage is that, with increasing carbide spacing, the wear resistance decreases [63, 64]. Only carbides of titanium TiC and niobium NbC, behave more favorably; they precipitate from the melt, unevenly distributed [64, 65].

Apart from the type, quantity and distribution, the *shape of the carbides* also has a great influence on the wear resistance; with the same type and quantity of carbide, the wear resistance decreases with the size of the carbide. This relationship is easy to understand. If the carbide has to be cut through during machining or during the wear process because, due to its size, it adheres to the base material, the wear reducing effect of the carbides is high. On the contrary, when the size of the carbides is exceptionally fine, their effect on the wear resistance is almost completely lost. This influence of the carbide size also explains the different behavior of tool steels produced by powder metallurgical and conventional methods. The grindability of powder-metallurgical steels is very much better, that is, they show a lower wear resistance than conventional steels whose carbides are generally considerably coarser. However, if micromachining is involved, as with abrasive wear, then the wear behavior approximates that of conventionally-produced steels because the carbides are thick in comparison to the chip cross-section in both materials.

The possibility of increasing wear resistance by varying the chemical composition or the heat treatment has its limitations, so efforts are being made to an increasing degree to *improve durability (life) by changes in the surface layer of tools.* Chemical coating processes, such as hard chromium plating and particularly thermochemical treatment processes in which elements such as carbon, nitrogen or boron diffuse into the surface zones of tools, have proven their value in increasing wear resistance [66, 67]. These diffusion processes form hard particles, such as carbides, nitrides or borides, and increase the wear resistance in the same way as hard particles inherent in the steel. The most important and reliable thermochemical processes are carburizing [68] and nitriding [69, 70]. Boronized surface layers are very hard but brittle [71–73] and, for this reason, are used only for simple tool shapes. In recent years increasing use has been made, to some extent with very good results, of coating processes in which wear-resistant layers of titanium carbide and titanium nitride have been applied to the work surfaces of tools. Coating of tools with titanium carbide and titanium nitride or combinations of the two can be done by chemical deposition (CVD method – Chemical *V*apour *D*eposition) or by physical deposition (PVD method – *P*hysical *V*apour *D*eposition) from the gaseous phase [74–81]. Great improvements in the wear resistance are brought about, in particular, by coatings of titanium nitride using the PVD-method. The efficiency of cutting and machining tools is more than quadrupled by this method [82–87]. The majority of surface treatment processes are carried out at temperatures of at least 500 °C, so only steels having a high retention of hardness can be used with these coating methods.

Wear resistance diminishes with increasing tool temperature because the hardness is reduced with increasing temperature. At higher temperatures, as at ambient temperature, the steel with the highest hardness (tensile strength) has the greatest wear resistance (low degree of wear). Figure C 11.28 shows this dependency [58] on the basis of test results for the steel X 32 CrMoV 3 3. Due to this relationship, all

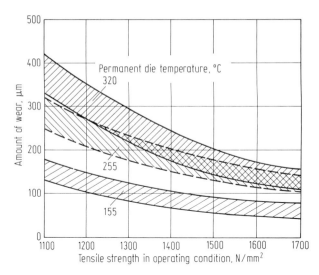

Fig. C 11.28. Relationship between die wear, tensile strength and operational temperature for the steel X 32 CrMoV 3 3 (see Table C 11.3). After [58].

factors which increase the hardness at elevated temperatures and the retention of hardness closely connected therewith, also have a positive influence on the wear resistance at elevated temperatures. The relationship between the effect of alloying elements on the hardness at elevated temperatures and wear resistance can be seen from Fig. C 11.29 [58]. As can be seen from the factors at the abscissa axis, vanadium has the greatest influence on the wear resistance at higher temperatures.

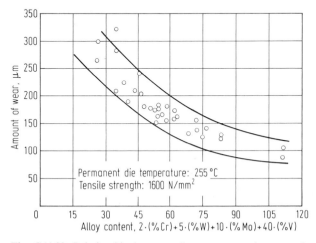

Fig. C 11.29. Relationship between alloy content and wear resistance of different hot work tool steels. After [58]. Evaluated range of composition: 0.30 to 0.44% C, 0.02 to 0.38% Si, 0.31 to 0.83% Mn, 1.18 to 3.03% Cr, 0.08 to 3.04% Mo, 0.26 to 1.92% V, and 1.42 to 5.85% W.

Tungsten and molybdenum increase the wear resistance in approximately the same way. However, due to its higher atomic weight, double the quantity of tungsten is required to produce the same increase in wear resistance.

C 11.2.1.8 Thermal Conductivity

1. As a matter of principle, *high thermal (heat) conductivity* is desirable both in the production of tools and in respect of their service behavior. With good thermal conductivity, locally introduced amounts of heat are quickly carried off so that large temperature and stress fields cannot build up. Good thermal conductivity reduces susceptibility to grinding cracks. Small changes in shape are to be expected during heat treatment. Tool steels with good thermal conductivity can be cooled with water during use with less danger of cracking. Good thermal conductivity can also be important to the functioning of a tool; for example, the time cycle of processing machines for plastics can be reduced if the temperature of the plastics is controlled by internal heating or cooling of the tools.

2. References were made in B 2 to the *bases of thermal (heat) conductivity* and its testing can be found. Therefore, only the vital effect of the chemical composition on the thermal conductivity will be mentioned here. On the basis of a statistical evaluation of many individual measurements, thermal conductivity can be determined to an adequate degree by the content of the alloy elements in at.-%. Figure C 11.30 [88] shows the relationships found.

3. When selecting a steel with respect to good heat conductivity, only the content of the alloy elements (at.-%) calculated from the chemical composition need be considered. For a given type of tool, the possibilities of choosing between various steel grades are very limited because the alloy content for adjusting other necessary tool properties such as retention of hardness, hardness at elevated temperatures, and wear resistance has a greater importance.

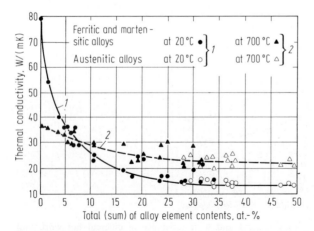

Fig. C 11.30. Influence of alloy content and microstructure on the thermal conductivity of steel at 20 °C and 700 °C. After [88].

C 11.2.1.9 Resistance to Thermal Fatigue

1. The surfaces of some tools are exposed to *periodic temperature changes*. This is significant for forging tools, dies for pressure casting, glass forming and plastics molds. The extreme temperature changes in dies for pressure casting and in tools for forging and glass forming and even the high temperature gradient during the different cycles of heating and cooling causes heat checking (see Fig. C 11.31). These cracks are of great importance in the failure of pressure casting dies. Different investigations explain the formation of heat checking in the following way [89–95]:

In contact with the molten metal the surface of the mould is suddenly heated (thermal shock) and therefore tends to expand. However, the colder region beneath the hot surface prevents the surface area from free expansion. Temperature stresses are thus built up that lie above the yield (flow) stress and cause elastic or even plastic compressive deformation of the surface. During the following cooling of the surface, tensile stresses are generated in the surface region, due to elastic and plastic deformations of the surface area compressed previously.

As with the generation of (mechanical) fatigue fractures, changes in the stress direction are the reason for cracking, beginning at points with maximum local stress or at failures of microscopic size. Because of the planar state of stress in such regions a planar crack field develops.

Following this conception, it is apparent that cracking will be delayed by the use of tool steels that are characterized by a *high yield (flow) stress* at elevated temperatures [95] as well as by a high plastic deformability, that means a high toughness at elevated temperatures. Recent investigations explain that the failure of material by heat checking is caused by the creep behavior of the tool steel [57]. Therefore a high thermal fatigue resistance is related to the resistance to the development of slight failures during the creep test.

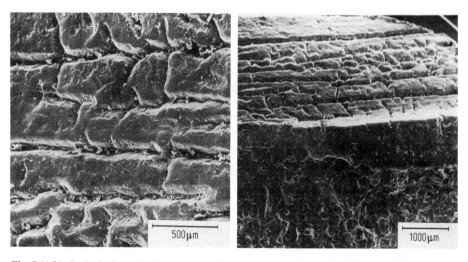

Fig. C 11.31. Reticular heat checking on the surface of a pressure casting die (left) and laid bare fracture surface of a heat check (right).

2. Numerous *investigations* into the problem of the reproducible development of heat checking and of the possibility of comparing experimental results with those of service conditions have been made under definite conditions [89–94]. One method to define the tendency of heat check formation is to measure the number of temperature cycles up to the time when the first cracks can be observed with the eye. In conducting these investigations attempts are made to simulate the distribution of temperature (the temperature field) and its periodic changing as exactly as possible, which requires a considerable expenditure on apparatus [90, 91, 94]. The results show that the tendency to heat check formation is influenced not only by the mechanical properties of the investigated material but also by the resistance to oxidation and the accompanying formation of scale under atmospheric conditions [91, 92, 94]. Therefore, the results of these investigations cannot be transferred directly to the conditions of pressure casting. Likewise, individual parameters do not have a uniform effect on crack formation which is delayed e.g. by a higher yield (flow) stress [95]. This statement was confirmed for the steel X 38 CrMoV 5 1 [96], whereas for the steel X 30 WCrV 5 3 an inverse result was observed [94]. These contradictions can probably be explained by the phenomenon of hot embrittlement that characterizes the hot work tool steels with 2.5% Cr and higher tungsten contents and that rises with increasing strength [97] (see C 11.2.1.6, too).

During recent years, tests have been made to try to establish a correlation between heat check resistance and the results of investigations of the microstructure and of the technological properties such as the behavior in impact tests. As an indirect result, these investigations give the degree of homogeneity, as measured by a homogeneous microstructure or by homogeneous and even high values of the energy absorbed in fracturing. Resistance to heat checking can be deduced from the degree of homogeneity determined by these tests.

3. As the propagation of heat checking is connected with oxidation processes, it is important to use *chromium-alloyed steels* since chromium improves the scaling resistance. At higher tool temperatures, as for glass forming molds, the chromium content should be higher.

Because of the relationships mentioned above the tensile strength should be as high as possible, but it should not exceed the tensile strength that results from the tempering effect under operating conditions of the tool. Therefore tools for pressure casting of aluminum should have a tensile strength of 1500 N/mm^2 to 1700 N/mm^2, for pressure casting of heavy metals 1300 N/mm^2 to 1500 N/mm^2, and for glass forming 700 N/mm^2 to 900 N/mm^2. It may also be concluded that the resistance to heat checking could be improved by using steels with high thermal conductivity. But calculations have shown [95] that this property does not have the expected importance because the range of surface temperatures of the possible hot-work steels shows only minor differences.

Recent investigations have been concerned with the influence of the degree of segregations and with the homogeneity of the microstructure. The results show that it is advantageous to use steels with a homogeneous, quenched and tempered structure [98]. Ingot segregations must be reduced by remelting and microsegregations must be diminished by homogenizing (diffusion annealing) (see Fig. C 11.32). Such steels contain fewer areas with unfavorable properties, where

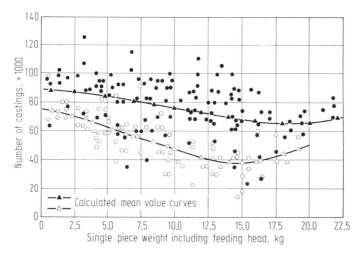

Fig. C 11.32. Life time of dies for cold chamber pressure casting of aluminum. After [98]. Influence of different steel melting processes (steels X 38 CrMoV 5 1 and X 40 CrMoV 5 1, see Table C 11.3). Life time of dies measured as number of castings up to the first main repair. (●: steel remelted by the electroslag process, ○: steel conventionally melted in the electric furnace.)

cracking may start. Consequently, the first visible heat checks originate significantly later in these materials. An improvement of the homogeneity and therewith the resistance to heat checking can also be expected if the carbon content of hotwork steels with about 5% Cr is reduced (see Table C 11.3). Investigations into the impact energy absorbed in fracturing these steels have shown that this energy is increased extensively in the transverse directon if the carbon content is below the eutectoid point at about 0.3% [100, 101].

C 11.2.1.10 Corrosion Resistance

1. *Special corrosion resistance* is seldom required during the application of tools. Among the few exceptions are tools for *plastics* molding. If faults in processing occur, for instance, if the temperature of the mold is chosen too high, the curing plastics release aggressive acids. As an example, during the processing of polyvinylchloride hydrochloric acid is released and that needs a corrosion-resistant tool steel. Processing of synthetic textile fibers also requires special corrosion resistance if corrosive agents are used or if the tools are cleaned in an aggressive salt bath. Tools for glass forming operate at high temperatures that need good scale resistant steels.

Corrosion resistant steels are necessary also in food manufacturing. For example, tools for sealing cans must be resistant to acids in fruits. Tools in pill presses should be corrosion resistant for the same reason.

During pressure casting and low pressure casting tools may be corroded by the molten material, especially those parts that are in direct contact with the liquid metal. Tool steels should be resistant to liquid zinc, magnesium and aluminum alloys which are known as aggressive.

2. With these possibilities of corrosion the conditions of surface attack are not so uniform and clear as they are with chemical process technology. Hence there are *no universal testing methods for evaluating the corrosion resistance* of tool steels that may be helpful in selecting certain steels. Estimating the behavior under corrosive service conditions by measurements of the current density and potential and by determination of the equilibrium rest potential are not useful with tool steels because of the non-predictable conditions. Thus, corrosion tests are only useful under operational conditions.

3. The corrosion resistance of cold-working tools for food processing is satisfactorily high, if the steels have a *content of free chromium* of about 12%. If the steels are to be used at a hardness of about 60 HRC – that means a carbon content of 0.6% to 1.0% – they must be alloyed with about 17% Cr. To improve the corrosion resistance, 0.5% to 1.0% Mo is added. These steels have the best corrosion resistance for use as cold-working tools if they are hardened or tempered only at low temperatures (Fig. C 13.3 [102] which applies also to the tool steel X 42 Cr 13). At tempering temperatures above 400 °C chromium, which maintains the corrosion resistance, is removed from the steel matrix by formation of carbides.

Tools for the *processing of aggressive plastics* do not need such high hardnesses. These steels are hardened and tempered to a tensile strength of only about 1000 N/mm^2 to improve the workability. Additionally, care should be taken to be sure that not too much chromium is fixed by carbon by forming carbides in the high temper condition. These steels, therefore, are alloyed with chromium contents of 12% to 17% and molybdenum additions of up to 1% and usually have carbon contents of less than 0.35%.

The good scaling resistance necessary for *glass forming tools* is attained by the use of ferritic steels with 17% Cr or by austenitic steels with 17% to 25% Cr and – to improve the scaling resistance – with about 1% Si. Nozzles for spinning tools are made from steels with similar compositions.

The influence of *corrosion by metals* has to be considered for each type of alloy individually. It depends on the mutual solubility, on the possible formation of compounds and on the temperature, which modifies such reactions. Corrosion is also influenced by the flow rate of the molten metal, which may wash out boundary layers [103]. Corrosion by metals is not influenced by the strength or the mechanical properties of the steel [104].

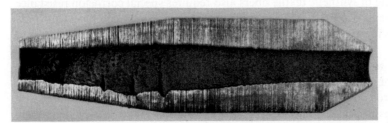

Fig. C 11.33. Corrosion by zinc in the pouring lip of a hot-chamber die-casting machine after 30 000 cycles. Tool material: X 32 CrMoV 3 3 (see Table C 11.3) (1/3 of the original dimensions). After [106].

Zinc alloys have only a small solubility for iron at pressure casting temperatures between 400 °C and 450 °C; effects during service conditions are negligible and there is no formation of iron-zinc compounds on the tool surfaces [105]. Materials are applied that reduce the contact between the tool and the liquid zinc, and the contact time during one casting cycle is very short. Corrosion effects have been observed on pouring lips, which are in permanent contact with the molten zinc alloy. In principle, iron and zinc form a protective compound layer, growing rather slowly only by the diffusion of zinc and iron atoms across this layer [105]. The protective efficiency may be reduced by the erosion of the layer, due to the flowing zinc and an accelerated dissolution of the tool may then result. Figure C 11.33 shows corrosion phenomena on a pouring lip of steel X 32 CrMoV 3 3, (see Table C 11.3) after 30 000 cycles, whereas the maraging steel X 3 NiCoMo 18 8 5 shows no visible corrosion after $1.2 \cdot 10^6$ cycles.

Magnesium can dissolve iron to a greater extent than zinc [107, 108]; but this effect will become a problem only at temperatures above 800 °C, that is at temperatures above the usual magnesium casting temperature of 600 °C to 660 °C. Steels that have low nickel levels are resistant to magnesium alloys. As shown by immersion tests in magnesium alloys at 700 °C, an alloy with 70% Ni has a 30 times higher dissolution rate, compared with the hot work steel X 38 CrMoV 5 1, containing 5% Cr (see Table C 11.3). The resistance of steels to liquid magnesium is improved by raising the chromium content. Additions of molybdenum, tungsten and vanadium also are favorable because of their low solubility in liquid magnesium [107].

Aluminum and iron form intermetallic Fe-Al-phases at temperatures above 600 °C [109]. To prevent corrosion by aluminum at pressure casting temperatures of 650 °C to 730 °C, aluminum alloys are manufactured only by cold chamber pressure casting, where the tools are not in permanent contact with the molten aluminum. However, corrosion phenomena by aluminum are observed during die casting at low pressures, where molten aluminum is poured into the tool mold and the contact with the liquid aluminum lasts longer. Under these conditions service tests with liquid aluminum containing 12% Si at 720 °C gave the following results: corrosion resistant steels, such as X 20 Cr 13, precipitation hardening nickel alloys, as e.g. X 3 NiCoMo 18 9 5, and austenitic steels, such as e.g. X 5 CrNi 18 9 show a far less favorable behavior than conventional hot-work steels, as e.g. X 38 CrMoV 5 1. The best corrosion resistance was observed with low alloyed steels, such as 25 CrMo 4. The retention of hardness and elevated temperature strength of these steels, however, are not sufficient to qualify it as a tool material. A satisfactory selection of materials for this application has not yet been found.

C 11.2.2 Steel Properties of Importance in the Manufacture of Tools

C 11.2.2.1 Dimensional Accuracy During Heat Treatment

1. It is difficult and expensive to machine tools in the hardened condition. Efforts, therefore, are made to produce the final tool shape by machining the steel in the

soft-annealed condition. Subsequent heat treatment processes change the dimensions and shape of a tool in a manner which cannot be accurately predicted. As a result, further machining of the hardened tool often becomes necessary, so the general requirement of *good dimensional accuracy* during heat treatment is understandable. Inadequate dimensional accuracy is also called distortion. According to DIN 17014 [23] and EURONORM 52-83 [110], distortion is the sum of all changes in dimension and shape as opposed to the initial state that appear after heat treatment.

To avoid dimensional and shape changes resulting from heat treatment, this work is often carried out prior to manufacturing of the tool. Low-stressed molds for plastics with hardness values of about 300 HB are normally machined in the hardened and tempered condition. With high hardness values spark erosion methods may be used to form the tool [111, 112].

2. No standard methods are used to *testing dimensional accuracy*, which is confined to the measurement of changes in dimension and shape of the tool after heat treatment.

3. Dimensional and shape changes during heat treatment cannot be avoided. *Three essential influencing variables in respect of dimensional accuracy* have become apparent from the various studies [25, 26, 50, 113–119]. These are: the differing specific volumes of the phases existing prior to and after heat treatment; the thermal stresses resulting from the temperature differences between surface and core which occur with each cooling process; and the transformation stresses – likewise caused by the temperature differences – which result from the time difference in structural transformation between surface and core. It should be noted that strong segregations can also cause transformation stresses. Volumetric changes during heat treatment lead to increases or decreases in all dimensions. Thermal and transformation stresses can influence the shape of the tool by changing the angular relationships.

Figure C 11.34 [119] shows, for mild steels with carbon contents up to 2%, the *order of magnitude of dimensional changes* as a result of structural influences. To reduce dimensional changes, efforts are made to balance out the positive volumetric changes by negative volumetric changes in other phases. This objective is achieved to a satisfactory degree with ledeburitic steels with about 12% Cr (X 210 Cr(W) 12 and X 155 CrVMo 12 1 in Table C 11.11). As can be seen from Fig. C 11.35 [114], selecting the appropriate hardening temperature allows the volume fractions of the structure constituents to be influenced in such a way that the volume increase through martensite formation compared to the original ferritic structure, is equivalent to the volume decrease attributable to the residual austenite. If dimensional changes are not to be determined empirically, but to be calculated, the volume fractions of ferrite, martensite and retained austenite as well as changes in the carbide quantities and their specific volumes must be taken into account.

Changes in shape occur when *thermal stresses are reduced through plastic deformations*. With slower cooling, the temperature differences and the related volumetric differences that develop over the cross-section, become less, so that large stresses cannot develop between the surface and core to influence the form.

C 11.2 Properties Required of Tool Steels   341

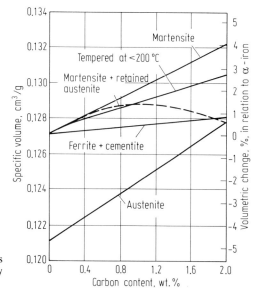

Fig. C 11.34. Volumetric change of mild steels due to the change of the microstructure by hardening. After [119].

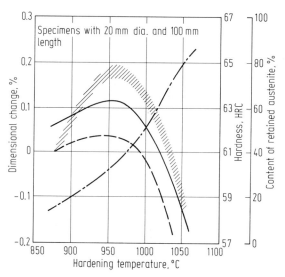

Fig. C 11.35. Influence of the hardening temperature on the dimensional changes of the steel X 210 Cr 12 (see Table C 11.11). After [114].

— Dimensional change in longitudinal direction
– – Dimensional change in transverse direction
////// Hardness, HRC
–·– Content of retained austenite

With steels having alloy contents that improve the depth of hardening (hardness penetration), and can be more slowly cooled from hardening temperature, the danger of changes in shape is less than with steels that due to their low alloy content need to be hardened in water. The tool steels alloyed with more than 2% Cr and with molybdenum show a transformation-free temperature range between

400 °C and 600 °C in the TTT diagram. The thermal stresses which develop in these steels during hardening can be very effectively reduced by cooling in steps with temperature equalization at 500 °C. The soaking time at about 500 °C has no influence on the structure or the properties of the hardened tool [120].

To reduce the *influence of transformation stresses* on shape changes, it is necessary to make the martensitic transformation take place over the entire cross-section as simultaneously as possible. This effect can be achieved by interrupting cooling of the tool during hardening at a few degrees above the martensite reaction temperature to equalize temperature between surface and core. However, this method requires a steel with good depth of hardening.

The influences on the dimensional accuracy of *irregular structural transformations in segregation zones* cannot be alleviated or eliminated during heat treatment but only by reducing the degree of segregation. In low- and medium-alloyed steels, ingot segregations can be reduced through production processes, and crystal segregations can be eliminated to a far-reaching degree by homogenization (*diffusion annealing*) [121, 122]. In ledeburitic steels in which the carbides are arranged in bands in the direction of deformation, this carbide distribution causes elongation of the tool in the deformation direction during hardening [123]. Changes in shape of this type, in plate-shaped bodies, can be prevented to a large extent by rolling processes which induce an irregular distribution of carbides (see Fig. C 11.36 [116]).

C 11.2.2.2 Hot Formability

1. The *hot formability* of tool steels is not generally of importance in the production of tools because tools are rarely produced by hot forming except, e.g. in rolling or

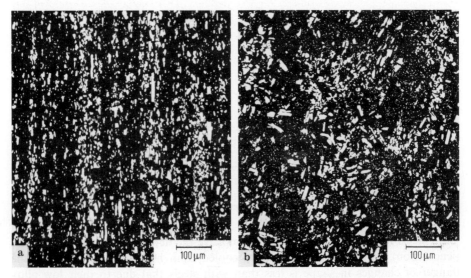

Fig. C 11.36. Distribution of carbides in steel X 210 Cr 12 according to Table C 11.11: **a** as conventionally rolled: banded arrangement of the carbides; **b** rolled according to special processes: irregularly arranged carbides.

pressing of twist drills. Of considerably higher importance to the steel producer is the hot ductility of tool steels in the rolling or forging of bars or disks. For all types of hot forming, the forming resistance should be as low as possible. This requirement cannot always be fulfilled by increasing deformation temperatures because they lead to undesirable changes in the microstructure. These changes include grain growth, dissolution of carbides and their subsequent precipitation at the grain boundaries and, in high-speed steels, formation of coarser carbides by coagulation at deformation temperature [6, 38] (see Figs. C 11.37 and C 11.38). An easily deformable material also requires that it can be deformed without fissuring within as wide a temperature range as possible.

2. *Methods for testing hot formability* are described in B 6. Essentially, tool steels are subjected to the torsion or compression tests, mainly for basic investigations [124–127].

3. *Hot formability depends* fundamentally on the homogeneity of the tool steel. Grain boundary deposits of hard and brittle carbide layers or of low-melting compounds can worsen the hot ductility considerably. Steels with a low sulphur content can be hot-formed more easily than steels having an intentional sulphur addition to improve the machinability. Tool steels with a ledeburitic composition have a particularly poor hot formability, but it is improved by a more homogeneous as-cast structure. Such a structure, for example, is achieved by remelting processes. Ledeburitic chromium steels with a carbon content of more than 3% cannot be hot-worked because they have very coarse, stalk-like carbides in the microstructure that are eutectically precipitated [128].

Introduction of greater variations in the chemical composition to improve the hot formability can usually not be taken advantage of because the composition is prescribed for other reasons. It has been shown, however, that small contents of tramp elements such as arsenic, copper or antimony can often worsen the hot formability in difficult deformation processes. This difficulty has sometimes been experienced in rolling of twist drill blanks from high-speed steel.

It is also important to maintain a good surface condition in difficult deformation operations; the first incipient cracks often emanate from roughly ground surfaces.

C 11.2.2.3 Cold Formability and Suitability for Hobbing

1. In connection with this section the term, "cold formability" is mainly used with the meaning given in B 7.1 and C 6.1. Tool steels are seldom cold-formed at lower temperatures, i.e. below the recrystallization temperature, e.g. by stamping or hobbing. *Cold formability is of importance* in the manufacture of tools if the form is to be produced by *hobbing*. This process may be used for tools of which a large number of pieces with the same contour are required, such as, for example, stamps for coins and screw heads, or for hollow-formed tools whose contours are simple to produce with a positively formed stamp. Deformation processes in the production of stamping tools are short. For this reason, the stamp material must have only a low deformation resistance and little deformability. For shapes which are to be deeply hobbed, apart from deformation resistance as low as possible hardness values of below 120 HB, a good deformability is also important.

344 C 11 Tool Steels

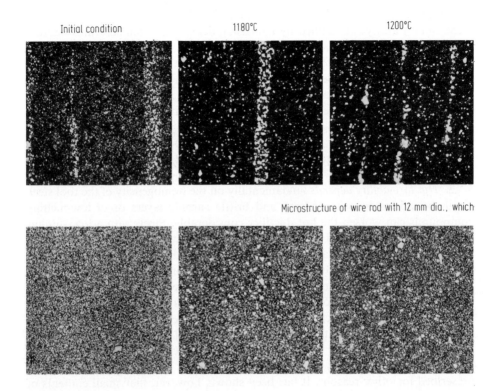

Fig. C 11.37. Influence of the temperature prior to rolling on the carbide formation in billets and in wire

2. The *methods for characterization of cold formability* are dealt with in B 7.

The suitability for *hobbing* of a steel is directly *dependent on its Brinell hardness*, which is easy to understand since the impression made by a steel ball in the surface during the hardness test is very similar to the hobbing process. According to various studies [129–131], the hobbing pressure P that characterizes deformation resistance, depends directly on the hardness, and its increase with increasing hobbing depth t on the ratio of t to the stamp diameter d (see Fig. C 11.39 [131]). Conformity of the cold formability of steels having the same hardness goes so far that the ratio of the hobbing pressure P to the Brinell hardness HB of the steel to be deformed is constant with a specific hobbing ratio t/d (see Fig. C 11.40 [131]).

3. *Cold formability is improved* by all procedures that lead to a *reduction of the hardness*. With a given composition, the cold formability can only be influenced by soft annealing to the lowest hardness values possible. In this condition tool steels have a microstructure with particularly large, well-spheroidized carbides. The shape of the ledeburitic carbides cannot, however, be appreciably influenced by soft annealing, so that the cold formability of tool steels with very high carbide content is limited.

holding for 3 h at a furnace temperature of:

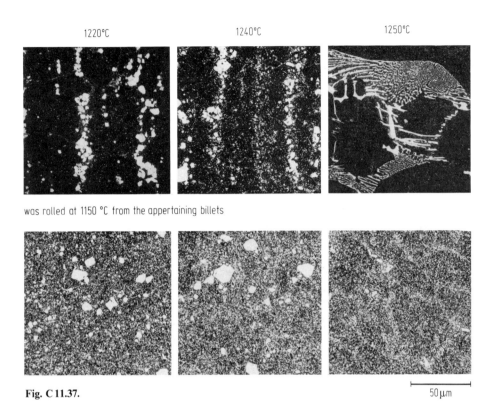

was rolled at 1150 °C from the appertaining billets

Fig. C 11.37. 50 μm

rod produced from them. Material: high-speed steel S 6-5-2 (see Table C 11.11). After [6].

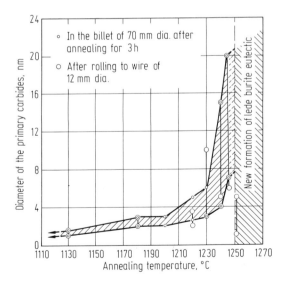

Fig. C 11.38. Formation of coarse carbides in high-speed steel S 6-5-2 according to Table C 11.11 through coagulation at high temperatures. After [6].

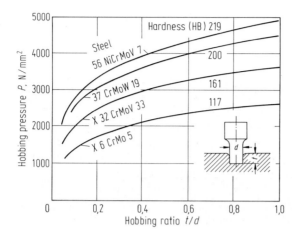

Fig. C 11.39. Dependency of the hobbing pressure P on the hardness of the steel to be deformed and on the hobbing ratio t/d. After [131]. (Concerning the chemical composition of the grades of steel examined, attention is drawn to the tables in C 11.3 and in DIN 17 350 [1].)

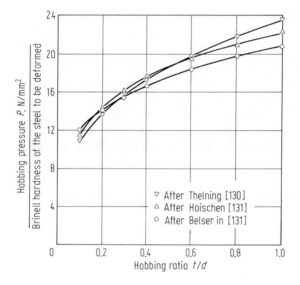

Fig. C 11.40. Relationship between the hobbing pressure in relation to the Brinell hardness P/HB and the hobbing ratio t/d (see Fig. C 11.39). After [131].

Cold formability can be influenced by the chemical composition if the content of alloy elements that increase the ferrite strength, such as carbon, silicon and phosphorus, is kept as low as possible. Care must also be taken that the content of elements that lower the transition point and thus have an adverse effect on the annealability, is low. Steels for which a good suitability for hobbing is required are usually alloyed with chromium, which has only a minor influence on the ferrite strength (see Fig. C 11.5).

C 11.2.2.4 Machinability

1. Machining processes are usually employed in the production of tools. The term machinability means all the properties of a material that play a role in shaping by

the use of cutting tools. *Tool steels* must *have good machinability*, i.e. they should possess properties that make it possible to remove as large a volume of chips as possible in a short time. At the same time, the energy requirement should be low, the edge life of the tool long, and the machined surface as smooth and even as possible.

2. The *methods for assessing machinability* are discussed in B 9. It must be emphasized here that none of the methods allows tool steels to be listed in order of preference according to their machinability. On the whole, machinability can only be compared in relation to specific machining conditions; further details are to be found in B 9. To assess the machinability of tool steels, methods similar to the machining conditions during tool production must be used.

3. Generally, those steels which have a particularly soft or hard *microstructure condition* are difficult to machine. Low-carbon tool steels with high ferrite content present difficulties because there is adhesion and smearing between tool and workpiece. For this reason, tool steels with low carbon content, e.g. case hardening steels, are not machined in the soft-annealed but in the normalized condition with a ferrite-pearlite microstructure. The tool steels with higher carbon contents, on the contrary, are always machined in the soft-annealed condition with spheroidized carbides because only in this heat treatment condition is the hardness low enough that cutting edge pressure and wear on the tools lead to satisfactory results. Even small amounts of carbide which, due to deficient soft annealing are not spheroidized, reduce machinability. The best machining results are obtained with hardnesses between about 180 and 230 HB.

Machinability is impaired by *hard, abrasive microstructure constituents* of alumina or silica which increase cratering on the cutting surface and abrasion on the top face. Hard special carbides in tool steels reduce machinability in the same way.

Additions of *sulphur* improve the machinability of tool steels in a manner similar to that of free-machining steels through their influence on chip formation [124, 132]. The sulphur content is generally limited to 0.1% so as not to reduce the transverse properties excessively.

The machined surface becomes smoother with *increasing hardness* of the tool steel being machined. This phenomenon is taken advantage of in special circumstances even if machining is difficult. Thus, high-speed steel milling cutters which must be backed-off, are backed-off with hardnesses of about 390 HB. As a result, a machined surface is obtained which is so smooth that (wet) grinding after heat treatment is not necessary. Tool steels which are backed-off at such high hardnesses are usually alloyed with 0.1% S.

C 11.2.2.5 Grindability

1. Tools that are preworked by machining are usually given their final shape by grinding after heat treatment. For this reason, tool steels should have a good grindability. *Good grindability* is understood to mean that a large amount of material can be removed by grinding in a short time, without the surface becoming damaged. Surface damage can occur as a result of too much heating in the surface

zone which may cause the surface to be tempered or even re-hardened (Fig. C 11.41). Stresses are connected with the volumetric changes occurring during such heating which can lead to so-called grinding checks (Fig. C 11.42).

2. Even less than machinability, grindability can be measured with absolute or relative values on the basis of a generally applicable *test method*.

In general, *three measuring values* are used to assess grindability, namely, abrasion on the workpiece, that means the removal of workpiece material, the abrasive wear on the grinding wheel, that means the trueness to form of the wheel, and the abrasion ratio of the abrasive wheel wear and workpiece abrasion. Good grindability can be characterized by a high workpiece abrasion, but it can also be measured on the basis of a high degree of trueness to form of the grinding wheel; both data are not dependent on each other. Grindability cannot be described satisfactorily by the relationship of the workpiece abrasion to the abrasive wear of the grinding wheel, which is called the abrasion ratio (see above). As Fig. C 11.43 [133] shows, with very differing abrasive wear of the grinding wheel and workpiece abrasion, the abrasion ratio can be the same. Grindability therefore can only be individually assessed on the basis of the combined action of a system with the principle factors tool, workpiece and machine, whose reciprocal influences are too little known.

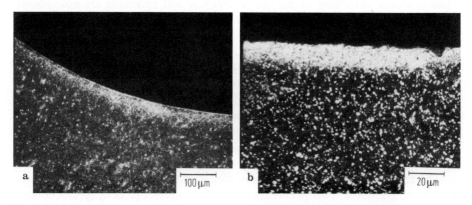

Fig. C 11.41. Damage to the surface of threading taps during grinding as a result of too much heating: **a** strong tempering effect and drop in the hardness to a depth of 0.08 mm; **b** high degree of rehardening to a depth of 0.03 mm.

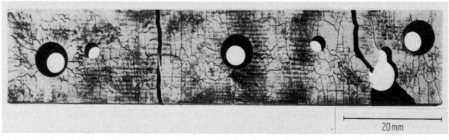

Fig. C 11.42. Grinding cracks made visible by magnetic particle testing.

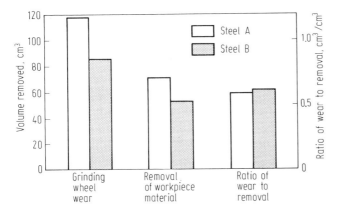

Fig. C 11.43. Example showing the difficulty of assessing the grindability of steels on the basis of grinding wheel wear, removal of workpiece material or ratio of wear to removal. After [133].

3. Grindability of a steel generally decreases with increasing *hardness*, with increasing *carbide content* and with increasing hardness of the carbides. Since hardness and carbide content are governed by the service properties of the material examined, grindability cannot normally be influenced by alloying measures or other heat treatment conditions.

Where composition and hardness are the same, a certain influence is exerted by the *size of the carbides*, i.e. grinding becomes more difficult with increasing carbide size (see also C 11.2.1.7). It is therefore important, for instance in high-speed steels, that the carbides are not coagulated [6]. Similar impairments of the grindability are found with all ledeburitic steels if the carbides are irregularly distributed and have accumulated in bands or in spots. The influence of the carbide size and distribution becomes very apparent if one compares the grindability of conventionally-cast high-speed steels and high-speed steels produced by powder metallurgy. Due to their fine and uniformly distributed carbides, high-speed steels made by powder metallurgy under the same conditions, a material wear rate about three times that of steels produced by conventional methods melting processes [134].

The *effects of differing hardness and composition* of the steel on the grindability become evident mainly when soft grinding wheel material is used. Hard abrasives such as boron nitride or diamond with very differing material conditions lead to almost the same abrasion of material. If the work can be flushed with coolant during grinding with boron nitride or diamond wheels, there will be little difference in the grindability of the two materials.

C 11.2.2.6 Polishability

1. Tools for stamping, forming and, in plastics processing, basic forming are often produced with polished surfaces. The steels used for these tools must be suitable for producing faultless, smooth, and as brightly finished surfaces as possible. The polishing agents used on these steels must uniformly remove the surface roughness

resulting from previous grinding and the polishing agent particles, which roll and slide over the surface during polishing, must level and smooth the surface through plastic deformations without forming localized holes or elevations in the plane surface [135].

2. Generally applicable *methods for assessing polishability* of a steel have not yet been introduced. However, a test method has been adopted by many steel producers for assessing the products from individual heats, in which a specimen is polished with a felt or cloth disk for a specified time with specified contact pressure using a polishing agent. With this type of stressing, where the disk always operates in the same direction over the specimen surface, poor polishability is easy to recognize from elongated holes. The polishability is assessed on the basis of size, depth and distribution of surface defects after polishing, allowing categorization from very good to not polishable.

3. It is understandable that it is more difficult to polish ledeburitic tool steels than other tool steels, due to the hard carbides contained in the ledeburitic steels.

Apart from this consideration, all studies have shown that polishing difficulties are purely the result of *inhomogeneities* in the steel. Therefore, examinations of the microscopic degree of purity provide a good indication of the polishability. In this regard, the size, number, and type of inclusions are significant. Particularly detrimental to the polishing result are oxidic inclusions such as silicates and alumina compounds. Ingot and crystal segregations, which can lead to hardness differences in plane surfaces and, thus, to differing degrees of abrasion during polishing, also influence the polishing result. Whenever particularly high requirements are placed on the polishability, steels are used that have been remelted by the electroslag process, in the vacuum arc or electron beam furnace to improve the microscopic degree of purity. This remelting not only limits the size and quantity of non-metallic inclusions, but reduces ingot and crystal segregations to such an extent that no disturbances are caused by hardness differences or accumulations of carbidic phases during polishing.

Polishability increases with the *hardness* of the tool to be polished. The possibility of increasing hardness cannot always be taken advantage of because the hardness or strength is prescribed by other demands placed on the tool.

C 11.3 Characteristic Steel Grades for the Different Fields of Tool Application

C 11.3.1 Steels for Plastics Molds

Using heat and pressure, plastics are formed in tools made of steel, by injection molding, pressing, extrusion, blowing, rolling, and other processes. The pasty or liquid plastics materials solidify in the tools, as they cool about 200 K or are heated to set (cure). The tools are subjected to stress through the closing forces applied by the processing machines and through the plastics or its additions. Tool temperatures do not, as a rule, substantially exceed 250 °C, which means that *no particular requirements are placed on retention of hardness or hardness at elevated temperatures* in respect of the mold steel. Steels that are commonly used and have proven reliable are listed in Table C 11.2. Strength requirements in respect of tools for forming of plastics are generally of minor importance, so that even very low tensile strengths of about 1000 N/mm^2 are sufficient. Only if hard fillers such as silica flour or reinforcing materials such as glass fibers are added to the plastics, do signs of wear occur. These materials may have such an influence on the mold that satisfactory life times can only be attained with very hard, high-carbide steels or special surface treatment processes.

Due to the often complicated *machining* and *surface* treatments used for plastics molds, the machining properties are *more important than the properties required of the tool in operation*. Particularly in the production of large molds, because of the high volume of machining involved, good machinability is decisive in the selection of the steel. Various steels with low to medium carbon contents are used. Because of the low degree of distortion in hardening, nickel-alloy steels of higher hardenability are preferred not only as case hardening steel (X 19 NiCrMo 4) but also as quenched and tempered steel (X 45 NiCrMo 4). To completely avoid distortion in hardening, however, 40 CrMnMo 7 and the more easily machinable 40 CrMnMoS 8 6 are frequently used. Both these steels can be quenched and tempered to a tensile strength of 900 N/mm^2 to 1200 N/mm^2, and are directly processed into finished molds without further heat treatment. For medium-sized and small molds, the low-alloy case hardening steel 21 MnCr 5 is very popular. The soft case hardening steel X 6 CrMo 4 is particularly recommended for producing multiple cavity molds by means of hobbing.

For plastics that may be abrasive molds are usually produced from the steels 90 MnCrV 8 and X 155 CrVMo 12 1, of which the former has a lower wear resistance but is more easily polished.

If there is a danger that decomposition products of the plastics will chemically attack the surface of the mold such as in processing of polyvinylchlorides, the *corrosion resistant steel* X 36 CrMo 17 has proven very successful.

Frequently, the *surfaces* of plastics molds are deliberately patterned by photochemical etching of the tool surface. The uniformity of the etching requires low segregation and low carbide steels and a low content of non-metallic impurities. For this reason, the steel 40 CrMnMoS 8 6 alloyed with sulphur is not suitable for molds that are to be etched because the sulphide inclusions may be recognizable in the reproduction of the etched image in the molded plastics.

The highest homogeneity and purity is required of mold steels used to produce transparent (crystal-clear) components of plastics. The requirements in respect of surface polish for such applications can often be only fulfilled by using remelted steels.

If operating temperatures in excess of 200 °C are encountered in the processing of plastics, hot work tool steels such as X 38 CrMoV 5 1 have proven reliable for the molds.

For molds of higher hardness and with high requirements for *dimensional accuracy*, the steel X 3 NiCoMoTi 18 9 5, which is machined in the solution-annealed condition and attains its hardness by means of a simple aging process at 480 °C, has also been used successfully.

Table C 11.2 Chemical composition, hardness characteristics, operational hardness and typical applications of tool steels for plastics molds

Steel grade Designation	Chemical composition								hardness, soft annealed HB max.	Reference data for		opera- tional hardness HRC	Frequent applications	Compa- rable AISI- steel[c]
	% C	% Si	% Mn	% Cr	% Mo	% Ni	% other	%P, %S[a]		hardening temper- ature °C	cooling medium[b]			
X 6 CrMo 4	≤0.7	≤0.2	≤0.2	3.8	0.5	–	–	2	108	860 … 900	Oil	62[d]	Hobbable molds, multiple cavity molds	~ P 4
21 MnCr 5	0.21	0.3	1.3	1.2	–	–	–	2	212	810 … 840	Oil	62[d]	Medium sized molds, case hardened	
X 19 NiCrMo 4	0.19	0.3	0.3	1.3	0.2	4.1	–	2	255	780 … 810 800 … 830	Oil Air	62[d]	Large molds, case hardened	~ P 6
X 45 NiCrMo 4	0.45	0.3	0.3	1.4	0.3	4.1	–	2	262	840 … 870	Oil	48 … 56	Large molds, well hardenable	
40 CrMnMo 7	0.40	0.3	1.5	2.0	0.2	–	–	1	230[e]	830 … 880	Oil	32	Large molds, pre-heat treated[h]	~ P 20
40 CrMnMoS 8 6	0.40	0.4	1.5	1.9	0.2	–	0.08 S	8	230[e]	830 … 880	Oil	32	Large molds, pre-heat treated[h], easily machinable	
X 36 CrMo 17	0.38	≤1.0	≤1.0	16.0	1.2	≤1.0	–	2	285	1000 … 1040	Oil	27 … 31	Processing of aggressive plastics	

Steel	C	Si	Mn	Cr	Mo	Ni	Other	P,S[a]	HB	Heat treatment[b]	HRC	Application	AISI
X3NiCoMoTi1895	≤0.03	≤0.1	≤0.2	≤0.3	4.9	18.0	9.3Co 1.0Ti	6	330[f]	[g]	50...53	Molds of highest dimensional accuracy	~O2
90MnCrV8	0.90	0.3	2.0	0.4	–	–	0.2V	2	229	790...820 Oil	60...62	Small molds, wear resistant, easily polishable	
X155CrVMo121	1.55	0.3	0.3	11.5	0.7	–	1.0V	2	255	1020...1050 O, HB, A	60...64	Very wear-resistant molds, well nitridable	D2
X38CrMoV51	0.39	1.1	0.4	5.2	1.3	–	0.4V	2	229	1000...1040 O, HB, A	40...45	Molds for temperatures above 300°C	H11

[a] In the tables of C11 the highest admissible phosphorus and sulphur contents are indicated by the following figures:
1 = ≤0.035% P and ≤0.035% S, 2 = ≤0.030% P and ≤0.030% S, 3 = ≤0.025% P and ≤0.025% S, 4 = ≤0.020% P and ≤0.020% S,
5 = ≤0.015% P and ≤0.015% S, 6 = ≤0.010% P and ≤0.010% S, 7 = ≤0.030% P and ≤0.015% S, 8 = ≤0.030% P,
9 = ≤0.020% P and ≤0.010% S, 10 = ≤0.020% P and ≤0.005% S

[b] O = oil, HB = Hot bath, A = air, W = water

[c] Steel grades according to Steel Products Manual of the American Iron and Steel Institute 1978 [145].

[d] On the surface

[e] These steels are usually supplied in the quenched and tempered condition with a hardness of about 300 HB

[f] Solution annealed

[g] Solution annealing at 800°C to 850°C, cooling in air, aging at 480°C to 550°C, cooling in air

[h] Quenched and tempered

Further information regarding tool steels for the processing of plastics can be obtained from the literature [136–144].

C 11.3.2 Steels for Pressure Casting Molds

During pressure die casting, molten alloys, e.g. aluminum, magnesium, zinc, lead, or copper, are injected under high pressure into steel dies. The tools are heated to about 500 °C, depending on the alloys to be processed and under high compression forces are exposed to high mechanical stresses and even to attack by erosion and – possibly – to metal corrosion by the flowing alloy. But the usual *causes of failure* of pressure casting dies are the reticular heat checks that are caused by severe temperature changes in each cycle. Heat checking cannot be prevented but its beginning can be delayed by increasing the operational hardness values of 44 to 46 HRC, which were usual in former times, to values of 46 to 48 HRC or more. This extra hardness necessitates special steels with little segregation and a martensitic hardness structure. If such a structure cannot be attained, the tool will fail more rapidly at a higher operating hardness. Likewise, the hardness of the tool should be adapted to the temperature stress of the tool surface, which is influenced by the wall thickness of the castings. Figure C 11.44 [98] shows the relation between the hardness HRC of the tool and the wall thickness S of aluminum castings. These empirical data are obtained from tools with a good life time. The relation between wall thickness and hardness may be expressed by $HRC = 56 \cdot S^{-0.14}$. Proved steels for pressure casting dies are presented in Table C 11.3.

The usual *steels* for pressure casting dies are the secondary hardening steels alloyed with chromium, molybdenum and vanadium X 38 CrMoV 5 1, X 40 CrMoV 5 1 and X 32 CrMoV 3 3. In these steels, the former, frequently applied, hot work steels have been superseded by steels with tungsten contents, because these molybdenum containing steels have higher toughness and improved thermal fatigue resistance compared with steels containing tungsten, so that heat checking is retarded. The hardenability also is improved so that distortion due to heat treatment is diminished. Additionally, machinability is better because of the lower temper hardness. The steels X 38 CrMoV 5 1 and X 40 CrMoV 5 1 are used predominantly for the processing of light metal alloys, the first-mentioned having a better hardenability and the second a somewhat higher wear resistance. Because of its better retention of hardness and its higher hardness at elevated temperatures the third steel, X 32 CrMoV 3 3, with a higher molybdenum content of about 3% is specially suited to the processing of copper alloys which causes a stronger heat stress of the tool, due to the higher melting point of the copper alloys.

Besides these three standard steels, some other *steels with a higher retention of hardness* are qualified for some parts of pressure casting dies, which are subject to especially high thermal stresses. In particular the cobalt-containing steels X 45 CoCrWV 5 5 5 and X 20 CoCrWMo 10 9, and the hot work steels X 30 WCrV 9 3 and X 30 WCrV 5 3 with a high tungsten content, have been used frequently for copper alloy die casting.

Maraging steels are sometimes employed for pressure casting dies because of their good toughness properties, their favorable strength and good thermal fatigue

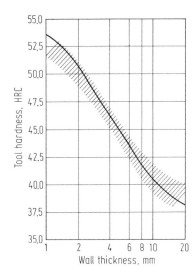

Fig. C 11.44. Advice for the favorable hardness of pressure casting dies for aluminum processing made of the tool steels X 38 CrMoV 5 1 and X 40 CrMoV 5 1 according to Table C 11.3. The given values are based on empirical results and give the relationship between favorable tool hardness and wall thickness of the aluminum casting [98].

resistance. In particular, the steel X 3 NiCoMoTi 18 9 5 should be mentioned in this connection [146]. This steel combines beneficial service properties with the simple heat treatment of aging at 480 °C to 550 °C after machining in the solution annealed condition. This simple heat treatment avoids non-desirable distortions.

Nitriding is used to improve the wear resistance of pressure casting dies and all the steels in Table C 11.3 are excellent for this purpose. However, no reliable experience is available on the success of this surface treatment.

The strongest wear and temperature conditions are experienced in the area of the inflowing molten metal. A remarkable improvement in the life of these tools can be obtained by roughening the surface by *spark erosion* [147, 148]. The effect of this surface roughening treatment is based on a reduction of the wettability and therewith of the heat transfer and not on an absorption of tungsten carbide from the spark erosion electrode, as is often supposed.

For further information about tool steels for pressure casting dies refer to [31, 99, 146, 149–157].

C 11.3.3 Steels for Glass Processing

Molten glass is formed in metallic permanent molds or by rolling to make sheet-glass. Production of hollow articles requires two steps and the tool stress is very similar to that encountered in tools for pressure casting. However, *the stress* caused by *alternating heating* by the relatively hot molten glass is higher. Therefore, *oxidation and corrosion phenomena* occur at the surface of the glass forming molds which destroy the mold surfaces. This mold surface effect is connected with a deterioration of the glass surface and the pressed parts may then stick in the mold. Hence, glass forming molds require a good resistance to thermal fatigue plus good

Table C 11.3 Chemical composition, hardness characteristics, operational hardness and typical applications of tool steels for pressure casting dies

Steel grade Designation	Chemical composition									Reference data for			Frequent applications	Comparable AISI-steel [145]
	%C	%Si	%Mn	%Cr	%Mo	%V	%W	%other	%P, %S[a]	hardness, soft annealed HB max.	hardening temperature[b] °C	operational hardness HRC		
X38CrMo51	0.39	1.1	0.4	5.2	1.3	0.4	–	–	2	229	1000...1040	42...46	Pressure casting dies	H11
X40CrMoV51	0.40	1.1	0.4	5.2	1.4	1.0	–	–	2	229	1020...1060	42...46		H13
X32CrMoV33	0.32	0.3	0.3	3.0	2.8	0.6	–	–	2	229	1010...1050	40...46	Pressure casting dies for heavy metals	H10
X45CoCrWV555	0.45	0.4	0.4	4.5	0.5	2.0	4.5	4.5 Co	3	250	1120...1160	44...48	Mold cores, cavity inserts, die inserts	H19
X20CoCrWMo109	0.20	0.3	0.5	9.5	2.0	–	5.5	10.0 Co	1	320	1050...1100	50...55	Mold cores, mold parts	–
X30WCrV93	0.30	0.2	0.3	2.7	–	0.4	8.5	–	1	250	1100...1150	45...50		H21
X30WCrV53	0.30	0.2	0.3	2.4	–	0.6	4.3	–	1	240	1050...1100	45...50		–
X3NiCoMoTi1895	≤0.03	≤0.1	≤0.2	≤0.3	4.9	–	–	18 Ni 9.3 Co 1.0 Ti	6	330[c]	[d]	50...53	Pressure casting molds for light metals with good resistance to thermal fatigue	–

[a] See notation [a] in Table C 11.2
[b] All steel grades (besides the last one) may be cooled from hardening temperature in oil, hot bath or air
[c] Solution annealed
[d] Solution annealing at 800 °C to 850 °C, cooling in air, aging at 480 °C to 550 °C, cooling in air

oxidation resistance. These requirements lead to a selection of materials for glass forming molds that are comparable to corrosion and heat resistant steels (see Table C 11.4). The formerly often used cast iron nowadays is only used for the production of small lots.

Steels for dies and stamps for the production of glass hollow-ware are *the heat treatable steels* (steels for quenching and tempering) X 21 Cr 13 and X 23 CrNi 17. These steels are suitable, if the demands on the glass surface are not too high and if the heat effect is low. Due to the extreme temperature stresses the tools are hardened and tempered to very low operational hardnesses of a maximum of 300 HB. Steel with a chromium content of 17% is often used in the soft annealed condition without additional heat treatment. Even if the tools are cooled to reduce the temperature of the mold, the temperature conditions (stresses) of the tool surface are extremely high. To improve the oxidation resistance, therefore, the chromium content of the first-named above steel (see also Table C 11.4) is often raised to values of 15%. But these steels are not suitable for the processing of glass with a high melting point. Higher operating temperatures require the use of austenitic heat-resistant steels, e.g. X 16 CrNiSi 25 20 or X 13 NiCrSi 36 16. Due to their excellent scaling resistance and the slight reaction of the molten glass with the surface of the mold, these steels tend to stick only slightly during processing. Hence, these steels will produce smooth and clear surfaces from high melting-point glasses over long processing periods.

In particular, it is pointed out that nickel alloys are used successfully for rolling of crystalline glasses, e.g. alloy NiCr 20 Co 18 Ti (the last one in Table C 11.4).

As a rule, tool steels for glass forming molds are air-melted. If a special surface finish of the glass products is needed as with TV-tubes or optical glasses, the applied steels should have good polishability. For these applications, therefore, it is recommended to use steels that have been remelted.

To improve their properties the heat-treatable steels containing about 0.2% C and 13% to 17% Cr are often surface treated with chromium or nickel.

For further information refer to [158–160].

C 11.3.4 Steels for Cold-forming Tools

Non-ferrous metals, unalloyed and alloyed steels, up to and including the stainless austenitic steels are frequently made into complicated molds by cold-forming. *Forming usually takes place at ambient temperature*, however, to improve the formability of the workpiece, it may also be heated to several hundred degrees C. The advantages of the cold-forming process as opposed to other forming methods are high dimensional accuracy and a good surface finish which frequently makes additional machining of the formed parts unnecessary. The forming processes are known as deep drawing, stamping, extrusion, bending, rolling, and drawing. A systematic classification of metal-forming technology processes is to be found e.g. in DIN 8580 [3]. The processes and terms are described extensively and in detail in the literature [161].

Table C 11.4 Chemical composition, hardness characteristics, operational hardness and typical applications of tool steels for glass processing

Steel grade	Chemical composition									Reference data for hardening or solution annealing			Frequent applications	Comparable AISI-steel [145]
	% C	% Si	% Mn	% Co	% Cr	% Ni	% other	% P, % S[a]	hardness, soft annealed HB max.	temperature °C	operational hardness HB			
X 21 Cr 13 {possibly chromium or nickel plated}	0.20	0.4	0.3	–	13.0	–		1	220	980 … 1010[b]	250 … 300	Glass forming molds for low-melting glasses, glass rolls	~420
X 23 CrNi 17	0.18	≤1.0	≤1.0	–	17.0	2.0		1	275	1000 … 1050[b]	220 … 300	Glass rolls comparable to X 21 Cr 13, but for higher numbers of pieces	~431
X 16 CrNiSi 25 20	≤0.20	2.1	≤2.0	–	25.0	20.0		1	223[c]	1050 … 1100[d]	rd. 200	Glass forming molds for high-melting glasses, glass rolls for higher output	~310
X 13 NiCrSi 36 16	≤0.15	1.8	≤2.0	–	16.0	35.5		1	223[c]	1050 … 1100[d]	rd. 200		
NiCr 20 Co 18 Ti	≤0.13	≤1.0	≤1.0	18.0	20.0	Bal.	1.5 Al 2.5 Ti	7	277[c]	1050 … 1080[d]	rd. 300	Glass rolls for crystalline glasses	

[a] See notation [a] in Table C 11.2
[b] Hardening with quenching in oil
[c] Solution annealed
[d] Solution annealing and cooling in air

Tools for cold forming are chiefly subjected to stress through pressure and friction. Therefore, as Table C 11.5 shows, steels of high hardness and high carbide content having the same compositions and hardness as those used as cutting steels (refer Table C 11.11) are mainly concerned here. If tension, shear or bending forces act on the forming tools, reinforcement by means of shrink rings is recommended.

Very high pressure and wear stresses are present in *extrusion tools*. Only high-speed steels[1], e.g. S 6-5-2 and S 10-4-3-10, are suitable for absorbing the compressive stresses which often exceed 3000 N/mm^2 in extrusion dies. The steel X 155 CrVMo 12 1 fulfils the requirements for compression stresses up to about 2000 N/mm^2. Due to the high friction stress and, in addition, the heating that results from the deformation work, press boxes are also produced from both these high-speed steels. To absorb the transverse forces, the press boxes are held by prestressed reinforcing rings for which the steels X 40 CrMoV 5 1, X 45 NiCrMo 4 and 56 NiCrMoV 7 have proven efficient.

Deep drawing tools are particularly subjected to stress through frictional forces. The ledeburitic chromium steels, such as X 155 CrVMo 12 1 already mentioned, provide good service behavior for drawing punches and drawing rings; to reduce cold welds the steel is often nitrided. The low-carbide 90 MnCrV 8, which is also used for ejectors, is adequate for the blank holders used in deep drawing tools.

The pressure and wear stresses encountered by *stamping tools* such as are used for the production of coins are much lower than with extrusion tools. However, the alternating (cyclic) stress is, on the other hand, considerable. Apart from the 90 MnCrV 8 steel the likewise low-carbide steel 60 WCrV 7 has proven reliable for these operating conditions. But ledeburitic steels, e.g. S 6-5-2 and the steel with 1.5% C and 12% Cr are also used. However, these steels must be carefully soft-annealed before hobbing the relief. If special demands are placed on the relief, the low-carbide nickel-alloyed steel 75 NiCrMo 5 3 3 is particularly suitable because it can be soft-annealed for hobbing, attains a high hardness due to its carbon content and, because of its low carbide content, can be polished very well. If a very high gloss is required, remelted steels should be used to improve the degree of purity and segregation.

Bending tools are subjected to the least pressure and wear stress of all cold-forming tools. Therefore, bending punches are often produced from the low-carbide but tough first two steels in Table C 11.5.

A great deal of *forming work* is performed *by rolling processes*. An example is the rolling of threads for which circular or flat rollers are used. For rolling large threads, a steel with 1.5% C and 12% Cr is predominantly used, for rolling small threads, high-speed steels such as S 6-5-2 prove better. Sheets, strips and similar products are produced by cold rolling using flat rolls. For working rolls in two-high and four-high stands, the low-alloy steel 85 CrMo 7 has good polishability, due to its low carbide content, and has proven efficient. For high-stressed working rolls in multiroll stands, on the contrary, high-carbide wear-resistant materials,

---

[1] The abbreviations denoting the high-speed steels state their content (wt.-%) of the characteristic alloying elements tungsten, molybdenum, vanadium and cobalt (rounded off) in this sequence.

Table C 11.5 Chemical composition, hardness characteristics, operational hardness and typical applications of tool steels for cold-forming tools

Steel grade	Chemical composition									Reference data for			operational hardness HRC	Frequent applications	Comparable AISI-steel [145]
	% C	% Si	% Mn	% Cr	% Mo	% V	% W	% other	% P, % S[a]	hardness, soft annealed HB max.	hardening temperature °C	cooling medium[b]			
90 MnCrV 8	0.90	0.3	2.0	0.4	–	0.1	–	–	2	229	790 … 820	Oil	60 … 64	Stamping tools, blank holders	~ O 2
60 WCrV 7	0.60	0.6	0.3	1.1	–	0.2	2.0	–	2	229	870 … 900	Oil	58 … 62	Stamping (coining) tools	~ S 1
75 NiCrMo 5 3 3	0.75	0.3	0.7	0.8	0.3	–	–	1.4 Ni	3	240	820 … 850	Oil	60 … 64	Stamping (coining) tools	
S 6-5-2	0.90	≤0.5	≤0.4	4.2	5.0	1.9	6.4	–	2	300	1140 … 1180	O, HB, A	62 … 65	Extrusion dies, press boxes, thread rollers, cold rolls	M 2
S 10-4-3-10	1.28	≤0.5	≤0.4	4.2	3.6	3.3	9.5	10 Co	2	300	1140 … 1180	O, HB, A	62 … 65		
X 40 CrMoV 5 1	0.40	1.1	0.4	5.2	1.4	1.0	–	–	2	229	1020 … 1060	O, HB, A	48 … 52	Shrinking/intermediate ring	H 13
X 45 NiCrMo 4	0.45	0.3	0.3	1.4	0.3	–	–	4.1 Ni	2	262	840 … 870	Oil	44 … 48	Shrink ring	
X 155 CrVMo 12 1	1.55	0.3	0.3	11.5	0.7	1.0	–	–	2	255	1020 … 1050	O, HB, A	59 … 62	Drawing punch, drawing rings, thread rollers, live (working) rolls in multiroll stands	D 2
C 60 W	0.60	0.3	0.7	–	–	–	–	–	1	231	800 … 830	Oil	[c]	Base plates, build-up components	~ W 1
85 CrMo 7	0.85	0.3	0.3	1.8	0.3	–	–	–	2	230	820 … 850 / 830 … 860	Oil / W	61 … 65	Cold rolls	
56 NiCrMoV 7	0.55	0.3	0.8	0.7	0.3	0.1	–	1.7 Ni	2	248	830 … 870	Oil	45 … 50	Shrinking rings	~ L 6

[a] See notation a in Table C 11.2
[b] See notation b in Table C 11.2
[c] Steel is generally used in the soft annealed condition

such as the steels with 1.5% C or with 6% W, 5% Mo and 2% V, already mentioned, have been used very successfully [162].

Further information in respect of cold forming tools can be found in the literature [163–171].

C 11.3.5 Steels for Forging and Pressing Dies

Steels and non-ferrous metals are hot formed by forging and pressing. The usual tool steels for this purpose are listed in Table C 11.6. Selection of steels for forming tools depends on the type of forming process.

Hammer-forging operations are characterized by a short time of contact between tool and workpiece. Heating of the tool is negligible, so that requirements for retention of hardness and hardness at elevated temperatures are not great. On the other hand, the high striking velocity requires a low risk of failure (fracture). These conditions are satisfied by the steels 55 NiCrMoV 6 and 56 NiCrMoV 7. As a rule, these steels are used for massive dies ("Vollgesenke") with an operational hardness of only 350 to 400 HB because at higher values the risk of failure is too high. The wear resistant, carbide-rich steel 145 V 33, is suitable for forging dies with flat cuts and is a surface (layer) hardening alloy with a good resistance to rupture. The hardness penetration of this steel can be regulated over a wide range by the temperature used for hardening, in contrast to the unalloyed surface (layer) hardening alloys. To reduce the wear on the surface, forging dies can be nitrided or hard chromium plated for flat cuts.

During *press* forging there is a longer time of contact between die and work piece and the die becomes heated to a greater extent. The hot-work steels X 38 CrMoV 5 1, X 32 CrMoV 3 3, and X 38 CrMoV 5 3 are therefore recommended. These steels have a high retention of hardness, good elevated temperature strength, and can be used at a hardness above 45 HRC because of the lower impact loads imposed. To reduce surface wear, which usually leads to a failure, still higher wear-resistant steels are in use. These steels are characterized by a higher chromium-, molybdenum- and vanadium content, e.g. the steel X 48 CrMoV 8 11.

A substantial *improvement in the life* of forging dies is obtained by the use of the precipitation hardening nickel alloys NiCr 19 NbMo and NiCr 19 CoMo [172]. But only specific temperature stressed pre-dies (rough dies, ("Vorgesenke") show a significant increase of output. These austenitic alloys are not suitable for press working with finishing dies ("Fertigteilgesenke") because the effective stresses lead to die deformation as indicated by the low yield stress of the alloys. Therefore, finishing dies are made from the chromium-molybdenum-vanadium steels previously mentioned.

As a rule, non-ferrous metals are forged only by press working. The most-used steels are X 38 CrMoV 5 1 and X 32 CrMoV 3 3. To delay the tendency to heat checking in press-working dies for the processing of copper alloys, it is important that the tool steels used have a martensitic hardened structure. Further improvements are possible by the use of steels that have been homogenized by remelting and diffusion annealing.

Table C 11.6 Chemical composition, hardness characteristics, operational hardness and typical applications of tool steels for forging and pressing tools

Steel grade Designation	Chemical composition									hardness, soft annealed HB max.	Reference data for hardening or solution annealing temperature °C	cooling medium[b]	operational hardness HRC	Frequent applications	Comparable AISI-steel [145]
	% C	% Si	% Mn	% Cr	% Mo	% Ni	% V	% other	% P, % S[a]						
55 NiCrMoV 6	0.55	0.3	0.8	0.7	0.3	1.7	0.1	–	2	248	830 .. 870	Oil	32 … 40	Hammer (forging) dies for steel	~L6
56 NiCrMoV 7	0.55	0.3	0.8	1.1	0.5	1.7	0.1	–	2	248	830 … 870 860 … 900	Oil Air	32 … 40	Massive dies	~L6
145 V 33	1.45	0.3	0.4	–	–	–	3.3	–	2	230	800 … 950	W	45	Hammer (forging) dies for flat cuts	
X 48 CrMoV 8 11	0.48	0.8	0.4	7.6	1.4	–	1.4	–	10	250	1040 … 1090	Oil	45	Wear resistant pressing dies with higher hardness at elevated temperatures	
NiCr 19 NbMo	0.06	≤0.4	≤0.4	19.0	3.1	52.5	–	0.6 Al 0.004 B 5.1 Nb 0.9 Ti	5		960[c]	Air	[d]	Pre-pressing dies with high efficiency for steel	
NiCr 19 CoMo	≤0.12	≤0.5	≤0.1	19.0	9.8	Bal.	–	1.6 Al 11.0 Co 3.1 Ti	9		1080[c]	W	[d]		
X 38 CrMoV 5 1	0.39	1.1	0.4	5.2	1.3	–	0.4	–	1	229	1000 … 1040	O, HB, A	42 … 46	Pressing tools	H 11
X 32 CrMoV 3 3	0.32	0.3	0.3	3.0	2.8	–	0.6	–	1	229	1010 … 1050	O, HB, A	40 … 46	Pressing tools	H 10
X 38 CrMoV 5 3	0.38	0.4	0.5	5.0	3.0	–	0.6	–	1	230	1050 … 1080	O, HB, A	45 … 50	Pressing tools	

[a] See notation [a] in Table C 11.2
[b] See notation [b] in Table C 11.2
[c] Solution annealing
[d] Tensile strength in the solution annealed condition: 1300 N/mm²

C 11.3.6 Steels for Hot Extrusion Press Tools

During hot extrusion heated metal billets are placed in a container and are pressed through the die opening by a punch. This method is used to produce profiles and tubes and involves a long contact period between the tool and the alloy to be shaped. However, the influence of alternating temperatures on the tool is considerably smaller than in other hot forming processes. The most important requirement for hot extrusion tools is an *elevated temperature strength*. The pressing temperature can vary between 200 °C for the processing of lead and 1200 °C for that of steel. The mechanical stresses imposed on individual tool components are high, due to the high pressing forces during hot extrusion. A review of proven tool steels for hot extrusion tools is given in Table C 11.7.

The forces within the extrusion *container* can be controlled only by prestrained assemblies of several parts. This concept leads to a beneficial stress distribution and material utilization. For an exact calculation of the shrinkage forces, the creep behavior of the tool steel must be considered [173–175] because of the high temperatures and stresses and the duration of loading of the tools. For the highly-stressed liners (containers) used for processing of light alloys, the first two steels of Table C 11.7 are mostly employed. These steels contain about 0.4% C, 5% Cr, 1.4% Mo and 0.4 or 1.0% V and are used with a tensile strength of about 1400 N/mm^2. A suitable steel for liners for the extrusion of heavy metals is the austenitic precipitation hardening steel X 6 NiCrTi 26 15. This steel retains a yield strength of 600 N/mm^2 even at a temperature of 700 °C. The liner holders (intermediate boxes, "Zwischenbüchsen") and intermediate mantles ("Zwischenmäntel") of the containers (recipients) are made of the steel X 38 CrMoV 5 1 or lower alloyed steels, e.g. 40 CrNiMo 7. These steels have a tensile strength of about 1200 N/mm^2 (for the liner holder) and 1100 N/mm^2 (for the intermediate mantle) and, therefore, their strength is lower than that of the liner.

Extrusion stems for average temperature conditions are made from the steel 56 CrNiMoV 7, but the steels X 38 CrMoV 5 1 and X 40 CrMoV 5 1 mentioned above are also suitable. The stems are used with a tensile strength between 1600 N/mm^2 and 1800 N/mm^2, considering the high compressive forces to be encountered.

Hot extrusion of tubes involves extreme temperature conditions of *the mandrels* which, therefore, are made from the more creep resistant steels X 32 CrMoV 3 3, X 45 CoCrWV 5 5 5 and X 20 CoCrWMo 10 9 or even nickel alloys, such as NiCr 19 CoMo [172]. Sometimes, the required high strength can only be achieved by strong internal cooling of the mandrels.

The *extrusion dies* are those parts of the tool which must sustain the strongest thermal and mechanical conditions. During the passage through the die, the billet reaches the highest temperature due to the inner friction. An additional, considerable frictional heat develops at the area of contact between the die and the billet. Extrusion of heavy metals leads to an especially strong heating of the die and causes a tempering effect. This tempering leads to higher abrasion as well as to deformations and a loss of dimensional accuracy.

For the processing of light metals, the first three steels of Table C 11.7 are adequate. Complicated extrusion shapes require the dies to be made from remelted

Table C11.7 Chemical composition, hardness characteristics, operational hardness and typical applications of tool steels for extrusion press tools

Steel grade Designation	Chemical composition									hardness soft annealed HB max.	Reference data for		operational hardness HRC	Frequent applications	Comparable AISI-steel [145]
	% C	% Si	% Mn	% Cr	% Mo	% V	% W	% other	% P, % S[a]		hardening or solution annealing temperature °C	cooling medium[b]			
X 38 CrMoV 51	0.39	1.1	0.4	5.2	1.3	0.4	–	–	2	229	1000 … 1040	O, HB, A	42 … 52	Extrusion dies containers (recipients)	H 11
X 40 CrMoV 51	0.40	1.1	0.4	5.2	1.4	1.0	–	–	2	229	1020 … 1050	O, HB, A	42 … 52	Press stamps, stems, press mandrels, press tools	H 13
X 32 CrMoV 33	0.32	0.3	0.3	3.0	2.8	0.6	–	–	2	229	1130 … 1160	O, HB, A	40 … 52	Press tools	H 10
X 45 CoCrWV 555	0.45	0.4	0.4	4.5	0.5	2.0	4.5	4.5 Co	3	250	1120 … 1160	O, HB, A	44 … 48	Press die	H 19
X 20 CoCrWMo 109	0.20	0.3	0.5	9.5	2.0	–	5.5	10.0 Co	1	320	1050 … 1150	O, HB, A	50 … 55	Press tools / Press dies / Press mandrels	
X 30 WCrV 93	0.30	0.2	0.3	2.7	–	0.4	8.5	–	1	250	1100 … 1150	O, HB, A	45 … 50	Dies	H 21
X 30 WCrV 53	0.30	0.2	0.3	2.4	–	0.6	4.3	–	1	240	1050 … 1100	O, HB, A	45 … 50	
X 38 CrMoV 53	0.38	0.4	0.5	5.0	3.0	0.6	–	–	1	230	1030 … 1080	O, HB, A	45 … 48	Parts of recipients (containers)	~H 13
X 6 NiCrTi 2615	≤0.8	≤1.0	1.5	14.8	1.3	0.3	–	26 Ni / 2.1 Ti	2	220[c]	980[d]	W	1000[e]	Liners (internal boxes) for the extrusion of copper-alloys	
NiCr 19 CoMo	≤0.12	≤0.5	≤0.1	19.0	9.8	–	–	1.6 Al / 11 Co / 3.1 Ti / Bal. Ni	9		1080[d]	W	1300[e]	Dies, mandrels for the extrusion of heavy metals	
X 50 NiCrWV 1313	0.50	1.4	0.7	13.0	–	0.7	2.2	13 Ni	1		hot-cold worked		35 … 40	Dies for the extrusion of heavy metals	
40 CrNiMo 7	0.40	0.3	1.5	2.0	0.2	–	–	–	1	230	830 … 880	Oil	30 … 35	Mantles of recipients (containers)	
56 NiCrMoV 7	0.55	0.3	0.8	1.1	0.5	0.1	–	1.7 Ni	2	248	830 … 870	Oil	48 … 52	Press stamps, stems	~L 6

[a] See annotation a in Table C 11.2. [b] See annotation b in Table C 11.2. [c] Solution annealed
[d] Temperature of solution annealing. [e] Tensile strength after solution annealing, N/mm²

steels with little segregation which are characterized by better transverse properties. Dies for the processing of light alloys in general are nitrided to reduce the frictional wear. For the extrusion of heavy metals the steel with 3% Cr and 3% Mo is often not suitable and steels with a higher retention of hardness and a better elevated temperature strength, containing 4.5% or 5.5% W and 4.5% or 10% Co, must be used.

Simple shaped die orifices – e.g. for production of wire – lead to the successful employment of hard metals based on cobalt. Precipitation-hardening high temperature nickel alloys can also be used, e.g. alloy NiCr 19 CoMo mentioned above. Another successful method to counter the high service temperatures of extrusion press dies is to use cold deformed austenitic steels, e.g. the steel X 50 NiCrWV 13 13. The austenitic dies strain hardened by forging at very low temperatures retain their additional strength due to strain hardening up to the temperature of recrystallization, which is above 850 °C.

C 11.3.7 Steels for Machining Tools

Simple forms of machining tools, particularly tools for turning, are chiefly produced from sintered carbide hard materials or from oxide-ceramic materials. These cutting materials are difficult to form and to machine and, in comparison to tool steels, also have less toughness, so the more complicated tool shapes susceptible to fracture used for drilling, milling, countersinking, reaming, broaching, sawing and turning, are produced from steel. Due to the high cutting speeds involved in machining and the resulting heating, high-speed steels are chiefly used, and these are listed in Table C 11.8.

Stressing of the machining tools leads to changes in the cutting edge geometry. The machining accuracy and the machining pattern are changed as a result of abrasive wear on the top face, the cutting edge is weakened by adhesive wear from the flowing chip, and the tool hardness can drop so much as a result of heating that machining becomes impossible. Therefore, the ability to maintain a good cutting edge (cutting ability) up to red-hot temperature of about 600 °C is essential to steels used for machining tools.

The large number of high-speed steel alloys used today can be reduced to *three almost equivalent basic alloys* as shown in Table C 11.9 [6]. Apart from these, the alloy IV, a lower tungsten variation of alloy I, is also used in low-temperature stress conditions. The three basic alloys I to III with a carbon content between 0.8% and 0.9% fulfill in approximately the same way the requirement of a minimum hardness of 65 HRC; the chromium content of 4% ensures adequate through-hardening, and equivalent tungsten and molybdenum contents have the effect that the retention of hardness and the hardness at elevated temperatures are of a comparable order. Since the three basic alloys have the same carbide quantities of about 8% and the same types of carbide M_6C and MC [176–179] after hardening from average hardening temperature, they also have almost the same wear resistance. Of these three basic alloys, alloy III has gained extraordinary significance and application. Due to the way in which it solidifies in the ingot, alloy III has finer

Table C 11.8 Chemical composition, hardness characteristics, operational hardness and typical applications of tool steels for machining tools

Steel grade Designation	Chemical composition[a]									hardness soft annealed	Reference data for hardening		operational hardness	Frequent applications	Comparable AISI-steel [145]
	% C	% Si	% Mn	% Co	% Cr	% Mo	% V	% W		HB max.	hardening temperature °C	cooling medium[b]	HRC		
S 6-5-2	0.90	≤0.5	≤0.4	–	4.2	5.0	1.9	6.4		300	1190 ... 1230	O, HB, A	64	Twist drills, saws	M 2
S 6-5-2-5	0.92	≤0.5	≤0.4	4.8	4.2	5.0	1.9	6.4		300	1200 ... 1240	O, HB, A	65	Milling cutters, screw taps	~M 41
S 6-5-3	1.22	≤0.5	≤0.4	–	4.2	5.0	3.0	6.4		300	1200 ... 1240	O, HB, A	65	Screw taps, countersinks, brotches	M 3
S 7-4-2-5	1.10	≤0.5	≤0.4	5.0	4.2	3.8	1.8	6.9		300	1180 ... 1220	O, HB, A	67	Tools for machining austenitic steels and titanium alloys	M 41
S 2-10-1-8	1.09	≤0.5	≤0.4	8.0	4.0	9.5	1.2	1.5		300	1170 ... 1210	O, HB, A	67		M 42
S 10-4-3-10	1.28	≤0.5	≤0.4	10.0	4.2	3.6	3.3	9.5		300	1210 ... 1250	O, HB, A	66	Turning tools	T 15
S 12-1-4-5	1.38	≤0.5	≤0.4	4.8	4.2	0.9	3.8	12.0		300	1210 ... 1250	O, HB, A	66	Milling cutters for shaping	
S 6-3-2	0.90	≤0.5	≤0.4	–	4.2	3.0	2.1	6.0		300	1190 ... 1230	O, HB, A	64	Twist drills	
S 3-3-2	1.00	≤0.5	≤0.4	–	4.2	2.7	2.4	2.9		300	1170 ... 1210	O, HB, A	64	Saws	
X155CrVMo121	1.55	0.3	0.3	–	11.5	0.7	1.0	–		255	1120 ... 1150	O, HB, A	60 ... 63	Highly stressed wood processing (machining) tools	D 2
115 Cr V 3	1.18	0.2	0.3	–	0.7	–	0.1	–		223	780 ... 810	W	61 ... 64	Twist drills, thread taps ≥12 mm dia.	L 2
											810 ... 840	Oil		Twist-drills, thread taps <12 mm dia.	
100 Cr 6	1.03	0.3	0.4	–	1.5	–	–	–		223	820 ... 850	Oil	60 ... 63	Tools for wood processing (machining)	~L3
105 W Cr 6	1.05	0.3	1.0	–	1.0	–	–	1.2		229	800 ... 830	Oil	61 ... 64	Twist-drills, threading dies	
75 Cr 1	0.75	0.4	0.7	–	0.4	–	–	–		238	810 ... 840	Oil	42 ... 48	Wood saws	
80 Cr V 2	0.80	0.3	0.4	–	0.6	–	0.2	–		248	810 ... 840	Oil	42 ... 48	Wood saws	
C 125 W	1.28	0.2	0.2	–	–	–	–	–		213	760 ... 790	W	63 ... 66	Files	~W112

[a] For all steel grades of this table: ≤0.30% P and ≤0.030% S
[b] See annotation [b] in Table C 11.2

Table C 11.9 Types of alloys forming the basis of the high-speed steels in commen use today

Basic alloy	Chemical composition				
	% C	% Cr	% Mo	% V	% W
I	0.80	4.0	–	1.0	18.0
II	0.90	4.0	9.0	1.0 ... 2.0	–
III	0.90	4.0	5.0	1.0 ... 2.0	6.0
IV	0.90	4.0	–	1.0 ... 2.0	12.0

carbides than alloy I, at high deformation temperatures its carbides are less likely to become coarse, and it also decarburizes less than the basic alloy II, which is alloyed with molybdenum [6]. In addition, the basic alloy III is less dependent on fluctuations in the availability of the individual elements. All other types of high-speed steel that are adjusted to specific stress conditions can be derived from these basic alloys, as is shown in Table C 11.10 using the basic alloy III as an example.

It should be mentioned that high-speed steels alloyed with molybdenum have advantages in respect of the density. The basic alloy I with 18% W has a density of 8.7 kg/dm^3, the basic alloy II with 9% Mo has a density of 7.95 kg/dm^3 and the basic alloy III has a density of 8.2 kg/dm^3. Thus, 5.7 to 8.6% more tools can be produced from the high-speed steels containing molybdenum than from the same quantity of high-speed steels alloyed with 18% W.

The basic alloy III – shown in Table C 11.8 as S 6-5-2 – is chiefly used for simple twist drills, circular metal saws and saw blades. If a higher hardness or hardness at elevated temperatures is required, a cobalt addition of between 5% and 12% has proven successful; for this reason, S 6-5-2-5 is often used for milling cutters and threading taps. In conditions of high wear stress, the vanadium content is increased to 3% to 5% as compared to the basic alloy, as the example of the S 6-5-3 for countersinks, broaches and threading taps shows. The high-speed steels of very high hardness – S 7-4-2-5 and S 2-10-1-8 – which have a higher carbon content than the basic alloys and are alloyed with cobalt, have proven efficient for the machining of very tough but not too hard materials. For types of tools where wear resistance *and* hardness at elevated temperatures are required, as with turning

Table C 11.10 Adaption of the basic alloy III according to Table C 11.9 to specific stress conditions

Requirement	Chemical composition					
	% C	% Co	% Cr	% Mo	% V	% W
(Basic alloy)	0.90	–	4	5	2	6
Higher hardness at elevated temperatures	0.90	5 ... 12	4	5	2	6
Highest hardness	1.15	5 ... 12	4	5	2	6
Higher wear resistance	1.15 ... 1.50	–	4	5	3 ... 5	6
Hardness at elevated temperatures and wear resistance	1.35	10	4	5	4	6

tools, alloys with more vanadium and additional cobalt as compared to the basic alloy – steel S 10-4-3-10 – have been found efficient.

For lower stresses, the lower alloy economy steels S 6-3-2 and S 3-3-2 are commonly used for do-it-yourself tools or hand saw blades. There is, however, a wide range of machining work and cutting tools for which the use of high-speed steels is not necessary and more simple tool steels suffice. Accordingly, twist drills, threadcutting tools, and milling cutters, among other things for machining metallic materials without particular requirements placed on the life of the tool, are produced from 115 CrV 3 and 105 WCr 6.

Unalloyed or low-alloy steels with very high carbon content, such as C 125 W, are used for files. Wood-working tools of the most varied kinds, such as knives and cutters, are produced for low stresses from the steels 115 CrV 3 and 105 WCr 6 and, for high wear and temperature stresses, from ledeburitic steels with 12% Cr – the X 155 CrVMo 12 1 mentioned above. For low-stressed circular wood saws and master saw blades, 75 Cr 1 is the usual steel, and for high-stressed wood saws, such as circular saws, multiple-blade saws and master saw blades 80 CrV 2 is used.

Machining tools are generally formed by machining in the soft-annealed condition and by grinding after heat treatment. Thin twist drills are also produced by hot rolling or are ground from solid hardened blanks. To improve the grindability and to reduce change of shape during heat treatment, *powder-metallurgically produced steels* are used on a small scale for thread taps, broaches, and pinion-type cutters among other things. However, powder-metallurgical steels of the same chemical composition show no improvement of the tool life.

Machining tools are sometimes *cast* to shape to reduce production costs. However, the considerably greater susceptibility to fracture in the as-cast condition limits the use of cast machining tools [180].

Further information on steels for use in machining tools is to be found in the literature: on the status of development of high-speed steels [181], on the significance of carbon [182], on the influence of nitrogen [183], on the effect of silicon [184, 185], on the influence of sulphur [124, 186], on the influence of cobalt [187, 188], on the influence of molybdenum [189], on the effect of titanium [190], vanadium [191] and niobium [192], on influences of production methods [6] and of heat treatment [193, 194].

C 11.3.8 Steels for Cutting Tools

The term cutting tools is applied to a large number of tool types for shearing off, blanking (cutting out), punching and burring or – in connection with forming work – for stamping. A comprehensive survey of the processes is to be found in the literature [195].

The *durability (life) of a cutting tool is determined by the shape stability of the cutting edges*. These edges are subjected to stress through pressure and wear and, with increasing cutting gap width, also through bending and shearing forces. The compression stresses on the cutting edges are very much higher than is calculated from the mean stamp load $P = sd\tau$, where P = stamp force, s = length of cutting line, d = thickness of material to be cut, and τ = shear strength of the material to be cut. The compressive forces concentrate themselves in the stamp (mandrel) and in the die body on a zone at the cutting edge. The width of this zone is approximately the thickness of the material to be cut, and the center of the cutting die is often free of pressure as is illustrated on the left in Fig. C 11.45. This example is a normal cutting tool with wide shear gap of 5% to 10% of the thickness of the material to be cut. Even if the compression stress is concentrated on the cutting edge zone, the limits of the compression strength of hard tool steels are, however, only attained if the thickness of the material to be cut approaches the diameter of the stamp. From Fig. C 11.46 it can be seen that the stamp stress in this critical example is about 2700 N/mm² when cutting a material with a tensile strength of 700 N/mm². Therefore, with material that, on the whole, is thin in comparison with the stamp cross-section or to parts of the die cross-section, it is not usually the

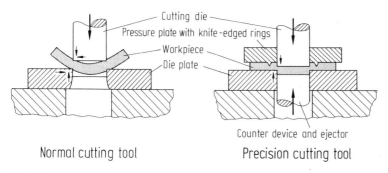

Fig. C 11.45. Operating principles of tools for normal blanking and precision blanking.

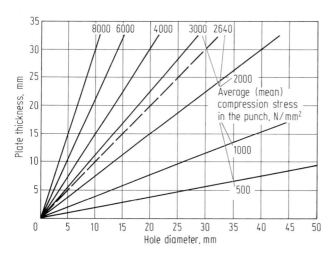

Fig. C 11.46. Mean compressive stress in a punch as a function of the die diameter (punch diameter) and thickness of the material to be cut, with a tensile strength of the material to be cut of 700 N/mm² (approx. St 60-2, see Table C 2.5).

compression strength but the wear resistance of the tool steel which is decisive for the life of a cutting tool.

For this reason, *tools for cutting thin material* up to about 6 mm thick are chiefly made from the high-carbide ledeburitic chromium steels X 210 Cr 12, X 210 CrW 12 and X 155 CrMoV 5 1 in Table C 11.11, which provides a survey of the steels commonly used for cutting tools. The latter steel is always the preferred choice if a high retention of hardness is required in respect of surface treatment to reduce wear. The ledeburitic chromium steels have very low and easily controllable dimensional changes during heat treatment, which is very important in cutting tools for perfect fit between punch and die (see C 11.2.2.1).

To an increasing degree, the high-speed steel types S 6-5-2, S 6-5-3 and S 10-4-3-10 are being used instead of chromium steels. With these steels, due to the possibly higher operational hardnesses of up to 65 HRC, the cutting edge wear is lower; in addition, just as with low-carbon steel with 12% Cr, due to their high retention of hardness these steels provide an excellent base on which to provide hard wear-resistant layers. High-speed steels having the same hardness, because of their fine carbides, are less susceptible to fracture than the ledeburitic steels with 12% Cr; this characteristic is of particular advantage with complicated tool shapes.

With *increasing thickness of the material to be cut*, the cutting edges of normal cutting tools, due to their wide shear gap, are subjected to more stress through bending and shearing forces. Because of the resulting increasing danger of fracture connected with increasing thickness of the material to be cut, the low-carbide, tougher steels from X 100 CrMoV 5 1 to 45 WCrV 7 in Table C 11.11 are used. The hardness of these steels for plate thicknesses above 12 mm must be reduced to about 48 HRC to further increase the toughness. However, this lower hardness reduces the cutting performance of the tools considerably. The necessary reduction of the operational hardness of the tools for cutting thicker plate also reduces their compressive stressability which should actually be increased for work with thicker plates. Thus there are inevitable limits to the plate thicknesses that can be cut and to their strength.

For *precision cutting tools*, the toughness concerning stress at the cutting edges is much lower due to the very small shear gap width. This gap width is usually distinctly less than 1% of the thickness of the material to be cut, due to the tool guide which is determined by the process (Fig. C 11.45). Therefore, even where material thicknesses exceed 12 mm, ledeburitic chromium steels and high-speed steels whose hardnesses must be reduced to between 58 and 60 HRC for the higher thicknesses, are still used for these tools. In this type of tool, however, the wear stress on the cutting edges through adhesive and abrasive wear is greater and the tool performance correspondingly lower.

For precision cutting tools, *powder-metallurgically produced cold-forming tool steels* and high-speed steels have also proven efficient [51]. Apart from the fact that very wear-resistant types of steel with high carbide content can be used that cannot be produced with conventional processes, powder-metallurgical steels offer more safety in tool production than those produced conventionally.

Because of the smaller shear gap width, which is often only a few µm, punches and die cutting stamps can only be produced with the necessary accuracy by spark

Table C11.11 Chemical composition, hardness characteristics, operational hardness and typical applications of tool steels for cutting tools

Steel grade Designation	Chemical composition									Reference data for			operational hardness	Frequent applications		Comparable AISI-steel
	% C	% Si	% Mn	% Cr	% Mo	% V	% W	% other	% P, % S [a]	hardness soft annealed	hardening temperature	cooling medium [b]		Type of tool	for sheet or plate thickness of mm	[145]
										HB max.	°C		HRC			
S 6-5-2	0.90	≤ 0.5	≤ 0.4	4.2	5.0	1.9	6.4	–	2	300	1100...1200	O, HB, A	62...65 61...63 59...61 58...60	{ Precision cutting tools { Cutting tools	≤ 3 ≤ 6 ≤ 12 > 12	M 2
S-6-5-2 S 10-4-3-10	1.22 1.28	≤ 0.5 ≤ 0.5	≤ 0.4 ≤ 0.4	4.2 4.2	5.0 3.6	3.0 3.3	6.4 9.5	– 10 Co	2 2	300 300	1180...1220 1210...1250	O, HB, A O, HB, A	62...65 61...63	Cutting tools Precision cutting tools	≤ 3 ≤ 6	M 3
X 210 Cr 12 X 210 CrW 12 X 155 CrVMo 121	2.05 2.13 1.55	0.3 0.3 0.3	0.3 0.3 0.3	11.5 11.5 11.5	– – 0.7	– – 1.0	– 0.7 –	– – –	2 2 2	248 255 255	940...270 950...980 1020...1050	O, HB, A O, HB, A O, HB, A	58...64 58...64 58...64 58...60	{ Cutting tools { Cutting tools { Precision cutting tools	≤ 3 ≤ 3 ≤ 6 > 12	D 3 D 6 D 2
X 100 CrVMo 5 1 90 MnCrV 8 105 WCr 6 60 WCrV 7	0.98 0.90 1.05 0.60	0.3 0.3 0.3 0.6	0.6 2.0 1.0 0.3	5.2 0.4 1.0 1.1	1.1 – – –	0.2 0.1 – 0.2	– – 1.2 2.0	– – – –	1 2 2 2	240 229 229 229	950...980 790...820 800...830 870...900	O, HB, A Oil Oil Oil	58...64 55...60 55...60 50...55	{ Cutting tools { Billet cropping tools	≤ 6 ≤ 12 ≤ 12 > 12	A 2 ~ O 2 ~ O 1 ~ S 1
45 WCrV 7 X 45 NiCrMo 4	0.45 0.45	1.0 0.3	0.3 0.3	1.1 1.4	– 0.3	0.2 –	2.0 –	– 4.1 Ni	1 2	225 262	890...920 840...870	Oil Oil	48...50 48...50	Cutting tools Cutting tools	> 12 > 12	S 1
C 105 W 1	0.05	0.2	0.2	–	–	–	–	–	4	213	770...800	W	56...60	Punches (piercing dies)	> 12	W 110

[a] See annotation [a] in Table C 11.2
[b] See annotation [b] in Table C 11.2

Table C 11.12 Steel selection for hand tools, their operational hardness and hardness characteristics

Tool	Steel grade Designation	Chemical composition							Hardness soft annealed	Reference data for hardening			Operational hardness
		% C	% Si	% Mn	% Cr	% Mo	% V	% P, % S[a]	HB max.	Hardening temperature °C	Quenching medium[b]	HRC
Bench (hand) hammers	C 45 W	0.45	0.30	0.70	–	–	–	1	190	800...830	W	56...58 at head and peen
Ball hammers	C 60 W	0.60	0.30	0.70	–	–	–	1	231	800...830	Oil	~ 60 at head and point
Axes	C 45 W	0.45	0.30	0.70	–	–	–	1	190	800...830	W	54...56 at the cutting edge
	C 60 W	0.60	0.30	0.70	–	–	–	1	231	800...830	Oil	
Scythes	C 60 W	0.60	0.30	0.70	–	–	–	1	231	800...830	W	42...46
Sickles	C 85 W	0.85	0.35	0.60	–	–	–	3	222	800...830	Oil	
Saws	80 CrV 2	0.80	0.35	0.40	0.55	–	0.20	2	220	800...830	Oil	
Shears	C 60 W	0.60	0.30	0.70	–	–	–	1	231	800...830	W	56...60
	75 Cr 1	0.75	0.35	0.70	0.35	–	–	2	220	810...840	Oil	
	85 Cr 1	0.85	0.40	0.65	0.40	–	–	1	225	800...830	Oil	
Nippers, pliers	C 45 W	0.45	0.30	0.70	–	–	–	1	190	800...830	W	40...46 at the pliers flange
	C 60 W	0.60	0.30	0.70	–	–	–	1	231	800...830	Oil	Cutting edges induction post-hardened to 50...60 HRC
	31 CrV 3	0.31	0.35	0.50	0.60	–	0.10	2	220	830...860	W	
	51 CrV 4	0.51	0.25	1.05	0.95	–	0.15	2	231	830...860	Oil	
Wrenches	31 CrV 3	0.31	0.35	0.50	0.60	–	0.10	2	220	830...860	Oil	50...56
	51 CrV 4	0.51	0.25	1.05	0.95	–	0.15	2	231	830...860	Oil	
Hand chisels	45 CrMoV 7	0.45	0.25	0.95	1.8	0.25	0.05	2	240	840...860	Oil	~ 54
Screwdrivers	61 CrSiV 5	0.61	0.85	0.75	1.15	–	0.10	1	220	850...880	Oil	56...60 for screwdrivers about 62 for screwdriver bits
	73 MoV 52	0.73	1.20	0.50	–	0.55	0.20	4	220	800...830	Oil	
Screw stocks	C 105 W 1	1.05	0.15	0.15	–	–	–	4	213	770...800	W	about 60
	145 Cr 6	1.50	0.25	0.60	1.40	–	–	1	234	830...870	Oil	

[a] See annotation [a] in Table C 11.12
[b] See annotation [b] in Table C 11.2

erosion [112]. In conventionally-produced steels the stresses in the hardened die material can be so high and so irregularly distributed that irregular changes in shape or even cracks can be caused through stresses released during eroding, making the tool components unusable. The stresses are lower in hardened powder-metallurgical die material and much more uniformly distributed, so that the difficulties described do not occur during eroding.

For *simple cutting (separating) tools* such as shearing knives or circular shearing knives [196, 197], depending on the thickness of the material to be cut, approximately the same tool steels and operational hardnesses are used as for normal cutting tools because the cutting stresses are comparable. For perforating (piercing) thicker plate of lower strength, the unalloyed tool steel C 105 W 1 has also proven efficient. Piercing punches made from this steel do not have a high compression strength; they do, however, have an adequate maintenance of the cutting edge and, due to the compressive stresses present in shell hardening steels, show a good resistance to fracture.

Cutting at high temperatures is not commonly practiced in the press-working of metals; it is only done in rolling mills when cutting (separating) rolled bars on billet shears. As the billet being at rolling temperature has a low strength, tool hardnesses of about 50 HRC are sufficient. However, due to possible heating of the tool, the steels 60 WCrV 7 and 45 WCrV 7 or even X 45 NiCrMo 4, which have higher hardness retention, are used.

C 11.3.9 Steels for Hand Tools

Hand tools are not only indispensible aids in the private and handicraft sectors but also in industrial-scale production processes. In the oldest field of application for steels, unalloyed steels are used for some types of tool, such as hammers, axes, pliers, saws and files, whereas for other types of tools, such as wrenches, screwdrivers, chisels or boring tools, low-alloyed tool steels are used, and for twist drills, even high-speed steels are used.

Apart from the service properties, hardness and resistance to fracture, a simple hardenability as well as good hot ductility are required of steels for hand tools because many hand tools are formed by forging or pressing. The most commonly-used steels and operational hardnesses for the different types of tool are shown in Table C 11.12.

C 12 Wear-resistant Steels

By Hans Berns

The annual cost of losses from wear in the Federal Republic of Germany is estimated at 5 billion DM [1]. A considerable part of this cost is to be allotted to wear-resistant iron alloys. *Abrasion, adhesion, surface fatigue and tribochemical reactions* are the main wear mechanisms (see B 10).

C 12.1 Key Properties and their Characteristics

In machine parts and tools for mineral extraction (mining) and processing, in metal working tools, gears, bearings, wheel/rail contact and so on, mixed modes of wear are usually encountered in which one of the wear mechanisms is prominent. *Wear resistance* is an important property in service of the steels to be discussed here but it is not solely a materials property, and it is therefore characterized best in service or simulated service tests. Basic studies on wear behavior use various laboratory model tests like the pin-on-disc, the blast wear or the four-ball test [2, 3]. In addition to wear resistance at the surface, bulk properties like strength and toughness are required to withstand the overall mechanical loading.

The machine parts and tools are manufactured by casting, hot or cold working, or powder metallurgy (PM) and may be clad by welding or spraying. Heat and surface treatments play an important role. All these production processes call for specific manufacturing properties like *deformability, weldability, machinability*.

C 12.2 Microstructural Considerations

The microstructure of wear-resistant alloys results from the chemical composition, production and heat treatment. *Carbon* in solution at hardening temperature determines the matrix hardness and the amount of retained austenite. Excess carbon is combined as carbides. Higher carbon content, therefore, increases the wear resistance via the matrix hardness and the carbide content and diminishes the toughness. The other *alloying elements* improve the hardenability of the matrix and, like chromium, molybdenum and vanadium, raise the carbide hardness above

---

References for C12 see page 787.

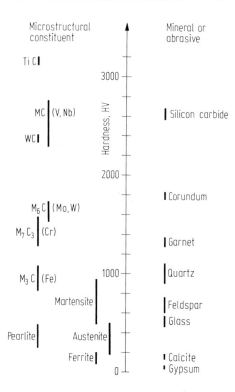

Fig. C 12.1. Comparison of mineral or abrasive hardness with that of microstructural constituents in iron-base alloys.

that of the iron carbide M_3C by forming M_7C_3, M_6C and MC carbides respectively (Fig. C 12.1).

With *production processes* like casting, hardfacing and powder metallurgy, the growing solidification rate reduces the sizes of primary and eutectic carbides almost down to 1 µm. Secondary carbides precipitated from solid solution are below 1 µm in thickness. Net-like carbide distributions are encountered in hypoeutectic castings and hypereutectoid steels. Hot working with low finishing temperatures shifts the carbide distribution towards a dispersion with some alignment of the primary and eutectic carbides (Fig. C 12.2). The size and distribution of coarse carbides is brought about by the production processes, and the strength of the matrix is achieved by *heat treatment*.

Normalizing reduces the interlamellar spacing and the thickness of the pearlite lamellae. A more pronounced strengthening results from quenching, which is usually followed by tempering to ensure a sufficient toughness. In steels with more than one percent of molybdenum and vanadium, precipitation hardening sets in at a tempering temperature of about 500 °C. This secondary hardening produces a higher hot strength in the matrix.

Abrasion is dominant in *grooving wear* [4]. To increase wear resistance the microstructural constituents should be harder than the attacking particles. The highest possible matrix hardness of about 900 HV is obtained in high carbon martensite. Unstable retained austenite proves beneficial as it transforms to hard

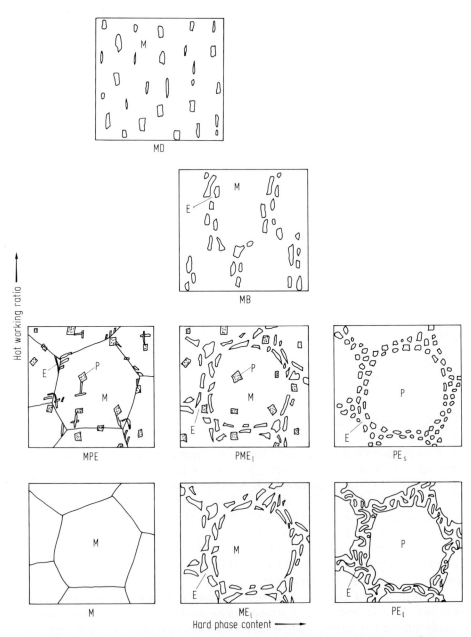

Fig. C 12.2. Schematic representation of the microstructure of wear-resistant iron alloys. M: metal dendrites, possibly with small secondary carbides (St 37, C 70, X 120 Mn 12); ME_1: metal dendrites surrounded by a net-like lamellar eutectic of carbides (X 210 Cr 12); PE_1: primary hard phases (carbides or borides) surrounded by a net-like eutectic of carbides (white cast irons, hardfacing alloys); MPE: metal dendrites containing primary carbides and little eutectic (X 120 Mn 12 + Nb or Ti); PME_1: as ME_1 with some primary precipitation of carbides (X 210 Cr 12 + Nb, Ti); PE_s: as PE_1 but with an interlocking eutectic skeleton (hardfacing alloys with > 4% B); MB: banded hypoeutectic carbides after a low hot-working ratio (X 210 Cr 12); MD: dispersion of aligned hypoeutectic carbides after a high hot-working ratio (X 155 CrVMo 12 1)

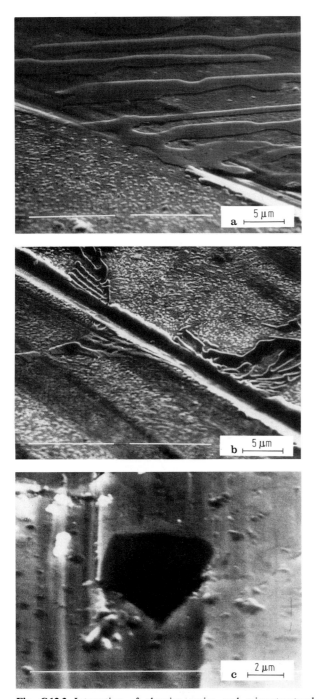

Fig. C 12.3. Interaction of abrasive grains and microstructural constituents in hardened steel X 210 Cr 12 + Ti: the matrix is scratched by the harder abrasive grains, but the carbides are harder than the abrasive grains (see Fig. C 12.1) and, therefore are not cut, if they are of sufficient size. **a)** M_7C_3 (slow solidification)/flint; **b)** M_7C_3 (fast solidification)/flint; **c)** TiC/Al_2O_3.

martensite causing compressive residual stresses in the wear surface. Several minerals like quartz and corundum are harder, though (Fig. C 12.1), which requires that even harder carbide phases are embedded in the matrix to block the scratching path (Fig. C 12.3). To be most effective the size of the carbides should exceed the groove cross section and they should be homogeneously dispersed. The stronger the matrix the better the carbides are supported (Fig. C 12.4).

Adhesion is a key mechanism in *sliding wear* of two metal components. The matrix hardness has no general influence but a higher hardness supports the asperities in a number of tooling applications. This unevenness keeps the real contact area small and may thus promote the formation of tribochemical layers due to higher spot temperatures. Carbides prohibit local metal contact thus reducing adhesion. Therefore in many applications microstructures similar to those used in grooving wear also proved useful in resisting sliding wear.

Surface fatigue plays an important role in wear by *rolling contact* and *cavitation*. In general a high matrix hardness suppresses microplasticity thereby prolonging the life of a component. On the other hand, coarse hard phases develop stress concentration thus initiating cracks and eventually pitting. Hardened steels of high purity with a lower content of fine, dispersed carbides are required to resist these conditions.

Tribochemical reactions may reduce wear by the formation of parting layers. These layers are best supported by a hard substrate. To summarize, a hardened matrix with carbides is certainly not the only, but a most common microstructure in wear-resistant parts.

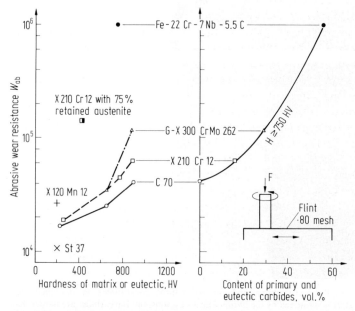

Fig. C 12.4. Abrasive wear-resistance in a pin-on-disc test [8, 13] depending on the content of sufficiently-coarse carbides, harder than flint, and on the hardness of the matrix or eutectic.

C 12.3 Applications of Steels and Iron-base Alloys

Steels subject to wear in service are selected from different groups of steels including high-strength weldable steels (see C 2), quenched and tempered, and surface-hardened steels (see C 5), tool steels (see C 11), and roller bearing steels (see C 26). In the narrow sense, wear-resistant steels and alloys are used for wear protection in the presence of hard granular, mainly mineral substances.

In sheet metal structures like bunkers, chutes, and cyclones, *weldability* prevails. Normalized low-carbon steels such as St 52-3 with a ferritic microstructure of low pearlite content may be partially coated by hardfacing welding (see Table C 12.1). Quenched and tempered high-strength steels like StE 690 give a somewhat higher hardness. Conveyor chains of steel 23 MnNiMoCr 6 4 are similarly heat treated (see C 30). In low-alloy dual-phase steels the hard dispersion of martensite in the soft ferrite is beneficial [5].

Low cost *quenched and tempered steels* like 42 CrMo 4 and 50 Mn 7 are used for dragline or grab buckets, caterpillar chains and their induction-hardened rollers. The dispersion of fine carbides in the tempered martensite or bainite provides a good combination of strength, toughness and wear resistance to withstand the severe mechanical loading [6].

Raising the carbon content to about 1% reduces the toughness but improves the wear resistance. Parts of complex shape are prone to distortion and, are therefore, normalized to a pearlitic microstructure containing a considerable amount of lamellar M_3C carbides (90 Mn 4, Table C 12.1). Hardening reduces the carbide content but increases the matrix hardness (100 Cr 6). Mill liners, rods and balls are made from these *hypereutectoid steels* as well as parts for scrapers, crushers and stirrers. Air hardening steels like X 100 CrMoV 5 1 are used for pump housings. The higher tempering temperature of this secondary hardening steel allows residual stresses to be reduced and the toughness improved for a given hardness. An exception among the steels with 1% C is the *Hadfield manganese steel* X 120 Mn 12. In this steel, after quenching from solution temperature the austenite is stabilized by the high manganese content. This exceptionally tough microstructure work hardens up to 600 HV in the wear surface when impact loading occurs as it does in crusher jaws. The resistance to grooving wear may be improved by a fine grain size [7], dispersed carbides [8] and a reduced manganese content [9]. These conditions were applied, for instance, to crusher cones.

In chromium steels and white cast irons with 1.5 to 3.5 per cent C, an increasing amount of coarse eutectic carbides precipitated from the melt, impede grooving wear [10]. The M_3C carbides of the pearlitic *white cast iron* and the *martensitic cast iron* G-X 330 NiCr 4 2 are transformed by added chromium to the harder M_7C_3 carbides found in martensitic cast iron G-X 300 CrMo 26 2. This material provides long life in ball mill liners used in cement grinding and of baffle plates used in impact rock crushing machines, especially when harder minerals like quartz are present (compare Fig. C 12.1). *Steels* with 12% Cr can also profit from the M_7C_3 carbides. Steel X 210 Cr 12, for example, is used for compacting tools in manufacture of bricks [11] and grinding wheels, and in powder metallurgy. The addition of niobium and titanium leads to even harder MC carbides [8]. Higher amounts of

Table C 12.1 Selection of iron-base alloys for wear parts (each material resembles a group of similar alloys)

Alloy	Chemical composition				Heat treatment	Hardness HV in service	Carbide type	Carbide content vol.%
	% C	% Mn	% Cr	% other				
St 52-3	≤ 0.2	≤ 1.5			n	160...180	M_3C	3
90 Mn 4	0.9	1.0			u, n	280...380	M_3C	14
StE 690	0.15	0.8	0.5	0.3 Cu, 0.5 Mo, 0.8 Ni, 0.06 V	h + t	215...240		
42 CrMo 4	0.42	0.7	1.0	0.2 Mo	h + t	250...400	M_3C	< 4
50 Mn 7	0.5	1.8			h + t	350...450	M_3C	< 4
100 Cr 6	1.0	0.4	1.5		h + t	600...800	M_3C	6
X 100 CrMoV 5 1	1.0	0.5	5.0	1.0 Mo, 0.2 V	h + t	500...750	M_7C_3	
X 210 Cr 12	2.0	0.4	11.5		h + t	600...800	M_7C_3	16
X 155 CrVMo 12 1	1.55	0.4	12.0	0.7 Mo, 1.0 V	h + t	600...800	M_7C_3	12
White cast iron	3.3	1.0			u	400...550	M_3C	48
G-X 330 NiCr 4 2	3.3	0.5	2.0	4.0 Ni	(h) + t	550...750	M_3C	42
G-X 300 CrMo 26 2	3.0	0.7	26.0	2.0 Mo	h + t	800...900	M_7C_3	30
[b]	5.5	0.7	22.0	7.0 Nb	u	750...800	M_7C_3, MC	52
X 120 Mn 12	1.2	12.0			s	180...220	–	0

[a] h = hardened, n = normalized, s = solution annealed, t = tempered, u = untreated
[b] Filler wire for hardfacing

retained austenite also may have a favorable effect [11, 12]. Steel X 155 CrVMo 12 1 will take secondary hardening and, therefore has a better hot strength. These qualities are required in handling hot materials, as in the crushing of sinter. Hot sinter sieves, however, are usually made from heat-resistant austenitic steels.

The highest carbon contents of up to 6% are encountered in hardfacing weld deposits. Excavator shovels in open-pit mining for example, are clad with the filler wires listed in Table C 12.1. A partial exchange of carbon by boron results in the formation of boride hard phases. A comparable wear resistance may be achieved with a lower alloy content [13].

C 13 Stainless Steels

By Winfried Heimann, Rudolf Oppenheim and Wilhelm Wessling

C 13.1 Required Design Properties of the Steels

C 13.1.1 Fields of Application for Stainless Steels

Stainless steels are used primarily under normal environmental conditions where resistance is required to attack by atmospheric oxygen, airborne moisture, aqueous solutions, river and industrial water, chloride-rich sea water and brackish water, and, secondarily, in the more aggressive media such as inorganic and organic acids or alkalis, where chemical resistance is required. Because of the range of corrosive attacks encountered, stainless steels, generally are required to have a chromium content of more than 12%; Further alloying elements are added to suit the required properties of use.

The principal industrial user in both historical and volume terms has been the chemical industry, which early recognized the advantages of this group of steels and has specified it increasingly ever since as the preferred metallic material for large-scale plant. Energy production and, more recently, new areas in marine and environmental engineering, represent further plant construction applications in which stainless steels play an important part. As durable and attractive material, stainless steel also has gained a number of uses outside the sphere of industrial engineering, e.g. in architecture, automotive accessories, cutlery, in domestic machines and household appliances, and in medical equipment.

C 13.1.2 Resistance to Various Types of Corrosion

The requirements these steels must meet in terms of corrosion resistance are derived from the media and reaction products involved in the chemical reactions and from process-related conditions such as temperature and mechanical stress. The interplay between these chemical, thermal and mechanical influences produces wide ranges of requirements which must be considered in choosing the correct material for the application in question.

In low-alloy steels the mechanical properties predominate, but when stainless steels with comparable mechanical properties are used, their *chemical resistance* constitutes *the main selection criteria* (see B 3). Further requirements, particularly

---

References for C13 see page 787.

with regard to the mechanical, technological and physical properties, may be satisfied by the relevant set of desirable performance characteristics.

The many and varied demands concerning chemical resistance can be summarized in a *fundamental requirement*: the predominant concern is to ensure resistance to local corrosion phenomena whereas general surface removal shall be uniform and tending to be minimal. A prerequisite for this resistance is that the steel must be used in its passive condition. In aggressive media such as hot hydrochloric acid or sulphuric acid, and in hydrofluoric acid, this condition is realized, if at all, only with a high alloy input. In these instances a limited degree of general attack may be permitted, if the surface removal during exposure remains uniform and thus calculable. Compensation for this removal can be provided in the form of corrosion allowances. However, even the slightest degree of surface removal must sometimes be avoided,.for example, when the product being manufactured or processed must remain completely free of any contamination from dissolved metal ions. The handling of pharmaceutical, photochemical and high-purity chemicals, and the food chemistry sector are examples. High demands are also made in medicine with regard to surgical instruments and implants for the human body, for instance.

Rather than this controllable uniform attack, the *local corrosion phenomena cause limitations in durability* and availability for large industrial plant and apparatus. Exposure to chloride-containing water and solutions can induce point corrosion in the form of pitting. Crevice corrosion occurs as a result of changes in concentration and a depletion of oxygen inside the crevice as opposed to ample oxygen supply to the metal outside. Galvanic or bimetallic corrosion occurs when a metal (the anode) has a positive current flowing from it to a less reactive metal (the cathode) with which it is in electrical contact in the presence of an electrolyte. Chromium depletion as a result of chromium carbide precipitation at the grain boundaries leads to intergranular corrosion (IGC). Stress corrosion cracking (SCC) is triggered by tensile stresses combined with corrosive chemical attack.

Exothermic reactions or externally supplied energy designed to accelerate chemical reactions for the sake of operational economy correspond with a higher intensity of attack on the materials used. Corrosion attack in aqueous solutions, for example, is promoted or may be initiated as temperatures increase. In mechanically stressed components, the temperature dependency of strength and toughness properties must also be taken into account.

C 13.1.3 Mechanical and Technologically Significant Properties

Requirements for *mechanical properties* generally refer to strength and toughness under the usual stress conditions as brought about by static, cyclic or shock loading in the normal environment, at low and elevated service temperatures. Thus internally pressurized piping and pump housings are subject not only to static loads but, if the medium pulsates, to cyclic stress. Surface wear can be brought about not only by material removal as a result of corrosion but also by mechanical attack in the form e.g. of abrasion or cavitation, such as occurs with pistons and impellers in pumps or agitators of mixing apparatus, and in piping, valves and pumps carrying fast-flowing liquids. There are close similarities with the requirements and service conditions that prevail for quenched and tempered steels in comparable applications.

Economy in the production and construction of industrial plant and equipment depends on the existence of *good processing properties* in the material to be used. These properties include machinability, sufficient cold and hot formability and the often indispensable weldability. Good deep drawing properties are an important prerequisite for the manufacture of domestic articles, and the heavy deformation process employed for the manufacture of threaded fasteners requires the material to have particularly good cold upsetting behavior.

Thus the *most important material properties* required in rust- and acid-resisting (stainless) steels can be summarized as follows:
- Corrosion resistance in gases, aqueous solutions and acids
- Adequate strength and toughness even under conditions of thermal loading
- Cold and hot formability, machinability
- Weldability

C 13.2 Characterization of the Required Properties by Test Values

C 13.2.1 Testing Corrosion Resistance

The *corrosion resistance* of steels is not verified in absolute terms. Rather, an indication must be provided of the conditions under which corrosion susceptibility or resistance generally is found in materials in relation to their chemical composition and as-treated condition. Thus, the corrosion resistance of a steel can only be described in relation to the specific requirements of an application. The conditions under which either the steel remains immune from corrosive chemical attack or the rate of corrosion stays at a permissible, calculable level must therefore be determined. Owing to the many types of corrosive attack, the existence of innumerable aggressive media, and the frequent involvement of additional mechanical and thermal loading, only a few selected tests will be described here.

The most frequent types of corrosion, some of which can occur side by side and as a result of tensile stress, are described in Fig. C 13.1 [1].

The *chemical resistance* of stainless steels *is due to the formation of a passive layer* on the surface (see B 3.6). Under reducing conditions – in aggressive media such as hydrochloric acid, sulphuric acid and phosphoric acid, for example – the build-up of this stable protective layer is inhibited or even completely prevented with the result that the steel undergoes uniform dissolution. The determining factors here are the temperature and concentration of the corrosive solution and the rate of surface removal. The even, general dissolution represents a linear rate of surface removal in relation to time and permits the use of results from short-term tests for extrapolation to predict long-term behavior (see Fig. C 13.2) [2].

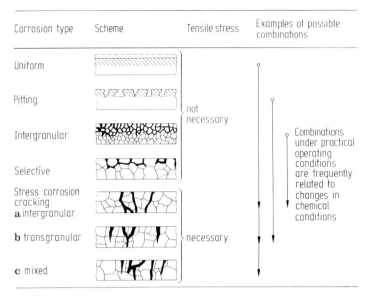

Fig. C 13.1. Schematic representation of common types of corrosion. After [1].

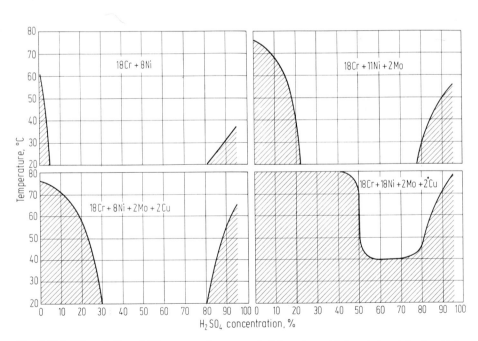

Fig. C 13.2. Effect of alloy additions on surface removal of chromium-nickel steels in sulphuric acid. In the hatched areas metal removal is less than 0.1 mm/year. After [2].

Removal rates of 1 g/m² h, for example, will reduce the wall thickness of a vessel by 1 mm/year. If the weight loss is not more than 0.3 g/m² h, a steel is regarded as being resistant for practical purposes. This rule of thumb may be applied across a wide range of applications, but must sometimes be adapted to special requirements, i.e., reduced. This statement is particularly true of steels for vessels used in the food processing industry and the pharmaceutical industry.

In addition to the chemical composition of the steels, their *microstructure* which can be influenced by heat treatment is also *of fundamental significance* for uniform, general corrosive attack. Stainless heat-treatable steels in their hardened condition such as, for example, the knife steel X 40 Cr 13 with about 0.45% C and 13% Cr, have a degree of corrosion resistance which is in keeping with their composition. Tempering causes the chemical corrosion behavior to undergo a considerable change (Fig. C 13.3 [3]). Owing to the precipitation of chromium carbide and the associated local reduction in the chromium content of the matrix, the weight losses sustained may increase several-fold, with the highest rates of surface removal occurring in the area of the greatest decrease in hardness.

Intergranular corrosion in ferritic and austenitic stainless steels is the result of chromium carbide formation at the grain boundaries [4]. The resistance of the steels to this form of preferential (selective) corrosion depends to a large extent on their carbon content. The carbon however, can be tied up (fixed) with titanium or niobium to form stable carbides and thus prevent this form of corrosion (see B 3.7.2). Tests to determine susceptibility to intergranular corrosion are conducted in solutions of sulphuric acid and copper sulphate [5], with results plotted in

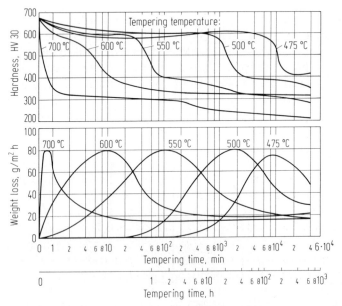

Fig. C 13.3. Influence of tempering steel X 40 Cr 13 on hardness and corrosion properties in boiling 5% acetic acid. The steel was tempered as indicated after 1030 °C 15 min/550 °C 1 min/air. After [3]. Tempering followed by cooling in air.

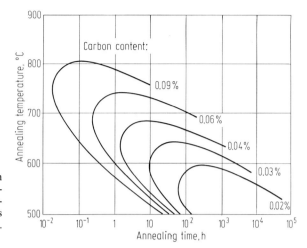

Fig. C 13.4. Influence of the carbon content on the position of the sensitivity range of intergranular corrosion in unstabilized austenitic steels with about 18% Cr and 9% Ni. After [6].

intergranular corrosion diagrams as a function of annealing temperature against time (Fig. C 13.4) [16]. DIN 50914 [5] states that intergranular corrosion is present when the attack at the grain boundaries exceeds a depth of penetration of 0.05 mm. Higher chromium contents render chromium depletion in the usual test (DIN 50 194) no longer detectable. With steels of this kind, sensitization is determined by more stringent tests such as those described in SEP 1877 [7]. For applications involving nitric acid the test specified in DIN 50921 [8] is recommended.

In chloride-containing aqueous solutions, rust and acid resisting steels can undergo point (local) attack in the form of *pitting corrosion*. At these points protection is no longer provided by a passive layer and the steel is locally active; local cells form, leading to the anodic dissolution of the active region (Fig. C 13.5 [9]). By plotting current density against potential measurements it is possible to ascertain the conditions under which this electrochemical process is activated. The indicative parameter to be determined is the electrical potential at which a steel undergoes pitting in an aqueous solution. The representation of the pitting potential as a function of the test temperature is often used to characterize resistance to pitting (Fig. C 13.6 [10]).

Pits are frequently the starting points for a further type of preferential corrosion, namely *crevice corrosion*. If there is a lack of oxygen needed for repassivation or if it cannot be supplied in sufficient quantities, progressive corrosion can be observed. Corrosion in small crevices is due to the same electrochemical processes discussed above. A reliable prediction of the resistance of a steel can only be made on the basis of long-term exposure to the corrosive medium concerned with an artificial crevice being created by the placement of a plastic disc on the steel specimen.

The interaction between corrosive chemical attack and mechanical tensile loading can lead to the dangerous phenomenon of *stress corrosion cracking* occurring in the steel [11]. Crack propagation is promoted not only by the

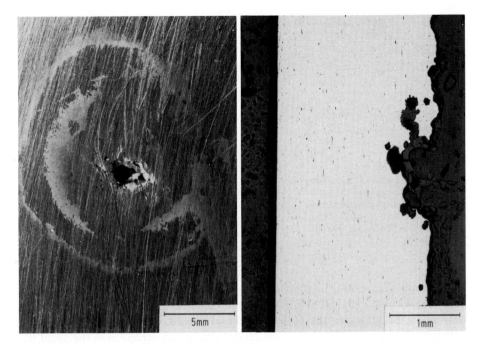

Fig. C 13.5. Pitting in a storage vessel of steel X 5 CrNi 18 9 (with about 0.05% C, 18% Cr and 10% Ni) for cola seed concentrate. After [9].

Table C 13.1 Chemical composition of the steels investigated according to Figs. C 13.6 and C 13.7

Steel grade Designation	Chemical composition					Microstructure
	≤ %C	%Cr	%Mo	%Ni	%others	
X 5 CrNi 18 9	0.05	18.5		9.5		⎫
X 2 CrNi 19 11	0.02	19		11		⎬ austenitic
X 2 CrNiMo 18 10	0.03	17	2.2	11.5		⎭
X 2 CrNiMoSi 19 5	0.03	18.5	2.7	4.7	1.7 Si	⎫
X 2 CrNiMoN 22 5 3	0.02	22	2	5	0.15 N	⎬ ferritic-austenitic
X 3 CrMnNiMoN 25 6 4	0.04	25.5	2.3	3.7	5.8 Mn, 0.37 N	⎭
X 2 CrNiMoCuN 17 16	0.03	17	6.3	16	1.6 Cu, 0.15 N	austenitic

mechanical stress but also by electrochemical reactions at the crack front as with crevice corrosion. In these conditions it is necessary to determine service life as a function of mechanical stress and electrochemical potential. Below a certain stress-dependent threshold potential it can be expected that stress corrosion cracking will no longer occur [12]. One common way to measure the resistance of stainless steels to SCC is to determine the time to failure at a stress value expressed as a percentage of tensile strength (Fig. C 13.7 [13]). Data must include the type of electrolyte used because transference of the results obtained in one test solution to another medium is possible only in exceptional circumstances and is generally not permissible.

In those steels stabilized with titanium or niobium, the relatively high level of thermal loading resulting from multi-pass welding may introduce susceptibility to

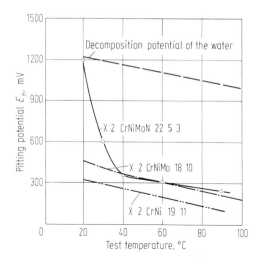

Fig. C 13.6. Pitting potential of three stainless steels in artificial sea water at 20 °C to 100 °C. For chemical composition of the steels, see Table C 13.1. After [10].

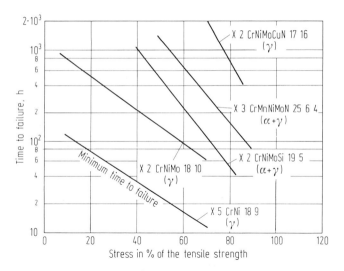

Fig. C 13.7. Resistance of five different steel grades (for chemical composition, see Table C 13.1) to stress corrosion cracking. After [13]. In the tests tensile specimens were electrically heated in a common-salt solution with 3% NaCl to such an extent that a salt crust formed on them at the water line.

intergranular corrosion along a narrow zone in the base material running close to the weld. This so-called *knife-line corrosion* is caused by formation of chromium carbide following the preceding dissolution of titanium or niobium carbides (see B 3.7.2) and is detected by means of a bending test.

Stress corrosion cracking occurs under static tensile loading, but *corrosion fatigue cracking* takes place under conditions of cyclic mechanical stresses. The fatigue limit of steels (fatigue strength for an infinite number of stress cycles) is usually determined by plotting Wöhler curves. Under corrosive service conditions

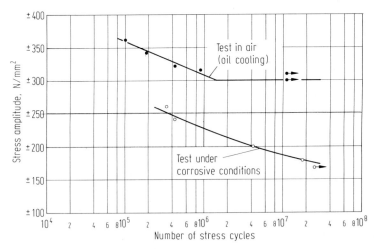

Fig. C 13.8. Wöhler curves for steel X 10 CrNiNb 18 9 (with 0.06% C, 18% Cr, 10% Ni and 0.7% Nb) for testing in air and under corrosive conditions (in aqueous solution with 44.7 g NaCl + 11.3 g FeSO$_4 \cdot$ 7H$_2$O/l, adjusted to $p_H = 1.7$ with H$_2$SO$_4$. After [14].

in which repassivation cannot take place, it is not possible to determine a final stressability value for the fatigue limit because the failure stress falls progressively as the number of cycles increases (Fig. C 13.8 [14]).

C 13.2.2 Parameters Characterizing Mechanical and Technologically Significant Properties

The mechanical and other technologically significant properties of stainless steels are determined by the same tests used for other steels. Thus the *tensile test* provides the designer with yield strength data such as the 0.2%-proof stress and tensile strength values. As toughness-indicators, elongation after fracture and reduction of area. In austenitic steels employed for construction of pressure vessels and similar apparatus, the 1%-proof stress is being used as the basis for design to allow for the relatively high level of formability exhibited by the austenites. For a number of stainless steels, however, the test is not limited to room temperature; instead, test temperatures ranging from 4 K (boiling point of helium) to the service limits for this group of steels at around 550 °C, are used.

The notched-bar impact test constitutes a further means of measuring the *toughness* of steels (Fig. C 13.9 [5]). It provides a reliable indication of the temperature at which the impact energy of steels with a bcc crystal lattice drops from the high range to the low range (impact transition temperature). In austenitic steels, the impact test offers a means of detecting changes in the material microstructure such as precipitation processes.

With regard to the *forming of sheet steel*, the tensile test provides certain characteristic data such as the yield stress to tensile strength ratio, and elongation before necking (elongation at maximum force). In austenitic steels, this uniform

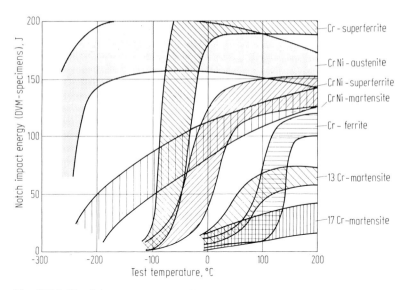

Fig. C 13.9. Notch impact energy-transition temperature curves for various grades of stainless steel. After [15].

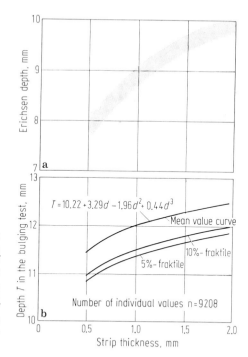

Fig. C 13.10. Relationship between thickness and deep drawing properties of cold wide strip of stainless steels. After [16]. **a** Scatter band of the Erichsen test depth values for strip of ferritic steels with about 0.05% C, 17% Cr (X 8 Cr 17, X 6 CrMo 17, X 8 CrNb 17, X 8 CrTi 17, and X 5 CrTi 12). **b** Relationship between depth T after cupping test and strip thickness d for the austenitic steel X 5 CrNi 18 9. Statistical evaluation.

elongation value might be seen as the limiting value for pure stretch forming. However, if an assessment must be made concerning stretch-forming behavior under multi-axis tensile loading, then the cupping test, similar to the Erichsen test, is more appropriate (Fig. C 13.10 [16]). The characterizing parameters of this test are the depth of draw achieved prior to cracking, and the deformation in the vicinity of the crack [17]. To assess deep drawing behavior it is useful to apply not only the anisotropic values r, Δr and r_m, but also the limiting drawing ratio as characterizing parameters [18] (for details, see B 8).

Welded joints must exhibit properties that are comparable with those of the base or parent material. Thus *assessment of the weld metal and heat-affected zone* again involves test methods such as the tensile and notched-bar impact tests. Unlike a solution-annealed and quenched base material, a weld does not have a homogeneous microstructure but rather a highly-varied structure in keeping with the influences of the welding heat. This condition is found particularly in multi-pass welds such as those needed for heavy gauge plate. The high level of heat input can lead to deleterious brittleness and reduced corrosion resistance. It is therefore necessary to ensure weld quality by careful selection of materials and comprehensive testing. Apart from the notched-bar impact test, the bending test has also proven a useful indicator in the evaluation of toughness, the parameters here being the bending angle achieved and the level of bending strain (Fig. C 13.11 [19]). Finally it should be mentioned that the weld metal and the heat-affected zone of the base material must be tested for cracking. Among the possible tests the X-ray and dye-penetration methods are particularly widely used.

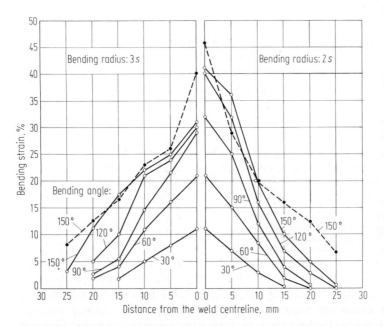

Fig. C 13.11. Results of toughness tests on welds between 8 mm thick plates. After [19]. The drawn curves refer to steel X 2 CrNiMoN 22 5 3, the dotted curve refers to X 2 CrNi 19 11 (s = plate thickness)

C 13.3 The Importance of Microstructure and its Chemical Composition for Corrosion Performance

C 13.3.1 The Decisive Importance of Chromium Content

The *corrosion resistance* of stainless steels is achieved by *the chromium content*, as explained in detail in B 3.3. With a chromium content of about 12% or more, a metallically-bright steel surface forms in air or in an oxidizing aqueous electrolyte the passive state which is characteristic for chromium metal. This chemical behavior which is also observed in noble metals is due to the formation of a submicroscopically – thin oxide layer – chromium oxide in stainless steels – and it is to this layer that such metals owe their chemical resistance. The formation and maintenance of this covering layer requires an adequate supply of oxygen. The electrochemical standard potential of iron as compared to the standard hydrogen electrode is increased by the addition of about 12% Cr, taking it into the positive range (Fig. C 13.12 [20]). As the chromium content is increased further, the metal wastage in various corrosive media is reduced to a negligible value.

From investigations into the causes of intergranular corrosion in stainless steels it has become apparent that corrosion resistance largely depends on the amount of *free chromium* in the steel – that is, chromium not tied up by carbon or nitrogen. Where the chromium is only bonded by carbon in the carbide $Cr_{23}C_6$, the amount of free chromium can be calculated, according to [21], as $Cr_{free} = \% Cr - 14.54 \cdot \% C$. In martensitic steels, in which the hardness and strength properties are of predominant importance and are largely determined by the carbon content, the free chromium content always lies in the region of the lower resistance limit. Adequate corrosion resistance is therefore obtained in steels with 13% Cr only in mildly aggressive media or under certain atmospheric conditions. A further prerequisite is a good surface finish as can be obtained by grinding or polishing. In the

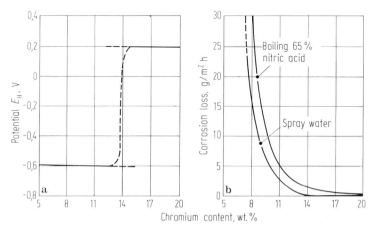

Fig. C 13.12. Influence of the chromium content of iron **a** on electrochemical potential as compared with the standard hydrogen electrode in normal iron sulphate solution, **b** on corrosion resistance (loss by uniform corrosion) in spray water and in boiling 65% nitric acid. After [20].

hardened and stress-relieved condition the content of free chromium is at its highest, thus providing for maximum corrosion resistance (Fig. C 13.13 [22, 23]). At temperatures from 400 to 600 °C the chromium-rich carbide Cr_7C_3 is precipitated with a correspondent detrimental effect on corrosion resistance; this tempering range must therefore be avoided. At the usual tempering temperatures between 650 and 780 °C precipitation takes the form of extremely finely distributed $Cr_{23}C_6$ and does not adversely affect corrosion resistance.

In the ferritic and austenitic steels in which the carbon is not required for strength purposes, the emphasis changes to *reducing the carbon content as far as is feasible* to achieve the proportion of free chromium required for adequate corrosion resistance. This measure has the effect of causing the intergranular corrosion ranges to shift to lower temperatures and longer incubation periods (see Fig. C 13.4). Austenitic steels with carbon contents of max. 0.03% are sufficiently resistant to intergranular corrosion for welding (even of thick-section workpieces), hot straightening work and stress relieving annealing of limited residence time.

In the austenitic chromium-nickel steels, it has been demonstrated that *nitrogen*, even in contents up to about 0.5%, does not promote intergranular

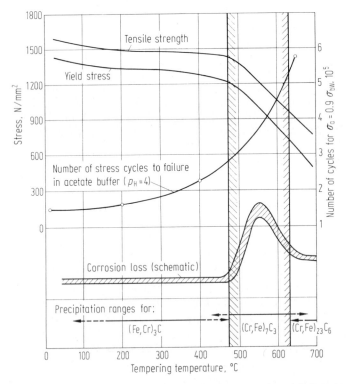

Fig. C 13.13. Influence of the tempering temperature of steel X 20 Cr 13 on the type of carbide precipitate, mechanical properties and resistance to uniform corrosion (latter schematic only). After [22, 23].

corrosion; and in austenitic steels with molybdenum, it even enhances resistance to pitting, entering the pitting resistance equivalent (PRE) ("Wirksumme") [24] with a factor of 30 [25]. This effect does not occur with the ferritic steels because the solubility of carbon and nitrogen in the ferrite is very low. To ensure that their resistance to intergranular corrosion is sufficient without the addition of stabilizing elements, the content of carbon and nitrogen together must be reduced to substantially below 0.01%. This reduction is possible only on an industrial scale if a sophisticated and expensive process of steel production under vacuum is employed.

C 13.3.2 Influence of Other Alloying Elements on Corrosion Behavior

Besides chromium *Nickel* is the most important alloying element encountered in stainless steels. In conjunction with a chromium content of 12% to 30% nickel particularly improves resistance to acids by bringing about a considerable reduction in the passivation current density. Where the nickel content is above 20% it even counters metal dissolution in the active condition if the formation of a protective passive layer should not be possible owing to a lack of any oxidizing capacity of the aggressive agent (i.e. under reducing conditions).

In the martensitic chromium steels, high content of nickel improves corrosion resistance because the attainment of martensitic structure becomes possible even with a reduced carbon content so that the proportion of free chromium practically corresponds to the chemical content. The effect of nickel on resistance to chloride-induced stress corrosion cracking varies. If purely chromium steels are already resistant to this particular form of corrosion, even small nickel additions are sufficient to substantially reduce the crack-initiating threshold stress [26]. On the other hand, ferritic-austenitic steels with 6 to 8% Ni again exhibit good resistance to stress corrosion cracking in neutral or weakly-acidic media. As the austenitic portion of the microstructure increases, resistance falls once more; only if the structure is purely austenitic and the nickel content exceeds 10% there is an improvement [27]. The low sensitivity of high-nickel steel to stress corrosion cracking in sulphuric acid solutions is well documented. As regards improving pitting, resistance nickel only has a synergistic effect in combination with molybdenum.

The addition of *molybdenum* not only extends the passivity range and promotes passivation, it also improves corrosion resistance in the active condition. Molybdenum contents up to about 4% in ferritic steels, and up to about 6.5% in austenitic steels, acting in combination with chromium, increase resistance to the preferential (selective) corrosion phenomena initiated by chlorides such as pitting and crevice corrosion. The influence on the pitting potential is demonstrated by the pitting resistance equivalent (% Cr + 3.3 · % Mo + 30 · % N) [25]. In the martensitic steels, molybdenum has the favorable effect – as already mentioned in the discussion on nickel – of ensuring that the martensitic transformation remains possible where the carbon content has been reduced.

The addition of *silicon and copper* can bring about a specific improvement in corrosion resistance for special applications. Silicon, for example, considerably reduces the tendency to general corrosion in superazeotropic nitric acid (highly concentrated acid, "Hoko-acid") [27]. Copper-alloyed chromium-nickel-molybdenum steels are particularly suitable for service in sulphuric acid [2].

Mention has already been made of the fact that carbon and nitrogen can be tied up by *titanium and niobium* to prevent intergranular corrosion, which is known to be caused by the formation of chromium carbide and chromium nitride at the grain boundaries. The amount of addition is generally larger than that required by the stoichiometry of the desired compounds. In the standards the minimum content for titanium (or niobium) specified for austenitic steels is $5 \cdot \% C$ ($10 \cdot \% C$), and for ferritic steels it is $7 \cdot \% C$ ($12 \cdot \% C$).

To improve machinability, *sulphur* is added to stainless steels – as it is to other free-machining steels – in contents up to 0.35%; a reduction in corrosion resistance is tolerated in such treatments. The addition of selenium has been discontinued owing to its toxicity.

The alloying measures described can be used, within certain limits, to counteract external electrochemical attack on steel. Apart from selecting the right chemical composition for the steel, however, it is also important to obtain, as far as possible, a *homogeneous microstructure*; imperfections such as intermetallic phases, sulphides or oxides with differing electrochemical potentials are, generally speaking, initiators of preferential (selective) corrosion.

C 13.4 Attainment of the Desired Microstructure by Chemical Composition and Heat Treatment

C 13.4.1 Dependence of Structure Type on Content of Ferrite and Austenite Formers. Phase Diagram for Stainless Steels

The most important alloying element for stainless steels – chromium – restricts the gamma field in the phase diagram with iron (Fig. C 13.14 [28]); in steels with 13 to 50% Cr and max. 0.1% C, a ferritic structure with a bcc lattice exists at room temperature. Molybdenum and silicon are likewise ferrite formers. The carbide and nitride forming elements vanadium, tungsten, titanium and niobium have a double stabilizing effect on the ferrite solid solution in that they are involved in the formation of the ferritic solid solution and also tie up (fix) the carbon and nitrogen, thus limiting the austenite-forming effect of the latter.

On the other hand, nickel forms with iron a continuous row of fcc solid solutions (Fig. C 13.15 [28]), and thus constitutes a strong austenite former. In steels with a sufficiently high percentage of austenite-stabilizing alloying elements – manganese, carbon, nitrogen – the gamma field is so much widened, and thus the A_3 transformation temperature so reduced, that an austenitic structure remains intact at room and even lower temperatures. The nickel content of such steels is between 8 and 30%, the actual content being a matter of coordination with the chromium content and the other ferrite-forming alloying elements.

Because carbon and nitrogen as well as nickel and manganese have an important influence on the extension of the gamma field in the iron-chromium

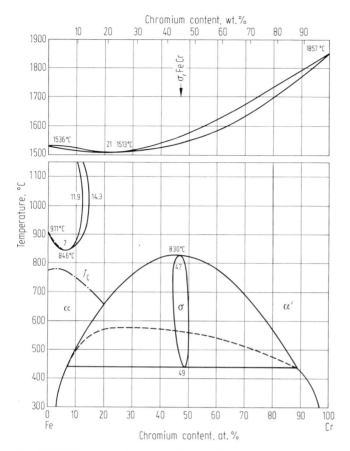

Fig. C 13.14. Iron-chromium phase diagram. After [28].

system (Fig. C 13.16 [29]), it is understandable that in the most common ferritic steels, isolated mixed structures comprising both ferrite and austenite can occur at high temperatures, and consequently traces of transformation structure can also be present at room temperature. Steels with carbon contents up to about 0.05% are therefore still only partially ferritic: above 850 °C, partial austenitization occurs which, on cooling the metal to room temperature, results in a martensitic transformation (Fig. C 13.17 [30]).

With respect to the difference in properties imparted to the stainless steels by their basic microstructure and the possibilities of influencing it by alloying, the next logical step was to attempt to present the involved relationships in a form which is easy to understand. Strauss and Maurer were the first to ascertain the structures which would result in chromium-nickel steels with about 0.2% C when cooled in air from 1100 °C (Fig. C 13.18 [31]). The consequential *classification into ferritic, martensitic, austenitic and ferritic-austenitic steels* according to the dominant struc-

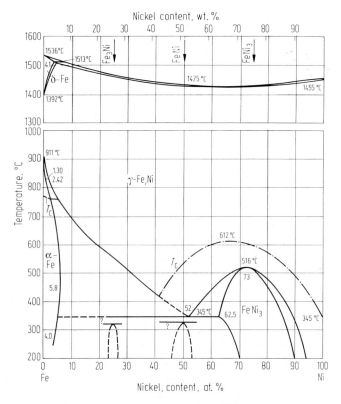

Fig. C 13.15. Iron-nickel phase diagram. After [28].

tural features has stood the test of time as many service properties are correctly determined by the structure. Given the fundamental significance of this phase diagram there has been no lack of attempts to combine the ferrite-forming and austenite-forming elements into respective equivalents ("Wirksummen") as chromium equivalent and as nickel equivalent. Here the effective factors of the individual elements represent their ferrite-forming or austenite-forming powers in comparison to the reference elements chromium and nickel. The best-known structure diagram of this type is the *Schaeffler diagram* (Fig. C 13.19 [32–34]). This diagram has been compiled for deposited weld metal of chromium-nickel steels and describes the condition following cooling from very high temperatures. The diagram is widely used because of its clarity and its importance for the welding of high-alloy steels. Thanks to improvements in the analytical technique, latest developments have included the influence of the nitrogen content on ferrite formation [34].

For the structure diagram in Fig. C 13.19 the chromium equivalent was calculated as being % Cr + 1.4 · % Mo + 0.5 · % Nb + 1.5 · % Si + 2 · % Ti, and the nickel equivalent was calculated to be % Ni + 30 · % C + 0.5 · % Mn + 30 · % N_2.

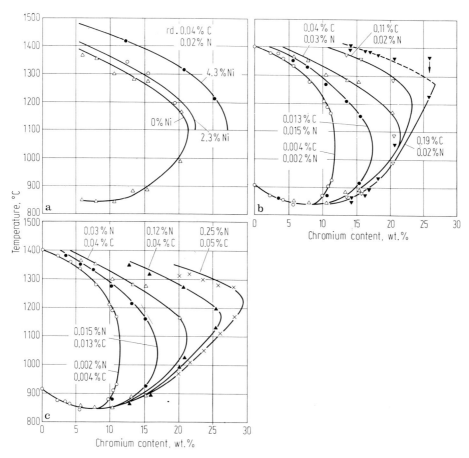

Fig. C 13.16. Shift in the $(\gamma + \alpha)/\alpha$ boundary line in the iron-chromium system by increasing contents of nickel, carbon and nitrogen. After [29].

C 13.4.2 Heat Treatment of the Characterizing Structure Groups

As already mentioned a *ferritic microstructure* can only be obtained in steels alloyed solely with chromium in the range 13% to 30% Cr where the carbon content does not exceed 0.1% C max. According the considerable influence exerted by carbon and nitrogen on the gamma field, higher carbon contents can be expected to trigger austenitization as the metal is heated. Cooling then leads to at least a partial transformation into martensite. To obtain a completely ferritic structure, even with chromium contents at the upper limit, up to 5% Mo must still be added, and the carbon content reduced to max. 0.015%; in addition it may also be necessary to tie up the carbon and nitrogen with niobium. Such *superferritic steels* are used for special applications.

On the other hand a completely austenitic structure at high temperatures and thus a *martensitic structure* following quenching to room temperature can only be

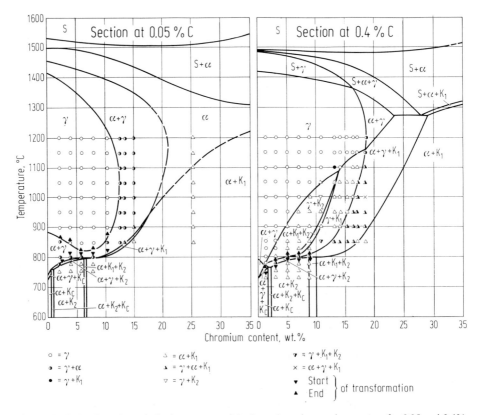

Fig. C 13.17. Sections through the iron corner of the iron-chromium-carbon system for 0.05 and 0.4% C. After [30].

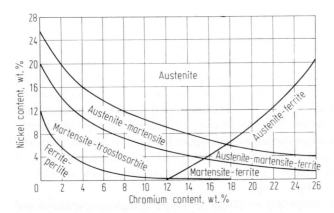

Fig. C 13.18. Structure diagram of the chromium-nickel steels according to Strauss and Maurer [31].

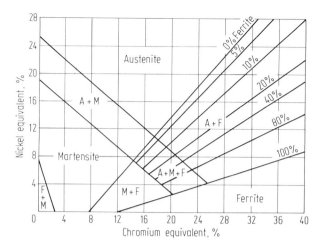

Fig. C 13.19. Structure diagram for stainless steels according to Schaeffler [32–34], plotted for deposited weld metal, i.e. for cooling from very high temperatures. The chromium and nickel equivalents were calculated as follows:
Chromium equivalent: % Cr + 1.4 · % Mo + 0.5 · % Nb + 1.5 · % Si + 2 · % Ti;
Nickel equivalent: % Ni + 30 · % C + 0.5 · % Mn + 30 · % N_2

expected in chromium steels where the carbon contents are relatively high. Steels with 13% Cr require more than 0.15% C and austenitization at min. 950 °C, and those with 17% Cr need 0.3% C and heating to 1100 °C (Fig. C 13.20 [20]). Steels with 12 to 17% Cr can therefore be rendered transformable or hardenable through the addition of carbon and by limited additions of nickel and nitrogen, as such measures extend the austenite field. Martensitic transformation is a prerequisite of a hardened or hardened and tempered structure. With carbon contents from 0.10 to 0.45%, these steels can be heat-treated for high strength accompanied by good toughness. Chromium steels are termed "hardenable" if their carbon content

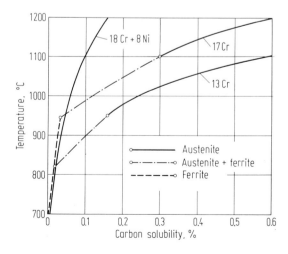

Fig. C 13.20. Maximum solubility of carbon in steels with 13% Cr and 17% Cr and also with 18% Cr and 8% Ni. After [20].

exceeds 0.40%. These materials are hardened for service and only stress-relieved at 200 to 350 °C. This treatment produces at least some toughness and still maintains a high level of hardness through removal of the tetragonal distortion of the bcc lattice in the hardened condition. The degree of increase in hardness depends on the content of free carbon in the steel during austenitization; because the solubility of the chromium carbides at hardening temperature decreases as the chromium content increases, the minimum hardness achieved with a given carbon content and hardening temperature is higher in steels with 13% Cr than in steels with larger chromium contents. In the annealed condition the achievable strength properties are likewise dependent on carbon content (Fig. C 13.21 [16]).

The desire to improve the weldability of the quenched and tempered steels with about 13% Cr and 0.2% C led to the idea of replacing part of the carbon content with an equivalent of nickel so that heat-treatability was retained and weldability was substantially enhanced as a result of reduced (HAZ) hardening. This approach also had the effect of extending the size range for high strength and good toughness properties to section diameters beyond 400 mm [35]. The *nickel martensitic group* thus developed was followed by other precipitation hardening steels on the basis of martensitic structures.

The process of *precipitation hardening* results from the ability of *nickel martensitic structures* under exposure to temperatures in the 400 °C to 600 °C range to precipitate intermetallic phases; steels which lend themselves to this process are therefore often referred to as "maraging" steels [36]. The greatest degree of precipitation hardening is obtained with titanium, followed in order of decreasing efficacy by aluminum, copper, niobium and molybdenum. The arrows in Fig. C 13.22 [37] indicate the point at which the content of the elements promoting precipitation hardening begins to contribute to the formation of either delta ferrite or austenite. The presence of these elements must be taken into account when determining the contents of carbon, chromium and nickel, because excessive ferrite or austenite is deleterious to strength, as is demonstrated by the hardness curves. In an alloy with 0.1% C and 17% Cr, maximum hardness is achieved with 4% Ni following solution annealing (Fig. C 13.23 [38]). It is also noticeable that, as the nickel content increases, the proportion of delta ferrite falls from 64% to about 5%; parallel to this drop the martensite point decreases to about 100 °C. Above 4% Ni, the occurrence of only partial (martensitic) transformation leads to a steep decline in hardness until the martensite point falls to 0 °C at about 7% Ni. A metastable austenitic structure now exists that can no longer transform completely into martensite even when cryogenic cooling is applied down to -78 °C.

The above example should demonstrate the extent to which the chemical composition of the precipitation hardenable stainless steels must be precisely coordinated: after martensitic transformation, no metastable austenite should remain in the structure as it will prevent full precipitation hardenability and any subsequent transformation under mechanical influences may lead to undesirable cracking (non-tempered martensite, hydrogen pick-up).

Steels from the *austenitic structure group* are used in the solution-annealed and quenched condition. Solution annealing temperatures vary with chemical composition between 1000 °C and 1150 °C. Following quenching, the steels with less than

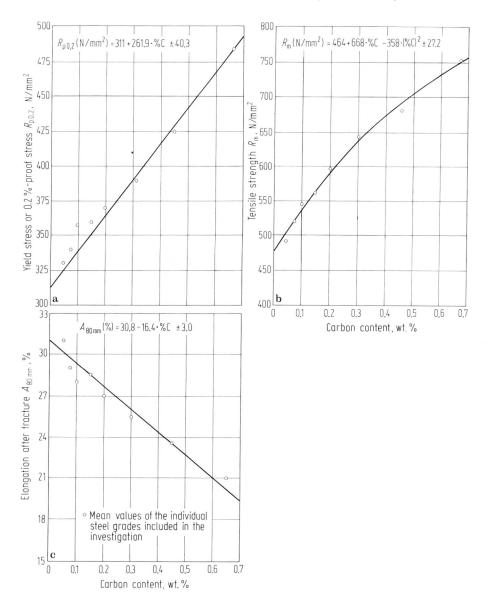

Fig. C 13.21. Influence of the carbon content on **a** yield stress or 0.2%-proof stress, **b** tensile strength, and **c** elongation after fracture, of annealed cold wide strip of steels with 13% Cr. After [16]. The steel grades X 17 Cr 13, X 7 CrAl 13, X 10 Cr 13, X 15 Cr 13, X 20 Cr 13, X 30 Cr 13, X 40 Cr 13 and X 60 Cr 13 were incorporated into the on-going test program.

16% Ni are in a metastable condition between room temperature and the temperature of liquid helium. Below the solution annealing temperature both diffusionless and diffusion-controlled structural changes can take place; thus unstable austenite can be transformed into α'-martensite by deep cooling and cold forming. An

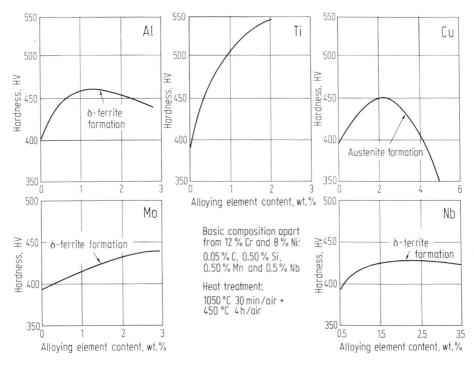

Fig. C 13.22. Influence of aluminum, copper, molybdenum, niobium and titanium on the aging properties of martensitic steels with 12% Cr and 8% Ni. After [37].

increase in the alloy content by which the ferrite-forming elements also stabilize the austenitic solid solution at low temperatures counteracts martensite formation. Saturation polarization measurements can be used to determine the amount of the magnetizable martensite within the paramagnetic austenite. Thus it is possible to define the influence of the various alloying elements on austenite stability (Fig. C 13.24 [39]). Martensite formation is associated with an increase in strength and a decrease in toughness. Detailed knowledge of the processes involved is therefore particularly important for chipless forming operations when, for example, alloying measures must be carried out to improve the deformability of sheet material. Thus austenitic steels which, owing to their chemical composition, possess a certain instability of structure, permit α'-martensite to a limited extent and exhibit good stretch-forming properties [40]. Martensite formation in force-transmission zones enhances forming behavior, but structural stability is needed in the deformation zones themselves as it reduces the degree to which work hardening can take place.

Ferrous alloys containing 5 to 10% Ni and more than 20% Cr, either by itself or interacting with other ferrite-forming elements such as molybdenum or silicon, do not transform completely into austenite following ferritic solidification (Fig. C 13.25 [41]); they therefore exhibit a *ferritic-austenitic structure*. Below 900 °C the ferrite precipitates sigma phase; this precipitation has the effect of increasing brittleness but can be suppressed by quenching. Carbon and nitrogen extend the existence of austenite while reducing the delta ferrite field (Fig. C 13.26 [42]). When cooling takes place from the A_4-temperature after complete ferritization these elements determine the point at which austenite formation commences. At a given cooling rate and a constant ratio between the chromium and nickel contents, carbon and nitrogen also influence the proportion of ferrite and austenite in the structure (Fig. C 13.27 [36]). Starting from the original ferrite/ferrite grain boundary, the austenite is formed in a Widmanstätten arrangement [42]. However, if a supersaturated ferrite solid solution that has been rapidly cooled from a high temperature is annealed at say 1050 °C, austenite precipitates within the initially ferritic grains. A structure with nearly equivalent proportions of ferrite and austenite results in a favorable balance between strength, toughness and corrosion resistance [10, 19].

Austenitic and ferritic-austenitic steels can also be produced as *precipitation-hardenable* steels; however, such austenitic steels are more likely to be used for their high strength at elevated temperatures than for their corrosion resistance. The particles precipitated at the grain boundaries during age-hardening often promote

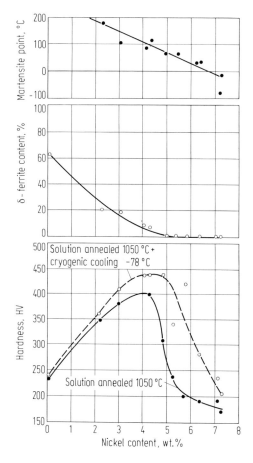

Fig. C 13.23. Influence of Nickel on microstructure, hardness and martensite transformation temperature of steels with 0.1% C and 17% Cr. After [38].

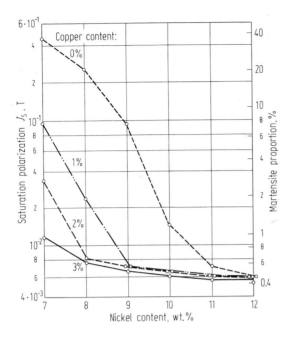

Fig. C 13.24. Influence of nickel and copper contents on the martensite portion of the microstructure of steels with 0.02% C and 18% Cr after cryogenic cooling to $-269\,°C$. After [39]. Heat treatment: $1050\,°C$ 10 min/water + $-196\,°C$ 10 min + $-269\,°C$ 10 min. Test temperature $+25\,°C$.

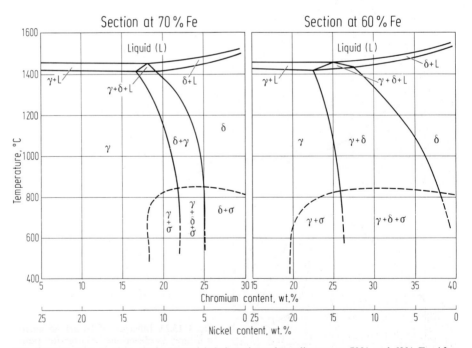

Fig. C 13.25. Sections through the iron-nickel-chromium phase diagram at 70% and 60% Fe. After [41].

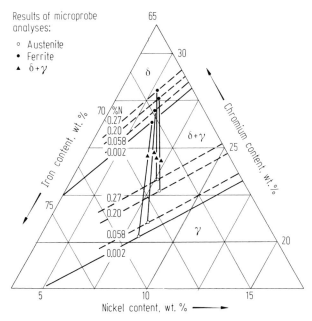

Fig. C 13.26. Section through the iron-nickel-chromium system at 1200 °C at the region where the steel composition is about 0.02% C, 25% Cr and 7% Ni (X 2 CrNiN 25 7) to investigate the influence of nitrogen. After [42]. The specimens were quenched in water after 52 h 1200 °C.

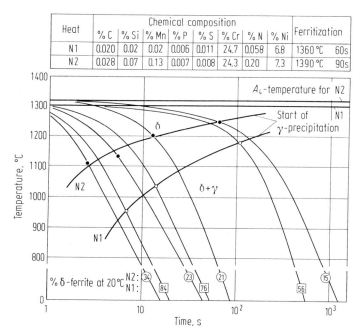

Fig. C 13.27. Time-temperature-transformation diagrams (continuous cooling) for two heats of steel X 2 CrNiN 25 7: Influence of small differences in the chemical composition and in the solution annealing temperature on the proportion of delta ferrite at room temperature. After [42].

intergranular attack under corrosive conditions if the carbon and nitrogen have not been stabilized as they are in a steel with 0.06% C, 15% Cr, 25% Ni and 2% Ti for example. In ferritic-austenitic steels, precipitation hardening adversely affects toughness so that such a process is suitable only for certain specific applications.

Concluding this section on the importance of heat treatment for stainless steels it is perhaps worth to mention that heating or cooling rates applied must be properly adjusted to the specific chemical composition of the material involved to obtain the best possible properties. *The final microstructure will only rarely correspond to the thermodynamic equilibrium.* It must therefore be remembered that at the high service temperatures allowing diffusion processes to take place the steel will tend to gain its equilibrium condition. This process can result in substantial changes of properties such as the development of brittleness by precipitation of intermetallic phases.

C 13.5 Typical Steels Proven in Service

The growing demand for stainless steels for a widely varied range of requirements has favored intensive development work in this field, the beginnings of which go back more than 70 years [43]. A *selection of typical steel grades that characterize their respective microstructure groups* and are used for a *wide range of applications* is offered in Table C 13.2; in accordance with the classification given in C 13. 4, the list is broken down into ferritic, martensitic, austenitic and ferritic-austenitic steels. The designatory code indicates the chemical composition of the steels with the symbols of the characterizing elements such as chromium, nickel, etc. together with their percentage contents (figures at the end) and carbon content (figure following X representing multiples of 0.01% C). Table C 13.3 contains the strength values for the same steels in their usual heat treated condition so that they do not always require mentioning in the following descriptions of a particular microstructure steel group.

In general, the *strength values at elevated temperatures*, decrease more or less uniformly as the temperature increases for all the steel groups under discussion. Where utilizable limit temperatures are indicated for yield stress, they will be below the so-called creep range in which creep limits and creep-stress rupture strength values must be included in the design calculations for structural elements; other temperature limits may have to be adopted where corrosive chemical attack and structural changes in long-term service conditions are expected.

C 13.5.1 Ferritic Steels

Sufficient *chemical resistance* to rusting in damp environments can be obtained from steels with chromium contents of about 12%. Simple household utensils and low-stressed apparatus in the petrochemical industry are manufactured, for

Table C 13.2 Characteristic alloy types of stainless steels

Steel grade Designation[a]	Chemical composition					Microstructure[b]
	% C	% Cr	% Mo	% Ni	Other elements	
Ferritic steels						
X 6 Cr 13	≤ 0.08	10.5...15	≤ 1.3	–		F (TF)
X 6 Cr 17	≤ 0.08	15.5...20	≤ 2.3	–	(Al, Ti)	F (TF)
X 1 CrNiMoNb 28 4 2	≤ 0.015	25...29	0.75...4	≤ 4.5	(Ti, Nb, S)	F
					Nb, Ti, (Zr, Al)	
Martensitic steels						
X 20 Cr 13	0.08...0.35	11.5...15	≤ 1.5	≤ 1.5		M_V (TM)
X 20 CrNi 17 2	0.10...0.43	15.5...18.5	≤ 1.3	≤ 2.5	(S)	M_V (TM)
X 40 Cr 13 (X 45 CrMoV 15)	0.25...1.15	12...16	≤ 1.5	–	(S)	M_H
X 105 CrMo 17 (X 90 CrMoV 17)	0.85...1.20	15.5...19	≤ 1.3	≤ 0.3	(Co, V)	M_H
					(Co, V)	
Nickel-martensitic steels[c]						
X 4 CrNi 13 4	≤ 0.05	12...15	≤ 2	3...5		NM_V
X 4 CrNiMo 16 5	≤ 0.05	15...17	0.8...1.5	4...6		NM_V
Maraging steels						
X 5 CrNiCuNb 15 5	≤ 0.07	14...17		3...5	Cu, Nb	NM_{AH}
X 7 CrNiMoAl 15 7	≤ 0.07	14...16	2...2.5	6.5...8	Al	AH_{AH}
Austenitic steels						
X 5 CrNi 18 10	≤ 0.07	17...20		8.5...10.5		A
X 5 CrNiMo 17 12 2	≤ 0.07	16.5...18.5	2.0...2.5	10.5...13.5		A
X 2 CrNi 19 11	≤ 0.03	18...20		8.5...11.5		A
X 2 CrNiMo 18 14 3	≤ 0.03	16.5...18.5	2...3	10.5...14.5	(Ti, Nb, N)	A
X 2 CrNiMoN 17 13 5	≤ 0.03	16...20	3...5	12.5...19	(Ti, Nb, N)	A
X 1 CrNiMoN 25 25	≤ 0.02	24...26	2...2.5	22...26	N	A
X 1 NiCrMoCu 25 20 5	≤ 0.02	19...28	3...7	24...32	N	A
					Cu, (N)	
Ferritic-austenitic steels						
X 2 CrNiMoN 22 5 3	≤ 0.03	18...28	1.3...3	4...7.5	N, (Si, Nb)	AF

[a] In the designation, the number immediately following the X gives the carbon content in units of 0.01%, followed by the chemical symbols for the characterizing alloying elements, the approximate percentage content of which is indicated in the same sequence by the figures at the end

[b] A = austenitic, AF = austenitic-ferritic (duplex), AH = precipitation hardenable, AM = austenitic-martensitic, F = ferritic, M = martensitic, M_H = martensitic (hardened structure), M_V = martensitic (quenched and tempered structure), NM = nickel martensitic, NM_{AH} = nickel martensitic (aged), NM_V = nickel martensitic (quenched and tempered structure), TF = partially ferritic, TM = partially martensitic

example, from grades similar to the basic X 6 Cr 13 steel with 11 to 15% Cr. For kitchen equipment such as sinks and washing machines, and applications in the automotive industry, steels with 17% Cr (such as X 6 Cr 17) are used. Steels with 18% Cr, 2% Mo, and low carbon and nitrogen contents stabilized with titanium and/or niobium are widely used because they exhibit a good corrosion resistance/alloy cost ratio which makes them economical. These steels are particularly popular in cooling systems that operate with chloride-containing river water. For very aggressive conditions in tubular heat exchangers, steels have been developed with 25 to 29% Cr, up to 5% Mo and 2 to 4% Ni. An example is X 1 CrNiMoNb 28 4 2 which, in spite of its low carbon and nitrogen contents, still must be stabilized with niobium [27, 45]. The steel group characterized by this grade is occasionally referred to as *superferritic*.

Like the fcc austenitic solid solution, the bcc solid solution of ferritic chromium steel can also be strengthened. Owing to the low level of solubility of nitrogen and carbon, however, only substitution solid solution hardening is possible. As the chromium and molybdenum contents increase, so do *hardness, yield stress and tensile strength*, with no adverse effects on elongation after fracture or reduction of area (Fig. C 13.28 [46]). Silicon also increases strength.

In the impact test, the tendency of ferritic steels towards *cold brittleness* usually manifests itself in the form of a steep (sharp) drop in energy values at around room temperature; increasing chromium contents and coarse grain formation has the effect of shifting the transition from ductile fracture to brittle fracture upwards to higher temperatures.

The embrittlement of ferritic steels is accelerated by chromium and even more by molybdenum. The supersaturated solid solution exhibits *two temperature ranges below the solution annealing temperature where the tendency towards embrittlement is considerably higher*: Up to about 550 °C this effect is due to the single-phase separation (segregation) of the alpha solid solution, a process generally known as 475 °C-embrittlement. Above 550 °C it is the result of precipitation of the chi- and sigma-phases (Fig. C 13.29 [46]). In addition brittleness must be expected in unstabilized ferritic steels as a result of carbide and nitride formation. The 475 °C embrittlement must be avoided and should not adversely affect the material over a period of at least 10^5 h thus defining an upper temperature limit for applications of such steels. Generally, this temperature falls as the chromium content increases. The service temperatures common in practice for ferritic steels in fact rarely exceed 250 °C so that the metallurgical behavior described above can largely be ignored.

The tendency towards embrittlement also leads to *restrictive measures in steelmaking and processing*. When there is a high alloy content, annealing treatments, for example, must be followed by sufficiently rapid cooling to obtain adequate toughness. Consequently the upper product section sizes must be limited [47].

Likewise *welding work* can only be carried out within the limits imposed by the metallurgical changes that occur. Welding the superferrites requires particular care as their considerable structural instability means that brittleness can be caused by the heat input. For this reason, high-alloy ferritic steels are weldable only up to a certain thickness, the limit of which is typical for the steel and must

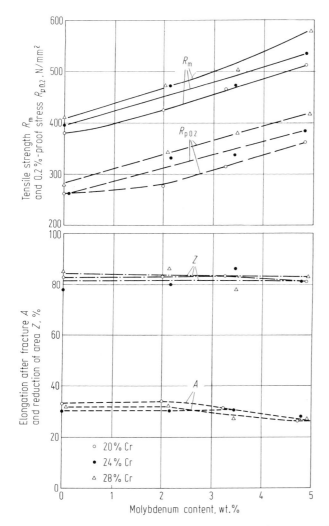

Fig. C 13.28. Influence of 2 to 5% Mo on tensile strength, 0.2%-proof stress, elongation after fracture and reduction of area of three steels with 20, 24 and 28% Cr. After [46]. Heat treatment: 1000 °C 30 min/water.

be determined for each steel grade. Hot-rolled plates in nickel-containing steel X 1 CrNiMoNb 28 4 2, however, are still weldable in the 10 to 20 mm thickness range.

C 13.5.2 Martensitic Steels

The corrosion resistance of these stainless steels with relatively high hardness and strength visibly depends on the microstructure obtained by heat treatment. The cutlery steels such as X 40 Cr 13 and X 105 CrMo 17 (see Tables C 13.2 and C 13.3)

Table C13.3 Mechanical properties of some stainless steels

Steel grade Designation	Head treatment condition	Yield stress (0.2%-proof stress) N/mm²	Tensile strength N/mm²	Elongation after fracture A^a % longitudinal	transverse	Hardness HB or HV	Notch impact energy (ISO-V-specimens) J longitudinal	transverse
Ferritic steels								
X 6 Cr 13	quenched and tempered	≥ 400	550...700	18	13			
X 6 Cr 17	annealed	≥ 270	450...600	20	18	< 185	60	45
X 1 CrNiMoNb 28 4 2	quenched	≥ 500	600...750	20	16	< 240		
Martensitic steels								
X 20 Cr 13	quenched and tempered, stage I	≥ 450	650...800	14	10		25	
	quenched and tempered, stage II	≥ 550	750...950	12	8			
X 20 CrNi 17 2	quenched and tempered	≥ 550	750...950	14	10		30	20
X 40 Cr 13 (X 45 CrMoV 15)	annealed		≤ 800 (≤ 900)			≤ 285		
X 105 CrMo 17 (X 90 CrMoV 17)	hardened					≥ 58 HRC		
Nickel-martensitic steels^b								
X 4 CrNi 13 4	quenched and tempered, stage V1	≥ 550	760...900	17	16	240...290	90	70
	quenched and tempered, stage V2	≥ 685	780...980	17	14	245...310	90	70
	quenched and tempered, stage V3	≥ 850	900...1200	14	11	275...340	80	50

Steel	Condition							
X 4 CrNiMo 16 5	quenched and tempered, stage V1	≥ 550	830...1030	16	14	260...325	90	70
	quenched and tempered, stage V2	≥ 685	850...1100	16	14	265...345	80	60
	quenched and tempered, stage V3	≥ 850	900...1200	14	11	280...385	70	40
Maraging steels								
X 5 CrNiCuNb 15 5	aged	790...1170	960...1310	9...12	5...9	311...451	20...34	14...20
X 7 CrNiMoAl 15 7	aged	700...1200	950...1350	5...15	4...8	311...451		
Austenitic steels								
X 5 CrNi 18 10	quenched	≥ 195	500...700	45	40		85	55
X 5 CrNiMo 17 12 2	quenched	≥ 205	510...710	40	35		85	55
X 2 CrNi 19 11	quenched	≥ 180	460...680	45	40		85	55
X 2 CrNiMo 18 14 3	quenched	≥ 190	490...690	35	30		85	55
X 2 CrNiMoN 17 13 5	quenched	≥ 285	580...800	35	30		85	55
X 1 CrNiMoN 25 23	quenched	≥ 255	540...740	40	30		70	40
X 1 NiCrMoCu 25 20 5	quenched	≥ 220	500...750	35	30	≥ 160	85	30
Ferritic-austenitic steels								
X 2 CrNiMoN 22 5 3	quenched	≥ 450	680...880	30	25		100	70

[a] The figures represent the minimum attainable values (in nickel martensitic steels the impact energy values are mean values), subject to product type and thickness (see e.g. DIN 17 440 [44])
[b] According to SEW 400 [43a]

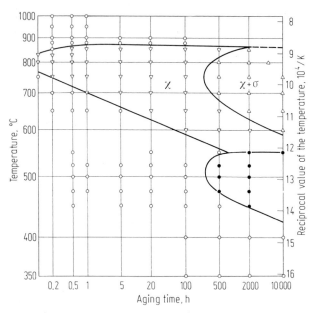

Fig. C 13.29. Time-temperature-aging diagram of an experimental steel with 34% Cr and 3.5% Mo. After [46].

are only hardened and stress-relieved, but X 20 Cr 13 or X 20 CrNi 17 2 are hardened and tempered for service. These cutlery steels are used not only for domestic purposes but also for needle valves, nozzles and antifriction bearings owing to the high level of hardness which they can achieve. The quenched and tempered steels X 20 Cr 13 and X 20 CrNi 17 2 are used for many machine components such as shafts, spindles, valves, fittings and steam turbine blades, or in plastics moulds.

In *resistance to chemical corrosion* the hardened and merely relieved steel offers a favorable microstructure because carbides are largely in solution. However, there is the danger cracking may occur where hydrogen is picked up from a partial cathodic reaction. As the temperature increases to 400 to 600 °C, chromium-rich carbides precipitate and the solid solution becomes depleted of chromium that is effective in chemical resistance. Tempering conditions that may induce intergranular chromium depletion should therefore be avoided (see Figs. C.13.3 and C 13.13). In soft annealing, too, care must be taken to prevent excessive chromium carbide precipitation in this group of steels.

For many important applications of the martensitic stainless steels a high level of hardness is essential. An improvement of *toughness* which is naturally low after quench-hardening is therefore only feasible by stress-relieving at low temperatures, say between 200 °C and max. 300 °C; in high carbon steels such as X 90 CrMoV 17 and X 105 CrMo 17, this treatment can produce a hardness of 50 to 60 HRC creating the conditions for adequate wear resistance in antifriction bearings and tools, and for the maintenance of the cutting edges of knives. If mechanical properties comparable with those of the higher-strength structural steels are required, they can be provided in the steels X 20 Cr 13 and X 20 CrNi 17 2 – possibly with a small addition of molybdenum – through hardening and tempering at 650 °C to 750 °C, as shown in Fig. C 13.30 [48] and indicated in Table C 13.3. The highest service temperatures of these hardened and tempered steels are generally 100 °C below their tempering temperature, i.e. around 550 °C. Their toughness below room temperature is low because the notch impact transition occurs at ambient temperatures.

The martensite of these heat-treatable chromium steels (with about 0.20% C) is hard and brittle, but chromium steels with carbon contents up to only 0.05% develop a tough, cubic martensite when alloyed with 3 to 6% Ni (X 4 CrNi 13 4 and X 4 CrNiMo 16 5 in Tables C 13.2 and C 13.3) on γ-α transformation [35, 49, 50]. Quench-annealing is done from temperatures between 950 and 1050 °C and the tempering range extends to about 620 °C. Between 500 and 600 °C these *nickel martensitic steels* form a stable austenite of finely dispersed distribution and they owe their remarkable toughness to this austenite (see Fig. C 13.31 [51]) accompanied by a high level of as-tempered tensile strength (Fig. C 13.32 [51]). Their low carbon contents exhibit better corrosion resistance in comparison to steels with 0.20% C and 13 to 17% Cr (see Fig. C 13.33 [35]). Their good toughness coupled with satisfactory weldability means that these steels are suitable for applications involving corrosive attack and alternating mechanical stress, such as for shafts and blades of water turbines, or pump casings. The through hardenability of nickel martensitic steels ensures uniform strength properties even in thick workpieces across the full cross-section. However, application temperatures must be limited to 300 to 350 °C max; extremely fine carbide precipitates form in these steels during long periods of exposure at higher temperatures and raise changes in toughness

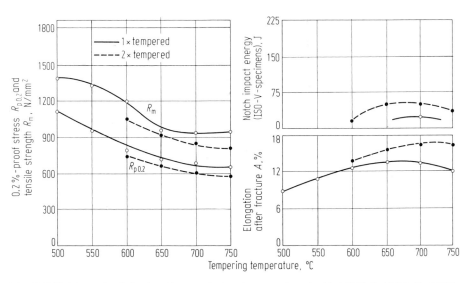

Fig. C 13.30. Tempering behavior of a quenched and tempered steel with 0.20% C, 16.5% Cr, 0.10% Mo and 1.7% Ni. Heat treatment: 1050 °C/oil, 2 h tempered. Longitudinal specimens. After [48].

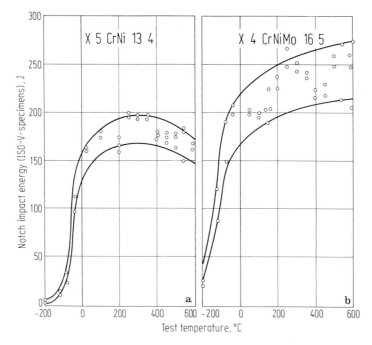

Fig. C 13.31. Notch impact energy-transition temperature curves for longitudinal specimens of the nickel-martensitic steels **a** X 4 CrNi 13 4; **b** X 4 CrNiMo 16 5. After [51]. For mechanical properties in the initial condition see Table C 13.4.

[52]. These steels do, however, remain tough at temperatures down to $-100\,°C$ (Fig. C 13.31).

The hardening and tempering treatment of the nickel martensitic steels can be supplemented by precipitation hardening if copper or aluminum are added to the alloy [49, 53]. The resultant materials make up the *maraging steels* subgroup which includes the grades X 5 CrNiCuNb 15 5 and X 7 CrNiMoAl 15 7 shown in Tables C 13.2 and C 13.3. These steels are used for components that must withstand high mechanical loading, e.g. in aerospace engineering applications. By applying a sequence of heat treatments, this steel group can be given a 0.2%-proof stress of about 1000 to 1200 N/mm², a tensile strength of about 1100 to 1300 N/mm² and an elongation after fracture of about 10%. Their upper service temperature limit is, however, around 300 °C to 350 °C as the aging reactions at these relatively low temperatures occur through diffusion.

C 13.5.3 Austenitic Steels

Chromium-nickel steels with an austenitic microstructure are suitable for many processing and application requirements. They have therefore become irreplaceable in all spheres of technology and daily life. By volume (quantity) they account for the largest proportion of total stainless steel consumption. With steels from the

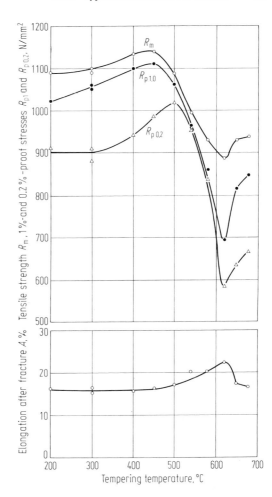

Fig. C 13.32. Proof stress values, tensile strength and elongation after fracture of the nickel-martensitic steel X 4 CrNiMo 16 5 as a function of tempering temperature. After [51]. Austenitization as in Fig. C 13.31 (Table C 13.4). Tempering time 2 h.

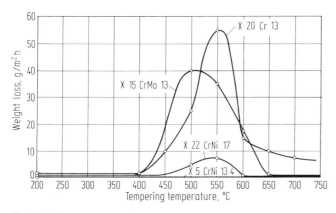

Fig. C 13.33. Comparison of the corrosion resistance of a nickel-martensitic steel (X 4 CrNi 13 4) with three other martensitic steels after tempering at 200 °C to 700 °C, for 8 h each. Test conditions: 2 h in boiling 20% (5% for X 20 Cr 13) acetic acid. After [35].

Table C 13.4 Mechanical properties of the steels dealt with in Fig. C 13.31 in their initial condition

Steel grade Designation	Initial condition	Mechanical properties				
		0.2%-proof stress N/mm²	1%- N/mm²	Tensile strength N/mm²	Elongation after fracture A %	Reduction of area %
X 4 CrNi 13 4	950°C 30 min/oil + 600°C 2 h/air	789	807	864	20	70
X 4 CrNiMo 16 5	950°C 30 min/oil + 600°C 2 h/air	763	812	909	20	76

basic grade X 5 CrNi 18 10 right through to the highly alloyed X 1 NiCrMoCu 25 20 5 (see Tables C 13.2 and C 13.3) it is possible to obtain high resistance to many forms of corrosive attack.

Applications for austenitic chromium-nickel steels include household kitchen equipment and internal and external building accoutrements. For external service and in automotive engineering applications, unfavorable conditions such as industrial environments, marine atmospheres, the presence of de-icing salt in winter, etc. should be countered by employing steels alloyed with molybdenum. The highest requirements of the chemical industry can be met with steels such as X 2 CrNiMoN 17 13 5 or X 1 NiCrMoCu 25 20 5 where applications involve organic acids or sulphuric acid.

Steels with a purely austenitic microstructure, tested at room temperature, exhibit a relatively low 0.2%-proof stress of about 200 to 250 N/mm², but have the advantage of good *toughness*. High values of elongation after fracture of about 50%, i.e. double the value of the ferritic and ferritic-austenitic steels (both about 25%), are a particular advantage offered by the austenitic steels; these high values, coupled with their low yield stress to tensile strength ratio of less than 0. 5, points to a high degree of plastic deformability (Table C 13.3).

To achieve high strength values, good use can be made of *strain-hardening the austenitic steels*. The increase in strength is due, once, to the strengthening of the austenitic solid solution through the usual mechanisms of physical metallurgy and, further, to the formation of α'-martensite (Fig. C 13.34 [54]). These characteristics are utilized particularly in the manufacture of stainless steel springs. In heavy deformation processes such as cold-heading employed in the manufacture of screwed fasteners, the high degree of strain-hardening in austenite, however, is disruptive. Copper-alloyed grades exhibit a more favorable behavior for such applications (Fig. C 13.35 [55]). The cold-formed condition of the austenitic steels, however, is such that welding must be ruled out.

It is possible to substantially increase both the proof stress and tensile strength values by substitution or interstitial dissolution of alloying elements (Fig. C 13.36 [56–59]). The greatest *effect* is achieved with carbon and *nitrogen* atoms lodged on interstitial sites. As high carbon contents have a detrimental effect on intercrystalline corrosion, nitrogen is used in preference; steels of a suitable composition, in the solution-annealed and quenched condition, can be alloyed with nitrogen contents up to 0.40% [57]. Without impairing the good toughness

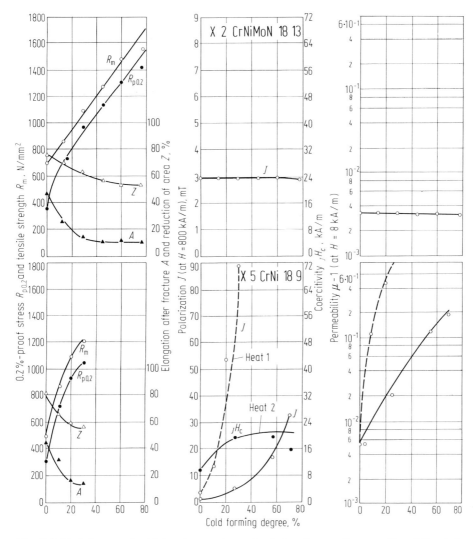

Fig. C 13.34. Influence of cold forming on the mechanical and magnetic properties of the steels X 5 CrNi 18 9 and X 2 CrNiMoN 18 13. After [54].

properties of the material, the 0.2%-proof stress can thus be raised to over 500 N/mm² at room temperature, and tensile strength is increased to values between 800 and 1000 N/mm². The favorable effect of nitrogen is also apparent at elevated temperatures [58]. Very high nitrogen contents up to about 1% can be obtained by processing the molten steel under pressure. As these contents do not correspond to the thermodynamic equilibrium – particularly at low temperatures – there is a danger of chromium nitride formation at service temperatures by diffusion processes and during welding relatively thick sections.

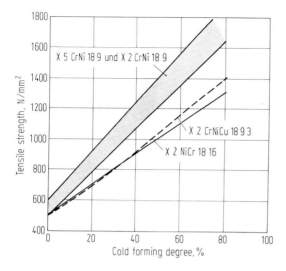

Fig. C 13.35. Strain hardening of the stainless cold upsetting steels X 2 NiCr 18 16 and X 2 CrNiCu 18 9 3 compared with the conventional austenitic steel grades X 5 CrNi 18 9 and X 2 CrNi 18 9. After [55].

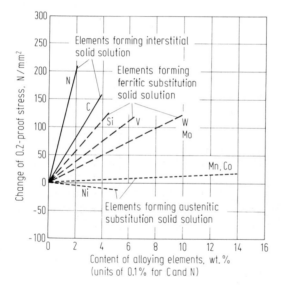

Fig. C 13.36. Influence of some elements forming interstitial or substitution solid solutions on the 0.2%-proof stress of steel. After [56].

Owing to their high alloy content the molybdenum-bearing austenitic chromium-nickel steels tend to be prone to the formation of intermetallic phases such as the Chi-, Sigma- and Laves-phases, with detrimental effects on corrosion resistance and toughness [59–62]. The addition of nitrogen retards the rates at which these phases form, and the phase space is shifted to lower temperatures (Fig. C 13.37 [63]). *Nitrogen* therefore *contributes* to the structural stability of the austenite and *higher chromium and molybdenum contents can be utilized* to improve corrosion resistance without increasing the tendency to intermetallic phase precipitation [63–66]. Examples for such developed steels are X 2 CrNiMoN 17 13 5 and X 1 CrNiMoN 25 23 contained in the two tables.

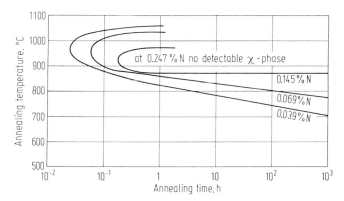

Fig. C 13.37. Influence of the nitrogen content on the start of chi-phase precipitation in steel X 5 CrNiMo 17 13. After [63].

The *maximum allowable service temperatures* for the austenitic steels are derived from the usual intergranular corrosion diagrams. According to these diagrams, all stabilized austenites are suitable for long-term service with an upper temperature limit of 400 °C. The low-carbon steels, also those alloyed with nitrogen, are generally used for continuous service at temperatures up to 400 °C, but with carbon contents of about 0.05%, the long-term application temperature should not exceed 300 °C. For the austenite group of steels, the general rule also applies that in the event of wet corrosive attack below the limit temperatures mentioned no sensitization (increasing susceptibility to intergranular corrosion) may occur within 10^5 h (see DIN 50914 [5]). If the austenitic steels are used in media which do not promote intergranular corrosion, higher service temperatures up to 550 °C can be permitted, when the relevant design parameter is the 1%-proof stress value. However, it must be remembered that as the alloy content increases, particularly with chromium and molybdenum contents, so does the tendency for intermetallic phases to form with the result that the limiting long-term service temperature may be lowered.

C 13.5.4 Ferritic-austenitic Steels

Thanks to the two-phase microstructure prevailing in the ferritic-austenitic steel group the 0.2%-proof stress, at over 450 N/mm², is almost twice that of the common ferritic and austenitic steels (see Table C 13.3). The structure of the readily weldable grade X 2 CrNiMoN 22 5 3 is made up of approximately equal proportions of ferrite and austenite.

Chromium and molybdenum produce a high level of *resistance to* pitting and other forms of *preferential corrosion*. Resistance to stress corrosion cracking in chloride-contaminated cooling water and organic acids is significantly better than in standard austenites. Owing to its good wear resistance under conditions of corrosion attack, these so-called duplex steels are also used where corrosion and abrasion combine. With high strength values accompanied by good corrosion resistance, these steels also offer the advantages of improved corrosion fatigue strength [67].

The *precipitation behavior* of ferritic-austenitic steels is determined by the two structural constituents. Owing to the ferrite content there is a tendency towards both the precipitation of intermetallic phases and the occurrence of 475 °C embrittlement. The addition of nitrogen in the steel grade X 2 CrNiMoN 22 5 3, however, does have the effect of delaying sigma-phase formation. The high solubility of carbon in the austenite prevents the formation of chromium carbides at the grain boundaries of the ferrite and consequently sensitization leading to intergranular corrosion, provided the cooling rate is sufficiently rapid [68]. Besides, nitrogen promotes the formation of austenite starting at the ferrite/ferrite phase boundary. This process must be particularly controlled during cooling of a weldment, as it is also necessary to obtain a weld metal structure that comprises roughly equal parts of ferrite and austenite, ensuring the requisite combination of strength, toughness and corrosion resistance. Thus, also in this group of steels, the addition of nitrogen constitutes a prerequisite for the reliable production of thick cross sections and for improved weldability [19]. The ferritic-austenitic steel, for the reasons mentioned above, has an upper *service temperature limit* of 280 °C, while regarding the toughness, it can be used at temperatures down to about -60 °C.

C 14 Steels for Use with Hydrogen at Elevated Temperatures and Pressures (High-pressure Hydrogen Resistant Steels)

By Erich Märker

C 14.1 Hydrogen-induced Damage to Steel Used in High-pressure Applications

According to the *definition* given in SEW 590-61 [1] *high-pressure hydrogen resistant* steels are steels that, under increased pressures and at elevated temperatures, are resistant to hydrogen-induced decarburization and to the associated embrittlement and formation of intergranular cracks.

Owing to its small atomic radius, hydrogen easily diffuses into metals, causing most varied types of damage. At room temperature and slightly elevated temperatures – up to about 200 °C – the following types of damage are known to occur in steel: flaking, fish eyes, pickling blisters, pickling cracks, hydrogen-induced stress corrosion cracking, underbead (weld) cracking, delayed cracking, hydrogen embrittlement, and fish scale formation.

The *type of damage to steel* occurring below 200 °C, known as "hydrogen embrittlement" or "hydrogen-induced stress corrosion cracking" (the latter being a rather recent term), is caused by a mechanism entirely different from those types observed at temperatures above 200 °C. The basics of hydrogen-induced damage to steel at temperatures below 200 °C and above all the special importance it assumes in the production, transmission, and purification of natural gas containing hydrogen sulfides, are treated in detail by Reuter [2].

Therefore, the following material will discuss only those types of damage that steel may suffer under the *effect of high-pressure hydrogen* at temperatures *above about 200 °C*. These effects have become known and important mainly since the chemical industry began to use high-pressure technology on a large scale for chemical processes. For example, in 1911, during the first large-scale attempt to synthesize ammonia from nitrogen and hydrogen by the Haber-Bosch process, two carbon steel tubes burst after a service life of 80 hours at 200 bar and an operating temperature of between about 500 and 600 °C [3]. Examination of the burst tubes revealed that the pearlite in the microstructure of the internal zone of the tube wall had disappeared. As a result, the steel had lost its cohesion so that even the unaffected external layer of the tube wall cracked. Obviously, under the combined effect of high-pressure hydrogen and high temperatures, the pearlite was decarburized in favor of methane formation.

References for C 14 see page 789.

Bosch's immediate counteraction was a special design solution: he protected the pressure bearing tube by an internal liner tube made from low-carbon iron. The hydrogen which had penetrated this liner tube could then escape through many tiny vent holes which had been bored into the pressure bearing tube wall. Despite this rather straightforward solution, extensive experiments with (unalloyed) carbon steels as well as with chromium and tungsten alloyed steels led to a number of patents that were granted to the Badische Anilin und Soda-Fabrik (BASF), as early as 1912, 1913, 1916 for the production and utilization of special steels for hydrogen applications involving high pressures and elevated temperatures.

But it was not until the twenties and thirties, with the further developed high-pressure processes following the Haber-Bosch process – such as the synthesis of methanol and higher valence alcohols or hydrogenation of coal and mineral oil to produce synthetic fuels – that high-pressure hydrogen-resistant steels of the type still being used today were developed and employed on an industrial scale. Damage as a result of hydrogen-induced decarburization also occurred in steam generators [4].

The failure of steels due to the effect of high-pressure hydrogen is a subject that has already been extensively discussed in literature. A list of the publications up until 1965 is shown in [5–9].

C 14.2 Basics of Physical Metallurgy to Attain High-pressure Hydrogen Resistant Steels

The *reactions* caused by *high-pressure hydrogen diffusing into steel* at temperatures above 200 °C, including the reactions with the carbon in the cementite, are discussed in detail in C 3.3. Here, it is sufficient to bear in mind that the carbides in the steel are decomposed into methane and ferrite, and that this decarburization reduces the cohesion between the crystallites in the microstructure of the steel. The methane accumulating in the affected discontinuities causes (an explosive) effect accompanied by crack formation because it cannot diffuse out of the steel.

Figure C 14.1 shows a micrograph of a carbon steel with discontinuities in its structure resulting from high-pressure hydrogen attack in an ammonia plant. However, as a result of the water reacting with the steel at high temperatures and pressures, hydrogen may generate and diffuse into the steel in steam generators as well [10].

In both examples, the damaging consequences of the chemical effect of high-pressure hydrogen at elevated temperatures leads to a reduction in all mechanical properties – tensile strength, elongation after fracture, reduction of area, bending capacity, and especially notch impact energy. Further phenomena are weight loss, changes in volume and of the steel's physical properties [6].

Figure C 14.2 is a schematic illustration of the *damage* generated in an *ammonia synthesis tube* with a wall thickness of 120 mm by decarburization as a result of *hydrogen attack* [6]. In more than two thirds of the wall thickness are decarburized, the hardness and tensile strength are increasingly reduced towards the attacked

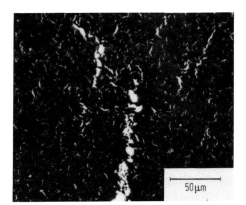

Fig. C 14.1. Appearance of hydrogen attack on unalloyed steel. After [10].

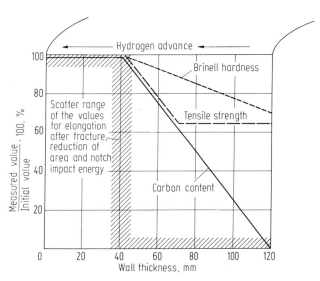

Fig. C 14.2. Schematic illustration of the change in carbon content and mechanical properties as a function of the wall thickness of a steel tube with 0.26% C and 3% Ni damaged by high-pressure hydrogen attack. After [6].

internal surface. Particularly remarkable is the rapid (steep) drop in elongation after fracture, reduction of area, and notch impact energy values as soon as decarburization commences.

Two methods can be used to avoid damage at the grain boundaries as a result of high-pressure hydrogen attack. The first method involves reduction of the carbon content or of the carbide activity in the steel. This reduction can be effected either by lining the pressure bearing steel tube with a very low-carbon steel ("soft iron") tube (which is the method chosen when the Haber-Bosch ammonia synthesis process was first carried out on an industrial scale) or by alloying the steel with elements that strongly promote the formation of carbides. Today, only alloying measures have maintained their importance in the production of high-pressure hydrogen resistant steels.

C 14.3 Conclusions Regarding the Chemical Composition

C 14.3.1 Resistance to High-pressure Hydrogen

The results of systematic investigations into the effect of alloying additions on the high-pressure hydrogen resistance of steels are schematically represented in Fig. C 14.3. The diagram shows the effect of various alloying elements on the hydrogen resistance of steels containing 0.1% C which were exposed to a pressure of 300 bar for a period of 100 h.

The elements *silicon, nickel and copper* do not form carbides, so they have no effect on high-pressure hydrogen resistance.

In contrast, *chromium, tungsten and molybdenum* bring about a strong and continuous increase in resistance. These elements initially dissolve in iron carbides, so their effect is based on the stabilization of the iron carbides. In chromium steels, depending on the carbon content a further discontinuous increase in resistance can be observed from a certain chromium content upwards. This increase is attributable to the replacement of the iron in M_3C by chromium and to transformation into the M_7C_3 special carbide. Figure C 14.3 [11] also indicates that, in practice, the steels with about 3% Cr assume particular importance.

After an initial minor increase of the high-pressure hydrogen resistance, *vanadium, titanium, zirconium and niobium* yield no immediate further improvement, and their effect in this range is surpassed by tungsten and molybdenum. The initial increase in resistance is due to the fact that the elements dissolve in the iron carbides. However, since their solubility is rather low, the resistance is not greatly improved in this range. After a certain level has been exceeded so that the iron carbides have disappeared completely and only special carbides are left, high-pressure hydrogen resistance suddenly undergoes a considerable (steep) increase. These elements form very stable carbides so extraordinarily high resistance levels can be obtained.

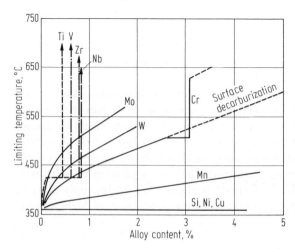

Fig. C 14.3. Effect of alloying elements on the high-pressure hydrogen resistance of steels (test duration 100 h in hydrogen at 300 bar). After [11].

The *combination of chromium and molybdenum* represents a suitable and economic basis in the production of steels intended for high-pressure applications. High-pressure hydrogen resistance and high-temperature strength can be even further increased through additions of tungsten and vanadium.

C 14.3.2 Consideration of the Processing Properties

To obtain optimal resistance to high-pressure hydrogen the steel intended for use must not only be purposefully selected with regard to its chemical composition, but also melted, rolled, or forged and heat treated with great care so that a largely defect-free and uniform microstructure with carbides is obtained that yields high resistance to hydrogen attack and excellent elevated-temperature strength. Therefore, when selecting the chemical composition of the steel, various properties must be purposefully balanced one against the other.

The supply of hollow bodies with wall thicknesses exceeding 200 mm, in lengths up to 18 m and ingot weights of *more than 300 t* represents *special problems* to be solved by the steel industry. The reliability of steel components for high-pressure applications is governed by the most varied processing parameters; for instance, the degree of deformation is just as important as the through-hardenability and the retention of hardness at the heat treatment. The manufacture of thick-walled high-pressure tubes from strongly air hardening chromium-alloyed steels using the pierce and pilger rolling process, must be considered as well.

Since cold deformation increases the speed of the hydrogen attack quite considerably, *high-pressure hydrogen resistant components* are almost exclusively installed in the *quenched and tempered or normalized and tempered condition*. High temperatures and long holding times are required for the tempering treatment following quenching, to obtain optimal high-pressure hydrogen resistance.

The microstructure in the as solidified condition of a *weld* is *considerably more susceptible* to hydrogen embrittlement than that of a forged or rolled component. Similarly, the martensitic microstructure after hardening of the type which may occur, for instance, in the heat-affected zone next to the welding zone of ferritic-pearlitic steels, will be less resistant than the evenly through-hardened and tempered microstructure of the base material. Depending on the alloying content, the high-pressure hydrogen-resistant steels must be preheated more or less before welding and, if possible, they should be given a complete heat treatment (quenching and tempering) after welding. In practice, owing to the component dimensions, this latter requirement can very rarely be fulfilled. However, the weld area at least should always be subjected to a tempering treatment at sufficiently high temperatures, and sufficiently long holding times.

C 14.4 Application-oriented Selection of the Steels

As already pointed out in C 14.3.1, the basis of physical metallurgy to make a steel resistant to high-pressure hydrogen attack is by alloying it with chromium and

Table C 14.1 Chemical composition of traditional high-pressure hydrogen resistant steels according to SEW 590, December 1961 edition [1]

Steel grade (Designation)	Chemical composition				
	% C	% Cr	% Mo	% Ni	% V
16 CrMo 9 3	0.12 ... 0.20	2.0 ... 2.5	0.30 ... 0.40		
20 CrMo 9	0.16 ... 0.24	2.1 ... 2.4	0.25 ... 0.35	≤ 0.80	
17 CrMoV 10	0.15 ... 0.20	2.7 ... 3.0	0.20 ... 0.30		0.10 ... 0.20
20 CrMoV 13 5[a]	0.17 ... 0.23	3.0 ... 3.3	0.50 ... 0.60		0.45 ... 0.55
X 20 CrMoV 12 1[b]	0.17 ... 0.23	11.0 ... 12.5	0.80 ... 1.2	0.30 ... 0.80	0.25 ... 0.35
X 8 CrNiMoVNb 16 13[c,d]	≤ 0.10	15.5 ... 17.5	1.1 ... 1.5	12.5 ... 14.5	0.60 ... 0.85

[a] For detailed information on the properties see Table C 14.2
[b] Further data on the properties are contained also in DIN 17 175 [12] and DIN 17 243 [13], as well as in SEW 670 [14] and SEW 675 [15]; see also C 9
[c] Furthermore 0.07% to 0.13% N and % Nb ≥ 10 · % C
[d] Further data on the properties are contained in SEW 670 [14] and SEW 675 [15]; see also C 9

molybdenum. Therefore steels alloyed in this way have long established themselves in the sphere of high-pressure applications. Table C 14.1 shows some examples taken from SEW 590 [1]. Tubes, fittings, vessels and other components for high-pressure applications in the temperature range between 200 °C and 400 °C, in connection with the hydrogen pressure of max. 700 bar, are manufactured from steels with chromium contents between 1 and 3%. These steels have proved themselves under operating conditions for many decades. In 20 CrMoV 13 5 steel, the maximum operating temperature in the specified pressure range is as high as 480 °C. This steel with about 3.25% Cr and 0.5% Mo and V each is the premium steel in the group of low-alloy high-pressure hydrogen resistant steels.

This steel is the only survivor of the traditional group of high-pressure hydrogen resistant steels after the latest revisions geared towards updating the specifications for chemical composition and other properties. With today's international interlocking in the construction of chemical processing equipment, all other steels of this group have lost their economic importance and have been replaced by steels used in the entire western world. These are chromium-molybdenum steels with chromium contents ranging from 1 to 9% and molybdenum contents between 0.5 and 1% (see also the standards issued by the American Society for Testing and Materials [16]). These steels are employed not only in chemical high-pressure processes, but also in energy-producing plants and in the processing of crude oil and natural gas. This usage has the advantage that even smaller lots are readily available for refitting measures, in compliance with current German or international standards for high-temperature steels, e.g. DIN 17 155 (plate and strip) [17], DIN 17 175 (pipes and tubes) [12], and DIN 17 243 (forgings and bar steel) [12]. These steels can be fusion welded, and their behavior during and after welding is well understood by plant constructors. This interchangeability is particularly important because flanged connections – previously standard in high-pressure plant engineering – has been largely superseded by welded connections.

The development of catalysts has led to changes and modifications in the process parameters; for example, the reaction temperatures have been lowered, which means that steels can be made resistant to high-pressure hydrogen with lower alloy contents.

SEW 590 [1] will be replaced by DIN-standards for the special products, e.g. DIN 17 176-Nahtlose kreisförmige Rohre aus druckwasserstoffbestandigen Stahlen (Nov. 1990). Of the large number of steels detailed in the above mentioned standards, the steels listed in Table C 14.2 are accentuated here as being specific high-pressure hydrogen resistant. This has been indicated in the August 1984 draft for the new edition of SEW 590 [1].

SEW 595, which deal with steel castings for crude oil and natural gas plants [1] contains also internationally accepted steels with 2.25%, 5% and 9% Cr and with 0.5% to 1% Mo (also in compliance with American ASTM standards) (see Table C 14.3). The high-alloy steel grades for castings described in this SEW are characterized by their higher resistance to the effect of high temperature corrosion typical in such plants.

The *operating limit curves for carbon and alloyed steels as a function of high-pressure hydrogen loads* established by Nelson [18] have gained substantial importance in practice with regard to the selection of materials suitable for a given hydrogen pressure and tube wall temperature. Most certainly better than individual curves would be the determination of scatter bands for the initiation of the attack so that the effects of differences in the chemical composition, grain size, microstructure, exposure time, etc. could be taken into account. However, a definite advantage of the representation method chosen by Nelson is its greater clarity. The "Nelson diagram" curves are adjusted at regular intervals to the relevant new findings [19]. Figure C 14.4 shows such a diagram. The relevant limit temperatures are plotted against the hydrogen pressure. The solid lines stand for the resistance limit and the dashed lines indicate the beginning of surface decarburization. The diagram also illustrates the lower resistance of welding seam of carbon steels.

Since Naumann published his work [11], it has been known that *titanium, vanadium and niobium* contents up to 0.1% have the same *effect as molybdenum* and that the effect of molybdenum is four times that of chromium. But under operating conditions the favorable behavior steel with 0.5% Mo seems to have been overestimated. In the 1983 edition of API-Publication 941 [18] eight failures were reported in steels with 0.5% Mo operating below the 1977 – curve for this steel in catalytic reforming units. Since then more failures have been experienced so that the curves for steels with 0.25% and 0.5% Mo will have to be revised (Fig. C 14.4). Until the reasons for these failures are better understood the steel with 0.5% Mo should not be used for conditions exceeding the limit curve for carbon steel.

In industrial chemical processes, hydrogen occurs together with other gases. In *hydrogen-nitrogen-ammonia mixtures*, ammonia-induced nitriding must be taken into account in addition to hydrogen-related effects. Comparative investigations concerning the nitriding (nitrogen pick up) of low and high-alloy steels and one nickel base alloy revealed that – contrary to low-alloy steels which during nitriding do not suffer weight loss but increases in hardness and embrittlement – high-alloy chromium steels at temperatures above 400 °C suffer a particularly strong corrosion attack owing to nitride layers peeling off. On the contrary, the nickel base alloy and the austenitic chromium-nickel steel form thin, crack-free nitride layers that assume a protective function [21].

Table C 14.2 Chemical composition and mechanical properties of high-pressure hydrogen resistant steels according to the August 1984 draft for the new edition of SEW 590 [1] for seamless tubes in the quenched and tempered condition as example

Steel grade (Designation)	Chemical composition[a]						Mechanical properties at 20°C						at 450°C 0.2%-proof stress	at 500°C 10⁴ h creep up to strength
	% C	% Si	% Mn	% Cr	% Mo	% others	Yield stress[b] N/mm² min.	Tensile strength N/mm²	Elongation after fracture A[e] % min		Notch impact energy[c,d] J min		N/mm² min	N/mm²
									l	tr	l	tr		
25 CrMo 4	0.22...0.29	≤ 0.40	0.50...0.90	0.90...1.2	0.15...0.30		345	540...690	18	15	48	27	185	176
12 CrMo 9 10	0.10...0.15	≤ 0.30	0.30...0.80	2.0...2.5	0.90...1.10	0.010...0.040 Al; ≤ 0.30 Ni; ≤ 0.20 Cu	355	540...690	20	18	64	48	275	191
12 CrMo 12 10	0.06...0.15	≤ 0.50	0.30...0.60	2.65...3.35	0.80...1.06		355	540...690	20	18	64	48	275	
12 CrMo 19 5	0.06...0.15	≤ 0.50	0.30...0.60	4.0...6.0	0.45...0.65		390	570...740	18	16	55	39	280	130
X 12 CrMo 9 1	0.07...0.15	0.25...1.0	0.30...0.60	8.0...10.0	0.90...1.10		390	590...740	20	18	55	34	295	215
20 CrMoV 13 5	0.17...0.23	0.15...0.35	0.30...0.50	3.0...3.3	0.50...0.60	0.45...0.55 V	590	740...880	17	13	55	34	420	186

[a] The phosphorus and sulphur contents are ≤ 0.030% each for all steels.
[b] If the yield stress cannot be determined, the values for the 0.2%-proof stress shall apply instead.
[c] Key to abbreviations: l = longitudinal specimen, tr = transverse specimen
[d] Determined on ISO-V specimens, average values of three specimens
[e] Tensile stress which leads to fracture after 10000 h (average values)

C 14.3 Conclusions Regarding the Chemical Composition

Table C 14.3 Chemical composition and mechanical properties of high-pressure hydrogen resistant steel grades for castings according to SEW 595 [1]

Cast steel grade (Designation)	Chemical composition			Mechanical properties					
				at 20 °C				at 450 °C	at 500 °C
	% C	% Cr	% Mo	0.2%-proof stress N/mm² min.	Tensile strength N/mm²	Elongation after fracture A % min.	Notch impact energy[a] J min.	0.2%-proof stress N/mm² min.	10⁴ h-creep rupture strength N/mm²
GS-12 CrMo 9 10	0.08 ... 0.15	2.0 ... 2.5	0.90 ... 1.1	345	490 ... 690	18	55	245	200
GS-12 CrMo 19 5	0.08 ... 0.15	4.5 ... 5.5	0.45 ... 0.55	410	640 ... 840	18	34	295	145
G-X 12 CrMo 10 1	0.08 ... 0.15	9.0 ... 10.0	1.1 ... 1.4	410	640 ... 840	18	27	295	175

[a] DVM-specimens (German standard specimen)
[b] Tensile stress which leads to fracture after 10 000 h (average values)

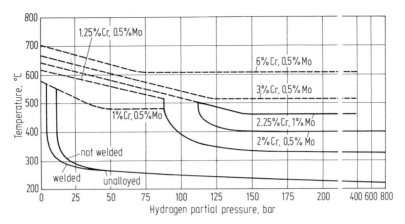

Fig. C 14.4. Operating limits for the resistance of steels to high-pressure hydrogen. After [18].

A further example that should be mentioned here is the *risk of sulfidation* involved in some petro-chemical processes.

Finally, when selecting a steel for a given application, besides the corrosion resistance, the *susceptibility to embrittlement* by sigma-phase precipitation must be taken into account. This susceptibility applies particularly to high-alloy chromium steels (see C 13).

Practical experience shows that the chemical high-pressure technology processes can be mastered by the application of steels which have proved themselves for decades.

C 15 Heat Resisting Steels

By Wilhelm Wessling and Rudolf Oppenheim

C 15.1 Required Properties of Use

Heat resisting steels are widely used for the manufacture of equipment and installations designed for high temperature operations in chemical, petrochemical, metallurgical and ceramics industries as well as in the technology of waste gas processing. *Applications* for example, include radiation tubes, supporting beams in furnaces, protective tubes for temperature measuring devices, soft annealing boxes and bright annealing muffles, dust cleaning at elevated temperature, conveyor belts, and enameling grids as well as automobile exhaust decontamination systems [1].

The designation, heat resisting and scale resisting steel by definition [2] is applied to steels that form, at temperatures above 500 °C, a tightly adhering surface oxide layer protecting the steel against damaging effects from hot gases, flue particles, liquid salts and metals. Construction elements that require heat resisting steels are generally heated by high environmental temperatures and are expected to exhibit not only a high resistance to scale formation – also called hot corrosion resistance – but must also be suited for longtime service under mechanical stresses. Sharp temperature changes may cause sudden linear expansions and for that reason a satisfactory service behavior can only be expected if the steel is also resistant to thermal shock. Embrittlement in longtime service also must be avoided to maintain plant safety.

The chemical attack at elevated temperatures is not limited only to oxygen or nitrogen from the surrounding atmosphere. One must also consider the steam content and the question whether the attack takes place under the influence of oxygen in a predominantly oxidizing or by the influence of carbon and hydrogen in a chemically reducing atmosphere. Bright annealing gases consisting of mixtures of nitrogen and hydrogen may have a nitriding effect but with the formation of nitrides one also must observe in carbon bearing gases the quantities of carbon-monoxide, carbon-dioxide and carbon-hydrogens, e.g. methane or propane, because these constituents govern the oxidizing, reducing or carbonizing character of the gas [4]. Similar distinctions are necessary for sulphur-containing gases which are characterized either by sulphur dioxide or hydrogen sulphide. Finally, attacks from halogens as for example chlorine, from salts, vitreous enamels, ceramics and liquid metals or low melting metal oxides such as vanadium and molybdenum oxides must be mentioned.

References for C 15 see page 789.

C 15 Heat Resisting Steels

The construction of reliable plant equipment also requires the steels to possess a satisfactory processing behavior i.e. hot- and cold-formability and weldability.

Essential properties when selecting heat-resisting steels and alloys are:
- resistance to scaling and hot corrosion
- high temperature strength and good long time behavior at elevated temperatures
- stability of the microstructure, that is, resistance to embrittlement
- weldability and deformability.

C 15.2 Characterization of the Required Properties by Test Values

According to agreement a steel is designated as *scale resistant* at a specified temperature if the scale formation at this temperature does not exceed about 1 g/m² h and at a 50 °C higher temperature about 2 g/m² h [5]. Testing commonly takes place during a 120 h test-run with 4 intermediate cooling cycles. However, as shown in Fig. C 15.1 the test may be extended to obtain more reliable results.

Such a cyclic test procedure makes possible an evaluation of the scale adherence. In removing the scale from the test pieces weight errors have to be avoided (for example by using a mordant of sodium hydroxide solution (NaOH) or sodium hydride (NaH) [6]. Table C 15.1 indicates for the chemical compositions of some heat resisting steel grades the limiting temperatures in air resulting from this agreed test procedures.

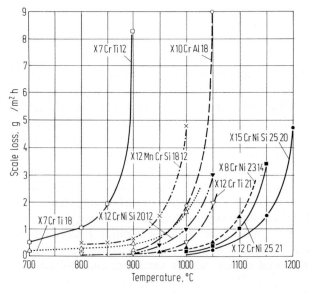

Fig. C 15.1. Curves of scale loss of heat resisting steels in still air for cyclic test procedure (13 × 24 h at the given temperatures each time cooled to room temperature). (The chemical composition of some heat resisting steels in common use that have been tested in this connection can be found in Table C 15.1.)

Table C 15.1 Chemical Composition of heat resisting materials and their limiting scale resistance temperatures in still air

Steel grade Designation	No.	% C max.	% Al	% Cr	% Ni	% Si	% Others	Scaling resistance in air up to
Ferritic steels								
X 10 CrAl 13	1.4724	0.12	1.0	13.0		1.0		850 °C
X 10 CrAl 18	1.4742	0.12	1.0	18.0		1.0		1000 °C
X 10 CrAl 24	1.4762	0.12	1.5	24.5		1.0		1150 °C
Austenitic steels								
X 12 CrNiTi 18 9	1.4878	0.12		18.0	10.0	1.0	0.4 Ti	850 °C
X 15 CrNiSi 20 12	1.4828	0.20		20.0	12.0	2.0		1000 °C
X 12 CrNi 25 21	1.4845	0.15		25.0	20.5	0.5		1100 °C
X 15 CrNiSi 25 20	1.4841	0.20		25.0	20.5	2.6		1150 °C
X 10 NiCrAlTi 32 20	1.4876	0.12	0.3	21.0	32.0	0.5	0.4 Ti	1100 °C
Non-ferrous alloys								
NiCr 15 Fe	2.4816	0.12	(0.1)	15.5	75		0.2 Ti, 8.0 Fe	1150 °C
CoCr 28 Fe	2.4778	0.10		28.0			48.0 Co, 20.0 Fe	1250 °C

The *strength properties* at room and elevated temperatures are also important in evaluating the material properties during processing and service.

Table C 15.2 contains information on *properties at room temperature*. For ferritic steels they apply only to thin gauges. Hot-formed products with larger cross-sections and hence larger grain sizes tend to exhibit cold brittleness, which makes it necessary to carry out forming operations at slightly elevated temperatures. Improved forming behavior of ferritic steels is found in cold rolled and hence fine grained flat products.

Figure C 15.2 demonstrates the differences in tensile strength and 0.2%-proof stress between ferritic and austenitic steels in dependence on the temperature and in particular in the application range above 600 °C. The common calculation strength value for designing long life installations is however the 1%-creep limit for 10 000 h (Fig. C 15.3) which in ferritic steels is mostly only slightly below the 100 000 h-creep-stress rupture strength. The difference is only slightly larger in austenitic steels (as to the test procedures see B 1).

Table C 15.2 Mechanical properties of heat resisting materials listed in Table C 15.1 at room temperature.

Material	0.2%-proof stress N/mm^2	Tensile strength N/mm^2	Elongation after fracture A %
Ferritic steels	210 ... 380	400 ... 700	10 ... 15
Austenitic steels	210 ... 330	500 ... 750	30 ... 35
NiCr 15 Fe	≥ 175	490 ... 640	≥ 35
CoCr 28 Fe	≥ 350	650 ... 900	≥ 5

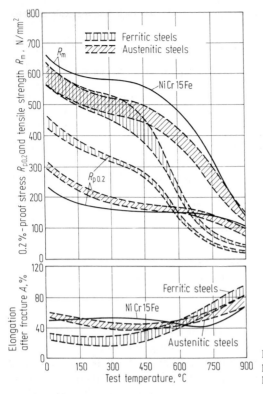

Fig. C 15.2. Relation between mechanical properties of heat resisting steels and temperature.

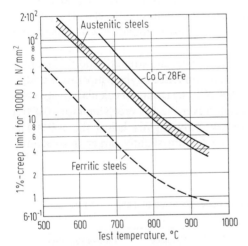

Fig. C 15.3. Stress resulting in a creep of 1% after 10 000 h (1%-creep limit) in dependence on temperature for heat resisting steels. After [2, 24].

Checking for possible embrittlement effects in long term behavior generally uses hardness measurements, notch bar impact testing at room or elevated temperatures and bending tests. The results indicate the technological behavior that may be expected. Additional metallographical and chemical investigations are necessary to analyze the causes of the embrittlement. It is assumed that the precipitation mechanism of intermetallic phases in ferritic chromium steels resulting in embrittlement is well known (see C 13). Embrittlement can be proved by an increase in hardness after long term annealing at different temperatures [7]. The stability of the microstructure of austenitic steels and their embrittlement behavior is also verified by notch impact testing after long term annealing [8]. Significant differences are due primarily to variations in the chromium content (Fig. C 15.4).

In connection with the hot strength properties attention must be given to thermal fatigue. Heat treatment installations are frequently subjected to sharp temperature changes that may cause sudden changes in length so that adequate *resistance to thermal shock* is required [10]. The required precondition is a high ductility at elevated temperatures in combination with a high yield stress at these temperatures. A low expansion coefficient together with good thermal conductivity to avoid overheating is also necessary to ensure sufficient resistance to thermal shock or alternating temperatures [1, 11].

C 15.3 Chemical Composition and Microstructure of Applicable Steels

According to B 3, corrosion resistance is due to the formation of a tight oxide layer which prevents or at least delays the diffusion of the attacking media into the steel thereby protecting the surface against chemical attack. The most important element for increasing the scale resistance is *chromium*, because in combination with oxygen, it forms a dense and tightly adhering layer of chromium-oxide. Increasing amounts of chromium increase the scaling resistance. Thus to provide sufficient scaling resistance at a given service temperature an appropriate amount of chromium is required. Other elements such as *silicon*, *aluminum* and *titanium* improve the effect of chromium, and can replace a certain amount of chromium in the protective layer. The adherence properties of the oxide layer can also be improved by additions of cerium [12]. Increasing amounts of *nickel* have a similar effect on austenitic steels because nickel reduces the different thermal expansion coefficients of the oxide layer and the base metal. Nickel alloys exhibit less thermal expansion than iron-chromium-nickel steels. Nickel-chromium spinels on iron-free

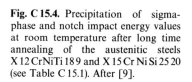

Fig. C 15.4. Precipitation of sigma-phase and notch impact energy values at room temperature after long time annealing of the austenitic steels X 12 CrNiTi 18 9 and X 15 Cr Ni Si 25 20 (see Table C 15.1). After [9].

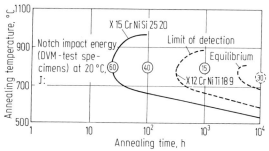

nickel-chromium alloys offer a more efficient protection against scaling than iron-chromium spinels.

Resistance to scaling is also influenced indirectly by nickel additions *in carbon- or nitrogen-rich atmospheres*. Diffusion into the more densely packed austenite lattice is hindered and the solubility for carbon and nitrogen decreases strongly with increasing nickel additions. Silicon also increases resistance to carburizing [4] and nitriding [13]. Experiments have also been performed to find ways to combine the diffusion of carbon into the steel with diffusion of niobium to render the carbon harmless.

The *scaling resistance determined in air* may not be generalized, as shown by these findings. One must also consider the behavior against the specific attack conditions of hot corrosion. If a protective oxide layer is not formed oxidation will continue, producing growing scale layers that eventually will burst off. An effective protection against oxidation therefore also requires exact temperature control to avoid overheating. Cracking of the scale layer as a result of local overheating often leads to total mechanical failure of the components.

If the requirements for the totality of the properties are not too high ferritic chromium steels are mostly used containing additions of silicon or aluminum (Table C 15.1). The amount of the alloying additions governs the sensitivity towards the 475 °C-embrittlement [14] or the precipitation of the sigma-phase [15] (Fig. C 15.5).

Aluminum and silicon additions to steels containing 13% Cr do not lead to an increase in hardness within 1000 h at 475 °C. This increase in hardness, however, is observed in steels containing 18 to 30% Cr with or without additions of these elements: aluminum favors the 475 °C-embrittlement and silicon exerts a somewhat retarding effect [17]. However silicon additions to steels containing 13 to 30% Cr shortens considerably the incubation time until the precipitation of the sigma-phase starts, whereas aluminum exhibits the opposite effect [15]. The ferritic

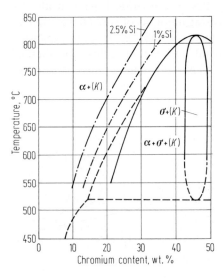

Fig. C 15.5. Range of the sigma-phase in pure iron-chromium-alloys and in chromium steels with 1% and 2.5% Si (K) = $M_{23}C_6$. After [16].

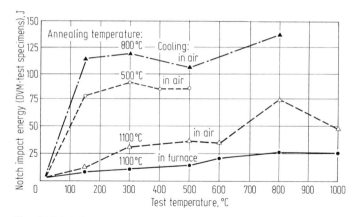

Fig. C 15.6. Influence of 1000 h annealing on the notch impact energy values of the ferritic steel X 10 CrAl 24 (see Table C 15.1) in dependence on the test temperature.

microstructure is adversely affected by carbide precipitations and grain growth after critical hot forming operations and high annealing temperatures resulting in cold embrittlement. However, rising service temperatures improve toughness properties to such a degree that even chains in ferritic steels exhibit satisfactory behavior at elevated temperatures (Fig. C 15.6). Intercrystalline micro-precipitations after the heat treatment must be avoided by cooling as fast as possible.

For high quality requirements use is made of the better quality properties of *austenitic chromium-nickel-steels* or nickel-chromium alloys. Precautions must be taken to ensure that the nickel content is sufficient to produce a complete austenitic microstructure because remanent fractions of delta-ferrite will result in embrittlement due to the precipitation of the sigma-phase. Such precipitation may lead to a premature carburizing because carbon diffusion takes place preferentially in ferrite [9]. Solution annealing at a sufficiently high temperature will produce a homogeneous recrystallized microstructure and to obtain an optimal creep rupture behavior the austenitic grain size should not be smaller than number 5 of DIN 50 601 [17a].

C 15.4 Characteristic Grades of Heat Resisting Materials

Table C 15.1 contains a list of heat resisting steels and non-ferrous alloys, classified according to the explanations in C 15.3, and information as to the limiting scaling temperature in air.

C 15.4.1 Ferritic Steels

Ferritic heat resisting steels recrystallize at 800 to 850 °C. Because of their sensitivity to grain growth they are mainly processed in hot rolled small cross-sections or as cold rolled strip [18]. Cold forming is possible with certain precautionary

Fig. C 15.7. Microstructure of the ferritic steel X 10 CrAl 18 (see Table C 15.1) **a** as delivered, **b** after carbon pick-up during service at medium temperatures.

measures and at somewhat elevated temperatures. In welding ferritic steels attention must be paid to preheating, limited heat input to avoid grain growth in the heat-affected zone and a suitable post-heat treatment. It is recommended that austenitic filler materials may be used, and only the final pass should be made with material similar to the base metal.

Ferritic steels have good resistance under oxidizing conditions in air and sulphur-containing gases. Additions of aluminum increase the resistance against the attack of sulphur (compounds), however, in gases with higher nitrogen contents *scale efflorescences* may form at high temperatures as a result of nitrogen pick up on account of formation of aluminum nitride. These scale efflorescences result in premature failure due to catastrophic scale cracks and fractures [19]. Nitrogen pick-up in the surface zones may even lead to formation of an austenitic microstructure. Aluminum-free steels containing 18 and 24% Cr and about 2.2% Si are less sensitive to this reaction, but tend to exhibit sigma-phase-embrittlement and therefore are now little used.

If the gas stream is chemically reducing in nature under *the influence of carbon monoxide or methane* the slightly restrained carbon diffusion in the bcc lattice results in a very quick precipitation of chromium carbide at the grain boundaries. As a consequence embrittlement and locally a depletion of free chromium will occur, decreasing the resistance to oxidation and sulphur attack (Fig. C 15.7). Ferritic steels therefore should not be used under carburizing conditions. Under reducing conditions in hydrogen sulphide the resistance against sulphides of ferritic steels is lower than for a sulphur attack under oxidizing conditions.

C 15.4.2 Austenitic Steels and Alloys

Austenitic steels in contrast exhibit good forming properties, less sensitivity to embrittlement, better high-temperature strength and good resistance to alternating temperatures (thermal shock). Steels containing more than 30% Ni are outside the constitution field of the sigma-phase, and steels with 15–20% Cr having good resistance against carburizing and nitriding may be employed in a wide temperature range without danger of embrittlement (see Fig. C 15.8).

Austenitic steels can be welded without technical difficulties using common welding techniques. Occasionally however, it is observed that full-austenitic chromium-nickel-silicon-steels show a tendency for *hot cracking*. This tendency can be controlled by a low heat input and by use of a line welding deposit (stringer bead). In complicated cross-sections and with high residual welding stresses is it good practice to use filler metal having a nickel basis with 15 or 20% Cr and additions of niobium [20], by which it will be possible to obtain welds without hot cracks, using a suitable heat input and intermediate layer temperatures.

Austenitic steels have a good resistance to scaling in both oxygen-rich and oxygen-poor atmospheres, the upper service temperature depends again on the chromium content. For service in overheated steam steel grades X 12 CrNiTi 18 9 and X 10 NiCrAlTi 32 20 are extensively employed and high temperature strength properties must be taken into account. Resistance to scaling is better in dry steam than in humid hot air; an addition of only 5% steam causes a higher oxidation rate. Nickel alloys with a chromium content of at least 15% give good service results in steam at temperatures above 760 °C.

Heat treatment installations that are operated alternately with reducing and oxidizing atmospheres may be affected by a particular kind of high temperature corrosion, which is known as *green rot* ("Grünfäule"). The material to be heat treated often carries residuals of greases and oils etc. into the installation which leads to a carbon contamination of the furnace chamber walls. If the furnace is

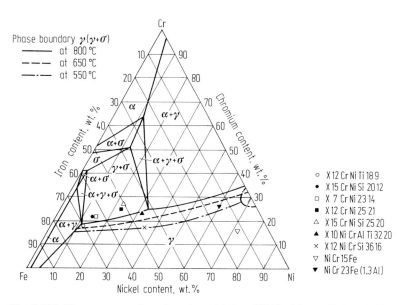

Fig. C 15.8. Ternary system iron-chromium-nickel at 800 °C. After [12]. (In addition see also Fig. C 16.1.)

operated under a chemically-reducing atmosphere carbon will diffuse into the material along the grain boundaries leading to the formation of chromium carbide. The local depletion of chromium near the grain boundaries decreases the resistance to oxidation so that under oxidizing conditions an *"inner oxidation"* can take place along the carbon enriched grain boundaries. By the same degree as the carbide formation proceeds the sensitivity for brittle fracture increases [21]. Resistance to this type of hot corrosion is found in particular steels with a high nickel content (for example X 12 NiCrSi 36 16). Examples of microstructures exhibiting carburizing and heavy scaling can be seen in Fig. C 15.9.

Nitrogen pick-up may also lead to local chromium depletion and thus to a decrease in scaling resistance. The danger of a certain cold brittleness must also be taken into account. For this reason nickel-rich steels are used for the installations of nitriding furnaces (Fig. C 15.10). Nitrogen pick-up at elevated temperatures leads to the formation of socalled *"false"* nitrogen pearlite (Fig. C 15.11).

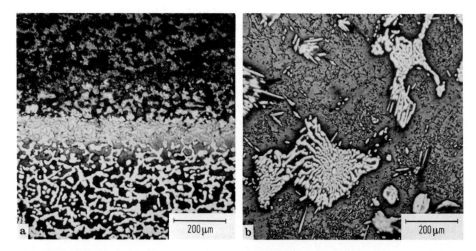

Fig. C 15.9. Microstructure of the austenitic steel X 15 CrNiSi 25 20 (see Table C 15.1). **a** After carbon pick-up; **b** after carbon pick-up and additional heavy scaling due to overheating.

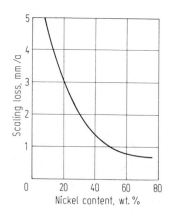

Fig. C 15.10. Scale loss of austenitic steels after 586 h at 540 °C to 580 °C in a nitriding furnace. After [22].

Fig. C 15.11. Microstructure of the steel X 15 CrNiSi 25 20 (see Table C 15.1) after nitrogen pick-up at very high temperatures: "false" nitrogen pearlite.

Bright annealing muffles for longtime operation at temperatures of about 1100 °C are manufactured from the nickel-chromium alloy NiCr 15 Fe (see Table C 15.1) because impairment of the scaling resistance, either of the hot gas side or at the bright annealing side is not admissible. The mechanical long term properties of the furnace material are also important for this type of application.

Nickel-containing austenitic steels and alloys are sensitive to *corrosion attacks by sulphur compounds at high temperatures*. Under oxidizing conditions the scale layer initially prevents sulphur absorption, but the sulphur pick-up in a reducing atmosphere proceeds very quickly with the formation of chromium-sulphide. This process is followed by the disruption (break down) of the steel by oxidation since the chromium depletion reduces the scaling resistance. Figure C 15.12 clearly shows the penetration of chromium sulphide deep into the steel which is followed by a strong inner oxidation. Nickel-containing steels can only be made more resistant to sulphur compounds by an increased chromium content, so a *nickel alloy with 50% Cr* was developed which exhibits outstanding resistance to *corrosion by oil ashes* [23].

Additions of aluminum also can improve the resistance to sulphides on a limited scale as the experience with NiCr 23 FeAl has shown. This alloy is particularly suited for automobile exhaust systems [1].

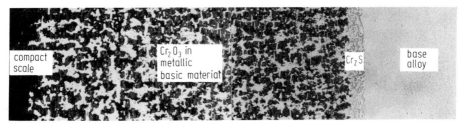

Fig. C 15.12. Scaling of the austenitic steel X 15 CrNiSi 20 12 (see Table C 15.1) after sulphur pick-up in the surface in a reducing annealing atmosphere.

Nickel-containing steels and alloys are generally only employed for atmospheres poor in sulphur which at most contain less than 0.02 vol.-% sulphur dioxide in an oxidizing atmosphere and 0.02 vol.-% hydrogen sulphide in a reducing atmosphere.

Manganese-austenite steels with low nickel contents (e.g. a steel with 0.10% C, 18% Mn, 12% Cr and 2% Si) may be considered for service in sulphur-containing gases at temperatures up to 900 °C, but furnace construction elements intended for extreme conditions need to be designed from only the *cobalt alloy* CoCr 28 Fe shown in Table C 15.1 [24].

C 15.4.3 Criteria for the Selection of Heat Resisting Material Groups

In conclusion it is recommended that selection procedures should primarily consider the most essential quality requirements for a given service condition. In most instances that will be the heat resistance, often in combination with the long term strength behavior (creep rupture strength). Material properties must not be endangered by microstructural instability at service temperatures. In other words, a type of steel should be selected which shows the least tendency for embrittlement at a given service temperature. If no special requirements need to be satisfied, for example, resistance to sulphides, a main criteria factor should always be the long term service behavior in the widest possible temperature range.

Ferritic steels exhibit good resistance to sulphur pick-up in neutral to oxidizing atmospheres, but they fail in reducing or carburizing atmospheres. They have little high temperature strength and are sensitive to sigma-phase-embrittlement. Poor cold forming properties and the tendency to grain growth pose problems during forming operations.

Austenitic steels and nickel alloys have good welding properties and are well suited for cold and hot forming operations. On the one side they show a certain sulphide sensitivity but on the other side they exhibit good resistance to carbon and nitrogen pick-up which increases with increasing nickel contents and they also have adequate high temperature strength properties. High stability of the microstructure in a wide temperature range and good behavior under mechanical loads are characteristic in particular for steels and alloys containing more than 30% Ni.

C 16 Heating Conductor Alloys

By Hans Thomas

C 16.1 Required Properties and their Testing

In electrothermic technology, heating elements are made, for instance, of platinum, molybdenum or of intermetallic compounds ($MoSi_2$). However, with regard to practicality and economy, alloys (solid solutions) with a high and only slightly temperature-dependent electrical resistance often are selected – to provide simple power supply and high resistance to oxidation for a long life.

The high *resistance* to scaling of the typical heating-conductor alloys is not based on a reduced affinity to oxygen. Instead, the resistance depends on the fact that at higher temperatures a dense, well adhering coat is formed on the surface, protecting the metal beneath against any progressive (further) reaction (see B 3 and C 15).

Due to their *electrical resistance*, heating conductor alloys are also suitable for resistance devices, e.g. in-line, control and braking resistors and potentiometers. With certain modifications some of the alloys can be used as materials for measurement- and precision-resistors.

Electrical resistance is measured by conventional methods (Wheatstone or Thomson Bridge, voltage comparison, digital ohmmeters). Temperature dependence is determined by heating the sample under a protective gas (to avoid changes in the cross-section due to oxidation) in an electric furnace. The temperature coefficient of materials for measurement and precision resistors is determined in liquid baths.

The *oxidation resistance*, or scaling behavior can be investigated by several well-known methods [1]. One of the methods more commonly used in practice is to set a wire sample, either a helice or a straight piece, under current and to heat and cool it alternately (each 2 mins) to a pre-given temperature, e.g. 1200 °C. The characteristic term for the "life time" is either the number of switching processes or the total time prior to failure. This method has the advantages that very little material is required and due to the extreme stress, results are obtained after a short period, giving comparative values not only for the oxidation rate but also for the adhesive strength of the scaling layer since the different coefficients of thermal expansion of metal and scale produce mechanical forces during each change in temperature. The tests can be conducted in air or in different gases or mixtures of gases [2].

References for C 16 see page 790.

The test gives reference values for the *highest permissible usable temperatures* and allows reference values to be derived for electrical *loadability*. The heat developed by a heating element under current is dissipated outwards through the surface. It follows that the specific surface load i.e. the electrical power per surface unit of the heating element is a suitable term for dimensioning. The electrical power per surface unit largely determines the temperature of the heating element under the projected conditions of use. The power per surface unit can be set the higher the more heat is accepted from and dissipated by the environment and the higher the permissible temperature of use, i.e. the higher the scaling resistance of the heating conductor alloy.

Mineral insulated heating conductors are a special design. The heating wire under current is embedded in mineral insulating powder (MgO, Al_2O_3, SiO_2) inside a metal tube. This sheath is electrically neutral which means that no other insulation is required providing the temperatures are not too high.

C 16.2 Relationship Between Physical Metallurgy and Chemical Composition

The technically important heating conductor alloys are based on nickel-chromium and nickel-chromium-iron (face-centered cubic crystal structure – austenitic) or iron-chromium-aluminum (body-centered cubic crystal structure – ferritic).

C 16.2.1 Austenitic Heating Conductor Alloys

In the nickel-chromium-iron system fcc solid solutions occur between 0 and 18% and up to 30% Cr and between 20 and 100% Ni depending on the iron content (Fig. C 16.1). Good scaling resistance is found only above 15% Cr [4], consequently for austenitic alloys only a relatively narrow zone is available in the phase diagram that covers the range from 20% Ni through to iron-free nickel-chromium-alloys. Technologically these alloys do not present any particular problems. They have a good strength at elevated temperatures that becomes noticeable in hot forming and in various applications. As with many other metallic materials tensile tests of these alloys show minima of elongation after fracture and of reduction of area between 500 and 700 °C [5, 6]. As a rule these changes result from the shearing off along grain boundaries that occurs in the relevant temperature range and leads to micro-cracks in the crystalline zones in their immediate vicinity, and later merge to macro-cracks [7].

The oxide layers that develop at high temperatures contain different metal-oxygen-compounds, i.e. spinelle ($NiCr_2O_4$, $FeCr_2O_4$ amongst others) and chromium oxide [8, 9] depending on the formation conditions and the layer thickness.

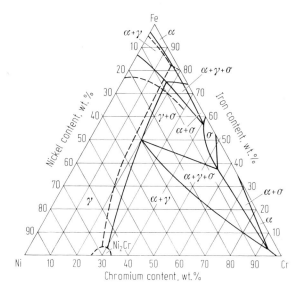

Fig. C 16.1. Phase diagram of the nickel-chromium-iron-alloys at 650 °C (dotted curves at 550 °C). After [3]. (Supplementary information see Fig. C 15.8).

C 16.2.2 Ferritic Heating Conductor Alloys

Ferritic heating conductor alloys are based on the iron-chromium-system whose phase diagram [10] includes the σ-phase, which is difficult to work because of its hard, brittle nature, and presents below 520 °C the possibility of the precipitation of a chromium-rich phase causing the so-called 475°-embrittlement [11, 12]. A silicon content accelerates the formation of the σ-phase and shifts the limit α/(α + σ) towards lower chromium contents [13], but additions of aluminum of more than 1% prevent the development of the σ-phase [14, 15]. It follows that in practice a transformation-free bcc solid solution exists with between 0 and 30% Cr and an aluminum content of more than 1% and less than 10% (Fig. C 16.2).

Heating conductors with particularly high scaling resistance and operating temperatures are found in the field of ferritic chromium-aluminum-steels. In aluminum-free steels a chromium content of more than 15% is required to attain adequate resistance to oxidation (Fig. C 16.3). On the other hand, high chromium

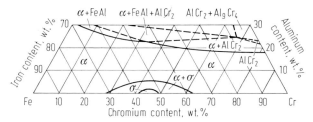

Fig. C 16.2. Phase diagram of iron-chromium-aluminum-alloys at 700 °C. After [16].

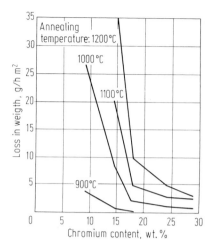

Fig. C 16.3. Scaling of chromium steels with 0.5% C during annealing for 220 h in air. After [17].

contents coupled with additions of aluminum cause certain technological difficulties, such as development of a coarse grain structure and transition to brittle fracture [18, 19], and further additions or impurities may cause a substantial shift in the transition temperature. For the above reasons, in practice, steels containing less than 25% Cr and 6% Al are preferred. In spite of this restriction, the tendency to cold brittleness calls for certain preventive measures in the production of the alloys and heating elements. The materials are easily formable at high temperatures due to their resulting relatively low strength, a fact that must be taken into consideration in later use. By heating to temperatures of more than 1000 °C in air, for longer periods, almost pure alumina (Al_2O_3) is developed on the surface, preventing the metal from further oxidation [20]. To ensure sufficient aluminum supply for the formation of this surface layer a minimum content of 4% Al is required.

C 16.2.3 Influence of Special Alloying Additions

Although the basic composition of the alloys is an important prerequisite for the oxidation behavior, the high scaling resistances now in use and the extremely high permissible application temperatures are made possible only by small additions of other metals. In nickel-chromium-alloys and -steels additions of about 1% Al or Si are effective, but additions of 0.01 to 0.3% Ca or Ce (cer-mixed metal) are far more effective in all heating conductor alloys. The additions considerably improve the adhesive strength of the oxidic surface layer, so that it does not flake off and even with temperature changes it can exercise its protective effect. However, the mechanism behind this improved adhesion has so far not yet been fully understood. As to the question of whether the small additions that improve the life-time actually develop their main effect in a metallic or oxidic form, there are different schools of thought [21–24].

C 16.3 Technically Established Heating Conductor Alloys

The heating conductor alloys are standardized [25]. Table C 16.1 lists several typical materials.

The highest permissible *operating temperatures* given apply to wires with a minimum diameter of 2 mm.

Apart from these highest permissible temperatures there are marked differences in the specific *electrical resistance* and *related temperature dependence* (Fig. C 16.4).

Table C 16.1 Typical heating conductor alloys (according to DIN 17470 [25] and to data of producers)

Material (Designation)	Chemical composition (main constituents)					Usable in air up to °C	Electric resistivity at 20 °C Ohm·mm²/m
	% Al	% Cr	% Ni	% Si	Fe		
			Austenitic materials				
NiCr 30 20	–	19 … 22	30 … 35	0.5 … 2	Remainder	1100	1.04
NiCr 80 20	–	19 … 21	77 … 80	0.5 … 2	≤ 2	1200	1.12
			Ferritic steels				
CrAl 15 5	4 … 5	13 … 16	–	≤ 1	Remainder	1050	1.25
CrAl 20 5	4.5 … 5.5	19 … 21	–	≤ 1	Remainder	1300	1.37
CrAl 25 5	5 … 6	21 … 25	–	≤ 1	Remainder	1350	1.44

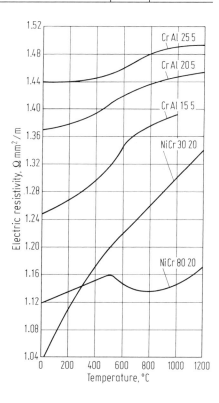

Fig. C 16.4. Dependence of electrical resistivity on temperature of the heating conductor alloys listed in Table C 16.1.

At room temperature the ferritic alloys that are ferromagnetic have an extremely high electric resistivity whose temperature dependence is very small in comparison with many other ferromagnetic materials. The temperature dependence of the resistance of non-ferromagnetic austenitic alloys also exhibits special features. The resistance versus temperature curve of NiCr 80 20 has a pronounced S-shape; with increasing iron content this unusual shape slowly disappears until, at about 45% Fe the curve for NiCr 30 20 is almost free of anomalies [26].

The small initial rise in the curve of the (almost) iron-free material NiCr 80 20 can be even further reduced by an addition of between 2% and 4% Al, so that by adding small amounts of other constituents – silicon, iron, copper – *materials* with a temperature coefficient of resistance in the range of $\pm 10 \cdot 10^{-6}/K$ are obtained, suitable for *measurement- and precision-resistances* [27].

The nickel-chromium-alloys and the low iron content nickel-chromium-iron-alloys just as the aluminum- and chromium-aluminum-steels [28] have the peculiarity that the electric resistivity at room temperature is reduced by several percent as a result of cold deformation after soft annealing and is relatively strongly increased by heat treatment at 300 °C to 500 °C [29]. This behavior must be taken into account when using the material [30].

Table C 16.2 gives the values for *other properties* of the alloys discussed so far. The mechanical properties apply to the state after soft annealing and naturally depend also on the microstructure. It follows that only ranges of values are given. A comparison shows that the austenitic alloys have better ductility (elongation after fracture) at room temperature. At higher temperatures their strength (characterized by the creep limit in a 1000 h test) is about two to four times higher than that of ferritic alloys. Use of the alloys discussed so far *in gases other than* air or in contact with liquid or solid materials may lead to different types of damage forcing the temperature of use to be reduced. For details please refer to the literature [31].

The *austenitic alloys* become increasingly *sensitive to* a pick up of *sulphur* as the nickel content rises resulting in the formation of low melting nickel/nickel-sulphide-eutectic and brittleness. The pick up of carbon lowers the melting point; molten zones in the microstructure then may lead to sudden failure of the heating elements.

The expression *"green rot"* ("Grünfäule") refers to a particular phenomenon that arises under special application conditions when only the chromium oxidizes.

Table C 16.2 Mechanical and physical properties at 20 °C of the two groups of heating conductor alloys listed in Table C 16.1 (according to DIN 17470 [25] and to data of producers)

Microstructure	Density g/cm^3	Tensile strength N/mm^2	Elongation after fracture A_{100mm} %	Specific heat J/g·K	Thermal conductivity W/cm·K	Coefficient of thermal expansion[a] 10^{-6}/K
austenitic	7.9 ... 8.3	600 ... 750	20 ... 35	about 0.46	about 0.14	17 ... 19
ferritic	7.1 ... 7.3	600 ... 800	12 ... 20	about 0.46	about 0.13	15

[a] Between 20 °C and 1000 °C

This condition leads to the formation of a sponge-like microstructure without any resistance to scaling.

If the protective Al_2O_3 surface layer of *ferritic alloys* is damaged, *nitrogen may penetrate* and combine with the aluminum to nitride, causing a drastic reduction of the scaling resistance.

C 17 Steels for Valves in Internal Combustion Engines

By Wilhelm Wessling and Friedrich Ulm

In internal combustion engines, the admission of the combustion mixture, the discharge of the exhaust gases, and the closing of the combustion chamber within the working cycle is made possible by closing and opening the inlet and exhaust valves. A typical valve and a description of its usual characteristics are shown in Fig. C 17.1 [1].

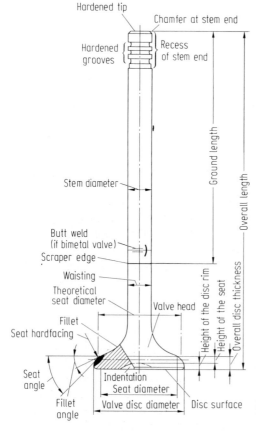

Fig. C 17.1. Designation of the parts of a valve. After [1].

References for C 17 see page 791.

C 17.1 Basic Properties of Valve Materials

C 17.1.1 Requirements Resulting from Service Conditions

The highest *temperatures* occur at the valve head, because this part, which is subject to pressure within the combustion chamber, does not benefit from the usual engine cooling. The head of the exhaust valve becomes particularly hot – in borderline cases, up to 900 °C – in the opening phase of the cycle by the burnt out gases [2]. On inlet valves, however, temperatures up to about 500 °C are reached. For this reason, the materials used for exhaust gas valves have to meet higher requirements than do inlet valve materials.

The level, distribution, and timewise changing of the temperature field in the valve head determine whether the chosen material will stand to the mechanical and corrosive stresses. The surface temperatures, which are decisive in valve head corrosion, were found, through metal vapor-deposition experiments, to be at least 950 °C in diesel engines [3].

Requirements for *fatigue strength* under pulsating stresses [4] arise from the changes in pressure during the working cycle and from the closing force of the valve spring. In addition, *alternating thermal stresses* occur at the rim of the exhaust valve, which is heated by the outflowing exhaust gases and cooled during closing against the valve seat [5]. Irregular operating conditions are caused by dynamic, impactlike phenomena such as excessive valve play, overspeed, and off-center guiding of the valve in the valve seat [5, 6].

In addition to nitrogen and water vapor, the exhaust gases contain varying quantities of carbon dioxide, oxygen, hydrogen, and hydrocarbons [5]. The strong *corrosive attack* is caused by additives to the gasoline and to the oil and by impurities in the diesel fuel. Gasoline contains an organic lead compound (antiknock) agent that forms lead dioxide on combustion. To remove the lead oxide rapidly, organic halogen compounds are sometimes added to the gasoline. These compounds produce lead chloride and lead bromide, which, when mixed with lead oxide, have low melting temperatures of around 500 °C [5, 7]. The main impurities in diesel fuel are sulphur and vanadium compounds that are converted into sulphur dioxide and vanadium pentoxide on combustion. Alkaline and alkaline-earth metals also appear in the combustion residue as sulphates or vanadate [8]. Lubricating oils, too, contain metallic additives such as calcium, whose products are corrosive [3, 9].

Scale-resisting steels (see B 3 and C 15) resist progressive oxidation by forming dense protective layers in gases containing oxygen; these layers are otherwise largely neutral. Combustion residues, which are solid or liquid at high temperatures, attack or even destroy these protective layers. A frequent cause of valve damage is "burn-through," which is produced at the seat area of the valve head by local overheating, leading to increased hot corrosion (Fig. C 17.2). In addition, the combustion of diesel fuel may lead to the bonding of chromium as carbide or sulphide and thus further restrict the stability of the valve material [4, 8].

The seating area of the valve head is exposed to *wear* by the solid combustion residues. The reduction of lead additives in gasoline has led to increased wear of the

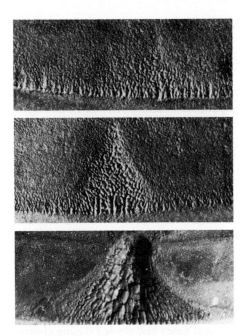

Fig. C 17.2. Development of a burn-through of an exhaust valve. After [19a].

sealing surfaces due to the reduced lubricating effect of the lead compounds [2, 10]. Wear also occurs at the stem end through the transmission of the opening forces and at the guiding part of the stem [2, 5].

The *physical properties*, such as thermal conductivity and thermal expansion and their effect on thermal stress, must be considered when selecting suitable valve materials. It is not possible, however, to derive these requirements from the steels themselves. Rather, the physical properties (see B 2) are taken into account in the design of the valves, by considering such properties as good high-temperature strength and good corrosion behavior.

C 17.1.2 Requirements Concerning the Processing Properties

B 6 and B 9 present the basic preconditions required for hot-formability of the valve heads, for hot extrusion of the valve stems, and for machinability. However, for machining, the high strength of the heat-treatable steels and the tendency to work hardening of the austenitic materials much be taken into account.

Steel bars used to produce valves are always supplied in the ground condition, sometimes with restricted dimensional tolerances.

C 17.2 Characterization of the Requirements by Test Values

Whether a valve material is acceptable for practical use can be qualified only by an extensive engine test. Because of the high cost of such a test, efforts have been made to establish requirements by using suitable experiments, to derive material characteristics from these experiments, and to compare different valve steels or their microstructures.

It is difficult to identify the stress exerted on the valve by internal and external forces and to derive a test value for the *mechanical properties*. It is easy, however, to determine the high-temperature strength by tensile and creep tests. During the last 15 years, fatigue tests at high temperatures have been recognized as being decisive [4, 5, 11, 12]. Previously, a high value for notch-impact energy [5] was considered important, but it was found that valve steels proving themselves reliable for years often had a low toughness [12].

It was also difficult to discover the mechanism of *hot corrosion*. Theories were formed about the complicated processes involved, and the results of individual test methods were rightly mistrusted.

One of the first tests developed was the *lead oxide test* [5, 13], after lead compounds were found in the residues [7]. This test determines the loss of mass after annealing for half an hour in lead oxide at 915 °C. This method is still used today to estimate the behavior of different steels in spark-ignition (Otto-cycle) engines [4].

In other procedures, exhaust gases and residues were also included in the experiments [5]. The results showed the strongly negative effect of sulphur dioxide on nickel alloys and the importance of chromium (Fig. C 17.3). To follow the aggressiveness of the corrosion on valve seats, valves with through-holes through the valve disc were fitted in experimental engines [7], thus imitating a burnthrough. The results are valuable, but the experiment is expensive. An insight into the course of corrosion in diesel engines was obtained with tests using microprobes [8]. By using similar work on the corrosion of turbine blades, it was found that the covering scale layer is interspersed with sulphur, and chromium sulphide is formed. The basic material is depleted of chromium, thus reducing the corrosion resistance. The process is maintained by oxidation of the chromium sulphide itself. After the chromium is largely combined, nickel sulphide is formed, which causes a considerable acceleration of the corrosion in nickel alloys. Sulphates and chlorides also accelerate the process. In another experiment [14], samples of valve materials are exposed to the hot exhaust gases of a diesel engine while also being stressed mechanically.

Of great importance in the corrosion behavior are the temperature and the air factor. In comparison, the influences of the diesel fuel's vanadium content and of the mechanical stress is small. In recent years, the discovery of "paving stone structures" and the "formation of smelt ball" (of burnt-out particles) has produced new indications that the progress of hot corrosion has not yet been explained sufficiently [14, 15].

The realistic *testing of wear resistance* (see B 10) is possible only within narrow limits because of the complicated interrelationships among the various wear-resistance parameters. By choosing the strength in the quenched and tempered condition and by producing wear-resistant layers by surface hardening, deposit welding, or hard chromium deposition, it is possible to produce practical wear resistance.

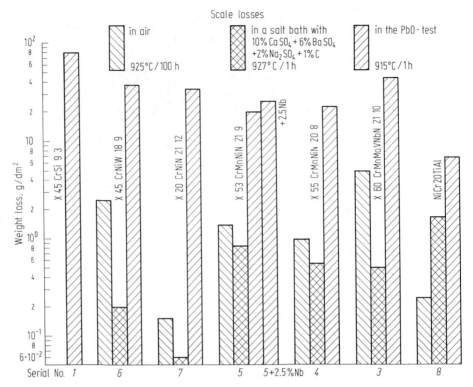

Fig. C 17.3. Resistance to oxidation, sulphidation, and lead oxide corrosion of some valve materials according to Table C 17.1. The serial numbers refer to that table. After [4].

C 17.3 Measures for Satisfying the Requirements

C 17.3.1 Measures of Physical Metallurgy

The most important service requirements for valve steels are good high-temperature strength and matching hot corrosion resistance. B 1 gives the basis for obtaining high-temperature strength, and B 3 for obtaining corrosion resistance.

With respect to *high-temperature strength* and operating temperatures up to 550 °C, microstructures of the quenched and tempered condition with a sufficiently high content of special carbides must be produced. Concerning corrosion resistance, heat-treatable steels with a higher chromium content must be considered. Above 550 °C, microstructures of the quenched and tempered condition are no longer useful, because recovery processes occur increasingly in the bcc lattice. It is then necessary to employ materials with a basic austenitic structure (fcc lattice) having also a considerably improved *corrosion resistance*.

Valve steels have the required *weldability*, compared with the usual high-temperature steels (see C 9), even with a higher carbon content. The hardenability of these steels is improved, and the proportion of the carbides in the microstructure, which affects the heat resistance, can be increased considerably. This results in better *wear resistance*. These correlations are similar in materials with a basic austenitic microstructure.

There are *three groups* of materials whose microstructures and chemical compositions fulfill the requirements for the properties:
1. *Heat-treatable steels* (steels for quenching and tempering) that, with special reference to corrosion resistance, are specially alloyed with chromium and silicon. For increased high-temperature strength and retention of hardness, these steels contain carbide-forming elements such as molybdenum, vanadium, and tungsten. With increasing carbon and chromium contents, the proportion of primary carbides in the microstructure increases.
2. *Austenitic* chromium-nickel and chromium-manganese (nickel) *steels* with high-temperature strength based on higher carbon and nitrogen contents the hot hardness of which is increased by carbide- and nitride-forming elements such as tungsten, molybdenum, vanadium, and niobium.
3. *Nickel alloys* with a basic austenitic microstructure which are precipitation hardened, for example, by the precipitation of titanium and aluminum as γ'-phase.

C 17.3.2 Measures for Shaping Valve Parts and Protecting Their Surfaces

It is not possible to influence the physical properties of valve materials by physical metallurgy per se. If the thermal conductivity given by the basic properties is insufficient, attention must be paid to the design. For example, it is possible to improve the heat flow toward the stem by converting the stem into a sodium-cooled hollow valve where the reduced temperature also means a better utilization of the high-temperature strength. For heat flow toward the cylinder wall, important factors are the size of the contact area, the temperature difference between valve seat and valve ring, the duration of contact, and the specific thermal conductivity of the hard-faced seat and the valve seat ring, as well as the extent of the oxide layers or other heat-insulating deposits on the seat surface. The fit of the valve seat and the guiding play of the stem also must be dimensioned carefully. For inlet valves of low-alloy heat-treatable steel, which in any case have a better thermal conductivity, the heat abstraction (leading off) can be improved by dip-aluminizing the surface. This makes it possible for the temperature to be reduced by about 100 °C [16].

The resistance to hot corrosion and wear of the valve materials is limited by the fundamentals of physical metallurgy. When these limits must be exceeded, special measures must be taken. To improve the hot corrosion and the wear resistance, hard facing is often applied to the valve seat and the valve disc. Solid and hollow valves are also produced as bimetals by joining the valve head made of an austenitic material to the stem made of a heat-treatable steel by friction welding, for example. In this way, it is possible to select a good high-temperature material for the part that is subjected to high temperatures and to impart the necessary sliding characteristics to the stem by surface hardening. At the same time, a good heat abstraction (leading off) through the stem is provided. The stem of austenitic valves is chromium plated to improve the gliding properties. The valve tip is hard-faced to resist wear by the valve tappet. However, if the stem is made of hardened and tempered steel, wear resistance is produced by surface hardening.

After friction welding, the bimetal valves much be temper-annealed. It is also necessary to carefully stress-relieve the weld-deposited hard facing on chromium heat-treatable steels.

C 17.4 Presentation of Standard Valve Materials

C 17.4.1 Heat-Treatable Steels (Steels for Quenching and Tempering)

For the less temperature-stressed *inlet valves*, low-alloy or even plain heat-treatable steels are selected, such as Ck 45 and 41 Cr 4 in accordance with DIN 17 200 [16a], heat treated to a tensile strength of 800 N/mm² to 1000 N/mm².

For *exhaust valves*, on the other hand, only highly alloyed steels and nickel alloys can be considered. Table C 17.1 gives a survey of the established materials from DIN 17 480 [17].

The *heat treatment* of the steels for quenching and tempering can be derived from the time-temperature-transformation diagrams for continuous cooling (see A 9 and B 4). For example, for a chromium heat-treatable steel, Fig. C 17.4a shows the transformation behavior of the steel X 45 CrSi 9 3. The chromium and silicon determine the position of the Ac_3 point. If the content of these two elements is high, there is a tendency toward a pre-eutectoidal ferrite precipitation, which can be suppressed by raising the austenitizing temperature to 1050 °C to 1100 °C. The final

Table C 17.1 Chemical composition of valve materials (values for wt.% given as typical only)

Serial No.	Material Designation	% C	% Si	% Mn	% Cr	% Mo	% N	% Ni	% Others
		\multicolumn{8}{l}{Heat-treatable steels (steels for quenching and tempering)}							
1	X 45 CrSi 9 3	0.45	3.0	0.40	9.0				
2	X 85 CrMoV 18 2	0.85	0.5	0.75	17.5	2.25			0.45 V
					Austenitic steels				
3	X 60 CrMnMoVNbN 21 10	0.60	0.15	10.5	21.0	1.0	0.45	(1.0)	0.85 V
4	X 55 CrMnNiN 20 8	0.55	0.5	8.5	20.5		0.3	2.3	
5	X 53 CrMnNiN 21 9	0.53	0.15	9.0	21.0		0.42	3.8	0.15 V
6	X 45 CrNiW 18 9	0.45	2.5	1.2	18.0			9.0	1.0 W
7	X 20 CrNiN 21 12	0.20	1.0	1.25	21.0		0.22	11.5	
8	X 12 CrCoNi 21 20	0.12	0.5	1.0	21.0	3.0	0.15	20.0	20.0 Co, 2.5 W
					Nickel alloy				
9	NiCr 20 TiAl	0.08	0.5	0.5	19.5			70.0	1.4 Al, 0.003 B, 2.2 Ti

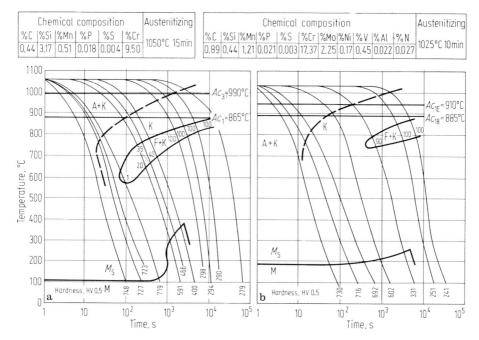

Fig. C 17.4. Time-temperature-transformation diagrams for continuous cooling of the heat-treatable valve steels, **a** X 45 CrSi 9 3; **b** X 85 CrMoV 18 2 (steels 1 and 2 according to Table C 17.1).

forming temperature must be in the same range, since the critical deformation of a undercooled austenitic microstructure or of part of the microstructure already transformed at about 750 °C will lead to considerable grain growth when reheating above that temperature. To refine the grain, a hardening temperature above the Ac_3 point is required. A hardening temperature that is too low increases the formation of coarse grains after critical deformation [18]. The pronounced pearlite field of the X 45 CrSi 9 3 steel presupposes a low nickel content. For a nickel content up to 0.6%, the pearlitic transformation is reached only after slow cooling. These melts exhibit a low Ac_{1B} point of about 800 °C, which must be taken into account when soft-annealing and tempering.

The chromium-molybdenum-vanadium steel X 85 CrMoV 18 2 (Fig. C 17.4b) also shows different transformation points that similarly depend on the chemical composition, especially on the chromium and carbon contents.

The chromium heat-treatable steels are hardened in oil between 1020 °C and 1050 °C and then tempered to a tensile strength of 900 N/mm² to 1100 N/mm² (Fig. C 17.5). With this oil hardening, the steels reach a hardness of 56 HRC to 60 HRC and, even with air cooling, the hardness obtained is still considerable. This causes a certain susceptibility to heat-treatment cracking, which makes careful handling of hot forming and heat treatment necessary.

Since the steel X 45 CrSi 9 3 has a noticeable temper brittleness with silicon contents above 3%, quenching in water is recommended after tempering or

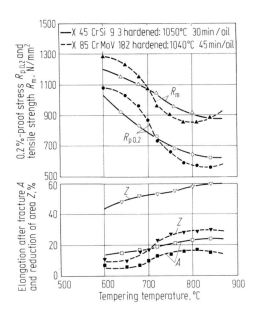

Fig. C 17.5. Tempering behavior of valve steels X 45 CrSi 9 3 and X 85 CrMoV 18 2 (steels 1 and 2 according to Table C 17.1). Tempering time 2 hours in each case.

soft-annealing. The temper brittleness can be avoided by the addition of molybdenum. For this reason, a steel containing 2.5% Si, 10.5% Cr, and 1.0% Mo (X 40 CrSiMo 10 2) is used occasionally if larger cross sections are to be quenched and tempered.

The high-temperature properties of the heat-treatable steels are compared with those of the materials of the other two groups (see C 17.3.1) in the final section, C 17.4.4.

C 17.4.2 Austenitic Steels

The austenitic steels (see Table C 17.1) are solution-annealed at 1000 °C to 1200 °C followed by quenching. In this condition, the tensile strength is about 800 N/mm^2 to 1000 N/mm^2. With a higher nitrogen content, the solution heat-treating temperature is raised (Fig. C 17.6). The increasing dissolution of precipitated phases becomes recognizable by an increase in elongation after fracture, reduction of area, and notch-impact energy. The tensile strength and the 0.2%-proof stress are only slightly reduced with an increasing solution annealing temperature.

The mechanical properties of the steels X 45 CrNiW 18 9 and X 20 CrNiN 21 12 are only moderately improved by subsequent artificial aging (at elevated temperatures) and, even for the X 53 CrMnNiN 21 9 steel, an increase in strength by precipitation is limited (Fig. C 17.7). However, the toughness characteristics of the steel X·53 CrMnNiN 21 9, with the higher nitrogen content of 0.4%, is affected markedly; after 5 minutes of aging, the toughness of the solution-annealed condition is reduced by more than 50%. In the temperature range around 760 °C, there is a general precipitation in the grains and at the grain boundaries (Fig. C 17.8a).

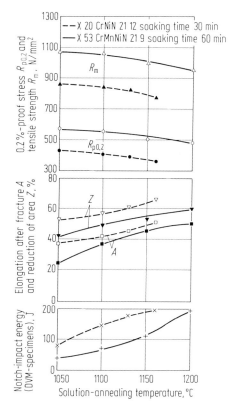

Fig. C 17.6. Effect of solution-annealing temperature on the mechanical properties after quenching in water of the steels X 20CrNiN 21 12 (0.22% C, 0.89% Si, 1.16% Mn, 20.8% Cr, 13.0% Ni, and 0.21% N) and X 53 CrMnNiN 21 9 (0.49% C, 0.14% Si, 8.58% Mn, 21.4% Cr, 3.9% Ni, and 0.39% N) (steels 7 and 5 according to Table C 17.1).

Above 900 °C, a laminar phase is found that is formed discontinuously, starting from the grain boundaries. After a longer holding time, spheroidization takes place and the microstructure becomes a very fine recrystallized austenite with interstitial granular pearlite (Figs. C 17.8b and 8c). The laminar phase, found also in steel castings of a similar composition, consists of alternately precipitated chromium carbide $Cr_{23}C_6$ and chromium nitride Cr_2N [19, 20]. X 45 CrNiW 18 9 steel exhibits only carbide precipitations after artificial aging at 800 °C (Fig. C 17.8d). The precipitations become very noticeable in the tensile test by affecting the elongation after fracture and the reduction of area or in the notch-impact test.

C 17.4.3 Nickel Alloys

As an established nickel alloy NiCr 20 TiAl is included in Table C 17.1. It is solution annealed at a temperature of 1050 °C (similar to the austenitic steel X 45 CrNiW 18 9), has a fine-grain recrystallization, and is precipitation hardened at 700 °C to a tensile strength of at least 1000 N/mm².

The material X 12 CrCoNi 21 20 can be regarded as a transition between the austenitic steels and the nickel alloys. Because of the addition of nitrogen and

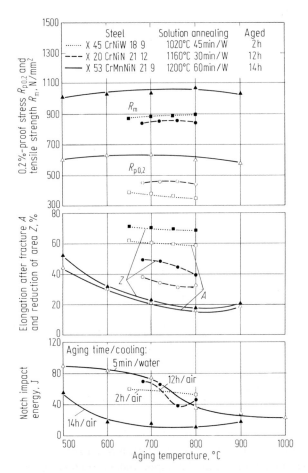

Fig. C 17.7. Effect of temperature and duration of precipitation hardening on the mechanical properties at room temperature of the steels X 53 CrMnNiN 21 9, X 45 CrNiW 18 9, and X 20 CrNiN 21 12 (steels 5, 6, and 7 according to Table C 17.1).

carbide-forming elements, it must be solution-annealed above 1120 °C; for aging, similar temperatures and times apply as for NiCr 20 TiAl.

C 17.4.4 High-Temperature Behavior of the Three Material Groups in Comparison

The high-temperature strength diagrams of the chromium steels for quenching and tempering (Fig. C 17.9) show a noticeable strength reduction above 500 °C; thus, the valve steels for quenching and tempering can hardly be used above 600 °C. In the as-hardened condition, the change in the high-temperature strength is still more sensitive; in this case, only working temperatures up to 400 °C appear possible (Fig. C 17.10).

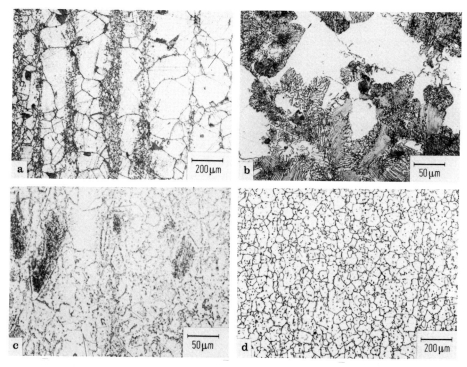

Fig. C 17.8. Microstructure of nitrogen-alloyed steel X 53 CrMnNiN 21 9 (steel 5 according to Table C 17.1) in different heat-treatment conditions compared with the austenitic-carbidic steel X 45 CrNiW 18 9 (steel 6 according to Table C 17.1). **a** X 53 CrMnNiN 21 9: Precipitation-hardened at 760 °C; **b** X 53 CrMnNiN 21 9: Nitrogen pearlite after a brief halt at 1050 °C; **c** X 53 CrMnNiN 21 9: Spheroidized nitrogen pearlite in an austenitic microstructure after 2 hr at 1050 °C; **d** X 45 CrNiW 18 9: Carbidic austenite microstructure after precipitation-hardening at 800 °C.

Above 500 °C to 600 °C, the austenitic valve steels can withstand a higher stress (Fig. C 17.11). By comparing the curves for the steels X 45 CrNiW 18 9, X 20 CrNiN 21 12, and X 53 CrMnNiN 21 9, the effect of the chemical composition on the high-temperature strength becomes apparent; carbon and nitrogen have a particularly favorable influence. Precipitation hardening has less effect than does the chemical composition, however. Solution-annealed valves are therefore aged only briefly and left to precipitation-harden fully in the engine.

Figure C 17.11 also shows, for the chromium-manganese-nickel steel X 53 CrMnNiN 21 9, the change of notched tensile strength in the temperature range from 500 °C to 900 °C. The ratio of the latter to the tensile strength of a smooth test specimen is clearly more than 1, which indicates that even after a longer operating time, no hot embrittlement is to be expected.

Figure C 17.12 gives long-time values for 1000 hours under steady (constant) and pulsating tensile loads for some valve materials. The pulsating tensile load was applied to notched specimens that were preloaded with a basic load of about 20 N/mm^2, and the creep strength was measured on plain (smooth) test pieces [12].

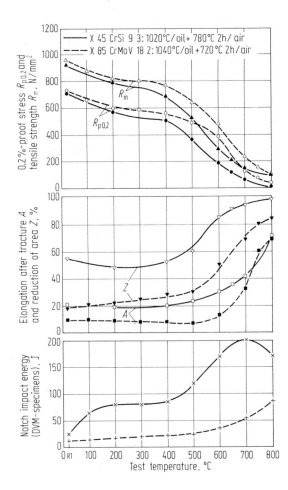

Fig. C 17.9. Mechanical properties at higher temperatures of the heat-treatable (quenched and tempered) valve steels X 45 CrSi 9 3 and X 85 CrMoV (steels 1 and 2 according to Table C 17.1).

The fatigue strength values throughout are above those for creep-stress rupture strength, so long-time values for steady tensile loads can serve as a guideline for the choice of a valve steel. Additional high-temperature strength values can be seen in Table C 17.2.

Since thermal stresses are decisive in the life expectancy of a valve, the physical properties of the selected materials (Table C 17.3) must be supervised to avoid harmful interrelations. Thus, when changing from a chromium heat-treatable steel to an austenitic material, the austenitic material's lower thermal conductivity and considerably higher expansion coefficient must be considered. These parameters cause higher demands to be made on the high-temperature properties. The difference in the thermal expansion of an austenitic steel compared with that of a cemented carbide (hard metal alloy, cermet) produce internal tensile stresses in the hard facing of the seat.

C 17.4 Presentation of Standard Valve Materials   465

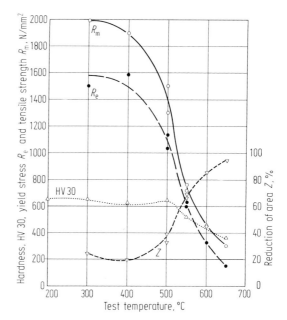

Fig. C 17.10. Mechanical properties of the steel X 45 CrSi 9 3 containing 0.43% C, 2.85% Si, 0.50% Mn, 9.5% Cr, and 0.24% Ni (steel 1 according to Table C 17.1) in the hardened condition at temperatures between 300 °C and 600 °C. The test pieces were hardened to 57–58 HRC (1000 °C 30 min/oil) and tested after soaking at test temperature for 1 h. The hardness was then measured at room temperature.

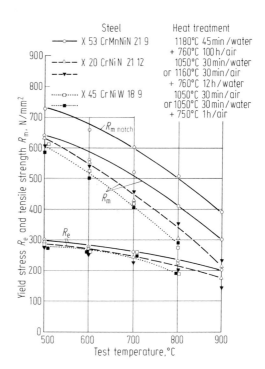

Fig. C 17.11. Yield stress and tensile strength of the austenitic steels 4, 5, and 6 according to Table C 17.1 at 500 °C and 900 °C after different heat treatments (see also Fig. C 17.7). X 53 CrMnNiN 21 9 containing 0.51% C, 0.12% Si, 8.8% Mn, 21.9% Cr, 4% Ni, and 0.39% N; X 45 CrNiW 18 9 containing 0.44% C, 2.4% Si, 1.3% Mn, 17.1% Cr, 8.8% Ni and 0.81% W; X 20 CrNiN 21 12 containing 0.22% C, 0.89% Si, 1.2% Mn, 20.8% Cr, 0.20% Mo, 13.0% Ni, and 0.21% N.

C 17 Steels for Valves in Internal Combustion Engines

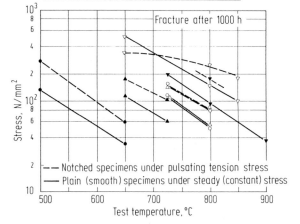

Fig. C 17.12. Stresses leading to fracture after 1000 h at 500 °C for six valve materials according to Table C 17.1. Both plain (smooth) test pieces under a steady (constant) stress and notched test pieces subjected to pulsating tension stress were tested.

Table C 17.2 Values for 0.2%-proof stress and tensile strength at elevated temperatures for some materials from Table C 17.1 in the precipitation-hardened condition (values given as typical only)

Material Designation	0.2%-proof stress at				Tensile strength at			
	500 °C	600 °C	700 °C	800 °C	500 °C	600 °C	700 °C	800 °C
	N/mm²				N/mm²			
X 60 CrMnMoVNbN 21 10	500	450	400	350	850	750	600	400
X 53 CrMnNiN 21 9	400	350	300	250	720	630	500	350
X 12 CrCoNi 21 20	260	240	210	150	550	520	450	350
NiCr 20 TiAl	550	550	540	350	900	850	680	400

Table C 17.3 Physical properties of the materials from Table C 17.1

Material Designation	Density at 20°C kg/dm³	Young's modulus at 20°C 10³ N/mm²	Linear expansion coefficient between 20°C and $10^{-6} K^{-1}$				Thermal conductivity at 20°C W/K·m	Heat capacity at 20°C J/K kg
			100°C	300°C	500°C	700°C		
			Heat-treatable steels (steels for quenching and tempering)					
X 45 CrSi 9 3	7.7	221	12.9	13.2	13.6	14.0	21	500
X 85 CrMoV 18 2	7.7	226	10.9	11.5	11.7	11.9	21	500
			Austenitic steels					
X 60 CrMnMoVNbN 21 10	7.8	215	16.1	17.2	18.0	19.0	14.5	500
X 55 CrMnNiN 20 8	7.7	215	14.5	16.7	17.8	18.4	14.5	500
X 53 CrMnNiN 21 9	7.7	215	14.5	16.7	17.8	18.4	14.5	500
X 45 CrNiW 18 9	7.9	205	15.5	17.5	18.2	18.6	14.5	500
X 20 CrNiN 21 12	7.8	210	15.3	16.2	16.9	17.7	14.5	500
X 12 CrCoNi 21 20	8.2	205	14.2	15.5	16.5	17.6	12.5	460
			Nickel alloy					
NiCr 20 TiAl	8.2	216	11.9	13.1	13.7	14.5	13	460

C 18 Spring Steels

By Dietrich Schreiber and Ingomar Wiesenecker-Krieg

C 18.1 Required Properties

The *function of springs* is to take up intermittent or pulsating loads (e.g. vehicle springs) or to accumulate energy under stationary loads (e.g. brake cylinders). Springs are also used to measure forces. In comparison to other materials steel possesses not only a high tensile strength, but also a high modulus of elasticity, especially after strain hardening and/or heat treatment. Therefore springs are preferably made of ferrous alloys. The *required properties* such as strength at elevated temperatures (valve springs), resistance to corrosion by aggressive media or non-magnetizability for applications to physical instruments, determine the kind and contents of alloying elements suitable for a given application. A number of influencing factors of general importance, independent of these aforementioned factors, are described extensively in the literature [1–3]. The most important are:
- a high limit of elasticity and high 0.2%-proof stress, required to provide the maximum possible elastic deformability, and keep the permanent setting of the spring within acceptable limits;
- good ductility (toughness), so that a sufficient degree of plastic deformability is maintained as a safeguard against fracture by overloading of the spring and in the forming process; and
- high fatigue strength for finite or indefinite life.

Steels for the industrial production of springs have *additional requirements* including adequate flexibility for forming operations, so that takes account of the design requirements. Cold-formed springs also need good formability for the preceding fabrication into wire or strip as well as for the final shaping of the spring. Ability to permit cutting at room temperature promotes cost-saving fabrication. In addition, the possibility for a thorough heat treatment – complete through-hardening to be more specific – is usually a prerequisite for achieving the calculated, required strength values.

References for C 18 see page 792.

C 18.2 Measures of Physical Metallurgy to Attain Required Properties

Selection of the chemical composition for the manufacture of springs requires the same considerations as for heat-treatable steels should the manufacturing process require quench hardening [4, 5] (see C 5). The carbon content predetermines the degree of hardenability as well as the elastic limit and the yield stress essential for springs in the cold-drawn and quenched and tempered condition. Alloying elements permit the required through-hardening to be achieved over the entire cross-section. They also ensure strength at elevated temperatures, toughness at low temperatures, and corrosion resistance or nonmagnetizability as required. In former years silicon was regarded as a special influencing factor in effecting an increase in the elastic limit and the yield stress. This view is no longer held, however, in view of the negative influence on decarburization.

When a martensitic *microstructure* is developed over the cross-section by hardening, uniform properties with favorable strength values under static and dynamic loads are produced after tempering. These favorable values are needed to avoid permanent set and to secure a long life. These conditions are valid in a similar way when homogeneous bainite is obtained in the lower transformation temperature range. The limiting dimensions for achieving through-hardening must therefore be given consideration when selecting particular steel grades for heat treated springs (see e.g. DIN 17 221). An increase in hardenability by means of alloying elements should be used only to the extent necessary, because excessively alloyed steels tend to intensified stress cracking.

Measurements of the *hardenability* may be made by determining the end quench hardening curve and the corresponding transfer to equivalent cross-sections [6]. Due consideration must be given to the hardening intensity or measuring the hardness in the core of all-round (all over) quenched samples, which allows conclusions to be reached on the martensite content in the core. In contrast to quenched and tempered constructional steels, for which normally a martensite content of 50% is regarded as a sufficient criterion for through-hardening, spring steels must have a content of at least 80% and even 100% where there are especially high stresses [7, 7a].

Spring steels are advantageously produced as fine-grained steels because of their good toughness properties. They offer a lower overheating sensitivity, thereby facilitating hot working and heat treatment, and they show a slighter tendency to distortion during hardening. The austenitic grain size should be determined after quenching.

Both *the mechanical properties* and the microstructure are criteria for the selection of the material [8]. A wide field opens up for targeted processes by a suitable heat treatment (quenching and tempering or patenting), by cold forming (drawing, rolling) and by a combination of both possibilities, e.g. for drawn spring wires [9] and spring steel strip.

The *tensile strength* of quenched and tempered springs lies normally in the range of 1000 N/mm^2 to 1900 N/mm^2 (Fig. C 18.1), and that of subsequently cold

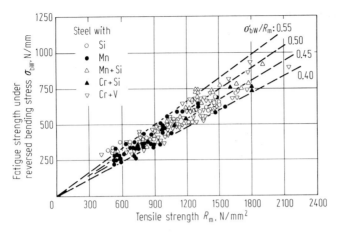

Fig. C 18.1. Fatigue strength under reversed bending stress of hardened and tempered spring steels in dependence on their tensile strength (compiled on the basis of publications in [5]).

formed springs as well as that of springs made of precipitation hardenable steels may rise to 3500 N/mm² in special circumstances. Raising the tensile strength reduces the *toughness* of the steel, however, whereas notch sensitivity [10] and the velocity of crack propagation both increase. In this connection, all forms of damage to the surface and to some extent nonmetallic inclusions within the surface zone must be regarded as notches. Therefore, for the highest tensile strength with a low toughness, and because of the required cleanliness and absence of segregations, spring steels should be melted by special processes, e.g. the vacuum or the electroslag remelting process. For the same reason, spring steels should be so alloyed as to have a sufficient plastic deformability to safeguard them against fracture when overstressed. The normal quenched and tempered spring steels are expected to have an elongation after fracture of at least 5% and a reduction of area of at least 25% in their final condition before use.

The most prominent characteristic of a resilient structural part is its *fatigue strength*. Extensive statistical investigations on structural steels with tensile strengths up to 1000 N/mm² have demonstrated a definite link between tensile strength, reduction of area and fatigue strength [5, 8].

The *surface quality* of a spring gains special importance among the above-listed prerequisites [11]. Steel for springs is hot rolled in the form of bars, wire rod and strip. During hot rolling an impairment of the surface is possible due to scratches, folds and cracks, by scale formation and rolling-in of scale and scars and by decarburization. A rolled material completely free of decarburization, cracks and scratches at this state of the art is impossible to produce. Nor can surface damage of a mechanical kind be excluded in spring manufacture, assembly and operation.

Chemical changes in the surface zone constitute a factor that seriously influences fatigue strength. A totally decarburized surface is especially detrimental. because quenching and tempering provides virtually no increase in strength. The loss of fatigue strength may be up to 40%, independent of the thickness of the ferritic zone

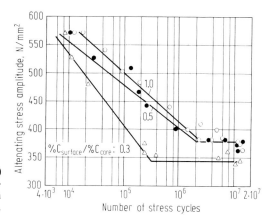

Fig. C 18.2. S/N curves (Wöhler-curves) for bars of steel 50 CrV 4 with varying decarburization. After [13]. Samples 5.8 mm dia. with a tensile strength of 1450 N/mm².

[12]. A partially decarburized surface may have a similar effect but to a substantially lesser extent (Fig. C 18.2) [13]. Silicon steels have a distinct tendency to partial and total decarburization. This problem must be watched closely when they are used e.g. for hot-rolled laminated springs without machining of the surface.

Diffusion of oxygen and internal oxidation along the grain boundaries underneath the surface may happen concomitantly with scale formation and decarburization and may result in a scabby and seamy surface [3]. Consequently, even a relatively minor decarburization may be accompanied by considerable deterioration [14]. Any improvement of the surface condition therefore means increased fatigue life [15], so highly-stressed helical springs are often submitted to surface machining at some stage before fabrication.

The surface factor m, shown in Fig. C 18.3 stands for the relation between the fatigue strength of samples with a predetermined roughness R_t and that of polished samples with a roughness $R_t \approx 1$ μm, depending on the value R_t [16]. The same close relationship is also found for a fairly large number of steel grades with different strength values for reversed tension-compression stresses as well as for rotating bending stresses. This negative surface influencing factor is countered

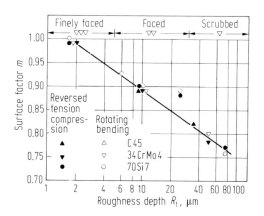

Fig. C 18.3. Surface factor m in fatigue tests under tension-compression (reversed) stresses and under rotating bending stresses for heat treated spring steels after various kinds of surface treatment. After [16]. m = Relation of fatigue strength of samples with predetermined roughness to fatigue strength of polished samples ($R_t \approx 1$ μm).

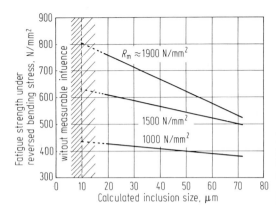

Fig. C 18.4. Summarized diagram of the size effect of nonmetallic inclusions on fatigue strength (rotating beam test) in steels with a tensile strength R_m between 1000 N/mm² and 1900 N/mm². After [26–28].

by shot peening, which causes residual compression stresses in the surface zones [14, 17–21].

The second important factor that influences the fatigue life of quenched and tempered springs of more than 1200 N/mm² tensile strength is to be seen in the content of nonmetallic inclusions [22–31]. Generally speaking, the oxide and silicate inclusions, especially those with sharp edges, on account of their different thermal expansion, are more detrimental than the elongated (stretched) thin manganese sulphides [32, 33]. The danger of initiating a fatigue crack increases, the smaller the distance to the surface becomes, assuming equal sizes of nonmetallic constituents. Cracks develop, for instance at a distance of only 0.4 mm below the surface with inclusions of about 25 µm dia. [34]; nondeformable globular inclusions are a detrimental factor when they reach diameters of 30 to 100 µm, depending on the distance from the surface [29–31].

The influence of the size of nonmetallic inclusions on the fatigue strength is shown in Fig. C 18.4 for different strength values. By inclusion-size in this connection [26, 27] is meant the effective cross-section of the inclusion measured at the site of crack initiation (length multiplied by width in the fracture plane). A comparison of the relationships shown in Fig. C 18.4 with those in Fig. C 18.2 and Fig. C 18.3 indicates that the life time of an oscillating spring is influenced to a much greater degree by the surface conditions (grooves, scratches, pores etc.) than by the size of nonmetallic inclusions present in spring steels melted by modern processes [30].

C 18.3 Spring Steel Grades Proven in Service

Fields of application, shapes and steel grades for springs are extremely numerous [35]. Details of standardized steels and their properties are contained e.g. in DIN 17221, 17222, 17223, 17224 and 17225 [36]. A summary giving chemical composition, tensile strength and applications of characteristic grades of each class of alloyed steels for springs can be found in Table C 18.1 [37]. For the selection of the spring steel best suited to the particular work pieces and applications, refer to [37a].

C 18.3.1 Steels for Cold Formed Springs

Unalloyed steels are employed largely for cold formed springs.

Springs for lower stresses or unspecified qualifications are made of wires drawn from wire rod e.g. according to DIN 17 140 [37b], or of cold rolled strip e.g. according to DIN 1624 [37c], with carbon contents from 0.05% to 0.50%. In selecting the steels it must be considered that the hardness of the hot rolled products influences greatly the processing by drawing and cold rolling as well as by cold coiling of the springs. A sufficient degree of deformability after drawing or cold rolling must be maintained for the subsequent shaping of the spring (see also C 4 and C 7). To increase the limit of elasticity, a short-term tempering of the springs at 200 °C to 300 °C is useful.

Unalloyed steels containing 0.40% to 1% C, mainly as patented and drawn or drawn and quenched, and tempered wires, are used for cold formed springs e.g. according to DIN 17 223. Thus steels with 0.60% to 0.70% C are used in the quenched and tempered condition for cold formed valve springs. Also roll springs and coiled springs as well as highly stressed engine springs are made from cold rolled strip or wire with 0.70% to 1.20% C in the quenched and tempered condition.

C 18.3.2 Steels for Quenched and Tempered Springs

Hot-formed quenched and tempered vehicle springs for low stresses are made from *unalloyed* steels containing 0.40% to 0.70% C.

Silicon alloyed steels with about 1% to 2% Si (Group A in Table C 18.1) are employed for the production of quenched and tempered springs for bending stresses, especially for vehicle springs in railroad and automobile construction. Frequently a small chromium addition of 0.20% to 0.40% is used for increasing the hardenability. Silicon raises the ratio of yield stress to tensile strength in the quenched and tempered condition as well as the retention of hardness in tempering, but it also greatly enhances the tendency towards decarburization. Therefore the use of these steels (examples are 55 Si 7 and 65 Si 7) is not appropriate for highly stressed springs. In helical and torsion bar springs the decarburized layer may be largely removed by machining. The steel grade 71 Si 7 is used as cold rolled strip for highly stressed watch springs and other horlogical (e.g. driving) springs.

Manganese alloyed spring steels, e.g. 50 Mn 7, 51 MnV 7 (Group B in Table C 18.1) and 60 MnSi 5, are still employed to a small extent for vehicle springs (mainly for leaf, cup or annular buffer springs). Whereas manganese certainly increases hardenability and through hardening even in bigger cross-sections, it does make the material sensitive to overheating and thus to crack formation. Vanadium and chromium are often added to counter this sensitivity and to

Table C 18.1 Chemical composition and examples for application of heat treatable alloyed spring steels (grouped according to alloy content) [36]

Group	Alloying element	Chemical composition							Steel grades (Designation)	Range of tensile strength N/mm²	Field of application
		% C	% Si	% Mn	% Cr	% Mo	% Ni	% V			
A	Si	0.35 to 0.55	1.50 to 2.0	0.50 to 0.80					38 Si 6, 38 Si 7, 46 Si 7	1200...1600	Support-, spiral-, buffer springs for railways
									50 Si 7, 51 Si 7		Spring washers, spring plates, vibration springs, cup springs, leaf- and cone springs
	Si(+ Cr)	0.55 to 0.75	1.50 to 2.0	0.60 to 1.10	(0.20 to 0.40)				55 Si 7, 60 Si 7, 65 Si 7	1300...1600	Vehicle springs (leaf springs)
									50 SiCr 7, 65 SiCr 7, 71 Si 7	1900...2400	Cup-, coil- and torsion bar springs Springs for watches and horological industry (e.g. driving springs)
	Si + Cr	0.52 to 0.72	1.20 to 1.80	0.40 to 0.90	0.40 to 0.60				67 SiCr 5	1500...1700	Seat and value springs for temperatures up to 300 °C, helical springs
									54 SiCr 6	1300...1700	
B	Mn	0.40 to 0.55	0.15 to 0.35	1.60 to 1.90				(0.07 to 0.12)	46 Mn 7, 50 Mn 7, 51 MnV 7	1200...1600	Vehicle springs (leaf springs)
C	Cr	0.50 to 0.65	0.15 to 0.40	0.70 to 1.00	0.60 to 0.90				55 Cr 3, 60 Cr 3	1300...1700	Highly stressed vehicle springs (helical-, torsion bar-, leaf springs)
	Cr + (Mo) + V	0.45 to 0.60	0.15 to 0.40	0.70 to 1.10	0.90 to 1.20	(0.15 to 0.30)		0.07 to 0.20	50 CrV 4, 58 CrV 4 (51 CrMoV 4), (50 CrMo 4)	1300...1700	
D	Ni + Cr + Mo	0.55 to 0.65	0.20 to 0.35	0.95 to 1.00	0.40 to 0.60	0.15 to 0.25	0.40 to 0.70		55 NiCrMo 2, 60 NiCrMo 2	1300...1700	Vehicle springs, support- and helical springs, torsion bar springs

increase hardenability. The pronounced longitudinal fibrous structure found, especially in manganese steels, does not yield advantages at high stresses, as shown by present experience, and in torsional stress it is even a disadvantage.

Chromium- and chromium-vanadium-steels (group C in Table C 18.1) are often preferred for helical and torsion bar springs in vehicle construction as well as for leaf and spiral spring that work under high stresses. These steels are used with advantage because of their high and uniform hardenability, they present no problems in heat treating and show a favorable behavior under bending and torsional stress. The range of sizes extends to 40 mm dia., and bigger sizes are alloyed with molybdenum. These two groups include the steel grades 55 Cr 3, 50 CrV 4, 58 CrV 4 and 51 CrMo 4. These materials as well as the silicon and chromium alloyed steels, such as 67 SiCr 5, may be selected because of their high retention of hardness in tempering and their high strength at operating temperatures of up to 300 °C, if so required. In Fig. C 18.5 the hardenability of these steels is compared to that of spring steels alloyed only with silicon.

Spring steels containing about 0.6% C, 0.5% Cr, 0.6% Ni and 0.2% Mo (group D in Table C 18.1), including the American steel grades AISI 8650 and 8660, are used for heavy leaf and helical springs, in sizes up to 45 mm dia., especially on account of their high hardenability and good toughness properties. At lower temperatures these steels permit higher toughness values to be achieved than with steels having no nickel and molybdenum additions.

C 18.3.3 Elevated-temperature Strength Steels for Springs

If strength requirements at temperatures up to about 300 °C are to be met, chromium-vanadium-, silicon-chromium- and silicon-chromium-molybdenum

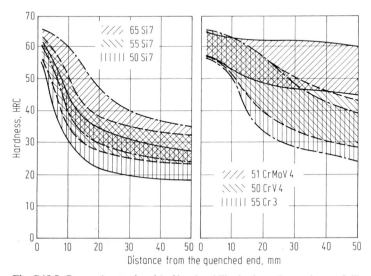

Fig. C 18.5. Ranges (scatter bands) of hardenability in the end quench test of silicon- and of chromium-(molybdenum-vanadium) spring steels. After [37].

steels may be considered. At higher temperatures chromium-molybdenum-vanadium steels and sometimes even tungsten alloyed hotworking steels are used, and austenitic stainless steels are employed if corrosion resistance is also required [38, 39]. These steels are necessary for springs to work at high temperature stresses in valves, seals, exhaust valves in engines (motors), and in locomotives, steam turbines, etc.

Steel selection [39a] depends, additionally on the properties at room temperature, on the yield stress at up to about 350 °C and at still higher temperatures on the creep limits. Accordingly, the working stresses must be reduced and the effects of higher temperatures must be considered in view of the fact that creep and relaxation increase whereas the modulus of elasticity (Young's modulus) and shear modulus, tensile strength and yield stress decrease considerably. High temperature steels for springs are employed comparatively frequently, especially the steel grade X 20 CrMoV 12 1 as well as the precipitation hardening steels X 7 CrNiMoAl 15 7 and X 5 NiCrTi 26 15, especially if corrosion resistance is also required. For very high temperatures, high temperature alloys, e.g. the material NiCr 20 TiAl are employed (chemical composition of such alloys see C 9).

C 18.3.4 Spring Steels with Good Toughness at Low Temperatures (Spring Steels for Low Temperature Service)

In the selection of cryogenic steels for springs with good toughness at low temperatures [40] attention must be paid to the fact that decreasing temperatures are accompanied by increasing tensile strength and notch sensitivity due to loss of toughness [33, 41–43]. Depending on service conditions and, above all, on service temperature, chromium-molybdenum- and chromium-nickel-molybdenum-quenched and tempered steels or austenitic chromium-nickel-steels are used.

C 18.3.5 Stainless Steels for Springs

Springs subjected to corrosive influences, e.g. in the chemical industry, require to be resistant to rust and acids. For these applications, stainless steels with at least 12% Cr, and, if required, at least 6.5% Ni (X 30 Cr 13 or X 12 CrNi 18 8) possess the necessary elastic deformability as a result of cold forming and/or heat treatment [44, 45].

Hardenable steels permit higher stresses but have a lower corrosion resistance. These are the steels X 30 Cr 13, X 20 CrMo 13 and X 35 CrMo 17, of which those containing molybdenum may be used at temperatures up to 400 °C and in bigger cross-sections.

Austenitic steels have better corrosion resistance but their elastic properties must be obtained by additional cold deformation (cold drawing or cold rolling). The attainable strength values are nevertheless much lower than those of the quenched and tempered steels. Austenitic steels also have lower elasticity and shear moduli. Steels representative of this group are X 12 CrNi 18 8 and X 5 CrNiMo 18 8 and, as a material with higher strength but less corrosion resistance, X 12 CrNi 17 7. After cold forming the strength values and the elastic limit may be raised by heat treatment at moderately high temperatures up to about 400 °C.

It is possible to combine the advantage of a good cold formability with a considerable increase of basic strength acquired from precipitation hardening by using *austenitic precipitation hardening steels*. The steel grade X 7 CrNiAl 17 7 is the main precipitation hardenable alloy [46]; the hardening effect being obtained by aluminum additions. To achieve the same target in other steels, titanium or niobium in particular are added.

The chemical composition of the above-named stainless steels: X 30 Cr 13: $\approx 0.3\%$ C and 13% Cr; X 12 CrNi 18 8: $\approx 0.12\%$ C, 18% Cr and 8% Ni; X 20 CrMo 13: $\approx 0.2\%$ C, 13% Cr and 1.1% Mo; X 35 CrMo 17: $\approx 0.35\%$ C, 17% Cr and 1.1% Mo; X 5 CrNiMo 18 10: $\approx 0.07\%$ C, 17% Cr, 2.2% Mo and 11% Ni; X 12 CrNi 17 7: $\approx 0.12\%$ C, 17% Cr and 7% Ni; X 7 CrNiAl 17 7: $\approx 0.09\%$ C, 17% Cr, 7% Ni and 1.2% Al.

C 19 Free-Cutting Steels

By Helmut Sutter and Günter Becker

C 19.1 Characteristic Properties

Shaping by chip removal (machining) offers the advantage of supplying a finished surface with very close tolerances and the possibility of producing complicated part shapes. In addition, machining is an economical method of producing parts in relatively small lots.

Machining can be automated for serial production if the following important conditions are met:
- High cutting speeds should be possible with limited tool wear, so that reliable operating tool life is obtained.
- Exact tolerances and smooth surface finish should be produced.
- Short breaking chips should be formed, with small chip volume, thus precluding any removal problem.

The metallurgy measures required to fulfill these conditions are mostly contrary to certain physical and technological material parameters that are decisive for the subsequent processing routes, heat treatments, and various kinds of stress. This problem was solved by a technical compromise. Free-cutting steels have been developed as a special metallurgical group of steel grades [1] that aim for the best machinability, but also meet the basic requirements of the common steel grades. In particular, free-cutting steels have to correspond to standard structural steels, case-hardening steels, and steels for quenching and tempering, as used in the mass production of cars, machines, and appliances in precision mechanics, in assembly technology, and so on. Free-cutting grades are also produced for various other special steel uses, e.g. as relay or stainless materials, but they will not be discussed here.

The examination and assessment of free-cutting steels are problematic because the machinability cannot be described by simple and generally valid characteristic parameters. Experience shows that the results of laboratory standard test methods (see B 9.1) must be supplemented by analysis and comparison of operating results. Characteristics that are especially important are surface quality, dimensional tolerance, and chip form, as well as cutting performance and tool life. It is preferred that long-time tests that are very close to the operating conditions be agreed upon by the producer and the customer. Test methods using shorter times and lower

---

References for C 19 see page 793.

C 19.2 Consequences Considering Physical Metallurgy

C 19.2.1 Nature of the Microstructure

Because of the conditions that are discussed in more detail in B 9, a favorable machinability depends on *the absence of hard and abrasive microstructural constituents*, which must be ensured by the metallurgical methods employed as well as by an *optimized combination of least possible strength with highest possible brittleness*. These criteria are clearly met in low-carbon free-cutting steels that cover most of the general demand. With a carbon content of about 0.10% – comparable to St 37 or SAE 1010 – these steels contain only a small quantity of cementite as a component of the strength-determining pearlite, while the ferrite toughness is reduced by a controlled proportion of embrittling elements such as nitrogen and phosphorus. These elements become particularly effective after the bright-drawing process that is necessary to improve the geometry of the hot-rolled initial material.

One or more *inclusion phases* of suitable nature, size, and distribution are the essential characteristic of all free-cutting steels. These phases exert *favorable influences* on *the chip-forming and chip-breaking mechanisms* and eventually on the *contact reactions between the tool face and the off-flowing chip* by forming lubricating and wear-retarding depositions on the tool face. At about the end of the 19th century, it was recognized that an increased number and size of sulfide inclusions favor machinability. From that time on, sulfide inclusions became the basis for the development of the free-cutting steels. These inclusions are still considered to be a classical characteristic, even though the modern free-cutting steel family also includes certain nonsulfurized types, with other alloy additions creating or influencing certain inclusion phases.

C 19.2.2 Effects of Special Alloying Elements on Free-cutting Steels

Sulfur contents of 0.10% to 0.40% have become usual, even in stainless free-cutting steels (see C 13). The microstructural and mechanical properties – in the hot-rolled as well as in the cold-drawn, case-hardened, or quenched and tempered condition – are affected in only a minor way by the increase of the sulfide inclusion content (Fig. C 19.1). The only exception is the toughness in the transverse direction, which is noticeably reduced by the banding structure resulting from a more or less pronounced elongation of the inclusions during the hot-rolling process into bar or wire rod. This reduction in transverse toughness is decisive when choosing the suitable sulfur content for a given application. Ultimately it limits the

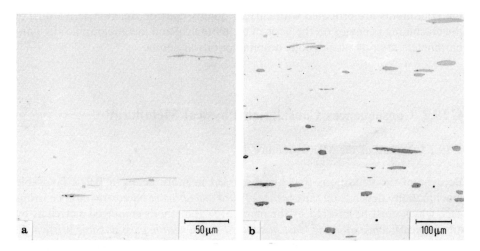

Fig. C 19.1. Sulphide shape. **a** In a steel with 0.020% S; **b** in a free-cutting steel with 0.25% S and a similar composition. After [17].

improvement in machinability that can be obtained by using elevated sulfur contents.

The *sulfide composition and precipitation mode* exert important influences. In free-cutting steel metallurgy, the manganese sulfide, MnS, created by addition of sublimed sulfur during ladle pouring, does not strictly follow the stoichiometry and may be described as a complex substitution compound such as $(Fe, Mn, Si, O)_x S$. The size, amount, and distribution of this sulfide compound, as well as the deformation resistance during hot rolling, determine the effect on machinability. The deoxidation state of the solidifying melt plays a decisive role in the development of three different sulfide types. Type 1, a globular oxisulfide with an average diameter of a few microns, develops at bath oxygen contents above about 200 ppm [2–10] and has lower plasticity during the hot-rolling process. Type 1 is desirable for free-cutting steels.

This sulfide type can be advanced and stabilized, beyond a suitable melt process conduction, by additions of a few different alloying elements, mainly *tellurium*. The resulting tellurides, especially MnTe, combine with the sulfides partly being built in, partly as an enveloping phase. They cause reduced sulfide plasticity during hot rolling [11] and, consequently, a conspicuously globulized inclusion shape (Fig. C 19.2). Usually, the amount of tellurium added is about 10% of the sulfur content, although tellurium's effects are perceptible at a much lower level. The presence of any low-melting telluride phases as $FeTe_2$, being liquefied at about 850 °C, on the grain boundaries is strongly detrimental and complicates the metallurgical and hot-forming operations on tellurium-bearing steels.

Finally, there are other sulfide-globulizing alloying elements, such as *selenium*, which is closely related to tellurium. *Cerium, zirconium, and calcium* also globulize sulfides, although the use of these elements is not economical in the production of free-cutting steels. For the applications and effects of these elements, see B 9.

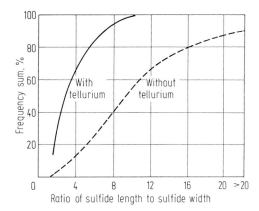

Fig. C 19.2. Influence of tellurium addition on the sulfide deformation during hot rolling of semikilled free-cutting steel round bar of 25 mm dia. After [1]. Chemical composition of the examined heat: 0.09% C, trace Si, 1.22% Mn, 0.067% P, 0.320% S, 0.012% N, and addition of 0.039% Te.

SEP 1572 [12] contains rating charts used to assess inclusions. These charts present a systematic classification of the sulfide shapes that occur in free-cutting steels. Attempts to derive machinability indices from metallographic parameters may be more successful because of the use of modern methods of identifying chemical inclusions and the use of quantitative optical metallography. This latter technique objectively describes the geometry of sulfides as well as oxides [13].

Lead is the second-most important inclusion-forming element after sulfur, in the production of free-cutting steels. Lead is soluble in steel at the steel melting temperature and appears as a finely dispersed metallic phase in the range of 0.15% to 0.30%. Practical alloying methods may guarantee an almost ideally homogeneous distribution, whose macroscopic appearance can be made visible by a simple chemical print procedure similar to the sulfur print. Using microscopy, the lead particles, whose average size is about 1 µm or less, can be observed either insulated in the matrix or – especially in the resulfurized steel grades – associated with the sulfides as lead appendices (Fig. C 19.3), which may also adhere to the oxides. Evidence of nonvisible protecting lead envelopes around the sulfides can be derived from the improved corrosion resistance that is occasionally observed [14]. If

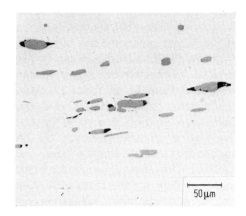

Fig. C 19.3. Deposits of metallic lead at manganese sulfides. After [25]. Chemical composition of the steel: 0.09% C, trace Si, 1.08% Mn, 0.080% P, 0.305% S, and 0.20% Pb.

tellurium is also present, an additional PbTe phase occurs with the same behavior as pure lead.

Bismuth is a close chemical relative of lead and is similar to it in the metallurgical process as well as in its influence on machinability. Additions of 0.06% Bi in various free-cutting steel types containing lead aim to increase lead efficiency; i.e., further improved machining performance and surface quality. Beyond this, recent experience shows that bismuth can be substituted completely for lead, in suitably increased bismuth contents of 0.2%. This need only be considered if high machining performance is required in cases where steel processing or application problems exclude the use of lead and/or tellurium. A synopsis of tellurium, selenium, and bismuth use exists in the literature [15].

Recently, *boron* has become more important as an inclusion-creating element in free-cutting steels. Under certain metallurgical conditions, it causes the precipitation of a complex oxide phase resembling the type $B_2O_3 \cdot MnO$. Small amounts of 30 ppm to 50 ppm are sufficient. The inclusions occur mainly as a fine dispersion of < 1 μm particle size in the matrix, eventually as manganese sulfide deposits. Of course, the precipitation of the hard boron-nitride phase must be avoided [16].

C 19.3 Consequences for the Production of the Free-cutting Steels

There is not much room to optimize the chemical compositions of the free-cutting steel types that are intended for heat treatment and special mechanical loads.

On the other hand, it is possible to influence the low-carbon free-cutting steels to improve their machinability and other technological properties. The *deoxidation process has particular importance* because it has to exclude hard oxide inclusion types. A higher oxygen level is necessary to produce the less ductile sulfide of Type 1. The effervescent melting practice meets these requirements. Since sulfur is the most segregating of all steel alloying elements, it must be considered that considerable horizontal and vertical differences in the structure and composition of the ingots results in very disadvantageous mechanical properties and nonuniform machinability. Furthermore, the ring of coarse blow-holes enclosed within the transition between the rimming zone and the segregated core leads to detrimental segregation peaks [17], eventually containing iron sulfide FeS. This sulfide is liquid at 988 °C and causes the risk of red shortness.

These two disadvantages can be avoided by melting *semikilled or "balanced" steel*; i.e., by *elevated additions of manganese* as a soft-acting deoxidizer that can be controlled. Contents of about 1% Mn guarantee a less segregated "quasikilled" solidification structure (Fig. C 19.4), do not disturb the conditions for favorable sulfide precipitation, and result in relatively harmless manganese oxides as the deoxidation product. However, the blow-hole-ring now situated closer to the surface may require more expensive surface inspection and cleaning. Because excessive manganese enhances the ferrite hardness and may impair machinability, an upper limit for the manganese content, depending on the sulfur content, is necessary. A maximum manganese/sulfur ratio of 4.5 (check analysis) can be

0,50%Mn　　　　　　0,64%Mn　　　　　　0,81%Mn

0,93%Mn　　　　　　0,99%Mn　　　　　　1,10%Mn

Fig. C 19.4. Influence of increasing manganese contents on the segregation pattern of free-cutting steels with about 0.10% C and 0.25% S. After [17].

specified, which is corrected to a maximum (CxMn)/S ratio of 0.55 *with regard to carbon contents higher than 0.12%*. Though these limits are not usually guaranteed directly, they do correspond to the actual technical standard, especially since higher sulfur contents themselves favor the "quasikilled" solidification [19] (Fig. C 19.5).

According to experience, the semikilled free-cutting steels react to the more or less inevitable contents of *silicon*. Silicon may act by influencing the corresponding oxide quantity in the microstructure, as well as by the detrimental influence of reduced bath (melt) oxygen contents on the sulfide shape. Therefore, the least possible silicon content is recommended.

On the other hand, since silicon and *aluminum* are the current strong deoxidants, they cannot be completely avoided during processing of the fully killed steels. In this case, it is desirable – considering the tool wear caused by silicates and alumina – to obtain a small content of those inclusions by suitable metallurgical measures.

All standard *melting processes* can be applied to the production of free-cutting steels, provided that the fundamental conditions of charging, adding, and post-treating materials as well as of melting and deoxidation practice are respected and,

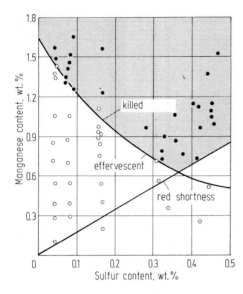

Fig. C 19.5. Influence of the manganese and sulfur content on the solidification process and the hot-rolling behavior of certain steels. After [19].

Chemical composition range of the examined steels

	% C	% Si	% Mn	% P	% S	% N	% O
effervescent	0.04/0.10	traces	0.40/0.68	0.044/0.061	0.182/0.250	0.007/0.017	0.027/0.039
killed	0.08/0.18	≤ 0.20	0.50/1.61	0.045/0.099	0.127/0.375	0.009/0.020	0.004/0.055
	partly with further additions of Al, Ca, Mg, or Zr.						

if necessary, specially adapted. Thus, the typical steel accompanying elements *phosphorus and nitrogen*, which favor ferrite-hardening and consequently chip brittleness and surface quality, were supplied in the Thomas basic-converter process in a convenient percentage; i.e., 0.050% to 0.090% P and 0.008% to 0.015% N. During the modern melting processes – especially the basic oxygen steel process – these elements are usually added as needed and adjusted in smaller ranges. Attention must be paid especially to the amount of the residual elements, above all chromium, that might come into the steel with scrap. Their amounts should be restricted below the standard limits to avoid any loss of machinability. Rather expensive ladle-alloying methods have been approved to allow the homogeneous incorporation of the inclusion-creating or -influencing additions. Large safety precautions for health and the environment are required because of the toxicity of elements such as lead and tellurium.

Since the *continuous casting process* is nearly universal in steel-making technology, this process is also used for free-cutting steels. Previous objections to this process, based on concerns about the influence of the solidification process – which is considerably different from the traditional ingot casting process – on sulfide shape and machinability, have not been borne out. It turns out that the continuous casting process even offers certain advantages based on the generally more homogeneous material structure [19a].

C 19.4 The Free-cutting Steel Grades of Common Use and Their Properties

C 19.4.1 Steel Grades Characterizing the Application Range

The remarkable *variety of commercial grades* reflects the steel-makers' and users' efforts to benefit from the economic advantages of improved machinability and to achieve a flexible compromise between the machining and service properties characteristic of the free-cutting steels.

The quality development can be observed primarily in the *low-carbon free-cutting steels* that dominate the current market. While the rimming steelmaking process represented by the 9 S 20 (comp. SAE 1109) was still standard up to 1965, it was quickly replaced by semikilled (or "balanced") steel practice, originally in the form of the grade 9 SMn 23 containing 0.20% to 0.27% S and 0.90% to 1.30% Mn (comp. SAE 1213). The semikilled steel process made possible the safe handling of lead and other additions favoring machinability, as well as increasing the sulfur level to obtain further improved machinability. The necessary type diversification is evidenced by the succeeding grades 9 SMn 28 containing 0.27% to 0.33% S and 9 SMn 36 containing 0.34% to 0.40% S. Derived from these, the grades 9 SMnPb 28 (comp. SAE 12L 14) and 9 SMnPb 36 are taking a larger share of the market.

Table C 19.1 specifies the most important free-cutting steel grades. These are primarily the sulfurized grades that are standardized, for example, in DIN 1651 [20] and also in the nearly congruent ISO 683, Part 9 [21]. Some standard grades for case-hardening and for quenching and tempering also are given. Considering the great number of special grades with different combinations of additions or "semi-free-cutting steels" such as cold-upsetting steel for nuts according to DIN 17 111 [22], such a list cannot be complete.

C 19.4.2 Machinability

The semikilled steel grades improve machining performance by about 20% to 30% for every 0.1% increase in sulfur content. Similar figures may also be stated for lead and tellurium in usual contents. Based on the grade 9 SMn 28, it is estimated that a 15% to 25% improvement in machinability results from adding lead and a 25% to 35% improvement results from adding tellurium. The advantage of adding both elements results in an estimated 35% to 65% machinability improvement. Based on these improvements, a number of *high-performance free-cutting steels* with different alloy combinations, such as S + Pb + Te, have been developed. The cost of the additives and possible output losses cause higher prices for these steels. Therefore, the use of special steel grades is decided by economic criteria. These special steel grades are consequently reserved for a limited application range.

To roughly quantify the specific machinability improvements obtained by means of various alloy additions and combinations, the typical low-carbon free-cutting steels are compared to an unalloyed steel with similar tensile strength in

Table C 19.1 Typical free-cutting steels

Steel grade (designation)	% C	% Si	% Mn	% P	% S	% Pb	% Others
Free-cutting steels not intended for heat treatment							
9 SMn 28	≤ 0.14	≤ 0.05	0.90–1.30	≤ 0.100	0.27–0.33		
9 SMnPb 28	≤ 0.14	≤ 0.05	0.90–1.30	≤ 0.100	0.27–0.33	0.15–0.30	
9 SMn 36	≤ 0.15	≤ 0.05	1.00–1.50	≤ 0.100	0.34–0.40		
9 SMnPb 36	≤ 0.15	≤ 0.05	1.00–1.50	≤ 0.100	0.34–0.40	0.15–0.30	
9 SMnPbTe 28	≤ 0.10		~ 1.20	0.040–0.060	0.27–0.33	0.15–0.30	~ 0.04 Te
9 SMnPbBi 28	0.05–0.12	Traces	0.85–1.25	0.040–0.080	0.27–0.33	0.15–0.30	~ 0.06 Bi
Case-hardening free-cutting steels							
10 S 20	0.07–0.13	0.10–0.30	0.50–0.90	≤ 0.060	0.15–0.25		
10 SPb 20	0.07–0.13	0.10–0.30	0.50–0.90	≤ 0.060	0.15–0.25	0.15–0.30	
C 15 Pb	0.12–0.18	≤ 0.40	0.30–0.60	≤ 0.045	≤ 0.045	0.15–0.30	
16 MnCrPb 5	0.14–0.19	≤ 0.40	1.00–1.30	≤ 0.035	≤ 0.035	0.20–0.35	0.80–1.10 Cr
Quenched and tempered free-cutting steels							
35 S 20	0.32–0.39	0.10–0.30	0.50–0.90	≤ 0.060	0.15–0.25		
45 S 20	0.42–0.50	0.10–0.30	0.50–0.90	≤ 0.060	0.15–0.25		
60 S 20	0.57–0.65	0.10–0.30	0.50–0.90	≤ 0.060	0.15–0.25		
C 35 Pb	0.32–0.39	≤ 0.40	0.50–0.80	≤ 0.045	≤ 0.045	0.15–0.30	
C 45 Pb	0.42–0.50	≤ 0.40	0.50–0.80	≤ 0.045	≤ 0.045	0.15–0.30	

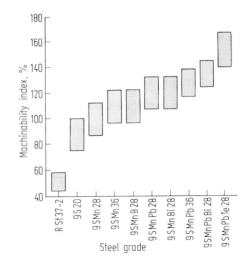

Fig. C 19.6. Machinability of various free-cutting steels. Machinability index: Chip volume removed within 8-hour tool life, based on 100% for 9 SMn 28 (see Table C 19.2).

Fig. C 19.6 (in connection with Table C 19.2). This diagram is based on laboratory and field experience with longitudinal turning operations with high-speed tools and about 8-hour tool life and corresponds to data in the literature [1, 14, 16, 24, 25]. The corresponding evaluation of free-cutting steels for quenching and tempering is shown in Fig. C 19.7 (in connection with Table C 19.3). The indicated ranges take into account the extremely diverse influences of the processes and materials.

C 19.4.3 Mechanical Properties

There is no significant difference of the mechanical characteristics measured on specimen in the longitudinal direction between the free-cutting steels and their comparable basic grades. The remarkable *anisotropy of the resulfurized types*, however, considerably reduces the ductility and impact values in the transverse direction [5]. Therefore, the minimum transverse elongation of 5% as required, for

Table C 19.2 Chemical composition of the steels examined according to Fig. C 19.6

Steel grade	Average chemical composition				Tensile strength N/mm²
	% C	% Mn	% S	% Others	
R St 37-2	0.10	0.45	0.045	0.15 Si	400
9 S 20	0.10	0.90	0.210		420
9 SMn 28	0.10	1.10	0.280		430
9 SMn 36	0.12	1.30	0.360		440
9 SMnB 28	0.10	1.20	0.280	0.0030 B	430
9 SMnPb 28	0.11	1.10	0.280	0.20 Pb	430
9 SMnBi 28	0.11	1.10	0.280	0.25 Bi	430
9 SMnPb 36	0.12	1.30	0.360	0.20 Pb	440
9 SMnPbBi 28	0.11	1.10	0.280	0.20 Pb, 0.06 Bi	430
9 SMnPbTe 28	0.11	1.10	0.280	0.20 Pb, 0.035 Te	430

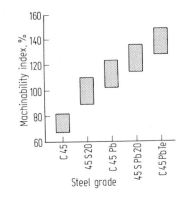

Fig. C 19.7. Machinability of various steel grades for quenching and tempering. Machinability index according to Fig. C 19.6, but based on 100% for 45 S 20 (see Table C 19.3).

Table C 19.3 Chemical composition of the steels examined according to Fig. C 19.7

Steel grade	Average chemical composition				
	% C	% Si	% Mn	% S	% Others
C 45	0.45	0.15	0.60	0.035	
45 S 20	0.45	0.15	0.75	0.200	
C 45 Pb	0.45	0.15	0.60	0.035	0.20 Pb
45 SPb 20	0.45	0.15	0.75	0.200	0.20 Pb
C 45 PbTe	0.45	0.15	0.60	0.035	0.20 Pb, 0.035 Te

instance, in nut production, can be guaranteed only under certain conditions if low-carbon free-cutting steels are used. A more globular sulfide shape, as attained especially by tellurium additions, makes possible decisive improvements [1] (Fig. C 19.8).

In comparison to sulfur, the same weight percentage of *lead* takes only 1/8th of the volume and consequently impairs the mechanical properties, in general, much less. This advantage guarantees the principal use of lead even for unalloyed and

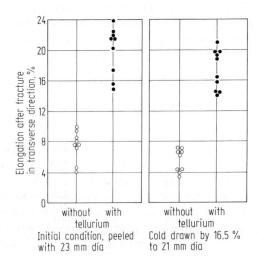

Fig. C 19.8. Improvement by tellurium of the elongation after fracture in transverse direction of a free-cutting steel in the peeled and the cold-drawn condition. After [1]. Same composition of the heat as in Fig. C 19.2.

alloyed heat-treatable steel intended for high mechanical loads and safety purposes, because the problem of the fine homogeneous lead distribution is solved by the actual processing and testing technologies. As far as known, only one usage restriction might be derived from a warm embrittling effect being observed between about 200 °C and 450 °C, with a maximum at the lead melting point (327 °C) [26].

C 19.4.4 Subsequent Processing and Heat Treatment

Because of the required dimensional accuracy and surface quality, the free-cutting steels are mostly machined in the *cold-drawn condition*. The resulting embrittlement by strain hardening favors machinability in the low-carbon grades. No improvement is generally possible in the steels for quenching and tempering – i.e., for steels with a carbon content over 0.3% [27] – because the optimal strength range is often exceeded.

But strength requirements are not the only factor. Due to the notch effect induced by the sulfides, excessive cold strain hardening may cause an increased risk of tension cracking, especially in the higher carbon grades such as 45 S 20 (comp. SAE 1144/1146) and 60 S 20. However, the residual stress can be limited if the cold-drawing process is suitably adapted [28].

Precautionary measures are also recommended for the *hot forging* of free-cutting steels – including the "quasikilled" grades – which is becoming more and more common. At any rate, attention must be paid to a sufficient forging temperature level, small broadside deformation (spread), and sufficient machining allowances. The risk of brittleness and cracking caused by the special microstructure must also be considered during hot and cold deburring operations, particularly if residue segregations may be forged into the burr and finally laid bare in deburring. Tellurium-bearing steels are normally not suitable for hot-forging operations, due to their hot-shortness tendency.

The fully killed free-cutting steels are intended for *case-hardening or quenching and tempering* applications. The achievable properties do not practically differ from those of the corresponding low sulfur grades, with the exception of the transverse properties. The case-hardening grades are often successfully replaced by the much more machinable semikilled types, if sufficient experience concerning heat-treatment conditions is available.

C 19.4.5 Other Properties

The *surface quality* of the free-cutting steels includes a rather high risk of defects, due to the intentionally heterogeneous microstructure and the embrittlement tendency. As a result, the steel grades standardized, for example, in DIN 1651 have been provided with auxiliary data for the allowable depth of longitudinal cracks in the peeled and the cold-drawn conditions. Also, the production of a "technical crack-free" condition by means of the common test methods must consider faked defects such as inclusion bands near or within the surface and so-called "ghostlines." Skin-embedded inclusions, occurring through elevated sulfur content, may impair the corrosion resistance, especially the homogeneous building up of a galvanic protection coating. Therefore, nonporous gloss coatings require relatively expensive grinding and polishing operations, as well as increased precautions during preliminary pickling treatments [29].

The requirements referring to a homogeneous *internal condition* are extending in practical use as far as to high density hydraulic applications – even concerning the semikilled steel grades. This amounts to a diameter limitation of all inclusion types to the microscopic order of size, which is considered problematic despite all the progress in the metallurgical and testing technologies. In the industrial production of machined parts of this kind, the generally more homogeneous structure of continuous-cast free-cutting steels has proved advantageous.

The free-cutting steels are generally not *weldable*. Nevertheless, a certain weldability is expected; for instance, for connection welding during wire drawing or assembling of prefabricated construction elements. Satisfactory results may be achieved by taking appropriate measures, especially with CO_2-shielded arc welding [30].

Low-remanence (soft) magnetic properties are occasionally required from free-cutting steels for use in electrical engineering. For this purpose, the semikilled types furnish sufficient values, even if they contain the usual special additions [1]. However, the strong strain hardening and aging tendency with corresponding increase of the "magnetic hardness" must be taken into account.

The development of free-cutting steels with improved quality and performance seems to be continuing. The low-carbon grades, however, may be reaching their limits; these grades are already diverse and are approved for numerous applications. Continuing improvement of machining technologies will probably lead to further progress in economical use of these steels. It is possible that special adaptation of microstructures and chemical compositions will lead to the combination of easy, noncutting methods of chip removal. Finally, better compromises between machinability and other processing and technological properties will probably be achieved by sulfurizing high-grade and special steels above the standard limits, combined with special additions that influence the sulfide shape [31].

C 20 Soft Magnetic Materials

By Ewald Gondolf, Fritz Assmus, Klaus Günther, Armin Mayer,
Hans Günter Ricken and Karl-Heinz Schmidt

C 20.1 Range of the Soft Magnetic Materials

As a rule, all materials that have *coercivities of less than 1000 A/m* are classified as soft magnetic [1]. Clearly distinguished from these are the hard magnetic materials (see C 21) with coercivities exceeding 10 000 A/m. The classification in the intermediate range is less clearly defined.

Recently, the International Electrical Commission (IEC) concerned itself with the *classification* of soft magnetic materials. It arrived at the following rough arrangement [2, 3]:

A: Pure iron
B: Low-carbon steel
C: Silicon steel
D: Other steels
E: Nickel-iron alloys
F: Cobalt-iron alloys
G: Other alloys
H: Soft magnetic ceramics

This section will follow this classification and will include steels with special requirements for magnetic properties, which often have coercivities between 1000 and 10 000 A/m. Soft magnetic ceramics will not be included.

Closely connected with the *requirement for low coercivity* for soft magnetic materials is the desire for *high permeability* and *high saturation flux density*. Additional requirements arise from technical demands and physical and economic conditions [4–8].

The great variety of requirements for the properties of the materials has engendered a large number of developments that have led to very different results and thus to very different soft magnetic materials. In particular, the "structure-sensitive" magnetic properties – such as coercivity, initial and maximum permeability, and core losses – have needed further development. To present a well-rounded representation of the individual groups of materials, they are discussed individually in C 20.3 to C 20.7.

For theoretical explanations of magnetism and the individual characteristics, see B 2 and a book by Kneller [9]. Bozorth [10] contains numerous details about the magnetic properties of iron and its alloys as well as information on measuring methods.

C 20.2 Characteristics for the Evaluation of Soft Magnetic Materials

Because of their historical development, the designations of the *magnetic properties* and especially their units are characterized by great variety, both nationally and internationally. This makes it difficult for a nonspecialist to enter this field. The worldwide application of the SI system has calmed the situation somewhat, and it must be hoped that the definitions and units published nationally in, for example, DIN 1325 [11], and internationally in IEC publication 50 (901) [12], will finally displace the old cgs units. Following these standards, the specifications and units given in Table C 20.1 are recommended for the most important magnetic characteristics. The designations of the magnetic characteristics of the hysteresis loop can be seen in Fig. B 2.8.

Table C 20.1 Designation of the most important magnetic characteristics, their units, and their relationships with each other

Term	Symbol	Definition	SI-unit	Note
Magnetic field strength	H	s. DIN 1325	A/m	$1 \text{ A/m} \triangleq \frac{4\pi}{10^3}$ oersted (Oe)
Magnetic flux density	B	s. DIN 1325	$\frac{\text{Vs}}{\text{m}^2} = $ tesla (T)	$1 \text{ T} \triangleq 10^4$ Gauss (G)
Magnetic polarization	J	$J = B - \mu_0 H$	$\frac{\text{Vs}}{\text{m}^2} = $ tesla (T)	
Magnetization	M	$M = \frac{J}{\mu_0}$	A/m	
Magnetic field constant (in void)	μ_0	$\mu_0 = \frac{B}{H}$	$\frac{\text{Vs}}{\text{Am}} = \frac{\text{henry}}{\text{m}} = \frac{H}{\text{m}}$	$1 \frac{H}{m} \triangleq \frac{10^7}{4\pi} \frac{G}{Oe}$
Permeability (in a substance)	μ	$\mu = \frac{B}{H}$	$\frac{\text{Vs}}{\text{Am}} = \frac{\text{henry}}{\text{m}} = \frac{H}{\text{m}}$	
Permeability coefficient (relative permeability)	μ_r	$\mu_r = \frac{\mu}{\mu_0}$	–	
Magnetic saturation	J_s	$J_s = B - \mu_0 H$	$\frac{\text{Vs}}{\text{m}^2} = $ tesla (T)	(after saturation)
Remanence	J_r	$J_r = B_r$ where $H \to 0$		
Core (power) loss per cycle	P	$P = \oint H \, dB$	$\frac{\text{Ws}}{\text{m}^3}$	
Core (power) loss at frequency f and density ρ	$P_{f\rho}$	$P_{f\rho} = \frac{f}{\rho} \oint H \, dB$	$\frac{\text{W}}{\text{kg}}$	
Coercitivity	$_J H_c$, $_B H_c$	H_c where $J = 0$ H_c where $B = 0$	A/m	for permanent magnetic materials for soft magnetic materials (after saturation)

The large number of important magnetic characteristics has led to highly developed measuring techniques to ensure the maintenance of the specified requirements [13–15]. In accordance with the desire for *narrow hysteresis loops*, the measurement of the *passage through zero of the magnetic polarization* – i.e., of the static coercivity – occupies an important place. This is one of the characteristic properties of magnetic materials – because, on the one hand, it is required as a magnetic property and, on the other, its value (or its change in value) can, under certain conditions, be used to form conclusions about hardness, grain size, tendency to aging, etc. This measurement is soon to be specified in a DIN standard. The passage through zero of the magnetic polarization $_JH_c$ is difficult to measure and differs only slightly from the passage through zero of the magnetic flux density $_BH_c$ for soft magnetic materials, which is generally given as coercivity. It is normally measured in an open magnetic circuit after sufficient saturation of the sample in a constant magnetic field. The passage through zero of B or J can be determined either with a moving coil or with external probes.

Another frequently determined characteristic value is the *magnetic flux density* (induction) as a function of magnetic field strength. For this value, ring or rod probes can be used that either are provided with a winding or are inserted in finished coils. In addition to the initial magnetization curve, the hysteresis loop with the remanence and coercivity can be determined in the same measuring setup. The permeability coefficient can be determined from the magnetic field strength and the magnetic flux density and is frequently used as a test quantity. Between the coercivity and the permeability coefficient, there is an empirical reciprocal relationship, which is often an inducement to substitute the coercivity measurement for the more complicated permeability measurement.

To determine the *stability with time of the magnetic properties*, the coercivity may also be measured after heating to $100\,°C \pm 5\,°C$ for 100 h. The percentage change compared with the nonaged condition is known as the *aging coefficient*.

In special cases, the *saturation polarization* is measured. This cannot be determined by the ring or rod measuring method, because it is not possible to produce the high field strength required. To measure the saturation value, it suffices to produce a sufficiently high field strength (whose value does not have to be known accurately) with a strong electromagnet, for example, and to measure the magnetic polarization of the sample with a suitable measuring instrument – e.g., inductively with a ballistic galvanometer – in comparison with a sample of known saturation.

If the material is used in an *alternating field*, suitable *tests* take the place of the static methods. Normally, an Epstein test frame is used, which is a measuring transformer in which the iron core is formed by the samples to be measured. The procedure is standarized, for example, in DIN 50462 [16]. The most important characteristics are the core (power) losses at 1, 1.5, or 1.7 T, the polarization at different field strengths, and the permeability, generally all at 50 Hz. Other characteristics, such as the specific electrical resistance, the density, the dependence of the magnetic properties on direction, etc., are determined in specific cases. Similarly, depending on the application of the material, the properties are also measured at frequencies other than 50 Hz.

C 20.3 Soft Iron: Applications, Properties, Production, and Grades

C 20.3.1 Basic Requirements

For the applications discussed here, soft iron is defined as a steel that has minimum unintended impurities and only small amounts of elements added for processing or to achieve the required magnetic properties. A compromise must be made between the optimum magnetic properties and economical manufacture.

Soft iron is used wherever constant magnetic fields must be produced, amplified, or shielded, such as in relays, small motors, pole shoes and tips, direct current DC magnets for atomic research, and shielding of magnetic DC fields.

Soft iron can be rolled, drawn, forged, sintered, or cast.

The following *requirements for the material properties* arise from the different applications of soft iron, which are partly interrelated: ease of magnetization to high values – i.e., high values of magnetic flux density (induction) – even with weak fields; high saturation polarization; high maximum permeability; low coercivity; constant magnetic properties with time (resistance to aging); magnetic homogeneity in parts of large magnet components, both within each part and among parts; and good machining and processing properties.

Because of its good electrical conductivity, soft iron can be considered where low core (power) losses are required – such as for electrical steel sheets – only if special measures are taken (e.g., sufficient reduction of sheet thickness and heat dissipation). Where applicable, silicon steels can be used in such cases (see C 20.4).

Corrosion resistance in soft iron must be obtained by using protective coatings. For stainless steels for relays, see C 20.5.3.

C 20.3.2 Findings of Physics and Physical Metallurgy Regarding Influences on the Required Properties

The magnetic characteristics of iron, and of all soft magnetic alloys, are subject to numerous influences. It is the task of the manufacturer to promote favorable influences and to keep the harmful ones to a minimum. To do this, it is necessary to recognize the interconnections between properties and internal material parameters. Some of the important ones will be dealt with briefly here (see also B 2 and [9]).

Saturation Polarization

In iron, the magnetic saturation polarization is reduced with the addition of virtually all elements except cobalt. This parameter is connected with the mean magnetic moments shown in Fig. B 2.18. Figure C 20.1 [1] shows the *effect of some elemental additions*. According to this figure, oxygen and carbon cause particularly strong reductions in the magnetic saturation polarization. These elements are present predominantly in inclusions or precipitations that are nonmagnetic or are less magnetic than the matrix.

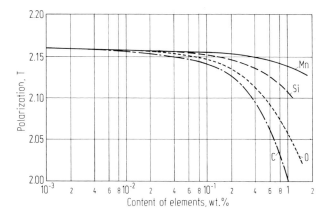

Fig. C 20.1. The effect of additions of carbon, oxygen, silicon, and manganese on the polarization of pure iron. After [1].

Coercivity and Bloch-wall Movements

The coercivity increases with defects in the ideal crystal structure. In iron, coercivity increases mostly with the volume fraction of nonmagnetic or, in comparison with the matrix, less magnetic *inclusions* or *precipitations*. The size of the inclusions is important, as shown in Fig. C 20.2 for iron with 0.02% C [17].

This behavior is connected to *Bloch walls* and their displacement (shifting) by a magnetic field (see B 2). The Bloch walls are held by certain *pin points* until they are torn away by an increasing magnetic field (irreversible wall displacements). In iron, it is primarily inclusions or precipitations that act as pin points, and their influence is strongest when they are roughly the same thickness as the Bloch wall (in iron, about 0.1 μm). Additional restrictions on the Bloch-wall movement are caused by *grain boundaries* or small grain size. Figure C 20.3 illustrates, with the

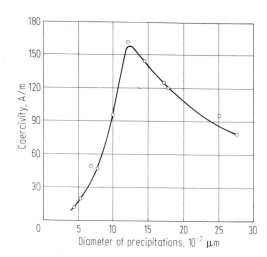

Fig. C 20.2. Dependence of coercivity on the mean diameter d of the precipitations in iron with 0.02% C at a constant volume fraction of the precipitations. After [17].

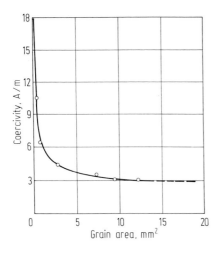

Fig. C 20.3. Dependence of coercivity on the mean value of the grain area for pure iron with 0.002% C and up to 0.07% Si. After [18].

example of an extremely pure iron with 0.002% C (both vacuum and zone melting), the relationship between coercivity and grain size. This relationship is approximately inversely proportional to the grain diameter [18]. In normal iron, this effect is small compared with that of inclusions.

In addition to the included defects, the Bloch-wall movement is also affected by *inhomogeneous stresses* and by plastic and elastic *deformations*. These forces act in much the same way as do the inclusions. However, it is, possible to produce positive effects by intentionally applying metered stresses (see B 2). Under the influence of stress, the saturation magnetostriction becomes a material parameter in the interrelationships.

Anisotropy Energies

The crystal anisotropy constant K_1 also enters these interrelations as a material parameter. It determines the major part of the anisotropy energy that is necessary to turn the polarization direction to one other than the preferred one (comp. B 2). This is especially true after the completion of the wall displacements during the remaining turning of the polarization into the direction of the field (turning processes). K_1 is particularly large for iron and its alloys as compared to, say, nickel and nickel alloys.

Analogously, the constants of the saturation magnetostriction λ also enter into these interrelationships. Together with the local stresses, these constants determine an additional anisotropy (see C 20.6.2).

Implications for Manufacturing

Without doubt, the best soft magnetic properties are obtained with *undisturbed ferritic microstructures* without dislocations that have the highest possible *degree of purity* and *large (coarse) grains*. So as not to impair the saturation polarization and the coercivity of the soft iron more than absolutely necessary, additives or impurities, as well as the unintentional picking up of foreign substances, must be kept to an absolute minimum. This applies particularly to elements that form precipitations or inclusions (e.g., carbon, nitrogen, oxygen, and sulfur). These elements must be removed as far as possible during melting or must be combined in a suitable form. Any remaining particles must be made less harmful by coagulation or dissolving.

In addition, to reduce the coercivity, the number of grain boundaries should be small; i.e., the grain size should be made as large as possible.

Internal (residual) stresses should be avoided as far as possible and should be reduced by annealing followed by slow cooling of sufficient duration.

C 20.3.3 Production Technique

To meet the requirements given in C 20.3.1 and C 20.3.2, the following two procedures must be followed, either singly or together.

Melting and Deoxidation

Soft iron produced by the well-known melting process, nowadays accomplished mostly by blowing with pure oxygen, always contains a number of the above-mentioned elements that are dissolved or are precipitated as compounds. The amount of effort that should be made to produce iron that is as pure as possible must be decided according to the application and economic factors. Using *vacuum metallurgy*, it is possible to maintain a content of $\leq 0.001\%$ C, $\leq 0.003\%$ N, and $\leq 0.001\%$ O in large-scale production.

Important *elements added for deoxidation* are manganese and aluminum. A content of up to about 0.20% Mn does not noticeably affect the magnetic characteristics, although the hot formability is improved increasingly. In the range of about 0.03%, aluminum combines with the nitrogen in the steel, preventing its combination with iron and thus magnetic and mechanical aging. If no aluminum is added to the steel, iron nitride particles go into solution during annealing and are precipitated again in a fine, and therefore magnetically very harmful, form, at a temperature between ambient and 200 °C. This can cause the coercivity to increase to a multiple of its initial value. As an example, Fig. C 20.4 shows the increase in coercivity at different temperatures of a soft magnetic steel without aluminum, as a function of the aging temperature. In this illustration, the highest measured value after aging at 100 °C is higher by a factor of five than the initial value. Aging resistance is tested according to DIN 17 405 [9], for example, by measuring the coercivity after heating to 100 °C and holding this temperature for 100 h.

Additional Processing and Heat Treatment

In addition to the metallurgical measures taken to achieve and maintain magnetic properties, further processing and heat treatment (such as coarse grain annealing, refining annealing, and stress-relief annealing) are very important. When choosing the temperature, the cooling rate, and the annealing atmosphere, attention must be paid to the phase relationships in the material, the gas equilibriums, and the kinetic relationships that occur during precipitation processes and gas reactions. These

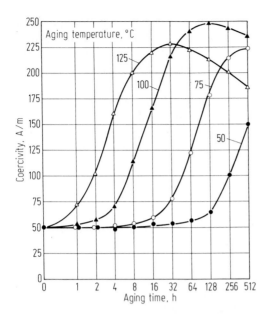

Fig. C 20.4. Changes in coercivity of an aluminum-free soft magnetic steel by aging.

factors all influence the achievement of a final microstructure with the largest possible grain, little precipitation, and low residual stress.

For *products of low thickness* (strip, rods, and wire), both melting and deoxidation can be used to obtain a ferrite microstructure without defects. In addition, these products are generally cold-formed by 8% to 12% to obtain a coarse-grained microstructure by recrystallization during the final annealing treatment, where values for coercivity of ≤ 50 A/m are reached. The final annealing treatment is a combination of refining annealing and coarse grain annealing that is carried out in hydrogen at 800 °C to 850 °C to remove carbon from the steel.

The example in Fig. C 20.5 shows the effect of annealing temperature and time on coercivity, where grain growth and refining on the one hand and α-γ-transformation on the other are opposing influences. The most favorable annealing

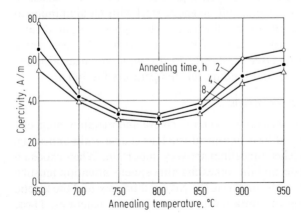

Fig. C 20.5. Influence of temperature and annealing time in hydrogen on the coercivity of a 2-mm-thick soft iron strip. Chemical composition of the strip: 0.025% C, 0.01% Si, 0.16% Mn, 0.008% P, 0.007% S, 0.025% Al, 0.02% Cr, and 0.02% Ni. Cold deformation of the strip: 10%.

temperature is 800 °C. To keep residual stresses – which have an unfavorable effect on the magnetic properties – low, slow cooling at least to 400 °C should take place at 20 °C/hr to 30 °C/hr.

Depending on the effectiveness of the annealing treatment, the microstructure exhibits a stalk-like (columnar) coarse grain growing from the edge, with a fine-grain remainder in the core, or – depending on the deformation, annealing temperature, and annealing time – continuous coarse grain over the whole cross section to a center line at which the grains from both sides of the strip have grown together.

For *large parts*, the only possibility is to apply as pure a melting process as possible to avoid unwanted elements and constituents, since refining annealing is not possible because of the long diffusion path. However, after hot forming, these parts also are annealed between 800 °C and 1000 °C to even out inhomogeneities across the cross section. These parts also must cool slowly after annealing – e.g., at no more than 20 °C/hr down to at least 300 °C – to avoid residual stresses.

C 20.3.4 Magnetic Properties of Important Soft Iron Grades

Relay Materials

Soft iron for relays (see DIN 17 405, for example) [19] is generally produced in the form of hot- or cold-rolled strip; forged, rolled, or drawn bars; precision forged parts; or castings, both annealed and unannealed. Further processing (bending, deep drawing, stamping) is generally done in the unannealed, fine-grain condition before recrystallization annealing, because processing difficulties and surface defects occur in the coarse-grain condition.

The seven soft irons in DIN 17 405 are graded according to the value of the coercivity from RFe 160 (with $H_c \leq 160 \text{ A/m}$) up to RFe 12 (with $H_c \leq 12 \text{ A/m}$). The lower the coercivity, the higher the values of the magnetic flux density (induction) at low field strengths. For all grades, the aging coefficient of the coercivity must be $\leq 10\%$. The magnetic values of DIN 17 405 apply by definition to the final annealed conditions.

The expenditure required for the fabrication of the individual grades varies. For instance, maximum permissible coercivity values of RFe 20 and RFe 12 can be reached only with a ferrite microstructure of the greatest purity after special annealing treatments in the coarse-grained and low-residual-stress condition.

If low core (power) loses are required for magnetic applications, it is necessary to change over to the silicon steels RSi 48 to RSi 12, which are also specified in DIN 17 405 [19], or to electrical steel sheet according to DIN 46 400 [20], for example.

Heavy Forgings

There are no special standards or material specifications for forgings of soft iron (e.g., for accelerating magnets), nor have there been many publications on this subject [21]. As a rule, the *properties to be attained* – namely, coercivity, magnetic flux density (induction), scattering of the flux density, resistance to aging, and mechanical properties – are agreed upon between the steel producer and the user. Of greatest importance is the homogeneity (measured by the scattering of the flux density) of all the parts of the yoke of a magnet, both in themselves and with each other, because asymmetries of the field would otherwise occur in the air gap. These field asymmetries can be corrected only with great difficulty [22].

As already mentioned, the purity of the material must be ensured by using selected burdens in the melt that have the lowest content of unwanted accompanying elements, if necessary with the aid of vacuum metallurgy [21]. A *degree of purity of the soft iron* of 99.5% to 99.9% is then obtainable. This purity is characterized by the following limiting contents of accompanying elements: 0.002% to 0.010% C, 0.002% to 0.010% Si, 0.03% to 0.20% Mn, $\leq$ 0.005% P, $\leq$ 0.005% S, 0.003% to 0.040% Al, 0.010% to 0.030% Cr, 0.010% to 0.030% Cu, $\leq$ 0.010% Mo, 0.003% to 0.006% N, 0.010% to 0.030% Ni, 0.001% to 0.003% O, and $\leq$ 0.010% Sn. For aluminum and manganese, the upper limit refers to a deliberate addition.

No refining annealing is possible. The inhomogeneities due to the casting of the ingot and the forging (irregular distribution of foreign substances, different grain sizes, etc.) are equalized by annealing at relatively high temperatures (between about 850 °C and 1000 °C) for longer periods. The required low-stress condition is obtained by cooling slowly at a rate of $\leq$ 20 °C/hr.

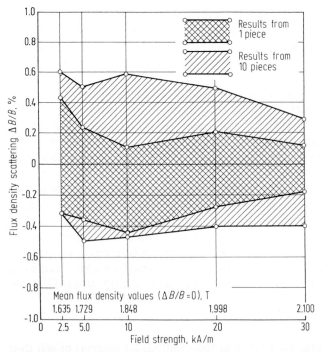

Fig. C 20.6. Flux density scattering in forgings from an 80 t ingot. Chemical composition of the examined individual ingot: top: 0.032% C, < 0.01% Si, 0.16% Mn, 0.008% P, 0.016% S, 0.023% Al, 0.006% N, and 0.001% O; bottom: 0.033% C, 0.025% Al, the rest the same as at the top. (Coercivity $\leq$ 60 A/m, yield stress 130 N/mm^2, tensile strength 260 N/mm^2).

The following figures are an example of the magnetic properties of forgings from 100 t melts of soft iron with a yield stress of ≥ 120 N/mm^2 [23].

Coercivity A/m	Magnetic flux density at a field strength of			
	2500 A/m	5000 A/m	10000 A/m	30000 A/m
	T			
45	1.667	1.756	1.871	2.106

As an example, Fig. C 20.6 shows the scattering of the flux density values at the poles of accelerator magnets from 80 t ingots that were measured on samples from the ingot taken over the cross section and over the length. With $\Delta B/B \leq \pm 0.5\%$, where $H = 2500$ A/m and $\Delta B/B \leq \pm 0.35\%$ at $H = 30000$ A/m, the scattering is very small; the specification $\Delta B \leq \pm 0.0175$ T, which in this special case was specified to ensure a symmetrical field distribution in the air gap, is exploited only by about one-half. It must be pointed out, however, that with field strengths of $H \leq 2500$ A/m, the range of the measure of variation ΔB is greater.

It can be seen from the two examples that, by taking special measures, it is possible to produce excellent soft magnetic properties even in heavy forgings made of soft iron.

C 20.4 Electrical Steel Sheet

C 20.4.1 Basic Requirements

Electrical machines and devices such as motors, generators, transformers, transducers, ballast transformers, and electromagnetic switches contain magnetic cores consisting of fine laminations. The thin sheets of silicon-alloyed or nonalloyed steel used for their production are summarized by the generic term *electrical steel sheet*.

With respect to industrial production, electrical steel sheet is the most important of the soft magnetic materials used in electrical engineering. Nonoriented electrical steel sheet is used for most production because it is to a large extent, magnetically isotropic. This sheet is applied to cores with variable direction of the magnetic flux (rotating machines). In addition, there is grain-oriented electrical steel sheet. It has a special crystallographic texture and particularly favorable soft magnetic properties, if it is magnetized so that the magnetic flux runs parallel to the direction of rolling. This type of steel sheet is applied mainly to the cores of transformers.

As a *core material*, electrical steel sheet must meet the following *requirements*. It must be easily magnetizable; i.e., it should reach the required polarization J or flux density (magnetic induction) B at a small field H to keep the currents in the windings low and to minimize the amount of material for the windings and the core. Furthermore, it should have a low core loss – i.e., only a small part of the electrical power should be transformed into heat – to achieve a high efficiency and to be able to carry off the heat easily.

For small machines, the most important *criterion for the choice* of the core material is the polarization. With increasing power and operation time, the loss of electrical energy – and thus the problem of carrying off the heat – becomes more and more important. In the case of big machines, therefore, the core (power) loss is the decisive criterion.

The requirements of easy polarizability and low core (power) loss mean that the *magnetization curves* $J(H)$ or $B(H)$, corresponding to the conditions of application, must rise steeply to a high level and the hysteresis loop must be as *narrow* as possible. The area of the hysteresis loop corresponds to the energy loss per cycle and unit volume. From a technical point of view, the energy loss is conveniently referred to the time unit and mass unit. This quantity is called the core (power) loss (see Table C 20.1 and Fig. B 2.8).

In addition to the requirements for the magnetic properties, the customer has further demands on electrical steel sheet for workability; e.g., with respect to punching. These aspects will not be discussed here (see [20]).

C 20.4.2 Considerations of Physics and Physical Metallurgy

Before it can be shown how these requirements can be achieved in practice, the parameters and processes that determine the properties of silicon-alloyed iron must be discussed.

Magnetization Processes and Eddy Currents

Because the addition of silicon does not basically change the lattice structure, the magnetization processes for electrical steel sheet are analogous to those for pure iron (see C 20.3.2). This applies to both the Bloch-wall displacement (shifting) and the domain rotation when approaching the saturation, which determine the steep part in the middle and the upper transitional part of the static hysteresis loop, respectively.

Electrical steel sheets, however, are normally exposed to alternating fields with technical frequencies. In this case, in addition to the *hysteresis loss* P_H from the static hysteresis loop, a *loss P_W due to the eddy currents* occurs, which broadens the static hysteresis loop (Fig. C 20.7). The *total core (power) loss* is the sum of P_H and P_W:

$$P = P_H + P_W = P_H + \eta P_{W,C}$$

$P_{W,C} = (\pi B f d)^2 / 6\rho_E \rho$ (where $d =$ the thickness of the sheet, $f =$ the frequency, $\rho_E =$ the resistivity, $\rho =$ the mass density, and $B =$ the flux density) is the "classical" core loss due to eddy currents [5, 6, 9]. In calculating $P_{W,C}$, the existence of the Bloch walls has not been taken into account. The actual loss due to eddy currents is usually larger, which is formally accounted for by the factor η (with $\eta \geq 1$), because the strong and fast change of induction due to the movement of a Bloch wall gives rise to larger eddy currents than does the smooth increase of induction assumed in

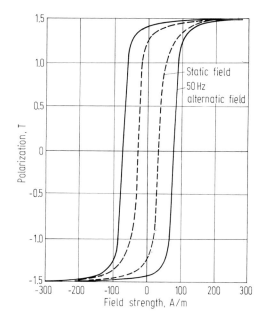

Fig. C 20.7. Dynamic and static hysteresis loop of a nonoriented electrical steel sheet made of steel with 3.2% Si (gauge 0.50 mm). The maximum value of the polarization J is 1.5 T.

the calculation. This additional loss is called *anomalous eddy-current loss* or *excess power loss*. The *loss ratio* η depends on the Bloch-wall displacements in a complicated way. It turns out that the larger the distance of the Bloch walls in comparison to the thickness of the sheet [24], the larger the loss ratio η. Large distances of the Bloch walls occur mainly in coarse-grained materials, such as grain-oriented electrical steel sheet.

Figure C 20.8 gives an impression of the orders of magnitude by which the *different parts contribute to the total loss*. The figue refers to a grain-oriented sheet.

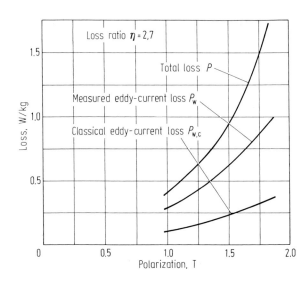

Fig. C 20.8. Constituents of the loss in a grain-oriented electrical steel sheet, grade 097-30-N5 (see DIN 46400 Teil 3 [20]). After [25].

The loss ratio is 2.7. The anomalous loss is also present in nonoriented electrical steel sheets. In this case, the loss ratio can reach an order of magnitude comparable to that in the grain-oriented sheets, although the size of the grains is relatively small. This is due to correlation effects between the Bloch walls, which increase their effective distance. These effects are not yet completely understood [26].

Chemical Composition

The main element for the alloying of electrical steel sheet is *silicon*. With iron, it forms substitutional solid solutions in the commonly used concentrations and gives rise to the following effects (among others):
- A narrowing of the gamma-region in the phase diagram such that, above 2% Si, transformation no longer occurs. Thus, transformation stresses after high-temperature annealing are avoided, and the chances for the formation of a texture are improved.
- An increase of the electrical resistivity (see Fig. B 2.17), leading to a decrease of eddy current loss.
- A decrease of the constant K_1 of the crystal anisotropy, which facilitates domain rotations [9].
- A lowering of the saturation polarization (compare Fig. C 20.1 and Fig. B 2.18).

Except for the last one, these effects are advantageous. A high degree of alloying is used if the requirement of a low core (power) loss is dominant (big machines); a low degree of alloying is used if a high polarization is the main requirement. For cores of small machines, nonalloyed electrical steel is also used.

Nowadays, silicon-alloyed nonoriented electrical steel sheet normally contains about 0.1% aluminum as an additional component of the alloy. The total aluminum content of the alloy is limited to about 4% for cold rolling.

The effects of *aluminum* with respect to the narrowing of the gamma region, etc., are similar to those of silicon. However, aluminum has a higher deoxidizing power and thus helps to improve the degree of purity (smaller number of disadvantageous oxide particles). Furthermore, aluminum is able to bind the nitrogen to create large particles of aluminum nitrides that are less disadvantageous to the magnetic properties.

Texture, Grain-oriented Electrical Steel Sheet

In cold-rolled electrical steel sheet with a relatively high silicon content, the crystallites can be orientated such that the direction of rolling coincides with a direction of easy magnetization (crystal axis $\langle 100 \rangle$). At the same time, the direction perpendicular to the direction of rolling within the plane of the sheet coincides with a face diagonal $\langle 110 \rangle$ of the cubic elementary cell (see Fig. C 20.9a). The $J(H)$ curves thus become similar to those of a single crystal in the respective directions (comp. Fig. B 2.10). This texture is called *Goss texture*, after its discoverer [27]. It is generated by a secondary recrystallization, during which the matrix of fine primary grains is absorbed by the Goss crystallites. To control the secondary grain growth, *inhibitors* are used, the most common of which is manganese sulfide (MnS) [28]. Grain-oriented electrical steel sheet with high permeability resulting

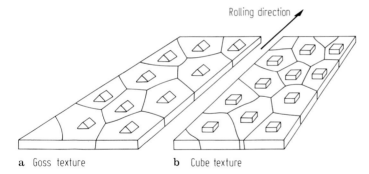

a Goss texture **b** Cube texture

Fig. C 20.9. Position of the elementary cubes of the crystal lattice: **a** for Goss texture and **b** for cube texture. After [8].

from recent further developments shows a particularly perfect Goss texture. This is achieved by an additional inhibitor (e.g., aluminum nitride [29, 30], antimony and manganese selenide [31], or boron nitride [32]). The new product shows an average deviation of the direction of rolling from that of the crystal axis of at most 3 deg, while in conventional electrical steel sheet with Goss texture, this deviation is up to 7 deg.

As a measure of the *degree of orientation*, the polarization in a magnetic field of 8 A/cm is used (J_8-value). The more perfect the texture, the closer the J_8-value comes to the saturation (Fig. C 20.10) [33]. The average J-value is around 1.8 T in conventional material and around 1.9 T in material with high permeability. The better orientation is accomplished at the expense of an increase of the average grain diameter from 2 mm to 5 mm in the conventional material and up to 10 mm to 15 mm in the highly permeable sheets [34]. Although highly permeable sheet shows a lower hysteresis loss in comparison to the conventional electrical steel sheet, the

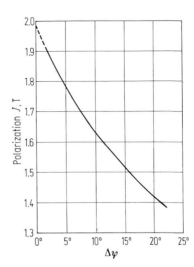

Fig. C 20.10. Dependence of the polarization J_8 on the average deviation $\Delta\psi$ of the cube orientation from the rolling direction for grain-oriented electrical steel sheet. After [33].

excess power loss is increased due to the enlarged Bloch-wall spacing caused by the larger grain diameter. A particularly low total loss is achieved by a special surface coating, which creates a stress that lowers the Bloch-wall distance (Fig. C 20.11 [35] and Fig. C 20.12 [36]). This stress also decreases the Bloch-wall distance in the conventional material, but this does not lead to a considerable decrease of the loss [37].

A *further improvement of the degree of orientation* in grain-oriented electrical steel sheet is reasonable only in connection with efforts to limit the anomalous eddy-current loss. Therefore, attempts have been made to increase the number of Bloch walls by applying local stresses to the material after the manufacturing process. In practice, this is done by mechanical scratching or laser scribing of the surface perpendicular to the rolling direction (Fig. C 20.12 [36]). The most effective refinemet of the domain structure is achieved by a combination of tension due to surface coating and local stresses due to surface scribing. In pilot plants, a material has recently been produced with this procedure and shows a core loss at 1.7 T of a full grade lower than the best grades in Table C 20.2.

The introduction of electrical steel sheet with Goss texture into the construction of transformers 40 years ago must be regarded as important progress in the reduction of power loss and core size. In stacked transformer cores, the limbs meet the yokes at a right angle. In this case, a texture with an orientation parallel as well as perpendicular to the rolling direction (Fig. C 20.9b) should be more favorable than the Goss texture. In the past, considerable efforts were undertaken to produce grain-oriented electrical steel sheet with a cubic texture. These efforts are reflected in a large number of patents. The practical results, however, have been disappointing. On the one hand, the fabrication of the sheets turned out to be very expensive; on the other hand, the material failed with respect to the two main criteria of application (see C 20.4.4); i.e., power loss and noise of the transformer cores.

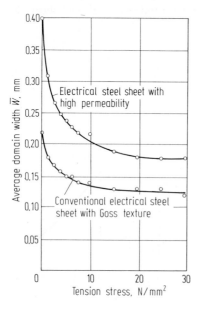

Fig. C 20.11. Average domain width $\bar{W}$ of 180-deg domains as a function of an external tension stress. After [35].

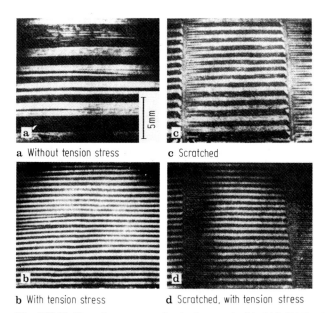

a Without tension stress
c Scratched
b With tension stress
d Scratched, with tension stress

Fig. C 20.12. Domain structure of a single crystal with (110) [001]-orientation. After [36].

Table C 20.2 Core (power) loss of the electrical steel sheets currently available on the market

Kind of material	Gauge mm	Number of grades	Maximum value of core loss at polarization W/kg		Standard
			1.5 T	1.7 T	
Nonoriented, unalloyed	0.50	3	6.60–10.50		DIN 46400
Semiprocessed[a]	0.65	3	8.00–12.00		Teil 2 [20]
Alloyed	0.50	4	3.40–5.60		DIN 46400
Semiprocessed[a]	0.65	4	3.90–6.30		Teil 4 [20]
Alloyed or unalloyed resp.	0.35	4	2.50–3.30		DIN 46400
Fully processed	0.50	11	2.70–8.00		Teil 1 [20]
	0.65	9	3.30–9.40		
Grain-oriented	0.27	1	0.89	1.40	DIN 46400
Conventional product[b]	0.30	1	0.97	1.50	Teil 3 [20]
	0.35	1	1.11	1.65	
With high permeability	0.30	2	1.10 and 1.20		

[a] The core loss values correspond to a reference state that results from specimen annealed according to the standard
[b] The core loss is fixed only for the polarization 1.5 T or 1.7 T, not for both values of polarization simultaneously

In the early 1970s, pilot manufacturing was stopped. The adjustment of the core construction to only one favored orientation has meanwhile ended interest in the cubic texture for sheets of power transformers.

The core sheets in rotating machines must be isotropic and soft magnetic at the same time. The second requirement means that, in the plane of the sheet (in which the magnetic flux is running), the most unfavored directions (the diagonals of the cube) must not occur; i.e., all crystallites must be oriented in such a way that the faces of the cube are oriented in the plane of the sheet. This *cube-on-face texture* is important for the further development of nonoriented electrical steel sheet. In the fabrication of electrical steel sheet from highly alloyed steel with optimal magnetic properties, suitable methods are sought to considerably increase the number of these grains with cube-on-face orientation.

Consequences for the Fabrication

For manufacturing of electrical steel sheets, the explanations given in C 20.3.3 on pure iron apply analogously. The following aspects should be mentioned in addition:
- The *silicon content* must be optimized to achieve the required magnetic properties and workability. A high content, although suitable for the coercivity as well as for the hysteresis and eddy-current loss, is disadvantageous with respect to the saturation polarization, cold rolling, and punching.
- For the *thickness of the final sheet*, a compromise also must be made. With decreasing thickness, the eddy-current loss is diminished, but the manufacturing expenses rise for the producer as well as for the customer (user). Increasing requirements on the magnetic properties imply a lower sheet gauge. The electrical sheets of highest grade within their class are those with the smallest gauge (Table C 20.2).
- The electrical sheet must be supplied with *a surface insulating coating*, so that the laminations of the magnetic core, which will be constructed of the sheet later, will be electrically insulated against each other to avoid additional eddy currents.
- In the production process of grain-oriented electrical steel sheets, particular attention must be directed to providing an undisturbed and complete *secondary recrystallization*. Nonrecrystallized regions must be avoided.
- For the production of nonoriented electrical steel sheets, the most important goal is to achieve a sufficiently *coarse-grained microstructure*. Therefore, particles in the steel, which hinder the growth of primary grains, are disadvantageous. On the other hand, the coarsening of the structure must not be pushed too far, because with increasing grain size, the anomalous eddy-current loss increases (see above). The superposition of hysteresis loss (H_c curve in Fig. C 20.13) and excess power loss results in an optimal grain diameter for the magnetic properties; this diameter must be adjusted.

C 20.4.3 Production Processes of Electrical Steel Sheets

The *production steps* for non-grain-oriented (NO) and grain-oriented (GO) electrical steel sheets are summarized in Fig. C 20.14. This figure shows that, to technically achieve the requirements on the grain-oriented material, a large number of processing steps are necessary. Of particular importance is the processing of the hot-rolled strip to the final product.

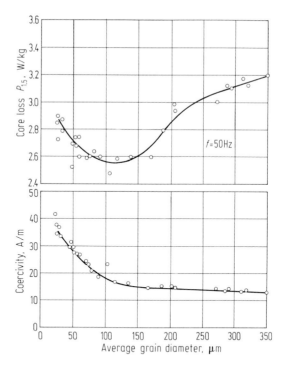

Fig. C 20.13. Dependence of the core (power) loss $P_{1.5}$ and the coercivity on the average grain diameter for nonoriented electrical steel sheet with about 3.2% Si.

One should, however, be aware that the soft magnetic properties of the final product are determined in large part by the hot band. This also applies to the production of high-grade NO electrical steel sheets without any restriction.

Steel-making and Hot Rolling

For the steel used for NO electrical sheets, as the requirements on the magnetic properties rise, an increasingly high *degree of purity* is necessary. For the highest grades, the elements oxygen, sulfur, and nitrogen must be eliminated from the steel, as much as possible, by metallurgical means, because they can later lead to unfavorable inclusions and precipitations that cannot be removed in the solid state. In the same way, the carbon content must be reduced as much as possible to limit the decarburization time in the subsequent cold processing. The reduction of carbon and oxygen today is performed in vacuum degassers and also in argon oxygen decarburization, (AOD) vessels to a lesser extent.

In the case of steel for GO electrical steel sheets, observing a *narrow specification for the chemical composition* is of most importance. This applies in particular to the inhibitor elements, the tolerable deviations of which are on the order of 0.001%. The oxygen content must be as small as possible, because oxide particles unfavorably influence the distribution of the inhibitor in the hot-rolled sheet. From the point of view of metal physics, the *hot rolling* must provide a favorable (i.e., finely dispersed) distribution of the inhibitors, which is achieved by dissolving the inhibitors when the slabs are heated up in the pushing furnace and reprecipitating them in the mill.

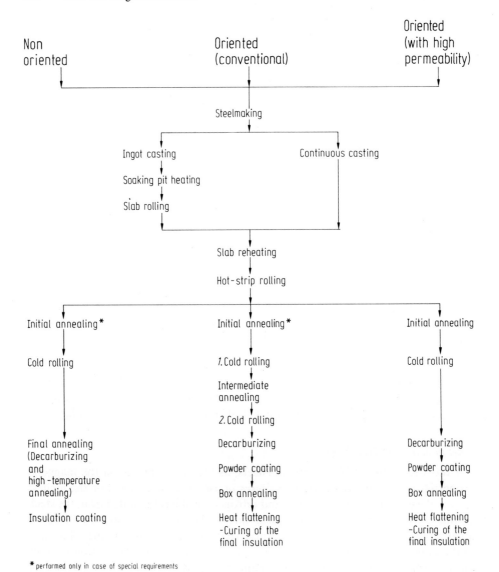

Fig. C 20.14. Production of fully processed electrical steel sheet. After [38].

Processing of the Hot-rolled Strip

The production of highly permeable GO electrical steel sheet requires a subsequent accurate *adjustment of the inhibitors*. This is done by annealing the hot-rolled strip in a special heating and quenching cycle. The annealing is followed by "heavy cold rolling" (directly down to the final gauge), which provides a sufficient number of crystallites in sharp Goss position later on (before secondary recrystallization). The second inhibitor is required to stop the primary grain growth, which is favored by the high degree of deformation caused by the heavy cold rolling, before the beginning of the secondary recrystallization. In the case of GO electrical sheet, the cold rolling is followed by *decarburization* to prevent magnetic aging. During this *annealing process*, the surface of the sheet is also oxidized to some extent. The thin oxide film improves the adhesion of the later insulation on the metal. In general, the decarburization annealing is connected with magnesia coating that prevents the windings from sticking together in the subsequent high-temperature annealing. During the heating cycle of the high-temperature annealing, the secondary recrystallization takes place, and a glass film (Forsterite layer) develops on the surface. To remove the inhibitor particles and residual oxides from the steel matrix, the material must be soaked at high temperature ($\approx 1200\,°C$) for many hours. As mentioned in C 20.4.2 (see also Fig. C 20.11), the surface coating must exert a permanent stress on the matrix. This is achieved in the last step of production. Here, a special phosphate coating is applied to the sheet surface and reacts with the Forsterite layer. The heat treatment is connected with a thermal flattening (leveling) which, by improving the flatness of the strip, can avoid compressive stresses in the later application of the sheet.

In the case of NO electrical steel sheet, the *final annealing* is the most important step of the processing. It mainly serves two purposes:

1. A further *decrease of the carbon content* [39] to prevent magnetic aging in the core material by carbide precipitations during the operation of the electrical machine. In the example of the steel sheet in Fig. C 20.15, the carbon content had to be reduced to about 0.0015%, for which an annealing time of 8.5 min was necessary. The residual carbon content of the final product that is tolerable with regard to magnetic aging depends on the degree of alloying (Fig. C 20.16): The smaller the silicon content, the more important the requirement for a complete decarburization.
2. *Adjustment of the grain size* in the optimal region according to Fig. C 20.13. In case of particularly high requirements, a hot initial annealing must be performed prior to the cold rolling. Thus, a larger amount of grains in the magnetically favorable cube-on-face position is achieved in the final product.

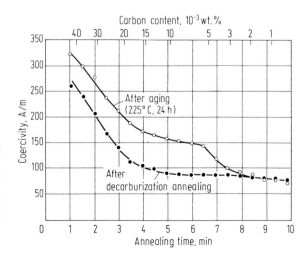

Fig. C 20.15. Influence of annealing time in the decarburization annealing on the coercivity and the magnetic aging of cold-rolled non-oriented electrical steel sheet with about 1% Si. After [25]. (Decarburization annealing at about 800 °C in a wet hydrogen-nitrogen mixture.)

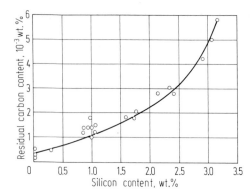

Fig. C 20.16. Tolerable final residual carbon content in nonoriented electrical steel sheet. After [25].

If the material is delivered in a not finally annealed state, the customer must anneal the punched laminations in special equipment. In this procedure, the annealing temperature has an upper limit because of the danger of shape deviations. To ensure that the customer can nevertheless achieve satisfactory soft magnetic properties, the producer must take measures to provide a sufficient grain size. As has been described in C 20.3.3 for strip and wire, these measures consist of an intermediate annealing and subsequent cold-rolling. An advantage of this procedure is that detrimental stresses, which occur in punching, are eliminated without further effort, and an insulating coating in the form of an oxide layer on the laminations is also gained without additional cost.

C 20.4.4 Electrical Steel Sheets Currently Available on the Market and Their Properties

The technical conditions for delivering electrical steel sheets are fixed in DIN 46 400 [20], for example, or the respective IEC standards [40]. For the measurement techniques, DIN 50 462 and 50 466, for example, apply [16, 41]. The various grades of electrical steel sheet are classified with respect to the core (power) loss and the sheet gauge. Table C 20.2 gives a survey over the ranges of values and the number of the standardized grades. For comparison, note that the measured data in the case of GO sheets refer to the favored (rolling) direction, while in the case of NO sheets, the samples used for measurement consist of an equal number of strips in the longitudinal and transverse directions.

The *NO electrical steel sheet* is available in two different types: finally annealed (fully processed) and not finally annealed (semiprocessed). This has already been mentioned in C 20.4.3. Within both types, there are both alloyed and unalloyed grades, although the alloyed grades predominate within the finally annealed type, and the unalloyed grades predominate within the not finally annealed type. The top grades are manufactured of iron-silicon alloys and are delivered in the fully processed state. Ten years ago, these materials were still produced as hot-rolled sheets. Their silicon content was about 4%. Only after several years of development has it become possible to produce NO electrical steel sheet of top quality by the now-common hot-rolling cold-rolling procedure, where the silicon content is limited to about 3%.

Figure C 20.17 shows the dependence of the core (power) loss on the polarization for two top grades of NO electrical steel sheets, for a conventional GO electrical steel sheet, and for a highly permeable grain-oriented sheet. The highly permeable GO sheet shows a particularly small loss above about 1.2 T. This sheet is particularly suitable for manufacturing transformer cores of high flux density. The loss reduction at high induction is not the only advantage of the more perfect orientation. It also leads to a decrease of the magnetostriction (in highly permeable materials, there are less 90-deg Bloch walls than there are in the conventional electrical steel sheets). Magnetostrictive oscillations of the transformer core are the main source of undesired transformer noise. Also with regard to the small magnetostriction (Fig. C 20.18), it is important to provide the surface of the grain-oriented electrical steel sheet with a coating that produces a high stress. Because there are unavoidable pressures connected with the core manufacturing, the insulation can reduce the magnetostriction considerably.

The *highly permeable electrical steel sheet* is clearly superior to the conventional GO electrical steel sheet in the main criteria for application: power loss, apparent power, and transformer noise. The only reason that highly permeable sheet has not yet replaced conventional steel sheet is its higher cost. Because economic considerations finally determine the material selected, both materials will continue to be used in the foreseeable future.

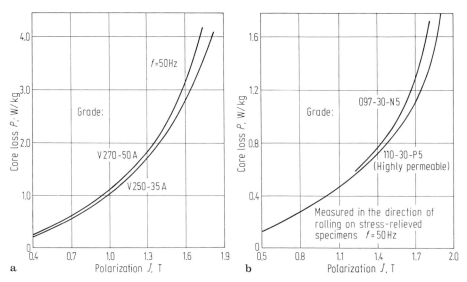

Fig. C 20.17. Core (power) loss: **a** of nonoriented electrical steel sheet, **b** of conventional and high-permeability grain-oriented electrical steel sheets. After [42]. In the grade notation (see also [20, 40]), the first number denotes 100 times the maximum value of the power loss $P_{1.5}$ (in W/kg) or $P_{1.7}$ (in the case of the highly permeable product), the second number is 100 times the gauge (in mm), and the third number is 1/10 the value for the frequency (in Hz).

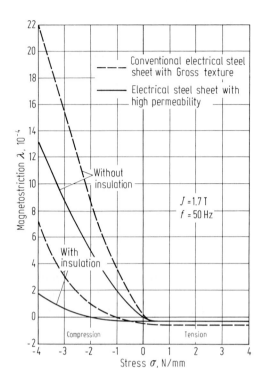

Fig. C 20.18. Magnetostriction of grain-oriented electrical steel sheet in an alternating field (maximum values). Samples with and without insulation under tension and compression stress. The measurements have been made parallel to the rolling direction at stress relieved Epstein samples [43].

C 20.4.5 Comments on Other Materials

Recent Developments

Amorphous Metals

Because of their different structures and new material properties, the amorphous metals (see also C 20.7.2) stimulate the interest of scientific institutes and others. In the United States, considerable efforts are undertaken to apply metallic glasses on the basis material of iron-boron-silicon (e.g., $Fe_{78}B_{13}Si_9$) in the production of distribution transformers with wound cores. After tempering in a magnetic field, the new material has the same coercivity as does highly permeable grain-oriented electrical steel sheet. Its core (power) loss at the same polarization, however, is considerably lower because of the strip thickness, which is smaller by one order of magnitude. A disadvantage of the amorphous metals in comparison to the silicon steels is their clearly smaller saturation polarization.

Silicon Steel with 6% Si

In the past, attempts have been made to lower the core (power) loss further by increasing the degree of alloying up to 6% Si [44]. In a procedure that was proposed a number of years ago, the cold-rollability is maintained by preventing the hot strip from forming superstructures that would lead to high brittleness. This is achieved by rapidly passing through the temperature interval from 650 °C to

420 °C in cooling down from the rolling temperature or by continuous deformation during the cooling process. On modern hot-strip mills, this procedure turned out to be impractical. However, with the development of the melt-spinning technique for the production of amorphous ribbon, steel with 6% Si again generated interest. By quenching the alloy from the molten liquid state, the crystallization is not excluded, but brittleness due to the formation of a superstructure is hindered. The produced strip, with a thickness of 100 μm, is ductile. Favorable soft magnetic properties can be obtained only by a high-temperature annealing with relatively long soaking time, where the cube-on-face texture is formed in the material. Besides the low saturation polarization, another drawback of this new development is that the high-temperature annealing probably must be performed at the final stage – i.e., by the customer – because the alloy can no longer be processed after the required slow cooling.

Silicon-alloyed Materials Besides Electrical Steel Sheets

Apart from electrical steel sheet, steels with silicon as the main alloying element are also supplied in various other forms; e.g., for relays and flux guides. These materials are used in applications where lower coercivity (compared to that of pure iron) is required or where a load with alternating magnetic fields or loads that act over a short period are applied. A lower saturation in these materials must be tolerated.

The explanations in C 20.4.1 to C 20.4.4 apply to the manufacturing and treatment of these products. Forms supplied include sheets of higher gauge and strips, wires, rods, and shaped products that are delivered in the semiprocessed or fully processed state. Table C 20.3 shows values for some magnetic properties after DIN 17 405 [19] for the silicon steels mentioned there.

C 20.5 Other Steels with Special Requirements for Their Magnetic Properties

C 20.5.1 Applications and Requirements for the Material Properties

In addition to soft iron and silicon steel, many other iron-based materials are used for their magnetic or other properties. Of the many applications, we will discuss, as examples, steels for generator shafts and pole sheets, steel castings for electrical apparatus, and stainless steels for relays. These steels, with coercivities of up to

Table C 20.3 Magnetic properties of silicon-alloyed relay material according to DIN 17405 [19]

Material (designation)	Silicon content (approx.) %	Coercivity A/m max.	Magnetic flux density for a field strength of		
			100 A/m	300 A/m	500 A/m
			T min.		
RSi 48	2.5–4	48	0.60	1.10	1.20
RSi 24		24	1.20	1.30	1.35
RSi 12		12	1.20	1.30	1.35

10 000 A/m, approach the upper limit of the materials considered to be soft magnetic. In practice, only the magnetic flux density (induction); the coercivity; and, in exceptional cases, the saturation are used to assess the magnetic properties needed for the specific application. The selection criteria vary with the respective application, as discussed below.

In the case of *generator shaft steels*, although the determining factors appear to be the mechanical properties demanded by the loads imposed by the design, the soft magnetic behavior of the materials is a precondition for their use. The requirements extend mostly to the highest flux densities possible at the lowest possible field strengths.

Soft magnetic steel castings are preferred in electrical engineering for complicated formed parts. Because of the absence of deformation and recrystallization, the magnetic requirements, in general, cannot be as high as for hot-formed steels, although the difference for large components – e.g., for lage magnets – is small. In addition to minimum values for the flux density at specified field strengths, certain mechanical strength characteristics are also required (see below).

Stainless chromium steels are used in a corrosive atmosphere when coated soft steel is not sufficiently resistant either mechanically or chemically. Compared with soft steels, however, certain compromises must be made in the magnetic properties of these stainless steels. For example, coercivity values up to 500 A/m are accepted; these values are considerably above those for soft iron. In addition, good machinability is required. This is in contrast to the requirement for lower coercivity, which requires lower mechanical strength characteristics that in turn impair the machinability. It is therefore necessary to make a compromise.

For the sake of completeness, we mention the *alloys of iron with aluminum* (up to 16%) and *with silicon* (up to 10%). Because of the permeability and coercivity that can be reached, these alloys are used in applications similar to those of the iron-nickel alloys and are therefore dealt with in C 20.7.

C 20.5.2 Relationships Between the Required Properties and the Microstructure

The basic processes for magnetizing these steels and the main influences on their magnetic properties are the same as those for the materials covered in C 20.3.2 and C 20.4.2. However, a much greater variety of microstructures must be achieved in the steels discussed here, because additional properties such as strength and corrosion resistance are required. Therefore, the chemical compositions, production processes, and microstructures, with their associated magnetic, mechanical, and chemical properties, are so interwoven that they must be discussed separately for each application.

C 20.5.3 Characteristic Examples of Other Steels with Special Magnetic Properties

Steels for Generator Shafts

The term *generator shaft steels* covers materials that are used for shafts in electrical machines because of their magnetic properties. These include steels for shafts with integral poles and with field windings in the slots of the rotor body (shafts in one piece or shafts for turbo-generators assembled from several forged prats, full and half speed). These steels are also used for multipolar shafts in salient-pole machines for hydraulic power stations and for the steel sheet of laminated poles.

Depending on the strength and toughness values required for the operating loads, unalloyed or alloyed steels that comply, for example, with SEW 550 and SEW 555 [45] are used for generator shafts. Other steels are also used, mainly outside Europe, but these will not be dealt with here. For pole laminations, steels according to DIN 17 100 [46] or SEW 092 [47] are possible. Table C 20.4 shows the most important steels to be considered, the maximum diameter that can be used (for two-pole machines), and the minimum values for the 0.2%-proof stress. There is no sharp dividing line between the different grades of steel, the diameters, and the 0.2%-proof stress. Table C 20.5 contains nominal values for the magnetic flux density (induction) of the most common steels.

The *plain carbon steels* containing about 0.35% and 0.45% C (Ck 35 and Ck 45), with their relatively low strength values (300 N/mm² or 350 N/mm² for the 0.2%-proof stress), are used only for low-stressed shafts with small diameters. Because of the high carbon content, the magnetic properties of these steels are only moderately good. It must also be borne in mind that, due to the low capability of these steels for through-hardening, differences in the magnetic properties between the surface and the core of the body are unavoidable.

The steels most often used for generator shafts are *alloy steels for quenching and tempering*, in particular the NiCrMoV steels listed in the Tables C 20.4 and C 20.5 that have varying quantities of nickel and chromium [48, 49, 50].

Table C 20.4 Survey of the most important steels that can be considered in Europe for generator shafts

Steel grade designation	Chemical composition					Maximum body dia. mm	0.2%-proof stress N/mm² min.
	% C	% Cr	% Mo	% Ni	% V		
Ck 35	0.35					500	300
Ck 45	0.45					500	350
28 NiCrMo 5 5	0.28	1.15	0.40	1.15	(0.10)	750	500
28 NiCrMoV 8 5	0.28	1.30	0.40	1.95	(0.10)	1000	600
26 NiCrMoV 11 5	0.26	1.50	0.40	2.75	0.10	1250	700
26 NiCrMoV 14 5	0.26	1.50	0.40	3.60	0.10	1250	750

Table C 20.5 Magnetic flux density of the generator shaft steels according to Table C 20.4

Steel grade designation	Magnetic flux density at a field strength (A/m) of						
	2500	5000	10 000	20 000	30 000	50 000	100 000
				T			
Ck 35	1.33	1.54	1.705	1.85	1.925		
Ck 45	1.23	1.49	1.65	1.80	1.89	1.97	2.07
28 NiCrMo 5 5	1.43	1.63	1.77	1.89	1.95		
28 NiCrMoV 8 5	1.47	1.66	1.80	1.91	1.97	2.03	2.11
26 NiCrMoV 11 5	1.53	1.70	1.83	1.94	1.99	2.05	2.13
26 NiCrMoV 14 5	1.57	1.74	1.86	1.965	2.015	2.07	2.145

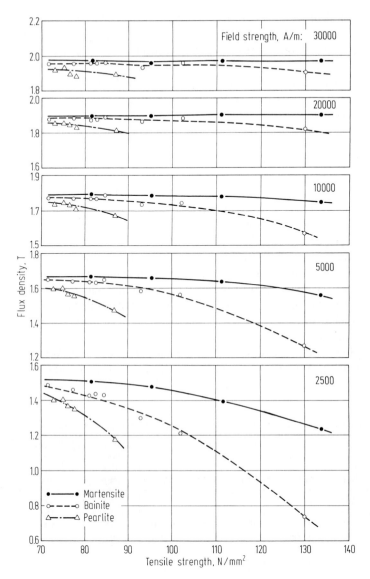

Fig. C 20.19. Magnetic flux density of the steel 28 NiCrMoV 8 5 in dependence on the tensile strength for field strengths from 2500 A/m to 30 000 A/m and for different microstructures. After [51].

The dependence of the magnetic properties on the microstructure [51] can be seen by example, in the steel 28 NiCrMoV 8 5 (see Table C 20.4). This dependence applies similarly to all the listed steels for quenching and tempering. Figure C 20.19 shows the dependence of the magnetic flux density on the tensile strength, with the kind of microstructure as the parameter for different field strengths.

The following conclusions can be drawn from the illustration:
- The magnetization of the microstructure developed from martensite by tempering is better than that of the microstructure of tempered bainite, which in turn is better than that of the annealed pearlite-ferrite microstructure.
- The magnetic flux density is reduced with increasing tensile strength for all conditions of the microstructures that were considered. This relationship is more pronounced for tempered pearlite-ferrite microstructures than for those of tempered bainite and martensite.
- The dependence of magnetic flux on tensile strength and the microstructure is far more marked at low field strengths than at high ones.

This behavior is conditioned by the basic magnetic processes. At high field strengths, in the saturation range, the flux density is determined primarily by the chemical composition; while at low flux densities, it is determined by wall displacements that, because of the homogeneous structure, take place more easily in the martensitic microstructure than in the bainite microstructure and even more easily in the pearlite-ferrite microstructure. The general reduction of the flux density with increasing mechanical strength is related to structural changes and internal (residual) stresses in the lattices that occur with the lower tempering temperatures necessary for higher strengths.

The different chemical compositions of the steels listed in Table C 20.4 also affect the *magnetic flux density*. Thus, by changing from the steel 28 NiCrMoV 8 5 to 26 NiCrMoV 14 5, the flux density is improved due to the lower carbon content (see Table C 20.5).

Because of its chemical composition, the steel 26 NiCrMoV 14 5 can be better through hardened, quenched, and tempered. Compared with the other steels in Table C 20.5, this steel reaches higher values for flux density, because larger cross-sectional regions are transformed into the more favorable bainite and martensite microstructures. In addition, as a result of the higher nickel content of 26 NiCrMoV 14 5 compared with 28 NiCrMoV 8 5, the flux density is improved even more, as shown in Fig. C 20.20. This illustrates the effect of nickel on the magnetic flux density. The results of further investigations into the influence of

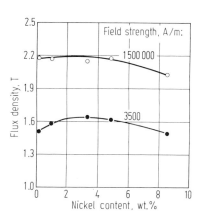

Fig. C 20.20. Effect of nickel on the magnetic flux density of steel with 0.10% to 0.16% C. After [52].

chemical composition on the magnetic properties of generator shaft steels have been published in [49, 52–55].

Pole sheets for generally slow-running shafts must have certain mechanical properties, depending on the loads to which they are subjected, as well as the best obtainable flux densities over the whole range of field strengths. In principle, the following two material groups can be used:
- Steels in accordance with DIN 17 100 [46], for example, such as St 37-3, St 52-3 or St 70-2, which have yield stresses of ≥ 235 N/mm^2, ≥ 355 N/mm^2, or ≥ 365 N/mm^2 in the normalized condition. The magnetic induction values attainable are not particularly high, due to the relatively high carbon content, but are sufficient in many cases.
- Steels according to SEW 092 [47], for example, which must be used in the thermomechanically treated condition. Depending on the steel grade, minimum yield stress values between 260 N/mm^2 and 550 N/mm^2 are attainable with very good magnetic properties. Due to the low carbon content of these steels (generally $\leq 0.10\%$ C), the same magnetic induction values can be obtained as for grades of soft iron with the same carbon content, but with considerably higher yield stress values.

Steel Castings for Magnetic Applications

For steel castings, the chemical composition, the heat treatment, and the resulting microstructure also play an important part in determining the magnetic properties. For magnetic applications, the choice is overwhelmingly unalloyed steel, so most available data concern the *effect of carbon* [56–58]. An increase in the carbon content impairs the magnetic flux density, as can be seen clearly in Fig. C 20.21 for about 60-mm-thick castings. The illustration also shows that normalizing with uncontrolled cooling of the casting in air does not alter the magnetic properties

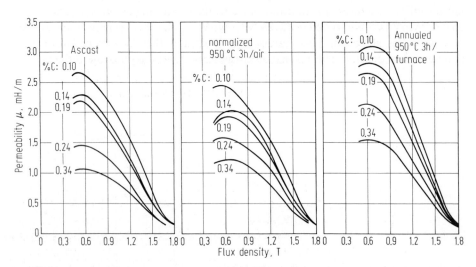

Fig. C 20.21. Permeability as a function of the flux density for unalloyed steel castings with a wall thickness of about 60 mm in the as-cast condition and after annealing. After [58].

significantly compared with the as-cast condition, whereas annealing followed by slow cooling in the furnace makes a marked improvement. This latter annealing causes a coarser formation of the pearlite with thicker ferrite and carbide lamellae, which has a positive influence on the magnetic properties.

C 20.3.2 referred to the large effect of the carbon content; it has the same importance for unalloyed steel castings that are used for electrical equipment and motors. Therefore, in addition to the usual four grades of unalloyed steel castings, for which the flux density is specified in DIN 1681 [59], castings made of pure iron have been developed [60] whose chemical composition differs only slightly from that of hot-formed soft iron (see above). To obtain the best magnetic properties, however, a suitable *heat treatment* is necessary [57, 58, 60]. It is then possible to obtain coercivities from 50 A/m to 100 A/m at maximum permeabilities up to 8.7 mH/m (6900 G/Oe) where the saturation magnetization is at about 2.15 T. If there is a requirement for better mechanical properties, steel castings with a carbon content above 0.15% must be chosen; two grades from DIN 1681 are listed as an example.

Stainless Chromium Steels for Magnetic Applications

The best magnetic properties of the steels containing about 13% Cr and 0.15% C are obtained by annealing above the transformation temperature and slow cooling to about 500 °C. From this temperature, cooling is optional. In addition to a reduction of internal stress, this annealing causes the austenite to decompose into ferrite and carbide, giving the most favorable magnetic properties. Figure C 20.22 shows the extent to which a steel containing 0.15% C and 13% Cr exhibits better coercivity, induction, and permeability properties in the annealed condition than in the hardened and tempered condition [61].

Under *corrosive conditions*, stainless ferritic steels containing 12% to 17% Cr could be used as soft magnetic materials. If the parts are highly stressed mechanically, such as armatures in solenoid valves, these steels are the only ones to consider, since the protective layer necessary for soft iron can easily be destroyed, allowing corrosion to start. For this application, steels with a low carbon content are chosen (see Table C 20.7), with additional sulfur added if machinability is important. Especially good magnetic properties are expected from the high-purity "super-ferritic steels" [62], but their chromium content must not be too high, or the induction is markedly reduced.

Table C 20.6 Magnetic and mechanical properties of steel castings for electric motors according to DIN 1691 [59]

Steel casting grade designation	Magnetic flux density at a field strength of			Tensile strength	Yield stress	Elongation after fracture A	Reduction of area
	2500 A/m	5000 A/m	10000 A/m	N/mm² min.	N/mm² min.	% min.	% min.
	T min.						
GS 38	1.45	1.60	1.75	380	190	25	35
GS 60	1.30	1.50	1.65	600	300	15	–

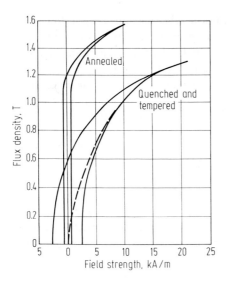

Fig. C 20.22. Magnetic properties of a stainless (free-cutting) steel with 0.15% C and 13% Cr in the annealed and in the quenched and tempered condition. After [61].

Table C 20.7 Magnetic properties of stainless chromium steels in dependence on the treatment. After [61]

Steel grade designation	Chemical composition		Condition	Coercivity	Maximum permeability
	% C	% Cr		A/m	mH/m
X 7 Cr 13	0.06	13	Annealed	400–700	0.5–0.8
			Drawn	600–800	0.4–0.6
			Quenched (hardened)	2500–3000	0.1–0.2
X 8 Cr 17	0.06	17	Annealed	400–600	0.5–0.9
			Drawn	600–800	0.3–0.5

For all stainless steels, a considerable impairment of the magnetic properties must be expected compared with those of soft iron. In addition, the processing condition (e.g., after cold forming) has a considerable effect. Table C 20.7 shows the magnetic properties for different treatment conditions [61].

In a "works report" [63], slightly better values are given for coercivity with a saturation magnetization of 1.6 T to 1.7 T and a remanence of 0.8 T to 1.0 T. By using a more expensive annealing method, even lower coercivities can be obtained. This type of treatment makes sense, however, only if there is no subsequent working. Where there are very severe requirements, it is essential to anneal the finished part in a protective atmosphere to obtain the best magnetic properties, as has always been usual for relay parts made from soft iron.

C 20.6 Soft Magnetic Nonferrous Alloys

C 20.6.1 Applications and Requirements

Applications in electrical engineering and electronics often require soft magnetic qualities that are superior to those of the iron-based materials. Such requirements include very low coercivity, very high initial permeability, extreme shape of the hysteresis loop, and low losses. These requirements can be fulfilled using alloys with high percentages of nickel or cobalt [3–6]. The total production of these materials is far smaller than that of soft magnetic iron and silicon-iron; nevertheless, alloys with high nickel or cobalt content are indispensable because of their unique properties.

Examples of *application* include highly sensitive relays, special transformers and chokes in power electronics, communication and data technology, magnetic cores for switching and storing, sensors and transducers for electric control systems, and magnetic shields for a low residual field strength.

The *requirements* depend on the specific application. Generally, they include static properties (e.g., high saturation and/or permeability, low coercivity, high or low remanence) and dynamic properties (e.g., low losses in the particular working frequencies or pulse conditions, short times of flux reversal). Additional requirements are high stability for long times and either a negligible dependence on temperature or a specific dependence. In some cases, a combination of magnetic softness and mechanical hardness is required. For details on applications and requirements, see [3, 5, 6].

C 20.6.2 Aspects of Physics and Physical Metallurgy

Phases and Transformations

The *nickel-iron* alloys crystallize in the face-centered cubic gamma-phase in the whole range from 100% down to about 30% Ni. Atomic ordering may occur, especially in the upper half of the range. Near the lower limit, the alloys may transform irreversibly to a body-centered cubic alpha-phase if cooled to low temperature; this changes the magnetic properties radically and should be avoided. Despite their single-phase character, the nickel-iron alloys are uniquely rich in specific magnetic properties because of a large variability of some physical constants in dependence on the nickel content and, above all, in the state of order of the alloys.

The *cobalt-iron* alloys of technical use (27% to 50% Co) exist in the body-centered cubic alpha-phase in the temperature range of the applications. At high temperatures, (about 805 °C to 900 °C), they transform to a nonmagnetic gamma-phase. In the alpha-phase, an ordering may occur that influences the magnetic properties.

For *general aspects*, see [9, 64]; for *special problems* of nickel-iron and cobalt-iron alloys, see [65].

Magnetic Constants and Ordering Processes

The basic magnetization processes resemble those in iron and silicon-iron (see C 20.3.2 and C 20.4.2). The magnetic properties are highly determined by a group of constants – the saturation polarization J_s and the Curie temperature T_C – and by two sets of constants expressing the magnetic anisotropy: the constants of the *magneto-crystalline anisotropy*, especially its first term K_1, and the constants of the *saturation magnetostriction* λ depending on crystal direction (orientation); the latter constants express the anisotropy by mechanical stress. In some metal systems, these constants pass zero at special compositions; in important cases, they depend on atomic order of the alloying elements in the common lattice and can be varied by heat treatment.

If the anisotropies by crystal structure and mechanical stress are small, a third anisotropy may take place: the *diffusion anisotropy* (expressed by a constant K_D), which is also called *uniaxial anisotropy* (expressed by a constant K_U). This anisotropy appears in binary or multicomponent systems after a heat treatment in a magnetic field, at high temperature but below the Curie point. This anisotropy is thought to be caused by a directional ordering of pairs (or groups) of atoms during the field tempering (e.g., Ni-Ni pairs in the iron-nickel system). The directional order can be preserved down to room temperature and causes the additional anisotropy. (Uniaxial anisotropy induced by other processes is not of interest here; compare [9].)

An extreme magnetic softness requires very low values (taken absolutely) of these three anisotropy constants. Depending on their height, the Bloch-wall movements are hindered more or less by inclusions, grain boundaries, regions of mechanical stress, etc. Important parameters of these obstacles are diameter and concentration, grain size, etc. (compare [66, 67]); the degree of hindrance typically passes a maximum at specific diameters (about at the Bloch-wall thickness).

All these constants are shown in Fig. C 20.23 for the iron-nickel system that is the basis of many commercial alloys. These alloys profit from the maximum magnetic saturation I_s (at 45% to 50% Ni); from the zero-passages of K_1 and λ, which depend on order (at 65% to 80% Ni); or from a high K_D (at 50% to 65% Ni). The commercial alloys of high nickel content mostly have ternary or quaternary components (especially copper, molybdenum, chromium) by which the zero-passages approach one another; furthermore, they improve the dynamic properties by a higher electric resistivity and by influencing the Bloch-wall structures (compare [9, 64–69]).

Specific Shapes of the Hysteresis Loop

The hysteresis loops of normal iron-based materials (N-loops) are smoothly rounded, more or less, with a remanence of about 50% to 85% of the saturation. Extreme values of the remanence can be reached in nickel or cobalt alloys, the values being considerably higher than 90% on the one side or less than 10% on the other. The corresponding hysteresis loops show a "rectangular" or a "flat" shape (Z- or F-loops, respectively).

A rectangular loop can be attained by the formation of a *cube texture* (Fig. C 20.9), with directions of the cube edges $\langle 100 \rangle$ in the rolling direction as well

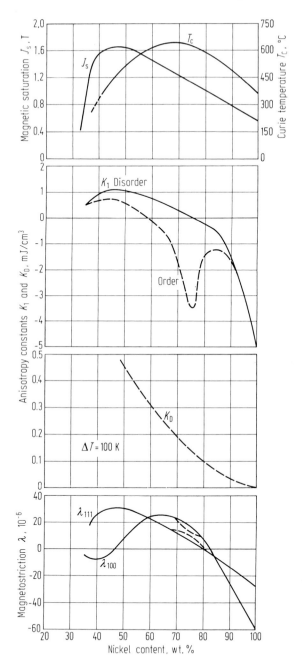

Fig. C 20.23. Magnetic constants in the iron-nickel system: J_s = magnetic saturation; T_C = Curie temperature; K_1 = constant of crystalline anisotropy (first term); K_D = constant of diffusion anisotropy after magnetic field-tempering at $T_C - 100$ K; λ_{100}, λ_{111} = saturation magnetostriction in the directions $\langle 100 \rangle$ and $\langle 111 \rangle$ (see also C 20.6.2).

as perpendicular to it in the plane of the sheet. If K_1 is positive (Fig. C 20.23), both are directions of a preferred magnetization; in these directions, the polarization approaches the saturation for low field strengths; compare the [100]-curve in Fig. B 2.10.

In another method, a *diffusion anisotropy* is induced by a heat treatment in a magnetic field. A rectangular loop results parallel to the direction of the field during the treatment, and a flat loop results perpendicular to it. In the first case, however, the anomalous eddy-current losses are often relatively high (compare C 20.4.2).

Pseudo-rectangular loops can also be attained without a strong anisotropy if the irreversible wall displacements all occur in a small range of field strengths. Up to its upper limit, the hysteresis loop appears rectangular, because a further increase of polarization requires much higher field strengths (rotational processes).

C 20.6.3 General Remarks on Processing and Products

The nonferrous alloys are basically processed in the same ways as are the iron and silicon-iron materials discussed in C 20.3 and C 20.4. Differences are caused by the smaller quantities and by the higher demands for purity and low levels of internal disturbances.

The alloys are produced, generally, in small electric melting furnaces (capacity of 0.1 to several metric tons). The higher quality alloys are often produced in vacuum induction furnaces; in special cases, powder metallurgy is used. The compositions are chosen according to the desired properties (see C 20.6.2).

The resulting ingots are rolled down to the final dimensions, including intermediary heat treatments. At the end of all mechanical processing, final heat treatments are required, often in several steps. In all these processes, the conditions should be chosen based on the required microstructure. Points of special attention are the grain size and purity, the parameters of residual inclusions, and the crystal textures and states of order. The latter two conditions can be desirable or not, depending on the particular case; often, the magnetic properties will react to small changes in these conditions.

The *steps* in these processes depend on the alloy and on the requirements for dimensions and properties. They cannot be treated here in detail; characteristic steps are mentioned in the individual sections.

The *main shapes* of supply are strip-wound cores and laminations, molded and solid parts, and shielding cans or tubes, most of which are delivered after the final heat treatment. Sometimes, further stages of manufacturing are included – e.g., winding and encapsulating (with or without other electronic elements) – to provide the user with components suitable for immediate operation [3]. Also available are semifinished products (rods, sheets, etc.) that require an appropriate heat treatment of the fabricated parts after mechanical processing. In exceptional cases, hard or semihard parts are applied (e.g., diaphragms for telephones).

C 20.6.4 Commercial Nonferrous Alloys for Magnetic Applications

The most important groups of alloys are shown in Table C 20.8, with some of their characteristic data. Within each group, the properties vary considerably, according

Table C 20.8 Survey of soft magnetic nonferrous alloys

Alloying components[a] wt-%	Main data			References to		Further data	
	Magnetic saturation T	Coercivity (static) mA/cm	Shape of hysteresis loop[b]	General features in section	Permeability flux density in Figure	Density g/cm³	Electric resistivity Ω·mm²/m
70 to 80 Ni, fc[c]	0.75–0.95	3–40	N, Z, F		C 20.24–C 20.26	8.7	0.55–0.6
50 to 65 Ni, (fc)[c]	1.25–1.50	10–100	N, Z, F		C 20.24–C 20.26	8.2–8.6	0.4–0.6
45 to 50 Ni[c]	1.50–1.60	50–200	N, Z, F	C 20.6.4	C 20.24 and C 20.25	8.2–8.6	0.4–0.5
35 to 40 Ni[c]	1.30–1.40	200–400	N		C 20.24	8.1–8.2	0.55–0.6
ca 30 Ni[c]	dependent on T				C 20.27	ca 8.15	ca 0.8
27 to 50 Co, (fc)[c]	2.3–2.4	200–2000	N, Z		C 20.29	ca 8.15	0.15–0.3
Fe + 16 Al	0.8–0.9	20–50	N, Z	C 20.7.1		ca 6.5	ca 1.4
Fe + 10 Si + 5 Al	ca 1.1	15–100	N, Z			ca 7	ca 0.6
Ni, fc	ca 0.8	5–100	N, Z, F				
Co, fc } metallic glasses	1.6 (to ca 1.8)	2–100	N, Z, F	C 20.7.2		ca 7.5	1.4–1.8
Fe, fc	ca 1.6	30–100	N, Z, F				

[a] fc: further components; ( ): partially
[b] N: normal (round), Z: rectangular, F: flat
[c] Residual components Fe and common additives

to the level of performance and the thickness and shape. Therefore, the table and the following figures give only a first survey; they all refer to materials heat treated appropriately, mostly about 0.05 mm to 0.5 mm thick. More detailed data are given in [3, 5, 6]. For minimum requirements see, the related standards [19, 70].

The following sections describe the most important facts about the typical groups of alloys. Some remarks are included concerning physical, physical metallurgical, or production specialities.

Iron-nickel Alloys with 70% to 80% Ni

These alloys of highest nickel content take advantage of the zero-passages of the magnetic constants K_1 and λ. *The main components* are 78% to 79% Ni and 4% to 5% Mo, or 75% to 77% Ni, 2% to 4% Mo and 3% to 5% Cu; molybdenum sometimes is replaced by chromium. The residual percentages are iron and common additives. The magnetic quality depends on an accurate choice of composition, purity, and manufacturing. In particular, care should be taken to avoid barriers of Bloch-wall movement and to introduce appropriate states of order. Figure C 20.24 shows characteristic curves (No. 1 and 2) for a high and a medium level of permeability.

In application, there are often *special requirements*; for example, for static or dynamic properties in given ranges of field strengths or flux densities or for a reduced temperature dependence of permeability [65, 66, 68, 69]. Another specific requirement is a high mechanical hardness, especially for magnetic heads. This requirement can be achieved by small additions (e.g., of niobium or titanium) that cause ultrafine segregations; they increase the hardness without a larger decrease of magnetic softness [71]. For other solutions, see C 20.7.1 and C 20.7.2.

All these alloys show a normal hysteresis loop (N-loop). *Rectangular or flat loops* (Z- or F-loops) can be produced by choosing specific compositions and treatments that ensure appropriate relationships among the relevant magnetic constants [69, 72]. Examples are given in Curve 1 of Figs. C 20.25 and C 20.26.

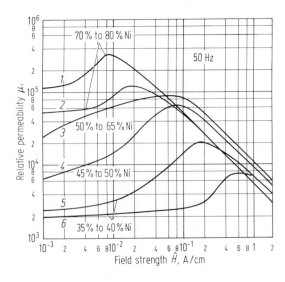

Fig. C 20.24. Permeability curves of iron-nickel alloys. After [3].

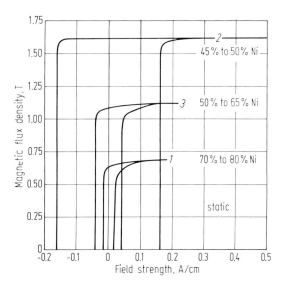

Fig. C 20.25. Rectanglar hysteresis loops of iron-nickel alloys. After [3].

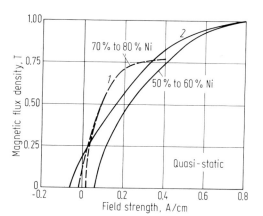

Fig. C 20.26. Flat hysteresis loops of iron-nickel alloys. After [3].

Iron-nickel Alloys with 45% to 65% Ni

The alloys of this medium range profit from the maximum saturation in the nickel-iron system. Table C 20.8 and Curve 4 in Fig. C 20.24 give data for a 50% Fe – 50% Ni alloy with a normal hysteresis loop. For a high saturation, all further components should be minimized; powder metallurgy may be applied to achieve this goal. Sometimes, however, a few percent of molybdenum or silicon are added to increase the electric resistivity, which is low with the binary alloys.

In the upper part of the nickel range, the permeability can be increased by a heat treatment in a magnetic field without losing the normal hysteresis loop [73], because of a specific relation of K_1 and K_D [74]. In permeability, these alloys approach that of 70% to 80% Ni; they are used frequently because of a higher saturation, despite their lower initial permeability. Curve 3 in Fig. C 20.24 gives an example.

Rectangular or flat hysteresis loops can be induced by the methods described in C 20.6.2. A nearly perfect cubic texture can be achieved by an appropriate recrystallization after very heavy cold rolling (see Curve 2 in Fig. C 20.25). A uniaxial anisotropy can be achieved by heat treatment in a magnetic field (see Curve 3 in Fig. C 20.25 and Curve 2 in Fig. C 20.26). Such materials are widely applied in nonlinear electronics and pulse technology.

Other special alloys combine good *glass-to-metal sealing* properties with magnetic softness, or with semihardness; they are applied in reed relays [3] (for details, see C 23).

Iron-nickel Alloys with 35% to 40% Ni

The alloys of this group are applied because of an economical relationship of low nickel content and magnetic softness. For properties, see Table C 20.8 and Curve 5 in Fig. C 20.24, which refer to an alloy with 36% Ni. The initial rise of the permeability curve can be made very small (see Curve 6 in Fig. C 20.24) for low-distortion transformer applications.

The electric resistivity of these alloys is rather high, and the thermal expansion is anomalously small (for the latter, see C 23).

Iron-nickel Alloys with Nickel Contents Near 30%

The Curie temperatures of these alloys reach down to about 100 °C and less (see Fig. C 20.23). Thus, the strong temperature dependence of polarization below the Curie point occurs just in the main regions of electrotechnical working. Usually, the alloys are applied to compensate for the temperature dependence of magnetic systems, chiefly of permanent magnets. The useful temperature range of each alloy depends on composition and, to a lesser extent, on treatment. Figure C 20.27 gives two examples. The curves of the *nickel-iron system* refer to alloys of about 30% Ni;

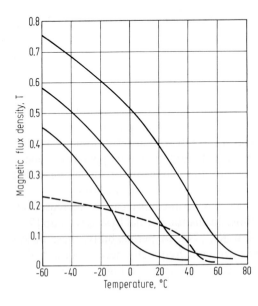

Fig. C 20.27. Temperature dependence of the flux density near the saturation in three iron-nickel alloys with about 30% Ni (solid) and in a nickel-copper alloy (dashed). Field strength 80 A/cm. After [5].

from one curve to the next, the nickel content varies by about 0.7%. The influences of small additions, especially of carbon, are very strong. For a possible transformation to the alpha-phase, compare C 20.6.2.

Similar effects can be obtained in other systems; e.g., in *nickel-copper alloys*. The dashed curve in Fig. C 20.27 refers to an alloy with 66% Ni, 30% Cu, and 2.2% Fe (compare [5]). These alloys are stable down to the lowest temperature; however, the working curve is relatively flat and, hence, the polarization is rather low.

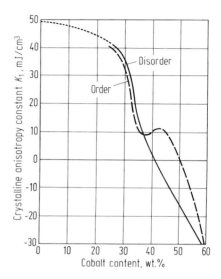

Fig. C20.28. Constant of crystalline anisotropy K_1 in iron-cobalt alloys. After [75].

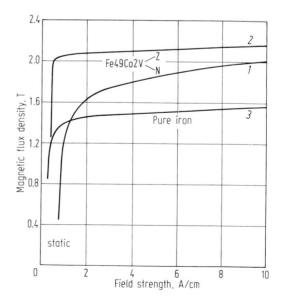

Fig. C20.29. Flux density curves of an iron-cobalt-vanadium alloy (1 and 2) compared to pure iron (3). After [3]. N and Z refer to materials with a normal (N) or with a rectangular (Z) hysteresis loop (the latter heat-treated in a magnetic field).

Iron-cobalt Alloys with 27% to 50% Co

The alloys of iron and cobalt offer the *highest flux densities* of all soft magnetic materials at normal temperatures. They surpass iron because of a higher saturation combined with a lower anisotropy constant K_1 (see Fig. C 20.28). The lower anisotropy constant causes an increase to high flux densities in rather low field strengths; see Fig. C 20.29 with the curve for pure iron for comparison [75] (saturation of the latter about 2.2 T!).

The alloys of practical use have a cobalt content of about 27%, 35%, or 50%. Further components are added for a higher specific electric resistance and a better ductility. A well-known alloy has about 49% Co and 2% V (besides iron); Fig. C 20.29 gives curves of flux density for an N-loop and a Z-loop material; see Curves 1 and 2, respectively. The Z-loop maerial is heat-treated in a magnetic field.

C 20.7 Remarks on Further Crystalline Special Alloys and on Metallic Glasses

Two groups of soft magnetic metals will be mentioned here briefly without going into details: an older one that has generated much interest but is applied only occasionally, and a new one with great future prospects.

C 20.7.1 Special Iron Alloys

An alloy of iron and *16% Al* approaches the iron-nickel alloys of medium nickel content in permeability and coercivity. Disadvantages are a lower saturation and great difficulties in plastic deformation because of a high hardness; the latter, however, favors an application for magnetic heads.

Alloys of iron and about *6.5% Si* show similar properties. They are recommended for higher frequencies because of a high electric resistivity (for other details, see C 20.4.5 and [76]).

Ternary alloys of iron, aluminum, and silicon come close in permeability and coercivity to the iron-nickel alloys with 70% to 80% Ni. Best known is the alloy with about 5.5% Al and 9.5% Si ("*Sendust*", compare [5]). Because the alloy is extremely hard, all processing is very difficult; small additives are said to be favorable. Recently, the alloy has been applied for magnetic heads.

C 20.7.2 Metallic Glasses

Unlike all the metals discussed before, the new metallic glasses are *amorphous*; their internal structures resemble those of the normal glasses. They are produced with a completely different method: an *ultrarapid cooling* of fine beams of liquid metal directed to a metal surface moving quickly. The metal glasses are formed directly in this strips, normally about 0.05 mm thick. Mostly, the composition is about $T_{80}M_{20}$ (up to $T_{70}M_{30}$); T refers to a transition metal (usually iron, cobalt, or nickel) and M to a metalloid (e.g., boron, carbon, silicon, or phosphorus). The elements of both groups may enter singly or combined. Examples are $Fe_{80}B_{20}$, $Fe_{40}Ni_{40}P_{14}B_6$, and $Co_{66}Fe_4(MoSiB)_{30}$.

The development and application of this new type of material are still in progress. For reviews, see [77, 78]. The following remarks give the present situation and some preliminary properties (see Table C 20.8).

In comparison to the usual crystalline alloys, the *saturation* is low; in nickel-iron alloys, it reaches up to about 1 T; in iron alloys, to about 1.6 T; and in iron-cobalt alloys, up to about 1.8 T. In some systems, *coercivity* and maximum permeability (and recently, the initial permeability) approach the properties of crystalline materials. The *hysteresis loops* show the same diversity (N-, Z-, or F-loops; see C 20.6.2).

The soft magnetic properties depend on the specific anisotropy constants that govern the magnetization processes (see C 20.6.2). The crystalline anisotropy is absent because of the amorphous structure. The saturation magnetostriction passes zero in some systems; thus, it can be brought to very small values (absolutely) by an appropriate choice of composition. A diffusion anisotropy can be induced by heat-treating in a magnetic field, just as in the crystalline alloys.

Due to their origin, the metallic glasses show good soft magnetic properties in the as-produced state; they can be further improved by a specific heat treatment with or without a magnetic field. However, the possibilities of heat-treating are restricted by the *instability* of the amorphous alloys, which crystallize at relatively low temperatures, often at only a few hundred degrees Celsius. This disturbs the magnetic quality completely. Even at temperatures below crystallization, internal processes may lower the magnetic quality over time; therefore, the composition and treatment should be chosen not only for good soft magnetic values, but also for good long-term stability at the highest temperatures that may occur during the application.

Besides their magnetic qualities, the metallic glasses show outstanding *mechanical properties* in the as-produced state. Because of a high hardness, they can be applied for magnetic heads; because of a high elastic limit and a high fatigue strength under bending stresses, they can fulfill the functions of both soft magnetic and spring materials. When heat-treated, however, they often are brittle.

As a whole, the soft magnetic metallic glasses appear very promising. The possibilities of application cannot be foreseen yet; the current discussions reach from big power transformers to minute elements in electronics [77–80]. Present applications are mostly in small electronic components.

C 21 Permanent Magnet Materials

By Helmut Stäblein and Hans-Egon Arntz

C 21.1 Required Properties

Permanent magnet materials, also referred to as hard magnetic materials, are required to maintain a magnetic flux in a working space without an external energy supply. In many cases, this flux must be as large as possible in relation to the volume of material required. Additional requirements may include high constancy, or at least reversibility, of the flux under changing ambient conditions (temperature, magnetic field, air-gap length etc.). Materials that can deliver constant magnetic flux over a prolonged period of time are used in a large number of statically or dynamically loaded magnetic systems or "magnetic circuits." Examples are given in the text, especially in C 21.4.

What physical properties make a material suitable for use as a permanent magnet?

With a given cross section A, a high magnetic flux Φ means that the flux density (induction) $B = \Phi/A$ in the magnetic material must be high. According to $B = \mu_0 H + J$, the flux density consists of the material-independent component $\mu_0 H$, emanating from the magnetic field strength H, and the material-dependent component of magnetic polarization J ($\mu_0 = 4\pi \times 10^{-7}$ H/m = magnetic constant). At a given field strength, therefore, the attainable flux density is given by the maximum polarization, called *saturation polarization* J_s. Permanent magnet materials *should always have a high saturation polarization*. In addition, the temperature dependence of saturation polarization should be low. This calls for a high Curie temperature T_C. Values of J_s and T_C for various materials are listed in Table C 21.1.

To maintain a magnetic flux – after magnetization up to the saturation point – without a continuous energy supply, the state of magnetization obtained should remain unchanged under fluctuating ambient conditions (see B 2.4.2). If changes do occur, they should be reversible – never irreversible. This means that the material must be "magnetically hard"; i.e., *its coercivity H_{cJ} must be high*. The state of magnetization can be changed either by the movement of Bloch walls or by rotational processes (i.e., rotating the polarization direction of a ferromagnetic domain). Both processes must be prevented, or at least impeded, and various mechanisms can be used to do this.

The *movement of Bloch walls* can be *inhibited by inhomogeneities* in the material; e.g., by precipitates. This mechanism is particularly effective if the precipitate

---

References for C 21 see page 796.

thickness is roughly equal to that of the Bloch wall. The magnetic hardness displayed by the high-carbon magnetic steels commonly used in the past results from this mechanism.

A second mechanism is based on *preventing the existence* of Bloch walls from the outset. This is achieved by dispersing small magnetic particles in a nonmagnetic matrix. The homogeneous polarization of such "single-domain particles" requires less energy than does a two-domain state with Bloch wall. In addition, measures must be taken to impede rotational processes, when the free energy depends on the direction of the atomic magnetic moments relative to the geometric or crystallographic preferred directions of the particles. In an elongated particle, the preferred direction is the longitudinal axis ("shape anisotropy"); in a hexagonally crystallizing substance with positive crystalline energy anisotropy, the preferred direction is the c-axis ("crystal anisotropy"). In both cases, rotating the atomic magnetic moments out of the preferred direction with the aid of a magnetic field requires energy and thus imparts magnetic hardness to the material. Shape anisotropy is effective, for example, in the AlNiCo, CrFeCo, and CuNiFe materials (see C 21.3.2.1, C 21.3.2.3, and C 21.3.2.4, respectively).

A third mechanism consists of *preventing the formation* of Bloch walls up to as *high a reverse field as possible*. Materials with high positive crystalline energy anisotropy, such as rare earth metal-cobalt and hard ferrite materials (see below), have a good chance of attaining a high degree of magnetic hardness in this way.

In practice, the mechanisms outlined above probably occur more or less together, rather than individually. Empirically, the effect of the occurrence of one or more of these mechanisms *is the attainment of high coercivity*. The coercivities H_{cJ} of various permanent magnet materials are given in Table C 21.1. These mechanisms must be effective over the entire temperature range required for the application; i.e., the materials must be relatively stable both structurally and chemically. Some permanent magnet materials do not have this stability: At room temperature, the microstructure is in a frozen, metastable state. At higher temperatures, the microstructure tends to change, adversely affecting the permanent magnet properties.

In summary, the requirement for high and constant magnetic flux can best be met by materials that have high saturation polarization, high Curie temperature, and high (shape or crystal) anisotropy energy.

Synoptic outlines of permanent magnet materials in general are given in [1–8], while the theoretical principles of magnetic hardness are dealt with in [2, 3, 9–12].

C 21.2 Permanent Magnet Characteristics

Of primary interest with permanent magnets is that part of the curve located in the second and/or fourth quadrant of the hysteresis loop, the *demagnetization curve* [2–8, 13]. This represents exactly the magnetic states existing in real magnetic systems, in which the magnetic field H is oriented antiparallel to the polarization J or flux density B. Possible notations are $B(H)$ or $J(H)$, which are basically

Table C 21.1 Magnetic and other characteristics of some permanent magnet grades[a]

Material[b]	B_r	$(BH)_{max}$	B_a	H_a	H_{cB}	H_{cJ}	$(B_pH)_{max}$
	mT	kJ/m³	mT	kA/m	kA/m	kA/m	kJ/m³
Cobalt steel[d]	950	7.2	600	12	20	21	10.4
ESD-magnet[e]	750	28	500	56	84	89	
AlNiCo 9/5 (isotropic)	550	9.0	350	25	44	47	15
AlNiCo 35/5 (anisotropic)	1120	35.0	900	39	47	48	44
AlNiCo 52/6 (anisotropic, columnar crystals)	1250	52.0	1050	50	55	56	62
AlNiCo 30/10 (anisotropic)	800	30.0	550	55	100	104	45
FeCoVCr 11|2	800	11.0	700	16	24	24	12.5
Cr-Fe-Co alloy (anisotropic)	1250	40.0	1050	38	48	50	
Cu-Ni-Fe (anisotropic)	500	10	350	28	40	41	
Hartferrit 7/21 (isotropic)	190	6.5	90	70	125	210	24
Hartferrit 25/14 (anisotropic)	380	25.0	177	141	130	135	53
Hartferrit 25/25 (anisotropic)	370	25.0	177	141	230	250	80
Hartferrit 3/18p (isotropic) (in plastic bond)	135	3.2	50	70	85	175	10
Hartferrit 9/19p (in plastic bond) (anisotropic)	220	9	75	110	145	190	25
SECo 112/100 (anisotropic)	750	112.0	375	290	520	1000	375
SECo 200 (commercial, typical values)	1030	200	500	400	750	>1200	740
Nd-Fe-B (commercial, typical values)	1220	280	600	465	860	940	1000
PtCo 60/40 (anisotropic)	600	60.0	290	210	350	> 520	190
Mn-Al alloys	600	50	360	140	230	270	100

[a] B_r remanent flux density, $(BH)_{max}$ static energy product, B_a flux density in $(BH)_{max}$ point, H_a field strength in the $(BH)_{max}$ point, H_{cB} coercivity of magnetic flux density, H_{cJ} coercivity of magnetic polarization, $(B_pH)_{max}$ dynamic energy product, μ_p recoil permeability, $(1/B)(dB/dT)$ temperature coefficient of flux density, $(1/H_{cJ})(dH_{cJ}/dT)$ temperature coefficient of coercivity H_{cJ}, T_C Curie temperature, J_s magnetic saturation polarization, R_m tensile strength, α coefficient of linear expansion, λ thermal conductivity, ρ resistivity, γ density

equivalent. For a state (B_1, J_1, H_1), Fig. C 21.1 shows how flux density B_1 breaks down into the "material" component J_1 and the "air" component $\mu_0 H_1$; i.e., $B = J + \mu_0 H$. After magnetization of the magnet, during which the magnetic saturation polarization J_s must be reached, at least approximately, a gradually increasing reverse field is applied and the magnetic states located on the demagnetization curve are traversed; in particular, the remanent flux density B_r at $H = 0$ and

C 21.2 Permanent Magnet Characteristics

μ_p	$\frac{1}{B}\frac{dB^c}{dT}$ %/K	$\frac{1}{H_{cJ}}\frac{dH_{cJ}^c}{dT}$ %/K	T_C °C	J_s mT	R_m N/mm²	α 10^6/K	λ W/m K	ρ nΩ·m	γ g/cm³
10			890	1500	1600	12		800	8.15
	−0.015		980	820					8.6
4.5	−0.03	−0.06	760	950	∼ 300ᶠ	13		630	6.9
4.2	−0.02	+0.03	890	1400	∼ 250ᶠ	12		470	7.3
2.3	−0.02	+0.03	890	1400				470	7.3
2.2	−0.02	−0.02	850		∼ 400ᶠ	10	∼ 22	500	7.2
4	∼ 0	∼ 0						600	8.2
3.5	−0.02								
4			410		800	12		180	8.6
1.2	−0.18	+0.40	450	460	∼ 30	9	∼ 5	> 10¹³	4.9
1.1	−0.18	+0.50	450	460	∼ 40	∥: 12	∼ 5	> 10¹³	5.0
1.1	−0.18	+0.27	450	460	∼ 40	⊥: 7	∼ 5	> 10¹³	4.8
1.1	−0.18	+0.40	450	460					3.3
1.1	−0.18	+0.40	450	460					3.8
1.1	−0.05	−0.3	724	1120	−	5.6	12	600	8.1
1.1	−0.035	−0.25	820	1.2	∼ 45	∥: 8 ⊥: 11	12	850	8.3
1.1	−0.13	−0.65	310	∼ 1.4	∼ 80	∥: 3.4 ⊥: −4.8		1440	7.4
1.1	−0.01	−0.35	500	720	12				15.5
1.1	−0.12	−0.4	300	710	290	17.8		800	5.1

ᵇ The materials identified by a combination of two numbers are dealt with in [1, 1a]. The values indicated in these examples for B_r, $(BH)_{max}$, H_{cB} and H_{cJ} are minimum values; this must be taken into account in comparison with other examples
ᶜ Valid in the 0 °C to 100 °C range at least
ᵈ Contains 1% C, 35% Co, 4.5% Cr, and 4.5% W
ᵉ Material containing 15% Fe, 10% Co, 66% Pb, and 9% Sb, anisotropic
ᶠ Bending strength; according to [88].

the two coercivities H_{cB} at $B = 0$ and H_{cJ} at $J = 0$. On the $B(H)$ curve, there is a pair of coordinates (B_a, H_a) with a (negative) maximum of the product of B and H, the $(BH)_{max}$ value. If, starting from a state (B_F, H_F) (i.e., point F), the reverse field strength decreases by ΔH, then the flux density increases by $\Delta B = \mu_p \mu_0 \Delta H$, with the relative recoil permeability μ_p depending only to a relatively small extent on the starting point (B_F, H_F) and ΔH. The flux density reached at $|\Delta H| = |H_F|$ (i.e., at

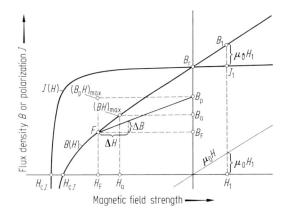

Fig. C 21.1. Demagnetization curves and characteristic points of permanent magnets.

$H = 0$), is termed permanence B_p. There is a pair of coordinates (B_F, H_F) for which the (negative) product of B_p and H_F has a maximum; this is referred to as $(B_p H)_{max}$. The $(BH)_{max}$ and $(B_p H)_{max}$ values are also referred to as the *static* and *dynamic energy product*, respectively, and *are two of the key characteristics* of a permanent magnet material; for measurement, see [14, 14a]. These are purely material characteristics.

By contrast, the *shape characteristics*, magnetic flux Φ and magnetomotive force θ, are obtained under homogeneous conditions from $\Phi = BA$ and $\theta = Hl$ and under inhomogeneous conditions from corresponding integral relations (A = cross section, l = length) [15].

All *magnetic material characteristics* are *temperature-dependent*. Some are reversible, some irreversible [16]. In addition to magnetic parameters, mechanical, thermal, and electrical characteristics must be taken into account (see Table C 21.1). For corrosion behavior, see [2].

C 21.3 Metallurgical Fundamentals and Production

C 21.3.1 High-Iron Materials

C 21.3.1.1 Steels

Permanent magnet steels contain approximately 1% C and up to 6% Cr, 6% W, and 40% Co. Shaping is done by casting or forging and hot rolling into strips, followed by cutting, stamping, drilling, etc. The heat treatment by which magnetic hardening is obtained consists, as a minimum, of quenching from 800 °C to 1000 °C to about room temperature, during which the face-centered cubic (fcc) austenite transforms into the rather unstable body-centered tetragonally distorted martensite (see A 6.4). Steels are hardly ever used nowadays because of their unfavorable characteristics and low structural stability. Table C 21.1 gives the characteristics of the highest grade 35% Co steel only. The magnets are isotropic.

C 21.3.1.2 Elongated Single Domain (ESD) Magnets

Elongated single domain magnets consist of submicroscopically small (up to several 10^{-5} mm thick), elongated, single-domain particles made of pure iron or iron containing 30% Co. The high coercivity of ESD materials is based on this shape anisotropy. These materials are manufactured by electrodeposition into mercury and heat treatment at about 200 °C. Succeeding steps include coating with a nonmagnetic material (tin, lead or antimony), removal of mercury, and pressing into final shape, where necessary, in a magnetic field to produce anisotropic magnets [17]. Magnetic quality is determined primarily by alloy composition and packing density. Because of the complex production process, the use of ESD magnets is limited in practice. Table C 21.1 gives the characteristics of an anisotropic high-remanence grade.

C 21.3.2 Medium-Iron Materials

C 21.3.2.1 Aluminum-Nickel and Aluminum-Nickel-Cobalt Alloys

Of main practical importance are alloys containing 27% to 60% Fe, 6% to 13% Al, 13% to 28% Ni, 2% to 6% Cu, up to 42% Co, up to 9% Ti, and up to 3% Nb. They are produced by melting and casting or by powder metallurgy.

In melting and casting, the starting materials are melted in open, medium-frequency furnaces and poured into sand or investment molds. Depending on the conditions used, globular or columnar crystals form upon solidification. Columnar crystallization gives optimum magnetic values [18]. In powder metallurgy, powders of iron, cobalt, nickel, and copper and prealloys of aluminum and titanium are thoroughly mixed and pressed into compacts at room temperature and a pressure of several kilobars. The compacts are then sintered in an oxygen- and moisture-free atmosphere at a temperature slightly below the fusion point, which lies between 1250 °C and 1400 °C, depending on composition. In the process, the linear dimensions of the compacts shrink by about 10% as density increases from about 5 g/cm^3 to 7 g/cm^3.

The *heat treatment* required depends on the formation kinetics of the possible phases and the temperature ranges where these phases exist. Table C 21.2 shows the phases of the AlNiCo 450 and AlNiCo 500 alloys as examples [19]. After homogenization in the region of the homogeneous α-phase, the formation of the detrimental γ_1-phase is largely prevented by sufficiently rapid cooling. The kinetics of this reaction are determined by the basic composition and the possible additions (e.g., silicon) and can be represented in a TTT curve [20]. The basis for magnetic hardening is the subsequent spinodal decomposition of $\alpha \rightarrow \alpha + \alpha'$ into a strongly magnetic phase and a weakly magnetic or nonmagnetic phase [18, 21–25]. In this process, elongated shape-anisotropic particles with high Fe or Fe-Co contents and transverse dimensions of some 10^{-5} mm are formed, along with a matrix rich in aluminum, nickel, copper, and titanium (see Fig. C 21.2).

In the high-grade AlNiCo alloys, the Curie temperature of the strongly magnetic phase is above the decomposition temperature. Here, *orientation can be*

Table C 21.2 Crystal structure of phases in AlNiCo 450 and AlNiCo 500 at temperatures between 1250 °C and 600 °C

Temperature °C	Existing phases		Crystal structure Type	Lattice constant[a] nm
		AlNiCo 450		
> 1250	α		α = bcc[a]	0.286
1250–845	$\alpha + \gamma_1$		α' = bcc	0.290
845–800	$\alpha + \alpha' + \gamma_1$		γ_1 = fcc	0.365[b]
< 800	$\alpha + \alpha' + \gamma_2$		γ_2 = fcc	0.359
		AlNiCo 500		
> 1200	α		α = bcc[a]	0.287
1200–850	$\alpha + \gamma_1$		α' = bcc	
< 850	$\alpha + \alpha'$		γ_1 = fcc	
600	$\alpha + \alpha' + \gamma_2$		γ_2 = fcc	0.356

[a] With superstructure
[b] At 800 °C

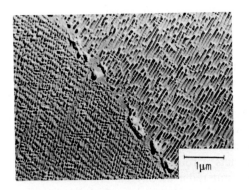

Fig. C 21.2. Electron micrograph of AlNiCo 450. A grain boundary can be seen across the center. The differing orientation of the two crystallites causes the difference in direction of the ($\alpha + \alpha'$) precipitates.

influenced by a magnetic field (of several hundred kA/m) applied *while the reaction is taking place*. The particles, with their long axes, then tend to precipitate in or near the $\langle 100 \rangle$ axes, which are adjacent to the magnetic field. An anisotropic magnet is formed with a macroscopic preferred direction that must be parallel to the later direction of magnetization. The desired orientation is best achieved with columnar microstructures because the columns extend parallel to the $\langle 100 \rangle$ axes. Spinodal decomposition is followed by single or multistage heat treatment between 650 °C and 550 °C, during which, through atom exchange, the difference in saturation polarization of α and α', and hence also coercivity, increase, and the form of the microstructure remains unchanged [26].

As a rule, AlNiCo magnets are hard and rather brittle in all stages of their manufacture. They *can be machined only by grinding*. Exceptions to the rule are hot-extruded magnets [27].

AlNiCo materials were *widely used commercially*. The main grades are standardized [1, 1a] or classified [28]. Table C 21.1 lists the characteristics of selected grades. Worthy of special note is the fact that high-cobalt AlNiCo alloys exhibit extremely good structural stability at temperatures up to at least 500 °C [29], which makes them particularly suitable for elevated-temperature applications.

C 21.3.2.2 Iron-Cobalt-Vanadium-Chromium Alloys

Common compositions lie in the range of 51% to 54% Co, 3% to 15% V, and 0% to 6% Cr [1, 1a, 28]. Coercivity increases with increasing vanadium and chromium contents. Shaping is done by casting, forging, hot and cold rolling, or drawing. Isotropic alloys are homogenized at approximately 1200 °C (in the γ condition), quenched, and tempered at about 600 °C (in the two-phase region). Anisotropic alloys require cold-working with a reduction in cross section of more than 90%, during which the face-centered cubic lattice gives way to a body-centered cubic (bcc) lattice, followed by tempering, during which the transformation is partially reversed. The macroscopic preferred direction after tempering lies in the direction of cold-working. Before tempering, the material is amenable to bending and stamping; afterward, it is very hard and can only be ground. The reason for the magnetic hardness may lie in the platelet shape of the particles of the $(\alpha + \gamma)$ two-phase microstructure [30–33]. Of scientific interest is the fact that the magnetic properties can be substantially increased by application of tensile stress [34, 35].

Iron-cobalt-vanadium-chromium alloys (often referred to as Vicalloy alloys) are used fairly widely, mainly for elongated miniature magnets and hysteresis applications [36], because of the scope they offer for cost-effective shaping. Table C 21.1 shows a grade with a relatively high coercivity. Particularly noteworthy is the high-temperature stability of the magnetic properties.

C 21.3.2.3 Chromium-Iron-Cobalt Alloys

Preferred alloy compositions are 10% to 23% Co, 23% to 33% Cr and, where appropriate, small additions of silicon, molybdenum, niobium, tantalum, aluminum, and titanium, among others [37–40]. After casting, forging, and, possibly, hot-working, heat treatment at 950 °C to 1300 °C gives a single-phase body-centered cubic α-phase. This phase is retained with rapid cooling. In this condition, the material is very amenable to cold-working. Shape-anisotropic single-domain particles are produced by subsequent multistage heat treatment. The process of spinodal precipitation (as for the AlNiCo magnets) is used. In this case (e.g., heating for 30 min to 600 °C to 650 °C), this process results in the formation of an iron-cobalt and a chromium-rich phase. This is followed by further heat treatments between 600 °C and 500 °C. These heat treatments, through atom exchange between the two phases, bring about a pronounced difference in the saturation polarization of the two phases and a corresponding increase in coercivity [41, 42].

Anisotropic materials are obtained in one of two ways: (1) spinodal precipitation in a magnetic field causes aligned shape-anisotropic, single-domain particles to form, or (2) spinodal precipitation produces spherical single-domain particles that are subsequently elongated and made shape-anisotropic by rolling and drawing [43]. The method chosen depends largely on the composition of the material.

Shaped parts can also be produced by powder metallurgy, with close dimensional tolerances being attained by sizing and re-pressing.

Table C 21.1 gives the characteristics of an alloy of this type.

C 21.3.2.4 Copper-Nickel-Iron Alloys

The composition of these materials lies in the region of 20% to 30% Fe, 50% to 60% Cu, about 20% Ni, and, where appropriate, up to a few % of Co. The alloys are homogenized at a temperature above 1050 °C; quenched; subjected to a high degree of cold-working (to produce anisotropy), where appropriate; and tempered at 550 °C to 600 °C. In the process, shape-anisotropic, FeNi-rich particles precipitate in a copper-rich matrix, causing magnetic hardening [44]. These materials are readily machined even after hardening.

Table C 21.1 shows characteristics of a heavily cold-worked anisotropic grade. Currently, only little use is made of this material.

C 21.3.2.5 Oxide Materials (Hard Ferrites)

Among the permanent magnet materials, a special position is occupied by the oxide materials, since they are sintered ceramic products rather than metallic materials. Today, these oxide materials, the hard ferrites, make up the predominant share of the market for permanent magnets (more than 90% on weight basis). They owe their magnetic properties to the phases $BaFe_{12}O_{19}$ or $SrFe_{12}O_{19}$, whose complex hexagonal structure [8] corresponds to that of the mineral magnetoplumbite [45]. The atomic magnetic moments of the ferric ions are coupled in part parallel and in part antiparallel to the crystallographic c-axis. This ferrimagnetic ordering is responsible for the relatively low saturation polarization of the hard ferrites (see Table C 21.1). The coupling is very strong, giving the high uniaxial crystal anisotropy on which magnetic hardness is based. The magnet materials have a polycrystalline microstructure with average crystallite sizes of about 2 µm [46]. If the c-axes of the crystallites are arranged randomly, the material is isotropic. The c-axes of crystallites in anisotropic hard ferrites, by contrast, tend to align in one direction [2].

In addition to these sintered ceramic materials, magnet materials in which *ferrite powder* is introduced isotropically or anisotropically into a *matrix of plastic or rubber* are also important.

Hard ferrites are *produced commercially only by powder metallurgy*, because the oxides decompose in air well below their melting point. The hard ferrite is formed by solid reaction from raw materials, usually barium or strontium carbonate and iron oxide [47–49]. Subsequent size reduction and fine-milling are used to ensure sintering activity and also, where appropriate, to obtain a single-domain particle size of < 1 µm and monocrystallinity of the particles [8, 46]. Shaping is done by pressing in a die; where appropriate, in a magnetic field to align the particles [50, 51]. Sintering in air at 1100 °C to 1300 °C gives compacts with 90% to 98% density. It is interesting that the degree of orientation of anisotropic magnets improves as a result of crystal growth [50–52]. Because of their hardness and brittleness, hard ferrites can be machined only by grinding, lapping, and honing [2, 53]. Magnet systems can assembled by bonding [54], fluidized bed coating [55], or plastic molding.

Table C 21.1 shows some typical grades. The 25/14 hard ferrite grade (preferably barium ferrite) is used mainly for loudspeaker systems, while the high-coercivity hard ferrite grade 25/25 (preferably strontium ferrite) is used primarily in motors. The relatively strong temperature dependence of both remanence and coercivity in hard ferrites must be taken into account. For irreversible losses due to cooling below room temperature, see [56].

A detailed description of hard ferrites, conventional and special production methods, and engineering properties can be found in [57].

C 21.3.2.6 Neodymium-Iron-Boron Materials

This class of permanent magnet materials, the most efficient and energy-rich so far encountered, became known in 1983. Their permanent magnet properties are based on a tetragonal $Nd_2Fe_{14}B$ phase, with the c-axis as the easy direction of magnetization. Based on the very high saturation polarization of this phase (J_s = about 1.6 T) and its relatively high crystal energy anisotropy, which corresponds to an anisotropy field strength of $H_A = 5.2$ MA/m (all values for room temperature), peak values of 360 kJ/m^3 for the static energy product $(BH)_{max}$ have been achieved with laboratory specimens. Comparison with the theoretical limit $J_s^2/4\mu_0$ of about 500 kJ/m^3 shows that further improvements can be expected.

This class of materials is also referred to as the third generation of the rare earth metal (RE)-3d transition metal magnets, after the $SmCo_5$- and Sm_2Co_{17}-base materials outlined in C 21.3.3.1 (the first and second generation). In addition to higher energy values, a further advantage of the new material class is its wider (i.e., more economic) raw material base: neodymium occurs many times more frequently in natural deposits than does samarium, and the replacement of cobalt by iron eliminates the need for this expensive and strategically vulnerable element. Another advantage is low brittleness and high resistance to fracture, which facilitate magnet handling. Disadvantages include the high negative temperature coefficients of remanence and, particularly, coercivity H_{cJ}, which currently limit applications to a maximum of 100 °C to 120 °C. Table C 21.1 shows characteristics of these materials.

Two manufacturing processes are used to produce these materials. The first closely resembles the powder metallurgy technique described for Re-Co materials [57a] (see C 21.3.3.1) and yields the magnetic properties mentioned above. The second approach [57b] uses the melt-spinning process developed for the production of amorphous or microcrystalline materials, in which a jet of molten metal strikes a rapidly rotating roll, where it solidifies very quickly in the form of thin strips or small flakes. Anisotropic magnets are formed by hot-forging a precompacted, flaky mass [57c]. [57d] gives a summary of the raw materials situation, technology, crystal chemistry, physical properties, and applications of the new class of materials.

C 21.3.3 Material Containing Little or No Iron

C 21.3.3.1 Rare Earth Metal-Cobalt Materials

Rare earth metals, notably samarium, form intermetallic compounds with cobalt [58, 59]. Compounds that have become known as permanent magnet materials are $RECo_5$ (RE = yttrium and the light rare earth metals lanthanum to samarium) and $RE_2(Co, M)_{17}$, in which M = copper, iron, zirconium, or similar metals. The hexagonal crystal structure of the $CaCu_5$ or Th_2Ni_{17} type causes a very high magnetocrystalline anisotropy. This is responsible for the very high coercivities attainable with these compounds. The parallel orientation of the atomic magnetic moments of the rare earth metal and cobalt atoms also results in relatively high remanence flux densities. The hexagonal c-axis is the easy direction of magnetization [59]. Permanent magnets with particularly good magnetic properties can be produced from aligned polycrystalline microstructures of these compounds [60–71].

To produce $RECo_5$ magnets, the metals are melted together or the mixture of oxides is reduced in the coreduction process [67]. The material thus obtained is crushed and fine-milled to an average grain size of about 5 μm [68]. Powders of this size can be readily aligned and allow high coercivity to be attained in the sintered part. Because of the high affinity of rare earth metals for oxygen, the RE-Co powders must be protected from oxidation [63, 64]; if a powder is sufficiently fine, it may ignite in air.

The powder is compacted, where applicable, in a strong magnetic field to crystallographically align the c-axes of the powder particles. The compact is sintered in a vacuum or under protective gas at about 1150 °C and then is subjected to controlled cooling. It is particularly important to avoid decomposition of the $SmCo_5$, which is stable above 800 °C, into the compounds Sm_2Co_7 and Sm_2Co_{17}, which are more stable at lower temperatures [63, 69–74]. The magnets produced in this way can be finished by wet grinding and cutting using diamond wheels. The material is brittle and hard. Its strong susceptibility to oxidation may necessitate protective coating [63], and the temperature of the application should not exceed 200 °C; otherwise, irreversible changes in the microstructure might occur. Table C 21.1 shows the magnetic and electrical properties of a standardized grade.

Permanent magnets based on the RE_2Co_{17} compounds are produced using a process similar to that used to produce magnets from $RECo_5$ materials. Compared with the latter, however, the RE_2Co_{17} materials offer a number of advantages: by virtue of their lower contents of rare earth metals and cobalt, their raw material base is more favorable; their saturation polarization J_s is higher, meaning that higher $(BH)_{max}$ values can be attained, currently up to 240 kJ/m^3 [75]; and, finally, they can be used at much higher service temperatures of 300 °C to 350 °C [71, 76]. Table C 21.1 gives typical figures for a high-quality commercial grade.

C 21.3.3.2 Platinum-Cobalt Materials

Above temperatures of about 825 °C, platinum-cobalt alloys near the equi-atomic composition PtCo exist as disordered face-centered cubic phase. With lower temperatures, the lattice becomes ordered to form the Cu-Au type (face-centered tetragonal) [77]. Materials quenched from above the ordering temperature display high remanence, which decreases with increasing tempering time below the ordering temperature, while coercivity increases substantially. The best magnetic properties are obtained in a partially ordered state, the optimum tempering temperatures being approximately 600 °C [2]. Coercivity is also influenced by the rate of cooling from the region of the high-temperature phase down to tempering temperature; a much higher coercivity can be achieved with slow cooling than with water quenching [78]. Table C 21.1 shows magnetic characteristics.

C 21.3.3.3 Manganese-Aluminum Materials

The magnets of Mn-Al alloys are based on the ferromagnetism of a metastable phase (τ phase) with an ordered tetragonal Cu-Au lattice. This phase is obtained under defined cooling conditions or by tempering from the ε phase that exists at high temperatures. The ε phase exists at 1000 °C in the 69% to 75% Mn composition range [79, 80]. The alloys display high remanence flux density but low coercivities [81, 82]. However, by introducing 0.5% to 1.5% C into the crystal lattice and by generating an easy direction of magnetization by means of mechanical working (rolling or extrusion) at temperatures from 600 °C to 700 °C, a pronounced anisotropy with a high $(BH)_{max}$ value is obtained [83–85]. Coercivity can also be improved by small additions of other metals; e.g., germanium [86] or titanium and zinc [87]. The material is ductile within limits and can be machined. The currently attainable magnetic, electrical, and mechanical properties of a high-quality magnet specimen are given in Table C 21.1.

C 21.4 Application Ranges

The main criteria for selecting materials for commercial permanent magnet applications are the cost of the magnet relative to the attainable magnetic energy, its resistance to demagnetizing influences, and the temperature dependence of its magnetic characteristics. *Barium* and *strontium ferrites* are the most cost-efficient permanent magnets because the raw materials are readily available and cheap and their manufacture is relatively straightforward. Therefore, these materials are used wherever there are no special requirements for the temperature dependence of magnetic properties. Barium and strontium ferrites also lend themselves to applications where components are not particularly small and service temperatures are neither high nor low. These materials are used preferentially in electro-acoustical transducers (e.g., loudspeakers). High coercivity, especially with strontium ferrite, and a straight demagnetization curve also make these materials suitable for dynamic applications such as small electrical machines (DC motors, synchronous motors, generators), couplings, and attachment systems.

If magnetic properties are required to change as little as possible with changing temperature, the more costly *aluminum-nickel-cobalt materials* are preferred. One example of their use is in electrical measuring instruments (electricity meters, moving-coil instruments). Another advantage of this group of materials is high flux density. For this reason, they are used in applications where, for example, a high air-gap flux density is required and no flux-concentrating pole-pieces can be used (e.g., core magnets in DC motors with iron-free rotors). In addition, AlNiCO materials can be used up to about 500 °C (permanent operating temperature).

Because of their high energy densities, straight demagnetization curve, very high coercivity, and high flux density, the *rare earth metal-cobalt materials* are widely used wherever demands for miniaturization and the highest magnetic performance must be met (e.g., clock magnets, miniature motors, aerospace, medicine), as well as in applications where the material is subjected to very high reverse fields (e.g., traveling-wave tubes, magnetic bearings, couplings, ion beam lenses). In the future, an increasing number of these applications will be filled by neodymium-iron-boron materials.

The *iron-chromium-cobalt-vanadium alloys* are particularly suitable for smaller magnets produced from strip or wire, because the material can be readily worked and displays relatively good and relatively temperature-insensitive magnetic properties. Major applications include moving-magnet measuring instruments, speedometers, hysteresis motors, compasses, and magnetogram carriers. In the future, the iron-chromium-cobalt-vanadium alloys used in these applications could be partially replaced by ductile iron-chromium-cobalt alloys, which offer lower raw materials costs and higher magnetic properties.

C 22 Nonmagnetizable Steels

By Wilhelm Wessling and Winfried Heimann

C 22.1 Required Properties

Steels are categorized as *nonmagnetizable* if their relative magnetic permeability μ_{rel} is totalized below 1.01 to 1.05 [1] in a field of 80 A/cm. As explained in B 2.3 on the physical properties of γ-iron solid solutions, this *requirement is fulfilled by the face-centered cubic (fcc) lattice structure of austenitic* chromium-nickel-(manganese) *steels*, provided a martensic transformation does not occur. It must be taken into account that tempering or annealing processes, cold-forming, and additional cooling to low temperatures may result in a γ-α transformation and, consequently, ferromagnetic behavior. In determining the appropriate amounts of the required alloying elements, it must also be considered that for steels with about 20 at.% Ni the conversion into the ferromagnetic basic state takes place in the range of absolute zero, and the Curie temperature grows rapidly with increasing nickel content.

Nonmagnetizable steels are required when disturbing interactions with the magnetic field lines of the environment must be avoided. These *applications* are found over a wide range of temperatures, from that of liquid helium (4.2 K) to above room temperature. Examples include clock housings, ship structures around the compass area, or in the construction of special ships; drill stems (boring bars) for magnetic directional control in deep drill holes of mining; water-cooled supporting (carrying) beams (arms) for electric cables or electrodes of electric furnaces; retaining rings (gaskets) for the ball ends of generator shafts to secure the exciting windings; cryotechnical devices of plasma physics in the field of particle acceleration and thermonuclear reactors, such as cryostats for superconducting magnetic field coils; and generators with superconductive exciting windings. In all these cases, the magnetic fields should not be impaired, and the eddy-current heating and hysteresis losses in the vicinity of high electric currents or magnetic shunts should be suppressed.

The fields of application reveal that, in addition to nonmagnetizability, high mechanical strength coupled with adequate ductility and corrosion resistance appropriate for the environmental conditions is required. Welded components also necessitate suitable filler metals and adequate weldability of the base materials.

References for C 27 see page 798.

C 22.2 Characterization of the Magnetic Properties by Test Results

The *determining characteristic property* of nonmagnetizable steels is *low magnetic permeability* or *low polarization* by high magnetic field strengths. The permeability can be tested by nondestructive magnetoinductive methods. The magnetic polarization is measured on specially taken test samples in yokes, either with the aid of vibration magnetometers or ballistically.

Naturally, a low magnetic polarization must be retained over a wide temperature range, depending on the application and independent of cold-forming. Under some circumstances, cold-forming can result in a microstructure that is unfavorable in this connection.

Figure C 22.1 gives, for four steels, the measured values of relative magnetic permeability, determined at 20 °C, in the fracture zone of tensile specimens after cooling them to low temperatures or after fracturing them at these temperatures. The results reflect the stabilizing influence of increasing alloy content on the austenite even at the lowest temperatures, despite a high degree of deformation. The slightly higher permeability values of X 5 CrNi 18 11 above room temperature are due to a small delta-ferrite content in the microstructure.

Figure C 22.2 presents the magnetic polarization J in dependence on the magnetic field strength for the steel X 3 CrNiMnMoN 19 16 5. These curves should be almost linear in nonmagnetizable steels; the arched run of the curves for 40 K, 30 K, and 4 K must result from a change in magnetic behavior at these low temperatures. The polarization of the steel X 5 NiCrTi 26 15 shows even more pronounced changes in dependence on the test temperature (see Fig. C 22.3). Antiferromagnetism which is known to occur in austenitic microstructures with higher nickel contents may be used to interpret these curves. According to both

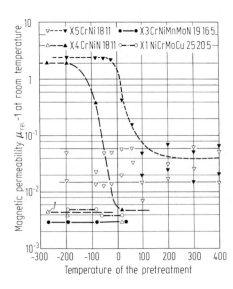

Fig. C 22.1. Magnetic permeability (at room temperature) of several nonmagnetizable steels after cooling to low temperatures (blank symbols), and after the tensile test at low temperatures (shaded symbols). (1 = specimen after 2000 stress cycles between 96 N/mm^2 and 676 N/mm^2 at 4.2 K tested at room temperature.) Chemical composition of X 1 NiCrMoCu 25 20 5: ≤ 0.02% C, about 25% Ni, 20% Cr, and 5% Mo; chemical composition of X 5 CrNi 18 11: ≤ 0.07% C, about 18% Cr, and 10% Ni; for X 3 CrNiMnMoN 19 16 5 and X 4 CrNiN 18 11, see Table C 22.1.

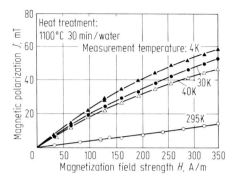

Fig. C 22.2. Magnetic polarization of the steel X 3 CrNiMnMoN 19 16 5 in dependence on the magnetic field strength at different temperatures (see Table C 22.1). After [2].

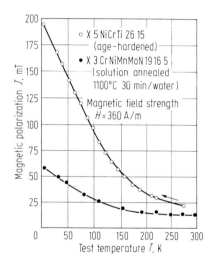

Fig. C 22.3. Magnetic polarization of the steels X 5 NiCrTi 26 15 ($\leq$ 0.08% C, about 15% Cr, 1.2% Mo, and 25% Ni) and X 3 CrNiMoN 19 16 5 (see Table C 22.1) in dependence on the temperature. After [2].

figures the magnetic polarization increases with decreasing temperature, although to much different extents, at room temperature, the magnetic polarization is about 50% higher for X 5 NiCrTi 26 15 than for X 3 CrNiMoN 19 16 5; however, at 4 K, it is about 200% higher. As Fig. C 22.2 shows, the polarization measurement values require details on the magnetic field strength for comparative reasons; in the experiments for Fig. C 23.3, the field strength was 360 A/m.

C 22.3 Conclusions on the Chemical Composition and the Microstructure

As discussed above, nonmagnetizable steels must retain a *stable austenitic microstructure* even at their lowest application temperature. This can be judged by the *martensite start temperature* M_s which indicates the beginning of the α'-martensite formation and is approximated as follows [3]:

$$M_s (°C) = 1305 - 1665(\%C + \%N) - 28 \cdot \%Si - 33 \cdot \%Mn - 41 \cdot \%Cr - 61 \cdot \%Ni$$

The above equation shows the *pronounced influence of carbon and nitrogen*, 0.1% of which has the same effect as about 2% to 2.5% Ni. The martensite point can be lowered by appropriate selection of the alloy constituents, so that, even at 4.2 K, the boiling point of helium, there is no further α'-martensite formation. Conversely, the onset of martensite formation can be shifted toward higher temperatures by cold deformation prior to or during cooling to low temperatures. Therefore, a steel with an unfavorable composition already becomes martensitic during cold deformation at room temperature, whereas other steels do not exhibit α'-martensite formation even during cold deformation in liquid helium right through to fracturing.

By increasing the content of austenite-stabilizing elements such as nickel, manganese, carbon, and nitrogen – and also of the ferrite-forming elements chromium, molybdenum, and silicon – the magnetizability is reduced, and the stability of the austenitic lattice under infuence of cold deformation and/or cooling to low temperatures is increased. The *existence range of nonmagnetizable steels* is shown in Fig. C 22.4 in a survey of the *iron-nickel-chromium system*. Alloys with a nickel content of over 30% are an exception; despite an fcc austenitic lattice, they are ferromagnetic at room temperature (similar to pure nickel). To establish the boundary between the homogeneous austenite field and the heterogeneous delta and gamma field in iron-chromium-multicomponent solid solutions, a knowledge of which is also necessary, a more detailed investigation on the phase boundary at 1050 °C was made, the results of which are presented in Fig. C 22.5. Using mathematical statistics as an aid, after having conducted investigations on steels

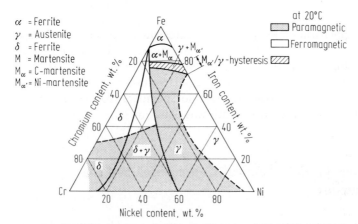

Fig. C 22.4. Magnetizability in dependence on the chemical composition in the iron-chromium-nickel system. After [4]. (Results on specimen with 0.03% C, 0.3% Si, and 1.2% Mn, after 150 hr at 1100 °C quenched in water.)

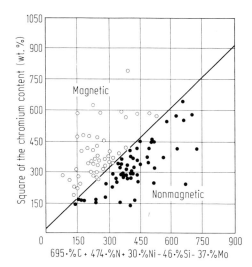

Fig. C 22.5. Distinction (division) of magnetic and nonmagnetic steels based on their chemical composition. After [5].

with 0.02% to 0.51% C, 0.08% to 2.74% Si, 0.5% to 24.8% Mn, 12.7% to 28.2% Cr, 0% to 3.5% Mo, 0.02% to 0.65% N, and 0% to 21.6% Ni, the following equation was obtained. It defines the admissible chromium content for a microstructure that is still ferrite-free:

$$(\% \, Cr_{admiss.})^2 = 695 \cdot \% \, C + 474 \cdot \% \, N + 30 \cdot \% \, Ni$$
$$- 46 \cdot \% \, Si - 37 \cdot \% \, Mo + 30$$

The percent data refer to wt. %.

It follows that the austenite-forming elements carbon, nitrogen, and nickel shift the field of the homogeneous austenitic solid solutions to higher chromium contents, and the ferrite-forming elements molybdenum and silicon shift the field to lower chromium contents. Manganese does not exhibit a recognizable influence here; apparently, it has no marked influence on the austenite formation at higher temperatures, but stabilizes the austenitic microstructure at lower temperatures [5, 6].

As a rule, austenitic chromium-nickel steels have relatively low *yield strength*. However, some applications require higher yield stress values. For such purposes, the mechanical properties can be improved in the solution-annealed state only by solid-solution hardening, essentially by additions of carbon and nitrogen, coupled, if necessary, with a limited addition of niobium [5, 7]. Along with a linear increase of the yield stress with increasing nitrogen content, 0.2%-proof stress values of over 500 N/mm^2 can be obtained. If the steels are to be resistant to intercrystalline corrosion in the welded state, the carbon content must be limited to a maximum of 0.03% [8, 9]. It follows that only unwelded pieces can use the strength-increasing effect of cabon.

Another possibility for *increasing the yield strength* is *cold deformation*. It is carried out on retaining rings (gaskets) by cold widening (Fig. C 22.6) or on the ends of drill stems (boring bars) by hot/cold strain hardening [11]. To prevent

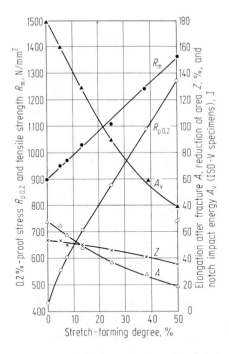

Fig. C 22.6. Values of the mechanical properties in dependence on the stretch-forming degree (cold work) for the steel X 55 MnCrN 18 5 (see Table C 22.1). After [10].

strain-hardening decay, an upper temperature limit of 300 °C may not be exceeded during processing or operation.

To meet requirements for *corrosion resistance* (e.g., in sea water), a "corrosion equivalent" ("Korrosionswirksumme") of % Cr + 3.3 · % Mo ≥ 26% [8, 9] is prescribed (also see B 3.5).

A precondition for the stability of the passive state of the steels alloyed with adequate chromium content is the metallic bright surface. However, it must be considered that even austenitic manganese (-nickel) steels with a chromium content below 12% can show restricted passive behavior [12, 13] in the metallic bright state and therefore must be protected during storage or application with a suitable surface coating (corrosion-resistant coat of paint, tinning, or silvering). If higher carbon content causes precipitation of chromium carbides at the grain boundaries which may also arise during age-hardening the steels must be handled appropriately for the degree of sensitivity to intercrystalline corrosion.

C 22.4 Characteristic Steel Grades Proven in Service

The *most economical method* of obtaining a nonmagnetizable steel is to add an adequate amount of *manganese* to a common unalloyed steel. With 0.5% C, at least 10% Mn must be added to reduce the martensite point below room temperature and obtain nonmagnetizability [15]. Additional requirements – such as a stable microstructure in the range of lowest temperatures and during cold forming, good machinability, resistance to rust, corrosion resistance in watery solutions, and high strength values – call for additional alloying measures. In the known fields of application, the individual requirements have been fulfilled by steels with different degrees of alloying additions (Tables C 22.1 and C 22.2).

Manganese steels such as X 45 Mn 18 are characterized by a tendency to extreme strain hardening as a result of their high content of interstitial carbon and as a result of a reduction in the stacking fault energy by manganese. Consequently, the deformability is limited and, under conditions similar to those found in drawing, is exhausted at about 40% to 50% cold deformation [4]. A positive aspect of the extreme strain hardening is the high wear resistance of manganese steels; on the other hand, this wear resistance impairs machinability. Small additions of chromium can improve the machinability and can also stabilize the microstructure, because the martensite point drops even further [15].

Using *chromium contents* of at least 12% can produce steels which are essentially rust resistant in air. If more stringent demands are made on the corrosion resistance, the chromium content must be increased further and the carbon content must be reduced to 0.05%. This alloying technique produced the steel X 5 CrMnN 18 18 K (Table C 22.1) which is used for generator retaining rings (gaskets) and is considered resistant to stress-corrosion cracking [13]. Its high strength of about 1200 N/mm^2 (Table C 22.2) is obtained by means of solid-solution hardening through 0.60% N and additional cold forming. The special demands on the toughness of retaining rings (gaskets) resulting from application on fast-running components call for a steel that is as homogeneous as possible. This requirement is fulfilled by electroslag remelting of the steel.

An increase in strength on larger rolling or forging cross sections via cold deformation is limited, due to the high force required, by the capabilities of the available equipment. Therefore, the steel X 5 MnCr 18 13 and the even more corrosion-resistant steel X 4 CrNiN 22 13 for drill stems (boring bars) and drill poles (boring rods) are "hot cold strain hardened" below the recrystallization temperature at temperatures of 700 °C to 1000 °C [14]. Along with good toughness properties, a tensile strength of about 900 N/mm^2 is obtained.

In the solution-annealed and quenched condition, the austenitic *chromium-nickel steels* without special additions are known to have only a low 0.2%-proof stress. Starting with the basic type of stainless steel, steels with appropriate properties were developed for special applications. With a view to application in the construction of special ships, the strength (Table C 22.2) and the resistance to corrosion in sea water of the steels X 3 CrNiMoN 18 14, X 3 CrNiMnMoN 19 16 5, and X 3 CrNiMoNbN 23 17 were brought to the required values by optimizing the analysis [16]. By using the combination of excellent toughness and nonmagnetizability over the entire temperature range down to 4 K, coupled with their structural stability, these steels are suitable for use in cryotechnology; e.g., for magnet systems and cryostats.

While the manganese-chromium steels with a carbon content of more than 0.03% are *suitable for welding* only with restriction, the austenitic chromium-nickel steels with less than 0.03 · % C can always be joined with fully austenitic steel and, in some cases, with filler metal of the same alloy type without loss in quality.

Table C 22.1 Chemical composition (mean values) of non-magnetizable steels and examples of application

Steel grade Designation	Chemical composition								Field of application
	% C	% Si	% Mn	% Cr	% Mo	% N	% Nb	% Ni	
X 35 Mn 18	0.35	0.4	18.0						Compass area in shipbuilding
X 55 MnCrN 18 5 K	0.52	0.7	18.0	4.5		0.10			Retaining rings (gaskets)
X 5 CrMnN 18 18 K	0.05	0.5	18.0	18.0		0.60		(1.0)	Retaining rings (gaskets), resistant to stress corrosion
X 15 CrNiMn 12 10	0.12	0.3	6.0	11.5				10.0	Anchor tie wire
X 5 MnCr 18 13	0.04	0.5	18.0	13.0	0.5	0.15		2.5	Drill stems (boring bars), drill poles (boring rods)
X 4 CrNiN 22 13	0.04	0.5	1.0	22.0		0.30		12.5	Drill stems (boring bars), drill poles (boring rods)
X 4 CrNiN 18 11	0.04	0.5	1.5	18.0		0.14		11.0	Cryomagnets
X 3 CrNiMoN 18 14	0.03	0.5	1.5	18.0	2.8	0.16		14.0	Cryostats, hydrogen bubble chambers
X 3 CrNiMnMoN 21 15 7 3	0.03	0.5	7.5	21.0	3.2	0.45	0.15	14.5	Special shipbuilding, drill stems (boring bars)
X 3 CrNiMnMoN 19 16 5	0.03	0.5	4.5	19.5	3.2	0.30	0.15	15.5	Special shipbuilding, cryo-equipment
X 3 CrNiMoNbN 23 17	0.03	0.5	5.5	23.0	3.2	0.40	0.20	16.5	Special shipbuilding, rope wires Cryo-generator rotors

Table C 22.2 Mechanical properties (reference data for minimum values) of the steels in Table C 22.1 at room temperature

Designation	Steel grade Condition	0.2%-proof limit N/mm	Tensile strength N/mm²	Mechanical properties Elongation after fracture A %	Reduction of area %	Notch-impact energy J	Test specimen
X 35 Mn 18	Quenched	245	700	30		70	DVM
X 55 MnCrN 18 5 [a]	Cold worked	1100	1250	25	40	55	ISO-V
X 5 CrMnN 18 18 [b]	Cold worked	1136	1197	24	63	100	ISO-V
X 15 CrNiMn 12 10 [c]	Cold drawn	1200	1400	(≥ 1.0)	–	–	–
X 5 MnCr 18 13 [d]	Hot/cold strengthened	760	910	40	70	120	ISO-V
X 4 CrNiN 23 13 [e]	Hot/cold strengthened	835	915	23	64	120	ISO-V
X 4 CrNiN 18 11	Quenched	270	580	40	50	85	ISO-V
X 3 CrNiMoN 18 14	Quenched	300	600	40	50	85	ISO-V
X 3 CrNiMnMoN 21 15 7 3	Quenched	500	850	35	45	70	DVM
X 3 CrNiMnMoN 19 16 5	Quenched	400	740	35	45	70	DVM
X 3 CrNiMoNbN 23 17	Quenched	500	850	35	45	70	DVM

[a] See Fig. C 22.6
[b] After [12, 13]
[c] Wire diameter 0.5 to 2.0 mm
[d] After [10]
[e] After [1]

C 23 Steels with Defined Thermal Expansion and Special Elastic Properties

By Hans Thomas and Herbert Haas

C 23.1 Properties and Tests

C 23.1.1 Determining the Coefficient of Expansion

The *characteristic property* of materials with defined thermal expansion is the *linear coefficient of thermal expansion (TEC)*. This property is isotropic in metals with cubic crystal lattice – i.e., is independent of the direction in the lattice – and, consequently, is independent of certain orientations or textures. With sufficient accuracy, it can be said that the thermal volume coefficient of expansion is three times the linear coefficient of expansion.

In the family of austenitic iron-nickel alloys, greatly varying coefficients of expansion are found, with values ranging between approximately 0 and $200 \cdot 10^{-7}/K$, depending on the composition. Measurement, control, and apparatus technology makes full use of this feature.

Table C 23.1 shows the *technically most important ranges of the expansion coefficient* and several examples of application.

In technical practice, the TEC is determined by measuring the increase in length of a sample relative to that of a quartz glass receptacle as a function of temperature; in the most straightforward case, in a dial gauge dilatometer between 0 °C and 100 °C. For wider temperature ranges, the TEC is measured by evaluating optically or electrically recorded expansion curves where the sample is either in an electric

Table C 23.1 Examples of application of materials with defined coefficients of thermal expansion

Range of the coefficient of thermal expansion $10^{-7}/K$	Examples of application
0–20	Measurement devices, geodetic measurement tapes, expansion controls, compensation limbs, thermobimetal components, cryotechniques
50–80	Expansion controls, thermobimetal components, sealings with hard glass, ceramic-to-metal sealings, core material of copper sheathed wire
80–110	Sealings with soft glass
180–210	Expansion controls, thermobimetal components

References for C 23 see page 798.

furnace or a cryostat. The results then must be corrected using the expansion of the quartz glass [1, 2].

As a rule, the usefulness of a material can be judged by the dilatometrically determined TEC. The usability of glass-to-metal sealing is checked by a more *sensitive* method; i.e., *by measuring* the elastic *stress in sample glass-to-metal seals* in polarized light [3–5]. Figure C 23.1 shows a comparison of the expansion curves of two different types of glass and corresponding metallic partners, as well as an example of a so-called polarimeter curve (the optical path difference of the two main beams as a function of temperature), from which the matching quality can be reliably determined.

Other properties important for particular applications, such as electrical resistance, tensile strength, yield stress and elongation after rupture, and thermal conductivity, are determined with conventional methods.

C 23.1.2 Determining the Modulus of Elasticity (Young's Modulus)

The group of alloys with exceptional thermal expansion includes steels with special elastic properties. Young's modulus is between approximately 130 kN/mm^2 and 200 kN/mm^2; with certain compositions, its temperature dependence almost vanishes. The fields of application of such constant-modulus alloys form two groups:
1. Static applications: primarily springs with temperature-independent force
2. Dynamic applications: mainly oscillators with temperature-independent inherent (natural) frequency

While the static determination of the elastic properties is relatively difficult, the dynamic measurement is an easy and reliable method (determination of natural

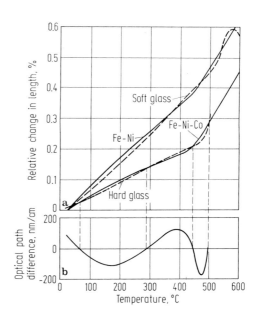

Fig. C 23.1. a Expansion curves of soft and hard glass and adapted iron-nickel and iron-nickel-cobalt alloys. **b** Polarimeter curve of the combination hard glass/iron-nickel-cobalt alloy

frequency, recording the resonance curve). Since constant-modulus alloys have a certain thermal expansion, it is important in practice that the temperature coefficient of the elastic modulus can be adjusted so that it just compensates for the changes in the spring force or natural frequency that are caused solely by thermal expansion.

C 23.2 Considerations of Physical Metallurgy

In the face-centered cubic (fcc) *austenitic* iron-nickel *alloys*, the spontaneous magnetization that arises during cooling from higher temperatures leads to an abnormally large, positive volume magnetostriction [6] (see B 2.3.2). To a greater or lesser extent, this compensates for the simultaneous thermal shrinkage, yielding a smaller TEC below the Curie temperature (Fig. C 23.2). Once spontaneous magnetization has been largely achieved, the TEC gradually increases again during further cooling.

Figure C 23.3 shows a section from the extensive systematic investigations carried out by Chevenard [7]. In the field of austenitic alloys, isotherms with fairly high values for less than 30% Ni pass through a pronounced minimum at approximately 36% Ni and subsequently approach the behavior of pure nickel. The TECs of the low-nickel-content ferritic alloys are close to the TEC of iron.

In the vicinity of room temperature, the TEC has a rather sharp *minimum at 35% to 36% Ni* ("Invar alloys") (see B 2.3.2). Impurities or additions of manganese, silicon, carbon, chromium, titanium, and others increase this minimum value for the TEC [8] (Fig. C 23.4) and shift the minimum to slightly different nickel

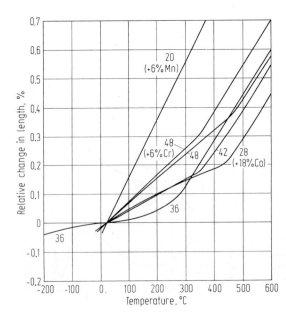

Fig. C 23.2. Expansion curves of different iron-nickel, iron-nickel-cobalt, iron-nickel-chromium, and iron-nickel-manganese alloys. The figures given in the graph represent the nickel content in wt.%.

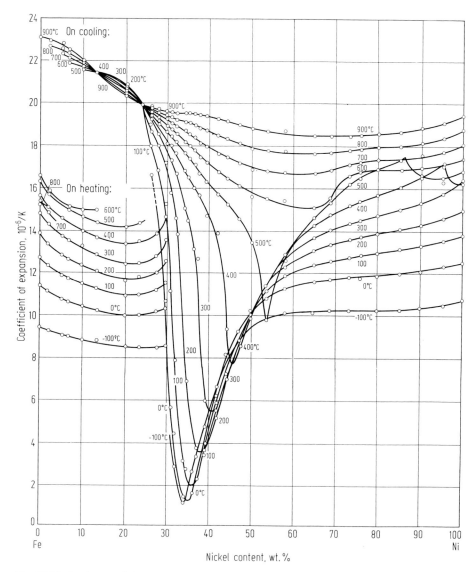

Fig. C 23.3. Isotherms of thermal expansion of pure iron-nickel alloys. After [7].

contents. To exclude, as far as possible, the influence of small carbon contents on the stability of the properties, a combined heat-treatment has been suggested [9].

Also in a pure iron-nickel alloy with 36% Ni, the *TEC is dependent on the pretreatment*. Rapid cooling from at least 800 °C or cold working decreases the TEC; combining the two treatments decreases the TEC even to close to zero. Moreover, unexpected time-dependent changes in the volume and TEC [10] should be mentioned here. Such instabilities can be eliminated in the soft or hard state by thermal aging between 300 °C and 500 °C; however, a slight increase in the

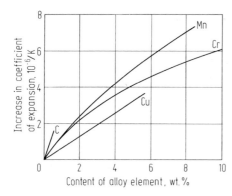

Fig. C 23.4. Influence of different elements on the coefficient of expansion of iron with 36% Ni. After [8].

TEC must be accepted. These phenomena have been traced back to lattice defects and rearrangements of atoms [11, 12].

The TEC can be brought completely to zero by replacing 4% to 6% Ni with the same amount of cobalt ("Superinvar") [13]. However, the TEC of this alloy is also substantially dependent on the pretreatment.

With *nickel contents of over 36%*, the TEC and Curie temperature rise. The expansion curve consists of a flatter part below the Curie temperature and a steeper part at higher temperatures. The flatter part can be extended into higher temperature ranges through increased additions of cobalt, which shift the Curie temperature upward (Fig. C 23.2) (see B 2.3.1). In this case, however, the stability limits of the fcc structure of iron-nickel-cobalt alloys must be observed [14], since exceeding them causes a martensitic microstructure with pronouncedly changed properties [15, 16]. An optimum is obtained with a cobalt content at which the Curie temperature attains its highest value for a defined TEC and where the martensitic temperature is below $-100\,°\text{C}$ [17].

In contrast to additions of cobalt, *chromium additions to iron-nickel alloys* lower the Curie temperature, yielding expansion properties that are particularly well adapted to those of certain types of soft glass.

The *iron-nickel system*, as such, exhibits a miscibility gap (a two-phase region), which strongly expands with decreasing temperature. However, this gap does not arise in alloys with more than 33% Ni, due to the extremely slow establishment of equilibrium conditions [18]. Below 33% Ni, instead of the two-phase equilibrium structure, we find a martensitic transformation (see B 2.3.1) with pronounced hysteresis ("irreversible alloys"), which is also particularly noticeable in the thermal expansion [19]. Extremely high TEC values are obtained with nickel contents between 14% and 25%, if the face-centered structure is stabilized by additions of manganese, molybdenum, or carbon [20] (also see B 2.3.2).

In the *iron-cobalt-chromium system* [21], there are alloys with very low thermal expansion that exhibit relatively good corrosion resistance ("Stainless Invar"). Other ferromagnetic materials with low thermal expansion are the iron-platinum alloys [22], where a negative TEC has even been found at 56% Pt [23], and the iron-palladium alloys, which exhibit the lowest TEC at around 50% Pd [24]. Several antiferromagnetic iron alloys (iron-manganese, iron-manganese-nickel, iron-manganese-chromium) also reveal a more or less pronounced "Invar Effect"

[25–28], which is primarily of scientific interest. Refer to B 2.3.2 and the comprehensive presentation in [29–31] for information on the fundamental physical-metallurgical properties.

There is a causal correlation between the thermal expansion and the temperature dependence of the elastic properties [29–31]. With positive volume magnetostriction, an increase in the spontaneous magnetization results in larger interatomic distances and, consequently, in a decrease of the elastic modulus (modulus defect ΔE_A). Elastic lattice distortion alters the spontaneous magnetization, which produces additional volume magnetostriction (resulting in ΔE_ω). Another modulus defect (ΔE_λ) arises as a result of the shape magnetostriction, where Weiß areas orient, under the influence of elastic stress, in such a way that additional elongation is obtained. It follows that the total modulus defect is composed of three parts:

$$\Delta E = \Delta E_A + \Delta E_\omega + \Delta E_\lambda ,$$

the first term being the most significant for technical applications. In certain alloys, the temperature dependence of the modulus defect is so great that the normal drop in the elastic modulus with increasing temperature is compensated (constant-modulus alloys) and even overcompensated.

The technically important *constant-modulus alloys* are derived from the iron-nickel alloys. Their fundamental behavior is presented in Fig. C 23.5 [32]. The drop in the elastic modulus below the Curie temperature represents the total modulus defect. This decreases when ΔE_λ is suppressed by high hardness or strong magnetic fields. This part of the curve can be straightened and raised to a horizontal course by additions of chromium [33] or molybdenum [34] (Fig. C 23.6).

To obtain good spring properties for static loads or good behavior under oscillating loads in such iron-nickel-chromium or iron-nickel-molybdenum alloys, further additions are necessary [35]. Some harden immediately (carbon, wolfram), and some enable precipitation-hardening (beryllium or titanium + aluminum), simultaneously effecting a fine adjustment of the temperature coefficient during the final heat treatment (Fig. C 23.7). In this case, the elastic properties are not isotropic; therefore, texture influences must be considered carefully [36].

Since the constant-modulus effect of these alloys is based on their ferromagnetism, the properties are influenced by external magnetic fields [37, 38]. To keep this influence to a minimum or to exclude it completely, it is tried, by an appropriate choice of the chemical composition, to have the Curie temperature as close to

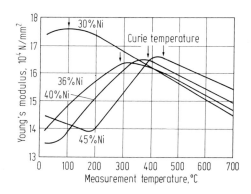

Fig. C 23.5. Temperature dependence of the elastic modulus of soft-annealed iron-nickel alloys. After [32].

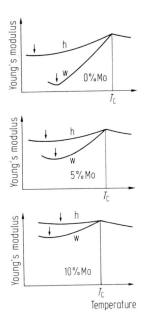

Fig. C 23.6. Influence of additions of molybdenum on the temperature dependence of the elastic modulus of iron alloys with about 40% Ni (schematically, after [34]). h = hard (cold worked), w = soft-annealed, T_C = Curie temperature. The arrows indicate the position of the relevant curve minima.

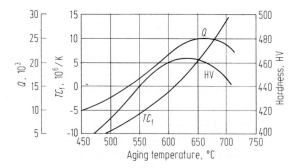

Fig. C 23.7. Example of the influence of final heat-treatment on the temperature coefficient of natural frequency (TC_f), the quality factor (concerning the behavior under alternating loads) (Q), and the hardness (HV) of a constant-modulus alloy. Aging time 30 min. After [38].

room temperature as possible so that a quasi nonmagnetic alloy is obtained. On the other hand, the use of antiferromagnetic alloys (iron-manganese, iron-manganese-nickel, iron-chromium-manganese, and others) can be considered [39–41]. However, processing difficulties stand in the way of large-scale technical use.

C 23.3 Technically Proven Materials

C 23.3.1 Materials with Special Thermal Expansion

Alloys made of iron with nickel, with nickel and cobalt, and with nickel and chromium, with defined thermal expansion, are widely used in technical engineering due to their excellent workability and the variety of their properties. Along with these alloys, the iron-cobalt-chromium alloys play a role because of their relatively good corrosion resistance. Furthermore, ferritic chromium steels are used for less sophisticated glass-to-metal sealings. The values given in Tables C 23.2, C 23.3, and C 23.4 are essentially based on data from handbooks [42], summarizing presentations [43–45], standards [46, 47], and company publications. The steels and alloys used on a large technical scale are categorized in accordance with the groups listed in Table C 23.1.

Apart from the occasional individual case, it is important to keep the *carbon content as low as possible*. This applies particularly to alloy 1 in the interest of the smallest possible TEC and the highest possible stability in the dimensions [48] and properties. The deoxidation additions should also be limited to ensure that they are kept just below the point where red-shortness occurs.

In the *austenitic alloys 4 to 6*, used for glass-to-metal sealings, too high a carbon content leads to the development of carbon dioxide during sealing and consequently to the formation of bubbles in the glass. Since these alloys are frequently

Table C 23.2 Commonly used alloys for defined ranges of the coefficient of expansion

Range of coefficient of thermal expansion $10^{-7}/K$	Consecutive numbers	Alloy composition[a]					
		% C	% Si	% Mn	% Co	% Cr	% Ni
0–20	1	< 0.05	< 0.5	< 0.5	–	–	35–37
50–80	2	≤ 0.05	≤ 0.3	≤ 1.0	–	–	41–43
	3	≤ 0.10	≤ 0.5	≤ 1.0	–	< 1	45–47
	4	< 0.05	≤ 0.3	≤ 0.5	17–19	–	28–30
80–110	5	< 0.05	≤ 0.3	≤ 0.6	–	–	50–52
	6	< 0.05	≤ 0.3	≤ 1.0	–	5.5–6.5	47–49
	7	< 0.1	< 1.0	< 1.0	–	24–26	–
180–210	8	0.55–0.65	≤ 1.0	6–7	–	–	12.5–14.5
	9	≤ 0.20	≤ 1.0	6–7	–	–	19–21

[a] Balance iron.

Table C 23.3 Coefficient of thermal expansion of technically used alloys after soft-annealing and slow cooling (approx. mean values)

Consecutive nos. from Table C 23.2	Coefficient of thermal expansion		
	Between 20 °C and 200 °C $10^{-7}/K$	Between 20 °C and 400 °C $10^{-7}/K$	Between 20 °C and 600 °C $10^{-7}/K$
1	22[a]	75	107
2	52	63	95
3	74	75	99
4	55	49	78
5	98	97	106
6	94	103	123
7	105	109	112
8	197	205	205
9	201	207	211

[a] Between 20 °C and 100 °C: $15 \cdot 10^{-7}/K$

Table C 23.4 Mechanical properties of the alloys according to Table C 23.2

Property	Condition			
	Soft-annealed		50% cold-worked	
	Austenitic alloys	Chromium steel 7	Austenitic alloys	Chromium steel 7
Tensile strength, N/mm^2	450–550	550–650	750–800	900–1000
Yield stress, N/mm^2	300–400	350–450	700–750	700–850
Elongation after rupture, %	30–35	15–25	5	5
Hardness, HV	130–150	140–160	200–250	200–250

used in vacuum equipment, they should also be largely free of gas. It follows that vacuum melting is an advantageous technique. On the other hand, powder metallurgical techniques (mixing, compressing, and high-temperature sintering) are also successful. If powder metallurgy is selected, the basic composition must be modified accordingly, to compensate for the lack of deoxidation additions and their influence on the properties and, in some cases, on the structural stability.

In the *ferritic chromium steel 7*, freedom from austenite stabilizers is of major importance, since even small amounts of austenite in the microstructure can alter the properties substantially.

The values given for the TEC in Table C 23.3 *apply to the soft state*; i.e., after annealing at about 900 °C and subsequent slow cooling. TEC values are several percent lower after substantial cold-deformation. The TEC of alloy 1 for the range between 20 °C and 100 °C can be reduced to about $5 \cdot 10^{-7}/K$ by cold-deformation. Alloys 1 to 7 are ferromagnetic. The Curie temperatures are indicated by the changes in direction of the expansion curves ("Kink temperature") (see Fig. C 23.2).

All the alloys listed in Table C 23.2 can be processed by hot and cold deformation without difficulties. Intermediate and final annealing is conducted between 800 °C and 1000 °C, with any type of cooling. If the annealing temperatures are too high or the annealing times are too long, the grain structure coarsens steadily, impairing the deep-drawing behavior.

The *mechanical properties* of the austenitic alloys are so similar that general data are sufficient (Table C 23.4). The values for the ferritic chromium steel 7 are listed separately.

The *elastic modulus* (Young's modulus) of the austenitic alloys shows a little greater variation. At 36% Ni, the Young's modulus has a minimum of about 135 kN/mm^2; for all other alloys, the values lie between 140 kN/mm^2 and 180 kN/mm^2. Young's modulus for the ferritic alloy is about 200 kN/mm^2. As a rule, Young's modulus can be increased by several percent by extreme cold-work.

The *electrical resistance* of the binary iron-nickel alloys has a maximum at 36% Ni [49]; at 20 °C, alloy 1 has a specific resistance of 0.76 Ω mm^2/m.

The *thermal conductivity* of this alloy is rather low (0.13 W/cm K) [50]. Together with the small TEC, this provides an alloy suitable for equipment and devices for use in cryotechnology; among other applications, as a material for containers in liquefied natural gas (LNG) tankers [51].

C 23.3.2 Materials for Thermobimetals

Steels with defined thermal expansion are also used in the production of thermobimetals. A layer of a material with low thermal expansion ("passive") is bonded to a second layer of a material with high thermal expansion ("active") by cladding and by rolling. Alloys 1 to 3 (depending on the working range of the thermobimetal) are used primarily as passive components, and alloys 8 and 9 (or a variant with 0.6% C, 22% Ni, and 3% Cr) as active components. When the temperature changes, the thermobimetal arches; the specific curvature is proportional to the difference in expansion of the two layers [52].

To complete the picture, it should be pointed out that the electrical resistance of the thermobimetals can be varied over a wide range through additional layers of nickel or copper. Furthermore, by using a manganese-rich manganese-copper-nickel alloy [53] with a TEC of about $280 \cdot 10^{-7}/K$ (preferably combined with alloy 1), a thermobimetal with an especially high specific curvature is obtained.

C 23.3.3 Constant-Modulus Alloys

The technical constant-modulus alloys on an iron-nickel base always contain other constituents (chromium, molybdenum, and others) and are often made precipitation-hardenable by additions of titanium and aluminum or beryllium. Table C 23.5 gives examples [35, 38].

As a rule, the final treatment to set the desired thermoelastic coefficients consists of two steps: cold-deformation (mostly by about 50%) and heat-treatment of varying duration at a defined temperature in the range between 400 °C and 700 °C [54]. In the highest quality alloys, the temperature coefficient of the force of a spring or the natural frequency of an oscillator can be adjusted to values of about $\pm 5 \cdot 10^{-6}/K$ (in the range between $-30\,°C$ and $70\,°C$).

The *specific electrical resistance* of constant-modulus alloys is approximately $1\,\Omega\,mm^2/m$, and the coefficient of thermal expansion is between $70 \cdot 10^{-7}/K$ and $80 \cdot 10^{-7}/K$.

Table C 23.5 Composition of common constant-modulus alloys

Consecutive numbers	Main constituents[a]					
	% Ni	% Cr	% Mo	% Ti	% Al	% Be
1	36	12	–	–	–	–
2	38	8	–	1	–	1
3	40	–	9	–	–	–
4	40	–	9	–	–	0.5
5	42	5.3	–	2.5	0.5	–

[a] Balance iron

C 24 Steels with Good Electrical Conductivity

By Karl Werber and Herbert Beck

C 24.1 Ranges of Application of the Steels and Expected Properties of Use

In addition to copper, bronze, aluminum, and aluminum alloys, steel is also used for the transmission of electrical currents. It is cheaper and stronger than these other metals, although its specific conductivity is lower. Steel used for this purpose must have as high an electrical conductivity as possible, because the energy losses arising in the conductor are reduced in direct proportion to the conductivity. Corresponding to modern usage, steels may be classified according to application requirements for good electrical conductivity.

In overhead lines for telecommunication and other signal transmission purposes, soft wires with an anticorrosive zinc coating and a minimum tensile strength, predominantly in the range of 300 N/mm^2 to 400 N/mm^2, are used. According to standard specifications, customer specifications, or suppliers' internal standards, minimum conductivity values of between 7 Sm/mm^2 and 9.3 Sm/mm^2 are required (Table C 24.1). Soft zinc-coated wires with a conductivity of ≥ 7 Sm/mm^2, which may also be coated with plastics, are also used for grounding and return lines of electrified railways [1].

For insulated *lines and cables in telecommunication and power installations*, wires of a low tensile strength are standardized. These wires consist of a copper jacket tightly joined to a steel core. The conductivity required for these wires is determined by the percentage of copper used: ≥ 14.7 Sm/mm^2 for 23% copper or ≥ 18.2 Sm/mm^2 for 30% copper [2]. Copper-plated steel wire possessing conductivity values of 15% to 80% of copper and a tensile strength of roughly 300 N/mm^2 to 1800 N/mm^2 is used for lead wires in electronic parts and for springs [2a].

In addition to these soft wires, *wires with a higher strength are used for overhead lines*. They may be zinc coated and have a minimum tensile strength in the range of 586 N/mm^2 to 1540 N/mm^2 and, accordingly, minimum conductivity values between about 8.3 Sm/mm^2 and 5 Sm/mm^2. These wires may be clad with copper, pure aluminum, or alloyed aluminum to serve in stranded conductors for power transmission. In these cases, minimum tensile strength values of the complete conductor from 579 N/mm^2 to 1370 N/mm^2 and minimum conductivity values between 23.2 Sm/mm^2 and 11.79 Sm/mm^2 are standardized. In this last group of wires, the steel core serves essentially as a reinforcing element carrying the electrical

---

References for C 24 see page 800.

Table C 24.1 Standards on steel wires with good electrical conductivity

Standard	Wire grade	Purpose	Minimum conductivity at 20 °C Sm/mm^2	Minimum tensile strength N/mm^2	Source
		Wires with low tensile strength			
DIN 48 300	Drawn, zinc coated 2–5 mm dia.	Telecommunication overhead lines	7	383	[5]
ASTM A 111-66	Drawn, zinc coated 2.1–6.05 mm dia.	Telephone and telegraph line wire	7.63–9.35[a]	314–353[b]	[6]
GOST 4231-48	Hot-rolled rod 5.5–7.0 mm dia.	Telegraph line wire	7.09 or 7.52	314	[7]
VDE 0203/XII.44	Copper-clad steel wires with 23% or 30% Cu	Insulated lines and cables in power and telecommunication installations	14.7 or 18.2	441 (max.)	[2]
		Wires with higher tensile strength			
ASTM A 326-67	Zinc coated 2.11–3.76 mm dia.	Telephone and telegraph line wire	6.86–8.27[a]	586–1330[b]	[8]
DIN 48 300	Zinc coated 2–5 mm dia.	Telecommunication overhead lines	5	677–1540[b]	[5]
DIN 48 200 Teil 7	Copper-clad steel wires 1–5 mm dia.	Stranded conductors	17.4 or 23.2[c]	579–1207[b]	[9]
DIN 48 200 Teil 8	Aluminum-clad steel wires 2–5 mm dia.	Stranded conductors	11.79	1080–1370[b]	[10]

[a] Depending on grade or strength classes, thickness of zinc coating, wire diameter, and percentage of copper
[b] Depending on grade or strength classes, wire diameter, and percentage of copper
[c] Conductivity 30% or 40% of that of wire of soft-annealed copper with same diameter and conductivity of 58 Sm/mm^2

conductor. The future development of these conductor materials is described in [3].

The processes for coating the steel wire, especially with aluminum, are also explained in detail [4]. For the drawability of steel wire rod, see C 7.

For these groups of wires, *standard specifications* of different countries and *specifications* of several producers and customers exist. Some of these are cited in Table C 24.1, together with data on minimum requirements for conductivity and tensile strength. In addition, the standards contain specifications for the ductility or the adherence of the coating; namely, elongation and behavior in the bending, wrapping, or torsion test.

In addition to the usual cold-drawn wires, hot-rolled products of steels with good conductivity are used as *third rails for electrified railways* [12]. These steel products have a conductivity at 15 °C of ≥ 8.5 Sm/mm^2, a tensile strength of ≥ 290 N/mm^2, and an elongation after fracture of $\geq 29\%$. *Bus bars for aluminum electrolysis* are also hot-rolled.

Heavy plates are used in chemical and mechanical engineering; e.g., for the construction of cells for the electrolysis of alkali chloride or for eddy-current brakes. The plates are supplied with a relatively high conductivity, minimum values of tensile strength and toughness, and good weldability. The plates are also used, for example, in large-scale magnets in nuclear research, because of their good magnetic properties. For the same purpose, forgings of the same steel grade may be supplied.

C 24.2 Determination of Electrical Conductivity

To determine the electrical conductivity – or its reciprocal, the specific electrical resistance – *current-voltage or bridge measurements* (e.g., Thomson bridge) are normally carried out. The measuring length (for the potential) is defined by knife edges or by welded-on wires in the case of higher temperatures. The cross section either is measured directly or is determined from the length, weight, and density of the sample [12, 12a]. In the case of third rails, the measurements are carried out on finished sections at least 4 m long.

The temperature of the sample, as long as it does not deviate too much from the reference temperature (usually 20 °C), is compensated for by calculating as follows: $R_T = R_t/(1 + \alpha_T(t - T))$, where R_t is the resistance at the sample temperature t, R_T is the resistance at the reference temperature T, and α_T is the *linear temperature coefficient*. With increasing amounts of impurities, the conductivity decreases and

Table C 24.2 Temperature coefficients of the electrical resistance

Material	Conductivity Sm/mm²	Reference temperature °C	Temperature coefficient 1/K	Temperature range °C	Source
Pure iron	10.3	0	0.00651	0–100	[13]
Low-carbon steel M2	9.585	0	0.0060	0–100	
Low-carbon steel M2	9.585	20	0.0057	20–100	[14]
Low-carbon steel M2	9.585	20	0.0051	18–22	
Telephone wire rod	9.45–9.49	20	0.0057	20–100	Own measurements
Third rails	8.5	15	0.005	Ambient temperature	DIN 17 122 [11] [12a]
ASTM B 193:					
Aluminum-clad steel		20	0.0036		
Copper-clad steel		20	0.00378		
Galvanized steel Grade EBB	8.85–9.35[a]	20	0.0056		
Galvanized steel Grade BB	7.63–8.46[a]	20	0.0046		
Galvanized steel Grade steel		20	0.0042		
Telegraph wire	7.09–7.52	20	0.00455	Room temperature	GOST 4231-48 [7]

[a] Depending on thickness of zinc coating, wire diameter, and percentage of copper according to ASTM A 111-66 [6]

C 24.3 Measures of Metallurgy and Physical Metallurgy to Attain a Good Conductivity

All deviations from the ideal lattice structure of a pure metal increase its electrical resistance, as explained in B 2. The most important *change* is caused by *the alloying and tramp elements*. With small additions, the resistance increases almost proportionally to the concentration of the foreign element. Figure C 24.1 shows how some elements as a single addition each increase the resistance of pure iron at room temperature (for which a lowest value of 0.0971 $\Omega \cdot$ mm^2/m is stated in the literature [13]).

As shown in Fig. C 24.1, the *carbon* dissolved in iron influences the electrical resistance most strongly. Its influence, however, depends very much on the *microstructure of the steel*. This is exemplified in Fig. C 24.2, which shows the resistance depending on the carbon content after a quenching treatment and for different kinds of cementite formation. After quenching from just below the A_1-temperature, the electrical resistance increases linearly up to the solubility limit in ferrite at 0.02% C by about 2.5% for each 0.01% solute carbon [17]. The resistance increases in the lamellar pearlitic condition by 0.046 $\Omega \cdot$ mm^2/m to 0.052 $\Omega \cdot$ mm^2/m and in the condition of spheroidized cementite by 0.024 $\Omega \cdot$ mm^2/m to 0.043 $\Omega \cdot$ mm^2/m for each wt-% of carbon [18–21]. If there is no normalizing annealing stage, cooling of the steel after rolling must be conducted slowly enough to obtain a ferritic-pearlitic structure in which the carbon is precipitated in the form of carbide as fully as possible.

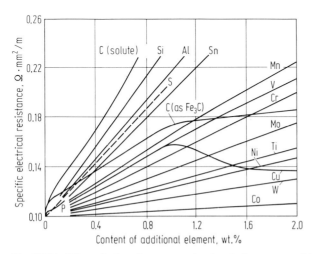

Fig. C 24.1. Effect of other elements on the electrical resistance of iron. After [15].

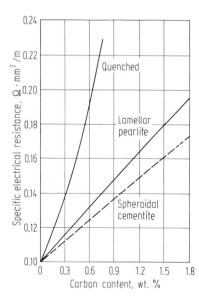

Fig. C 24.2. Effect of carbon and microstructure on the electrical resistance of unalloyed steel. After [16].

In the usual case where more than one additional element is present, the individual increments in resistance may be added as a first approximation, if the total contents are low. But there may be larger deviations; e.g., by precipitations of chemical compounds in the form of particles that are larger than the single lattice atoms by several orders of magnitude and that therefore contribute less to the scattering of the electrons. Thus, in the statistical analysis of steels, a negative contribution of sulfur to the total resistance is found, because manganese sulfides are formed and precipitated. Results of such calculations are enumerated in the literature [22, 23], in which the *simultaneous influences of several elements* were investigated, assuming a linear relationship to the total resistance. The magnitude and the significance of the coefficients for the different elements, however, depend very strongly on the particular steel groups that are submitted to the calculation. Using these coefficients for steels with different chemical compositions and production conditions, therefore, must be done with caution.

The *effect of the cold deformation* in wire drawing *on the electrical conductivity* is of minor importance, secondary to the influence of the chemical composition. The electrical resistance of pure iron increases with the degree of cold working by about 6% up to the highest deformation [24]. When the carbon content is raised, the increase of the resistance after cold deformation is reduced [19, 25] and may not be found at all in steels with medium or high carbon contents [26]. The increase of the electrical resistance caused by the growth of the number of lattice structure defects and the lattice stresses is counteracted by a decrease of resistance due to the rearrangement and straightening of ferrite and cementite lamellae in the direction of the deformation, in a sense switching their individual resistances in parallel. This effect at higher carbon contents may balance the increase of the resistance caused by cold deformation and may even lead to an increase of the conductivity [27, 28].

Table C24.3 Composition and application of steels with good electrical conductivity

| Material | Chemical composition | | | | | Conductivity | Source | Examples of application |
	% C	% Si	% Mn	% P max.	% S max.	Sm/mm²		
Low-carbon steel M2	≤ 0.015	trace	≤ 0.08	0.020	0.015	≈ 9.2	Own values	Electromagnetic parts
Telephone wire	≤ 0.03	trace	≤ 0.25	0.015	0.015	≥ 9.3	Own values	Zinc coated wire 0.4–6 mm dia.
STRS	≈ 0.10	trace	≈ 0.25	0.030	0.040	≥ 8.5		Third rails
Ck 7	≤ 0.08	≤ 0.15	≤ 0.50	0.035	0.035	≈ 6.4	[32]	Larger forgings and heavy plates for electromagnets
Ck 5	≤ 0.07	≤ 0.10	≤ 0.35	0.025	0.025	≈ 7.2		Heavy plates for electrolysis of alkali chloride,
Special structural steel SHL	≤ 0.07	≤ 0.10	≤ 0.35	0.020	0.020	≈ 6.7		eddy-current brakes

Aging may have an effect on the electrical resistance of a steel after extended storage at room temperature or heating to temperatures up to 300 °C, especially after cold deformation. In this process, the anchoring of carbon or nitrogen atoms at dislocations or the precipitation of fine carbides or nitrides reduces the electrical resistance [17, 27, 29, 30].

An increase of the resistance due to cold deformation may be reversed by *recovery and recrystallization* in a subsequent *annealing*, the best temperature normally being in the range 500 °C to 600 °C [24, 27, 31]. The conductivity can be increased further by tempering, which causes spheroidization of the carbides; this is more pronounced at higher carbon contents. When annealing a cold-deformed structure with lamellar cementite, the increment of the conductivity increase will be raised with the degree of cold deformation, since cold deformation causes the cementite lamellae to break up and thereby produces favorable conditions for the formation of granular cementite [20].

A *decarburizing anneal* in moist hydrogen is another way to obtain a higher conductivity.

C 24.4 Characteristic Steel Grades

As indicated above, it is necessary to keep the contents of impurities and of alloying elements within certain limits to satisfy the requirements for minimum electrical conductivity values. Table C 24.3 lists the limiting values for some characteristic elements. In producing the steels, special care must be taken to start with selected raw materials when a higher conductivity is required. The *most favorable method* is seen in the production of a *rimming* steel with *low carbon and manganese* (and silicon and aluminum) contents. To ease production and working of heavy plates and forgings, however, certain restrictions may have to be observed in preference to the conductivity. The lowering of the contents of carbon and other accompanying elements will have to be limited, or even excluded, if certain minimum values are needed to give higher tensile strengths.

C 25 Steels for Line Pipe

By Gerhard Kalwa and Konrad Kaup

C 25.1 Required Properties

The property requirements to be met by line-pipe steels depend on the type of material to be transported, the environmental and operating conditions of the pipelines, and safety considerations.

Pipelines are required to transport, over long distances, the primary energy sources oil and gas, as well as various chemical products, sewage, industrial and potable water, and solid materials such as coal and ores. Solid materials are normally tansported as water suspensions. Natural gas is transported either in the gaseous state or in the fully or partially liquefied state.

Pipelines are operated both on land, where they pass through temperate zones (down to $-10\,°C$) and arctic zones (down to $-60\,°C$), and in off-shore installations. Both welded and seamless pipes are used in pipeline construction, but welded pipes can be made with a larger diameter than can seamless pipes. In modern pipe mills, longitudinally welded pipes, for example, are produced with outside diameters reaching up to 1600 mm.

For economic reasons, high operating pressures and high diameter-to-thickness ratios are aimed for in pipeline construction. This results in a requirement for steels with high *yield stress and tensile strength* values. The required yield stress values are the values determined on the finished pipe. Therefore, any possible reduction in the yield stress determined on the flat plate or strip steel used to make pipe – resulting from the pipe manufacturing process (Bauschinger effect) – must be taken into account when selecting appropriate pipe materials.

Under certain conditions, critical cracks, once initiated, can propagate unstably with constant velocity over large lengths of gas transmission lines. For safety reasons, the steels used to make high-pressure transmission lines must possess high *toughness* to ensure adequate resistance to crack initiation and propagation. Crack initiation at various defects can be prevented even when the steels exhibit only relatively low notch-impact energy values. But if a crack of overcritical length is formed by external factors, high notch-impact energy values will be required, in addition to full material ductility at the service temperature, to prevent propagation of the crack [1, 2].

In pipe mills, large-diameter line pipes are usually welded using single or multiwire submerged arc welding (SAW). The weld metal is deposited in two passes, one for making the inside seam and the other for making the outside seam. The steels used for line pipe applications should have proper *weldability*. The pipes are either longitudinally welded or spiral-welded. Adequate strength properties, freedom from defects, and proper weld seam geometry are the important criteria for *production welding* of the pipes. During pipe laying, the adjacent pipes are positioned so that their production seam welds are misaligned. Thus, an unstable crack, running through the heat-affected zone (HAZ) of a pipe, for example, encounters the highly tough base material in the adjacent pipe. Therefore, there is no need to place severe toughness requirements on production welds (weld metal and HAZ).

Field welding of pipeline girth welds is preferably carried out using vertical-down cellulose-coated stick electrodes. However, there are certain limitations to the use of these electrodes for welding higher strength steel grades (X 80). In this case, the use of relatively low-strength cellulosic electrodes to make root and hot passes and basic vertical-down electrodes to make filler and cap passes is an economical solution that also complies with the more stringent requirements. In addition, fully or partly mechanized gas metal arc welding is increasingly employed. Field welding is usually carried out by depositing the weld metal in multiple passes from the outside of the pipe. The steel must be capable of being welded without difficulty employing welding methods and consumables commonly used for the particular application. The steel's weldability must comply with this requirement. Depending on the method of operation, welding conditions, and steel chemistry, high cooling rates, high stresses, hard low-temperature transformation products (hardened microstructure), and high hydrogen contents can occur. These factors must be taken into account when evaluating weldability (see B 5). In addition, the material's tendency to harden must be estimated based on the carbon equivalent (see C 25.2).

Pipelines may suffer *corrosion* damage in service unless appropriate countermeasures are taken. To avoid corrosion on the outer surface, the pipes are generally coated and cathodically protected. Pipelines may also undergo internal corrosion caused by the materials being transported through them. Corrosion damage in the form of hydrogen-induced cracking may occur even in the unpressurized condition before service, if, for example, fluids containing hydrogen sulfide react with the steel. In service, hydrogen-induced stress-corrosion cracking may occur in pressurized pipelines under certain conditions of mechanical stress and corrosion [3]. Both forms of corrosion can be combated most effectively by removing the corrosive species from the material being transported and by drying the pipeline and the fluid (gas). Inhibitors also may be added to combat corrosion. When it is not possible to adopt these preventive measures, appropriate design and material selection must be used to make the pipeline resistant to corrosion.

C 25.2 Characterization of the Required Properties

The *strength properties* of the material are characterized by the yield stress and tensile strength values determined in *tensile tests* (see B 1). The specimens are generally taken from the pipe so that the length of the specimen is in the circumferential direction of the pipe; i.e., in the direction of the main operating stress. Straightening of the test specimens can result in a significant alteration of the yield stress owing to the Bauschinger effect (see B 1.2.1). To determine the strength properties, it is prudent to perform the testing on full-size ring specimens using ring-testing presses. But this test method is too expensive to use routinely. A reasonable alternative specimen form for determining the yield stress is an unstraightened tensile specimen with circular cross section removed from the pipe wall with the specimen axis transverse to the pipe axis.

The *toughness* of the material is characterized by the notch-impact energy values determined in *impact toughness tests* according to the German standard DIN 50 115 [3a], for example, using ISO V-notch specimens with different notch axis orientations. Additionally, the Battelle drop-weight tear (BDWT) test [4–6] is used in many cases for testing large-diameter pipes. In this test, unwelded full-thickness (plate thickness) specimens containing a sharp notch are broken at different temperatures. The percentage of the shear fracture area on the total fracture surface is then measured. Usually, the temperature at which the fracture surface of the broken specimen exhibits a specified percentage of shear fracture – area e.g., 50% or 85% – is given as the characteristic value. Figure C 25.1 shows the fracture surfaces of four BDWT test specimens broken at different temperatures. The increasing percentage of shear fracture area with increasing test temperature is clearly evident.

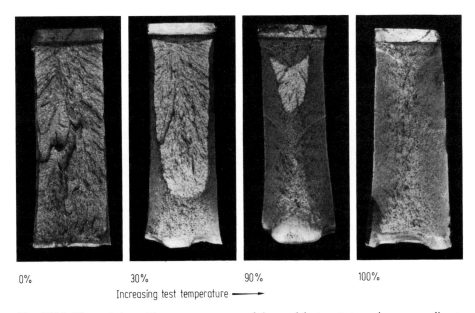

Fig. C 25.1. The variation of fracture appearance of drop-weight tear test specimens according to Battelle [4–6] tested at different temperatures. The figures shown below the photographs represent the percentage of the shear fracture area on the total fracture surface of the respective specimen.

To investigate the *fracture behavior* of pipelines, very expensive tests are carried out on specimens similar to actual components and on full-size sections of a pipeline. Experiments are conducted on test lines over 200 m long made of pipes of the original size to determine the minimum toughness required to prevent crack propagation and, hence, the formation of long, running cracks. In these tests, a crack is initiated in a test line that is under pressure, the velocity of crack propagation through the test pipes is measured, and the fracture behavior is evaluated. Full-scale tests carried out on large-diameter pipes [7] at various centers show that, with increasing stresses on the pipe wall, increasing impact energy values are needed to prevent the occurrence of long, running cracks (Fig. C 25.2). Since this figure is intended only to show the trend, details of the various equations used for calculating the curves will not be dealt with here, and the reader is referred to the published work [7]. Though the trend is the same, the quantitative individual test results differ quite significantly. To explain this discrepancy, full-scale tests on large-diameter pipes have been carried out by the European Pipeline Research Group (EPRG) [8–10].

The carbon equivalent, which is an index of a steel's tendency to form hard microstructures, can be used to estimate *weldability*, especially the weldability of steels in field welding, which is of primary concern here. The level of attainable hardness is determined primarily by the carbon content of the steel. The other elements have little or no influence. The carbon equivalent formulas commonly used to date – e.g., the IIW formula – do not accurately represent the effect of various alloying elements such as manganese, molybdenum, and nickel for the range of cooling rates encountered in the field welding of line-pipe steels [11]. The cooling rates are generally represented by the term $t_{8/5}$, which gives the time it

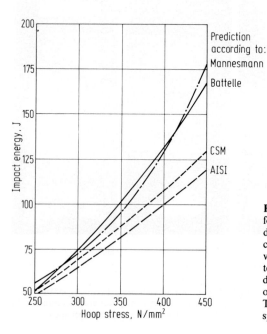

Fig. C 25.2. Comparison of various curves for predicting the fracture behavior of large-diameter gas transmission pipelines. The curves enable the notch-impact energy values required for fracture arrest to be determined. Pipe size: 48-in. (about 1220-mm) diameter, 20-mm wall thickness. For details of various prediction equations, see [7]. Type of specimens: ISO V-notch transverse specimens.

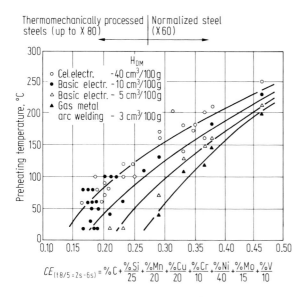

Fig. C 25.2a. Preheating temperature required to ensure crack-resistant welds as determined by implant tests as a function of the carbon equivalent. Applied stress in the implant test was equal to the yield stress. Heat input was 8 to 9 kJ/cm.

takes for the material to cool through the temperature range of 800 °C to 500 °C. The effect of individual alloying elements on the tendency of the steel to form hard microstructures is estimated differently by various investigators in their carbon equivalent formulas. The discrepancies can be attributed to differences in the welding conditions and the steels used to derive the formulas. Extensive evaluation of the available data led to the following new formula for the carbon equivalent [11, 12]:

$$\text{CE}(\%) = \%\,\text{C} + \frac{\%\,\text{Si}}{25} + \frac{\%\,\text{Mn} + \%\,\text{Cu}}{20} + \frac{\%\,\text{Cr}}{10} + \frac{\%\,\text{Ni}}{40} + \frac{\%\,\text{Mo}}{15} + \frac{\%\,\text{V}}{10}$$

This formula adequately represents the hardening tendency of the low-carbon line-pipe steels dealt with here for the $t_{8/5}$ values of 2s to 6s that are typically encountered in welding operations associated with pipe laying. The most stringent requirements on the behavior of the line-pipe steels during field welding are placed on the deposition of the first weld pass; i.e., the root pass. An unfavorable combination of welding conditions and the steel's chemical composition can result in cold cracking. The susceptibility to cold cracking can be determined in various tests; for example, the implant test [11a, 13]. Figure C 25.2a shows the preheating temperature required to avoid cold cracking in the implant test as a function of the carbon equivalent of the steel and the hydrogen content of the weld metal (welding process).

To determine the resistance of the material to *hydrogen-induced cracking*, rectangular specimens are usually exposed to aqueous solutions containing hydrogen sulfide for 96 hours, with no external stress applied to the specimens. The specimens are then metallographically examined for cracks, caused by hydrogen entering the steel, at nonmetallic inclusions [14]. The number, length, and distribution of the cracks correlate well with the microstructural inhomogeneity of the steel and the volume fraction, shape, and distribution of nonmetallic inclusions, especially sulfides [15, 15a].

To avoid *hydrogen-induced stress-corrosion cracking*, low-stress constructions should be aimed for. During design – i.e., when selecting the pipe diameter and wall thickness – care should be exercised to keep the stresses resulting from the internal pressure below a critical value. This critical stress level depends on the material being carried through the pipe or the nature of the hydrogen-sulfide-bearing condensate (pH value) forming from it. The extent to which steels can be stressed relative to their yield stress was found to lie between 20% and 60%, depending on the type of environment and the inherent susceptibility of the steels. Low stress levels can also be achieved by taking proper measures during manufacture – e.g., expanding or heat-treating (quenching and tempering) the pipes – which help reduce the residual stresses in the pipes [16].

Although both *hydrogen-induced cracking* and *hydrogen-induced stress-corrosion cracking* are caused by hydrogen, one should not overlook the fundamental differences between the two types of corrosion. It should not be assumed that a steel resistant to hydrogen-induced cracking is invariably resistant to hydrogen-induced stress-corrosion cracking [17].

C 25.3 Measures of Physical Metallurgy to Attain the Required Properties

It is possible to adjust a steel's *microstructure* and, consequently, its properties to comply with the requirements placed on the line pipe by varying the chemical composition and the type of treatment, such as normalizing, thermomechanical rolling, or quenching and tempering. To do this, several steel strengthening mechanisms are used. These mechanisms are dealt with in B 1 and B 4 and include solid-solution hardening, strengthening through grain refinement, precipitation hardening, and dislocation hardening.

Thermomechanical rolling has attained a special significance in producing the plate and strip of low-carbon steels used for making large-diameter line pipe [18, 19]. In these steels, it is possible to produce transformation products ranging up to bainite, which are characterized by a very fine grain size. Figure C 25.3 shows schematically how the physical metallurgical phenomena are used in thermomechanical rolling [20]. The cast structure of the slab is homogenized through dynamic recrystallization in the initial rolling stage. Static recrystallization leads to grain refinement in the intermediate rolling stage, and precipitation processes occur simultaneously. In the finish rolling stage, dislocation density is increased,

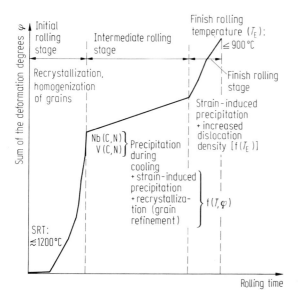

Fig. C 25.3. Physical metallurgical processes occurring during thermomechanical rolling as a function of the sum of the deformation degrees φ, rolling time, and rolling temperature T in various stages of rolling. SRT = slab reheating temperature. After [20].

strain-induced precipitation of fine particles takes place, and a further grain refinement is achieved. Figure C 25.4 shows the quantitative increase in yield stress with decreasing finish rolling temperature and transformation temperature for a low-carbon steel microalloyed with niobium [21]. Finally, a further significant improvement in the properties of steels with leaner chemical compositions can be achieved by accelerated cooling directly from the finish rolling temperature.

Refinement of the microstructure can also be achieved by a *quenching and tempering* treatment. The low-carbon, low-alloy, heat-treating steels developed for line-pipe applications transform to bainite during quenching. The bainite laths are smaller in width than the grain diameter of the thermomechanically rolled steels. The yield stress of the steel is also increased by the numerous lattice defects formed simultaneously during quenching and by the solid-solution hardening through the alloying additions of manganese, molybdenum, chromium, and nickel. A further increase in the yield stress can be achieved by the precipitation of intermetallic compounds and carbonitrides during the tempering that follows the quenching. The quenching and tempering heat treatment can be carried out at different stages of production. Quenching and tempering of the welded pipe, rather than the steel plate, has the advantage of also heat treating the weld and homogenizing and reducing the residual stresses in the pipe. Direct quenching of the plate from the finish rolling temperature is also employed in mill practice.

Although the yield stress and ductile-brittle transition temperature of the notch-impact energy are favorably influenced by grain refinement, the upper shelf energy, especially of the transverse specimens (relative to the rolling direction), is significantly improved by *desulfurization*; hence, the anisotropy of the toughness is

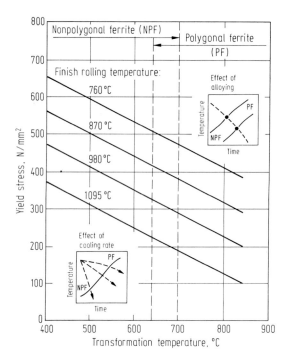

Fig. C 25.4. Dependence of yield stress on the finish rolling temperature and on the transformation temperature, which was varied by the addition of alloying elements and changes in cooling rates for a low-carbon steel microalloyed with niobium. After [21].

reduced, irrespective of the type of steel and its treatment (Fig. C 25.5) [20]. If the sulfur content is below 0.003%, hardly any elongated sulfides are found metallographically in the steel matrix.

Desulfurizing the steel, binding the unremoved sulfur to sulfides of reduced plasticity, preventing the occurrence of inclusions arranged in planes, and improving the basic cleanness of the steel are also appropriate measures to avoid the occurrence of sites where the atomic hydrogen entering the steel can recombine to

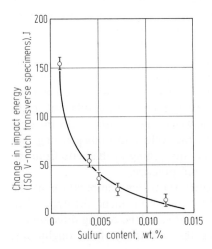

Fig. C 25.5. Improvement of the upper shelf notch-impact energy of transverse impact specimens by reduction of the sulfur content in fine-grained microalloyed steels. After [20].

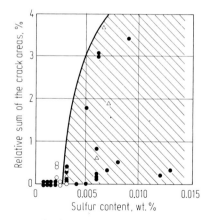

Fig. C 25.6. Effect of sulfur content on the susceptibility of steels to hydrogen-induced cracking. The ordinate represents the sum of the crack areas relative to the sum of the areas of the specimen cross-section examined. After [14].

△ Continuously cast steel with sulfide shape control
● Continuous casting
○ Ingot casting

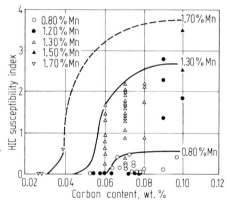

Fig. C 25.6a. Effect of carbon content on the resistence to hydrogen-induced cracking (HIC) of line-pipe steels with various levels of manganese in an aqueous solution saturated with H_2S, pH = 3. After [15a].

hydrogen molecules and lead to the formation of cracks. These measures will significantly increase the steel's resistance to hydrogen-induced cracking (Fig. C 25.6) [15].

Preventing the formation of hard, low-temperature transformation products in the otherwise softer microstructures leads to a further improvement of the resistance to hydrogen-induced cracking. This can be achieved by reducing the carbon and manganese contents of the steel (Fig. C 25.6a) [15a] and by taking proper measures during steelmaking to reduce macrosegregation.

C 25.4 Characteristic Steel Grades Proven in Service

Until the end of the 1950s, manganese steels in the conventionally rolled or normalized condition were used to make seamless and welded line pipes. The minimum yield stress that could be achieved with these steels was 360 N/mm^2. Later, by microalloying with the elements vanadium and/or niobium, the yield stress of these steels could be increased even in the normalized condition. These microalloying elements were used for grain refining and precipitation hardening. Minimum yield stress values of up to 420 N/mm^2 were achieved in this way (Fig. C 25.7) [22].

The ever-increasing strength and toughness requirements could no longer be met by conventionally rolled or normalized steels because further increases in their alloy content adversely affected weldability. The thermomechanical rolling of steels then emerged as a new method capable of producing higher yield stress and toughness values with reduced levels of carbon and alloying elements. The BDWT-transition temperature (see C 25.2), which is used to verify the suitability of a steel for the lowest operating temperature of a pipeline, is considerably reduced for thermomechanically rolled steels as compared to normalized steels. Because of the low carbon content, the field weldability of these thermomechanically processed steels is also significantly improved.

It is possible to produce steels up to grade X 80 with a minimum yield stress of 552 N/mm^2 using thermomechanical processing. To achieve an optimum combination of high yield stress and low ductile-brittle transition temperature, the following

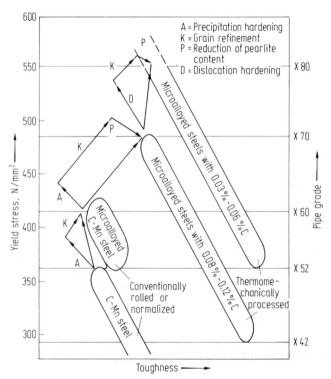

Fig. C 25.7. Development of steels for line pipe. (No numerical values are shown on the abscissa because the measured toughness values depend on the testing method, which is selected out of several possible methods.)

influencing factors must be carefully balanced [21]: the chemical composition of the steel, the slab reheating temperature, the temperature schedule during and after rolling, and the sequence of deformation steps (see C 25.3).

If necessary, a further increase in yield stress, along with good toughness properties, can be achieved by a quenching and tempering treatment of the large-diameter pipes or of the plates used to make these pipes. The steels are alloyed primarily with manganese, molybdenum, and nickel, together with additions of microalloying elements; for example, niobium.

Details of the chemical composition, the mechanical properties, and technological properties of the seamless or welded pipes made from these steels can be found in the German standard DIN 17172 [23], for example. Reference also should be made to standards 5L, 5LS, 5LX, and 5LU of the American Petroleum Institute (API) [24]. However, users quite often place demands on the pipes that exceed the requirements laid down in the standards mentioned.

C 26 Ball and Roller Bearing Steels

By Klaus Barteld and Anita Stanz

C 26.1 Required Properties

Bearings are used in machine parts, where they are stressed by reversed tension, compression, and shear loads. The loading capacity of bearings is a function of the material's *hardness* and *limit of elasticity*. In addition to the material properties, several other factors determine the working life of bearings. These include the quality and construction of the bearing and the external service conditions, such as lubrication, dirt accumulations, wear, corrosion, and violence or overstressing. If the bearings are correctly dimensioned and maintained, their useful life is determined by the *fatigue strength* of the material. If the fatigue strength is exceeded, incipient cracks occur and result in pitting during continuing loading. The smallest defects in bearing steels may be starting points for cracks at high stress. Therefore, the uniformity of the micro*structure* and, especially, the cleanness of nonmetallic inclusions (degree of purity) must satisfy high requirements. Through permanent cold deformation of the surface, small particles of the bearing track are rubbed off; these particles abrade the bearing components and reduce the service life by wear. Insufficient tensile strength, high loads, and poor lubrication also expedite the abrasion. A martensitic microstructure with inserted spheroidized carbides is the most desirable.

Temperature stress also must be taken into account. This stress can be caused by friction and elastic deformation in bearings, as well as by high external temperatures, such as in bearings of furnaces and engines.

When bearings are used in corrosive environments, such as salt water, acids, or alkaline solutions, *susceptibility to corrosion* must be considered. Residual stresses from heat treatment, from machining, or (especially) from rolling loads in service also influence the service life of the material [1]. For these reasons, bearing steels must have high hardness, a microstructure with only slight variations in regions of high stress, hardenability corresponding to constructional dimensions, resistance to distortion, and good wear resistance. Good resistance against corrosion and a high hardness at elevated temperatures is required for special bearings.

Attention must be paid to the *processing properties*. For cutting delivered steel bars or billets into pieces for hot forming, the sutability for cold shearing given by keeping a maximum strength is sufficient. For cold forming, a microstructure with spheroidized carbides is necessary. Also, machining using automatic lathes is

---

References for C 26 see page 801.

possible only with completely spheroidized carbides in the microstructure. Machinability can be further improved by higher sulfur contents in the range of 0.025% to 0.100%, additions of selenium, tellurium, or lead can be partly substituted for sulfur. It is controversial, however, whether a higher content of sulfides as a type of nonmetallic inclusions in the microstructure increases the danger of premature failure of bearings by pitting [2].

C 26.2 Characterization of the Required Properties by Test Values

As described above, the most important properties of interest for ball and roller bearing steels are hardness, fatigue strength under reversed stresses, wear resistance, and microstructure.

The *hardness* is generally examined by the Rockwell method.

Two procedures are currently used to determine the *fatigue strength* under reversed stresses. The evaluation of fatigue tests under reversed bending stresses on steel specimens with the high rotation speed of 6000 revolutions per minute (rpm) is shown in Fig. C 26.1 [3]. A more practical method is the statistical failure diagram developed by Weibull (Fig. C 26.2) [4]. For this method, finished bearings are tested under special conditions and the so-called B_{10}-value is determined. This value represents the lifetime at which 10% of the tested bearings have failed.

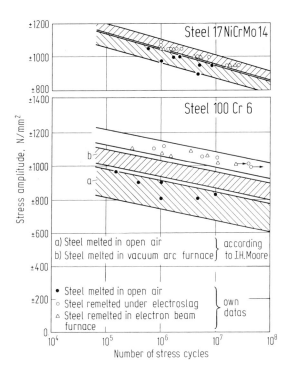

Fig. C 26.1. Influence of the melting process on the fatigue strength under reversed bending stresses of the steels 100 Cr 6 and 17 NiCrMo 14 according to Table C 26.1 (frequency of cycles: 6000 min^{-1}). After [3].

To date, there are no standardized methods for testing *wear resistance*. But it is well established that high hardness and high toughness reduce wear (see B 10 and C 12).

Essential for ball and roller bearing steels is an *examination of the microstructure*. Microcleanness (degree of purity) [5], distribution of carbides [6], decarburization [7], and the type of microstructure are tested by metallographic methods. Therefore, standard series of reference photomicrographs for metallographic testing have been developed for unified application [8]. For rating microcleanness, DIN 50 602 [5] classifies nonmetallic inclusions as sulfides and oxides according to distribution, type, and size. Under a microscope, all counted inclusions of a fixed size and larger within a particular area can be expressed as a characteristic value by applying area proportional factors to the sum of commeasurable inclusions [9]. This value is proportional to the sulfur and oxygen content of the steel. Specification of a maximum permissible value is contained in the standard DIN 17 230 for ball and roller bearing steels [10]. To assess carbide distributions, SEW 1520-78 has been established [6]. According to this standard, the following factors can be evaluated numerically: carbide size of spheroidized cementite, residual portions of laminated pearlite after spheroidizing, constitution of carbide network, and carbide bandings in a compact or noncompact shape [11]. According to the recommendations, limiting values for the results of metallographic tests can be stipulated on a uniform basis, where the size and the steel grade of respective products have a decisive significance.

The *hardenability* is determined by the Jominy test and by measuring the hardness over the cross section of uniformly dimensioned specimens under assessed conditions of heat treatment [12].

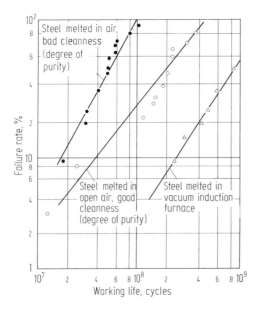

Fig. C 26.2. Weibull diagram of bearing steel 100 Cr 6 melted by different methods. (Radial load 800 kg, frequency of cycles 2750 min^{-1}.) After [4].

C 26.3 Measures of Physical Metallurgy to Attain the Required Properties

Chemical composition, melting process, and heat treatment are decisive for the microstructure at working conditions and for the required properties of ball and roller bearing steels. The requirement for a hardness from 50 HRC to 65 HRC and for a good wear resistance by a straight martensitic microstructure with well-distributed fine carbides results in the chemical composition of the *through-hardening steels* with 1% C and 1.5% Cr. Depending on the size of the bearings and the necessity of through-hardening of the material, the chromium contents are varied and supplemented by manganese and molybdenum additions. Figure C 26.3 demonstrates the influence of these elements on the through-hardenability of some bearing steels. It follows that higher alloyed steels must be used for larger dimensioned bearings.

The *melting process* today is generally done in electric furnaces or by basic oxygen steelmaking in combination with ladle metallurgy and vacuum treatment. Using these methods results in a cleanness of nonmetallic inclusions that is necessary to guarantee the lifetime of ball and roller bearings. It must be mentioned, however, that the content of nonmetallic inclusions also depends on the chemical composition of the steel [14]. Figure C 26.4 demonstrates the frequency distribution of the characteristic value $K\,4$ for oxides in different steel types. The influence of the carbon content on the characteristic value of the oxides can be seen. Corrosion-resistant bearing steels and hot-work tool steels, melted in open air, contain substantially more oxidic inclusions, because of their compositions.

For the highest requirements of microcleanness, remelting of regular melted steel *under electroslag or in vacuum* is necessary. The dependence of the characteristic inclusion value on the oxygen and sulfur contents after different melting or

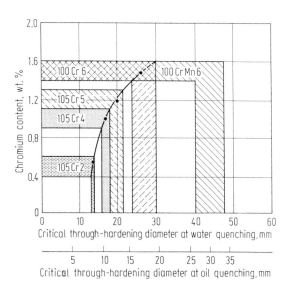

Fig. C 26.3. Through-hardenability of some bearing steels. After [13].

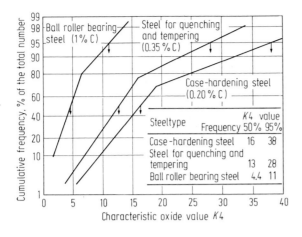

Fig. C 26.4. Cumulative frequency curves for the characteristic inclusion value $K4$ according to DIN 50 602 [5], determined on 90-mm to 130-mm square billets of case-hardening, quenching- + tempering, and ball and roller bearing steels. After [14].

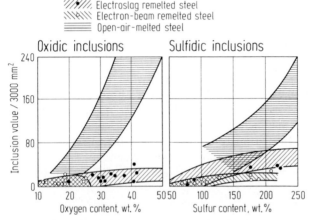

Fig. C 26.5. Relation between the characteristic inclusion value according to DIN 50 602 [5] and the oxygen and sulfur contents for steel 100 Cr 6 melted by different methods. After [3].

remelting processes is shown in Fig. C 26.5 [3]. At corresponding oxygen contents, open-air-melted steel has substantial higher characteristic inclusion values, compared to steel remelted under electroslag. These higher values result from a rougher inclusion distribution and the fact that finer parts of inclusions are not counted. The finer parts are not necessarily harmful. A steel remelted in an electron beam furnace has a lower oxygen content and a still lower inclusion value. The quicker solidification in remelted ingots is responsible for the finer distribution of nonmetallic inclusions, which have less time for growth and coagulation. Also, the inclusion size is reduced by hot and cold forming [15].

Influencing the *carbide distribution* is a further metallurgical measure that can be taken to increase the lifetime of bearings. Carbide segregation appears in hypereutectoidic steels and is especially marked in ledeburitic steels; it is caused by slower solidification in the center parts of ingots or continuous cast billets, by segregation at increasing carbon contents, and by precipitation caused by falling below the solubility temperature of carbon. This segregation may involve hypereutectoidic or ledeburitic carbide lines (bandings) at the subsequent hot forming. Homogenization at high temperatures improves the microstructural conditions, but raises the surface decarburization and also increases the cost of carbide removal. Regarding the carbide lines (bandings), a further improvement is obtained by using remelting processes with substantially faster and differently directed solidification. Carbide networks can develop at a slower cooling rate after hot forming (Fig. C 26.6). Carbide network formation can be reduced by special measures, such as lowering the final rolling temperature or cooling the rolled bars or wire more quickly. Figure C 26.7 shows the holding time and temperature necessary for the dissolution of carbides [13, 16]. Recent investigations report an improvement of carbide distribution and an increase of yield stress by thermomechanical treatment [18]. Carbide lines and networks are inhomogenities and remain spurious, but they are not totally avoidable.

C 26.4 Characteristic Steel Grades Proven in Service

Table C 26.1 shows ball and roller bearing steel grades that were considered effective in the first edition of DIN 17 230 [10]. Since the recommendations of the International Organization for Standardization (ISO) [20] and the results of

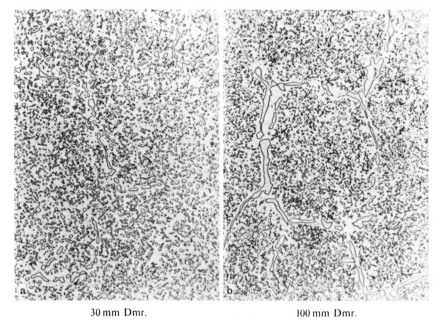

Fig. C 26.6. Dependence of the carbide network of 100 Cr 6 on the final size after hot rolling.

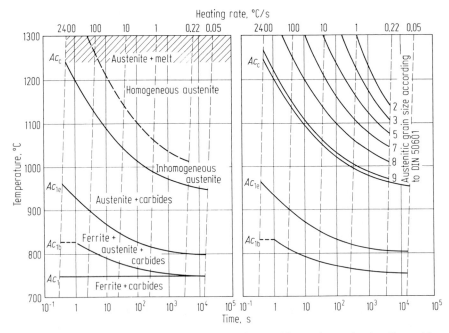

Fig. C 26.7. Time-temperature-dissolution diagram of 100 Cr 6 for continuous heating. (Austenitic grain size according to Euronorm 103-71 [17] or DIN 50601 [17].)

negotiations for a standard of the European Communion of Coal and Steel [19] are taken into account in this table, Table C 26.1 can be considered a synopsis of current steel grades in worldwide application. In Europe and especially in Germany, *through-hardening steels* are mainly used, whereas in the United States, the *case-hardening steels* are preferred. The advantage of case-hardening steels is that they have a soft center under a hard and wear-resistant surface, by which shocks can be better absorbed. But the processing is more intricate and more expensive because of the case hardening needed for a larger thickness of the carburized layer than it is obtained by the normal processing (hardening) of the through-hardening steels. The higher the carbon content, the lower the oxygen content, as well as the casting temperature; this results in better cleanness of steels melted in open air (see Fig. C 26.4). As shown in Fig. C 26.3, through-hardenability changes with increasing alloy content. This means that larger dimensioned bearings must be produced from higher alloyed steels.

Also, *steels for quenching and tempering* are used for larger parts as tracks for balls and rollers (e.g., for crane constructions), both in the quenched + tempered and in the surface-hardened condition. The *stainless steels* mentioned in Table C 26.1 are not fully rust-resistant because of their higher carbon content; they have a decreased capacity to withstand stresses because of a maximum obtainable hardness of 55 HRC to 58 HRC.

Table C 26.1 Chemical composition of ball and roller bearing steels most frequently used (ladle analysis). After [10]

Steel grade Designation	%C	%Si	%Mn	%P max.	%S max.	%Cr	%Mo	%Ni	%V	%W	%Cu max.
Through-hardening steels											
100 Cr 2	0.90–1.05	0.15–0.35	0.25–0.45	0.030	0.025	0.40–0.60	–	max. 0.30	–	–	0.30
100 Cr 6	0.90–1.05	0.15–0.35	0.25–0.45	0.030	0.025	1.35–1.65	–	max. 0.30	–	–	0.30
100 CrMn 6	0.90–1.05	0.50–0.70	1.00–1.20	0.030	0.025	1.40–1.65	–	max. 0.30	–	–	0.30
100 CrMo 7	0.90–1.05	0.20–0.40	0.25–0.45	0.030	0.025	1.65–1.95	0.15–0.25	max. 0.30	–	–	0.30
100 CrMo 73	0.90–1.05	0.20–0.40	0.60–0.80	0.030	0.025	1.65–1.95	0.20–0.35	max. 0.30	–	–	0.30
100 CrMnMo 8	0.90–1.05	0.40–0.60	0.80–1.10	0.030	0.025	1.80–2.05	0.50–0.60	max. 0.30	–	–	0.30
Case-hardening steels											
17 MnCr 5	0.14–0.19	max. 0.40	1.00–1.30	0.035	0.035	0.80–1.10	–	–	–	–	0.30
19 MnCr 5	0.17–0.22	max. 0.40	1.10–1.40	0.035	0.035	1.00–1.30	–	–	–	–	0.30
16 CrNiMo 6	0.15–0.20	max. 0.40	0.40–0.60	0.035	0.035	1.50–1.80	0.25–0.35	1.40–1.70	–	–	0.30
17 NiCrMo 14	0.15–0.20	max. 0.40	0.40–0.70	0.035	0.035	1.30–1.60	0.15–0.25	3.25–3.75	–	–	0.30
Steels for quenching and tempering											
Cf 54	0.50–0.57	max. 0.40	0.40–0.70	0.025	0.035	–	–	–	–	–	0.30
44 Cr 2	0.42–0.48	max. 0.40	0.50–0.80	0.025	0.035	0.40–0.60	–	–	–	–	0.30
43 CrMo 4	0.40–0.46	max. 0.40	0.60–0.90	0.025	0.035	0.90–1.20	0.15–0.30	–	–	–	0.30
48 CrMo 4	0.46–0.52	max. 0.40	0.50–0.80	0.025	0.035	0.90–1.20	0.15–0.30	–	–	–	0.30
Stainless steels											
X 45 Cr 13	0.42–0.50	max. 1.00	max. 1.00	0.040	0.030	12.5–14.5	–	max. 1.00	–	–	0.30
X 102 CrMo 17	0.95–1.10	max. 1.00	max. 1.00	0.040	0.030	16.0–18.0	0.35–0.75	max. 0.50	–	–	0.30
X 89 CrMoV 18 1	0.85–0.95	max. 1.00	max. 1.00	0.045	0.030	17.0–19.0	0.90–1.30	–	0.07–0.12	–	0.30
Steels with high hardness at elevated temperatures											
80 MoCrV 42 16	0.77–0.85	max. 0.25	max. 0.35	0.015	0.015	3.75–4.25	4.00–4.50	–	0.90–1.10	–	–
X 82 WMoCrV 6 5 4	0.78–0.86	max. 0.40	max. 0.40	0.030	0.030	3.80–4.50	4.70–5.20	–	1.70–2.00	6.00–6.70	–
X 75 WCrV 18 4 1	0.70–0.78	max. 0.45	max. 0.40	0.030	0.030	3.80–4.50	max. 0.60	–	1.00–1.20	17.5–18.5	–

[a] For further details, see DIN 17 230 [10]

If good tempering properties (retention of hardness) are required, a steel of the group with *high hardness at elevated temperatures* must be used. With these steels, the application of bearings is possible up to 540 °C, whereas the steel 100 Cr 6 allows bearing applications only up to 150 °C. Higher temperatures would involve loss of hardness by martensite decomposition, and premature failure would follow.

C 27 Steels for Permanent-Way (Track) Material

By Wilhelm Heller and Herbert Schmedders

Permanent-way material includes steels for rails, switches, and sleepers, plus connecting and fastening items. From the technical and economic points of view, the rails are of paramount importance. Therefore, this chapter is centered on rail steels.

C 27.1 Required Properties of Use

The most important properties of rail steels are governed by the functions that the rail must perform as a carrying element and runway [1].

The forces transferred by the wheel to the rail – such as axle loads, track guidance, acceleration, and deceleration – result in very high dynamic stresses, large deformations, and strain hardening of the rail steel within the contact zone (Fig. C 27.1). Thus, the most important properties of the rail steels within the rail contact zone are the *mechanical properties*, including the yield stress, the tensile strength, and the fatigue strength, plus strain-hardening behavior and deformability.

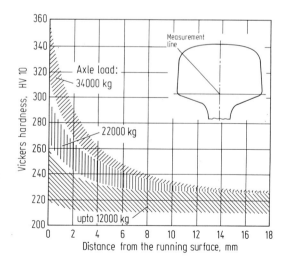

Fig. C 27.1. Strain hardening (characterized by the Vickers hardness HV 10) beneath the running surface of tracks under service conditions (steel grade UIC 70 according to Table C 27.1). For an axle load of 34 and 22 t: UIC 60 shape; for axle loads of up to 12 t: grooved rails.

References for C 22 see page 802.

The amount of wear caused by friction between the wheel and the rail generally determines the service life of the rails. Therefore, the *wear resistance* of the rail steel is an important property.

The need to provide maximum track reliability, particularly for high-speed trains, results in a requirement for sufficient fracture resistance of the rails used [2]. Also, to prevent fatigue failures due to stress to which the rails are exposed as supporting elements, sufficient *strength depending on shape* must be achieved.

Since rails today are usually welded together without any gap between them, they must possess sufficient *weldability*.

Rails must be *resistant to stress cracks* that might propagate rail failure (oval flaws) while in service, and they should possess good cleanness to prevent edge spalling of the running surface [3–5].

As the traveling speeds increase, the rail *straightness* is another requirement, which can be met only by means of roller straightening under carefully controlled conditions.

Corrugations (roaring rails) – i.e., a periodic unevenness of the running surface after a specific service life under operating conditions – are considered system-inherent phenomena that are hard to control by changing the rail steel [1].

C 27.2 Characterization of the Required Properties

This section deals only with properties that are typical of the rail steels and rails and whose characterization requires particular test methods. For other properties mentioned in C 27.1, and their determination, see Part C; for instance, B 1 describes the *mechanical properties*.

The *strength depending on shape* is determined on rail sections about 2 m long using the test equipment shown in Fig. C 27.2 [6].

To determine the fracture resistance, the specifications stipulate an impact test to be performed on whole rail sections. This is a technological test resulting in qualitative findings [7]. More recent tests have shown that the fracture resistance of rails under service conditions can be described quantitatively by the laws of fracture mechanics [8]. Illustrated in a simplified way in Fig. C 27.3, the interrelationship is characterized by the following:

– The fracture toughness K_{Ic} of the rail steel, about 1000 N/mm$^{3/2}$ for pearlitic steels

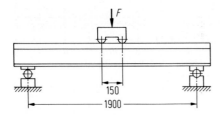

Fig. C 27.2. Equipment used for the fatigue test under fluctuating bending stresses.

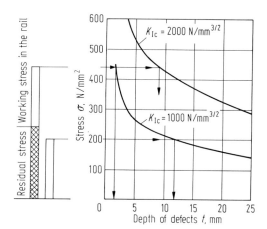

Fig. C 27.3. Fracture resistance of rails as a function of stress and depth of defects (K_{Ic} = plane-strain fracture toughness of the rail steels).

- The working stress within the rail σ, and additionally the residual stresses resulting from roller straightening
- The defects existing in the rail (for instance, fatigue cracks) having a depth t

With a high working stress within the rail of 200 N/mm² superimposed onto a residual stress of 240 N/mm², the rail will already fracture from a 2-mm-deep defect. The higher the fracture toughness and the lower the residual stresses, the rail will, under the same working stress, withstand larger – and thus more readily detectable – defects

Since the rail wear conditions vary greatly with respect to time and space, and the wear mechanisms are likely to occur in different ways and at different intensities, there is not yet a standard test method to determine the *resistance to wear* (see B 10).

C 27.3 Measures of Physical Metallurgy to Attain the Required Properties

The most important tested mechanical property of rail steels with which other test properties and services properties of use are closely associated – such as yield strength, fatigue strength, and wear resistance – is the *tensile strength*, which for rail steels ranges from 700 N/mm² to 1300 N/mm².

Most of the rails used in the as-rolled condition are made from plain-carbon or low-manganese steels that may also contain chromium and minor percentages of vanadium and molybdenum.

Under normal conditions of cooling on a hot bed, rail steels undergo transformation in the pearlite stage, presenting a *ferritic-pearlitic microstructure* within the tensile strength range from 700 N/mm² to 900 N/mm², and a *pearlitic microstructure* for a tensile strength above 900 N/mm². The main properties of rail steels are governed by the fraction (percentage) of each respective microstructure and their morphological formation (shape).

For ferritic-pearlitic steels, the tensile strength is governed mainly by the percentage of pearlite resulting from the carbon content (Fig. C 27.4). The tensile strength is also influenced by the ferrite grain size, the solid-solution hardening, and the lamellar pearlite spacing [9, 10].

In pearlitic steels, it is the *pearlite formation (shape)* that governs the mechanical properties, to a large extent [10, 11]. The yield strength (Fig. C 27.5) and the tensile strength increase as the lamellar spacing decreases.

The *toughness* increases as the lamellar cementite thickness (Fig. C 27.6) and the austenite grain size (Fig. C 27.7) decrease. In addition, another minor effect is exerted by silicon; it causes the tensile strength and the notch-impact energy transition temperature to increase.

Because the lamellar spacing produces different effects on the tensile strength and on the yield strength, the yield/tensile ratio rises from about 0.5 for a tensile strength of 900 N/mm² to more than 0.6 for a tensile strength of 1200 N/mm² (Fig. C 27.8).

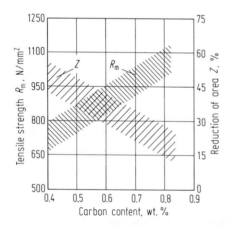

Fig. C 27.4. Tensile strength and reduction of area of a steel containing 1.2% Mn, as a function of the carbon content.

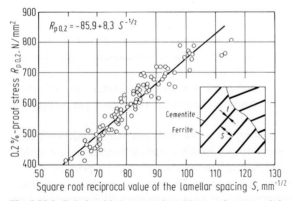

Fig. C 27.5. Relationship between the 0.2%-proof stress and the lamellar cementite spacing in pearlitic steels.

Fig. C 27.6. Relationship between the notch-impact energy transition temperature T_{tr15} and the lamellar cementite thickness in pearlitic steels (T_{tr15} = temperature at which the ductile fracture appearance of a DVMF-specimen amounts to 15%).

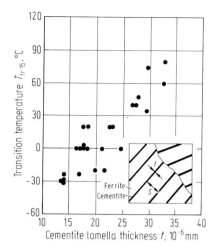

Fig. C 27.7. Relationship between the notch-impact energy transition temperature $T_{tr10.8}$ and the austenite grain size ($T_{tr10.8}$ = temperature at which the notch-impact energy on ISO V-notch transverse specimens amounts to 10.87). After [12]. The austenite grain size of the tested steel with 0.51% C, 0.17% Si, 0.87% Mn, 0.018% P, and 0.013% S was varied by different heat treatments.

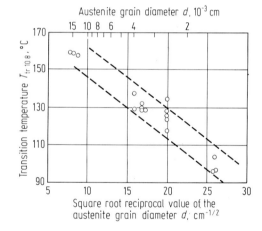

Fig. C 27.8. Relationship between the 0.2%-proof stress and the tensile strength of rail steels.

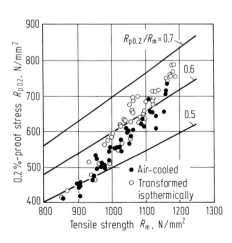

Based on the physical metallurgy interrelationships, it is recommended that the desired strength be attained by starting with a steel of small austenitic grain size and producing a fine-grained ferrite-pearlite or pearlite microstructure with a small lamellar spacing and a small cementite lamella thickness in the pearlite [13].

The *resistance to wear* increases as the tensile strength increases. Judging by the results of laboratory tests and service trials in curves of tracks pertaining to railway companies or industrial plants, a strength increase of 200 N/mm^2 causes the wear resistance to increase by roughly 100% in the rail steels considered here (see C 27.4) [14, 15].

Figure C 27.9 shows that the *fatigue strength* under reversed bending stresses improves as the tensile strength increases. An increase in tensile strength from 700 N/mm^2 to 1200 N/mm^2 will result in a nearly 70% improvement of the fatigue strength under reversed bending stresses [1].

As shown in Fig. C 27.10 for the fatigue strength under fluctuating bending stresses, the strength depending on shape of rails is less dependent on tensile strength than on the rail profile and the surface finish of the rails [6]. It can be seen at the bottom end of the figure that the rail profile exerts an influence only on the section modulus of the rail. If specific requirements must be met with respect to the strength depending on shape, these requirements will have to be taken into account when selecting the respective profile.

Considering the high carbon contents of the rail steels, their *weldability* can hardly be influenced by the steelmaking. For particular rail steel grades, higher contents of manganese, chromium, and other elements that considerably retard the pearlite transformation should be avoided to prevent formation of a martensitic structure subsequent to welding. Figure C 27.11 displays time-temperature-transformation diagrams for typical rail steels containing 1% and 1.5% Mn or 1% Mn and 1% Cr. The welding conditions must take into account the transformation behavior by specifying adequate preheating or preheating plus reheating to make sure that transformation takes place in the pearlitic stage [1, 16, 17].

So far, heat treatment of rails is important only for specific service conditions, such as for heavy-duty haulage. A fine lamellar pearlitic microstructure is attained by causing the transformation in the pearlitic stage to be shifted toward lower

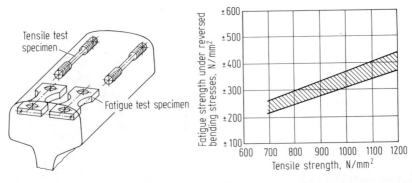

Fig. C 27.9. Fatigue strength under reversed bending stresses in rail steels of different tensile strength.

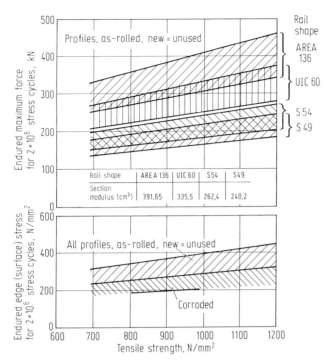

Fig. C 27.10. Strength depending on shape of rails of different shape and surface finish under fluctuating bending stresses (the edge or surface stress relates to the bottom side of the rail base).

temperatures as a result of accelerated cooling [18]. The aforementioned interrelationships of the physical metallurgy hold true here. The service behavior of such heat-treated and as-rolled rails of identical strength is largely the same. When welding such rails, care must be taken to ensure that the amount and extent of softening of the heat-affected zone are limited.

C 27.4 Characteristic Steel Grades and Their Application

The as-rolled steel grades shown in Table C 27.1 have been gaining ground internationally as rail steels. The tensile strength of the heat-treated rails briefly described in C 27.3 is above 1150 N/mm^2 in the zone of the running surface (determined in hardness tests) and gradually drops toward the core of the rail.

For *main line rails* with low axle loads, such as those encountered in suburban traffic, and with low daily runs ($\leq$ 10,000 GRT/day), frequent use is made of the UIC 70 steel grade.

For higher daily runs, in particular on lines with a great number of curves, the use of such steel grades as UIC 90 A, UIC 90 B, or AREA/ASTM-A-1 is more economical. For high axle loads ($\geq$ 25 tons), high average runs per day, and tight

Steel	Chemical composition								
	%C	%Si	%Mn	%P	%S	%Cr	%N	%Al$_{total}$	%Al$_{sol.}$
1	0.71	0.32	1.00	0.016	0.019	0.06	0.006	0.009	–
2	0.64	0.35	1.50	0.020	0.020	0.02	0.006	0.003	–
3	0.71	0.47	0.98	0.018	0.022	1.00	0.003	0.003	0.003

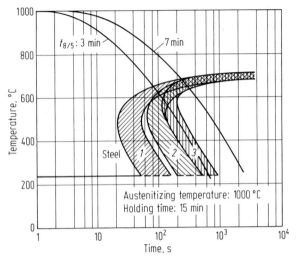

Fig. C 27.11. Time-temperature-transformation diagrams for continuous cooling of three rail steels.

Table C 27.1 Tensile strength and chemical composition of rail steels

Steel grade[a]	Tensile strength N/mm²	%C	%Si	%Mn	%Cr	%V max.	%Mo max.
UIC 700	680–830	0.40/0.60	0.05/0.35	0.80/1.25	–	–	–
UIC 900 A	880–	0.60/0.80	0.10/0.50	0.80/1.30	–	–	–
UIC 900 B	880–	0.55/0.75	0.10/0.50	1.30/1.70	–	–	–
AREA/ASTM-A-1	850–1000[b]	0.67/0.82	0.10/0.23	0.70/1.00	–	–	–
Special steel 1100	1080–	0.60/0.82	0.30/0.90	0.80/1.30	0.70/1.20	0.2[c]	0.1[c]

[a] UIC = Union Internationale des Chemins de Fer (International Railway Association)
[b] AREA = American Railway Engineering Association. Approximate figures calculated on the basis of the chemical composition
[c] Subject to prior arrangement

curves (radii ≤ 600 m), the use of rails made from the 1100 special steel is recommended.

If, under given service conditions, an increase in wear, squeezing, or edge spalling of the running surface is noted on the rail head, the use of high-strength rail steel is usually advantageous.

The steel grades shown in Table C 27.1 are also used for the construction of *switches*. Switch components that are exposed to particular stress, such as crossing frogs for the German Railways, are also heat-treated to obtain a fine pearlitic microstructure (see C 27.3) with a tensile strength from 1100 N/mm^2 to 1350 N/mm^2 [19]. It is also possible to use quenched and tempered steels containing about 0.5% C, 1% Cr, and 0.2% Mo or 0.15% V (50 CrMo 4 or 50 CrV 4) with a tensile strength from 1200 N/mm^2 to 1400 N/mm^2. The straight manganese steel containing about 0.7% C and 14% Mn (X 70 Mn 14) is also used in stressed switch components.

Steel sleepers are usually made from structural steel grades, preferably from the St 37-2 grade [20]. If the ribbed soleplates are welded onto the steel sleeper, adequate weldability is a prerequisite [21].

For *fishplates*, two strength grades are proposed for as-rolled steels under the terms of the UIC specifications [22]. The ISO recommendation under preparation also proposes a heat-treated grade (see Table C 27.2).

Soleplates are supplied in three strength grades under the terms of UIC specifications [23], as shown in Table C 27.3.

The physical metallurgy measures to be applied to obtain the properties required for these permanent-way material items are discussed in C 2 and C 5.

Table C 27.2 Mechanical properties of steels used for fishplates

Tensile strength N/mm^2	Elongation after fracture A % min.
470–550	20
550–640	18
≥ 710[a]	–

[a] Heat-treated

Table C 27.3 Mechanical properties of steels used for soleplates

Tensile strength N/mm^2	Elongation after fracture A % min.
360–440	24
410–490	23
470–540	20

C 28 Steels for Rolling Stock

By Klaus Vogt, Karl Forch, and Günter Oedinghofen

C 28.1 General Remarks

The category "steels for rolling stock" includes materials used in the production of shafts, wheel discs, and wheel tires, as well as solid wheels (internationally, also called "monobloc-wheels").

Several decades of railway development have led to similar delivery conditions in several European countries. Thus, the specifications of the International Railway Organization (UIC) in Paris are valid in Europe and beyond. These specifications also apply to steel grades, production processes, and properties of the steel products. It is hoped that a further worldwide standardization of conditions can take place within the scope of the International Organization for Standardization (ISO).

C 28.2 Required Properties of Use

For all steels used in the components mentioned in C 28.1, there are requirements for the *mechanical properties* (tensile strength, fatigue strength, and toughness) based on function and service stress. In steels used for wheel tires and solid wheels, there are additional requirements for *resistance to wear* and to tread damage, which are related to the *heat crack susceptibility*. These requirements are difficult to meet, because these properties conflict with each other. References [1–4] discuss this important complex of tread damage of wheel tires and solid wheels.

Independent of the steel, the *profile shape* of the wheel rim and rail and the running properties of the vehicles have a considerable influence on the attainable running performance (efficiency). To obtain realistic load parameters in the design (shape) and dimension of wheel set components, new processes have been developed in recent years that enable favorable stress distributions to be calculated and used in component design [6–8].

References for C 28 see page 802.

C 28.3 Characterization of the Required Properties

The *mechanical properties* of steels for wheel discs, wheel tires, solid wheels, and shafts are determined by *tensile test* and by *notched-bar impact test* (see B 1). Because of the peculiarities of these products, the tensile and impact specimens are taken in certain specific positions and directions. In addition, other tests may be needed: the preparation of sulfur prints with limiting values, hardness tests over the running rim or wheel section, fatigue strength tests, the drop-weight test, and ultrasonic and magnetic powder tests. A typical distribution of hardness over the cross section of a heat-treated (quenched and tempered) rim of a monoblock wheel (see below) with about 0.5% C, 0.3% Si, and 0.8% Mn is given in Fig. C 28.1.

To characterize the susceptibility of the steels to *crack formation by temperature changes*, "thermoshock" tests are carried out on specimens whose surface zones are inductively heated to about 600 °C to 900 °C and then cooled immediately afterward in water. This test simulates a possible heat stress arising during block-braking or caused by a high slip between wheel and rail. The number of thermocycles up to the occurrence of flaws (cracks) within the sample surface is measured.

It is not yet possible to clearly define the best chemical composition, microstructure, and selection of steel grades for operating practice [9].

C 28.4 Measures to Attain the Required Properties

To meet the required mechanical properties, measures are taken that are based on the principles of the physical metallurgy, treated in detail in B 1. To obtain *tensile strength* and *toughness*, one of the following microstructures is aimed for: (1)

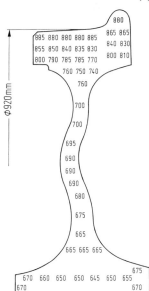

Fig. C 28.1. Tensile strength (N/mm^2) calculated on the basis of the measured Brinell hardness and its distribution over the cross section of a rim-heat-treated solid wheel of a steel with about 0.5% C, 0.3% Si, and 0.8% Mn.

a *microstructure* of a ferritic matrix with insertions of pearlite in proportion to the carbon content, B i.e., a microstructure that is available in unalloyed steels with a relatively low carbon content after hot forming; (2) a more uniform and fine-grained microstructure obtained after normalization; or (3) a quenched and tempered microstructure. For the different components, the following paragraphs apply.

For *wheel discs* for wheels with tires, which are manufactured from forged bodies with a straight or corrugated disc part (blade), as well as for wheel stars made of cast steel for electrical and steam locomotives, steels with a microstructure corresponding to the hot-formed or the normalized condition are used.

For a weight-saving design, such as a double-corrugated light wheel disc for wagons, low-alloy steels for quenching and tempering in a condition of the corresponding microstructure may be considered.

Wheel tires are used in the lower range of the required tensile strength (see below) in the rolled condition, or the microstructure of the normalized condition is adjusted. Wheel tires in higher strength ranges are heat-treated to obtain a quenched and tempered microstructure.

Solid wheels in the lower strength range (see below) are normalized or remain in the as-rolled condition. In the case of unalloyed steels, a nearly complete pearlitic-ferritic microstructure results. In the higher strength range and in the case of alloyed steels, a heat treatment of the whole solid wheel can be performed by oil quenching and tempering. For unalloyed steel, a microstructure of bainite and martensite in the areas near the surface results. For alloyed quenched and tempered steel, a microstructure of bainite and martensite results over deeper ranges of the cross section, while in the core zone, pearlite with amounts of ferrite is present. A special heat-treatment procedure, the so-called *wheel rim heat treatment*, has been developed and is often used for solid wheels of unalloyed and alloyed steels with low carbon content. In general, the wheel is positioned horizontally in a special hardening machine; then, under slow turning of the wheel, the rim area and the flange are gradually cooled from austenitizing temperature by means of circumferentially adjusted nozzles. The wheel disc itself is excluded from the water attack, so that in this area a near normalizing takes place. After this, the whole wheel is tempered (i.e., according to the chemical composition), in general, at least at 500 °C. This heat treatment aims to adjust distinctly lower strength values and thus achieve a better toughness of the disc, while the strength values of the rim zone of the wheel are high, which is favorable with regard to the wear behavior. Moreover, this heat treatment offers the possibility of combining high strength with good toughness in the rim. The wheel-rim heat treatment is far reaching and has been introduced internationally as a standard heat treatment for solid wheels of unalloyed steels with higher strengths.

In addition to the usual manufacturing in Europe that is done by prepressing in a die under a forging press and finish rolling by grooved rolls, solid wheels are also manufactured by a pressure-pouring process technology developed in the United States. This finish is used especially for wagons in North America.

In the case of wheel tires and solid wheels, special attention must be paid to the contradictory demands for *wear resistance* and *resistance to heat cracking* (see

C 28.2). For these steels and the kind of wear stress in question, wear generally decreases with higher material hardness and higher tensile strength (see B 10 for details). Because hardness and tensile strength depend essentially on the carbon content, steels with higher carbon content generally show a lower wear under traffic. The carbon content can be compensated for to a certain degree by alloying elements. Depending on the increase of tread strength by the application of higher carbon contents, on the other hand, a higher susceptibility to formation of certain tread damage is observed. Friction heat, especially quick and local heating caused by braking or, at high slip, by temperatures over the austenitizing temperature [10], may cause changes in the microstructure in the outer zone of the tread. Because of the quick heat flow into the direction of the cold part of the wheel, a martensitic zone of high hardness and brittleness forms. This excessive thermal hardening causes fine and mostly short surface cracks, and traffic is the source of material fracture up to fatigue fracture.

Thus, it is necessary to achieve the best possible combination of these contradictory influences. The carbon content and the carbide balance cannot be fixed only by considering the abrasive wear decrease. The unfavorable effect of high carbon contents on the resistance to thermoshock and on the martensitic formation by friction also must be considered. Therefore, attempts have been made to develop a substitute for the conventional manganese steel with carbon contents between about 0.40% and 0.70% C. The substitute steels should have lower carbon contents, and special carbide-forming elements or alloying elements forming solid solution are added to increase strength. In a project to investigate the limits of the wheel-rail system, unusual materials (for instance, relatively low-carbon steels containing about 0.25% C and austenitic steels) are being tested. Tests are being conducted on a steel with a reduced carbon content of below 0.10%. By using low-alloy steels instead of plain manganese steels, the possibility also exists to increase the impact values and, thus, the resistance to brittleness, combined with high strength.

In general, a fine, streaky pearlite is regarded as *a favorable microstructure in the tread area* of wheels and wheel tires. The above-mentioned wheel rim heat treatment of manganse steels leads to a structure that consists largely of such fine, streaky pearlite, in addition to the same amounts of bainite. With the appropriate heat treatment of alloyed low-carbon steels, a bainite-martensite microstructure is formed that is converted by tempering to a microstructure of the quenched and tempered condition. Results concerning the influence of the microstructure on the amount of the abrasive wear are not yet available.

The same microstructure is desirable if other wear types, such as crack initiation by thermoshock or fatigue and crack propagation under the existing cyclic stress, are considered. Recent results [11] show that, with the same chemical composition, the crack propagation da/dN (a = crack length, N = number of stress cycles) in the range of mean velocities does not depend on the type of the microstructure. The extension of the remaining area of fracture is lower the higher the toughness of the steel.

An improvement of the notch-impact energy cannot be attained on the basis of an unalloyed steel under retention of a fine pearlitic structure. Tests are being run to increase the toughness of such steels by developing a special, fine-grain microstructure using additions of special carbide-forming elements. In all, the development of low-alloy steels for quenching and tempering seems to be at least promising.

Wheelset shafts are produced either by forging or rolling or by rolling with subsequent forging, depending on the delivery conditions and the existing production facilities. With regard to the properties of steels, a microstructure of the normalized or the quenched and tempered condition is adjusted. In the case of driving shafts, only a quenching and tempering of the steels can be considered.

For wheelset shafts, the fatigue strength under reversed bending stresses is of great importance. It depends not only on the mechanical properties of the steel, but also on the shape of the shafts. Therefore, this constructional shape must be especially observed. It is one of the measures to ensure endurance under the service conditions of the traffic.

C 28.5 Characteristic Steel Grades Proven in Service

For *wheel discs*, unalloyed steels with carbon contents between about 0.22% and 0.45% (e.g., steel grades Ck 22, Ck 35, and Ck 45) are generally applied. In the normalized condition, these steels have tensile strength values of about 400 N/mm^2 to 750 N/mm^2. For these products material development is neither necessary nor advisable.

For the double-corrugated light wheel discs, as mentioned in C 28.3, a steel with about 0.46% C, 1% Mn, and 0.8% Si (steel grade 46 MnSi 4), quenched and tempered to tensile strength values between 750 N/mm^2 and 850 N/mm^2, is used.

The unalloyed steels used mainly for the manufacture of *wheel tires* are characterized by carbon contents between 0.4% and 0.7% and manganese contents up to about 1%. In special cases, conventionally alloyed steels for quenching and tempering – e.g., with about 0.5% C, 1% Cr, and 0.20% Mo (steel grade 50 CrMo 4) – or low-alloy, low-carbon steels for quenching and tempering – e.g., with chromium, molybdenum, silicon, and also boron additions – are applied in the normalized or quenched and tempered condition (such as steel grade 26 MnMoB 64 with about 0.26% C, 1.5% Mn, 0.4% Mo, and 0.003% B).

In the production of *solid wheels*, essentially the same steel types as for wheel tires are applied. These are unalloyed steels with carbon contents between about 0.4% and 0.7% and manganese contents up to about 1%. Occasionally, steels with higher carbon content, up to 0.77% – e.g., for wheels of railways in North America – are also applied. In special cases, alloyed steel grades – e.g., with about 0.50% or 0.58% C, 1% Cr, and 0.25% Mo (steel grade 50 CrMo 4 or 58 CrMo 4) – are chosen. Depending on the heat treatment and microstructural condition, tensil strength values of 600 N/mm^2 to 1200 N/mm^2 are achieved. In addition, in large-scale service, different quenched and tempered alloyed steels with partially reduced carbon content between about 0.2% and 0.3% are applied during traffic and test drive (see C 28.3). In this case, the necessary strength is attained by alloying additions of, for instance, chromium, manganese, and silicon. The present state of development is not yet sufficient to substitute new materials for the conventional steels [12–15].

For *traversing wheel shafts* with diameters of about 150 mm, unalloyed steels with about 0.35% or 0.45% C (e.g., Ck 35 or Ck 45) usually are applied. These steels are used in the untreated, normalized, or quenched and tempered condition. According to the respective microstructure, the tensile strength lies in the range of about 500 N/mm^2 to 700 N/mm^2, with notch-impact energy values up to about 30 J (determined with ISO-U-notch longitudinal specimens at 20 °C).

For the more highly stressed driving wheel shafts with diameters up to about 250 mm, alloyed steels for quenching and tempering are applied. Steel grades with about 0.25% C, 1% Cr, and 0.20% Mo (25 CrMo 4); with about 0.34% C, 1.5% Cr, 0.20% Mo, and 1.5% Ni (34 CrNiMo 6); and others have been introduced and have withstood the tests. The heat treatment is performed by quenching in liquids and tempering. The chemical composition of the heat-treatable steels for driving wheel shafts guarantees, in the quenched and tempered condition, a nearly ferrite-free microstructure of bainite and martensite over the whole cross section. The unalloyed steels for the carrying wheel set shafts show, in all heat-treated conditions, a structure of pearlite and ferrite. Depending on the microstructure, tensile strength values are in the range of about 650 N/mm^2 to 1000 N/mm^2, with values for the notch-impact energy of about 30 J up to 60 J (with ISO-U-notch longitudinal specimens at 20 °C).

Further data, generally valid for most of the above-mentioned steel grades, can be found in DIN 17 200 [16]. For details of the steels and products, reference should be made to the international regulations of the ISO and UIC (see C 28.1) [17, 18].

C 29 Steels for Screws, Bolts, Nuts and Rivets

By Klaus Barteld and Wolf-Dietrich Brand

C 29.1 Required Properties of Use

In steels that are used for the cost-efficient manufacture of large series of such mass-produced pieces as screws, bolts, nuts, and rivets, and also for special fabrications to meet extremely high requirements, *cold formability* (for definition and details, see B 7) is the main property of use upon which correspondingly high demands are placed [1]. At the same time, high demands are placed on the homogeneity of the microstructure and on the deformability, as well as on the surface quality of the steel products (merchant bars or wire [2]). Figure C 29.1 provides an impression of the steps involved when cold forming the steels into bolts of different shapes.

For large components, good *hot formability* (see B 6) is also required, depending on the manufacturing technique.

For certain modes of component manufacture, forming involves machining; for example, when processing the steels into smaller numbers of pieces, or when the mechanical properties of fasteners made of steels must meet high requirements (see below). Consequently, demands are additionally placed on the *machinability* (see

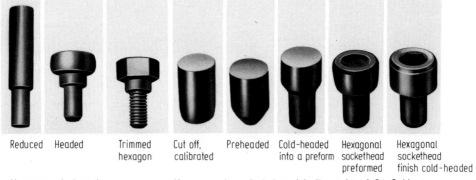

Fig. C 29.1. Characteristic examples of the steps involved in the cold forming of steel into screws and bolts (41 Cr 4: about 0.41% C and 1.1% Cr; 34 CrMo 4: about 0.34% C, 1.1% Cr, and 0.25% Mo).

References for C 29 see page 803.

B 9). If these requirements can be met only by free-cutting steels (see C 19), attention must be paid to the fact that they are alloyed with substances that, either alone or in the form of compounds, have a chip-breaking effect. This means that the service properties of the steels and of the fasteners manufactured from them can be adversely affected. For this reason, free-cutting steels are allowed only a limited use for screws, bolts, and nuts by certain standards (e.g., in DIN ISO 898, Teil 1 [3]).

The required *mechanical property* values, upon which the design of the fastener is based, must be achieved within tight scatter ranges, with service life being of decisive importance. Screw joints are very sensitive to fatigue failure as a result of the extremely high notch effect of the thread. To improve the service life of screw joints, constructional measures should be taken [3a]. Where the steels designed for making screws and bolts are concerned, DIN ISO 898, Teil 1, is recognized in many instances as the basis for the requirements to be met by the mechanical properties. [3, 4].

To attain the required mechanical properties in parts manufactured by cold forming, strain hardening must be taken into account and suitably combined with the other properties being considered (see B 7) [5].

High-strength screws, bolts, and nuts are subjected to heat treatment after forming. The materials therefore must display great uniformity of hardening throughout the cross section of the fastener and must fulfill corresponding *hardenability* requirements [6]. Screws involving only small deformations (e.g., long-shank screws and long-shank ball pins) increasingly are being made from quenched and tempered wire rod or merchant bar material.

To do this, no further heat treatments are necessary, but quenching and tempering values within a narrow scatterband are required throughout the entire length of the wire. The case hardening steels used for special screws, bolts, and rivets must be suitable mostly for quenching directly from the carburizing operation and, therefore, must be fine-grained (with grain size indexes > 5, see DIN 50 601 [7]) [8, 9].

For stainless steels used in the manufacture of fasteners, *corrosion resistance* to attack by different corrosive media and forms of corrosion is required, in addition to certain mechanical property and cold formability values [10].

Unalloyed steels are processed in considerable quantities into weld nuts; e.g., for automobile body construction. These steels must be suitable for *resistance welding*. Such suitability is displayed by the low-carbon steels generally used for screws, bolts, nuts, and rivets. The high thermal stresses generated locally by resistance welding, and their effects on the mechanical properties of the work-hardened material, must be considered.

Different demands are placed on the *surface quality* and *dimensional accuracy* of the products made of the steels under consideration here, depending on what parts are to be manufactured. For example, for the manufacture of screws, bolts, nuts, and rivets, use is made of wire rod or rolled bars having the dimensions and admissible dimensional tolerances as specified, for example, in DIN 1013, 59 115, or 59 130 [11–13]. Low-carbon steels are processed in the untreated condition; heat-treatable steels are processed after a spheroidizing annealing (GKZ) and, in special cases, in the quenched and tempered condition; and transformation-free steels are processed after solution annealing and subsequent quenching, if necessary, in the descaled or coated condition for better lubricant adhesion. When high dimensional accuracy is required, use is made of bright steel that has a drawn, peeled, ground, or ground and polished surface with correspondingly restricted, standardized tolerances, and, where applicable, is coated and redrawn. Wire rod and bright wire are mainly cold-formed, while merchant bars are mostly hot-formed or machined.

C 29.2 Characterization of the Required Properties

The basic criteria for characterizing the *cold formability* of steel can be found in B 7. For the steels mentioned in this section, the *tensile strength* and *reduction of area*, as determined by *tensile testing*, have proven suitable for a simple characterization of cold formability. A low tensile strength and a high reduction of area are aimed for [14]. Although good deformability is mostly achieved in this way, extremely low strength values make shearing of the wire lengths difficult on the cold-forming machines. Here, however, a remedy can be provided through a slight strain hardening of the region near the surface. This calibration additionally improves the dimensional accuracy of the feedstock.

Since steels having a higher pearlite proportion in their microstructure – i.e., steels containing more than about 0.2% C – are mostly spheroidized for cold-forming (see Fig. C 29.2), the *microstructure* is also frequently taken as a means of assessing cold formability. The microstructure is assessed by means of metallographic techniques (see, e.g., SEP 1520 [15]).

For the characterization of *hot formability*, reference should be made to B 6. Details concerning the characterization of the *machinability* of steels can be found in B 9.

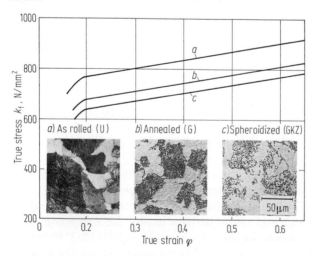

Fig. C 29.2. Change of the true stress-true strain curve of an unalloyed steel containing about 0.35% C (Cq 35) used for cold-forming brought about by annealing to a maximum admissible tensile strength (G) and by spheroidizing (GKZ).

Apart from the tensile strength following forming and, where applicable, following heat treatment, the *yield stress* and its relation to the tensile strength are important material characteristics where the *mechanical properties* are concerned. This also includes the toughness of the steels, because the manufactured pieces are notched many times as a result of the thread and of the change in section from shank to head. An indication of the toughness is given by the *elongation after fracture* determined, as already mentioned, in tensile tests. In many cases, the *notched bar impact energy* of the steels is also determined in *notch-impact tests*. Because of its importance for the service behavior of the screws and bolts, the toughness in the finished part is also ascertained: The *angular tensile test* (e.g., in accordance with DIN ISO 898, Teil 1 [3]) is a suitable technique for testing the service behavior of screws and bolts with regard to toughness and also tensile strength (Fig. C 29.3).

The *hardenability* is ascertained, if necessary, by the *end-quench hardenability test* (see B 4.6.1).

During cold-heading operations, the surface of the steel being processed must not split open. The *surface quality* of the steel products is examined by means of the *compression test*, in which a section with a length 1.5 times the diameter is compressed to 0.5 times the diameter (see also Fig. C 6.3). Cracks, rolling fins, pores, and scars on the surface, as well as nonmetallic inclusions close to the surface, can lead to splits and must therefore be avoided as far as technically and economically possible [16]. For smaller steel product diameters (e.g., < 5 mm), the *alternating torsion test* is selected instead of the cold compression test. In the alternating torsion test, specimens with a length 50 times the diameter are twisted in alternating directions (see, e.g., DIN 51 212 [17]). To characterize the surface quality, use can also be made of *magnetic-particle* or *ultrasonic methods*. The admissible depth of surface discontinuities is classified according to the method of processing (e.g., cold-heading or machining), taking into consideration any necessary heat treatment.

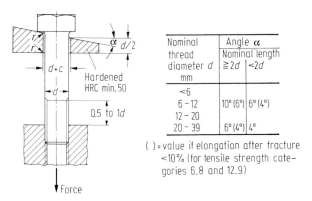

Nominal thread diameter d mm	Angle α Nominal length $\geq 2d$	$<2d$
<6		
6–12	10° (6°)	6° (4°)
12–20		
20–39	6° (4°)	4°

( ) = value if elongation after fracture <10% (for tensile strength categories 6.8 and 12.9)

Fig. C 29.3. Angular tensile test for testing the toughness and tensile strength of screws and bolts (in accordance with DIN ISO 898, Teil 1 [3]).

A further characteristic of the surface quality of the steel products used in screw, bolt, nut, and rivet manufacture is the *surface decarburization*. To ensure minimum surface decarburization, *metallographic inspection techniques* are used. As long as none of the surface is removed, all steels undergo more or less decarburization during hot-forming into wire, rods, or bars. Therefore, values for the admissible degree of decarburization must be fixed in each case (see, for example, the values for the degree of decarburization admissible for heat-treatable steels in DIN 1654, Teil 1 [18]). For instance, particularly in the case of screws and bolts with a rolled, fine pitch thread, the load capacity of the thread in a decarburized surface layer would not meet the given tensile strength category.

The *dimensional accuracy* of the rolled or drawn steel products (wire or merchant bars) exerts an important influence on their suitability for being worked into fasteners, as well as on their mechanical properties after cold-forming. Dimensional fluctuations lead to nonuniform work hardening and underfilling or overfilling of the dies, which can result in tool fracture. The dimensional standards stated in C 29.1 are adequate for most fields of application.

C 29.3 Measures of Physical Metallurgy to Attain the Required Properties

To achieve the described properties in the steels used for fasteners, special care must be taken during the steelmaking operation and during the subsequent processing into wire rod, bars, and bright steel. The chemical composition and cleanness must be optimized to achieve the required properties (e.g., tensile strength, yield stress, elongation after fracture, cold formability, or machinability).

In steels used for the cold fabrication of parts where *cold formability* is important (see B 7), a *microstructure* consisting of ferrite and small proportions of pearlite – i.e., an unalloyed steel with a low carbon content – should be aimed for. Such steels can be cold-formed without any special treatment, in the as-rolled condition. The higher the carbon and alloy contents, the lower the deformability and the higher the "flow stress" (the true stress for a given true plastic strain, see B 7). In addition to carbon, manganese and such tramp elements as phosphorus, sulfur, and nitrogen impair the deformability and reduce the toughness; these elements must be limited or reduced in the steel.

In contrast to a coarse-grained microstructure, a fine-grained microstructure leads to better toughness. Consequently, fine-grained steels are preferred for the manufacture of fasteners despite their somewhat lower deformability and hardenability. Grain fineness is achieved by adding elements that lead to the formation of precipitates during appropriate cooling from the rolling or annealing temperature; these precipitates bring about fine-grained recrystallization or impede grain growth in the austenite.

In steels having an elevated carbon and alloy content, the lamellar *carbides* of the pearlite must be *spheroidized* prior to cold-forming in order to achieve low flow stress (see above) and better deformability (see Fig. C 29.2). Depending on the chemical composition, an optimum annealed microstructure is achieved by choosing suitable annealing conditions, with allowances made for the initial microstructure in the as-rolled condition. For example, rapid cooling after hot rolling improves spheroidizability. At the same time, the characteristic features of carbide precipitation and of spheroidizability over normal annealing periods must not be overlooked. Generally, prolonged spheroidizing ("GKZ") is practiced, with a spheroidized pearlite proportion of more than about 70%, distributed uniformly in the structure, being adequate (Fig. C 29.4). Higher degrees of spheroidizing are achieved if, beforehand, cold-forming is performed to break carbide lamellae; e.g., by drawing. In addition to the spheroidization of the pearlite, the *distribution of the carbides* is important. The carbides must be distributed, not mainly at the grain boundaries, but uniformly in the structure.

Stainless, austenitic steels are solution-annealed and quenched. This treatment dissolves the precipitations in the microstructure that impede formability.

For the steels of the higher tensile strength categories used for manufacturing screws, bolts, and nuts, *hardenability* is of central importance with regard to the required mechanical properties. Generally, refer to B 4 and C 5. Particularly in steels mentioned in this chapter, boron has proved very useful for increasing hardenability, as shown in Fig. C 29.5. Despite the improvement in hardenability brought about by boron, the formability is hardly impaired (see the curves for steels 2 and 3 in Fig. C 29.6). The addition of boron renders unalloyed steel grades, also in thicker wire dimensions, suitable for oil hardening, while without boron they could be hardened only in water [19–21]. The true stress-true strain curve characterizing

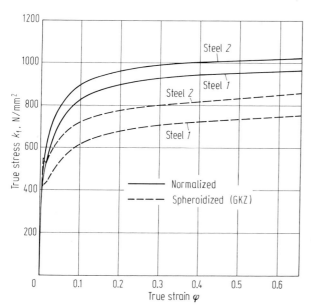

Fig. C 29.4. Influence of a long-time spheroidizing (GKZ) on the true stress-true strain curve of two case-hardening steels (steel 1 – 16 MnCr 5: about 0.16% C, 1.2% Mn, and 1% Cr; steel 2 – 20 MoCr 4: about 0.20% C, 0.8% Mn, 0.4% Cr, and 0.45% Mo).

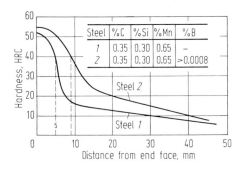

Fig. C 29.5. Hardenability, characterized by end-quench hardenability curves (mean values) of two steels for quenching and tempering used for cold-forming and having identical carbon contents but different boron contents (steel 1: Cq 35; steel 2: 35 B 2).

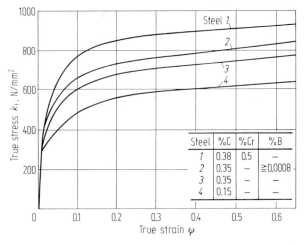

Fig. C 29.6. True stress-true strain curves of two steels (1 and 2) having roughly identical hardenability but different boron contents. (Additionally plotted as a comparison are the curves of two unalloyed steels used for cold forming (3 and 4)). All the steels were tested after spheroidizing (GKZ) (steel 1: 38 Cr 2; steel 2: 35 B 2; steel 3: Cq 35; steel 4: Cq 15).

the cold formability of a steel containing about 0.38% C and 0.5% Cr (38 Cr 2) is appreciably higher, and therefore worse, than that of a boron steel with comparable hardenability and an identical carbon content (35 B 2) (Fig. C 29.6). Plotted as a comparison are the true stress- true strain curves for a steel not alloyed with boron but containing about 0.35% C (Cq 35) – i.e., with the same basic composition but lower hardenability – and for a lower carbon unalloyed steel containing about 0.15% C (Cq 15).

Bolts and fasteners are traditionally manufactured from medium-carbon steels that sometimes contain alloying elements such as chromium, molybdenum, and nickel. As is well known, low-carbon microalloyed steel grades can be used for fasteners. In addition to steel grades containing boron, a niobium-boron microalloyed steel was developed with properties for bolts of class 10.9. This steel is produced without any heat treatment [21a, 21b].

The measures taken to obtain the other important properties (see C 29.1) of the steels discussed in this chapter are identical to those described for comparable steels in other chapters (see, for example, B 1 and B 9).

C 29.4 Characteristic Grades Proven in Service

As discussed in C 29.3, steels with a ferritic matrix and only small proportions of pearlite (in other words, mainly *low-carbon unalloyed steels*) present good cold formability without any special treatment and are therefore particularly well suited for manufacturing screws and bolts in cold condition. Table C 29.1 lists two examples of such steels – namely, the two grades UQSt 36 and UQSt 38 – which are also shown in Table C 6.1 but have been mentioned again here because they are two unalloyed steels typical in screw and bolt manufacture that can meet the most stringent surface and deformability requirements. For screws, bolts, or rivets of a simple shape and limited thickness that do not have to comply with such high surface requirements, the USt 36 steel grade listed in Table C 29.1 may be used. Since the case hardenability of these unkilled steels is impaired by the inherent segregations, use is frequently made of the killed RSt 36 and RSt 38 steel grades or, where particular demands are placed on the finished pieces, of steels such as Cq 15 (see Tables C 29.2 and C 6.2), which are better suited for heat treatment and have good cold upsetting characteristics. The low-carbon steels, as well as specially killed steels (e.g., QSt 32-3 in Table C 6.1), undergo relatively little strain hardening during cold forming, with the result that it is possible to achieve high degrees of forming with them, such as in the case of screws, bolts, or rivets that have a particularly flat and wide head.

Tables C 29.2 and C 29.3 present a selection of the *case hardening steels, steels for quenching and tempering* and *stainless steels* [23, 24, 26] standardized in DIN 1654, Teil 3 to 5, together with their characteristics for the two states of delivery: (1) spheroidized ("GKZ") for the steels having a γ/α transformation, and (2) quenched for the transformation-free materials. The mechanical property values given here are not limiting values specified in the standard, but exemplary reference values that have proven successful and been adopted in practice. This has led in individual cases to differences compared with the values in Tables C 6.2 to C 6.4, in which some of these steels are already mentioned yet have been listed again here because of their importance for screw and bolt manufacture. Tables C 29.2 and C 29.3 contain information about the use of each steel grade; this information is not complete, but is intended to facilitate the proper coordination and choice of steel to achieve the tensile strength categories required by DIN ISO 898, Teil 1 [3].

Of increasing importance in the manufacture of screws, bolts, and nuts, though not in terms of the quantities used, are *creep resisting* (elevated temperature) *steels* (see, for example, the steels specified in DIN 17 240 [28] and DIN 267, Teil 13 [29]). For the most part, these steel grades are supplied in the form of merchant bars, which are hot-formed and machined to make the fasteners. In the few cases in

Table C 29.1. Chemical composition and mechanical and technological properties of low-carbon, unalloyed, non-heat-treatable steels used for the manufacture of screws, bolts, nuts, and rivets (for further details, see [22]).

Steel grade (designation)	Chemical composition					Mechanical and technological properties in the hot-rolled condition[a]				
	% C max	% Si	% Mn	% P max	% S max	Tensile strength N/mm^2	Yield stress N/mm^2 min	Elongation after fracture A % min	Notched-bar impact energy (ISO-V-specimen) mean value J min	at °C	Compression test $h_1:h_0 = 1:3$ at °C
U St 36	0.14	traces	0.25–0.50	0.050	0.050	330–430	205	30	27	+20	900
UQ St 36	0.14	traces	0.25–0.50	0.040	0.040				27	+20	20
R St 36	0.14	≤0.30	0.25–0.50	0.050	0.050				27	+10	900
U St 38	0.19	traces	0.25–0.50	0.050	0.050	370–460	225	25	27	+20	900
UQ St 38	0.19	traces	0.25–0.50	0.040	0.040				27	+20	20
R St 38	0.19	≤0.30	0.25–0.50	0.050	0.050				27	+10	900
U 7 S 6	0.10	traces	0.30–0.60	0.050	0.04–0.08	(310–440)	(205)	–	–	–	–
U 10 S 10	0.15	traces	0.30–0.60	0.050	0.08–0.12	(340–470)	(225)	–	–	–	–

[a] The values in brackets are intended for information purposes only

Table C 29.2 Chemical composition, mechanical properties, and hardenability of case-hardening steels and steels for quenching and tempering used for the manufacture of fasteners [23, 24]

Steel grade (designation)	Chemical composition[a]						Mechanical properties in the spheroidized condition (GKZ)[c]		Hardenability		Examples of application[b]
	% C	% Si	% Mn	% Cr	% Mo	% Ni	Tensile strength N/mm² max	Reduction of area % min	Core hardness HRC min	Diameter[d] mm max	
Cq 15	0.12–0.18	0.15–0.35	0.25–0.50	–	–	–	440	65	–	–	Rivets, sheet metal and drilling screws
16 MnCr 5	0.14–0.19	0.15–0.40	1.00–1.30	0.80–1.10	–	–	550	62	–	–	Rivets, nuts, special screws
Cq 35	0.32–0.39	0.15–0.35	0.50–0.80	–	–	–	550	60	40	8	Nuts for tensile strength categories 8 and 10; 8.8 screws up to M 8
35 B 2[e]	0.32–0.40	0.15–0.40	0.50–0.80	–	–	–	530	62	40	18	Nuts for tensile strength categories 10 and 12; 8.8 screws up to M 20
34 Cr 4	0.30–0.37	0.15–0.40	0.60–0.90	0.90–1.20	–	–	580	60	42	24	8.8 screws up to M 24 10.9 screws up to M 18
37 Cr 4	0.34–0.41	0.15–0.40	0.60–0.90	0.90–1.20	–	–	600	60	44	24	10.9 screws up to M 20 12.9 screws up to M 8
41 Cr 4	0.38–0.45	0.15–0.40	0.50–0.80	0.90–1.20	–	–	610	59	45	26	10.9 screws up to M 26 12.9 screws up to M 8
34 CrMo 4	0.30–0.37	0.15–0.40	0.50–0.80	0.90–1.20	0.15–0.30	–	600	60	45	22	10.9 screws up to M 24 12.9 screws up to M 16
42 CrMo 4	0.38–0.45	0.15–0.40	0.50–0.80	0.90–1.20	0.15–0.30	–	630	59	48	28	10.9 screws up to M 30 12.9 screws up to M 24
34 CrNiMo 6	0.30–0.38	0.15–0.40	0.40–0.70	1.40–1.70	0.15–0.30	1.40–1.70	680	62	48	30	12.9 screws up to M 32 14.9 (special) screws up to M 12
30 CrNiMO 8	0.26–0.33	0.15–0.40	0.30–0.60	1.80–2.20	0.30–0.05	1.80–2.20	700	62	48	36	12.9 screws, 14.9 (special) screws up to M 16

[a] The phosphorus and sulfur content is max. 0.035% in each case.
[b] The tensile strength categories of the screws and bolts and their characteristic codes are laid down in DIN ISO 898, Teil 1[3] and DIN 267, Teil 3[25], where the first digit signifies 1/100 of the nominal tensile strength of the screws or bolts, and the second digit indicates 10 times the ratio of the nominal yield stress (or nominal 10.2% proof stress) to nominal tensile strength of the screws or bolts; the last digit together with M (e.g. M 8) signifies the thread diameter in mm.
[c] GKZ spheroidized
[d] Diameter up to which a core hardness of at least 40 to 48 HRC (see adjacent column) is attained after hardening in oil.
[e] 0.008% to 0.0050% B.

Table C 29.3. Chemical composition and mechanical properties of stainless steels used for the manufacture of fasteners [26]

Steel grade (designation)	Micro-structure	Chemical composition[a]				Mechanical properties[b]				Example of application[c]
						In condition GKZ[b]		quenched		
		%C	%Cr	%Mo	%Ni	Tensile strength N/mm² max	Reduction of area % min	Tensile strength N/mm² max	Reduction of area % min	
X10Cr13	Martensitic	0.08–0.12	12.0–14.0	–	–	600	60	–	–	Screws and nuts C1
X22CrNi17	Martensitic	0.15–0.23	16.0–18.0	–	1.5–2.5	850	55	–	–	Screws and nuts C3
X12CrMoS17	Martensitic	0.10–0.17	15.5–17.5	0.20–0.30	–	650	55	–	–	Screws and nuts C4
X8Cr17	Ferritic	≤0.10	15.5–17.5	–	–	–	–	570	63	Screws and nuts for special uses
X1CrMoNb182[d]	Ferritic	≤0.015	17.0–19.0	1.8–2.3	≤0.25	–	–	550	70	Screws, nuts, rivets for special uses, also in place of A2 and A4
X5CrNi1911	Austenitic	≤0.07	17.0–20.0	–	10.5–12.0	–	–	680	55	Screws and nuts A2
X5CrNiMo1810	Austenitic	≤0.07	16.5–18.5	2.0–2.5	10.5–13.5	–	–	680	55	Screws and nuts A2
X10CrNiMoTi1810[e]	Austenitic	≤0.10	16.5–18.5	2.0–2.5	10.5–13.5	–	–	680	55	Screws and nuts A4

[a] In addition for the martensitic and ferritic steels: max. 1% Si, max. 1% Mn, max. 0.045% P and max. 0.030% S (except for steel X12CrMoS17 for which a maximum of 1.5% Mn and 0.15% to 0.35% S apply), for the austenitic steels: max. 1% Si, max. 2% Mn, max. 0.045% P and max. 0.030% S.
[b] "GKZ" = spheroidized.
[c] the abbreviations relate to DIN 267, Teil 11 [27].
[d] $Nb \geq 15 \times (\%C + \%N) \leq 30 \times (\%C + \%N), \%C + \%N \leq 0.025$.
[e] $Ti \geq 5 \times \%C$.
[f] Such a possibility of exchange has not as yet been set down in standards.

which these steels are also intended for cold-forming, the same characteristics apply as for heat-treatable steels (steels for quenching and tempering).

Figure C 29.7 shows examples of the true stress-true strain curve of typical *stainless steels*. Included are the steeper curves for *austenitic steels* 1 to 3 in the quenched condition – especially, for example, for the frequently used steel X 5 CrNiMo 18 10 (for steel 2, see Table C 29.3) compared with low-alloy steels (see Fig. C 29.6). This illustrates the greater strain-hardening tendency of the austenites. Allowances must be made for the creep susceptibility of these steels [30]. The titanium-alloyed steel X 10 CrNiTi 18 9 (steel 1 in Fig. C 29.7), by comparison, displays the highest "flow stress" (the true stress for a given true plastic strain, see B 7) during deformation and is consequently not used for manufacturing fasteners because no welding is performed on screws and bolts. Where higher degrees of forming are required, use can be made of ELC (extra-low-carbon) steels; i.e., steels that have a particularly low carbon content, e.g., the steel X 2 CrNi 18 9 containing about 0.02% C, 18% Cr, and 11% Ni. The new steel grade X 1 CrMo 18 2 containing about 0.01% C, 18% Cr, and 2% Mo, which was developed as an ELA (extra-low-additions) ferrite (i.e., a stainless ferritic steel having especially low levels of alloying elements, extremely low nitrogen contents, and a very low carbon content) presents further advantages in respect of cold formability (steel 4 in Fig. C 29.7). As well as possessing good corrosion resistance, the amount of strain hardening of this steel grade when subjected to elevated degrees of forming is much less than that of austenitic steels.

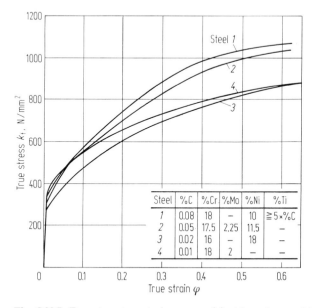

Fig. C 29.7. True stress-true strain curves of ferritic and austenitic stainless steels in the quenched condition (steel 1: X 10 CrNiTi 18 9; steel 2: X 5 CrNiMo 18 10; steel 3: X 2 NiCr 18 16; steel 4: X 1 CrMo 18 2).

If the pieces are manufactured by *machining* rather than by cold forming, it is possible to use, for example, the unkilled steel grades U 7 S 6 or U 10 S 10 mentioned in Table C 29.1. These steels have only limited cold formability, yet possess good machinability because of their elevated sulfur content; these can also be produced as killed steels upon request. They are used mainly for nuts [31] and also, to a limited extent, for screws and bolts manufactured by machining.

On a very limited scale, the unalloyed 6 P 10 steel grade containing about 0.06% C and 0.10% P is used in the manufacture of hot-pressed nuts. The high phosphorus content appreciably improves the flow properties of the steel at temperatures above 1000 °C. It also leads to short brittleness of the chips, which is beneficial for thread cutting.

Information regarding the use of the steels discussed in this section (C 29.4) for manufacturing screws, bolts, and fastening elements, and also regarding surface treatments, can be found in [32–40].

C 30 Steels for Welded Round Link Chains

By Hans-Heinrich Domalski, Herbert Beck and Helmut Weise

C 30.1 Required Properties of Use

The requirements applicable to steels for the production of round link chains are determined by the fabrication method and the properties of the finished chains [1–4].

Chains as finished products have essentially load-carrying or force-transmitting functions. Therefore, *strength properties* are the major requirements for chain steels. But it must be taken into account that chains are used not only at ambient temperatures, but also at lower and higher temperatures, and that working loads in service are not only static, but also may be impacting or pulsating. Thus, there may be special requirements placed on the mechanical properties of chain steels.

The range of the required strength depends on the kind of chains. Chains without any special quality requirements are used whenever low working loads are to be expected and whenever a chain break involves only a small risk [3].

Quality chains are used in standard grades or, if necessary, as quenched and tempered, high tensile strength, or wear-resistant grades. If necessary, the chains may have to have special physical or chemical material properties. These grades are submitted to special tests [5–7] (see, e.g., DIN 685 [8]). The strength values of the chains, as demanded by the chain manufacturers (see Table C 30.1), are defined by load-carrying capacity (working stress limit), test (proof) stress, and breaking stress. A chain factor of about 0.7 has evolved as a rule of thumb in chain manufacturing for steels with a ferritic-pearlitic microstructure or for steels in the quenched and tempered condition. This means that the breaking load of the *chain* amounts to about 70% of the breaking load as calculated from tensile strength and a cross section of the *steel* bar [4, 13]. A chain factor of 0.8 may be used for welded round link chains of fully austenitic steels because of their higher strain-hardening capacity. The reason for the difference between the breaking load of the chain and that of the steel bar may be seen in the stress distribution in the chain link: Fig. C 30.1 shows stress concentrations occurring at certain points that are several times higher than the normal stress. Large differences are observed in the distribution and extent of stresses, depending on the shape of the chain (width, pitch, stud) [4, 5, 14, 15]. Quality chains are normalized or quenched and tempered after welding, depending on steel grade, to obtain the desired mechanical properties.

References for C 30 see page 804.

Table C 30.1. Working stress limit of chains in comparison to tensile strength of chain steels

Type of chain[a]	Standard specification	Quality grade	Chain Working stress limit σ_{Tr} max.	Proof-stress σ_{Pr} min. N/mm²	Breaking stress σ_{Br} min.	$\dfrac{\sigma_{Tr}}{\sigma_{Br}} \leq$	$\dfrac{\sigma_{Pr}}{\sigma_{Br}}$ min. %	Steel Required tensile strength[b] $\sigma_{Br}/0.7$ N/mm²
H_H	DIN 766[9]	3[c]	80	200	320	1:4	62.5	460
H_M		5	106	315	530	1:5	59.4	760
H_H			125	315	530	1:4.2	59.4	760
H_M	DIN 5684[10]	6	125	400	630	1:5	63.3	900
H_H			160	400	630	1:4	63.3	900
H_M		8	160	500	800	1:5	62.5	1140
H_H			200	500	800	1:4	62.5	1140
F	DIN 22252[11]	1	about 341[d]	487	670 about	1:2	72.7	960
F		2	about 421[d]	602	800 about	1:2	75.3	1140
S		K1	—	depending on bar diameter, according to [12], Table 10.6			70.0	400[g]
S	GL[e]	K2[f]	—				71.4	490[g]
S		K3[f]	—				70.0	690[g]

[a] H_M = Load chain, engine driven
  H_H = Load chain, hand driven
  F = Conveyor chain.
  S = Ship chain, anchor chain.
[b] Informal data given as nominal only.
[c] According to DIN 766[9]: Heat treated.
[d] Informal data only.
[e] Specification of the Germanischer Lloyd [12].
[f] For the quality grades K 2 and K 3 of the ship-anchor-chain classification [12], notched-bar impact-energy values on ISO-V-notch test specimens of min. 27 J for K 2 and min. 59 J for K 3, each at 0 °C, are required.
[g] Value required in the specification for chain steel.

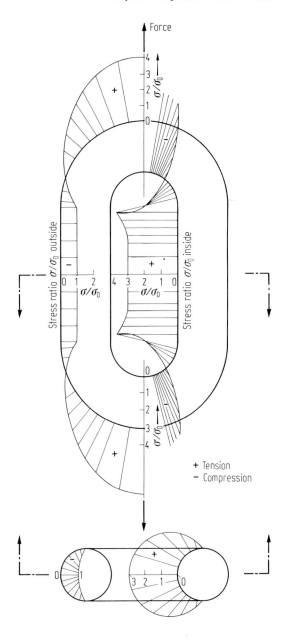

Fig. C 30.1. Distribution of stresses in a chain link with straight sides. After [14].

In addition to the strength properties, quality chains must be able to work safely. Brittle fractures must be prevented, which means that the steels must have a good *formability* and *ductility* and must be *nonsusceptible to aging*. A satisfactory formability of the chain material in service must ensure, on the one hand, that the formability reserves can cope with overloads that might lead to forced ruptures without an elongation. Excessively stretched chain links, on the other hand, are bent by faulty interaction with the pocket or sprocket wheel; this might lead to brittle fractures, especially in the region of the weld, if the steel is unsuitable or has been processed incorrectly [16]. Therefore, the formability of the steel used for motor-driven chains should not exceed certain values, in order to maintain the uniform pitch despite the elongation under load and to safeguard smooth running over the sprockets [5]. In testing the chain with the predetermined testing load, a specified total elongation must not be exceeded. The aforementioned bending and also the crushing damage to chain links caused by loads in service may cause strain hardening. Steels must possess sufficient resistance to such a loss of ductility (toughness).

It may be that the required mechanical properties of the chains can be attained only by heat treatment, depending on steel grade, the diameter of the steel product, and the intended application of the chain. This gives rise to requirements for the *hardenability* of the steels.

In conjunction with the load-carrying capacity of the chains, a certain amount of *wear resistance* of the chain steels is required. One particular wear mechanism is surface abrasion of chain material, which causes weakening (see B 10). Another mechanism seems to be the formation of martensite on the surface of the chain links as a consequence of the localized pressure on the asperity contacts at higher sliding velocity and pressure. This wear mechanism is observed mainly in steels with higher carbon contents, which normally show a higher resistance to abrasion. Cracks may form in the martensite due to the low formability of the structure; these cracks may represent the starting point of fatigue cracks or brittle fractures [17, 18]. Other causes of wear and fractures are corrosion, unsuitable design or construction of the transportation system, or defects in the chain fabrication [19–22].

As already mentioned, the steels must possess certain *physical properties* or *corrosion resistance*, depending on their intended applications.

Steel for chains is hot-rolled as wire rod or straight bars and, depending on required tolerances, is also cold drawn. Chain steels are used in sizes between 1 mm and 160 mm in diameter. Since it is then cut to length by shearing (bigger sizes by sawing), the steel must show suitability for *cold shearing*. After cutting, the ends are prepared for welding, bent and welded. Steels must fulfill requirements for *formability* by cold or hot bending and especially for *weldability*. Depending on the chain fabrication method, weldability may be limited to the suitability for flash butt welding and pressure butt welding, which are applied to smaller sizes (e.g., to special steels below 20 mm in diameter) [23].

C 30.2 Characterization of the Required Properties

The *mechanical properties* (yield stress and tensile strength) are measured in the tensile test (see B 1). Normally, the test is made at room temperature. The tensile test, together with its values for elongation after fracture and reduction of area, may indicate the formability and the toughness as well. If only indicative values are needed, the strength may be measured by a hardness test instead of by the tensile test. This is frequently done to characterize the soft-annealed condition and, especially, the state after annealing to spheroidize the carbides.

The *toughness* may be tested, depending on the steel grade, by the *notched-bar impact test* (see B 1). This test may also be conducted at temperatures below or above room temperature, because chains are sometimes used at such temperatures (e.g., in lifting assemblies or sling chains). Investigations have been conducted that allow a comparison to be made with the properties of other steels (see, e.g., C 2), in which the behavior of chain steel was tested at temperatures up to about 400 °C [16, 24, 25]. Figure C 30.2 shows the influence of temperature on the notch-impact energy of a steel containing about 0.21% C, 1% Mn, and 0.035% Al (21 Mn 4 Al).

The *susceptibility to strain-aging* of chain steels is also tested in the notched-bar impact test by using artificially aged test specimens. This artificial aging consists of a combination of cold working (by reduction of the cross section by about 5% to 15% relative to the initial cross section) and aging (normally 30 min at 250 °C). The difference between the notch-impact energy values in the nonaged and the aged condition is taken as the characteristic. Figure C 30.3 shows this difference for two commonly used steels. The largest deviation between the two conditions in these cases is on the order of 20%.

Hardenability may be tested, depending on the steel grade, by the end-quench test. To derive the hardness over the cross section of the chain steel product from the test results, reference should be made to the literature [26]. The hardenability characteristic, together with the curves of yield stress and tensile strength versus temperature after tempering, is the basis for fixing the procedure for the heat treatment of the chain. Thus, the load-carrying capacity and the proof-test load are given if the appropriate steel grade has been selected and – with some reservations – the breaking stress value is given, too. The importance of deformability before a possible break is again pointed out in accordance with the explanations in C 30.1; the limited allowable elongation of the chain must be borne in mind, however. A brittle fracture without deformation must be avoided in any case. In this respect, fracture mechanics (see B 1.1.2.6) are also an important factor [27].

The characterization of the *wear resistance* is difficult. If special instances such as the case hardening of chains are disregarded, wear resistance is accepted as

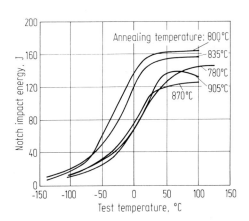

Fig. C 30.2. Notch-bar impact energy (DVM-keyhole test specimens) of a steel containing about 0.21% C, 1% Mn, and 0.035% Al (21 Mn 4 Al) annealed at different temperatures in dependence on the test temperature.

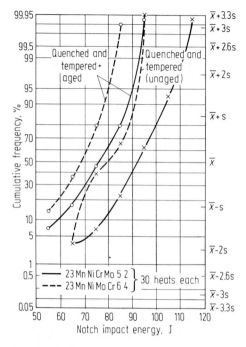

Fig. C 30.3. Cumulative frequency curves for the notch-bar impact energy (DVM-keyhole test specimens) of steel grades 23 MnNiCrMo 5 2 and 23 MnNiMoCr 6 4 (see Table C 30.2), quenched and tempered and quenched, tempered, and aged.

a result of the other governing properties. Hardness, with certain restrictions, gives an indication (see B 10).

It is possible to judge the suitability for cold shearing by taking the values for tensile strength and hardness, Brinell hardness usually being measured. Tensile strength and hardness also give a clue to the formability of the steels into chain links. Furthermore, *formability* may be characterized, in addition to the above-mentioned property values derived from the tensile test, by technological tests, e.g., by the behavior in the bending test.

Basically, the *weldability* is judged by the carbon content of the steels. The influence of alloying elements is taken into account by calculating a carbon equivalent (see B 5).

C 30.3 Measures to Attain the Required Properties

Two more or less contradictory properties must be taken into account when considering the kinds of steels and their requirements. On the one hand, the steels must possess a specified, sometimes high, strength, or permit the development of this property in the chain. The easiest and most economical way to achieve this is via an appropriate carbon content in steels either in the normalized or in the quenched and tempered condition, together with the appropriate microstructure. On the other hand, the steels must have sufficient weldability, and the surest way to achieve this property is a low carbon content, because it entails a low hardenability and hence a low susceptibility to weld cracking in the heat-affected zone. The demand for weldability thus means a limitation of the carbon content. If the required mechanical properties cannot be obtained – whether with or without quenching and tempering – the steel must be alloyed for a given chain thickness to achieve a microstructure with the desired properties (see B 1) to increase the strength. The other requirements (e.g., securing sufficient formability or preventing susceptibility to strain aging) are generally subordinate to these basic considerations.

More specifically, the above-outlined contradiction of demands necessitates that, in the interest of *weldability*, the carbon content be held below about 0.25% (for details, see C 30.4). To achieve the desired behaviour in welding, the aim is also to attain the lowest possible phosphorus and sulfur contents.

For *low strength* requirements, the steels are employed after normalizing (i.e., with a structure of ferrite and evenly distributed patches of lamellar pearlite), the various parts of the structure being more or less finely developed (see below). The diameter is taken into account by small alloying additions to produce solid-solution strengthening and a partial refinement of the microstructure by a shift of the transformation temperature and time. A sufficient *hardenability* must be provided by an appropriate alloying of the steels, if *high strength* values are demanded. In this case, heat treatment of the chains can produce a microstructure of the quenched and tempered condition, with possibly high strength and good toughness values in comparison to the strength, thus resulting in the most advantageous use of the alloying elements employed.

The *susceptibility to aging*, whether after deformation and storage for any length of time (strain aging) or after quenching and storage (quench aging), is characterized by a loss of toughness (see C 30.2). Aging is caused mainly by nitrogen (but also by carbon), with its low solubility in the ferrite. Iron nitrides, which may have precipitated (e.g., during cooling after the hot deformation), reduce the ductility (toughness) of the steels. Quench aging is also possible during cooling after welding. Thus, the tendency toward aging can be reduced primarily by lowering the nitrogen content to prevent the formation of the injurious iron nitrides. Another, perhaps better, way would be to bind the nitrogen (e.g., to aluminum), with the additional advantage of producing a fine grain structure. Adequate resistance to aging is normally attained with an aluminum content of at least 0.020% in conjunction with a heat treatment adapted to the steel. The decrease in notch-impact energy values by aging is generally on the order of 15% if the contents of aluminum and nitrogen are in a good ratio to each other (see Fig. C 30.3); decreases of up to 30% are possible. The notch-impact energy values of the investigated steels with tensile strengths of about 1200 N/mm^2 also characterize the usual scatter range of these steels after testing with keyhole-impact test samples (DVM-type). Low carbon contents contribute to a reduction in the tendency towards aging.

Suitability for *cold shearing* is influenced by the tensile strength or the hardness; under normal circumstances, it is satisfactory at tensile strengths of up to 850 N/mm². In certain cases, steels with strength values up to 1050 N/mm² can be sheared, depending on the dimensions of the steel products and the available shear forces. At higher strength values, the steel must be spheroidizing annealed before cutting. More or less the same applies to *cold bending*, which is possible up to a maximum tensile strength of about 700 N/mm². This limit again depends on the dimension of the steel product and the processing conditions. Steel grades showing a higher strength level must always be spheroidizing annealed. The annealing permits faster production in modern welding machines by reducing the spring-back of the chain links after cold bending. This enables the welded link to be released from the bending and contacting tool immediately after the welding process. The time-temperature-transition diagrams (see C 4) of the alloyed steels show that the time required until complete transformation is, in any case, longer than the time needed in the welding machine. The transformation of the austenite takes place during air cooling from the welding temperature after removal from the machine. The low ductility (toughness) of the newly formed microstructure increases the susceptibility to cracking under the tensile stress that is simultaneously present. An excessive spring-back would mean that the steel's yield stress at elevated temperatures was exceeded in the region of the weld seam. Therefore, from a diameter of about 25 mm steels for high-tensile strength chains are nowadays bent not in the cold condition, but after resistance heating, depending on available equipment.

Finally, the quality of the welded chain depends directly on the homogeneity of the material and on the fibrous structure of the product. The elements phosphorus and sulfur, which tend to segregation, are kept below limits that are lower than those in similar steels for other applications, in order to increase the homogeneity. This is especially important for the manganese-containing steels to avoid, as far as possible, a banded structure. The results of microstructure tests on welded chain links show a direct connection between the chemical composition and the degree of homogeneity in the distribution of different microstructural phases. The course of structure fibers in the weld seam is deflected to a greater or lesser degree, depending on the contact pressure during welding.

C 30.4 Characteristic Steel Grades Proven in Service

Unalloyed and low-alloy steels are applied as chain steels, a selection of which is given in Table C 30.2. The first listed steel is basically qualified for non-heat-treated chains without further quality specification (see C 30.1), but may be used for the purposes listed in Table C 30.2 for this and other steel grades as well. The decisive factor for the choice of the alloying elements and their combination is, in addition to economic factors, their influence on weldability on the one hand and on hardenability on the other; molybdenum is also used for the suppression of temper brittleness.

Other countries (e.g., the United States, Great Britain, and Japan) have laid down details on the steels for chains in their standards for chains [29–31]. In

Table C 30.2. Selection of steel grades and their chemical composition for the manufacturing of welded round link chains; for further details see DIN 17115 [28]

Steel grade	Chemical composition [a]									Application and classification	Final heat-treatment condition of the chain
	% C	% Si	% Mn	% P	% S	% Cr	% Mo	% Ni	% Al		
RSt 35-2	0.06–0.12	≤ 0.25	0.40–0.60	0.035	0.035					Anchor chains	Normalized [d]
15 Mn 3 Al	0.12–0.18	≤ 0.20	0.70–0.90	0.035	0.035				0.020–0.050	Sling chains	
21 Mn 5	0.18–0.24	≤ 0.25	1.10–1.60	0.035	0.035				0.020–0.050		Hardened and tempered
21 Mn 4 Al	0.18–0.24	≤ 0.25	0.80–1.10	0.035	0.035				0.020–0.050		
27 MnSi 5	0.24–0.30	0.25–0.45	1.10–1.60	0.035	0.035				0.020–0.050	High-tensile-strength chains for mining	
20 NiCrMo 2 [b]	0.17–0.23	≤ 0.25	0.60–0.90	0.020	0.020	0.35–0.65	0.15–0.25	0.40–0.70	0.020–0.050	Sling chains and lifting assemblies	
20 NiCrMo 3	0.17–0.23	≤ 0.25	0.60–0.90	0.020	0.020	0.35–0.65	0.15–0.25	0.70–0.90	0.020–0.050		Hardened and tempered
23 MnNiCrMo 5 2	0.20–0.26	≤ 0.25	1.10–1.40	0.020	0.020	0.40–0.60	0.20–0.30	0.40–0.70	0.020–0.050	High-tensile-strength chains for mining application, sling chains, and lifting assemblies	
23 MnNiMoCr 5 4 [c]	0.20–0.26	≤ 0.25	1.10–140	0.020	0.020	0.40–0.60	0.50–0.60	0.90–1.10	0.020–0.050		
15 CrNi 6 [b]	0.12–0.17	≤ 0.25	0.40–0.60	0.020	0.020	1.30–1.60		1.30–1.60		Fire and skid resistant chains	Case hardened
X 10 CrAl 18	≤ 0.12	0.70–1.40	≤ 1.0	0.040	0.030	17–19			0.70–1.20	Heat-resistant chains; e.g. for rotary furnaces	Annealed
X 20 CrNiSi 25 4	0.10–0.20	0.80–1.50	≤ 2.0	0.040	0.030	24–27		3.5–5.5	0.020–0.050		
X 15 CrNiSi 25 20	≤ 0.20	1.50–2.50	≤ 2.0	0.045	0.030	24–26		19–21	0.020–0.050		Quenched

[a] Ladle analysis.
[b] Attention must be paid to differences in chemical composition, especially in phosphorus, sulfur, and aluminum contents, as compared to case-hardening steel-grades in DIN 17210 [28a].
[c] Formerly 23 MnNiMoCr 64.
[d] Untreated, if circumstances permit, depending on application.
[e] Normalized, if circumstances permit, depending on application.

addition, in the ISO standards for chains [32], data on the corresponding steels also can be found. In all these standards, besides carbon steels, there are alloy steels with nickel, chromium, and molybdenum as the most important alloying elements. In the foreign standards, neither a range nor an average value for these elements is given. The decision to accept a chain is made considering the mechanical properties of the actual chain and the ranges prescribed. Three interdependent characteristics are listed:
- Load on the chain in service in relation to the mechanical properties, influenced by
- Chemical composition, and
- Heat treatment of the welded chain

The Japanese standard G 3105 specifies basic elements. Data are given for chemical composition and mechanical properties of bars for manufacturing welded round-link chains. No reference is made to application and heat treatment of the chain.

Table C 30.3 shows a comparison of a carbon steel classified for general use in the different standards and a nickel-chromium-molybdenum alloyed steel for high-tensile-strength chains. The best ductility (toughness) is realized by quenching and tempering welded chain links.

There is no chance of finding agreement in the instructions given in the standard specifications of the leading chain-making countries.

In regard to the cold shearing and bending properties, which are important for chain making, the steels are supplied in the following conditions, depending on their chemical composition and the microstructure resulting after cooling from hot rolling:
1. Untreated, if the steels' chemical-compositions are such that their ferritic-pearlitic microstructure is sufficiently soft after cooling down from the rolling temperature
2. Capable of being cold sheared, which normally means, for chain manufacturing, that the tensile strength does not exceed 1050 N/mm^2
3. Spheroidizing annealed for globular cementite, with microstructure and maximal tensile strength as the most important characteristics; this spheroidizing anneal is employed especially on multiple alloyed steels. If a particular microstructure is not specified, tensile strength must be held below a certain maximum figure by a soft annealing between 600 °C and 680 °C for 2 h.

In none of these three cases should the cementite be present in a coarse state, because of the welding behavior.

After manufacture, the chains are normalized or quenched and tempered. A combination of both processes may also be employed, depending on the steel grade. Of the many possibilities of heat treatment in fabrication, an example is the triple-alloyed martensite-hardening nickel-chrome-molybdenum steels, which have a good ratio of tensile strength to ductility (toughness) (see Fig. C 30.4). For these steels, a tensile strength of 1200 N/mm^2 after heat treatment is usual. This can be achieved by using the following typical procedure:
- Normalizing anneal between 860 °C and 940 °C, followed by cooling in still air
- Austenitizing at 870 °C to 890 °C; after reaching this temperature range in the

Table C 30.3 Steel grades for the same application according to international standards (a general survey for only two steel grades)

Standard specification	Steel grade	Alloying elements	Application/classification	Final heat-treatment conditions of the chain
DIN 17115 [28]	RSt 35-2		Anchor/fastening chains	Normalized
British Standard 6405 [30]	Grade 30-class 2		General duties	As-rolled or normalized
ASTM A 413-80 [29]	Grade 28		Proof coil chains	As-rolled
JIS-G 3105 [31]	S BC 31-class 1		No recommendation	As-rolled condition of the bars before chain-making
DIN 17115 [28]	23 MnNiMoCr 5 4	See Table C 30.2	High-tensile-strength chains for mining purposes, sling chains, and lifting assemblies	Hardened and tempered
British Standard (3114) [30]	Grade 80	Ni, Mo, Cr, minimum values specified	Calibrated load chain in pulley-blocks	Hardened and tempered
ASTM 01.05 (A 391-85) [29]	Grade 80	Ni and Cr or Mo, compulsory addition	Calibrated load chain in pulley-blocks	Hardened and tempered
JIS-G 3105 [31]	SBC 70-class 3	Ni, Mo, Cr, and V can be added, values not specified	No recommendation	Recommended: Bars may be quenched and normalized before chain-making

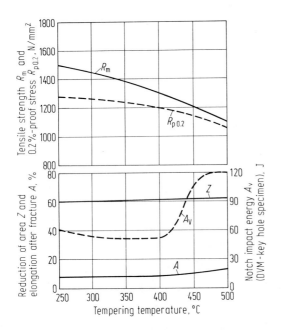

Fig. C 30.4. Tempering diagram (time 2 h) for the steel 23 MnNiMoCr 6 4 (see Table C 30.2) for lifting appliances or mine chains. Heat treatment: 880 °C/air + 880 °C/water; bar steel of 26.5-mm diameter.

core and complete carbide dissolution, a water quench follows
- Tempering between 450 °C and 550 °C for 2 h

The combination of these process steps permits optimal results to be obtained with the above-mentioned alloying elements. Deviations from this procedure, especially a reduction in tempering temperature, reduce the quality of the heat-treated parts [33, 34]. The important influence of the tempering on the mechanical properties, especially on the notch-impact energy, is shown in Fig. C 30.4. The curve for the notch-impact energy rises rapidly from temperatures of 420 °C upward, thereby showing a good ratio of strength to toughness above 450 °C. This minimum tempering temperature must be adhered to when using this processing combination for this group of high-tensile-strength, heat-treatable steels. The correlation between temperature and tempering time is of special importance (refer to the relevant discussion in B 4); an arbitrary shortening of the tempering time does not yield optimal properties of the chain, even at elevated temperatures.

Part D
The Influence of Production Processes on the Properties of Steel

Part D
The Influence
of Production Processes
on the Properties of Steel

D 1 The Influence of Production Processes on the Properties of Steel and Steel Products–An Overview

By Alfred Randak

In parts A to C of this handbook the *correlation between steel properties and microstructure* and the microstructure's dependence on *chemical composition*, heat treatment, and shaping have been explained. This picture would be incomplete without an overview of the methods used to produce, reliably and cost efficiently, a steel that has the chemical composition required for its intended application. Raw materials – like iron ore and coal, but most of all scrap – as well as energy sources and additives introduce undesirable or even harmful constituents. During production, steel is frequently exposed to the atmosphere, where the high working temperatures result in oxygen, nitrogen, and hydrogen pickup – again usually to the detriment of the desired propeties; in addition, shaping operations affect the steel's microstructure. Part B therefore covers briefly the production processes that affect the steel's physical properties and surface finish.

Heat treatment, often used to obtain a specific micro-structure, is discussed here only as it applies to obtaining properties desired; other heat treatments – for example, as used in subsequent fabrication – are described, in fundamental terms at least, in parts B 4 and C.

A knowledge of the principal relationships among production processes and the properties of steel and steel products is essential for all engineers engaged in the steelmaking and steel processing industries. In the following chapters the most important processes used in melting, casting, and hot and cold working are discussed. Heat treatment is mentioned to the extent that it is applied by steelmakers to impart desired properties to their products. Obviously this means that *to specify the properties of a particular steel product is to specify the production operations*, with all their associated expense and energy requirements. This inherent link among properties, production conditions, and cost makes it necessary for the steel's properties, quality, and grade to be specified and selected after consultation between producer and user so that the best solution is found, both technically and economically, for the particular application.

Steelmaking technology and the conditions under which steel is processed and treated are evolving. Moreover, the users demands for improved steel products and the technological developments and advances in steelmaking are closely related. Developments in the steel industry are always geared toward producing and processing steel as cost efficiently as possible under the constraints imposed by the available raw materials and energy sources and toward attaining the steel products' specified properties as accurately as possible. A number of examples are cited to highlight the impact that advances in process technology have had on production and hence on the steelmakers' ability to obtain specific properties.

Current methods of steelmaking and hot working became possible only with the advent of processes for *making steel in the liquid state*. Until the 1850s steel was produced almost exclusively by puddling, in which only small quantities of iron were refined per batch and worked into a pasty mass interspersed with slag. Steelmaking processes utilizing air, patented by H. Bessemer and latter by S. G. Thomas, and the open-hearth process, patented by the brothers W. and F. Siemens and E. and P. Martin (father and son), yielded more uniform steels of higher purity and offered better opportunities to control the chemical composition of the melts. Indeed, it was the use of air in steelmaking together with the development of the basic-lined converter that made refining iron smelted from high-phosphorus ores possible and hence provided the basis for modern steel production. Further advances steadily increased the volume of the melts, permitting rolled products of increasingly larger sizes and weights to be supplied. The open-air-blowing processes, such as the Bessemer, for producing steel with better service properties and a wider range of applications; the poorer results of the air-blowing processes were caused by nitrogen pickup from the blast. Refining with pure oxygen in order to remove excess carbon, phosphorus, silicon, and manganese was not economical until advances made by R. Linde and M. Fränkl appreciably reduced the cost of extracting pure oxygen from the air. Since 1950, the basic oxygen process has not only completely supplanted air-blowing practices but has also caused the open-hearth process to become less important; oxygen steel has better service properties for major applications because of its lower nitrogen content.

Continuous casting is rapidly supplanting conventional ingot casting, with the impetus for change coming not only from economic and ergonomic advantages of continuous casting but also from the qualitative improvements in product properties that are obtainable. Vacuum degassing and gas purging, generally known as ladle metallurgy or *secondary steelmaking*, represent additional major advansces and have made it much easier to adjust chemical composition, keep hydrogen content low, control nonmetallic inclusions, and control the temperature and composition of the melt.

The introduction of *further refining* and *remelting processes* has made it possible to produce steels meeting stringent requirements of homogeneity and freedom from nonmetallic inclusions.

Advances in *hot and cold working* and *heat treatment* have also brought about major improvements in product properties. The introduction of continuous processes is an example of these advances, and facilities such as wide-strip hot-rolling mills, cold -rolling tandem mills, and continuous furnaces for sections and flats deserve special mention. These state-of-the-art processes have contributed greatly toward improving the uniformity of important product properties. Heat treatment from the rolling temperature and thermomechanical processes should also be noted.

Everyday, engineers have to translate their knowledge of the relationships between production conditions and product quality into an operational setup that turns out the specified quality both accurately and reliably. To this end it has long been the practice in the steel industry to allot time for *quality-assurance analysis* in the production schedule. Reflecting the importance of quality assurance, the chapters outlining the influence of production conditions on properties are followed by a chapter dedicated to that subject; this chapters highlights the efficiency of quality assurance in meeting the demands imposed by exacting quality specifications.

It is important that steel users and producers agree on realistic quality characteristics that can be achieved economically. While the manufacture of flawless products is possible with the processes and facilities available today, it would not make economic sense to do so in most situations. The standardization of steel grades has for years contributed greatly toward this goal, and Germany can look back on a long tradition of applications-oriented standardization. Based on this tradition, efforts should be made toward a constant review of the quality characteristics of specific steel products to ensure that technical advances are incorporated by steel producers and users alike, if necessary by amending existing specifications.

The following chapters illustrate the technologies for making, shaping, treating and finishing steel. It will be evident that there are many interrelated factors involved in the production of steel, which is why in some process steps there are uncertainties as to which set of parameters will yield a product that will meet specifications. It is the task of the engineers involved to select from a variety of feasible solutions those that are technically and economically appropriate for the particular application. The following chapters are intended to underscore the need for this approach and to indicate, with reference to the latest state of the art, the most important relationships between manufacturing conditions and product properties.

D 2 Crude Steel Production

By Hermann Peter Haastert

The goal of steelmaking is to produce a steel of defined chemical composition, as homogeneous as possible, and free of deleterious impurities. Hence the production method used (Fig. D 2.1) is influenced by the demands imposed on steel properties and by the availability of adequate raw materials.

D 2.1 Raw Materials for Steelmaking

Elemental iron is not found in nature. To become steel iron has to be extracted from its ores by reduction and then has to be further refined and alloyed while in a liquid state.

High-grade ores are used almost exclusively as raw materials. Their reduction in a *blast furnace* with coke generates molten *pig iron* – hot metal – containing about 3.2% to 4.5% C, 0.2% to 1.2% Si, 0.2% to 1.5% Mn, 0.02% to 0.12% S, and, depending on the phosphorus content of the ore, either 0.06% to 0.30% P or 1.5% to 2.2% P. Other elements, for instance copper and titanium, vary in concentration depending on the composition of the raw materials; their concentration in the pig iron can be controlled only to a limited extent by the blast furnace process. Only trace levels of these impurities may be present if high-purity materials are used, but these may not always be available.

In order to improve the steel's quality, or to ease both operations it is customary to treat the hot metal while it is being transported between the blast furnace and the steelmaking vessel [1]. This treatment may be a desulfurization by addition of soda ash during a reladling operation, where dissolved silicon and nitrogen are also extracted from the hot metal [2–4]; by pneumatic injection [5]; by mechanical stirring [6, 7]; or by adding desulfurizing reagents based on calcium carbide, magnesium, or lime. Sulfur concentrations of less than 0.002% can be attained using these methods [8]. Other treatments remove silicon and phosphorus. In the dephosphorization treatment, however, the large amount of reagents like soda ash or lime-fluorspar-iron oxide mixtures required may cause problems [8, 9].

Processes of direct reduction of iron ores using solid or gaseous reductants produces solid *sponge iron*. The temperatures at which the ores are reduced vary

---

References for D 2 see page 806.

I. Reduction step
Production of metallic iron materials from iron ores

Raw materials: Additions:	Iron ores, coal, oil*, gas* Limestone, gravel*			
Reduction process:	Blast furnace process	Direct reduction processes		
Raw materials:	Iron-bearing materials: Screened ore Sinter* Pellets* Scrap*	Fuels and reductants: Coke Oil* Gas* Pulverized coal*	Iron-bearing materials: Screened ore Pellets Fine ore	Fuels and reductants: Coal Gas Oil
Process alternatives	Blast furnace	Shaft furnace	Rotary kiln	Fluidized bed reactor
Reduction unit:				
Intermediate product	Pig iron (hot metal)	Sponge iron materials		

II. Refining step
Crude steel production Melting and refining

Raw materials Additions and fluxes Oxygen carriers	Pig iron, scrap, sponge iron*, ore* Burnt lime; fluxes: deoxidizing and alloying materials Air, commercial pure oxygen, ore			
Charge preparation	Desulfurization (desiliconization, dephosphorization) with addition of desulfurizing or dephosphorizing reagents			
Pig iron (hot metal) Process alternatives	Reladling	Injection	Immersion	Stirring
Scrap	Preparation by classifying, compacting, shredding			

Fig. D 2.1. (continued)

D 2 Crude Steel Production

1. Process alternatives for crude steel production										
Type of melting and refining process	Bottom air refining processes	Oxygen blowing processes				Hearth and electric processes				
Type of vessel	Bessemer/Basic Bessemer converter	Top blowing converter	Bottom blowing converter	Combined top + bottom blowing converter	Side blowing converter	Open hearth furnace	Electric arc furnace	Induction furnace		
Deoxidizing and alloying, aditionally if required homogenizing	Additions of deoxidizing and alloying materials into furnace and ladle					Gas rinsing/stirring in the ladle				
Product	Crude steel for simple grades or for special applications					Intermediate product for aftertreatment				

2. Process alternatives for the aftertreatment of crude steel melts							
Added materials	Inert gases (argon*, nitrogen*), oxygen, calcium, magnesium-alloys, basic slag compounds, deoxidizing and alloying materials						
Steel after-treatment processes	Gas rinsing/stirring	Injection treatment	Heating		Vacuum treatment		
Treatment vessels	Bottom or lance rinsing, inductive stirring	Powder injection or cored wire feeding	Ladle furnace	Pouring stream	Ladle treatment	Arc heating	Vacuum refining	Partial mass-type processes circulation / lifting
Product	Crude steel for special applications and high demands						

*Alternatives

Fig. D 2.1. Process routes in steelmaking

between 700 °C and 1200 °C, depending on the ore and process characteristics; however, these temperatures are kept below the softening temperature of the ore. As a relatively clean iron-bearing material, sponge iron constitutes an important raw material for steelmaking [10, 11], especially in countries without coking coal deposits. Of its total iron content, more than 80%, and usually 90% to 95%, is in the metallic state. As opposed to pig iron, sponge iron's carbon concentration is comparatively low and controllable, ranging between 0.5% and 2.5%. The gangue content of the ore is unaffected during direct reduction, is not separated as slag as in the blast furnace, and can only be removed in the steelmaking operation. Sulfur concentrations below 0.010% can be attained in sponge iron if low-sulfur reductants are used. Similarly, undesirable elements in the iron can be kept at a level close to their detection limit if high-purity ores are used.

Other iron-bearing raw materials used in steelmaking for melting or cooling are *steel scrap, cast iron scrap, and iron ore*. The first two constitute circulating or waste materials, which are segregated into classes by origin and nature. Scrap of unknown origin, especially collected light weight classes, can introduce undesired impurities into the furnace. To improve scrap quality, more extensive preparation can be done, for instance, using shredding machines or scrap mills.

Additional raw materials for steelmaking are gaseous oxygen and oxidizing materials for refining; various, mostly inert gases for melt stirring; fluxes for slag forming; and deoxidizing and alloying materials.

D 2.2 Influence of the Charged Materials in Steelmaking

The concentrations of accompanying elements and nonmetallic impurities introduced into the steel melt by the raw materials are to be adjusted metallurgically, and if these additional elements are deleterious, they must be removed so that the lowest concentrations compatible with the economy of the operation are reached. The attainable concentrations of the dissolved elements are determined by their solubility limits, which depend on composition, temperature, and, occasionally, pressure, and by their reaction equilibria, especially with oxygen [12]. Together with the alloying material added, these elements and impurities establish the steels composition, which determine, in combination with the conditions existing during solidification, deformation, and heat treatment, the microstructure and, consequently, the properties and the surface quality of the steel product.

Undesired elements are those that strongly impair the steel's properties. Oxygen, sulfur, phosphorus, nitrogen, hydrogen, and some nonferrous metals belong to this group, when they are not added to attain specific steel properties. Oxygen and sulfur, for instance, determine almost exclusively the steel's purity and generally affect the steel's properties negatively; in some cases, however, their presence is desired (as in "free-cutting" steel, i.e., easily machinable steels). Both elements also strongly impair hot deformability (causing hot shortness), and reduce toughness, surface quality, and weldability. Phosphorus increases susceptibility to brittle fracture and causes temper brittleness, whereas nitrogen can cause aging phenomena. Owing to their tendency to segregate during solidification, phosphorus, sulfur, oxygen, and nitrogen can produce local defects, even if individually their concentrations are below "danger levels." Hydrogen, if present in sufficient amounts, may precipitate as a gas on fast cooling from forming temperatures, mainly at the grain boundaries, causing localized stress cracks ("flakes"). An example of the damaging effect of non ferrous metals is the intercrystalline surface cracking caused by excessive amounts of copper and tin (solder brittleness).

Besides the unwanted elements, nonmetalic impurities entrapped by the steel subsequently may cause a variety of defects due to insufficient purity. These impurities may come from the raw materials, the slags, the refractory linings of furnaces, or the ladles – so-called exogenous inclusions – or precipitate within the steel for instance, during deoxidation – so-called endogenous inclusions.

D 2.3 Steelmaking

In the production of steel, the hot metal is converted into a steel melt of predetermined composition by refining, alloying, and heating to the correct temperature for subsequent casting. For this purpose, the process steps outlined in Table D 2.1 are followed, according to the steel grade to be produced.

The elements associated with the raw materials can be classified according to their behavior during refining as follows:
1) elements leaving the system in the gaseous state: carbon, zinc, lead, hydrogen, nitrogen, sulfur (partly);

Table D 2.1 Process steps in steelmaking and their metallurgical aims.
(Combinations of ladle metallurgy processes are used as determined by the desired metallurgical result; also, specific operating or economic reasons are often determining factors)

Process step		Metallurgical aim
1. Melting and refining	standard process steps	Melting and adjusting to the required temperature. Removal of carbon, silicon, manganese, phosphorus, sulfur, nitrogen, hydrogen, chromium, zinc, vanadium, titanium, aluminum, magnesium, tantalum, nobium, tungsten, lead.
2. Deoxidizing and alloying		Removal of oxygen. Control of alloying elements
3. Ladle metallurgical after treatment processes a) gas rinsing/stirring treatment	special treatments	Homogenizing, improving the steel's cleanliness, "fine-tuning" alloying, sulfur removal
b) injection treatment		Removal of oxygen and sulfur, inclusion modification, homogenizing, improvement the steel's purity, "fine-tuning" alloying
c) heating by electricity		Steel heating, alloying, homogenizing "fine-tuning" alloying, improving the steel's purity[a], removal of sulfur[a]
d) vacuum treatment		Removal of hydrogen, carbon, oxygen, nitrogen, improving the steel's purity, "fine-tuning" alloying, removal of sulfur[a]

[a] by participation of reactive slag mixtures.

2) elements transferred completely to the slag as oxides: silicon, aluminum, titanium, boron, magnesium, tungsten, tantalum, niobium;
3) elements distributed between molten metal and liquid slag: manganese, phosphorus, sulfur (partly), chromium, vanadium;
4) elements remaining completely in the metal melt: copper, nickel, tin, arsenic, bismuth, antimony, selenium, tellurium, cobalt, molybdenum.

D 2.3.1 Melting and Refining

Once the charged iron-bearing materials have been melted and reach the required temperature, oxidizing materials, which contain oxygen in gaseous or combined form for *refining*, and fluxes are added to the liquid iron melt in order to remove elements with high oxygen affinity. The gaseous oxides generated leave the system with the flue gases or, if liquid, are transferred to the slag, which they form with the added fluxes. This part of the refining operation is mainly for decarburization, but also removes other undesirable elements, typically phosphorus, sulfur, silicon, manganese, hydrogen, and nitrogen.

The *decarburization* of the molten metal bath involves formation of iron oxide and carbon monoxide. The process parameters determine the rates of decarburization and the attainable carbon concentration. Carbon concentrations of 0.02% are customary in bottom blowing converter processes, whereas about 0.05% C are common in the top oxygen blowing process (BOF or BOP). Owing to the closer approach to the thermodynamic equilibrium of the slag-metal reactions inherent in the bottom blowing process and the resulting enhanced and more economic decarburization, oxidizing and/or inert gases are also injected through the converter bottom in the more commonly used top oxygen blowing process. This family of processes are known as combined top-bottom blowing converter processes. – In the open hearth and electric arc steelmaking processes, low carbon concentrations are difficult to attain. Higher concentrations can be attained in all processes by the so-called catch-carbon practice, whereby the oxidation of carbon is interrupted when the desired concentration has been reached.

Controlling the carbon concentration also determines the related *oxygen concentration* [13, 14]. The oxygen concentration found after refining in the converter, the open hearth, and the electric arc processes are shown in Fig. D 2.2. At low carbon concentration the oxygen concentration greatly increases because of the corresponding high solubility for oxygen (*Vacher-Hamilton relationship* [13]). To avoid defects in the steel, the oxygen concentration must be lowered by deoxidation (see D 2.3.2).

Phosphorus is removed by oxidation and is stably bound as phosphate in a lime-rich slag. Dephosphorization is enhanced by increasing the slag volume, slag basicity, and iron oxide content, and by decreasing the temperature. With a low-phosphorus charge, final phosphorus concentrations in steel of $\leq 0.02\%$ are customary; concentrations $\leq 0.01\%$ can be achieved by reducing the phosphorus and/or manganese input and also by a two-slag practice in the refining vessel.

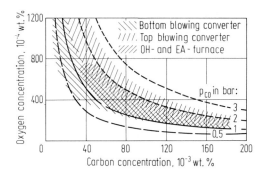

Fig. D 2.2. Relationship between oxygen and carbon concentrations in different steelmaking processes. After [15].

The *desulfurization* of iron melts is aided by maintaining the oxygen potential as low as possible in the steel-slag system; high temperatures and large slag volumes with a high basicity ratio are also helpful. Therefore, the sulfur removal conditions are not favorable in the oxidizing refining processes, which are characterized by high oxygen potentials. Generally, not more than a 50% sulfur removal is attained. Diffusion-governed sulfur transfer to the slag with combination to lime predominates, whereas a smaller proportion escapes in gaseous form. By processing low-sulfur raw materials a sulfur concentration of about 0.003% can be attained in the converter processes. In the open hearth and electric arc furnace processes, with their larger scrap input, concentrations of 0.015% to 0.075% S are reached, depending on the amount of charged sulfur and the sulfur content of the fuel. In the basic electric arc furnace, however, sulfur concentrations of $\leq 0.005\%$ can be obtained by a final refining with basic low-iron-oxide reducing slags. In all cases, sulfur concentrations as low as $< 0.003\%$ – under special conditions $< 0.001\%$ – can be attained by means of a secondary metallurgical treatment in the ladle after tapping (see D 2.3.3).

All *silicon* charged is oxidized and transferred to the slag during refining. *Manganese*, on the other hand, is only partly removed from the liquid metal – until the partition equilibrium between metal and slag is reached.

Nitrogen and hydrogen are removed by the floating up of the precipitated phases, e.g., nitrides, or by diffusion into the carbon monoxide bubbles that rinse the metal bath. Their final concentrations depend on the solubility of the gases in the liquid metal, their input with charged materials and refining gases, the intensity of the boiling action, and the concentrations in the furnace atmosphere and the slag [16, 17]. In the oxygen refining processes, the nitrogen concentrations at the end of the blow are about 0.002% and the hydrogen concentrations are below 0.0002%. In the electric arc process, the final nitrogen concentration of the steel before tapping is higher because of the favorable conditions for nitrogen pick-up present in the electric arc.

The commonly observed increase in the contents of gaseous elements during tapping, aftertreatment, and casting can be controlled by shielding with inert gas, vacuum treatment, and avoiding exposure to air by using shielding devices e.g., ceramic tubes. The hydrogen concentration can be lowered by a subsequent vacuum treatment (see D 2.3.3).

D 2.3.2 Deoxidizing and Alloying

After refining, the oxygen concentration of the liquid steel is generally too high. To avoid steel defects the oxygen content must be reduced by deoxidation:
– with slags having a low FeO-activity – deoxidation by diffusion – or
– by reaction with elements having a high affinity for oxygen and favorable equilibrium conditions – deoxidation by oxide precipitation – for instance, making use of the pressure-sensitive carbon deoxidation or the pressure-insensitive reaction with deoxidizing metals (Fig. D 2.3).

The deoxidation products generated in the steel should be removed as far as possible before solidification. Any remaining oxides will affect the cleanliness and hence the quality of the steel. These part of deoxidants not used for deoxidation are to be accounted for as alloying components and to be considered as such when adjusting the required steel composition.

In order to meet narrow composition ranges and when small amounts of alloys are added – e.g., in microalloying – aftertreatment processes are frequently applied (see D 2.3.3).

Deoxidizing and Alloying in the Steelmaking Vessel

When producing steels with a high alloy content in the electric arc furnace, it is common practice to remove the highly oxidized slag after refining and to deoxidize the steel melt by forming a new refining slag with a low iron oxide concentration ($\leq 1\%$ FeO). The reducing potential of this slag is obtained by using lime with additions of aluminum, ferrosilicon, calcium-silicon, and/or carbon-bearing materials. The oxygen concentrations attained are roughly between 0.004% and 0.008% for a carbon concentration of 0.05%. This oxygen removal is usually completed by

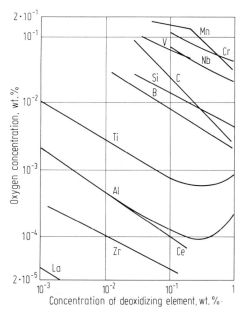

Fig. D 2.3. Deoxidizing potential of different elements at 1600 °C. After [18].

precipitation deoxidation in the ladle or by vacuum treatment – vacuum deoxidation – after tapping. The oxide and sulfide concentrations are lowered considerably in the reducing refining period owing to the high absorption capacity of the lime-active slag for suspensions. Intensive stirring in the melt or intensive mixing with the liquid refining slag during tapping into the ladle – *Perrin effect* – enhance the separation of undesired suspensions.

The alloying of steels with high concentrations of alloying elements is carried out mainly in the reducing refining period. To adjust the chemical composition, after treatment processes are frequently applied. By charging alloying materials (e.g., carbon, ferroalloys, pure metals, or mischmetal) and slag-forming fluxes, an increase in the nitrogen and hydrogen concentrations usually occurs. As mentioned before, the hydrogen concentration can be lowered to below the danger level by vacuum processes.

Deoxidizing and Alloying in the Ladle

Unalloyed and low-alloy steel grades are usually tapped from the steelmaking vessel into the ladle without prior deoxidation or after a weak furnace deoxidation with manganese. For a thorough deoxidation, aluminum, ferrosilicon, calcium-silicon, or deoxidizing mixtures as well as the alloying materials required for attaining the required steel composition are added in the solid or liquid state into the ladle. If oxidizing furnace slag is poured with the steel into the ladle, it reacts with the deoxidized steel causing losses of deoxidants and alloys as well as a phosphorus reversion. Attaining narrow composition ranges during alloying is a difficult task; it is common practice to adjust the chemical composition in subsequent stirring and rinsing operations.

D 2.3.3 Ladle Metallurgy Processes for Steel Aftertreatment

Steels for uses requiring special properties, especially high purity, may be subjected to aftertreatments in the ladle, as, e.g., rinsing or stirring of the melt, the injection of gases or solids heating and vacuum treatments, in separate or combined process steps [19, 20]. These ladle treatments can also be used for overall process optimization. Practical application examples of the process engineering units as concentrated in the model reactor of Fig. D 2.4 are depicted in Fig. D 2.1. The corresponding metallurgical goals are indicated in Table D 2.1.

Gas Rinsing/Stirring Treatment

Gas rinsing is the simplest form of aftertreatment. It now constitutes a standard process step in steelmaking. Inert gases (argon or nitrogen) are introduced deep into the melt through gas-permeable ceramic elements or through an immersed lance. The effects are to enhance the homogeneity of the composition and the temperature of the melt and its cleanliness. A precise adjustment or "trimming" of the chemical composition of the steel becomes possible if the alloys are added into the "eye" or naked steel spot on its surface where the gas emerges. In this operation, the measurement of the oxidation degree of the melt by the electrochemical determination of the oxygen activity (e.m.f.-sensor units) based on the principle of oxygen ion conductive solid-state electrolytes [21] can be useful.

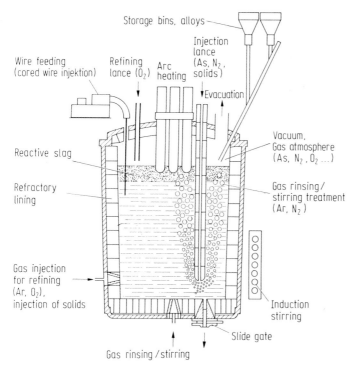

Fig. D 2.4. Process and operating elements for steel aftertreatment (model reactor). After [8].

Stirring by means of induction coils mounted outside the ladle serves the same purpose.

Deoxidized melts may be desulfurized by addition of lime-fluorspar mixtures; nondeoxidized melts may be dephosphorized to a limited extent by mixtures of iron oxide, lime, and fluorspar.

Injection Treatment

An extensive removal of oxygen and sulfur, and also of quality-impairing nonmetallic inclusions, from deoxidized steel melts is achieved by adding calcium- [22] or magnesium-containing materials or basic slag-forming mixtures. Owing to the high vapor pressures of calcium and magnesium, the addition is made by means of pneumatic injection, by shooting-in larger particles or by feeding in a hollow steel wire filled with reagent ("cored wire"). In particular, injecting powdered solids, entrained in argon, through a lance immersed in the liquid steel in the ladle (e.g.: TN-process, for Thyssen Niederrhein) [22–25] has proven to be useful. The oxygen and sulfur concentrations can thus be lowered to the detection limit – about 0.001%. Suspended nonmetallic particles that are easily deformed during subsequent hot rolling, as, e.g., manganese sulfides and aluminum silicates, are transformed during the calcium treatment into globular inclusions that are nondeformable during hot deformation, as, e.g., calcium aluminates. These globular inclusions are also separated more easily from the liquid steel than the former suspended particles. The result is improved toughness and a markedly improved isotropy in the mechanical properties, which is of great importance, especially for flat-rolled products. The overall cost-effectiveness of steelmaking can also be improved by transferring metallurgical tasks from the melting and refining vessel into the ladle, making use of aftertreatment processes.

Hydrogen pick-up caused by basic lime-slag additions can be counteracted with a vacuum treatment or hydrogen release by means of diffusion in the course of slow cooling-down technique after hot rolling, for those grades that are sensitive to hydrogen.

Heating

When subjecting a steel bath to a ladle treatment, additional heat losses occur that depend on the type and length of the treatment; these losses have to be compensated for by a corresponding overheating of the melt in the refining vessel or in the ladle. By heating in the ladle, for instance with the electric arc of a so-called ladle furnace (LF), the economical and metallurgical disadvantages of overheating in the steelmaking furnace are avoided. Moreover, quality can be improved by combining heating with ladle refining under active slags and an inert atmosphere: very low oxygen and sulfur concentrations in the steel can be attained, and the hydrogen and nitrogen concentrations do not increase or can be lowered by a subsequent vacuum treatment; in addition large amounts of alloying materials can be added without temperature losses.

Vacuum Treatment

The exposure of the melt to reduced pressure [26] is done to remove hydrogen, decarburize to low levels, deoxidize, and alloy precisely. Side effects are a homogenizing and purifying of the steel melt. Vacuum systems combined with arc heating (VAD-processes = Vacuum Arc Degassing) allow the addition of large amounts of alloys or performing metal-slag reactions. Of practical importance are pouring stream, ladle, and, especially, partial mass treatment processes; among the latter the most commonly used are the circulating (RH = Ruhrstahl-Heraeus) and the lifting (DH = Dortmund-Hoerde) processes. In order to reduce the hydrogen concentration in the liquid steel to values below the critical value for flake formation, typically 0.0002%, pressures of around 1 mbar are used.

Under vacuum, dissolved oxygen reacts to form gaseous carbon monoxide and is thus removed from the melt without leaving a residue, oxidic inclusions are partly reduced, and favorable conditions for the removal of other nonmetallic inclusions are created. Hence, very low oxygen concentrations and high purity can be attained. In the production of heavy forgings, the *vacuum-carbon-deoxidation* practice causes less segregation during solidification [27] (see also D 6.).

The pressure sensitivity of the decarburization reaction allows the carbon concentration of steel melts to be lowered to 0.001%, aided by oxygen addition if necessary. Uses of vacuum decarburization are both the so-called "light treatment," by which the last part of the decarburization of BOF-heats is done under weak vacuum as well as the production of low carbon and/or high-chromium corrosion-, acid-, and heat-resisting steels. Several processes are commercially available for *vacuum refining* [28–30], for example, VOD (Vacuum Oxygen Decarburization), VODC (Vacuum Oxygen Decarburization Converter), RHO (Ruhrstahl Heraeus Oxygen), RH-OB (Ruhrstahl Heraeus-Oxygen Blowing). For stainless steels, the required low carbon concentrations ($\leq 0.03\%$) can be reached by these processes at normal temperature levels and with a minimum of chromium loss by oxidation. The same results are achieved by reducing the partial pressure of gaseous oxygen in oxygen converter refining processes [31] (AOD = Argon Oxygen Decarburization, CLU = Creusot Loire Uddeholm), in which the gases are injected below the surface of the steel melt in a *side blowing converter*. Gas mixtures of oxygen with argon or oxygen with water vapor are used in this processes.

D 3 Casting and Solidification

By Peter Hammerschmid

D 3.1 Characteristics of Common Methods of Casting

Molten steel may be cast as an ingot or as a continuous strand depending on the available facilities and the subsequent processing steps. *Ingot steel* is top poured or bottom poured into molds; the resulting ingots weigh from 100 kg to 500 t (Fig. D 3.1). *Top pouring* is favored for heavier ingots (> 20 t) and where high purity is required. *Bottom poured* ingots have a better surface than top cast ingots, because in bottom pouring the steel rises smoothly in contrast to top pouring where the molten steel may splash against the mold walls. To avoid splashing, more rapid casting or splash-protection devices can be used.

The molten steel solidifies in the mold as *rimmed, semikilled,* or *fully killed* steel depending on the steel's dissolved oxygen content. Lower oxygen content means there is less oxygen available to react with the carbon in the steel and, therefore, the solidification proceeds more smoothly, i.e., without formation of bubbles or "boiling action." The oxygen content in molten steel is determined by the amounts of carbon, manganese, silicon, and aluminum present. In addition, gas generated during solidification causes blowholes or pipe (Fig. D 3.2). In contrast to rimmed ingots, fully-killed ingots are practically free from blowholes, but may show shrinkage cavities, so-called "pipe". If a heat-generating, exothermic hot top is used, the pipe or shrinkage cavity will not form in the big-end-up ingot; the big-end-down ingot may have only a secondary pipe, which will be welded completely shut during rolling or processing of the ingot. The degree of segregation increases from killed big-end-up to killed big-end-down to semikilled to rimmed ingots.

In contrast to casting of ingots, *continuous casting* of steel in a strand has the following advantages: solidification is faster, subsequent less deformation work is needed, one or two forming steps may be saved, and yield is higher.

Different requirements and designs during the development of continuous casting led to different types of *continuous casting machines.* The most important machines are the vertical type, the vertical bending type, the bow type with straight or curved mold, and the horizontal type (Fig. D 3.3). The current trend is toward lower-height casting machines, because of reduced capital costs. Thus, vertical-type machines are seldom used or built today and, they are used or built, only when extreme purity is required or the steels to be cast are sensitive to cracking. In

---

References for D 3 see page 806.

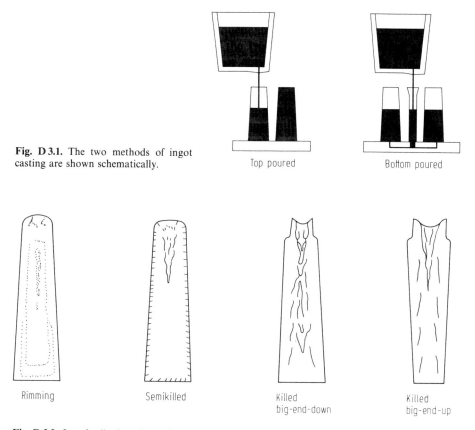

Fig. D 3.1. The two methods of ingot casting are shown schematically.

Fig. D 3.2. Longitudinal sections of rimming, semikilled, and killed steel ingots.

bow-casting machines, a distinction is made between continuous casting machines with straight molds and subsequent progressive bending from the vertical line into a circular arc; and continuous casting machines with curved molds.

Horizontal continuous casting machines can produce cross sections up to 250 mm square and 300 mm diameter and are preferred for higher alloyed steel grades. In spite of the sensitivity of this process and higher mold costs, horizontal continuous casting is a lower cost process than ingot casting, especially for small steel plants [1a, 1b].

D 3.2 Casting and Solidification of Steel in Ingot Molds

D 3.2.1 Reoxidation, Flow, and Superheat

During pouring the steel stream comes into contact with air and, depending on the shape of the pouring stream, *reoxidation* and *nitrogen pickup* may occur. These

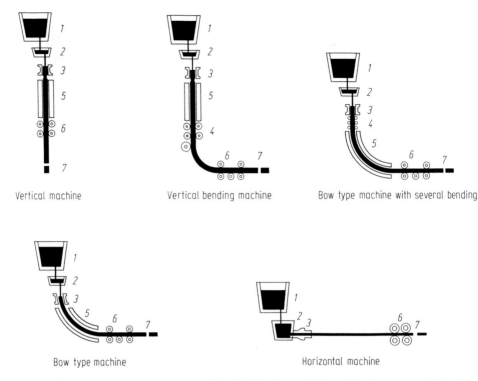

Fig. D 3.3. Types of continuous casting machines (1: ladle; 2: tundish; 3: mold with primary cooling; 4: bending zone with secondary cooling; 5: straightening with secondary cooling; 6: driving and straightening; 7: flame cutting).

effects may be mitigated partly or avoided totally by shrouding the pouring stream and by shielding with inert gas [1]. Compared with continuous casting complete shielding from air in ingot casting is expensive; therefore, only steel of the highest purity is shrouded when cast in ingots.

The fluid *flow* pattern of the steel in the mold during casting is produced by the flowing stream, and after casting it is influenced by free convection, by suction processes due to shrinkage during freezing, and by frozen steel crystals sinking in front of the solidification front [2].

In bottom pouring the direction of flowing steel may be inclined considerably from the mold axis, thus producing turbulence near the mold wall. This turbulence may cause weakness and overflowing of the solidified steel skin [3]. During top pouring the steel splashes against the mold wall when the stream falls on the bottom plate. These splashes form scabs, which may lower the surface quality [4].

Superheat is the difference between the pouring stream temperature and the liquidus temperature of the steel. The superheat should not fall below a fixed value if excessive inclusions are to be avoided. When argon ladle purging is used before the pouring starts and the ladle is covered with a lid, the temperature will fall only slowly during pouring. Excessive superheat, which causes surface cracks and increased segregation, can be diminished by argon ladle purging and, if needed, with simultaneous cooling by addition of scrap. Lower superheat allows a largely equiaxed solidification structure to be produced.

D 3.2.2 Crystallization Process

The crystallization process in ingots and continuous castings differs only in the rate of cooling, so the following statements are valid for both methods.

The melt is supercooled at the mold wall, so that crystallization starts spontaneously during casting and an equiaxed surface layer forms with a thickness of 5–10 mm. High superheat cannot prevent nucleation at the wall, but slows the growth rate (Fig. D 3.4).

After supercooling is balanced by the released heat of crystallization, the formation of a fine-grained equiaxed surface skin is followed by further solidification directed toward the interior, i.e., the zone of columnar crystals [5]. The constitutional supercooling of the melt increases toward the center of the ingot, so the original plane solidification front becomes unstable and a heterogeneous layer of aligned columnar crystals and melt develops [6]. Because of the low heat conduction of the mold, columnar crystals may also grow directly from the surface so that the normal fine-grained equiaxed quench layer does not form and surface cracks develop (Fig. D 3.5).

In the center of the ingot a distinct dendritic structure develops. During further solidification, free movable crystals may finally be formed by heterogeneous nuclei or by the melting and breaking away of the dendrite arms. These crystals may sink down in the remaining melt. Owing to convection within the melt, collisions between free movable crystals may occur giving rise to clusters, (Fig. D 3.6).

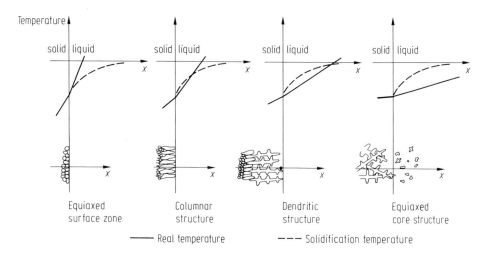

Fig. D 3.4. Schematic graph of the solidification structure. After [6].

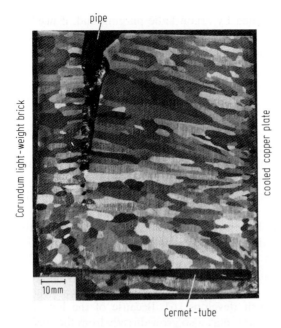

Fig. D 3.5. Solidification structure of a ferritic iron alloy with 2.85 per cent V after cooling under vacuum. Longitudinal section of the upper ingot half. After [7].

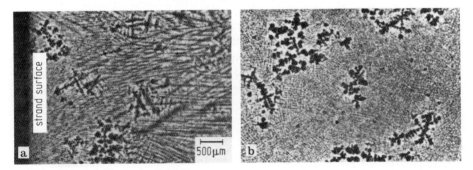

Fig. D 3.6. Dendritic and equiaxed floating crystals in a steel melt with about 0.37 per cent C and 1.25 per cent Mn: *a* transverse section 1.5 mm from the strand surface; *b* horizontal longitudinal section. After [8].

D 3.2.3 Heat Transfer and Solidification

Heat transfer and solidification are closely linked. Most of the heat given up by the ingot accumulates in the mold. The main resistance to heat transfer comes at the interface between ingot and mold. The thickness of the solidified shell is smaller than is formally described by the $\sqrt{t}$ law because of incomplete thermal contact due to microroughnesses, the uneven temperature distribution in the ingot surface, and the faster heat flow into the ingot corners. During the course of solidification, deviations from the $\sqrt{t}$ law manifest themselves as shorter *solidification times* for square ingots and for slabs with side ratios < 1.7. These variations are caused by increases in the ratio of surface to volume during solidification, which result in faster removal of the residual heat content (Fig. D 3.7).

In addition, the heat transferred to the mold is diminished by the gap between the hot steel and the mold wall that develops during solidification. However, fast and hot pouring, which increases the ferrostatic pressure, may force the skin back into contact with the mold wall after the metal had initially contracted away from it at the start of solidification. This sequence of events may also deform and burst the ingot corners, which had not moved.

D 3.2.4 Formation of Segregations

Ingot Segregation

Inverse segregation may occur at the bottom of *killed ingots*, which has a lower concentration of alloying and tramp elements than the molten steel [10]. The residual melt, thus enriched in these elements, is separated during sedimentation from the equiaxed crystals it formerly surrounded and is displaced to the top of the ingot. A normal segregation, partly due to shrinkage, develops in this melt, resulting in a V-shaped structure after complete solidification (Fig. D 3.8). Big-end-up ingots exhibit lower normal segregation than big-end-down ingots. Segregation in big-end-up ingots is marked less and the yield is higher if the ingot is slender and has a large taper [11]. Why "A"- (inverted "V"-) ghostlines develop in

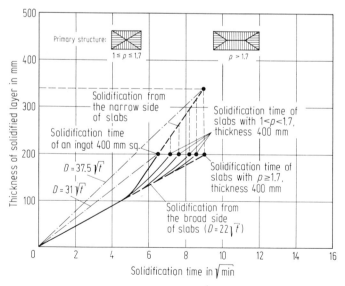

Fig. D 3.7. Solidification of square ingots and slabs with a side ratio $p > 1$ from the long and short sides. After [9] (D = thickness of the solidified layer; p = ratio of broad to narrow side of slabs; t = time in min).

big ingots is not yet clearly understood, but seem to depend on chemical composition, extent of supercooling, ingot size, and shape [11a].

Rimmed steel exhibits more segregation than killed steel; this segregation is caused principally by the drag flow due to gas bubbles evolved at the solidification front [12]. The enriched area preceding the solidification front is distributed by the fluid flow into the melt that is not yet solidified and raises the concentration of accompanying elements, leading to normal segregation. The extent of segregation in rimmed steel also depends on chemical composition, ingot size and shape, and duration of boiling (Fig. D 3.9).

Microsegregation

Crystal segregation during solidification develops because of the lower solubility of the alloying elements in the solid state as opposed to the liquid state. When equiaxed crystals solidify, the zones of highest segregation coincide with the grain boundaries; in dendritic solidification, however, the interdendritic spaces are enriched with segregated elements [13]. During cooling through the γ–α transformation of hypoeutectoid steel grades, a secondary structure is formed by a band-like precipitation of ferrite and pearlite, which may impair the steel's quality, i.e., its machinability and cold workability.

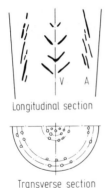

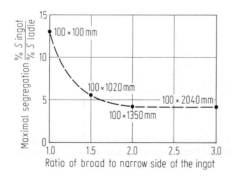

Fig. D 3.8. Schematic graph of A-(inverted V-) and V-segregations. After [12].

Fig. D 3.9. Influence of the ingot shape on maximal segregation in rimming ingots. After [12]. (Initial thickness 680 mm. Dimensional details = sizes after rolling.)

D 3.2.5 Formation of Oxide Inclusions

Endogenous oxidic inclusions in the solidified steel are distinct from exogenous inclusions. *Endogenous* inclusions may form by the reaction of deoxidizing elements dissolved in the steel with oxygen dissolved by supersaturation in the pouring stream or preceding the solidification front, or both. Oxygen supersaturation may result from pick-up in the pouring stream, decrease of solubility due to decreasing temperature, or accumulation during solidification. The inclusions that develop during solidification differ from those formed primarily by deoxidation. Thus, in the deoxidation of steel having, for example, a high oxygen content with aluminum, coarse spherical herzynite ($FeO \cdot Al_2O_3$) inclusions and alumina dendrites are formed. On the other hand, during solidification at low oxygen contents, "corals" of alumina are formed (Fig. D 3.10). These corals are captured in the solidification front and, therefore, cannot rise within the ladle or the ingot like those formed during deoxidation. During rolling, so-called alumina bands are formed and they impair the steel's properties. These bands can be diminished by ensuring low oxygen content in the liquid steel and by avoiding reoxidation.

Exogenous inclusions are formed by reaction of the steel with refractory materials or slags. In large forging-grade ingots most of the macroscopic inclusions come from refractory fragments and from the reaction of steel tramp elements with the refractories washed into the mold during pouring [15]. During the very fast solidification at the bottom plate, these inclusions are concentrated in a zone in the ingot bottom (Fig. D 3.11) and in addition these inclusions together with accumulations of dendrites may sink in the still liquid residual melt. Inclusions can be

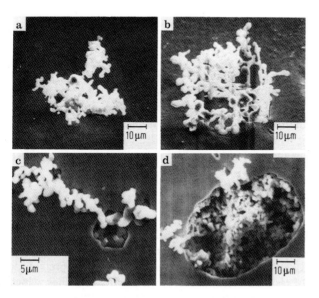

Fig. D 3.10. Shapes of alumina inclusions in a 7-t ingot, steel grade RSt38. **a**, ingot border; **b** to **d**, columnar zone. After [14].

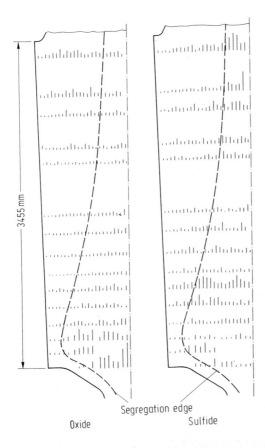

Fig. D 3.11. Distribution of oxide and sulfide inclusions in a 105-t ingot. Composition: C 0.21, Si 0.27, Mn 0.85, P 0.011, S 0.008, Cr 0.46, Mo 0.69, Ni 1.27, V 0.03, Al 0.021 per cent. After [15].

reduced by low initial oxygen content, suitable pouring techniques, protection against reoxidation, and high-quality refractories.

Ingot scum on rimming steel, formed mainly of manganese and iron oxides, may be sucked in by circulating fluid flow in the boiling steel and may be detrimental in steel processing. These defects can be avoided by adjusting the temperature, matching carbon and manganese contents, and skimming off the scum [16].

D 3.2.6 Formation of Sulfide Inclusions

At high oxygen and relatively low manganese activities liquid (Fe, Mn) oxides form; when sulfur is present, liquid (Fe, Mn)-oxisulfides form (Type I, Fig. D 3.12). With increasing manganese activity and diminished oxygen activity, sulfides precipitate in the interdendritic spaces of residual melts as corals with many branches, which appear in metallographic specimens as sulfide "chains" (Type II). When oxygen is

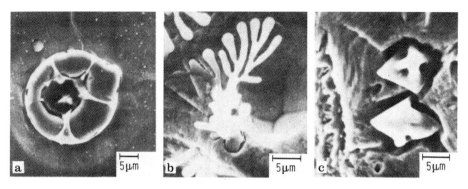

Fig. D 3.12. Scanning electron microscope photographs of samples with S 0.1 per cent. *a*, Manganese oxide bordered by oxisulfide at 0.4 per cent Mn (Type I); *b*, coral like sulfide of Type II, 2.1 per cent Mn; *c*, sulfide of Type III still incomplete in the centers of the faces Mn 13.2 and Al 0.16 per cent. After [17].

kept to very low residual amounts, sulfide dendrites form, with shapes varying from plane surfaces to compact plane surfaces. (Type III). These various shapes have different effects on the properties of the processed steel. Type I is desirable for good machinability, and Type II impairs formability and toughness [17, 18]. Generally, sulfides diminish transverse toughness and formability especially when present in high concentrations [18]. Even after desulfurization in the ladle lowers the concentration, the remaining sulfur still frequently accumulates to the saturation limit because of segregation and it may precipitate during solidification as a mixed sulfide [19]. By adding elements like zirconium, rare-earth metals, calcium, titanium, or tellurium, sulfides or mixed sulfides can be formed instead of manganese sulfides; these sulfides and mixed sulfides have low hot formability, which greatly improves the transverse toughness [20].

D 3.2.7 Formation of Blowholes

During solidification the carbon, which accumulates at the solidification front of the melt, reacts with oxygen forming bubbles of carbon monoxide. Depending on the amount of deoxidation the initial concentrations of oxygen result in different quantities of carbon monoxide and the steel solidifies with rimming, semikilled, or killed structure (Fig. D 3.13). The position of the blowholes trapped during solidification depends on the intensity of the carbon monoxide development. Blowholes close to the surface may be exposed during heating in the soaking pit and cause seams in rolling. If too little carbon monoxide is developed in semikilled steel, pipe is formed owing to insufficient compensation of shrinkage. Also, killed steel may show subsurface blowholes even at aluminum concentrations of 0.020 per cent, owing to the formation of carbon monoxide at rough mold walls; these blowholes may be avoided by smoothing the surface of the mold [22].

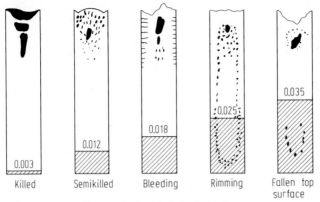

Fig. D 3.13. Solidification structure of steel ingots with different oxygen concentrations (schematically). After [21].

D 3.2.8 Formation of Pipe

During solidification of steel a sudden decrease of about 4 per cent of the specific volume occurs; with killed steels particularly, this shrinkage leads to pipe formation when the free flow of liquid steel for feeding during solidification is blocked. Heat radiated from the top surface is a special problem. Pipe formation depends on the ingot shape, i.e., on the ratio of side to height, and on the type of taper (big-end-up or -down). Pipe may be avoided by preferred axial heat flow directed from the bottom of the ingot to the top; by an insulating hot top; by heating or additional exothermic covering materials; and by control of casting temperature and rate. Pipe is harmful to processing and the operational characteristics of finished parts. Pipe should be avoided because it presents weak spots containing inclusions, which in processing or use may cause early failure of the finished part.

D 3.2.9 Treatment of Cast Steel

Ingots are placed in soaking pits or pusher-type furnaces as soon as possible after casting. The minimum duration of cooling in the mold depends on the degree of deoxidation and the ingot size. To avoid shrinkage, killed steel with a hot top has to solidify completely before being placed in the soaking pit. Slabs and ingots, which cannot be charged directly, are cooled slowly in air depending on alloy content. Improper cooling may cause clinks leading to internal and surface defects which cannot be removed by hot forming.

D 3.3 Processes in Continuous Casting

D 3.3.1 Reoxidation, Flow, and Superheat

In continuous casting the killed molten steel is *reoxidized* when the pouring stream picks up oxygen from the surrounding air between ladle and tundish as well as between tundish and mold. Moreover, tundish covering material and lining or slag from casting fluxes may reoxidize the steel [22a]. Reoxidation is indicated especially by oxidation losses of aluminum in killed low-carbon steel and may result in loss of up to a quarter of the ladle aluminum content when the ladle stream is unfavorably shaped [22b]. The molten steel can be protected from reoxidation by a submerged refractory tube or a pressurized chamber – between ladle and tundish – and a submerged nozzle or by shielding with argon or liquid nitrogen – between tundish and mold (Fig. D 3.14). These measures, especially the shrouding tube between ladle and tundish, greatly improve the purity [23, 23a]. Moreover, the purity is improved by using basic refractories with a low iron oxide content for shrouding tubes between ladle and tundish and for tundish linings [22a].

The inclusion content is considerably reduced when alumina–graphite submerged nozzles instead of silica nozzles are used between tundish and mold.

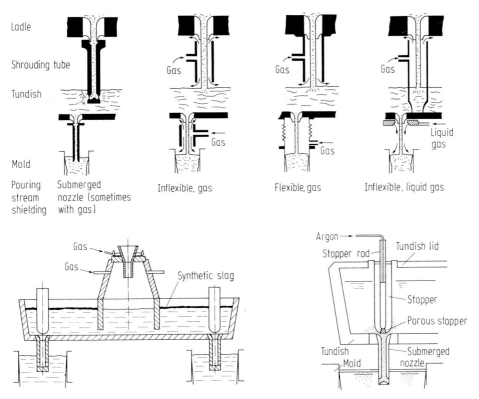

Fig. D 3.14. Measures for shielding the pouring stream to prevent reoxidation during continuous casting.

However, the ideal immersion nozzle for aluminum killed steel would be made of pure alumina, because use of graphite reduces alumina, yielding carbon monoxide, which is dissolved in liquid steel as carbon and oxygen forming alumina inclusions [23a].

The modification of the inclusions formed by reoxidation – spheres, dendrites, corals, plates, or compact shapes – is influenced by the *mass flow densities* of oxygen and deoxidation elements. It is possible to infer the source of reoxidation from the inclusion type and to deduce methods that will improve the process [24].

The *flow* in the tundish, the mold, and the still-liquid portion of the strand greatly influences the purity. In the tundish, the steel flow field should be uniform and the residence time should be long. Flowing along the surface helps to absorb both the slag carried over and the oxides in the covering slag. This flow pattern is attained by a suitably-shaped tundish (T-tundish) and by dams at its bottom [23]. The increased width of the tundish promotes removal of inclusions because of the longer residence time and the more uniform flow [24a]. Slag carry-over is caused by vortex formation and may be suppressed by breaking up the vortex or by a sufficiently high steel level in the tundish [24b, 25].

Because of its momentum while casting, the steel flowing into the mold can penetrate so deeply into the still-liquid portion of the strand that rising inclusions in bow-type machines may be trapped in the upper inclined solidification front resulting in an oxide band, which may cause laminations (double draw) and cracks during the deep drawing of cold strip [26].

In slab and bloom casting machines it is possible to pour the molten steel through a submerged nozzle with lateral orifices. By this means and by the injection of argon through the tundish stopper, the penetration depth is reduced so that the inclusions can be retained by the fluxed slag in the mold without reaching the inclined solidification front. The lateral orifices in the submerged nozzle directed upwards or horizontally produce a uniform temperature distribution in the casting flux layer, which leads to a better strand surface [26].

Superheat is measured in the tundish and may be lower than in casting ingots. When superheat is low, there is no danger of remelting the strand shell or of breaking through. Center segregations and internal cracks also are diminished by low superheat, because the equiaxed solidification structure has a greater extension in the strand center [25, 27]. Very low superheat may lead to freezing of the submerged nozzles. Therefore, superheat in continuous casting is adjusted to about 15–30 degrees C. In ingot casting the superheat range is held at 40–70 degrees C (bottom pouring) and 30–50 degrees C (top pouring). By using the ladle furnace, superheat variation is smaller and can be held in a narrow range [27a].

Regarding steel crystallization, continuous casting differs from ingot casting only by the intensity of the heat flow and, thus, by the solidification rate. Therefore, the statements of Section D 3.2.2 are also valid for continuous casting.

D 3.3.2 Heat Transfer During Solidification

Owing to more intensive cooling, in continuous casting more heat is extracted from the solidifying steel per unit of time than in ingot casting. This greater heat extraction rate is demonstrated by the higher *solidification constant* for continuous casting, an average of 26 mm min$^{-1/2}$ [28] in contrast to approximately 22 mm min$^{-1/2}$ for ingot casting (Fig. D 3.7).

The heat transfer in the mold depends on the film thickness of the mold slag layer and on the extent to which the strand is in contact with the mold. This contact determined by the *taper* of the mold, which depends on the casting rate. Because of the high absolute shrinkage of the broad face in continuous casting of slabs, it is important to keep the taper of the narrow faces of the mold very uniform.

The rate of *heat extraction in the secondary cooling zone* must not be too high for steels that are sensitive to cracks due to excessive temperature stresses. Mist cooling by water and compressed air cools the whole area of the strand between the pinch rolls, also in the arch spandrel between roll and strand. Mist cooling also avoids stress peaks because its uniformity prevents incipient cracks present in the mold from growing larger [29]. The heat transfer coefficient is the same for spray water as for air mist cooling; it depends on the water quantity per unit of time [29a].

D 3.3.3 Formation of Segregations

Macrosegregations in continuous castings are different from those in ingots because of the particular conditions of cross section and solidification.

Center segregation may occur in continuously cast slabs owing to flow during the final solidification in the liquid crater; bulging of the strand between the pinch rolls is the principal cause of the flow in the liquid crater [30, 31] (Fig. D 3.15). If there is no bulging, only a little segregation will develop by flow due to solidification shrinkage [32]. A larger core zone with a globulitic structure leads to an even distribution of the residual melt, i.e., to a moderate center segregation.

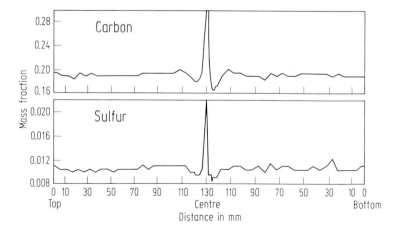

Fig. D 3.15. Distribution of carbon and sulfur through the cross section of a continuously cast slab. Steel grade: St 52-3, with columnar dendritic solidification. After [27].

Columnar core crystallization, however, leads to almost planar solidification of the residual melt giving rise to a distinct center segregation [27].

Center segregations may be controlled by low superheat; by proper cooling and avoiding roll eccentricity, which keeps down bulging or by pressing solidification fronts together. Additionally, electromagnetic stirring and smaller diameter split rolls (corresponding to a smaller roll distance) will reduce center segregation [32].

Segregations develop in continuously cast *billets* and *blooms* because the cross section is periodically cut off due to local solidification (Fig. D 3.16). Thus, so-called "mini-ingots" are formed, which means that segregation develops between cut-off areas as in killed ingots. Low superheat and electromagnetic stirring reduce segregation. Mean segregation values of high-carbon steel grades (0.55–0.85% C) fall when intensive secondary cooling is applied.

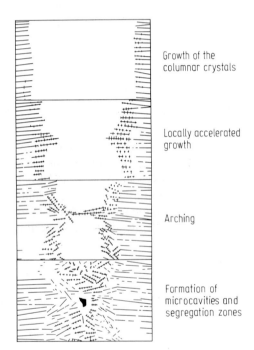

Fig. D 3.16. Formation of "mini-ingots" during continuous casting of billets and blooms. After [34].

D 3.3.4 Influence of Electromagnetic Stirring on Segregations

Stirring with line frequency (50 Hz) is not successful in *continuous slab casting*. However, stirring with an *electromagnetic traveling field* produces an equiaxed center solidification in structural and heat-treatable steels even in machines with a small arc radius, thus improving center segregation. Indeed, the quality of strand support (e.g., bulging, roll eccentricity) has more influence on segregation than electromagnetic stirring. The advantage of electromagnetic stirring is that it is not dependent on the level of superheat. Therefore, small defects in the strand support are not critical.

However, mechanical properties, resistance to hydrogen-induced cracking, weldability, and other properties of the formed material are not improved by electromagnetic stirring. Electromagnetic stirring may lead to an agglomeration of inclusions, thereby lowering the purity, but it is advantageous for strongly columnar solidifying material such as *stainless and heat- and acid-resisting steels*, which tend to form ribs during processing [34a, 36]. "White bands" are zones of inverse segregation that form between the heterogeneous solidification front and the forced flow in the liquid crater near the effective field of the stirrer. These bands have no effect on the properties of the materials rolled from stirred slabs [35].

In *continuous casting of billets and blooms*, stirrers with rotating electromagnetic field are installed in and below the mold and in the area of the secondary cooling zone up to shortly before the end of the liquid crater with the same field or in combination with an electromagnetic travelling field. An equiaxed crystal solidification structure may be attained and the "mini-ingot" formation in continuous billet casting restrained. Thereby segregation may be avoided to a large extent [35]. Furthermore, it is possible to diminish the intensity and frequency of the main defects such as porosity, inclusions, slaggy patches and subcutaneous blowholes, and, thereby, to yield a better surface and workability. The formation of "white bands" may be avoided by a system of several electromagnetic stirrers in which one stirrer operates with a lower intensity [35].

Increased casting rates and an improved strand quality in bloom machines with large cross sections have been secured, but statements about the success of electromagnetic stirring in slab machines should be weighed carefully considering the limiting conditions [37].

D 3.3.5 Formation of Internal Cracks

In continuous casting stresses and strains are caused by diverse deformations of the strand shell, which may induce cracks in the solidification front if critical values are reached. The cracks are filled by the residual enriched melt and thus are healed. The formation of internal cracks depends on the material properties at high temperatures, on the installation and its condition. Figure D 3.17 shows the *effect of several alloying elements* and their contents [38]. The composition can only be held within close limits to the steel alloy required. Reduction of sulfur to the lowest content ($\leq 0.010\%$ S), however, gives the best chance of avoiding internal cracks [39].

The different frequency of internal cracks in slabs of stainless steel and heat resistant steel grades can be explained by the high temperature properties material of these materials. Crack sensitivity can be reduced by lowering the phosphorus and sulphur levels [39a].

Problems with the formation of internal cracks are diminished by the *design of multiple bending machines* with closely-spaced split rolls. By the so-called "compression casting," the pressure of driven pinch rolls lowers the stress in the strand shell before the straightening point is reached and reduces internal cracks. Because of these advantages, "compression casting" is in use in several bow type machines [40].

D 3.3.6 Surface Defects

Figure D 3.18 shows schematically the most important surface defects and the defects occurring in the strand shells of slabs, blooms and billets [41].

Seams can be caused by ferrostatic pressure, excessively-temperature gradients and over-high frictional forces. Longitudinal cracks are located preferably in the central third of the broad side, and transverse cracks mainly close to the edges of the broad sides. Furthermore, corner cracks and longitudinal cracks near the corner must be recognised. These cracks occur more frequently in smaller cross sections. Star cracks are formed mainly by infiltration of copper. Blowholes directly

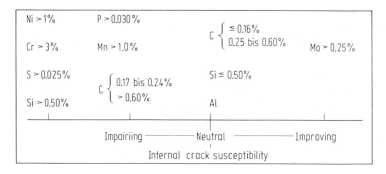

Fig. D 3.17. Effect of different alloying elements on the internal crack susceptibility of continuously cast steel. After [38].

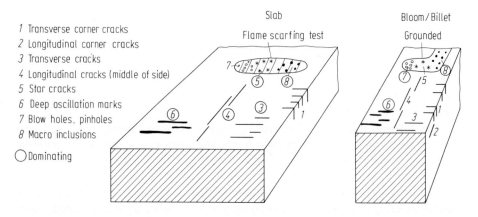

Fig. D 3.18. Possible surface defects on continuously cast (steel). After [41].

underneath the surface (pin holes), pores and slag inclusions also may be discovered. Depth and shape of oscillation marks depend on oscillation height and frequency, negative strip-time and viscosity of mould slag. If these marks are very strong they may result in transverse cracks [42, 42a, 42b].

D 3.3.7 Significance of Casting Fluxes

Casting fluxes are pulverised synthetic slags used to cover the steel in the mold. Figure D 3.19 shows the behavior of casting flux in a continuous casting mold [43]. The flux should melt at the temperature of the steel in the mold and act as lubrication at the interface between mold and strand shell. In addition, the material protects the steel against reoxidation by air, acts as flux for oxides precipitating out of the steel, prevents marked heat losses and influences heat transfer in the mold.

The choice of composition of casting flux to be used depends on steel grade, type of continuous casting machine, strand cross-section and casting rate. Melting and solidification behavior of the flux, viscosity, heat conductivity, spreading and similar properties are used to judge flux suitability [44]. No complete theory is yet available to predict the action of casting fluxes. So their behavior during casting is used as a final acceptance test.

Results of research published recently, help identify factors governing flux selection: crystallization during cooling of phases such as Cuspidin ($3CaO-2SiO_2-CaF_2$) and Nephelite ($Na_2O-Al_2O_3-2SiO_2$) greatly affects the efficiency of mold lubrication and hence the occurrence of break-outs caused by sticking [45]. The

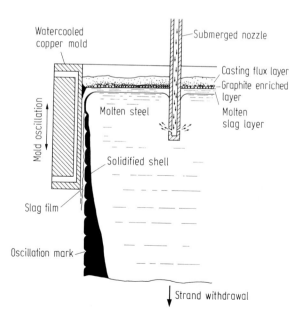

Fig. D 3.19. Behavior of casting flux in continuous casting (schematically). After [43].

morphology of carbon controls the powder melting rate and causes the unusual occurrence of longitudinal cracks in low-carbon high-alloy steel [46].

D 3.3.8 Hot Charging and Hot Direct Rolling of Continuous Cast Steel

To *save energy*, slabs are seldom cooled but are fed directly into the walking beam furnace. Still more recently, feeding of the strand from the caster directly into the mill, bypassing any furnace, has been developed recently to save more energy.

High-speed casting, manufacture of high-temperature slabs and quality improvements are indispensable technologies for direct hot rolling. Various measures are used to produce high temperature slabs including: mist spray cooling, and an insulating zone behind the secondary cooling zone resulting in slabs hotter than 1000 degrees C at the exit and slab edge heaters [47, 48]. Because of the higher temperature, the solidified shell is reduced in strength and is more susceptible to surface or internal cracks. This situation led to application of compression casting [48]. By these and other measures 50–80 per cent of the cast slabs were directly hot rolled. The hot on-line crack detection on slabs is now used to detect all or nearly all cracks on the surface [49].

D 3.4 Comparison of Ingot Casting and Continuous Casting

Continuously-cast steel is suitable for the whole range of steel grades and uses and is of at least equal in quality to ingot steel [50]. There are only a few exceptions in which ingot casting is still preferred. For instance, rimming steel may be necessary due to its outer skin being low in carbon and tramp elements (e.g. for enamelling) and ingot steel is still preferred in some special alloyed steels (e.g. some high alloy tool steel).

Continuously-cast steel in contrast to ingot steel has other specific possibilities of defects, e.g. in its segregation behavior (cf. D 3.3.3), formation of internal cracks (cf. D 3.3.5) and longitudinal and transverse surface cracks (cf. D 3.3.6). However, it is possible to avoid these defects by various technological processes, allied with suitable measuring and controlling equipment.

D 3.5 Outlook: Continuous Casting Approaching Final Shape

Reducing the work required to achieve the required form by continuous casting of steel to near the shape of the rolled product is of great economic interest. For hot strip thickness materials, probably only casting machines with movable molds are suitable. Heat flow seems to be no problem at higher casting rates but feeding the molten steel into the narrow entrance of the mold presents some difficulties [51, 52].

Several new processes have been developed such as the twin belt Hazelett machine [53] and the wheel caster [54]. Recently a process having high operating rates, using a tapered stationary mold has been introduced and is capable of reducing the thickness from 170 mm to 30 mm – 50 mm [55]. This process has clearly reached the stage of first industrial application, in conjunction with compact hot rolling mills technology. This has opened the way to the first mini-mills for flat products, in the carbon as well as in the stainless steel areas. Structure and surface quality of the hot strip was reported to be good [56].

Several processes have been proposed for casting a strip which has only to be cold rolled. These processes include casting with one or two rolls, and spray casting. The starting material quality must be equal to conventionally-cast steel. This condition may be secured by a fine solidification structure with a reduced dendrite arm spacing and smaller size of sulfide inclusions [51]. The various strip casting technologies need more time to mature into fully industrial processes, because of the higher level of difficulties which have to be solved such as the liquid metal feeding and level control, the side confinement, the problem arising from the welling and contact of liquid metal onto the travelling mold and the necessity of obtaining a good surface and a good geometry directly from the mold. However the result announced are optimistic [56].

D 4 Special Methods for Melting and Casting

By Hans Vöge

Because of physical laws and for economic reasons the possibilities of influencing steel properties by process engineering in conventional steel production methods are limited. These limitations apply to chemical composition (including gas contents), and freedom from non-metallic inclusions and segregations (homogeneity) and therefore to mechanical properties, also.

The high demands that are frequently made on steels for certain applications such as aircraft and spacecraft, and special forgings for power machinery, often cannot be met by means of steel production methods that are normally appropriate and economically adequate. For such applications it is of prime importance that the contents of oxygen, nitrogen, hydrogen and undesirable non-metallic inclusions should be kept as low as possible. To reduce these inclusions it is necessary to utilize remelting and/or vacuum-melting methods. Owing to the high costs involved, these methods can be used only in the production of high-quality steels, where conventional production techniques cannot provide satisfactory properties.

D 4.1 Remelting Methods

Remelting today is understood to refer to all methods whereby an oxygen converter or electric furnace ingot possessing essentially the final chemical composition is remelted by continuous fusion under exclusion of air (vacuum, slag), and resolidified in a water-cooled mold to ingots of the desired form and size [1]. If the fusion rate is chosen correctly a primary structure grows in the direction of the heat flow in the mold without primary segregations or local concentrations of non-metallic inclusions. Furthermore, the process reduces the unavoidable microsegregation compared to the conventional cast ingot, owing to the avoidance of the globulitic crystallization zone in the ingot center. For industrial steel production (Fig. D 4.1) the *Electro-Slag-Remelting* (ESR), *Vacuum-Arc-Remelting* (VAR) and the *Electron beam processes* (EB) are established. A new variant of the VAR-process is the *Vacuum Arc Double Electrode Remelting* (*VADER*) consisting essentially of fusing the consumable electrode below the liquidus temperature (Subliquidus melting). By this method, the steel is refined to the extent that the ingot structure is similar to that of ingots produced by powder metallurgy [2a].

References for D 4 see page 808.

D 4.1 Remelting Methods

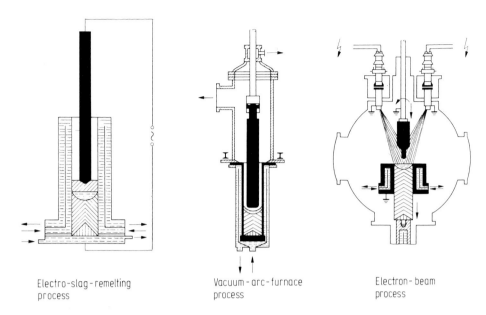

Electro-slag-remelting process

Vacuum-arc-furnace process

Electron-beam process

Fig. D 4.1. Schematic representaion of the principles of design and operation of some remelting methods. After [2]. Electro-Slag-Remelting-Process, Vacuum-Arc-Furnace-Process, Electron-Beam-Process.

Fig. D 4.2.

	Impaired	Same	Improved	Strongly improved
Ingot condition: surface, porosity and density, yield				
Chemical composition: basis metals, hydrogen, oxygen, sulphur, trace elements				
Cleanness: microscopical, macroscopical				
Ingot structure: ingot segregations, crystal segregations				
Mechanical properties: tensile strength, yield strength, ductility, isotropy				

The condensed graph in Fig. D 4.2 shows the influence of the above-mentioned remelting methods on various steel properties in comparison to steel properties of conventionally-produced steel.

The *chemical steel composition* is influenced in various ways by variations in the remelting process. While in the ESR, owing to slag reaction or by the flotation of highly oxygen-affinitive elements, such as oxygen, sulphur, aluminum and titanium are removed from the melt. The vacuum remelting process leads to the reduction of metallic inclusions and iron-affinitive elements possessing high gas pressure as well as of gases. The lower pressure in the electron beam furnace greatly favors reduction of undesired tramp elements, not only such elements as copper, antimony and tin, but also manganese, so that meeting the specified manganese

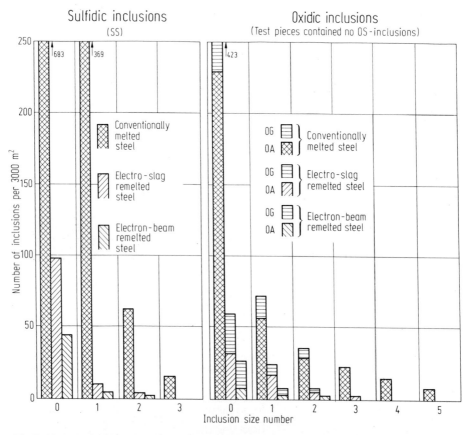

Fig. D 4.3. Nonmetallic inclusions in steel produced by various methods 100 Cr 6, measured in 240 test pieces from 60 heats. Evaluation was made in accordance with Stahl-Eisen-Prüfblatt 1570–71. Microscopic examination of special steels for non-metallic inclusions by means of picture series [4a]. After [5]. (SS = Sulphides, line form; OS = Oxides, line form (siliceous type); OA = Oxide, dissolved form (alumina type); OG = Oxide, dissolved form (alumina type); OG = Oxide, (globulitic form). (Chemical composition: 0.95 to 1.10% C, 0.15 to 0.35% Si, 0.25 to 0.45% Mn, $\leq 0.030\%$ P, $\leq 0.030\%$ S, 1.35 to 1.65% Cr) Sulphidic inclusions (SS), oxidic inclusions (test pieces contained no OS inclusions), number of inclusions per 3000 mm^2, conventionally melted steel, electro-slag-remelted steel, electron-beam-remelted steel.

content is rendered difficult. Remelting under vacuum results in such low hydrogen contents that flake development is avoided. In the ESR process the hydrogen content may increase compared with that of the electrode material. The correct selection of the working slag and its careful preparation also enables flake formation to be avoided, in the ESR process [4].

The *cleanness* of steels from *oxidic and sulphidic inclusions* is substantially improved by remelting. Plotted by way of example in Fig. D 4.3 are the data on the cleanness of ESR- and EB-remelted steels in comparison with conventionally-produced steels. In the ESR-process the reaction with the working slag results in an extensive slagging and separation of oxides and sulphides. Oxidic and sulphidic inclusions remaining in the steel are present in a finely distributed form as a result of the crystallization conditions peculiar to all remelting processes. In all methods, contact of the molten heat with the oxygen of the air and the refractory materials is excluded, so that macroscopic inclusions cannot form, as they do in conventional steel production methods.

The continuous slow growth of the ingot and the resulting as-cast structure in all remelting processes lead to a greatly reduced *ingot segregation* which does not vary over the ingot length, nor does the macrostructure. By appropriate control of the remelting processes the formation of *crystal segregation* also is greatly reduced

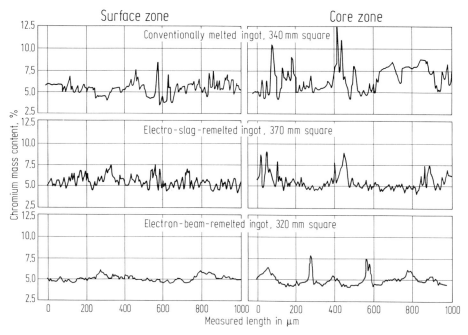

Fig. D 4.4. Comparison of crystal segregations of inclusion size number ingots of hot-working steel produced by various methods X 40 CrMoV 5 1, as – cast structure. After [5]. (Chemical composition: 0.37 to 0.43% C, 0.90 to 1.20% Si, 0.30 to 0.50% Mn, ≤ 0.030% P, ≤ 0.030% S, 4.8 to 5.5% Cr, 1.2 to 1.5% Mo, 0.90 to 1.1% V). Chromium mass content %, Measured length in μm. Surface zone, core zone, conventionally melted ingot, electro-slag-remelted ingot, electron-beam-remelted ingot. 340 (370, 320) mm square.

Table D 4.1 Segregation behavior of as-cast ingots of hot-working steel produced by various methods X 40 CrMoV 5 1. After [5].

Production method and size	Average chemical composition									Position of sample	Degree of segregation Max. Percentage of element i Min. Percentage of element i (Averages)			
	% C	% Si	% Mn	% P	% S	% Cr	% Mo	% V			Mn	Cr	Mo	V
conventionally cast ingot 340 mm square	0.42	1.10	0.35	0.020	0.006	5.23	1.41	1.08		surface	1.8	2.1	11.0	6.4
										center	2.3	2.7	13.0	18.4
electro-slag-remelted ingot 370 mm square	0.41	1.00	0.35	0.020	0.003	5.21	1.40	1.06		surface	1.6	1.6	3.8	4.1
										center	1.6	1.9	4.5	6.7
electron beam remelted ingot 320 mm square	0.42	0.95	0.06	0.008	0.008	5.03	1.35	0.95		surface	1.5	1.4	3.0	2.0
										center	1.8	1.8	7.5	3.6

due to the smaller dendritic zone, when compared with ingot casting. Concentration curves for chromium content, for example, established by the microprobe (Fig. D 4.4) demonstrate, the reduced microsegregation of remelted ingots in the hot working steel X 40 CrMoV 5 1. Manganese, molybdenum and vanadium also show smaller segregations (Table D 4.1).

Values for elongation at fracture and reduction of area of high-strength structural steels of the grade X 41 CrMoV 5 1 produced conventionally as well as by the ESR-process are shown in Table D 4.2. It will be recognised that considerable improvements are obtained, especially in the transverse direction.

Due to the greater structural homogeneity and the reduction of non-metallic inclusions, steels produced by the remelting methods possess improved properties. Advantages of remelted steels are improved hot workability, reduced minimum degree of deformation, lower sensitivity in heat treatments, lower tendency to distortion, better polishing properties and improved mechanical properties. Remelting methods may be utilized in various ways for the various steel grades and applications.

Stainless steels produced by the ESR-process also possess very much improved *resistance to pitting* and corrosion in general because of their extensive freedom from sulphidic inclusions [7].

D 4.2 Special Melting Methods for Heavy Forgings

The increasing capacity of power plants, especially of nuclear power plants, results in increased dimensions and weights of the components. Production of forgings for these components requires ingots that are guaranteed to be fault-free to ensure satisfactory service behaviour of large components in regard to segregations, absence of center porosity and non-metallic inclusions [8, 9]. To fulfill these demands several special metallurgical methods have been developed. Ingots with diameters of up to 2300 mm maybe produced by the ESR-method [10]. The advantage of the shallow depth of the melted pool with the resulting more uniform and faster solidification, size and distribution of the non-metallic inclusions, in both light and heavy ingots, is that macroscopic segregations are practically eliminated.

A modified ESR-method is also used for producing heavy conventionally-cast forgings for ingot top heating (*Ingot Top Remelting Process*). Methods developed to operational maturity include the BEST (Böhler Electroslag Topping) [11] and TREST (Terni Refractory Electroslag Topping) [12] which help to solve the problems arising during ingot solidification (shrinkage, segregation and formation of large inclusions) by remelting an electrode in the ingot top. These processes use a method in which the molten steel is covered by a premelted slag and the electrode is introduced into the slag pool. The metal melting away from the electrode, as in the ESR-process, mixes during solidification with the residual melt of the cast ingot. The slag is kept in a molten state until the final solidification with the result that shrinkage cavities can be avoided completely, ingot segregation and segregation lines can be reduced and cleanness substantially improved.

Table D 4.2 Mechanical properties of quenched and tempered forgings of 230 to 350 mm diameter, of steel remelted by various methods X 41 CrMoV 5 1[a]. After [6]

Position of sample	Production method	Elongation at fracture A		Reduction of area Z		Ratio of values transverse to longitudinal	
		long. %	trans %	Long. %	transv. %	A	Z
Surface	Conventionally-melted ingot (el arc furnace)	12	6	45	12	0.5	0.27
	Electro-slag remelted ingot	12	12	45	43	1.0	0.97
1/2 Radius	Conventionally-method ingot (el arc furnace)	10	2	34	5	0.2	0.15
	Electro-slag remelted ingot	10	10	40	38	1.0	0.95
Center	Conventionally-method ingot (el arc furnace)	10	2	30	4	0.2	0.13
	Electro-slag remelted ingot	10	9.5	40	36	0.95	0.90

[a] Range of tensile strength about 1550 to 1700 N/mm^2; Averages of about 200 heats

Another means of obtaining an almost faultless core zone in heavyforging ingots is presented by the *Core-Zone-Remelting process* [13]. In this process, a conventionally-cast ingot is heated to forging temperature, then pierced by a press, so that with center punching the largest part of the V-segregations is eliminated. The hollow body thus produced serves subsequently as a melting receptacle, as in the normal ESR-process, in which an electrode is melted. It is planned to produce ingots up to a weight of 300 tons by this method.

In forging ingots weighing more than 300 tons the enrichment in the segregated core is equalized by "after pouring" of a heat with a lower carbon content. In the "Multi-Pouring"-process, heats from several furnaces are poured one after the other to make one big ingot, the carbon content being reduced from heat to heat. This technique proved successful in keeping the differences in the carbon content between top and bottom relatively small [14].

D 4.3 Vacuum Melting Methods

High-temperature alloys and high-strength steels, e.g. for application in jet and gas turbine engines containing higher amounts of oxygen-affinitive elements, particularly aluminum and titanium, generally are melted in vacuum. The *Vacuum-Induction Furnace* is widely accepted as an economical melting method.

Metallurgy in the vacuum-induction furnace is defined essentially by the vacuum prevailing in the system as well as by the kinetic influence of the inductive bath movement and the surface condition of bath and crucible on the course of the reaction [3]. The vacuum usually reached at the end of the melt-down period of 10^{-2} to 10^{-4} mbar suffices for the extensive removal of hydrogen to below 0.0001% (1 ppm) and of nitrogen to contents of below 0.0020%, with a simultaneous carbon monoxide formation favoring this reaction process. By means of the carbon monoxide reaction in low-alloy steels, even in the range of 0.001% to 0.003% C, oxygen contents of approx. 0.001 may still be achieved. Although phosphorus can only be removed with great difficulty on account of the low oxygen contents in the vacuum-induction furnace, the removal of sulphur is relatively easy by way of the gas phase.

Of practical importance, particularly for the selection of the metallic charge materials in this process, is the partial vaporization of during melting, metallic impurity elements such as lead, arsenic, antimony, tin and copper. [15, 16].

It is not possible to influence the ingot structure in the vacuum-induction furnace. For this reason, vacuum-melted steel is normally remelted again (ESR, EB, VAR), if it is to be used for high-grade applications.

To make sure that possible impurities caused by ceramics of the furnace lining are separated, new methods are being developed for super refining in so-called cold-hearth melting and refining furnaces such as the Electron Beam Cold Hearth Refining (EBCHR) and the Plasma Cold Hearth Refining (PCHR) processes [16a].

D 5 Hot Rolling

By Klaus Täffner

D 5.1 Hot Rolling Process

Apart from steel castings, all primary and semi-finished products produced by casting molten steel are further processed by hot working. This process of hot working consists almost exclusively of hot rolling, as this operation is very economical. By hot rolling it is intended
- to give the products a definite shape within the dimensional tolerances as specified in the technical standards;
- to compress the as-cast structure in order to improve its internal and external state;
- to modify the microstructure as well as the mechanical and technological properties by controlling the rolling and cooling conditions.

The process of working takes place in the gap between the two rotating rolls. The frictional forces acting between the surface of the product and the rolls force the material continuously through the gap. The stress conditions in the roll gap will result in the material being elongated and spread laterally or perpendicular to the direction of rolling, i.e. the material becomes wider. The extent of this widening depends on several factors, of which the reduction of the thickness is most important. If material is rolled without restriction to its width, for instance in flat rolling, the product undergoes changes in length (increase), thickness (reduction) and width (increase).

Even in the rolling process with grooved rolls, spreading has to be taken into account in the roll pass design. Overfilled roll passes may be the cause of severe rolling defects.

Normally, final dimensions are achieved in several rolling passes. For single-stand mills, the direction of rolling has to be changed after each pass: this process is known as *reversing rolling*. Mill stands with only two work rolls are called two-high stands. Due to the high rolling forces, it is sometimes necessary to support the work rolls against bending by installing back-up rolls. So-called four-high stands with two back-up and two work rolls each are frequently used. Two-high ingot-slabbing mills and four-high plate rolling mills are typically representative of such heavy reversing mills.

Continuous tandem mills comprise several stands arranged in series with the steel being rolled in a single pass through each stand. The product being rolled is in

---

References for D 5 see page 808.

contact with the rolls of several stands at the same time, and it is necessary that in each stand the same product volume per unit of time is passed through. Examples of such designs are the finishing stands of hot wide-strip tandem mills or continuous wire rod mills.

The processes of hot rolling follow the steel making and casting steps described in the previous chapters. The processes differ depending on whether the initial material is ingot-cast or continuously-cast steel and whether flat products or sectional steels are to be produced.

For the production of *flat products*, either ingots or continuously-cast slabs are used. The unit weights usually amount to about 40 tons. Ingots have first to be roughed down into slabs, whereas continuously-cast slabs can be fed immediately into the finishing mills (strip and plate mills).

For *sectional steel products*, including wire rod, a variety of processing methods is available. On the one hand, a distinction has to be drawn between ingot-cast and continuously-cast steel as initial material and, on the other hand, between finished products rolled with and without reheating. Because of the often very small final cross-sections, both ingot-cast and continuously-cast steel of smaller initial cross sections must be used, resulting in unit weights being generally lower than 10 tons. Normally, ingot-cast billets for small final cross-sections (wire rod, steel bars < 50 mm diameter) are rolled with reheating. If continuously-cast steel is used, the first step of the hot-working process can be eliminated. Large final cross-sections (rails, beams, sheet piling, steel bars > 50 mm diam.) may also be rolled with or without reheating if the initial material is ingot-cast steel.

The manufacture of finished products having critical requirements for surface quality and homogeneity from ingot-cast steel may also necessitate rolling with reheating.

Table D 5.1, shows schematically the different operations used in the hot rolling of *flat products*, *semi-finished products* and *sectional steel products*.

In the following chapters, the influence of heating, hot-working and finishing to obtain the required properties of the products are described in more detail.

D 5.2 Heating

D 5.2.1 Heating Conditions

The hot material supplied by the melting shop is normally loaded into *soaking pits* to homogenize out the temperature differences arising during cooling in the mold and during transportation of the ingot, and to achieve the required temperature for rolling. During this step, the following conditions have to be taken into account:
1. To avoid material separation in the ingot head which will solidify last, the temperature of this part of the material must be below a certain level during rolling.
2. The required reductions in cross section must be achieved without exceeding the

Table D 5.1 Processing sequences in hot rolling of steel

	Flat products		Semi-finished and sectional steel products, steel bars and wire rod					
			Continuously-cast steel			Ingot-cast steel		
	Continuously-cast steel	Continuously-cast steel	Without reheating	With reheating	Direct rolling of semi-finished products	Without reheating	With reheating	
Initial material								
Heating	scalling Pit furnace			Pusher furnace, walking beam furnace		Soaking Pit furnace	Soaking Pit furnace	
Roughing	Slabbing mill (Automatic slab scarfing)		Pusher furnace, walking beam furnace] Roughing mill for semi-finished products, heavy section mill	Roughing mill for semi-finished products		Blooming mill (automatic hot scarfing)	Roughing mill for semi-finished products, blooming mill (automatic hot scarfing)	
Conditioning	Cold scarfing, grinding, flamecutting ultra-sonic testing		Levelling, crack testing, ultra-sonic testing, surface conditioning, annealing, release test				Levelling, crack testing ultra-sonic testing, surface conditioning, annealing, release test	
Heating	Pusher or walking beam furnace		Pusher or walking beam furnace	Pusher or walking beam furnace			Pusher or walking beam furnace	
Finish rolling	Strip mill, plate mill		Section, bar, rod mill	Section, bar, rod mill		Rolling mill for semi-finished products and heavy sections	Section, bar rod mill	
Conditioning	Descaling, levelling, ultra-sonic testing, surface finishing, annealing (plate), release test		Levelling, crack test, ultra-sonic testing, surface finishing, assorting, annealing, peeling					
Processing	Cold rolling, annealing, forming, welding etc.		Forming, forging, welding, drawing					

strain limits of the rolling mill. Therefore, the coldest areas in the ingot – normally in the core of the ingot bottom – must not fall below certain levels of temperature.
3. The stresses generated during cooling and reheating of the ingots must be kept within certain limits to avoid cracking.

Pusher or walking beam furnaces are normally used for heating of continuously-cast steel and ingot-cast special steel, as well as semi-finished products for final rolling. The main criterion in choosing and designing the process to be used is to reduce the heat energy required to a minimum. In this regard, a special problem is the elimination of stress cracks associated with differential stresses between surface (compressive) and core (tensile) generated by steep temperature gradients during heating. These relationships have been described in model equations (Fig. D 5.1).

Surface characteristics and the microstructures of the rolled products are substantially influenced by heating: Scale formation, surface decarburization,

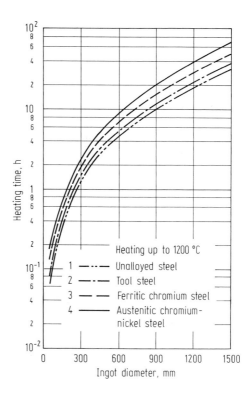

Fig. D 5.1. Shortest possible heating time for ingots of different diameters and steel types with heating from all sides. After [1].

Chemical composition of the tested steel types

Steel	% C	% Si	% Mn	% Cr	% Ni	% W
1	0.08	0.08	0.31	–	–	–
2	1.22	0.16	0.35	0.11	0.13	–
3	0.13	0.17	0.25	12.95	0.14	–
4	0.08	0.68	0.37	19.11	8.14	0.60

D 5.2.2 Scaling

The extent of scaling in the heating furnaces of hot rolling mills is determined by the migration speed of the iron/wüstite phase boundary which follows a $\sqrt{\text{time}}$ law [2] (see also B 3).

Figure D 5.2 shows that the scaling rate of steel D 75-2 of about 0.75% C is definitely lower than that of the steel grades D 9 and 9 S 20 with about 0.09% C. With increasing reheat temperature, higher humidity of the combustion gas and with an increasing amount of excess air, the scaling rate will increase [3].

During the rolling process it is important that the scale layer which has been produced can be easily removed by working (in the first upset pass) and/or by pressurized-water descaling. Strongly adhering scale – "sticking scale" – will be rolled in and will promote surface defects. The presence of some alloying elements, such as sulphur [3] and nickel [4], promotes adhesion between scale and base material and makes descaling more difficult.

D 5.2.3 Surface Decarburization

Decarburization and descaling of the surface are closely related, as both reactions will proceed with the participation of carbon dioxide. The rate-controlling process

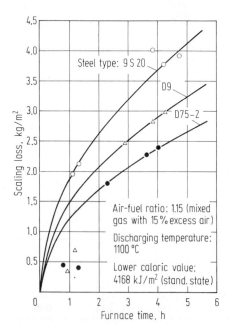

Fig. D 5.2. Scaling loss as a function of furnace time. After [3]. (○: 9 S 20, △: D 9, ●: D 75-2).

is the diffusion of oxygen to the metallic phase through the resulting iron oxide layer in which oxygen is consumed for the oxidation of both iron and carbon [2]. The profile of the carbon concentration in the layer with surface decarburization is determined by the diffusion rate of the carbon in the γ-iron. As with descaling, the decarburization reaction also follows a $\sqrt{\text{time}}$ law and is determined by the time of annealing above 900 °C.

Alloying elements may influence surface decarburization in different ways. Their influence on the activity of carbon, on the diffusion constant of carbon in iron and on the scaling rate are of major importance. Generally, decarburization is reduced by the presence of nickel and chromium, whereas a definite increase is caused by silicon [5], raising problems e.g. in the production of springs from such steel alloys.

If the furnace atmosphere contains humidity, additional carbon dioxide is generated which considerably accelerates decarburization.

Conditioning of the billet surface before running the billets through the reheating furnace also influences the extent of surface decarburization. By grinding the billets, for instance, material in which decarburization would otherwise occur during the first rolling pass can be removed before processing. As shown in Fig. D 5.3, a slightly lower decarburization will be found as compared to flame-scarfed or sand-blasted billets. For reasons of cost, grinding is justified only in the production of high-quality steels.

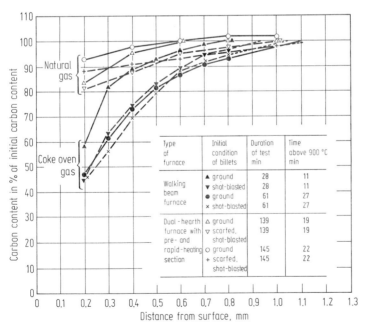

Fig. D 5.3. Surface decarburization of steel 41 Cr 4 as a function of the distance from the surface relative to duration of test, time above 900 °C, type of fuel and pre-treatment of the billet surface. After [3].

Surface decarburization will produce non-uniform and inadequate hardness levels after heat treatment of finished products. With surface decarburization, the fatigue strength of structural parts which have to endure dynamic loads may be reduced considerably. Therefore, depending on the application and on the type of load, the *maximum allowable depth of decarburization* must be defined. According to DIN 17 221 [5a], for example, total decarburization is not allowed for springs, while partial decarburization is limited to about 1% of the cross section.

D 5.2.4 Hot Shortness

In scaling, oxidation of iron and of the less noble elements silicon, manganese, aluminum and the like occurs. The nobler elements *copper, nickel and tin*, which are either added to the steel as alloying elements or are present as impurities, for example from scrap, are not subject to oxidation, and they, therefore, concentrate in the areas near the surface. Because of low diffusion rates of these elements in iron these concentrations are reduced by diffusion only slowly so that their solubility limits may be exceeded at the surface. When the limits are exceeded, liquid phases will be produced at grain boundaries resulting in the intergranular decohesion [6]. This "hot shortness" results in a substantial deterioration of the surface quality.

Copper, nickel and tin have an opposite influence with regard to their solubility in γ-iron and to the melting points of the resulting liquid phases. Whereas the solubility of copper in γ-iron is increased by nickel, it is substantially reduced by tin.

The maximum permitted contents of copper and tin in the steel to prevent hot shortness can be determined approximately according to the following formulae: % Cu + 8·% Sn $\leq$ 0.4% or % Cu + 6·% Sn = $9/E$, with E representing an enrichment factor [7]. Alloying with nickel will permit higher copper contents.

D 5.2.5 Influences on the Microstructure

The *austenitic grain size* before rolling increases with increasing reheat-soaking temperature and time. There will be a particularly strong grain growth if precipitates which existed before reheating are dissolved [8]. Extremely high reheating temperatures may result in the breaking up of the grain structure, especially if liquid films such as sulphides occur on the grain boundaries. Steel in this condition is said to be *"burned"*. A material which has not been properly heated will tend to be very brittle during hot rolling, showing another type of hot shortness.

If hot rolled products are to be supplied without subsequent specific heat treatment (i.e. in untreated (U), normalized (N) or thermomechanically treated (TM) condition), special attention should be paid to the influence of reheating on the microstructure and the mechanical-technological properties. For instance, for products to be treated thermomechanically, the maximum heating temperature must be limited because a small austenitic grain size is an important requirement [9]. On the other hand, the desired precipitation hardening can occur only if sufficient amounts of elements to be precipitated are first dissolved, and this dissolution requires the achievement of minimum temperatures. For thermo-mechanically treated niobium-alloyed pipe steels after DIN 17172 [9a], the specified yield point values may not be achieved if the temperatures in the reheating furnaces of the flat rolling mills fall below the temperatures at which dissolution occurs. Aluminum-killed special deep-drawing steel will have excessively high yield point values as a consequence of the fine grain structure, if nitrides are not completely dissolved.

D 5.2.6 Hot Feeding and Direct Rolling

Since 1981, the improved technology used for continuous casting and the need for saving energy have led to the development of hot feeding or direct rolling [10]. By means of these technologies it is possible to feed the hot continuously-cast semi-finished products without any conditioning into the reheating furnaces of the rolling mills – hot feeding – or to roll directly with the residual casting heat using selective supplementary heating (such as inductive edge heating) – direct rolling. To use these methods successfully, there must be a defect-free continuous casting process and a close program coordination between the melting shop and the rolling mill. These processes lead to substantial energy savings. Moreover, the associated almost complete elimination of scaling losses increases the metal yield by up to 1%, compared to conventional reheating [10].

5.3 The Rolling Processes

The rolling process chosen depends on the requirements, which may be: flat rolling, rolling with breaking-down passes, rolling with shaping grooves, double- or triple-roll passes, rolling with or without longitudinal tension and the like [11]. But within the context which of interest here between production conditions and steel properties, the different methods of rolling can all be discussed under the following headings.

D 5.3.1 Deformation Resistance

Hot forming is generally carried out above the recrystallization temperature (see B.6), so that during and immediately after the forming step, *work-hardening and recovery* will take place *simultaneously* [12]: Below a limiting value of the deformation φ_{ver}, the material is subject to work-hardening. In this area, the amount of work-hardening will decrease with an increasing deformation φ and will reach zero

686　D 5 Hot Rolling

at $\varphi = \varphi_{ver}$. Above φ_{ver}, the dynamic rate of recovery will increase. As a consequence of these processes, a generally low, but still quite different dependence of the flow stress k_f on the amount of deformation φ is observed in the hot forming of various materials. High-carbon steels, micro-alloyed steels and steels with phase transformation will be subject to more severe work-hardening than single-phase or low-carbon steels.

Besides the final shape of the product *the hot rolling mill design* will depend on the *deformation resistance* k_w [11]. This deformation resistance k_w may be calculated from the yield stress k_f taking geometric and friction conditions between product and roll into account. Figure D 5.4 shows some results from various rolling mill designs and the deformation resistance as a function of the roll gap ratio l_d/h_m.

Various relationships are important in the design of rolling mills and they also affect the operations to be carried out. If the rolling mill is not strong enough with reference to the steel products to be rolled, considerable mechanical and electrical trouble may be experienced. A requirement for holding close dimensional tolerances as specified by the technical standards is that such relationships are allowed for by the process computers in the automatic setting of the rolling mills.

D 5.3.2 Formability

Formability means the degree of deformation to which the material may be subjected before its capacity for deformation is exhausted. This degree of deformation depends on the type of product to be rolled and on the forming conditions, such as temperature, stress conditions and rate of deformation [11] (see also B 6).

Whereas the flow stress as a function of the forming conditions can be described mathematically [12, 13], it is necessary to rely on graphic representations for

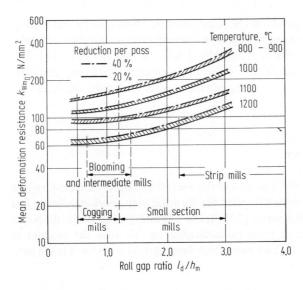

Fig. D 5.4. Deformation resistance as a function of the roll gap ratio at different temperatures for a product containing 0.1% C, 0.47% Mn, 0.063% P, 0.026% S. After [11]. (l_d = pressed length, h_m = mean opening of roll gap).

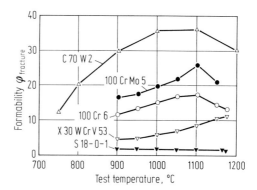

Fig. D 5.5. Formability (determined by the torsion test) as a function of temperature for alloyed steels. After [14]. (Chemical composition of the steels see also C 11).

formability. Different materials are evaluated largely by the hot torsion test [14]. Figure D 5.5, shows the formability of several alloyed steels as a *function of the temperature*. Generally, the formability is reduced with decreasing temperatures. High-alloy steels often show very poor values. Other parameters of important or unfavorable influence have been found as follows:
- the as-cast condition as compared to the pre-rolled condition [14];
- precipitates of aluminum nitride at around 1000 °C [15];
- low ratios of manganese to sulphur at around 900 °C [16];
- rolling in the dual-phase range of steels containing 1.5% Si [17] and of stainless steels [18];
- certain stress conditions, such as tension stresses with free lateral spread, as compared to pressure acting from all sides with closed passes [19, 20].

If, in hot rolling, the total deformation exceeds the formability of the material, defects of various kinds may occur. The most frequent defects are *edge breaks* and *edge cracks* in slab and strip rolling as well as *transverse cracks* in pass rolling of steels with low formability.

D 5.3.3 Control of Microstructure and Metal Properties

Depending on the later processing of the hot rolled products – hot pressing, hot treatment, cold rolling, cold drawing, cold forming etc. –, the microstructure as well as the mechanical and technological properties must meet special requirements. Subsequent heat treatment can normally be performed without restrictions. If later processing consists of cold rolling or drawing, specific microstructure conditions can be obtained by controlling the process parameters: For example, fine pearlite can be adjusted in unalloyed steels by controlling the cooling rate in strip and wire rod rolling mills, and in special killed deep-drawing steel ST 14 by avoiding the precipitation of aluminum nitride and using low coiling temperatures in strip rolling mills.

Often, the *mechanical and technological properties* of hot rolled products, *are directly* in structural components: Unalloyed structural steel in the untreated condition, reinforcing steel, normalized fine-grain structural steels and thermomechanically formed large pipe steels. For unalloyed structural steels, the specified mechanical properties can be adjusted by a "normal" control of the rolling process. The production of reinforcing steel by the so-called Tempcore process is based on the provision of well coordinated cooling conditions for adjusting the microstructure of ferrite/pearlite in the core and of tempered martensite on the surface [21]. In normalized forming, normalizing is replaced by maintaining certain forming rates in the final passes, using a specific final rolling temperature [22]. Finally, it is possible in thermomechanical forming to achieve improved properties as compared to other processes [22] by an appropriate chemical composition of the materials and by extremely accurate control of several rolling parameters (heating temperature, degree of deformation and temperatures for each rolling pass, cooling rate and perhaps also coiling temperature) (see also C 2 and C 25). All processes require highly developed rolling mill technology and process control, as any deviation from the closely limited process parameters will produce material that will not be suitable for further processing.

The *prevention of flakes* represents a special task in cooling hot-rolled products of large cross-sections – for instance in semi-finished products. If hydrogen contents during cooling exceed certain critical levels, separations of the microstructure, or so-called flakes, may result. If it is not possible to reduce the hydrogen content below preset limits by taking suitable measures during steel production, the danger of flake formation must be eliminated by a delayed cooling of the rolled product in special equipment. Delayed cooling is also necessary for materials with a tendency to develop stress cracks.

D 5.3.4 Improved Surface Characteristics

Surface characteristics can be influenced considerably by the rolling process. Although defects originating from the initial material are stretched and their

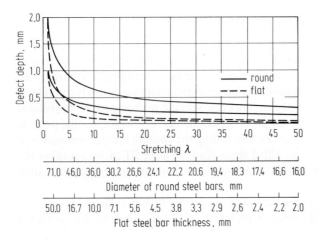

Fig. D 5.6. Examples of reduced defect depths in rolling flat and round steel bars. After [23].

absolute depth is reduced [23] (Fig. D 5.6), new defects, such as edge cracks or laminations, may be produced by the rolling process. To avoid such defects it often is desirable to include in the rolling process an operation by which surface defects will be eliminated – *automatic hot scarfing*. This process involves some loss of material but will safely remove surface defects up to a depth of 2–4 mm which might appear before or during rolling.

For steels which cannot be passed through automatic hot scarfing because of their chemical composition, it may be necessary to prepare the surface of the ingot or of a pre-rolled product by *grinding*.

In hot rolling, special care must be paid to any *scale* occurring before, during or after the rolling process. This need for care applies in particular to the production of sheet and tinplate where high quality surface finish is required. An unfavorable formation of the scale will lead to severe surface defects. The formation of scale may be controlled by the selection of suitable rolling temperatures, by roll cooling, high pressure water descaling and roll surface control [24].

Finally, it should be pointed out that the *surface of the rolls* is subject to mechanical and thermal wear as well as to mechanical damage. The resulting surface defects (marks, rough areas) are minimized by using proper roll materials and by appropriate roll change intervals.

D 5.3.5 Yield

The yield is determined by the percentage of the production which meets the required specification for material properties, surface characteristics and dimensional tolerances in relation to the input. A high yield is of vital importance for *profitability*. Losses are caused by scaling, down-grading due to unsatisfactory material and surface characteristics and, above all, by not meeting specifications for geometry, i.e. due to cropped ends, off-size lengths and off-tolerance material.

In rolling heavy ingots into slabs without reheating especially, the *formation of head and tail ends* is of substantial importance for an economical result. Due to the sequence of horizontal and vertical passes, so-called fish-tails and laminations occur at the ends of the slab and must be eliminated by crop-shearing (Fig. D 5.7).

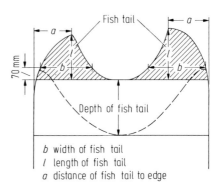

Fig. D 5.7. Break-down of process scrap by: "fish-tails" due to vertical passes and "laminations" due to horizontal passes. After [25].

b width of fish tail
l length of fish tail
a distance of fish tail to edge

Numerous attempts have been made to reduce such material losses; a detailed description of the possibilities offered by the rolling technology is given in [25]. In section rolling bars may show ends that cannot be used any further, but whose weight percentages are generally far below those occurring in slab rolling.

Specific requirements as to *shape and flatness* have to be met by *hot rolled wide strip*. A Report in [26] details the amount of measurement and process technology needed for meeting such requirements.

D 5.4 Finishing

Some work is necessary on semi-finished products as well as on finished products to find and to eliminate material and shape defects, to provide the desired condition of the material for shipment and to release the material, if it meets the specifications.

Since it is not possible to produce zero-defect products economically, as far as *surface, internal and shape defects* are concerned, producer and customer must agree for each individual order on the size and the number of defects that will be tolerated. As an example, delivery conditions may be specified for the surface finishes of heavy and medium plate and of wide flats [27], of hot-rolled steel bars, of wire rod [28] and of semi-finished products for die forging [29].

Inspection for defects is made automatically by means of crack detectors, ultrasonic test instruments, thickness gauges, width- and shape-meters operating on-line or separately in inspection lines which require considerable personnel [30–32]. Surface defects may be eliminated (within the permitted dimensional tolerances), by automatic scarfing equipment, grinding and peeling machines, and manual scarfing equipment. Product sections having internal defects must be cut off and scrapped. Some shape defects can be eliminated by straightening or skin-pass rolling. In case of material which is outside the specified dimensions, the order can be re-scheduled to an alternative order, with the penalty of some loss of material.

The *desired delivery condition* to a large extent is determined by the customer. The following examples of such conditioning should be mentioned: Surface finish – descaled by pickling or sand-blasting, peeled or drawn, with corrosion protection; form – side-trimmed, slit, sheared, in coils or sheets; plus marking for identification, packaging and means of shipment.

D 6 Hot Forming by Forging

By Hans Günter Ricken

D 6.1 Aims of Forging

Forging of steel has to perform the following work:
- give the forged part the desired form
- densify the material by closing the cavities present in the ingot
- create a favorable structure as a preliminary to the subsequent heat treatment or for further processing in general.

D 6.2 Forging Methods

A distinction is made – disregarding details – between *open-die forging* using tools open on all sides with free flow of material and *closed-die forging* (*swaging*) with closed tools and material flow limited on all sides. In open-die forging, tools are also used for certain forgings that are more or less closed, or processes are applied that are geared to specific forging forms.

The choice between *open or closed-die forging* depends on technical and economic considerations. Open-die forgings are produced singly or in series with gross weights ranging from some kilograms up to several 100 tons. The limits for closed-die forgings lie between several grams and several tons with certain minimum numbers being required for economic efficiency. Open-die forgings are used in almost all sectors of engineering, where parts are needed with large cross-sections, heavy weights and high requirements for properties. Closed-die forgings are widely used in the automobile and mechanical engineering fields, for instance.

D 6.3 Material for Forging

Ingots weighing from about 1 t up to 500 t [1] and rolled semi-finished products are used for open-die forgings. Closed-die forgings use rolled semi-finished steel and rolled bars; for components of greater thicknesses, forged semi-finished preliminary sections are used; and for special applications, strand-cast semi-finished material may be preferred.

References for D 6 see page 809.

Conditions prevailing in *steel melting* and *casting*, and hence in *solidification*, are of prime importance in the production of *forging ingots* [1–3] (cf. C 5 and D 4). Furthermore, the *geometric shape* of the ingot is of great importance. Forging ingots normally have octagonal or polygonal cross-sections; in the range of smaller weights, square ingots are also used. For special work round ingots are in use. Octagonal and polygonal cross-sections are preferred because they tend less to crack formation in the first forging passes than other ingot shapes, because columnar crystallization is interrupted at the ingot edges.

Forging ingots are generally of the big-end-up type, but some may be of the big-end-down type. Ingot taper ranges between 6 and 15 per cent, the range of 8 to 10 per cent being preferred. The ratio of height to diameter ranges normally between 1.0 and 1.8 in big ingots. The greater the ingot diameter the more a value of 1.0 is aimed at for this ratio. The hot top volume varies greatly and depending on the kind and geometry of the hot tops (consisting of refractories, insulating or exothermic material) and the technique of ingot hot top treatment – the ratio ranges between 10 and 25 per cent of the total ingot weight.

Generally, a certain part of the top and foot of each ingots is cropped. There are no generally-valid figures for the extent of the *yield*, which is determined by qualitative and economic considerations; ingot size, steel grade, kind and form of the planned forgings, and demands for freedom from defects appearing in volume and surface testing are decisive. With a given ingot form (shape of cross-section, height-diameter-ratio, taper, hot-top volume), the yield is influenced by the conditions in the ingot foot ("sand") and in the ingot top (segregations, chemical inhomogeneity). Yield is generally higher in small ingots than in big ones; for simple products made from small ingots the yield may be between 75 and 90 per cent, for high-quality parts made from heavy ingots yield is only about 50 to 75 per cent, and the foot cropping may be about 5 to 15 per cent, with top cropping of about 15 to 30 per cent. (Included in the yield estimates is a figure of 5 per cent for unavoidable material losses by scale formation on the ingots during the forging process (scaling loss); (variations in the scaling loss figures depend on the number of reheatings and the holding time at high temperatures.) Too little cropping of the top and foot scrap leads to qualitative risks, while too much scrapping is uneconomical. The yield can be improved using the methods for ingot top treatment mentioned in C 5.

D 6.4 Working Conditions in Forging

D 6.4.1 Heating

In heating ingots to forging temperature the dimensions of the cross sections must be taken into account. Low *heating rates*, especially in the low-temperature range and long soaking times are required; times necessary for heating the ingot core thoroughly are determined either empirically or by calculations.

The *temperature range for forging*, to be carried out in the γ-range, is limited for several reasons. The top limit of this range is the melting temperature of low-melting phases; in the lower temperature range the increasing yield stress (deformation resistance) must be overcome. Qualitative aspects may require the avoidance of high forging temperatures, a measure that is also backed by economic reasons; with a smaller grain size after forging, the heat treatment effort required to achieve good mechanical properties is lower. This relationship is especially important for transformation-free steels because forging parameters and forging temperatures or final forging temperatures determine the austenitic grain size remaining in the forged piece.

D 6.4.2 Forging Conditions

Effects on the General Quality Properties

The most important task of forging besides shaping the part is the closing of cavities and discontinuities present in the ingot. It is generally safe to say that these inhomogeneities are located in the core of the ingots and are mostly limited to about 60 per cent of the ingot height and up to 13 per cent of the ingot diameter [4]. On the question of closing cavities by forging and generally of the behavior of ingot defects in forging much experience has been gained in many decades, and the results of numerous plant and laboratory investigations are also available [4–6].

Some aspects on *closing cavities* will now be presented, without going into the details listed in the literature cited. The ratio of reduction by forging is taken as a measure for the deformation carried out on a forged piece by stretching, upsetting or a combination of the two. As to the definition of the ratio of reduction in combinations of methods, reference is made to the pertinent literature [6].

For closing cavities a certain minimum ratio of reduction is required, depending on the material, the data of ingot and forging and the forging conditions, so that no generally valid figure exists for this ratio. Therefore, a twofold ratio of reduction may be sufficient for the closing of cavities whereas for other examples a fourfold ratio may be needed. For remelted ingots and other ingots produced by a special process, twofold or smaller ratios of reduction are usually sufficient. For the dependence of mechanical properties on the ratio of reduction/stretching see below.

The major part of all forgings is stretched only; a small part, for reasons of forming or for quality reasons also, is upset or forged using a combination of both methods. The *application of the various die forms* (flat die, V-die, swage die) and die widths depends on the form and dimensions of the forging, on the material and its formability properties and on the size of the forging machine available.

Generally speaking, wide *flat dies* are better for a good through-forging of the core zone and for closing cavities than narrow dies. With too large a die width, however, cracks due to material displacements in combination with the stresses created in the transverse direction are all the more likely to develop, with greater ratios of bite width s_b to initial height h_o of the forging [7]. To obtain a good through-forging and closing of cavities as well as to prevent cracks, a ratio of S_b/h_o of 0.35 to 0.50 is favorable; a ratio of > 0.50 improves through-forging only a little, though it will encourage the appearance of interior cracks, whereas too small a ratio results in inferior through-forging. Therefore, for forging down large round dimensions to small diameters and to obtain and maintain the best forging ratios during forging, one or several tool changes are unavoidable. Because the deformations in the core zone decrease very considerably from the bite middle to its border, the forging procedure must be so controlled that in a secondary forging step – also after turning the forged piece – the middle of the bite is set at about the same spot where the die-edge was located in the preceding forging step. Table D 6.1 contains a number of recommendations on the selection of die width s_b and bite width s_0 together with the reason for their application.

Another means of obtaining a *good through-forging* and avoiding discontinuities and cracks in the core, is by forging round and polygonal cross-sections into an intermediate square dimension. This method is recommended for steels of low formability [7]. In big ingots with (frequently) fairly well-pronounced center porosity, a good through-forging can be obtained by cooling down the outer zones before forging [7, 8].

Forging by means of V-dies has often been investigated. In many reports the deficient depth effect of V-dies and the combination of flat and V-dies is pointed out. It has been shown that by applying V-dies with a small V-angle, e.g. 90 deg., through-forging in the core is imperfect. 135-deg. V-dies are prefered because they enclose octagonal ingots by combining a V-die with a flat die on three sides, or by applying two V-dies on four sides.

The working methods outlined for open-die forging cannot apply pressure simultaneously in all directions to the total volume of the forged piece. *Forging in a closed die*, however, does apply all-round pressure on the total forging during one single pressing action or hammer-blow. In closed-die forging, however, the closing of internal cavities is attempted only in exceptional instances, since solid semifinished material is used to start with. The aim here is rather to fill the die completely by causing the steel to flow under the applied pressure.

Effects on the Mechanical Properties

The mechanical properties of *open-die forgings* depend not only on the physical condition and the chemical composition of the ingot, but are influenced also by the deformation and its conditions.

Open-die forgings have a distinct *directionality of their properties* (anisotropy) despite the low ratios of reduction. This directionality is allowed for when taking test samples by choosing sampling locations with different directions related to the main direction of forging. Only by the use of extremely clean steels, notably steels with especially low sulphur contents, can the directionality be influenced favorably in terms of the material used. In Fig. D 6.1 the dependence of the mechanical properties on the ratio of stretching $R = F_0/F_1$ is plotted schematically for longitudinal and transverse test specimens and for intermediate sample positions. The graph shows that a 2- to 3.5-fold ratio of deformation produces favorable properties in multi-axially stressed forgings. Research and practical experience show that this ratio of deformation is also sufficient to close cavities within the forged piece, when the correct forging technology is used [9, 10].

Table D6.1 Die or bite width compared with the dimensions of the forging. Cf. [7].

Literature[a]	Material	Die- or Bite-width Forging dimension[b]	Reason
P. M. Cook	—	$s_b/h_0 = 0.25$ at $\varepsilon_h = 5\%$ $s_b/h_0 = 0.33$ at $\varepsilon_h = 15\%$	Uniform distribution of local deformations
I. Ja. Tarnowskij et al.	if φ_{Br} small, s/h greater and vice versa	$s/h = 0.5 + 0.7$ (Flat dies-square)	$s/h < 0.4$ more uniform distribution of deformations, but danger of crack formation by longitudinal tension stresses; $s/h > 1.0$ longitudinal cracks by shear stresses;
M. Kroneis and T. Skamletz	Steel	$s_b/h_0 \approx 0.4$ (Flat dies-square)	small $s_b/h_0 \rightarrow$ tension stresses and insufficient through-forging
A. Chamouard	φ_{Br} small (like Duralumin) Bronze (with nickel or iron)	s_b/h_1 or $s_b/d_1 \approx 0.33$ (135°-Vee-dies) $s_b/h_0 > 0.75$; $s_b/_1 < 1.33$ Flat dies $s_b/h_0 > 0.66$; $s_b/h_1 < 1.5$	ε_h small $\rightarrow s_b/_1$ bzw. $s_b/d_1 \approx 0.5$ s_b/h_0 too small $\rightarrow$ center cracks; s_b/h_1 too great $\rightarrow$ surface cracks
P. F. Ivanuškin	Steel	$s_b/h_0 > 0.6$; $s_b/h_1 < 2$ Square $s_b/d_0 = 0.5 + 0.8$ (Swage-dies)	s_b/h_n greater $\rightarrow$ better through forging $s_b/d_0 < 0.5 \rightarrow$ cracks $s_b/d_0 > 1$ too great shear deformation
B. E. Vactanov and Ja. M. Ochrimenko	—	s/d bzw. $s/h \approx 0.8 + 1.0$ (Hammer) s/d bzw. $s/h \approx 0.3 + 0.5$ (Press)	only small ε_h possible } Welding of internal defects ε_h is 20% possible
H. Heßler G. Richter and H. G. Lotze	Chromium and Chromium-Nickel-Steels	$s/d_1 \approx 0.5$	—
L. Jilek and B. Sommer	Steel	$s/d_0 \approx 0.5$	heavy pieces
		$s_b/h \approx 0.5 + 0.7$ (Square)	almost no transverse tension, stresses, longitudinal defects are closed; heavy pieces
A. Witte		$s/d_0 > 0.5$	> 100 t; $d_0 \approx 2300$ mm diameter 30-MN-Press } Good through-forging
H. Rothäuser J. G. Wistreich and A. Shutt		$s/d_0 \approx 0.25 + 0.33$	
M. Kroneis et al.,		$s_b/h > 0.33$ $s/h_n \approx 0.33$	— Discs $l_n/d_n < 1 \rightarrow$ Stretching

[a] Sources for the literature cf. [7]
[b] s = Die width, s_b = bite width, h_0 = initial height, h_1 = final height, d_1 = final diameter of forged piece.
d_0 = initial diameter of forging,

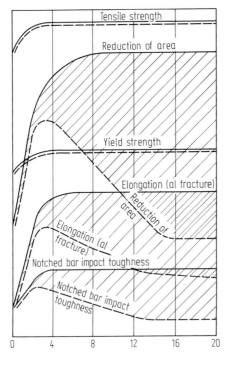

Fig. D 6.1. Mechanical properties related to the ratio of stretching $R = F_0/F_1$ (schematically). Cf. [6]. (F_0 = initial cross section, F_1 = final cross section). Tensile strength, reduction of area, yield strength, elongation (at fracture), notched bar impact toughness. Parallel (longitudinal) to fiber direction, normal (transverse) to fiber direction, directions between longitudinal and transverse.

The dependence of the mechanical properties on the application of *combined upsetting and stretching operations* has also been investigated many times [7, 11–15]. Several papers have shown that there is no decisive advantage in the application of upsetting. In addition, it has been found [16, 17] that upsetting improves mechanical properties only when the forging is stretched before and after upsetting. As regards the closing of cavities, however, the danger exists that with too small a degree of upsetting, cavities are not only not closed but even enlarged [9]. Irrespective of these considerations, the aim of upsetting is to attain definite forms and dimensions of forgings, e.g. of discs.

Closed-die forgings possess a special kind of *directionality* in their mechanical properties, especially in ductility properties. The longitudinal deformation of the semi-finished material in rolling and the material flow into the cut of the die produces a fibrous structure which is adapted to the form of the forging and to the working stresses. The deformation in forging is generally applied in the transverse direction, normal to the existing longitudinal deformation, which is limited to only a small volume in the fin. Non-metallic inclusions, notably manganese sulphides, are included in this volume, sometimes impairing the mechanical properties vertical to the transverse direction, depending on the amount of the inclusions.

The favorable *fibrous structure* resulting from the flow of the material within the die is also found in open-die forgings such as crankshafts, when they are formed with individual strokes using suitable tools ("in the die"). This procedure results in a marked improvement in strength [18].

D 6.4.3 Cooling from Forging Heat

Cooling after forging is important in two respects:
1. After cooling, the finished piece must be free from cracks and defects.
2. Cooling must result in a complete γ-α transformation if heat treatment is to be successful.

Cooling, therefore, is usually carried out by controlling furnace temperatures in the pearlitic range or by controlled air- and furnace-cooling in the bainitic range. The process selected depends on the chemical composition, on the degassing process used and on the cross-section of the piece. The process is controlled in such a way that excessive temperature- and transformation stresses are avoided.

Non-degassed steel must be cooled down slowly after forging – almost independently of the steel grade – in order to let the hydrogen diffuse outwardly, so as to minimize structural stresses, and thereby avoid flake formation. Because of the slow cooling, temperature differences in the forging are reduced and the danger of tension cracks is minimized.

In *vacuum-degassed* steel the reduced hydrogen content generally does not pose any danger of flake formation. However, care must be taken that tension cracks do not form. Several types of steels will cool down in air without problems when they are vacuum degassed. During this cooling, unalloyed steels are transformed in the pearlitic range; higher-alloyed steels for hardening and tempering, especially those used for heavy forgings transform predominantly in the bainite range, depending on the diameter.

It is *important* to obtain a *complete transformation even in the segregated portions* of the forging in the first cooling process after forging. After forging, *hypereutectoid steels* must be cooled rapidly in air to below the temperature range of the pearlite transformation. Thus carbide precipitations at the grain boundaries, which embrittle the steel and are not removable in the course of subsequent heat treatment, are avoided. Other detrimental precipitates are formed in some steels by slow cooling and cannot be dissolved by subsequent heat treatment. These steels must be treated similarly to the hypereutectoid steels.

D 6.5 Defects

Basically, a test report on the internal condition of a forging is desired at the earliest possible time during the production process. This report shows whether cavities are present or not, and the size and location of non-metallic inclusions and cracks. Many forgings can be tested by ultrasound to obtain initial information

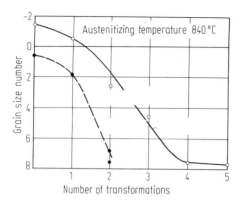

Fig. D 6.2. Grain size depending on number of transformations of steel 26 NiCrMoV 14 5 containing 0.22% to 0.32% C, ≤ 0.30% Si, 0.15% to 0.40% Mn, ≤ 0.015% P, ≤ 0.018% S, 1.2% to 1.8% Cr, 0.25% to 0.45% Mo, 3.4% to 4.0% Ni, 0.05% to 0.15% V. Cf. [19]. Grain size number, austenitizing temperature 840 °C, transformation in pearlitic, bainitic range, coarse grain annealed. Number of transformations.

after cooling from forging temperature, with defects sufficiently identifiable at this point. Smaller numbers of forgings – especially heavy forgings with large cross-sections – require one or several transformation treatments for grain refinement. This grain refinement reduces ultrasound attenuation, and makes defects more easily identifiable, as shown by means of simulated heat-treated samples in Fig. D 6.2 [19]. Defects of a circular sheave-reflector type of 1 mm diameter and below can be detected, by means of ultrasonic frequency and attenuation of ultrasonic waves, provided the surface has been treated properly. For further information on testing by ultrasonics and the present state of testing techniques cf. [20].

It is basically not possible to avoid defects such as non-metallic inclusions in large forgings. Because of this difficulty, connections which may exist between macroscopic defects and service properties are intensively researched, especially in making heavy-duty forgings for electrical generator rotors, for instance. Properties such as fracture sensitivity, ductility, defects caused by cyclic stresses and factors influencing creep resistance are particularly important.

D 7 Cold Forming by Rolling

By Jürgen Lippert

D 7.1 Definition and Scope of Cold Rolling

Cold rolled products are defined by Euro-Standard 79–82 [1a], as flat products which have undergone a reduction in cross-section of at least 25 per cent by cold rolling without prior reheating. Cold rolling allows thinner sheet and strip to be produced with closer dimensional tolerances, a higher-class surface finish and, by applying a suitable annealing process, better deformation behavior than is possible by hot rolling.

D 7.2. Production Sequences in the Cold Rolling Mill

The *types of equipment and plants* commonly used for production of the various categories of cold-rolled flat products are shown in Fig. D 7.1a–d. The application of this equipment depends on the particular working requirements and specific quality characteristics of the final product. *Fundamentals of the cold rolling process* are detailed in [1].

Material to be cold rolled has first to be *descaled*. Descaling is effected by pickling in acids with or without additional mechanical scale breakers.

Ferritic steels with a high chromium content are commonly *soft annealed* before cold rolling.

While standard cold rolled grades, electrical steels and tin-plate base metal are rolled in reversing mills or in tandem mills, stainless steel grades requiring higher roll forces are preferably rolled in *multiple-roll mill stands* (e.g. Sendzimir mills) which also allow better control of strip profile.

The rolling process and the auxiliary materials used (e.g. rolling emulsions) must be controlled continuously in order to detect the causes of possible quality losses in time to prevent them. This procedure is particularly important to avoid strip marks caused by damaged rolls or to avoid stains originating from emulsion residuals on the strip surface. It also helps to measure changes in the concentration and cleanliness of the rolling emulsion or rolling oil which may cause deviations from strip thickness and flatness specifications resulting from a non-optimal roll pressure.

References for D 7 see page 810.

700 D 7 Cold Forming by Rolling

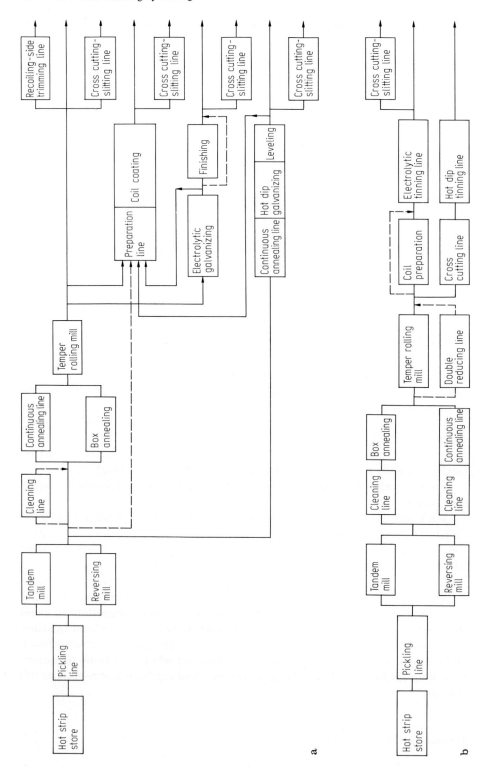

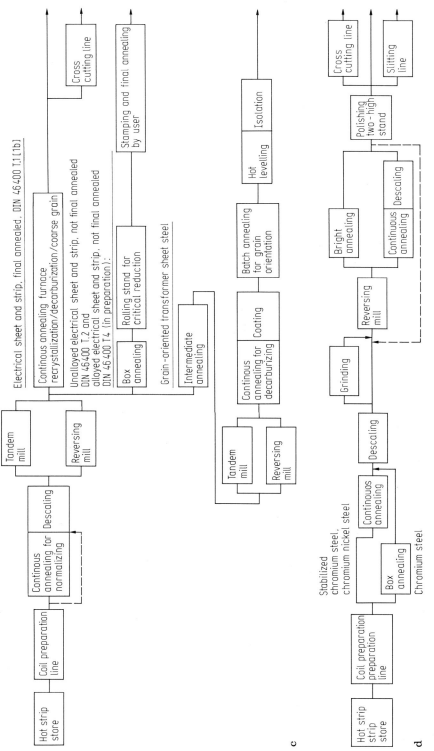

Fig. D 7.1. Sequences for the production of cold strip intended for: *a* Uncoated sheet, galvanized and color-coated sheet; *b* Tin plate; *c* Electrical sheet and strip; *d* stainless steel strip.

If strip is recrystallization annealed after cold rolling the following factors may contribute to the possible appearance of annealing stickers. These factors include the *annealing temperatures and times selected*, and the strip cross sectional profile in combination with the coiling tension and the surface finish (roughness). Optimal conditions must be established empirically for each individual cold rolling mill. The likelihood of annealing stickers decreases according to a common rule with increasing strip hardness and thickness.

The final dimensional accuracy (including flatness of the strip) is largely influenced by the correct choice of the *combination of the rolling emulsion or oil* with the "used" roll pressure and the coiling tension, by the rolling batch size and combination of steel grades to be rolled; and by the surface finish of the "hot" rolled strip.

D 7.3 Effect of Process Variables in Cold-Rolling on Material Properties

Users wish of an "ideal" cold strip with uniform *thickness* over the strip length and width in addition to a perfect *flatness*-maintained during further processing – will never be fully satisfied. As long as hot strip scheduled for cold rolling exhibits certain thickness variations across the strip width which are unavoidable because of the nature of the hot rolling technique, a compromise must be found between requirements for uniform thickness and for flatness (see also Fig. D 7.2). Applicable specifications for permissible dimensional and shape deviations are contained in specialized dimensional standards (e.g. in DIN 1541 for unalloyed steels [2a] and in DIN 59382 for stainless steels [2b]). For steel grades in the higher tensile strength range, larger thickness and flatness tolerances must be accepted.

At first sight cold strip may appear to have an almost undetectable *waviness vertical to the rolling direction*. This waviness may be caused by several factors. The most common causes are vibrations in the mill stands during rolling, and vibrations in the electric motors and working- or support-rolls ground to a non-round polygonal shape. This last cause has its origin in vibrations that affect grinding equipment used for finishing the rolls.

The inner structure of the steel and consequently the properties for use of the finished cold strip can be *influenced considerably by the cold rolling process*. Increasing reductions in rolling passes cause the microstructure to become finer after recrystallization annealing, as may be seen from the different recrystallization diagrams. This refinement results in an increase of tensile strength. In aluminum-killed unalloyed steel grades, changes in texture also increase the *vertical anisotropy*

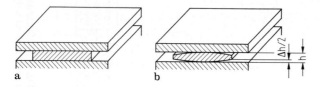

Fig. D 7.2. Definition of rolled strip flatness *a* ideal flatness *b* producible flatness (Δh = thickness deviation).

(r-value), thereby enhancing deep drawing properties but also increasing chances for earing.

D 7.4 Heat Treatment of Cold Strip

D 7.4.1 General Information

Cold strip is commonly *recrystallization annealed* to diminish work-hardening effects and to regenerate formability. Exceptions to this rule are stainless steel grades of strip which are subjected to *solution annealing*, strips for hot dip galvanizing by the Sendzimir process which requires a pretreatment of the strip surface by annealing above the Ac_3-temperature, and strips for the manufacture of electrical sheet. For electrical steels the annealing cycle should have a decarburizing effect or, depending on the chemical composition of the steel, should generate a certain texture.

The danger of a decarburization effect in recrystallization annealing [3] can be prevented almost completely by effective control of the protective gas atmosphere. Introduction of humidity by the material to be annealed also must be avoided as far as possible because humidity would cause a rise of the dewpoint in the annealing furnace, leading to undesirable surface coatings. To lower the dewpoint in box annealing furnaces it is common practice to change the protective gas partly or completely during the initial phase of the annealing cycle.

Cooling after annealing must also be controlled. Removing the inner covers too early may lead to oxidation coloring at the strip edges (annealing edges). The same effect may occur if the furnace base or the piping system leaks, allowing air to enter.

D 7.4.2 Continuous Annealing

Continuous annealing of aluminum-killed extra deep-drawing grades of steel under low-cost conditions requires special control of chemical composition and the hot rolling process. Development in this field is not yet complete.

Base metal for tinplate production is mainly annealed in continuous annealing lines, although from a quality point of view could be done just as well in box annealing furnaces.

Stainless steel grades must be recrystallization annealed in continuous furnaces after cold rolling because fast quenching is required after annealing to suppress the precipitation of carbides which otherwise would impair corrosion resistance. Annealing can then be effected by *bright annealing in a protective gas atmosphere* (hydrogen + nitrogen) or in an *open annealing process* followed by pickling. To avoid undesirable effects on the microstructure and surface finish of the strip, operating conditions (strip speed, temperature) must be varied to suit the particular strip gages.

Annealing of *electrical steels* is designed to produce special characteristics such as grain orientation as well as reducing work hardening effects due to rolling. Both continuous and box annealing furnaces are therefore employed (Fig. D 7.1).

Since open annealing requires adequate descaling, care must be taken to avoid over-pickling by choosing the appropriate speed for both continuous furnace and pickling line. Also in the continuous annealing process the bright annealing demands lowest values of the dew-point of the protective gas atmosphere in order to avoid temper colors.

D 7.4.3 Box Annealing

Cold strip in soft unalloyed steel grades and constructional steel grades still are predominantly recrystallization annealed in box annealing furnaces. The influence of temperature control on textures and on mechanical and technological properties has been described in numerous publications [4]. Aluminum-killed steel grades will possess optimal texture for deep drawing operations if the steel is heated slowly (about $\leq 40\,°C/h$). Grain size increases with higher temperatures and longer annealing times. The upper annealing temperature is limited for tight coils by the steel's sensitivity to sticker marks and finally by the A_1-point. The changes in structure which start above A_1 should be avoided with strip that is scheduled for severe deep-drawing operations.

D 7.5 Temper Rolling

Recrystallization annealing is commonly followed by temper rolling to improve flatness, to remove the yield point elongation so as to avoid stretcher strains in forming operations and to produce a specific surface structure (roughness). However temper rolling lowers the n-value to some degree (Fig. D 7.3).

Temper rolling of *unalloyed* and *microalloyed steel grades* and under certain conditions also of *ferritic chromium steels* first lowers the yield point as reduction increases to a minimum, then raises it again. The distinct yield-point elongation disappears simultaneously. These phenomena are due to the formation of new

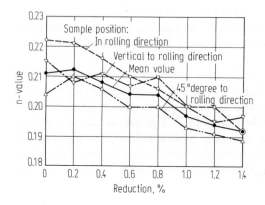

Fig. D 7.3. Influence of reduction in temper rolling on strain-hardening exponent in aluminum-killed cold-rolled steel. After [5].

dislocations which initially have low stress resistance; with increasing reduction, however, work hardening increases again. If *steel* (see B 8 and C 4) does not exhibit a distinct yield point because the carbon and nitrogen atoms are completely combined with niobium or titanium. Small amounts of straining cause a steady increase in yield strength values. *Austenitic chromium-nickel steel grades* exhibit a similar behavior (Fig. D 7.4).

Semi-finished *electrical steel grades* (not final annealed) are given a critical deformation by temper rolling, yielding a special microstructure in the final annealing process.

The *surface structure* (roughness) is of particular importance for *cold strip intended for deep drawing operations and for surface coating*. Existing roughness criteria (e.g. R, CLA) indicate only the overall depth of the surface structure and do not characterize the complete roughness structure. The surface fine structure of cold strip derives from the roughness imparted by cold rolling and by temper rolls. For this reason the roughness profile may vary widely [6]. Results are also influenced by reduction (Fig. D 7.5), by the strip gauge, roll diameter, rolling speed and strip tension in temper rolling. Another factor to be taken into account is that the roll roughness wears off during the rolling process so that the strip roughness is steadily reduced.

Strip flatness is affected in temper rolling as it is in cold rolling by the condition of the original strip, the roll finish and the roll pressure.

Analyzing the complexity of the interdependencies makes clear that changes in quality characteristics such as n- and r-values, yield strength, surface finish (roughness) and flatness cannot be changed individually without affecting the influence of the remaining characteristics.

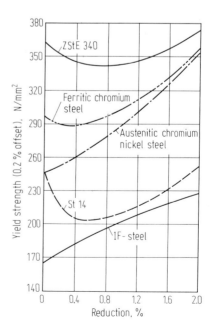

Fig. D 7.4. Influence of reduction in temper rolling on the yield strength (0.2% offset) (for steels cf. C 4).

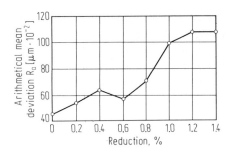

Fig. D 7.5. Influence of temper rolling on surface roughness of aluminum killed steel. After [5].

D 7.6 Finishing for Dispatch

Additional finishing operations may improve certain characteristics such as strip flatness, and rigid inspection helps to screen out unavoidable material defects. In addition the material finished must be made ready for shipment.

In processing the strip in slitting-, inspection- and cross-cutting lines care has to be taken to avoid damage to the product. Technological, physical and chemical tests ensure that material properties satisfy customer requirements. At this stage only the strip flatness may still be improved by a second leveling operation.

Inspection must detect non-permissible surface damage from earlier production steps. Possible surface defects are of widely different kinds [7].

The majority of cold rolled coils or sheets are coated with a corrosion protection oil prior to dispatch.

D 8 Heat Treatment

By Hans Vöge

Processing and service properties are required in quality and special steels and often can be obtained only by separate heat treatment after rolling or forging. The heat treatment to be applied depends on the steel grade (carbon and alloy content), and is selected to produce the structure and strength range suited to the particular application – e.g. machining, or cold forming. Accordingly, *iron and steel works and rolling mills* produce steel in the following forms [1]:
– untreated (U)
– stress relieving annealed (S)
– soft annealed (G)
– spheroidize annealed (GKZ)
– normalized (N)
– normalizing formed (N)
– thermomechanically formed (TM)
– heat treated for ferritic-pearlitic-structure (BG)
– treated for cuttability (C)
– heat treated for definite tensile strength (BF)
– heat treated for improved machinability (B)
– hardened and tempered (V)
– quenched
– hardened + tempered (H + A)
– warm work hardened and aged (WK + AL)
– solution annealed (L)
– precipitation hardening (PH).

Heat treatment and its influence on the microstructure and properties of steel have been described extensively in the preceding parts of this book. At this point, therefore, certain side effects are indicated only by way of example; for instance, inhomogeneities in structure and properties, high-temperature oxidation (scaling) (cf. also D 5.2.2), or surface decarburization (Cf. also D 5.2.3), which may result from some of the plant heat treatments. An intensive study of operational heat treatment effects on steel products literature has been referred to earlier [2, 3]. Data on heat treatments, especially on the effect of *stress-relieving heat treatment, normalizing* and *air and water quenching and tempering* on properties of *flat and sectional steel products made of normal and high tensile structural steels* are to be found in C 2.

References for D 8 see page 810.

The aim of *soft annealing* (cf. also B 4), carried out at temperatures of around A_1, with subsequent slow cooling is to produce a condition that is soft enough for the particular application. *Spheroidize annealing* (cf. also C 5, C 6, C 29), at only a little below or above A_1 with subsequent defined cooling, produces a microstructure which facilitates deformation – notably at room temperature. The yield stress (deformation resistance) is thus reduced and the formability is increased as far as possible. A microstructure is required that consists as far as possible of the ductile ferrite, with the hard constituents dispersed in a spheroidized form, so that they have little effect on the deformation.

The strength properties required for further processing necessitate an annealing process that must be controlled more carefully with increased carbon and alloy contents are in the steels. For instance, for the hypereutectoid ball and roller bearing steels containing about 1.0% C and 1.5% Cr are annealed by Connert [4]. Requirements for an extremely exact control of annealing conditions for the total production lot can be obtained only in continuous annealing furnaces. To spheroidize the carbides as completely as possible often requires such a long annealing time that the concomitant surface oxidation and decarburization may become excessive. Annealing of wire made from ball-bearing steels is carried out in a protective gas atmosphere, since its surface layer is not removed by machining.

The object of *normalising* (cf. also C 4) is to refine the microstructure and to make it as homogeneous as possible. The removal of inhomogeneities in the structure requires the steel to be heated to temperatures above A_3, with subsequent γ–α-transformation. The same effect may be achieved by *normalizing forming* (cf. also C 2.4.2) in which the final shaping takes place in the range of the normalizing temperature accompanied by complete recrystallization of the austenite. Cooling from the austenitizing temperature normally takes place in air outside the heat-treating furnace. The result is that not only a ferritic-pearlitic microstructure may result, but also a ferritic-bainitic or even martensitic structure, depending on the dimensions and chemical composition of the heat-treated material. In austenitizing hypoeutectoid steels the upper transformation temperature should not be exceeded by more than 30 to 50 °C if the formation of coarse secondary grains is to be avoided.

Austenitic grain growth is controlled in fine-grained steels with well-defined contents, e.g. of aluminum by grain growth checking particles, such as aluminum nitride. Local grain growth may happen occasionally, however, even with close control of normalizing conditions, when aluminum nitrides are present in an amount and size that are ineffective for grain growth checking because of the history of the processing batch, e.g. due to unfavourable hot-forming conditions.

Case hardening steels are usually *heat treated*, to obtain *a ferritic-pearlitic structure* (cf. also C 6) to improve machinability, by austenitization followed by an isothermal transformation in the range of the shortest transformation period in the pearlite range. This heat treatment develops a strongly-banded structure in steels alloyed with molybdenum and nickel but reduces machinability when cutting parallel to the structure bands. A banded structure due to microsegregation occurring during solidification of ingots usually cannot be removed even by an expensive homogenizing annealing at economic cost [5]. The banded structure may be countered, for instance, by carrying the ferrite-pearlite transformation to an incomplete stage accepting bainite formation, and subsequently tempering the steel, or by austenitizing at such high temperatures that grain coarsening results.

All heat treatments requiring austenitization result in a more or less considerable surface oxidation and decarburization so that descaling by sand blasting or surface machining to remove the decarburization may become necessary. Use of a protective gas atmosphere in heat treatment of billets and bars can be applied only in exceptional cases, since the construction of suitable continuous heat treatment plants is very expensive.

Still higher heat-trreatment temperatures may be applied for example in homogenizing annealing to equalize concentration differences or to a small extent for grain-coarsening annealing (e.g. at 950 to 1000 °C). These high temperatures produce such extensive oxidation that a considerable thickness loss and hence occasionally dimensional variations must be accepted, if surface machining by scalping or grinding is not scheduled.

Steels produced for fabrication (e.g. machining, cold forming, cold cutting) in a defined strength range – *treated for cuttability* (cf. e.g. C 30) or *heat treated for definite tensile strength* (see also C 30) – are tempered at temperatures of 500° to 700 °C, so that there are no problems with decarburization or oxidation. The same is true for *stress relieving annealing* (cf. also B 4, B 5) used for reduction of residual stresses, caused by for instance uneven heating, forming or cooling in straightening, machining and cold forming. Stress-relieving annealing must be carried out at temperatures which do not impair property values.

Hardening and tempering (cf. also B 4, B 5), particularly of steels for construction of vehicles and mechanical engineering, are carried out after hot deformation in stationary or in continuous furnaces usually before further manufacturing operations if the heat-treated condition permits such processing. Important information may be obtained from time-temperature-austenitization diagrams for the austenitized condition showing the austenitic grain size to be expected (cf. A 9). The TTT-time-temperature-transformation) diagrams (cf. also A 9) provide information on the expected microstructures in surface and center at the given cooling and piece dimensions, under the prevailing austenitization conditions.

Sufficient through-hardening and tempering assuming the correct steel has been selected, requires a uniform austenitization before quenching which must be performed at temperatures that are neither too high nor too low [6]. An excessive hardening temperature (overheating) or too long a heating time in the austenite range (overtiming) lead to austenitic grain coarsening and, after quenching, to coarse martensite that is particularly brittle and tends to from cracks in hardening. Too low quenching temperatures may occasionally not suffice for complete transformation to austenite or for complete dissolution of carbides, such low temperatures may give rise to non-uniform hardness levels (soft spots). Similar effects originate in incomplete hardening due to insufficient heat conduction in quenching caused by scale or formation of steam bubbles on the surface of the material being hardened.

In *quenching* for hardening purposes the selection of the quenching medium and a preferably uniform cooling are vital, and there must be (constant movement in the quenching medium). Hardening media which is too rapid in its action and non-uniform cooling both lead to increased distortion and to the danger of cracking. To avoid tension cracks the surface of the hardening stock must be smooth and without large flaws.

The liability of a steel tc develop hardening cracks depends essentially on its chemical composition; carbon in this respect is of the greatest importance among the other *hardenability* promoting elements. The literature contains several formulas for calculating hardenability and will enable hardening crack liability [2, 7, 8] to be estimated.

After heating in an oxidizing atmosphere, as with the other heat treatments at temperatures above A_1, decarburization of the surface zone may occur and may lead to a reduced strength in the zone of the stock to be heat treated.

The steel composition must also be considered in the specification of cooling conditions after *tempering*. Manganese, chromium and nickel favor temper brittleness if molybdenum or tungsten are not included. These such steels possess low ductility if they cool too slowly after tempering; therefore, they should be cooled in water or at least in oil or air [2]. The most dangerous temperature range lies between 450 and 550 °C so that a tempering treatment in this range should be avoided.

Heat treatment of austenitic chromium-nickel or chromium-nickel-molybdenum steels consists of *solution – or recrystallization annealing*. In treating unstabilized steel grades care should be taken that, depending on the carbon content the temperatures applied are outside the range of grain disintegration [9]. In order to prevent such precipitation effects will lead to intercrystalline corrosion unless quenching is made as precipitous as possible. However, with such treatment, distortion cannot always be avoided completely (cf. also C 12).

Influences of heat treatment on properties of cold-rolled sheet are described in C 4.

D 9 Quality Management in the Production of Steel Works Products

By Walter Rohde, Richard Dawirs, Friedrich Helck and Karl-Josef Kremer

D 9.1 Concept of Quality Management

The preceding sections have described the many possibilities of controlling the properties of steel works products. Knowledge of these relationships allows the producer to specify the methods which will result in a product of the required quality using the facilities available.

The sum of all the actions taken to secure product quality under given economical conditions is called *quality management*. The organizational implementation of these actions leads to the *quality system*. (Terms and definitions for quality management used in this text follow largely DIN 55350, Part 11, which deals with concepts in quality management and statistics [1a]. Some of these concepts are also to be found in ISO-standards [1b]).

This chapter deals with measures of quality management, beginning with charge materials for the melting of steel through the delivery of the specified steel works product [1–8], with quality planning, quality inspection and quality control being of prime importance. Questions concerning
– proving the qualitative state
– documentation of quality management measures
– organization of the quality system
– quality system review
– qualification of personnel
– improvement of quality

are discussed only as far as explanation seems to be necessary; for further details, literature references are given [9–11].

D 9.2 Measures for Quality Management

D 9.2.1 Quality Planning

Quality management begins with quality planning (Fig. D 9.1) and is concerned with the following tasks:
– definition of product properties to be achieved (quality aim),

---

References for D 9 see page 811.

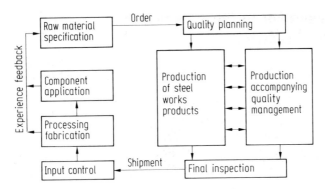

Fig. D 9.1. Typical loop of quality management in producing and working of steel works products.

- setting production parameters necessary for production of required quality,
- definition of quality-securing measures,
- specification of required documents on the qualitative product state and the implementation of measures for securing quality (documentation).

The extent of quality planning stems from the agreements made with the customer and is normally laid down by the staff of the quality department; any risks that may be linked to production and use of product must be taken into account.

Basic requirements for making an objective judgment on the quality of a product are the *quality characteristics*, which should be measurable if possible in numbers and by agreements on threshold values. Both numbers and threshold values are generally specified in delivery agreements, on which orders are based. If the standards governing material and dimensions quoted therein do not contain sufficient data, it is suggested that specific agreement should be reached between customer and producer on the required quality characteristics and their threshold values for delivery. Properties of most steel plant products are changed in processing by hot and cold deformation, welding and if necessary additional heat treatment. Quality-characteristic values of steel products in the as-delivered condition may, therefore, be different from those of the original material. If sufficient experience is not available, a preliminary test run at the customer's plant may be advisable so that an estimation can be made of the extent of possible property changes.

D 9.2.2 Quality Inspection

The term quality inspection comprises all test and inspection measures used to determine the product characteristics.

Tests during production provide information as to whether the applied production measures have led to the expected results. These tests are designed to show up deviations from the planned quality levels and to facilitate definite and immediate decisions on the continuation of production. The test results also serve to optimize production. *Testing of the final products* shows whether specified quality characteristics have been attained.

Quality inspection also focusses on results obtained from taking test samples produced under specified inspection procedures which are aligned with the product and its manufacturing stages.

Testing of samples from the material being processed during melting and ladle treating of the molten steel as well as during casting, and remelting if necessary, provides virtually the only possibility of obtaining data on the qualitative state. Inspections on sampled test pieces often must be performed outside the production plants, using properly equipped testing facilities. Sampling, delivery of test specimens, manufacture of test pieces and carrying out tests are time consuming. When the test results must be in hand before production-related decisions are taken, production delays are frequently unavoidable.

Development of modern testing methods has led to process-technological solutions for steel production and especially for the further-processing stages, which facilitate rapid determination of properties, and decisions on the continuation of the manufacturing process.

D 9.2.3 Quality Control

Comparison of the results of the quality inspection with the production conditions in use provides the basis for economic quality planning and effective quality control.

That the *application of statistical evaluation methods* to such comparisons may contribute considerably to increasing economic efficiency in industrial production was noted by Karl Daeves in Germany in 1922.[12] His work has since proven very fruitful in the area of quality management in steel works. Statistical evaluation methods are considered to be indispensable for quality management today, [13–20].

Generally, the normal distribution is applied in the statistical treatment of test results. If underlying model concept comes near to the existing circumstances, the test results plotted on the probability graph paper used for normal distribution yield a linear relation for the *cumulative frequency curve* and are in good approximation even with a relatively small number of values (Fig. D 9.2). On the basis of such a linear relationship, threshold values may easily be estimated and will be within small degree of probability. With regard to the confidence intervals to be used, refer to [16].

If the test results plotted on the probability graph paper for normal distribution do not yield a straight line, the disparity may be because the test results originated in a value lot that was of mixed distribution, and based on different production or testing conditions.

Frequently the cumulative frequencies do not yield a straight line on the probability graph paper either, even if the test results come from one homogeneously-composed value-population. Deviations are sometimes large and easily de-

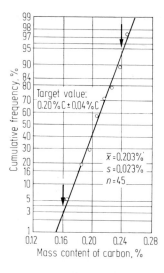

Fig. D 9.2. Cumulative frequency curve of carbon content in 45 heats of steel grade 21 Mn 4 containing 0.16 to 0.24% C, 0.10 to 0.25% Si, 0.80 to 1.1% Mn, ≤ 0.040% P, ≤ 0.040% S.

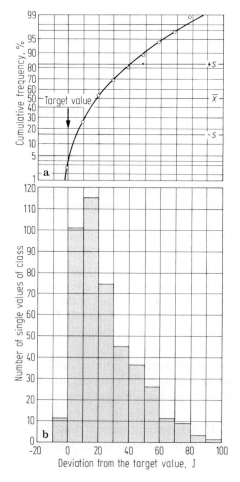

Fig. D 9.3. Cumulative frequency curve of deviations from the target value for values of notch impact energy (ISO-V-notch test pieces = CVN test pieces) at −60 °C in a fine-grained structural steel. *a* cumulative frequency curve for test results; *b* histogram of test results, both in classes of 10 J each.

tectable (Fig. D 9.3). At other times the detection of specific deviations is possible only when a large number of test results exist (Fig. D 9.4). The causes for such relatively frequently observed deviations from a normal distribution may be:
1. The total distribution contains parts which do not follow a random distribution but are influenced by deliberate assignment and production control procedures.
2. The production process used leads to a threshold value for the characteristic to be examined (Fig. D 9.5).
3. The target variable is linked to the influencing random variables in a non-linear manner.

This last-mentioned cause may be explained by considering the yield strength of unalloyed steels. The yield strength of unalloyed steels depends among other things on the carbon content as well as on the cooling rate in the temperature range of the γ-α transformation.

Increases in carbon raises the yield strength of unalloyed steels. Above a certain point however, the effect becomes weak with increasing carbon contents. Assume that the carbon content is normally distributed in a large number of heats which are of the same chemical composition (Fig. D 9.6a).

The non-linear connection between yield strength and carbon content shown in Fig. D 9.6b gives a frequency distribution for the yield strength values depicted in Fig. D 9.6c. The cumulative frequency curve (Fig. D 9.6d) then takes a downward

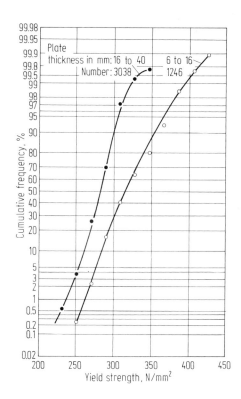

Fig. D 9.4. Cumulative frequency curves for yield strength values of steel St 37-3 in two thickness ranges.

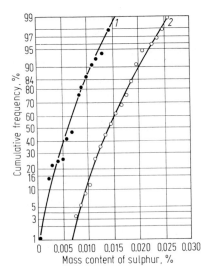

Fig. D 9.5. Cumulative frequency curves for values of sulphur contents in fine-grained structural steels (acc. to DIN 17102 [20a]. Influence of production process on the distribution of test results. 1 = Steels of cold-ductile special grade with $\leq 0.015\%$ S, 2 = steels of cold-ductile grade with $\leq 0.025\%$ S.

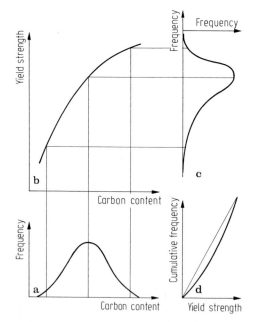

Fig. D 9.6. Relationship between yield strength and carbon content (schematically). **a** frequency distribution of carbon content; **b** relationship between carbon content and yield strength; **c** frequency distribution of yield strength values; **d** cumulative frequency of yield strength values.

deviating course from the normal distribution (right-hand side steep). The higher the cooling rates, the greater are the yield strength values, other conditions being equal. With a decreasing ratio of surface to volume the cooling rate decreases. When the surface to volume ratio depends only on the product thickness then a decrease of the cooling rate and therefore the yield strength are decreased with increasing product thickness. With a normal distribution for the product thickness, within a certain thickness range (Fig. D 9.7a), and from the nonlinear connection between yield strength and product thickness shown by Fig. D 9.7b, a cumulative frequency distribution curve results for the yield strength values as plotted in Fig. D 9.7c. The cumulative frequency curve then follows a course that rises from the normal distribution (left-hand side steep).

The distribution curve in any particular instance depends on which of the effective variables (carbon content, product thickness or cooling rate) influences the evaluation result more strongly.

With small product thicknesses the cooling rate generally has more effect than the carbon content, so that for a population with normally-distributed carbon content and product thickness a left-hand side steep cumulative frequency curve results for the yield strength values. With greater product thicknesses the influence of the carbon content predominates, producing, a right-hand side steep cumulative frequency curve for the yield strength values.

The aim of quoting these examples is to show that *mathematically-defined model distributions* can *aid* statistical evaluations. Their strict validity, presumed mostly for simplification of evaluations, is clearly present only rarely. If mathematically-defined model distributions are not used, the evaluation is more complicated, but gives an unbiased study of the test result distribution and provides more reliable conclusions that take account of metallurgical and physico-metallurgical links.

When realistic threshold values are to be specified on the basis of a statistical evaluation, the test value distribution which has been recognized as being charac-

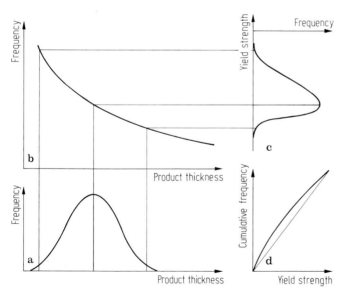

Fig. D 9.7. Relation of yield strength to product thickness in evaluations of greater thickness ranges (schematically). **a** frequency distribution of product thickness; **b** relation between product thickness and yield strength; **c** frequency distribution of yield strength values; **d** cumulative frequency of yield strength values.

teristic for the particular condition should generally be preferred instead of the model distribution, even when the statistical adaptation test has been proven successful mathematically.

Modern electronic data-processing facilitates effective statistical linking of quality inspection results to pertinent production parameters [21]. Thus, connections between production conditions and product properties may be derived quantitatively and weak points may be identified. Production operations and products may also be optimized from the viewpoints of qualitative and economical values achieving required quality characteristics within specified ranges in the making of steel products can thus be optimized. Advance information obtained from data processing may increase the value of the final product inspection in identifying faults [22, 23].

In addition to maintaining correct and suitable production conditions, quality control is used to compensate for random deviations and to make sure that specified product quality characteristics are attained with the greatest possible margin of safety. The final product inspection then serves essentially to confirm this last expectation.

D 9.3 Quality Management in Steel Making

In melting and teeming the crude steel, it is necessary to adjust the basic properties in such a way that the envisaged product properties can be attained accurately in subsequent processing [21]. This adjustment requires:
- achieving the specified ranges for chemical composition
- controlling precipitation and separation of non-metallic phases
- producing a solidification structure as homogeneous as possible and low in segregations
- avoiding internal and external flaws as far as they depend on melting, teeming and solidifying.

Quality management with the above requirements needs special attention to selection of charge materials, steel ladle treatment and casting and remelting – if special demands on the solidified structure and cleanness are to be met.

Extensive measuring and analytical techniques are available [24]. For the determination of *charge materials* properties and *heats* as well as for the control of metallurgical processes. As usual, the results of the analysis are vital it raw materials are to be transformed into steels of a desired chemical composition. The established methods for sampling, preparation of samples and their analysis are extensively described in the literature [25].

Ladle treatment of the molten steel requires rapid analysis to determine chemical compositions [26, 27] and temperature measurements, electrochemical measurement of the oxygen potential of the heat [28] and computer-controlled alloy additions.

An exact control of the oxygen potential of the heat is important in adjusting the contents of oxygen-affinitive alloying elements. It is also decisive for the controlled formation, precipitation and separation of oxidic and sulphidic inclusions.

Alloy calculation aims at bringing chemical composition into the specified range, while taking melting losses into account. Insofar as reliable regression equations exist for statistical links between the chemical composition and steel properties, a nominal composition may be calculated – starting from the heat composition measurements – permitting the specified steel properties to be calculated. This method is used, for example, in the production of case-hardening and heat-treating steels for which the target value of "hardenability" is included in the calculations as a measure of the capacity of the material to be heat treated.

Conditions in the various casting methods, depend on the solidification behavior of the steels as well as on the kind and dimensions of the product and on the product requirements. The aim is to:
- keep inhomogeneities in the structure caused by the solidification process to a minimum especially segregations.,
- avoid formation of pipe, center cavities and other internal imperfections,
- attain a uniform distribution of non-metallic inclusions,
- produce a surface that is low in flaws.

Special attention must be paid in *ingot casting* (cf. D 3.2) to keep to the specified casting temperature and rate, the hot top treatment and the cooling of ingots or slabs. Inspection must usually be limited to visual checks in which coarse surface imperfections may be recognized and the top and bottom formation may be judged. Dressing may be used to cut and remove surface inperfections by scarfing, grinding or machining; this work is expensive, however, and will reduce profits, so care should be taken to maintain a quality standard appropriate to the specified steel grade by careful preparation of the casting system and strict observation of casting specifications.

In *strand casting* (cf. D 3.3) it is important to use specified casting temperature and speed and especially to avoid reoxidation. Other important factors are steel level control, addition of casting powders, cooling in the mold and of the strand, electromagnetic stirring treatment if available, strand guidance and cutting of starting and finishing pieces. Essential quality-influencing casting parameters may be continuously monitored by means of modern computer technology and utilized which may also serve for quality control.

An essential and critical point is the *crude steel assessment* which governs the release of the steel (Fig. D 9.8) for further processing (hot deformation). The assessment is based on the results of analysis and controls as well as on adjustments of process parameter deviations that have a strong influence on quality. Such irregularities may trigger additional quality-management measures such as production of a prototype or additional tests on the semi-finished or finished product.

D 9.4 Quality Management in the Deforming of Crude Steel

The product to be delivered is made by a series of hot (cf. D 5 and D 6) and cold forming operations (cf. D 7) as well as additional heat treatments, if necessary, from as-cast raw products, such as ingots or slabs. The main steel works production lines

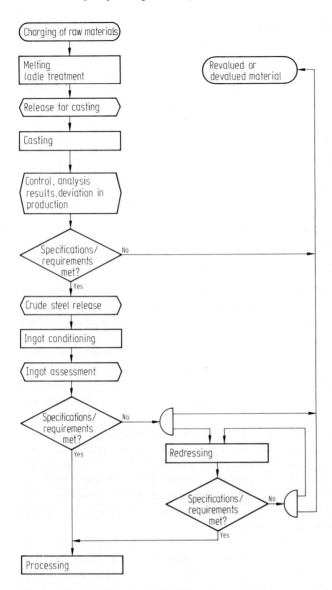

Fig. D 9.8. Production steps and critical points for measurements of quality management in steel making

manufacture semi-finished stock, flat product, sections and open-die forgings. The crucial point of quality control lies in the specification and the monitoring of quality-deciding parameters which are used to adjust the product properties.

Quality characteristics, which are indispensable to production control, include:
– structure,

- macroscopical and microscopical cleanness
- surface decarburization
- hardenability
- mechanical and technological properties, and
- chemical and physical properties

These characteristics are determined by tests [29, 30].

Sampling should not stop production but this is not always possible. Special stimulus is thus offered for development of non-destructive test methods to allow the state and properties of products to be determined. Such methods are now used to determine the form and dimensional accuracy of the product, as well as the external and internal condition [31, 32]. The results of the tests may be utilized immediately for automatic sorting by specified selection criteria.

One important task of quality management consists in maintaining the *material identity* during the whole production run, thus avoiding material mixups. For this purpose, both organizational and technical measures are necessary. Organizational measures include an information system which makes sure that the important product identification data, such as the number of pieces in a production lot or the characteristic dimension, accompany the product in its development. The information is transferred by accompanying cards (e.g. by pneumatic tubes) or by telecommunication installations, and thus maintain an unmistakable identification in each process step. Clear marking of the products is extremely important. The delivery lot ready to be shipped may be marked by [33]:

- Hot, cold or paint stamping
- paint marking
- weather-resistant tags which are fastened to the coils and usually contain the material characteristics as well as further coordination criteria, such as the number of the order/invoice, item number or customer's order number.

For immediate identification of each unit to be delivered, e.g. for sheet and coil of alloyed high-temperature structural steel made in accordance with DIN 17155 [33a], each individual piece is tested for possible material mixups and marked. These markings may even show the original position of the material in the ingot or strand. Within the framework of the operational identity testing procedure, semi-quantitative test methods are used. These methods include the grinding spark, [34] and spectroscopic or electromagnetic methods and are quick simple to carry out. In addition, movable plant spectrometers are available to determine the chemical composition, which make the material identification more certain [35].

Quality management measures adapted to the production line are described below by way of example for two important product groups – steel bars and plates.

D 9.5 Quality Management in Steel Bar Production

Values measured during steel bar production are plotted in Fig. D 9.9 as an example of quality management of sectional products [6].

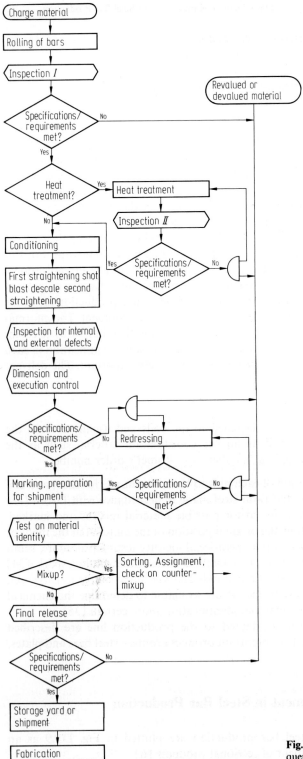

Fig. D 9.9. Production and testing sequence in the production of bar steel.

The ingot steel is released to the rolling mill. In rolling, the heating and deformation conditions must be controlled continuously, because they are chosen to suit the steel grade as well as the charge and the final cross-section. The aim is to obtain a uniform structure over the whole length of the rolled bar and the rolled lot by control of the final rolling temperature and the rate of cooling from the rolling temperature.

The possibility for producing desired structures directly from the rolling temperature decreases with increasing bar diameters, so a subsequent heat treatment frequently becomes necessary. The heat treatment conditions must be carefully controlled.

The adherence to permissible limits for deviations in form and dimension must be watched continuously during rolling. Attention must also be paid to rolling imperfections (cracks or slivers). Non-destructive test methods are available which facilitate testing the rolled strand in the hot state.

Tests made during the rolling or heat treatment represent fixed points in production, which is concluded when the material is released for further processing. At this point all samples are taken that are normally necessary for the determination of quality characteristics listed above, such as:
– structure
– macroscopical and microscopical cleanness
– surface decarburization
– hardenability
– mechanical and technological properties
– chemical and physical properties.

If required, finishing of steel bars, is mostly carried out on production lines. After a first straightening, shot blast descaling and final straightening, tests are carried out continuously to find internal and external discontinuities and a length control is performed with sorting yardsticks. Defective bars are sorted and internal and external defects are marked automatically by paint spraying at high throughput speeds. The defective bars are taken to a dressing station.

Internal inspection is effected by automatic ultrasonic testing in which the aim is to achieve the greatest possible volume ratio. Testing for surface flaws is carried out by an electromagnetic probe method, in which the defect signals are detected by rotating probes. The minimum defect depth to be detected by such test methods depends on the surface condition and the noise background which it causes.

Special demands for flawless surfaces as well as on form and dimensional accuracy may make it necessary to machine bars by scalping or grinding with a consequent loss of material. After rolling and heat treating the production line undertakes straightening, scalping (or grinding), testing on internal or external defects and checks on form and dimensional accuracy (cf. also Table D 5.1).

Only descaled and machined steel bars are checked during the final control procedures for product execution and marking. The last quality control measure before final release is normally a check on material identity.

The *final release* includes a check on the results of all control procedures used. Production lots that meet the specification can be released for shipping. If specified at the time of ordering, the plant-internal final release is followed by an acceptance inspection by representative of the customer or an independent organisation. This acceptance inspection may comprise an additional check for specified quality characteristics depending on conditions for delivery or special test and control measures.

Acceptance inspections, therefore, are no longer exclusively an integral part of the producer's quality management but represent rather an *incoming inspection at the customer's plant that has been transferred back into the producing plant*. Certificates on material testing listed in DIN 50049 [35a] may be provided to confirm results of inspections. When certificates are issued, they are based on conditions laid down in the relevant DIN standard as being required for the issue of *test certificates*. The kind of certificate to be issued generally is agreed upon with the customer.

D 9.6 Quality Management in Plate Production

As an example of quality management on flat products Fig. D 9.10 demonstrates the quality management measures for plate production [5]. The figures in parenthesis in the following explanations refer to the steps in the processes for the order listed in the figure.

As with all products, before beginning production, the producer must receive, a technically clear and complete order whose feasibility has been checked by the quality planning department (1). The production planning department (2) states the delivery conditions, the product types, permissible ranges of chemical composition and the sequence of testing operations. The order then given to the steel plant for the crude steel material must contain that part of the specifications necessary for making the steel and other details.

Requirements for surface condition of plates are mostly high, since the rolled surface will be approximately identical with the surface of the finished parts fabricated by the consumer. Removal of surface defects is less expensive and more successful the earlier the defects are detected. At least visual checks, therefore, are made during production of the crude steel slabs, after each production step, and in the slab yard (3). These checks form the basis on which the decision is made to release material for rolling.

During slab heating in a pusher-type furnace, (4) identification markings are lost. During the time interval between pushing into the furnace and finishing rolling the material identity must therefore, be safeguarded by means of a careful material-flow control system.

Plates are mostly welded during fabrication and frequently only heat treated to relieve stress. For the plate producer this procedure means that definitive values for mechanical properties usually are made part of the delivery specification and these target values have to be attained with a chemical composition that is especially favorable for weldability. In modern plate mills specified property values can be adjusted by temperature-controlled rolling (5) and subsequent controlled cooling.

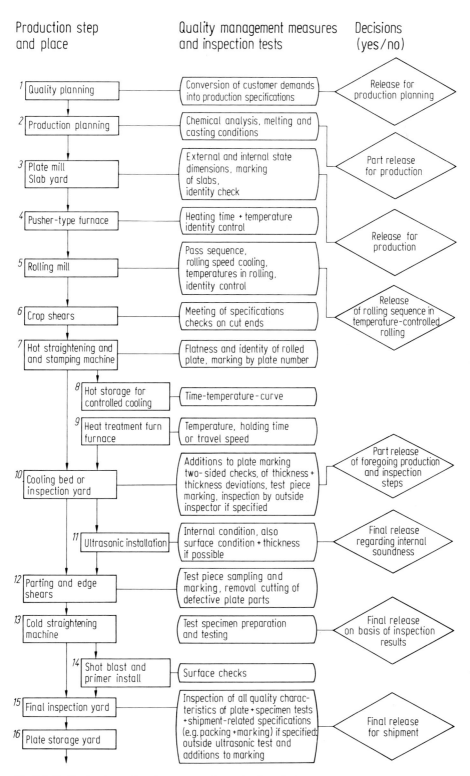

Fig. D 9.10. Production and testing sequence in production of plate.

If these methods are used, a comprehensive and very intensive control of rolling is required.

The rolled plate is generally hot adjusted after passing the crop shears (6) and is marked with a number derived from the material flow control data. If necessary, controlled cooling (8) follows.

Heat treatment (9) is carried out if required on the basis of the advance information specifying the desired quality characteristics or if it is called for in the delivery agreement.

The real product inspection begins on the cooling bed (10). Internal defects are detected by ultrasonic testing installations (11) which use in the plate mill the impulse echo technique. These installations may also be used to find surface defects and for control of product thickness [36].

At the dividing and edge shears (12) sample cuttings are taken, for determining the mechanical properties. Since test specimen preparation and testing do not run synchronously with production is completed before the test results become available. The production process is completed with final adjustment (13) and a subsequent surface treatment (14) if required. After final inspection (15) the plates are made ready for shipping, and taken to the plate storage yard (16).

The *release for shipping* follows after conclusion of the planned inspections (see production step 12) and satisfactory test results. The inspections required often depend on the intended final use of the plates. In plate made of unalloyed or low-alloy steels, mechanical and technological properties are of prime importance. An important part of the plate production is fabricated in installations for which safety requirements are laid down. The frequency of tests of such products is accordingly high.

With the release for shipment the product-related quality control measures are concluded. Data emerging during production are also utilized for quality control of follow-up orders.

Table 1. Comparison of the steel grades mentioned in Volume 1 and Volume 2 of "Werkstoffkunde Stahl" and identified here by their designations according to German standards (DIN Standard) with corresponding grades according to international, European and some other national standards.

Preliminary remarks: This comparison of steel grades can only serve for general reference purposes. Only rarely are the grades absolutely identical. In almost all cases, the steel grades printed side by side in a single line in Table 1 feature more or less significant differences in the details of the requirements on the decisive properties specified in the respective standards. The reference table cannot be used therefore as a basis for ordering or for delivery conditions. However, in view of the task set for this book the comparability of the grades is ensured to such an extent that the reader can easily understand the

Steel grade referred Designation	Werkstoff Nr.	Norm (DIN)	International Standard Designation	Standard ISO	European Standard Designation	EURO-NORM
097-30-N5	1.0861	IEC	097-30 N5	IEC 404 -8-5	FeM 097-30 N	107
9 S 20		1651 ex	9 S 20	683/9	10 S 22	87
9 SMn 28	1.0715	1651	11 SMn 28	683/9	11 SMn 28	87
9 SMn 36	1.0736	1651	12 SMn 35	683/9	12 SMn 35	87
9 SMnPb 28	1.0718	1651	11 SMnPb 28	683/9	11 SMnPb 28	87
9 SMnPb 36	1.0737	1651	12 SMnPb 35	689/9	12 SMnPb 35	87
10 CrMo 9 10	1.7380	17 155 17 175	13 CrMo 9 10 TS 34, P 34	9328-2 2604/4	10 CrMo 9 10	28
10 Ni 14	1.5637	17 280	12 Ni 14 (P 44)	9328-3	FeE 355 Ni 14	129
10 S 20	1.0721	1651	10 S 20	683/9	10 S 20	87
10 SPb 20	1.0722	1651	10 SPb 20	683/9	10 SPb 20	87
11 MnNi 5 3	1.6212	17 280			FeE 285 Ni 2	129
12 CrMo 19 5	1.7362	SEW 590 17 176	P 37	2604		
12 Ni 19	1.5680	17 280			FeE 390 Ni 20	129
13 CrMo 4 4	1.7335	17 155 17 175	14 CrMo 4 5 TS 32, P 32	9328-2 2604	14 CrMo 4 5	28
13 MnNi 6 3	1.6217	17 280			FeE 355 Ni 2	129
14 MoV 6 3	1.7715	17 243	P 33, F 33	2604		
14 Ni 6	1.5622				FeE 285 Ni 6	129
14 NiCr 14	1.5752					
14 NiCrMo 13 4	1.6657	AECMA*	FePl 71	AECMA*	14 NiCrMo 13	84
14 NiMn 6	1.6228	17 280	15 NiMn 6	9328-3	FeE 355 Ni 6	129
15 CrNi 6	1.5919	17 210			14 CrNi 6	84
15 Mn 3 Al	1.0468	17 115	2	(DP 4959)		
15 Mo 3	1.5415	17 155	16 Mo 3	9328-2	16 Mo 3	28
16 MnCr 5	1.7131	17 210 1654	16 MnCr 5 B 11 16 MnCr 5 E	683/11 4954	16 MnCr 5 16 MnCr 5 KD	84 119
17 Cr 3	1.7016	1654			15 Cr 2 KD	119
17 CrNiMo 6	1.6587	17 210	18 CrNiMo 7	683/11	17 CrNiMo 7	84
17 MnCr 5	1.3521	17 230	10	683/17	16 MnCr 5 F	94
17 NiCrMo 14	1.3533	17 230	16	683/17	18 NiCrMo 14	94
18 CrNi 8	1.5920	17 210				
19 9	1.4302	8556	19.9	3581		
19 9 L	1.4316	8556	19.9 L	3581		
19 12 3	1.4430	8556	19.12.3 L	3581		
20 MnCr 5	1.7147	17 210	20 MnCr 5	683/11		

*Association Européenne des Constructeurs de Matériel Aérospatial, Paris

general principles and correlations of physical metallurgy described in the book on the base of certain steel grades according to the German standard (DIN Standard) by referring to the reference steels in Table 1 with which he is familiar.

German-standard steels mentioned in the book, for which there are no comparable grades on an international level, are not listed in Table 1, but generally identified throughout the book by their chemical composition.

| USA | | UK | | France | | Japan | |
Designation Grade	Standard ASTM	Designation	Standard BS	Designation	Standard NFA	Designation	Standard JIS-G
		30 M 5	601/2	FeM 097-30	C28920		
1212	A 29	220 M 07	970			SUM 21	4804
1213	A 29	230 M 07	970	S 250	35-561	SUM 22	4804
1215		(230 M 07)	970	S 300	35-561		
12 L 13	A 29			S 250 Pb	35-561	SUM 22 L	4804
12 L 15	A 29			S 300 Pb	35-561	SUM 24 L	4804
P 22	A 335	622	1501/2	10 CD 9.10	36-206	SCMV 4	4109
T 22	A 213	622	3604	TU CD 9.10	49-213	STBA 24	3458
D	A 203	503	1501/2	3,5 Ni-355	36-208		
1109 S	A 29	210 M 15	970/1	13 MF 4	35-562		
11 L 08 S	A 29	210 M 15 lead	970/1	XC 10 q Pb	35-551		
060	A 27			0,5 Ni-285	36-208		
P 5, T 5	A 335/213	625	3604	Z 12 CD 5.05	49-213	SCMV 6	4109
5	A 387	(B 5)	(3100)				
	(A 645)			5 Ni-390	36-208	SFL 3	3205
12	A 387	620	1501/2	15 CD 4.05	36-205	SCMV 2	4109
T 12	A 213	620-440	3604	TU13CD 4.04	49-213	STBA 22	3458
				0,5 Ni-355	36-208		
				1,5 Ni 285	36-208		
		655 H 13	970/1	14 NC 11	35-551		
				1,5 Ni 355	36-208		
(E) 4320	A 304	815 M 17	970/1	(16 NC 6)	35-551	SNC 815	4102
	A 391	60	3113		36-206		
B	A 204	243	1501/2	15 D 3	36-602		
5120 H	A 304			16 MC 5	35-551		
5120 S 4	A 322	590 H 17	970/1	16 MC 5 FF	35-564		
6118 S 4	A 322	527 H 17	970/1	18 C 3 FF			
		822 M 17	970/1	(18 NCD 6)	35-551		
		(527 M 17)	970/1				
E 3310	A 534			16 NCD 13	35-565		
		822 M 17	970/1	20 NC 6	35-552		
308	AWS/ASME			Z6CND 20.10, 308 F 91	35-583		
308 L	AWS/ASME			Z2CND20.10, 308 F 92	35-583		
316 L	AWS/ASME			Z2CND19.12, 316 F 92	35-583		
				20 MC 5	35-552	SMn 420	4106

Table 1 (*Continued*)

Steel grade referred Designation	Werkstoff Nr.	Norm (DIN)	International Standard Designation	Standard ISO	European Standard Designation	EURO-NORM
20 MnMoNi 5 5	1.6311	(SEW 550)		2604/8		
20 MoCr 4	1.7321	17 210 1654	18 MoCr 4 B 31	683/11 4954	20 MoCr 4 20 MoCr 4 KD	84 119
20 NiCrMo 2	1.6522	17 115	6	(DP 4959)		
20 NiCrMo 6		Chap D 5	14	683/17	20 NiCrMo 7	94
21 CrMoV 5 7	1.7709	17 240				
21 Mn 4 Al	1.0470	17 115	3	(DP 4959)		
21 NiCrMo 2	1.6523	17 210 1654	20 NiCrMo 2 B 41	683/11 4954	20 NiCrMo 2 20 NiCrMo 2 KD	84 119
22 B 2	1.5508	1654	E 1 (CE 20 BG1)	4954	C 22 B KD	119
22 CrMoS 3 5	1.7333	17 210				
23 MnNiCrMo 5 2	1.6758	17 115				
24 CrMoV 5 5	1.7733				25 CrMo 4 KD	119
25 CrMo 4	1.7218	17 200	25 CrMo 4	683/1	25 CrMo 4	83
27 MnSi 5	1.0412	17 115				
28 Mn 6	1.1170	17 200	28 Mn 6	683/1	28 Mn 6	83
28 NiCrMoV 8 5	1.6932	SEW 550				
30 CrNiMo 8	1.6580	1654 17 200	31 CrNiMo 8 31 CrNiMo 8 E	683/1 DIS 4954	30 CrNiMo 8 30 CrNiMo 8 KD	83 119
31 CrMo 12	1.8515	17 211	1	683/10	31 CrMo 12	85
31 Mn 4	1.0520	21 544	A 1	1005/3		
32 CrMo 12	1.7361	SEW 550			32 CrMo 12	83
34 Cr 4	1.7033	1654	34 Cr 4 E (C 14)	4954	34 Cr 4 KD	119
34 CrAlNi 7	1.8550	17 211				
34 CrMo 4	1.7220	17 200 1654	34 CrMo 4 34 CrMo 4 E	683/1 4954	34 CrMo 4 34 CrMo 4 KD	83 119
34 CrNiMo 6	1.6582	17 200 1654	36 CrNiMo 6	683/1	35 CrNiMo 6 35 CrNiMo 6 KD	83 119
34 NiCrMo 16	1.2766					
35 B 2	1.5511	1654	C E 35 B (E 5)	4954	C 35 B KD	119
35 S 20	1.0726	1651	35 S 20	683/9	35 S 20	87
37 Cr 4	1.7034	1654	37 Cr 4 E (C 15)	DIS 4954	37 Cr 4 KD	119
37 Cr B 1	1.7007		37 Cr B 1 E (E 10)	4954	38 Cr B 1	119
37 CrMoW 19						
38 Cr 2	1.7003	1654	37 Cr 2 E (E 12)	4954	38 Cr 2 KD	119
38 Cr 4	1.7043	17 212			38 Cr 4	86
41 Cr 4	1.7035	17 200 1654	41 Cr 4 E (C 16)	683/1 DIS 4954	41 Cr 4 41 Cr 4 KD	83 119
42 CrMo 4	1.7225	17 200 1654	42 CrMo 4 42 CrMo 4 E (C 32)	683/1 4954	42 CrMo 4 42 CrMo 4 KD	83 119
43 CrMo 4	1.3563	17 230	(42 CrMo 4)	683/1	(42 CrMo 4)	83
45 Cr 2	1.7005	17 212			45 Cr 2	86
45 S 20	1.0727	1651	46 S 20	683/9	45 S 20	87
45 SPb 20	1.0757	1651	(46 S 20 Pb)			
45 WCrV 7	1.2542	(17 350)	45 WCrV 2 (13)	4957	45 WCrV 8	96
46 Si 7	1.5024				45 Si 7	89
49 CrMo 4	1.7238	17 212				
50 CrMo 4	1.7228	17 200 17 221	50 CrMo 4	683/1		

Table 1

USA		UK		France		Japan	
Designation Grade	Standard ASTM	Designation	Standard BS	Designation	Standard NFA	Designation	Standard JIS-G
GrB	A 533						
8620	A 304	808 M 17	970/1				
8620 S 4	A 322	805 M 20	970/1	18 CD 4 FF	35-564		
(E) 8620	A 304	80	3114	20 NCD 2	35-566	SNCM 220	4103
4320	A 304			20 NCD 7	35-565	SNCM 420	4102
				20 CDV 5.07	35-558		
				20 M 5	35-566	SBC 50	3105
8620	A 304	805 H 22	970/1	20 NCD 2	35-551		
8620 S 4	A 322	805 H 20	970/1	20 NCD 2 FF	35-564		
(Typ 2)	A 449			21 B 3 FF	35-564		
				18 CD 4q	35-551		
				23 MNCD 5	35-566		3105
		708 A 25	970/1	25 CD 4	35-552	SCM 430	4105
4130	A 304	9/1	3111				
				25 MS 5	35-566	SBC 70	3105
1330	A 304	150 M 28	970/1	35 M 5	35-552		
Class 5	A 288						
				30 CND 8	35-552		
				30 CND 8 FF	35-564		
		722 M 24	970/1	30 CD 12	35-552		
				30 CD 12			
				32 C 4 FF	35-564	SCr 430	4104
(Class B/C)	A 355			(40CAD6.12)	35-552		
4137 S 4				34 CD 4	33-557	SCM 435	4105
	A 322	708 A 37	970	34 CD 4 FF	35-564		
4340	A 304	817 M 40	970/1	30 CND 8	35-552		
4340 S 4	A 322			30 CND 8 FF	35-577		
Class 7	A 469	835 M 30	970/1	35 NCD 16	35-552		
		10/1	3111	38 B 3 FF	35-564		
1139	A 29	212 M 36	970/1	35 MF 6	35-562		
5135 S 4	A 322	530 A 36	970/1	38 C 4 FF	35-564	SCr 435	4104
H 12	A 681						
5135 S 4	A 322	(120 M 36)	970	38 C 2 FF	35-564		
5135	A 304	530 A 36	970/1	38 C 4	35-563	SCr 440	4104
5140	A 304	530 M 40	970/1	42 C 4	35-552	SCr 440	4104
5140 S 4	A 322	530 M 40	970/1	42 C 4 FF	35-564		
4142	A 304	708 M 40	970/1	42 CD 4	35-552	SCM 440	4105
4142 S 4	A 322			42 CD 4 FF	35-564	(SMB 7)	4107
				42 C 2p	35-563	SMn 443	4106
1146 S	A 29	216 M 44	970/1	(45 MF 5)	35-562		
11 L 44	A 29	216 M 44 lead	970/1				
				46 S 7	35-571		
4147	A 322			42 CD 4 TS	35-563	SCM 445	4105
(4150)	A 322	708 A 47	970/1			SUP 13	4801
4150	A 689						

Table 1 (*Continued*)

Steel grade referred Designation	Werkstoff Nr.	Norm (DIN)	International Standard Designation	Standard ISO	European Standard Designation	EURO-NORM
50 CrV 4	1.8159	17 200	50 CrV 4	683/1	50 CrV 4	83
		17 221	51 CrV 4	DIS 683/14	50 CrV 4	89
51 CrMoV 4	1.7701	17 221	52 CrMoV 4 (14)	DIS 683/14	51 CrMoV 4	89
51 CrV 4	1.2241	17 350	51 CrMnV 1 (12)	4957	51 CrMnV 4	96
51 Si 7		(17 221)	4	683/14	50 Si 7	89
55 Cr 3	1.7176	17 221	55 Cr 3 (8)	683/14	55 Cr 3	89
55 NiCrMoV 6	1.2713	17 350	55 NiCrMoV 2 (H 2)	4957	55 NiCrMoV 7	96
55 Si 7	1.0904	17 222			55 Si 7	89
55 SiCr 7	1.0958					
56 NiCrMoV 7	1.2714	17 350	55 NiCrMoV 2	4957	55 NiCrMoV 7	96
60 S 20	1.0728	1651			60 S 20	87
60 Si 7	1.0909		59 Si 7	DIS 683/14	60 Si 7	89
60 SiCr 7	1.0961	17 221	61 SiCr 7	DIS 683/14	60 SiCr 7	89
60 WCrV 7	1.2550	17 350	60 WCrV 2 (15)	4957	55 WCrV 8	96
70 Si 7	1.2823		60 SiMn 2	4957	60 SiMn 7	96
80 MoCrV 42 16	1.3551	17 230	30		80 MoCrV 40 16	94
90 MnCrV 8	1.2842	17 350	90 MnV 2 (18)	4957	90 MnV 8	96
100 Cr 2	1.3501	17 230				
100 Cr 6	1.2067	17 350	100 Cr 2 (16)	4957	102 Cr 6	96
	1.3505	17 230	1	683/17	100 Cr 6	94
100 CrMn 6	1.3520	17 230	3	683/17	100 CrMn 6	94
100 CrMo 7	1.3537	17 230	4	683/17	100 CrMo 7	94
100 CrMo 7 3	1.3536	17 230	5	683/17	100 CrMnMo 7	94
100 MnCrW 4	1.2510	(17 350)	95 MnCr W 1 (19)	4957	95 MnWCr 5	96
105 WCr 6	1.2419	17 350	105 WCr 1	4957	107 WCr 5	96
110-30-P5	1.0881	IEC 68(CO)	110-30-P5	IEC 404-8-5		
AlNiCo 9/5	1.3728	17 410	AlNiCo 9/5	404-8-1		
AlNiCo 30/10	1.3758	17 410	AlNiCo 31/11	404-8-1		
AlNiCo 35/5	1.3761	17 410				
AlNiCo 37/5			AlNiCo 37/5	404-8-1		
AlNiCo 38/11			AlNiCo 38/11	404-8-1		
AlNiCo 52/6	1.3759	17 410	AlNiCo 52/6	404-8-1		
AlNiCo 450		see	AlNiCo 37/5	404-8-1		
AlNiCo 500		see	AlNiCo 38/11	404-8-1		
BSt 420 S	1.0428	488	RB 400(W)	6935-2	Fe B 400	80
BSt 500 M	1.0466	488				
BSt 500 S	1.0438	488	RB 500(W)	6935-2	Fe B 500	80
C 15	1.0401	17 210	C 15	683/11		
C 15 Pb	1.0403	17 210	C 15 ES (Pb)	683/11		
C 35	1.0501	17 200	C 35	683/1	1 C 35	83
C 35 Pb	1.0502	17 200	(C 35 Pb)	683/1	1 C 35 Pb	83
C 45	1.0503	17 200	C 45	683/1	1 C 45	83
C 45 Pb	1.0504	17 200	(C 45 Pb)	683/1	1 C 45 Pb	83
C 70 W 2	1.1620	17 350	TC 70	4957	CT 70	96
C 80 W 1	1.1525	17 350	TC 80	4957	CT 80	96
C 85 W	1.1830	17 350	TC 80/TC 90	4957		
C 105 W 1	1.1545	17 350	TC 105	4957	CT 105	96
C 125 W	1.1663	(17 350)	TC 120	4957	CT 120	90
Ck 15	1.1141	17 210	C 15 E 4	683/11	2 C 15	84
Ck 22	1.1151	17 210	C 20	683/18		
Ck 35	1.1181	17 200	C 35 E 4	683/1	2 C 35	83

Table 1

| USA | | UK | | France | | Japan | |
Designation Grade	Standard ASTM	Designation	Standard BS	Designation	Standard NFA	Designation	Standard JIS-G
6150	A 29 A 689	735 A 50	970/5	50 CV 4	35-571	SUP 10	4801
				51 CDV 4	35-571		
6150	A 322	735 A 50	970/5	50 CV 4	35-571		
9255	A 29	Z 50 A 53	970/5	51 S 7	35-571		
5155	A 689	527 A 60	970/5	55 C 3	35-571	SUP 9	4801
L 6	A 681			55 NCDV 7	35-590	SKT 4	4404
		250 A 53	970/5				
9254	A 689			56 SC 7	35-571	SUP 12	4801
L 6	A 681			55 NCDV 7	35-590	SKT 4	4404
1151 S	A 29						
		250 A 58	970/5			SUP 6	4801
9254, 5160	A 401 to A 689			61 SC 7	35-571	SUP 12	4801
S 1	A 681			5 SWC 20	35-590		
6490	AMS			800 CV 40	35-565		
02	A 681	B 02	4659	90 MV 8			
E50100	A 29			100 C 2	35-565		
L 3	A 681	BL 3	4659	Y 100 C 6	35-590	SKC 11	4410
52100	A 295	534 A 99	970	100 C 6	35-565	SUJ 2	4805
2	A 485	(535 A 99)	970/1	100 CM 6	35-565		
3	A 485			100 CD 7	35-565		
5120	A 534						
		B 01	4659	90 MWCV 5	35-590	SKS 3	4404
07	A 681			105 WC 13	35-590	SKS 2	4404
		30 M 6	601/2				
AlNiCo 8	Alloy						
AlNiCo 5	Digest						
400	A 615	M 460/425	4449	Fe E 400	35-016		
W	A 185	460	4482				
	A 82			Fe E 500	35-016		
1015	A 29	08 M 15	970/1	XC 12	35-552	SC 15	4051
10 L 15	A 29	08 M 15 lead	970/1	XC 12 q Pb	35-552		
1034	A 29	80 M 36	970/1	XC 38	35-552	SC 35	4051
10 L 34	A 29	080 M 36 lead	970/1	XC 38 q Pb	35-552		
1045	A 29	080 M 46	970/1	XC 48	35-552	SC 46	4051
10 L 45	A 29	080 M 46 lead	970/1	XC 48 q Pb	35-552		
	A 686			Y 3 65/Y 1 70	35-590	SK 7	4401
W 1-7						SK 6	4401
W 1-8	A 686	BW 1A		Y 3 65/Y 1 80	35-590	SK 5	4401
W 5	A 686	BW 1B	4659	Y 1 105	35-590	SK 3	4401
W 1/12	A 686	BW 1C	4659	Y 2 120	35-590	SK 2	4401
		080 M 15	970/1	XC 12	35-551	S 15 CK	4051
		055 M 15	970	XC 25	35-552		
		080 M 36	970	XC 32	35-552		

Table 1 (*Continued*)

Steel grade referred Designation	Werkstoff Nr.	Norm (DIN)	International Standard Designation	Standard ISO	European Standard Designation	EURO-NORM
Ck 45	1.1191	17 200	C 45 E 4	683/1	2 C 45	83
Ck 60	1.1221	17 200	C 60 E 4	683/1	2 C 60	83
Cm 45	1.1201	17 200	C45 M 2	683/11	3 C 45	83
Cq 15	1.1132	1654	B 2-CE 15 E4	4954	C 15 KD	119
Cq 22	1.1152	1654	C 1-CE 20 E4	4954	C 21 KD	119
Cq 35	1.1172	17 240			C 35 KD	119
		1654	C 3-CE 35 E4	4954	C 35 KD	119
Cq 45	1.1192	1654	C 6-CE 45 E4	4954	C 45 KD	119
CrAl 25 5	1.4765	17 470				
D 15-2	1.0413	17 140 T1		8457-2	2 CD 15	16
GS-12 CrMo 9 10	1.7380	SEW 595	C 34 AH	DP 4991		
GS-12 CrMo 19 5	1.7363	SEW 595	37 AH	DP 4991		
GS-17 CrMo 5 5	1.7357	17 245	C 32 H	DP 4991		
GS-17 CrMoV 5 11	1.7706	17 245	C 35 (B) H	DP 4991		
GS-18 CrMo 9 10	1.7379	17 245	C 34 B H	DP 4991		
GS-22 Mo 4	1.5419	17 245	C 28 H	DP 4991		
GS-C 25	1.0619	17 245	C 23-45 A	DP 4991		
GS-38	1.0416	1681	20-40	3755		
GS-60	1.0553	1681	30-57	3755		
G-X8CrNi 12	1.4107	17 245	C 39 CNi H	DP 4991		
G-X12CrMo 10 1	1.7389	SEW 595	C 38 H	DP 4991		
G-X22CrMoV 12 1	1.4931	17 245	C 40 H	DP 4991		
Hardferrite 25/14	1.3649	17 410	Hard Ferrite 25/14	IEC 404-8-1		
Hardferrite 3/18 p	1.3614	17 410	Hard Ferrite 3/18 p	404-8-1		
Hardferrite 7/21	1.3641	17 410	Hard Ferrite 7/21	404-8-1		
Hardferrite 9/19 p	1.3616	17 410	Hard Ferrite 9/19 p	404-8-1		
Ni 36	1.3912	1715				
NiCr 15 Fe	2.4816	17 742	NW 6600	6208		
			NW 6600	9722		
NiCr 20 TiAl	2.4952	17 742	NW 7080	9722		
		17 480	NiCr 20 TiAl(11)	683/15	Ni-P 95-HT	AECMA
QSt 32-3	1.0303	1654	A 1 Al-CC 48	4954	CB 4 FF KD	119
QSt 34-3	1.0213	1654	A 2 Al-CC 8 A	4954	CB 7 FF KD	119
QSt 36-3	1.0214	1654	A 3 Al-CC 11 A	4954	CB 10 FF KD	119
QSt 38-3	1.0234	1654	A 4 Al-CC 15 A	4954	CB 15 FF KD	119
QStE 500	1.8957	SEW 092	HSF 490	5951		
QStE 500 TM	1.8959	SEW 092	FeE 490 TM	6930		
R 10 S 10	1.0703	17 111				
RSt 35-2	1.0208	17 115				
RSt 37-2	1.0038	17 100	Fe 360-B	630	FeE 235 BFN	25
RSt 38	1.0223	17 111	A 4 Si	4954	Fe 360-BFN	25
S 2-10-1-8	1.3247	17 350	HS 2-9-1-8	4957	HS 2-9-1-8	96
S 6-5-2	1.3343	17 350	HS 6-5-2	4957	HS 6-5-2	96
S 6-5-2-5	1.3243	17 350	HS 6-5-2-5	4957	HS 6-5-2-5	96
S 6-5-3	1.3344	17 350	HS 6-5-3	4957	HS 6-5-3	96
S 7-4-2-5	1.3246	17 350	HS 7-4-2-5	4957	HS 7-4-2-5	96
S 10-4-3-10	1.3207	17 350	HS 10-4-3-10	4957	HS 10-4-3-10	96
S 12-1-4-5	1.3202	17 350	HS 12-1-5-5	4957	HS 12-1-5-5	
S 18-0-1	1.3355	(17 350)	HS 18-0-1	4957	HS 18-0-1	96
S 18-1-2-5	1.3255	17 350				
S 18-1-2-10	1.3265		HS 18-0-1-10	4957	HS 18-0-1-10	96

USA Designation Grade	USA Standard ASTM	UK Designation	UK Standard BS	France Designation	France Standard NFA	Japan Designation	Japan Standard JIS-G
(1547), NV 2	SAE	080 M 46	970/1	XC 45	35-552	S 45 C	4051
1060		080 A 62	970/1	XC 60	35-553	S 48 C	4051
		080 M 46	970/1	XC 48 H 1q	35-552		
1015	A 545			XC 12 FF	35-564		
				XC 25 FF	35-564	SWCH 20 K	3507
(Typ 1)	A 449	080 M 36	970/1	XC 32 FF	35-564	SWCH 35 K	3507
(11 B)	B 603						
	A 510	0/3	3111/1		35-049		
WC 9	A 217						
C 5	A 217						
WC 6	A 217	B 2	3100	15 CD 4.05	36-602	SCPH 21	5151
9	A 356					SCPH 23	5151
10	A 356	B 3	3100			SCPH 32	5151
WC 1	A 217					SCPH 11	5151
WCA	A 216	A 1	3100			SCPH 1	5151
U-60-30	A 27					SCW 42	5101
80-50	A 148					SCW 56	5102
		410 C 21	3100				
C 12	A 217						
T 36	B 753			Fe-Ni 36	54-301		
N 06600	B 160	N 14					
Inconel/60							
HEV 5	SAE	2 HR 1	970	NC 20 TA	35-579		
1006				XC 6 FF	35-564	SWRCH 6 A	3507
				XC 10 FF	35-564	SWRCH 8 A	3507
				XC 12 FF	35-564	SWRCH 10 A	3507
				XC 18 FF	35-564		
1110	A 29			13 MF 4	35-562		
		40	1663			SBC 31	3105
		40 B	4360				
				E 24-2 NE	35-501		
M 42	A 600	BM 42	4659	2-9-1-8	35-590	SKH 59	4403
M 2	A 600	BM 2	4659	6-5-2 HC	35-590	SKH 51	4403
M 41	A 600	BM 2	4659	6-5-2-5 HC	35-590	SKH 55	4403
M 3-2	A 600			6-5-3	35-590	SKH 53	4403
M 41	A 600			7-4-2-5	35-590		
		BT 42	4659	10-4-3-10	35-590	SKH 57	4403
T 15	A 600	BT 15	4659	12-1-5-5	35-590	SKH 10	4403
T 1	A 600	BT 1	4659	18-0-1	35-590	SKH 2	4403
T 4	A 600	BT 4	4659	18-1-1-5	35-590	SKH 3	4403
		BT 5	4659	18-0-2-10	35-590	SKH 4 A	4403

Table 1 (*Continued*)

Steel grade referred Designation	Werkstoff Nr.	Norm (DIN)	International Standard Designation	Standard ISO	European Standard Designation	EURO-NORM
SECo 112/100	2.4135	17410	RECo...	IEC 404-8-1		
St 02 Z	1.0226	17162	02	3575	Fe P 02 G	142
St 03 Z	1.0350	17162	03	3575	Fe P 03 G	142
St 04 Z	1.0355	17162	04	3575	Fe P 04 G	142
St 14	1.0338	1623 T1	Cr 4	3574	Fe P 04	130
St 33	1.0035	17100	Fe 310-0	630	Fe 310-0	25
St 37-2	1.0037	17100	Fe 360-B	630	FeE 235 B	25
St 37-2G	1.0037 G	1623 T2	CR 220	4997		
St 37-3 N	1.0116	17100	Fe 360-D	630	Fe 360-D/FeE 235 D	25
St 37-3 U		17100	Fe 360-C	630	Fe 360-C/FeE 235 C	25
St 44			Fe 430-A		Fe 430-A	
St 44-2	1.0044	17100	Fe 430-B	630	Fe 430-B/FeE 275 B	25
St 44-3G	1.0044 G	1623 T2	CR 250	4997		
St 44-3 N	1.0144	17100	Fe 430-D	630	Fe 430-D/FeE 275 D	25
St 44-3 U	1.0144	17100	Fe 430-C	630	Fe 430-C/FeE 275 C	25
St 50-2	1.0050	17100	Fe 490	1052	Fe 490-2	25
St 52-3	1.0570	17100	Fe 510 C (D)	630	FeE 355-C (D)	25
St 52-3G	1.0570 G	1623 T2	CR 320	4997		
St 52-3 N	1.0570	17100	Fe 510-D	630	Fe 510-D/FeE 355 D	25
St 52-3 U	1.0570	17100	Fe 510-C	630	Fe 510-C/FeE 355 C	25
St 60-2	1.0060	17100	Fe 590	1052	Fe 590-2	25
St 70-2	1.0070	17100	Fe 690	1052	Fe 690-2	25
St 1080/1230	1.8862		wire rod	6934	rod Fe 1230	138/7
St 1470/1670	1.0662		wire 1670	6934	wire Fe 1670 D	138/4
St 1570/1770	1.0664		wire 1770	6934	wire Fe 1770 D	138/4
					strand Fe 7 SD 1770	138/6
StE 255	1.0461	17102	P 255 TN	9328-4	Fe E 255 KG	113
StE 285	1.0486	17102	P 285 TN	9328-4	Fe E 285 KG	113
StE 315	1.0505	17102	P 315 TN	9328-4	Fe E 315 KG	113
StE 320 Z	1.0250	17162	32 Z	4998	Fe E 320 G	147
StE 320.7	1.0409	17172				
StE 355	1.0562	17102	E 355 DD	4950	Fe E 355 KG	113
			P 355 TN	9328		
StE 360.7 TM	1.0578	17172				
StE 380	1.8900	17102	E 390 DD	4950	Fe E 380 KG	113
			PL 390 TN	9328		
StE 385.7 TM	1.8970	17172				
StE 415.7 TM	1.8973	17172				
StE 420	1.8902	17102	E 420 DD	4950	FeE 420 KG	113
StE 445.7 TM	1.8975	17172				
StE 460	1.8905	17102	E 460 DD	4950	Fe E 460 KG	113
StE 480.7 TM	1.8977	17172				
StE 500 V	1.8909	E 500 E	E 500 E	4950	Fe E 500 V	137
StE 690 V	1.8928	E 690 E	E 690 E	4950	Fe E 690 V	137
TStE 355	1.0566	17102			Fe E 355 KT	113
U 10 S 10	1.0702	17111				
UQSt 36	1.0204	17111	A 3 R-CC 11 X	4954	CB 10 FU KD	119
UQSt 38	1.0224	17111	A 4 R	4954		
USt 37-2G	1.0036 G	1623 T2	CR 220	4997	Fe E 235 BFU	25
USt 37-2	1.0036	17100	Fe 360 B	630	Fe 360-BFU	25

Table 1

USA Designation Grade	Standard ASTM	UK Designation	Standard BS	France Designation	Standard NFA	Japan Designation	Standard JIS-G
	A 527	Z 2	2989	GC	36-321		3302
	A 528	Z 3	2989	GE	36-321		3302
	(A 525)	Z 4	2989	GE	36-321		3302
DQAK	A 620	CR 1	1449/1	ES	36-401		
A	A 283			A 33	35-501	SS 330	3101
33, C	A 283, 570	40 B	4360	E 24-2	35-501		
	A 611						
(D)	A 284	40 D	4360	E 24-4	35-501		
		40 C	4360	E 24-3	35-501		
(40)	A 570	43 B	4360	E 28-2	35-501		
				E 28-4	35-501		
(42)	(A 572)	43 D	4360	E 28-4	35-501		
(42)	(A 572)	43 C	4360	E 28-3	35-501		
1025	A 576			A 50-2	35-501		
D	A 611			E 36-4	35-501		
50 W	A 709	50 D	4360	E 36-3	35-501		
50	A 572	50 C	4360	E 36-3	35-501		
1035	A 576			A 60-2	35-501	(SM 58)	3106
1045	A 576			A 70-2	35-501		
(i) wire	A 648	RR-1230	4486				
BA/WA	A 421	wire-1670-D	5896				
III	A 648	wire-1770-D	5896				
	A 416	1770-D	5896/3				
		40 DD	4360			SM 41	3106
A	633	43 DD	4360				
	A 808						
	(A 446)	(Z xx)	2989	C 320	36-322		
X 46	API 5 L			E 355	36-201	SM 50	3106
C	A 633						
X 52	API 5 L			E 375	36-201		
X 56	API 5 L						
X 60	API 5 L						
E	A 633			E 420	36-201		
B	A 678						
X 65	API 5 L						
65	A 812			E 460	36-201	SM 58	3106
X 70	API 5 L						
C	A 678			E 500 T	36-204		
80	A 812			E 690 T	36-204		
		50 E	4360				
1110	A 29			13 MF 4	35-562		
						SWRCH 12 A	3507
		RCR 3723	1449	E 24-2	35-501		
C	A 283				35-501		

737

Table 1 (*Continued*)

Steel grade referred Designation	Werkstoff Nr.	Norm (DIN)	International Standard Designation	Standard ISO	European Standard Designation	EURO-NORM
				IEC		
processed						
V 250-35 A	1.0800	46 400 T1	250-35-A5	404-8-4	FeV250-35 HA	106
V 270-50 A	1.0806	46 400 T1	270-50-A5	404-8-4	FeV270-50 HA	106
V 350-65-A	1.0810	46 400 T1	350-65-A5	404-8-4	FeV350-65 HA	106
alloy						
VE 340-50	1.0841	46 400 T4	340-50-E5	404-8-2	Fe340-50 HE	165
VE 390-65	1.0846	46 400 T4	390-65-E5	404-8-2	Fe390-65 HE	165
nonoriented						
VH 660-50	1.0361	46 400 T2	660-50-D5	404-8-3	(FeV280-50)	126
VH 800-65	1.0364	46 400 T2	800-65-D5	404-8-3	(FeV330-65)	126
grain oriented						
VM 97-30 N	1.0861	46 400 T3	097-30-N5	404-8-5	FeM 097-30 N	107
VM 111-30 P	1.0881	46 400 T3	110-30-P5	404-8-5	FeM 111-30 S	
VM 111-35 N	1.0856	46 400 T3		404-8-5	FeM 111-35 N	107
VM 140-30 S	1.0862	46 400 T3		404-8-5	FeM 140-30 S	
X 1 NiCrMoCuN 25 20 5	1.4539	SEW 400	A 4	683/13		
X 2 CrNi 19 9	1.4316	8556	19.9 L	3581		
X 2 CrNi 19 11	1.4306	17 440	10	683/13	X2CrNi 18 10	88
		17 441	X2CrNi 18 10	9328 5		
X 2 CrNiMo 17 13 2	1.4404	17 440	19	683/13	X2CrNiMo 17 13 2	88
(X 2 CrNiMo 18 10)		17 441	X2CrNiMo 17 12	9328.5		
X 2 CrNiMo 18 14 3	1.4435	17 440	19a	683/13	X2 CrNiMo 17 13 3	88
			X2CrNiMo 17 13	9328-5		
X 2 CrNiMoN 17 13 3	1.4429	17 440	X2CrNiMoN 17 13	9328-5	X2CrNiMoN 17 13 3	(88)
(X 2 CrNiMoN 18 13)			19 aN	683/13	X2CrNiMoN 18 13 3 KD	119/5
X 2 CrNiMoN 17 13 5	1.4439	17 440			X2CrNiMoN 17 13 5	88
X 2 NiCoMo 18 8 5	1.6359	aero			FE-PA 95	AECMA
X 2 NiCr 18 16	1.4321				X6NiCr 18 16 KD	119
X 3 CrNiMoN 18 14	1.6967				X2CrNiMoN 17 13 3	88
X3 CrNiN 18 10	1.6907				X2CrNiN 18 10	88
X 4 CrNi 13 4	1.4313	SEW 400				
X 4 CrNiMo 16 5	1.4418	SEW 400				
X 5 CrNi 18 10	1.4301	17 440	(D 21) 11	683/13	X5CrNi 18 10	88
			X5CrNi 18 9	9328-5		
X 5 CrNi 18 12	1.4303	17 440	(D 23) 13	683/13	X8CrNi 18 12	88
		1654	X5CrNi 18 12 E	DIS 4954	X8CrNi 18 12 KD	119/5
X 5 CrNi 19 9	1.4302	17 145	19.9	3581	X6CrNi 20 10 KE	144
		17 440	20	683/13	X5CrNiMo 17 12 2	88
X 5 CrNiMo 17 12 2	1.4401	17 441	X5CrNiMo 17 12	9328-5		
		1654 T5	X5CrNiMo 17 12 E	DIS 4954	X6CrNiMo 17 12 2 KD	119
X 6 Cr 13	1.4000	17 440	1	683/13	X6Cr 13	88
X 6 Cr 17	1.4016	17 440	8	683/13	X6Cr 17	88
		1654 T5	X6Cr 17 E(D1)	4954	X8Cr 17 KD	119
X 6 CrAl 13	1.4002	17 440	2	683/13	X6CrAl 13	88
(X 7 CrAl 13)						
X 6 CrMo 4	1.2341	(17 350)	5 CrMo 4 (24)	4957	5 CrMo 16	96
X 6 CrNi 18 11	1.4948	17 460	P 48	2604/4		
X 6 CrNiMo 17 13	1.4919	17 460	P 63, X5CrNiMo 17 13	2604/4		
		17 440	X6CrNiMoTi 17 12	9328-5	X6CrNiMoTi 17 12 2	88
X 6 CrNiMoTi 17 12 2	1.4571		(D 30) 21	683/13	X6CrNiMoTi 17 12 2 KD	
formerly X 10 ...		1654 T5	X6CrNiMoTi 17 12 E	DIS 4954		119

Table 1

USA		UK		France		Japan	
Designation Grade	Standard ASTM	Designation	Standard BS	Designation	Standard NFA	Designation	Standard JIS-G
(36 F 320 M)	A 677 M	250-35-A5	6404 − 8.4	FeV250-35HA	C28-900		
(47 F 370 M)	A 677 M	270-50-A5	− 8.4	FeV270-50HA	C28-900		
(64 F 459 M)	A 677 M	350-65-A5	− 8.4	FeV350-65HA	C28-900		
(47 S 392 M)	A 683 M	340-50-E5	− 8.2	FeV340-50HE	C28-926		
(64 S 428 M)	A 683 M	390-65-E5	− 8.2	FeV390-65HE	C28-926		
		660-50-D5	− 8.3	FeV660-50HD	28-925		
		800-65-D5	− 8.3	FeV800-65HD	28-925	65 A 800 C	2552
30 G 128 M	A 665 M	30 M 5	601/2	FeM 097-30 N	C28-920		
				FeM 111-30 S	C28-920	30 P 110 C	2553
35 G 146 M	A 665 M	35 M 6	601/2	FeM 111-35 N	C28-920		
30 H 183 M	A 725 M			FeM 140-30 S	C28-920	30 G 140 C	2553
				Z1NCDU 25 20	35-584		
			1501/3	Z2 CN 20.10	35-583		
304 L	A 167	304 S11	1449/2	Z2 CN 18.10	35-574 35-573	SUS 304 L	
316 L	A 240	316 S11	1449/2	Z2 CND 17.12	37-574		
316 L	A 167	316 S 61	1501/3				
317 L	A 167 A 240	316 S 13	1501/3 1449/2	Z2 CND 17.13 Az	35-574	SUS 317 L	
316 LN	A 167	316 S 12	1449	Z2 CND 17.13 Az	35-582	SUS 316 LN	
317 L	A 240	316 S 63	1501/3				
6512, 6520	AMS	5212	DTD	EZ2 NKD 18		AIR 9160/6	
XM-15	A 167				35-573		
316 LN	A 276	317 S 12 (N)	970/4	Z2CND17.13A2	35-582		
304 LN	A 276	304 S 12 (N)	970/4	Z2CN18.10A2	35-582		
F 6 NM	A 182			Z6 CND 16.04	35-573		
	A 240	304 S 15	1449	Z6 CN 18.09	35-574		
304	A 276	304 S 15	970/4	Z5 CN 18.09	36-209	SUS 304	
	A 240	305 S 19		Z4 CN 18.12	35-577	SUS 305 J1	4308
						SUS 305 J1	4315
(F 304)	A 182			Z5 CN 18.09	(36-209)		
	A 240			Z6 CND 17.11	35-574		
316		316 S 31	970/1		35-573	SUS 316	
	A 167			Z6 CND 17.11	35-577		
403	A 276	403 S 17	1449/2	Z6 C 13	35-574	SUS 403	
430	A 176 A 167	430 S 17	1449/2	Z8 C 17	35-577 35-574	SUS 430	
405	A 176	405 S 17	1449/2	Z6 CA 13	35-574	SUS 405	
F 304	A 182			Z6 CN 18.09	35-574		
F 316	A 182			Z6 CND 17.12	35-574		
				Z6 CNDT 17.12	35-574		
316 Ti				Z6 CNDT 17.12	35-574		

Table 1 (*Continued*)

Steel grade referred Designation	Werkstoff Nr.	Norm (DIN)	International Standard Designation	Standard ISO	European Standard Designation	EURO-NORM
X 6 CrNiNb 18 10 formerly X 10 ...	1.4550	17 440	X6CrNiNb 18 10 (50)	9328-5	X6CrNiNb 18 10	88
X 6 CrNiTi 18 10 formerly X 10 ...	1.4541	17 440	X6CrNiTi 18 10 15	9328-5 683/13	X6CrNiTi 18 10	88
X 8 Cr 17	1.4015	8556				
X 8 CrNiNb 16 13	1.4961	17 459	TS 56	2604/2		
X 8 CrNiMoVNb 16 16	1.4988	17 459	TS 67	2604/2		
X 8 CrNiTi 18 10	1.4941	17 460			X10CrNiTi 18 10	95
X 8 Ni 9	1.5662	17 280	X8Ni 9 (P 45)	9328	FeE 490 Ni 36	129
X 8 NiCrAlTi 32 21	1.4959	17 460	H 18	4955	X10NiCrAlTi 32 20	95
X 10 Cr 13	1.4006	17 440 1654	3 X12Cr 13 E (D10)	683/13 4954	X10Cr 13 X12Cr 13 KD	88 119
X 10 CrAl 13	1.4724	SEW 470	H 3	4955	X10CrSiAl 13	95
X 10 CrAl 18	1.4742	SEW 470	H 5	4955	X10Cr 18	95
X 10 CrAl 24	1.4762	SEW 470	H 6	4955	X10CrAlSi 24	95
X 10 CrNiMoTi 18 10: X 6 CrNiMoTi 17 12 2		see				
			X8NiCrAlTi 32 21 TQ	9328-5	X10NiCrAlTi 32 20	95
X 10 NiCrAlTi 32 20	1.4876	SEW 470	H 18	4955	X10NiCrAlTi 32 20	95
X 11 CrNiMo 12	1.4938				FE-PM 37	AECMA
X 12 CrCoNi 21 20	1.4971		X12CrCoNiMoWNb 21 20 20 (12)	683/15	X12CrCoNiMoWNb 21 20 20	90
X 12 CrMo 9 1	1.7386	17 176	TS 38	2604/2		
X 12 CrMoS 17	1.4104	17 440	9a	683/13	X14CrS 17	88
X 12 CrNi 17 7	1.4310	17 224			X12CrNi 17 7	151
X 12 CrNi 25 21	1.4845	SEW 470	H 15	4955	(X6CrNiSi 25 20)	95
X 12 CrNiTi 18 9	1.4878	SEW 470	(H 11)	4955	X10CrNiTi 18 10	95
X 15 CrNiSi 20 12	1.4828	SEW 470	H 13	4955	X15CrNiSi 20 12	95
X 15 CrNiSi 25 20	1.4841	SEW 470	H 16	4955	X15CrNiSi 25 20	95
X 20 Cr 13	1.4021	17 440	4	683/13	X20Cr 13	88
X 20 CrMoV 12 1	1.4922	17 175 17 243	TS 40 F40	2604/2 2604/1		
X 20 CrNi 17 2	1.4057	17 440 1654	9 b D12, X19CrNi16 2E	683/13 4954	X21CrNi 17	88
X 21 Cr 13	1.2082	(17 350)	20 Cr 13 (27)	4957	X21 Cr 13	96
X 30 WCrV 5 3	1.2567		30 WCrV 5 (H 7)	4957	X30 WCrV 5 3	96
X 30 WCrV 9 3	1.2581		30 WCrV 9 (H 8)	4957	X30 WCrV 9 3	96
X 32 CrMoV 3 3	1.2365	17 350	30 CrMoV 3 (H 4)	4957	30 CrMoV 12 11	96
X 32 NiCoCrMo 8 4	1.6974					
X 38 Cr 13	1.4031	17 440	(6)		X40Cr 13	88
X 38 CrMoV 5 1	1.2343	17 350	35 CrMoV 5 (H 5)	4957	X37 CrMoV 5 1	96
X 40 CrMoV 5 1	1.2344	(17 350)	40 CrMoV 5 (H 6)	4957	X40CrMoV 5 1 1	96
X 41 CrMoV 5 1	1.7783					
X 42 Cr 13	1.2083	(17 350)	40 Cr 13 (29)	4957	X41 Cr 13	96
X 45 CoCrWV 5 5 5	1.2678					
X 45 Cr 13	1.3541	17 230	20	683/17	X45Cr 13	94
X 45 CrMoV 15	1.4116	17 440				
X 45 CrNiW 18 9	1.4873	17 480	X45CrNiW 18 9 (6)	683/15	X45CrNiW 18 9	90

Table 1

USA		UK		France		Japan	
Designation Grade	Standard ASTM	Designation	Standard BS	Designation	Standard NFA	Designation	Standard JIS-G
347	A 240	347 S 31	1449/2	Z6 CNNb 18.10	36-209 35-574	SUS 347	
321 F 321	A 240 A 276	321 S 12	970/4	Z6 CNT 18.10	35-574	SUS 321	
430	A 176 A 240	430 S 15	1554	Z8 C 17	35-583		
	A 353 A 553	509	1501/2	9 Ni-490 Z8 NC 32.21	36-208 35-578		
410	A 240 A 176	410 S 21	1449	Z12 C 13	35-573 35-577	SUS 410 SUS 410	4304 4315
(405) (430)	A 240 A 240	405 S 17 (430 S 15)	1449/2 1449	Z6 CA 13 Z10 CAS 18 Z10 CAS 24	35-573 35-578 35-578		
N 08800	B 407 B 408	NA 15		Z8 NC 32.21	35-578		
HEV 1	J 775						
T 9 C 12	A 213 A 217	629	3604	TUZ 10 CD9	49-213		
310 321	A 240 A 240	(310 S 24) 321 S 20	1449/2 1449 4942	Z12 CN 17.07 Z12 CN 25.20 Z6 CNT 18.12 Z15 CNS 20.12 Z12 CNS 25.20	35-587 35-578 35-578 35-578 35-578		
420	A 314	420 S 29	970	Z20 C 13	35-574	SUS 420 J1	
616 Ev 4 (21-12 N)	A 565 SAE	762	3604				
431	A 314	431 S 29	970	Z15 CN 16.02	35-574	SUS 431	4319
S 42010		420 S 29	970	Z20 C 13	35-557		
H 21 H 21	A 681 A 681	BH 21	4659	Z32 WCV 5 Z30 WCV 9	35-590 35-590		
H 10 6524, 6526	A 681 AMS	BH 10		T32 DCV 28 Z40 C 14	35-590 35-575	SKD 7	4404
H 11	A 681	BH 11	4659	Z38 CDV 5	35-590	SKD 6	4404
H 13 6485, 6487	A 681 AMS	BH 13	4659	Z40 CDV 5 E40 CDV 20 Z40 C 14	35-590 AIR9160/6 35-590 35-595	SKD 61	4404
H 19	A 681						
EV 9	SAE	331 S 40	970/4		35-579		

Table 1 (*Continued*)

Steel grade referred Designation	Werkstoff Nr.	Norm (DIN)	International Standard Designation	Standard ISO	European Standard Designation	EURO-NORM
X 45 CrSi 9 3	1.4718	17 480	X45CrSi 9 3 (1)	683/15	X45CrSi 8	90
X 45 NiCrMo 4	1.2767	17 350	40 NiCrMoV 4 (H 1)	4957	40 NiCrMoV 16	96
X 46 Cr 13	1.4034	17 440			X45Cr 13	88
X 50 CrMnNiNbN 21 9	1.4882	17 480	X50CrMnNiNbN 21 9	DP 683/15		
X 53 CrMnNiN 21 9	1.4871	17 480	X53CrMnNiN 21 9	683/15	X53CrMnNiN 21 9	90
X 55 CrMnNiN 20 8	1.4875	17 480	X55CrMnNiN 20 8	DP 683/15		
X60CrMnMoVNbN21 10	1.4785	17 480				
X 75 WCrV 18 4 1	1.3557	17 230	32	683/17	X75WCrV 18 4 1	94
X 82 WMoCrV 6 5 4	1.3553	17 230	31	683/17	X80WMoCrV 6 5 4	94
X 85 CrMoV 18 2	1.4748	17 480	X85CrMoV 18 2 (3)	683/15		
X 90 CrMoV 18	1.4112	SEW 400				
X 100 CrMoV 5 1	1.2363		100 CrMoV 5 1 (20)	4957	X100 CrMoV 5 1	96
X 102 CrMo 17	1.3543	17 230	21	683/17	X100CrMo 17	94
X 105 CrMo 17	1.4125	SEW 400				
X 155 CrVMo 12 1	1.2379	17 350	160 CrMoV 12 (21)	4957	X160 CrMoV 12 1	96
X 210 Cr 12	1.2080	17 350	210 Cr 12 (22)	4957	X210 Cr 12	96
X 210 CrW 12	1.2436	17 350	210 CrW 12 (23)	4957	X210 CrW 12 1	96
ZStE 260	1.0480	SEW 093	FeE 275	6930	FeE 275 HF	149
ZStE 300	1.0489	SEW 093				
ZStE 340	1.0548	SEW 093	FeE 355	6930	FeE 355 HF	149
ZStE 420	1.0556	SEW 093	FeE 420	6930	FeE 420 HF	149

Table 1   743

USA		UK		France		Japan	
Designation Grade	Standard ASTM	Designation	Standard BS	Designation	Standard NFA	Designation	Standard JIS-G
HNV 3 S65007	J 775	401 S 45	970/4	Z 45 CS 9	35-579	SUH 1	4311
				Z 50 CMNNb 21.09	35-579		
EV 8	SAE	349 S 54	970/4	Z 52 CMN 21.09	35-579	SUH 35	4311
EV 12 S63012	SAE					SUH 35	4311
XEV-F	SAE						
T 1	A 600	BT 1	4659				
M 2 (r.C)	A 600	BM 2	4659	Z 85 WDCV 6	35-565		
				Z 85 CDMV 18.02	35-579		
440 B	A 314						
A 2	A 681	BA 2		Z 100 CDV 5	35-590	SKD 12	4403
440 C	A 276			Z 100 CD 17	35-565		
440 C	A 314			Z 100 CD 17	35-575		
D 2	A 681	BD 2		Z 160 CDV 12	35-590	SKD 11	4403
D 3	A 681	BD 3		Z 200 C 12	35-590	SKD 1 (2)	4403
Gr 45 cl 2	A 607		1449	E 275 D	36-203		
		SC 40 F 30	1449/1	E 335 D	36-203		
Gr 50 cl 2	A 607	SCU 3 F 35	1449/1	E 390 D	36-203		
Gr 60 cl 2	A 607			E 430 D	36-203		

Table 2. Brief definition of the scope and contents of the DIN Standards mentioned in Volume 1 and Volume 2 of "Werkstoffkunde Stahl" and listed in column 1 of Table 1.

DIN	Part.	Ed.	Short Title
A) *German Standards*			
267	1	08.82	Fasteners
	11	01.80	–, hot Dip Galvanized
	13	03.80	–; Corrosion Resistant (see 898 DIN-ISO, too)
488	1	09.84	Reinforcing Steels
	2	06.86	–; Bars, Dimensions
	3	06.86	–; Bars, Testing
	4	06.86	–; Fabries, Wire
	5	06.86	–; Fabries, Testing
	6	06.86	–; Quality Control
	7	06.86	–; Welding
685	1	11.81	Round Steel link Chains
	2		–; Safety Requirements
	3		–; Testing
	4		–; Marking, Certificate
	5		–; Utilization
766		01.86	Round Steel link Chains
779		12.80	Shaped Wires for locked Coil Ropes
898 DIN-ISO	1	08.85	Fasteners, Bolts, Screws, Studs
1013	1	11.76	Steel Bars, Round, Hot Rolled, Dimensions
	2	11.76	–; for Special Purposes
1045		07.88	Concrete, Reinforced Concrete, Design, Construction
1324	1	05.88	Electromagnetic Field
	2	05.88	–; Material Quantities
	3	05.88	–; Electromagnetic Waves
1599		08.80	Identification Markings
1614	1	03.86	Hot rolled Sheet and Strip
	2	03.86	–; Mild unalloyed Steels for immediate Cold Forming
1615		10.84	Welded Circular Tubes, unalloyed
1616		10.84	Tinplate and Blackplate, see EN 10203, too
1623			Cold rolled Sheet and Strip
	1	02.83	–; Unalloyed Steels for Cold Forming, see EN 10130
	2	02.83	–; General Purposes Structural Steels
	3	01.87	–; Mild Unalloyed Steels for vitreous Enamelling
1624		06.87	Cold rolled narrow Strips of Mild Unalloyed Steels
1626 to 1630			Circular Unalloyed Steel Tubes
1626		10.84	–; Welded; Special Requirements for (see EN 10217, too)
1628		10.84	–; Welded; High Performance
1629		10.84	–; Seamless; Subject to Special Requirements (see EN 10216, too)
1630		10.84	–; Seamless; Subject to Special Requirements
1651		04.88	Free Cutting Steels
1652			Bright Steels
	1	11.90	–; General
	2	11.90	–; For General Structural Purposes

Table 2 (*Continued*)

DIN	Part.	Ed.	Short Title
	3	11.90	–; Cuse Hardening Steels
	4	11.90	–; For Quenching and Tempering
1654			Cold Heading and Cold Extruding Steels
	1	10.89	–; General
	2	10.89	–; Killed Unalloyed Steels, not for Heat Treatment
	3	10.89	–; Case Hardening Steels
	4	10.89	–; Steels for Quenching and Tempering
	5	10.89	–; Stainless Steels
1662		–	CrNi Steels; Standard from 1932, withdrawn. See DIN 17 440 and further
1681		06.85	Cast Steels with improved Weldability: See DIN 17 182
1715	1	11.83	Thermostat Metal Sheet
	2		–; Specific Thermal Curvature
1736	1	08.85	Nickel Alloys; Welding Filh Metals
	2	08.85	–; Sample, Test Pieces
1910	1	07.83	Welding Processes Classification
	2	08.77	–; Welding of Metals
1912	1	06.76	Graphical Representation of Welded joints
	2	09.77	–; Working Positions, Slope, Rotation
1913	1	06.84	Covered Electrodes, Low Alloyed, Classification
2078		03.78	Steel Wires for Wire Ropes
2088		07.69	Helical Round Wire Springs
2089	1	12.88	Helical Round Wire Springs; Compression
2310	2	11.87	Thermal Cutting; Cut Surface Quality
3051			Steel Wire Ropes
	1	03.72	–; Characteristics
	2	03.72	–; Types of Ropes
	3	03.72	–; Calculation
	4	03.72	–; Technical Delivery Conditions
4099		11.85	Welding of Reinforcing Steels
4227			Prestressed Concrete
	1	07.88	–; Structural Members
	2	05.84	–; Partially Prestressed Members
4987		03.87	Indexable (throw away) Inserts
5684			Round Steel link Chains for Lifting
	1	05.84	–; Grade 5, Calibrated/Tested
	2	05.84	–; Grade 6, „
	3	05.84	–; Grade 8, „
6580		10.85	Terminology of Chip Removing (ISO 3002/3)
6581		10.85	Terminology of Chip Removing (ISO 3002/1)
8505	3	01.83	Soldering and Brazing; Classification of Process
8524			Defects in Metal Welds, Classification
	1	07.86	–; Fusion Welds
	2	03.79	–; Pressure Weldes Joints
	3	08.75	–; Cracks
8528	1	06.73	Weldability; Definitions
	2	03.75	–; Fusion Welding of Structural Steels

Table 2 (*Continued*)

DIN	Part.	Ed.	Short Title
8556	1	05.86	Stainless Steels; Filler Metals for Welding
8580		07.85	Manufacturing; Production Processes, Definitions, Classification
8583	2	05.70	Manufacturing Processes; Forming under Compressive Conditions
8588		06.85	Severing Processes
15 020	1	02.84	Lifting Appliances; Principles for Rope Drives
17 014	1	08.88	Heat Treatment of Ferrous Materials
17 021	1	02.76	Heat Treatment of Iron and Steel; Material Selection
17 100		01.80	Steels for General Structure Purposes – Withdrawn, see now DIN EN 10025
17 102		10.83	Weldable Structural Steels, normalized, Fine Grain
17 103		10.89	Forgings of Weldable Fine Grain Steels
17 111		09.80	Unalloyed Steels for Bolts, Nuts and Rivets low Carbon
17 115		02.87	Weldes Round link Chains, Steels for Steel Tubes for Structural Network, welded
17 115		06.84	–; Cold formed, Rectangular
17 120		06.84	–; Round
17 121		06.84	–; Seamless, Round
17 122		03.78	Conductor Contact Rails for Electric Traction
			Round Steel Tubes for Engineering:
17 123		05.86	–; Fine Grain, Welded
17 124		05.86	–; Fine Grain, Seamless
17 125		05.86	–; (Hollow Sections), Square and rectangular
17 140	1	03.83	Wire Rod for Cold Forming, Base and Carbon Steels
17 155		10.83	Steels for Elevated Temperature Service; Plate and Strip, see EN 10028 too
			Flat Products; Hot-Dip Zinc coated Strip and Sheet
17 162	1	09.77	–; Mild Unalloyed Steels – Withdrawn, see DIN EN 10142
17 162 E	2	04.88	–; Structural Steels – see DIN EN 10147
17 163		03.88	Flat Products; Cold Rolled Strip and Sheet, electrolytically Zinc coated
17 172		05.78	Pipes for Pipelines
17 173		02.85	Round Tubes for low Temperatures, –; Seamless
17 174		02.85	–; Welded
17 175		05.79	Tubes for Elevated Temperatures, –; Seamless
17 177		05.79	–; Electrically Welded
17 178		05.86	Round Tubes of Fine Grain Steels, –; Welded
17 179		05.86	–; Seamless

Table 2 (*Continued*)

DIN	Part.	Ed.	Short Title
17 182		06.85	Steel Castings with improved Weldability and Toughness
17 200		03.87	Steels for Quenching and Tempering, see EN 10 083
17 210		09.88	Case Hardening Steels
17 211		04.87	Nitriding Steels
17 212		08.72	Flame and Induction Hardening Steels
17 221		12.88	Hot Rolled Steels for Quenched and Tempered Springs
17 222		08.79	Cold Rolled Steel Strip for Springs
17 223	1	12.84	Round Spring Steel Wire –; Unalloyed, Patented Cold drawn
17 223	2	09.90	–; Quenched and Tempered; Unalloyed and alloyed Steels
17 224		02.82	Wire and Strip for Stainless Spring Steels
17 230		09.80	Ball and Roller Bearing Steels
17 240		07.76	Steels for Bolts and Screws for Elevated Temperature Service
17 243		01.87	Forgings and Bars of Weldable Steels, for Elevated Temperature Service
17 245		12.87	Ferritic Steel Castings, for Elevated Temperature Service
17 280		07.85	Steels for Low-Temperature Service, Plates, Sections, Bars and Forgings
17 350		10.80	Tool Steels
17 405		09.79	Soft Magnetic Materials – see IEC 404-8-6
17 410		05.77	Permanent Magnet Materials
17 440		07.85	Stainless Steels –; Plates, Wire, Bars, Forgings
17 441		07.85	–; Cold Rolled Strip and Sheet
17 443		04.86	–; Surgical Implants
17 455		07.85	–; Welded Circular Tubes
17 456		07.85	–; Seamless Circular Tubes
17 457		07.85	–; Welded Circular Tubes, Austenitic
17 458		07.85	–; Seamless Cirular Tubes, Austenitic
17 470		10.84	Ni-Cr and Fe-NiCr-Alloys for Electrical Heating
17 480		01.92	Valve Materials
17 745		01.83	Ni-Fe Alloys with additional Elements
21 254	1	04.80	Winding Ropes
22 252		09.83	Steel Chains in Mining
32 525	1	12.81	Testing Filler Metals –; Arc Welded Pieces
	2	08.79	–; Test Pieces with Low Heat Input
41 301		07.67	Electrosheet; Magnetic Materials for Transformers
41 302	1	05.86	Laminations for Transducers and Inductors (IEC 740 modified)
	2	05.86	–; Small Transformers and Clokes
43 101	1	02.75	Rolling Stock for Electric Traction

Table 2 (*Continued*)

DIN	Part.	Ed.	Short Title
46 400	1	04.83	Magnetic Steel Sheet and Strip –; Cold Rolled, non oriented (IEC 404-8-4)
	2	01.87	–; Cold Rolled, unalloyed, non-oriented, Semi-processed
	3	04.89	–; Grain oriented (IEC 404-8-7)
	4	01.87	–; Alloyed steels, Cold Rolled, nonoriented, Semi-processed
48 200	7	04.81	Steel Wires for Stranded Conductors –; Copper Clod
	8	04.77	–; Aluminium Clod
48 300		04.81	Wires for Telecommunication Lines
50 021		06.88	Corrosion; Sodium-clorid Spray Test
50 049		08.86	Documents on Materials Testing –; see EN 10021, too
50 100		02.78	Materials Testing; Continuous Vibration Test
50 103	1	03.84	Rockwell Hardness Test –; (C, A, B, F Scales)
	2	03.84	–; (N and T Scales)
50 106		12.78	Metallic Materials Testing; Compression
50 110		02.62	Grey Cast Iron; Transverse Bending Test
			Metallic Materials Testing
50 115		02.75	–; Notched Impact Test
50 118		01.82	–; Creep Rupture Test
50 125		04.91	–; Tensile Test, see DIN EN 10002 too
50 133		02.85	–; Vickers Hardness Test; Range HV 0.2 to 100
50 145		05.75	–; Tensile Test
50 150		12.76	–; Steel; Conversion Tables for Hardness and Tensile Strength
50 190	1	11.78	Determination of Effective Hardening Depth; –; Carburized and Hardened Cases
	2	03.79	–; After Flame or Induction Hardening
	3	03.79	–; After Nitriding
50 191		09.87	Hardenability Test by End Quenching
50 192		05.77	Determination of Depth of Decarburization
50 320		12.79	Wear Processes; Systematic Analysis
50 321		12.79	Wear Quantities
50 351		03.85	Brinell Hardness Test
50 462	1	09.86	Determination of Magnetic Properties –; General
	2	08.86	–; Total Losses
	3	08.86	–; Magnetic Polorisation and Flux Density
	4	07.79	–; Total Loss; Bridge Method
	5	07.79	–; Commutation and Remanent Induction
50 465		09.75	Determination of Induction and Permeability
50 466		09.75	Determination of Complex Permeability
50 470		09.80	Determination of Demagnetization Curve –; Induction Method

DIN	Part.	Ed.	Short Title
50 471		09.80	–; Magnetostatic Method
50 472		03.81	Determination of Magnetic Flux, Permanent Magnet
50 601		08.85	Determination of Ferritic or Austenitic Grain Size of Steels
50 602		09.85	Nonmetallic Inclusions; Microscopic Examination Using Standard Diagrams
50 914		06.84	Intergranular Corrosion –; Stainless Steels; Strauss-Test
50 921		10.84	–; Austenitic steels; local corrosion Huey-Test
51 212		09.78	Wire; torsion testing
51 226		12.77	Testing modines for creep test/tensile stress
59 115		11.72	Steel wire rod for bolts and nuts; dimensions
59 130		09.78	Steel bars for bolts and rivets; hot rolled; dimensions

B) European Standards

EN 10002		04.91	Metallic materials, tensile testing; Part 1: Method of testing (at ambient temperature)
EN 10025		01.91	Hot-rolled products of non-alloy structural steels; technical delivery conditions
EN 10028		1991	Flat products made of steels for pressure purposes Part 1: General requirements Part 2: Unalloyed and alloy steels with specified elevated temperature properties Part 3: Weldable fine grain steels, normalized
EN 10045		04.91	Metallic materials; Charpy impact test; Part 1: Test method
EN 10083		1991	Quenched and tempered steels; Part 1: Technical delivery conditions for special steels; Part 2: Technical delivery conditions for unalloyed quality steels
EN 10113		1991	Hot rolled products in weldable fine grain structural steels; Part 1: General delivery conditions Part 2: Delivery conditions for normalized steels Part 3: Delivery conditions for thermomechanical rolled steels
EN 10130		1991	Cold rolled low carbon steel flat products for cold forming; technical delivery conditions
EN 10142		03.91	Continuously hot-dip zinc coated mild steel sheet and strip for cold forming; technical delivery conditions
EN 10147		1991	Continuously hot-dip zinc coated structural sheet and strip; technical delivery conditions
EN 10203		1991	Cold reduced electrolytic tinplate
EN 10205		1991	Cold reduced black plate
EN 10207		1991	Simple pressure vessels; plates, strips, bars
pr EN 10208		1991	Steel pipes for pipelines
pr EN 10210		1990	Hot finished structural hollow sections of non alloy and fine grain structural steels; Part 1: Technical delivery conditions
pr EN 10213		1991	Steel castings for pressure purposes

Table 2 (*Continued*)

DIN	Part.	Ed.	Short Title
pr EN 10216		1991	Seamless steel tubes for pressure purposes; technical delivery conditions; Part 1: Non-alloy steels with specific room temperature properties
pr EN 10217		1991	Welded steel tubes for pressure purposes; technical delivery conditions; Part 1: Non-alloy steels with specified room temperature properties

Table 3. International and European standards as well as national standards of some non-German countries which are comparable with DIN. (The standards are identified just by their number; for abbreviated titles of the German standards refer to Table 2).

Preliminary remarks: EURONORM's (EU) and European national steel standards are replaced by European standards (EN). These are numbered 10 000 and higher and edited in English, French and German versions.
DIN 1616 is replaced by DIN EN 10203.
DIN 1623 is to be replaced by DIN EN 10130.
DIN 1626 is to be replaced by DIN EN 10217.
DIN 1629 is to be replaced by DIN EN 10216.
DIN 17100 is replaced by DIN EN 10025.
DIN 17102 is to be replaced by DIN EN 10113.
DIN 17155 is to be replaced by DIN EN 10028.
DIN 17120 and DIN 17121 are to be replaced by DIN EN 10210.
DIN 17123 to 125 are to be replaced by DIN EN 10210.
DIN 17162 part 1 is replaced by DIN EN 10142.
DIN 17162 part 2 is replaced by DIN EN 10147.
DIN 17172 is to be replaced by DIN EN 10208-2.
DIN 17200 is to be replaced by DIN EN 10083.
DIN 17245 is to be replaced by DIN EN 10213.
DIN 50115 is partly replaced by DIN EN 10045 part 1.
DIN 50125 is partly replaced by DIN EN 10002 part 1.
DIN 50145 is replaced by DIN EN 10002 part 1.

DIN	International ISO	EURO EU, EN	USA ASTM	GB BS	France NF A	Japan JIS-G
267 T3	898 T 1					
267 T11	3506					
267 T13	898					
488 T1	6935	10 080	A 615	4449	35-015	3112
488 T2	6935	80, 82	(A 706)	4461	35-016	
488 T3	6935	82				
488 T4	6935					
488 T5	6935					
488 T6		80				
488 T7	6935	80				
		10 080	A 648			
		10 138	A 821			3538
685 T2	1834					
685 T3	1834					
685 T4	1834					
685 T5	7592					
766						
779						
898 T1 DIN-ISO	898					
1013 T1	1035	60		4/1	45-001	3191
1013 T2				4360	45-003	3192
						3193
						3194
1045						
1324	ISO 35/1					
	IEC 50 (121)					
1599						
1614 T1	3576	46, 111	A 569	1449/1	36-102	3131
1614 T2	3573		A 621		36-301	
	6317		A 622			

Table 3 (*Continued*)

DIN	International ISO	EURO EU, EN	USA ASTM	GB BS	France NF A	Japan JIS-G
1615	(2547)		A 53 A 134		49-141	3452
1616	1111/1	10 203 (10 205)	A 624 A 625 A 626	2920	36-150 36-152	3303
1623 T1	3574	10 130	A 619 A 620	1449/1	36-401	
1623 T2	4995, 4997		A 366			3141
1623 T3	5001	10 209	A 424			3133
1624	6932	139	A 424	1449/1	36-401 37-501	3141
1626	559, 2604/3 9330	10 217	A 135 A 139 A 211	3601, 3602	49-142	3454 3455
1628	2604, 9330		A 513 A 587	778		3457
1629	2604/2, 9329-1	10 216	A 519 A 524	3601 3602/1	49-112 49-117	3454 3455
1630	2604/2 9329-1		A 524	778		3455
1651	683/9	87	A 29 A 582	970/1	35-561 35-562 47-410	4804
1652	683/13					
1654 T1	4954	119/1	A 545 A 548	3111	(35-053)	3507 3539
1654 T2		/2	A 549		35-564	3539
1654 T3		/3	A 576			3539
1654 T4		/4	A 354	1449	35-577	3539
1654 T5		/5	A 439			
1681	3755 9447		A 27	3100		5101
1715 T1			B 388 (B 389)			
1736 T1						
1736 T2						
1910 T1	857					
1912 T1	2553					
1912 T2	6947					
1913	2560					
2078	2232					
2088						
2089 T1						
2310 T2						

Table 3 (*Continued*)

DIN	International ISO	EURO EU, EN	USA ASTM	GB BS	France NF A	Japan JIS-G
3051 T1						
3051 T2						
3051 T3						
3051 T4						
4099						
4227 T1						
4227						
5684 T1						
5684 T2						
5684 T3	3077					
6580	3002/3					
6581	3002/1					
8524	6520					
8528 T1	581					
8556 T1	3581					
8580						
8588						
15 020 T1	4308/1	FEM				
17 014 T1	DP 4885	10 052		6562/1	02-010	
(17 100)	630, 1052 6316	10 025	A 6 A 242 A 283 A 284 A 570	4360 10025	35-501 10025	3101 3106
17 102	4950 9328-4	10 113 10 028	A 633 A 678 A 812	4360	36-201 36-204 36-207	3106 3115 (3126)
17 103			A 226 A 350			
17 111	4954	(80, 81)	A 31, A 502 A 544		35-053	3104
17 115	DP 4959	(62)	A 391 A 413 A 467	3113 (4942)	35-566	
17 119			A 500			
17 120	DP 630/2	10 210	A 134 A 135 A 139 A 501	10 210	10210	3444
17 121	DP 630/2	10 210	A 501	10210	10210	3444
17 123		10 210	A 618	10210	10210	3444
17 124						
17 125						
17 140 T1	8457/2	10 016	A 510 A 680 A 684	38 933 proj. 86 10 028	03-49 35-051	

Table 3 (Continued)

DIN	International ISO	EURO EU, EN	USA ASTM	GB BS	France NF A	Japan JIS-G
(17 155)	2604/4 9328-2	10 028	A 202 A 204 A 285 A 302 A 387 (A 299) A 455	1501/1 1501/2 (3602) (3604)	EN10028 36-205 36-206 (35-578)	3118 3119 3120 4109
17 162 T1	3575	10 142	A 361 A 446 A 525 A 526 A 527	2989 10 142 2989	36-321 10142	3302
17 162 T2	4998	10 147	A 525	3083	36-322	3302
17 163	5002	10 152	A 591 A 161 A 200	6687	36-160	3313
17 172	3183	10 208	A 333 API 5L	(3603)	(49-117)	
17 173	2604/2 DP 9329-3		A 333 A 334			3460 3464
17 174	2604/3 DP 9330-3		A 333 A 334 A 409	3603 3605		3460 3468
17 175	DP 9329-2		A 106 A 209 A 210 A 335	3602 3604 3059	49-211 49-213 49-214 49-215	3461 3462
17 177	DP 9330-2		A 178 A 214 A 226 A 250	3604	49-243	3456 3458
17 178						
17 179			A 192			
17 182	DIS 9477		A 148 A 216	3100		5111
17 200	683/1	10 083	A 29 A 304	970/1	35-552 35-553 35-554	4051 4052
17 210	683/11	84	A 304	970/1	35-551	4051 4101 4106
17 211	683/10	85	A 355	970/2	35-551	4102 4106
17 212		86	A 576	970/1	35-563	
17 221	683/14	89	A 689	970/5 1449/1	35-571	4801
17 222	4960	132	A 227 A 231 A 401 A 679 A 682		37-502	

Table 3 (*Continued*)

DIN	International ISO	EURO EU, EN	USA ASTM	GB BS	France NF A	Japan JIS-G
17 223 T1	8458-2	150	A 227 A 229 A 230	4637 4638 5216	47-301	3521 3560 3561
17 223 T2	DP 8458/3	150	A 229	2803	47-301	
17 224	6931	151		2056	35-587 35-588	
17 230	683/17	94	A 295 A 485 A 534 A 535	970/1	35-565	4805
17 240	4955 TR 4956	95	A 193 A 194 A 354 A 679	970/1	35-578 35-577	
17 243	2604/1 DP 9327		A 336	1503		3203
17 245	DIS 4991	10 213	A 216 A 217 (A 356)	3100	32-055	5151
17 280	2604/4 7705 9328-3	129 88 T3	A 203 A 516 A 553 A 662	1501/2 1501/3	36-208 (36-209)	3127
17 350	4957	96	A 597 A 600 A 681 A 685 A 686	970 4659	35-590	4401 4403 4404
17 405	IEC 404-8-6					
17 410	IEC 404-8-1		A 801			
17 440	683/13 2604/1 2604/4 DP 9327 9328-5	88	A 167 A 176 A 240 A 276 A 314	970 1449/2 1501/3	35-573 35-574 35-575 35-576 35-577 (35-578) 36-209	4303 4304
17 441	683/13	88	A 167 A 176 A 240 A 480	1501/3		4305
17 443	5832/1		F 55 F 56 F 342 F 354			
17 445			A 217 A 297 A 351 A 352 A 743 A 744	3100		5121

Table 3 (*Continued*)

DIN	International ISO	EURO EU, EN	USA ASTM	GB BS	France NF A	Japan JIS-G
17 455	2604/2		A 249	3605		3459
17 456	2604/5		A 268			3463
			A 269			
			A 358			
			A 376			
17 457	2604/5		A 358			
17 458	2604/2		A 376			
	DP 9329-4					
	DP 9330-4					
17 470			B 344			
			B 603			
17 480	683/15	90	A 182	970/4	35-579	
			(A 194)			
			A 232			
			SAE J775			
17 745						
21 254 T1	3154					
22 252	610					
32 525 T1						
32 525 T2	6847					
41 301	IEC 404-8-6-1					
41 302 T1	IEC 740					
41 302 T2						
43 107						
46 400 T1	IEC 404-8-4	106	A 677	601/1	C 28-900	
				6404		
46 400 T2	IEC 404-8-3	126		(601/1)	C 28-925	
				6404		
			A 665			
46 400 T3	IEC 404-8-7	107	A 725	(601/2)	C 28-920	
46 400 T4	IEC 404-8-2	165	A 683	6404	C 28-926	
				(601/1)		
48 300						
50 021	1456					
50 049	404	10 204		(42 276)	A 03-116	
	10474	(10 021)		proj. 86		
50 103 T1	6508	10 004		891/1	03-153	
				427/1		
(50 103 T2)	R 1024	10 109		4175/1	03-153	
(50 115)	148, 7705	10 045		113/3	03-156	
50 118	(R203-206)	123		3500...	03-355	
(50 125)	6892	10 002/1		18/2	03-151	
				1610		
50 133	6507, 409	5		427/1	03-154	

Table 3 (*Continued*)

DIN	International ISO	EURO EU, EN	USA ASTM	GB BS	France NF A	Japan JIS-G
(50 145)	6892	10 002/1		18/2 1610	03-151	
50 150	4964	circular 3 + 4	E 140	860	03-173	
50 190 T1	2639	105		6479 6481	04-202	
50 190 T2	3754	116		6481	04-203	
50 190 T3	3754	116		6481	04-203	
50 191	642	23	A 255	4437	04-303	
50 192	3887	104		6617/1	04-201	
50 351	6506 (410)	3		240/2	03-152	
50 462 T1	IEC 404-2	118		4175/2 6404/2	C 28-911	
50 462 T2	IEC 404-2					
50 462 T3	IEC 404-2					
50 462 T4	IEC 404-2					
50 462 T5	IEC 404-2					
50 465						
50 466						
50 470	(IEC 404-1)			6404/1		
50 471	(IEC 404-1)					
50 472	(IEC 404-1)					
50 601	643 2624	103	B 390 E 112	4490	04-102	
50 602	4968		E 45			
50 914	3651/2	114	E 708	5903	05-159	
50 921	3651/1	121	A 262-C		05-160	
51 212	7800		E 558			
51 226	204 206					
59 115		108	3111			
59 130		65		3111	45-075	

	International ISO	EURO EU, EN	USA ASTM	GB BS	France NF A	Japan JIS-G
Stahl-Eisen Werkstoffblätter und Lieferbedingungen des VDEh						
SEW 092	5951	149		1449	36-203	
SEW 093	6930	10 131		1449	36-203	
SEW 094						
SEW 095						
SEW 310						
SEW 390						
SEW 400						
SEW 470	4955	95	A 240	1449/2	35-578	
SEW 471						
SEW 500						
SEW 550						
SEW 590						
SEW 595						

Glossary of Repeatedly Used Symbols and Abbreviations

Generally, the symbols and abbreviations used in this book are explained on the pages on which they occur, with direct reference to the respective text. Symbols and abbreviations which are more commonly used are explained separately in the following glossary.

Latin characters

A	austenite
A	percentage elongation after fracture (for a gauge length of $L_0 = 5.65\sqrt{S_0}$ or of $L_0 = 5d_0$ in tensile test)
A	quenched (heat-treated condition)
A	tempered (heat-treated condition)
A_{gl}	uniform elongation
A_v	notch impact energy
a	activity of an element
a	lattice parameter (crystal lattice)
B	bainite (microstructure)
B	magnetic induction
b	Burgers vector (dislocations)
c	concentration
c	specific heat
D	diffusion coefficient
E	Young's modulus
F	ferrite (microstructure)
F	force (in general)
G	Gibbs' energy
G	shear modulus
G	(soft) annealed (heat-treated condition)
H	hardened (heat-treated condition)
H	magnetic field strength
H	enthalpy
HB	Brinell hardness
HRC	Rockwell hardness (cone)
HV	Vickers hardness
J	J-integral (fracture mechanics)
J	magnetic polarization
K	work hardened

K_I	stress-intensity factor (fracture mechanics)
K_{Ic}	fracture toughness (fracture mechanics)
K_L	pre-factor of Ludwik's equation (eq. B 1.17)
kfz	fcc (face-centered cubic) (lattice characterization)
krz	bcc (body-centered cubic) (lattice characterization)
k	Boltzmann's constant
k_f	yield stress
L	solution annealed (heat-treated condition)
L	life-time
L	path length of a dislocation
L	instantaneous gauge length (of a tensile-test specimen)
L_0	original gauge length (of a tensile-test specimen)
L_u	final gauge length after fracture
M	magnetization
M	martensite (microstructure)
M	momentum
N	normalised (heat-treated condition)
N	number of stress cycles to fracture (fatigue strength test)
n	frequency of stress cycles (fatigue strength test)
n	hardening exponent
n	number (for example: of dislocations)
P	pearlite (microstructure)
p	partial pressure
p_H	concentration of hydrogen ions
Q	activation energy
R	universal gas constant
R	ratio σ_M/σ_O (fatigue behaviour)
R	killed (killed steel)
RR	fully killed (fully killed steel)
R_e	yield stress
R_{eL}	lower yield stress
R_{eH}	upper yield stress
R_p	proof stress (in general)
$R_{p0.2}$	0.2% proof stress
R_{p1}	1% proof stress
R_m	ultimate tensile strength
RT	room temperature
r	radius in general (coordinates)
S	entropy
S	stress ratio of fatigue strength σ_U/σ_O
S	area, especially instantaneous cross-sectional area of the gauge length (of tensile specimen)
S_0	original cross-sectional area of the parallel length (tensile specimen)
S_u	minimum cross sectional area after fracture
SRK	stress corrosion
s	distance; length, in general

Glossary of Repeatedly Used Symbols and Abbreviations

s	stress ratio σ_U/σ_O (fatigue behaviour)
T	temperature
$T_ü$	impact transition temperature (of notch impact energy)
T_s	solidification temperature
TM	thermomechanical treatment
t	time
U	energy
U_w	interaction energy
U	untreated (heat-treated condition)
U	unkilled (unkilled steel)
V	quenched and tempered (heat-treated condition)
v	rate (of reaction)
W	section modulus
WEZ	HAZ (heat affected zone)
Z	percentage reduction of area after fracture (tensile test)

Greek letters

α	linear thermal expansion coefficient
α	ferrite (microstructure)
α_k	stress concentration factor
γ	austenite (microstructure)
γ	shear
γ	interfacial energy, surface energy, also energy in connection with fracture processes
Δ	changing amount of a variable
δ	crack opening (fracture mechanics)
δ	δ-ferrite (microstructure)
ε	strain, in general
κ	multi-axiality value
Θ	cementite
λ	thermal conductivity
λ	wave length
μ	permeability
ν	Poisson's ratio
π	degree of multi-axiality
ρ	dislocation density
ρ	specific electrical resistance
σ	stress, in general
σ_f	yield stress (see k_f)
σ_f^*	microscopic cleavage fracture stress
τ	shear stress
φ	deformation degree (logarithmic strain)

References

C 2 Normal and High Strength Structural Steels

1. Gräfen, H., K. Gerischer u. E.-M. Horn: Werkst. Techn. 4 (1973) S. 169/86.
2. Fiehn, H., W. Laber u. K. Nagel: Schweißen u. Schneiden 28 (1976) S. 346/48.
3. Feinkornbaustähle für geschweißte Konstruktionen. 2 Aufl. Düsseldorf 1974. Hrsg. von der Beratungsstelle für Stahlverwendung. (Merkblätter über sachgemäße Stahlverwendung Nr. 365.)
4. Jäniche, W.: Stahl u. Eisen 96 (1976) S. 1207/19.
5. Tauscher, H.: Dauerfestigkeit von Stahl und Gußeisen. Leipzig 1969.
6. DASt-Richtlinie 009. Empfehlungen zur Wahl der Stahlgütegruppen für geschweißte Stahlbauten. Ausg. April 1973.
7. Panknin, W., u. E. Marke: In: Werkstoff-Handbuch Stahl und Eisen. Hrsg.: Verein Deutscher Eisenhüttenleute. 4. Aufl. Düsseldorf 1965. S. E 5-1/5-7.
8. Rapatz, F., u. F. Motalik: In: Werkstoff-Handbuch Stahl und Eisen. Hrsg.: Verein Deutscher Eisenhüttenleute. 4. Aufl. Düsseldorf 1965. S. E 35-1/35-9.
8a. DIN 2310 Teil 2. Thermisches Schneiden; Ermitteln der Güte von Brennschnittflächen. Ausg. Mai 1976.
9. Degenkolbe, J.: In: Schweißen von Baustählen. Hrsg.: Verein Deutscher Eisenhüttenleute. Düsseldorf 1983. S. 1/15.
10. Degenkolbe, J.: Schweißtechn. Wien, 29 (1975) Nr. 2, S. 17/24.
11. Farrar, J.C.M., u. R.E. Dolby: Lamellar tearing in welded steel fabrication. The Welding Institute, Abington, Great Britain (1972).
12. Degenkolbe, J., u. B. Müsgen: Stahl u. Eisen 93 (1973) S. 1218/21.
13. Klöppel, K., R. Möll u. P. Braun: Stahlbau 39 (1970) S. 289/98.
14. Klöppel, K., u. T. Seeger: Zeit- und Dauerfestigkeitsversuche an Voll- und Lochstäben aus hochfesten Baustählen. Darmstadt 1969. (Veröffentlichungen des Instituts für Statik und Stahlbau der Technischen Hochschule Darmstadt. H. 7).
15. Degenkolbe, J., u. H. Dißelmeyer: Schweißen u. Schneiden 25 (1973) S. 85/88.
16. Bott, G., W. Neuhaus u. H. Schönfeldt: Schiffbau und Meerestechnik. Düsseldorf 1976. (VDI-Reihe Kleine Stahlkunde.) S. 29.
17. Minner, H.H., u. T. Seeger: Stahlbau 46 (1977) S. 257/63.
18. Minner, H.H., u. T. Seeger: OERLIKON-Schweißmitt. 36 (1978) Nr. 83, S. 13/23.
19. Müsgen, B.: Stahl u. Eisen 103 (1983) S. 225/30.
20. Braumann, F.: Stahl u. Eisen 83 (1963) S. 1356/63.
21. Degenkolbe, J., u. B. Müsgen: Mater.-Prüf. 12 (1970) S. 413/20.
22. Degenkolbe, J., u. B. Müsgen: Z. Werkst.-Techn. 3 (1972) S. 130/33.
23. Degenkolbe, J., u. B. Müsgen: Mitt. Dt. Forsch.-Ges. Blechverarb. u. Oberflächenbehandl. 22 (1971) S. 136/45.
23a. Erste Allgemeine Verwaltungsvorschrift zum Bundes-Immissionsschutzgesetz (Technische Anleitung zur Reinhaltung der Luft-TA Luft-). Heider-Verlag Bergisch Gladbach 1976.
24. Schmidt, W.: Blech Rohre Profile 23 (1976) S. 90/98.
24a. Müsgen, B., u. H.-J. Kaiser: Thyssen Techn. Ber. (1984) S. 116/23.
25. Degenkolbe, J., u. B. Müsgen: Arch. Eisenhüttenwes. 44 (1973) S. 769/74.
26. Haneke, M., u. W. Middeldorf: In: DVS-Ber. Nr. 33. 1975. S. 153/65.
27. Gulvin, T.F., D. Scott, D.M. Maddrill u. J. Glen: J. West Scotl. Iron Steel Inst. 80 (1972/73) S. 149/75.
28. Haneke, M.: Estel-Ber. 1974, S. 32/46.
29. Degenkolbe, J., u. B. Müsgen: Stahl u. Eisen 94 (1974) S. 757/63.
29a. DIN 50601. Metallografische Prüfverfahren. Ermittlung der Austenit- und der Ferritkorngröße

von Stahl und anderen Eisenwerkstoffen. Ausg. Juni 1985. – EURONORM 103. Mikroskopische Ermittlung der Ferrit- oder Austenitkorngröße von Stählen. Ausg. Nov. 1971.
30. Müsgen, B., u. J. Degenkolbe: Bänder Bleche Rohre 19 (1978) S. 56/62.
31. Haneke, M., u. G. Juretzko: Estel-Ber. 1974, S. 57/68.
32. Hofe, H. von, u. H. Wirtz: In: Fachbuchreihe Schweißtechnik Bd. 23. 1962. S. 95/104.
33. Stahl-Eisen-Werkstoffblatt 088-76. Schweißgeeignete Feinkornbaustähle; Richtlinien für die Verarbeitung, besonders für das Schweißen. Ausg. April 1987.
33a. DIN 85 28. Teil 1. Schweißbarkeit; metallische Werkstoffe, Begriffe. Ausg. Juni 1973.
34. Schweißen unlegierter und niedriglegierter Baustähle. 3. Aufl. Hrsg. von der Beratungsstelle für Stahlverwendung. Düsseldorf 1980. (Merkblätter über sachgemäße Stahlverwendung Nr. 381.)
35. Bersch, B., K. Kaup, u. F. O. Koch: Hoesch-Ber. 7 (1972) S. 36/51.
35a. DIN 17 172. Stahlrohre für Fernleitungen für brennbare Flüssigkeiten und Gase. Ausg. Mai 1982.
36. Degenkolbe, J., u. B. Müsgen: Schweißen u. Schneiden 17 (1965) S. 343/53.
37. Mück, G. H.: Über den Einfluß der Schweißbedingungen und der Werkstoffzusammensetzung auf die Gefügeänderung und Änderung der mechanisch-technologischen Eigenschaften in der wärmebeeinflußten Zone. Darmstadt 1971. (Dr.-Ing.-Diss. Techn. Hochsch. Darmstadt.)
38. Piehl, K.-H.: Stahl u. Eisen 93 (1973) S. 568/77.
39. Degenkolbe, J., u. D. Uwer: J. Soudure 5 (1973) S. 117/33.
40. Uwer, D., u. J. Degenkolbe: Schweißen u. Schneiden 27 (1975) S. 308/06.
41. Uwer, D.: In: Schweißen von Baustählen. Hrsg.: Verein Deutscher Eisenhüttenleute. Düsseldorf 1983. S. 85/113.
42. Wallner, F.: Berg- u. Hüttenm. Mh. 118 (1973) S. 295/304.
43. Degenkolbe, J.: Techn. Überwachung 10 (1969) S. 259/69.
44. Hornbogen, E.: Z. Metallkde. 68 (1977) S. 455/69.
44a. Stahl-Eisen-Werkstoffblatt 082. Begriffsbestimmungen zur thermomechanischen Behandlung von Stahl. Ausg. Juni 1989.
45. DIN 17 100. Allgemeine Baustähle. Ausg. Jan. 1980.
46. VdTÜV-Merkblatt 1263. Werkstoffe. Ausg. März 1976.
47. Meyer, L., u. H. de Boer: Thyssen Techn. Ber. (1977) S. 20/29.
47a. de Boer, H.: Blech Rohre Profile 30 (1983) S. 485/88.
47b. Stahl-Eisen-Werkstoffblatt 083. Schweißgeeignete Feinkornbaustähle, thermomechanisch umgeformt. Technische Lieferbedingungen für Blech, Band und Breitflachstahl. Ausg. Dez. 1984.
47c. Stahl-Eisen-Werkstoffblatt 084. Schweißgeeignete Feinkornbaustähle, thermomechanisch umgeformt. Technische Lieferbedingungen für Formstahl und Stabstahl mit profilförmigem Querschnitt. Ausg. Aug. 1988.
47d. Degenkolbe, J., J. Mahn, B. Müsgen u. H.-J. Tschersich: Thyssen Techn. Ber (1987) S. 41/55.
48. Forch, K.: Grundlagen der Wärmebehandlung von Stahl. Düsseldorf 1976.
48a. Lotter, U., B. Müsgen u. H. Pircher: Thyssen Techn. Ber. (1984) S. 13/23.
49. Tetelmann, A. S., u. A. J. McEvily, jr.: Fracture of structural materials. New York, London, Sydney 1967.
50. Tither, G., u. J. Kewell: J. Iron Steel Inst. 208 (1970). S. 686/94.
51. Duckworth, W. E., R. Phillips u. J. A. Chapman: J. Iron Steel Inst. 203 (1965) S. 1108/14.
52. Irani, J. J., D. Burton, J. D. Jones u. A. B. Rothwell: Strong tough structural steels. London 1967. (Spec. Rep. Iron Steel Inst. No. 104.) S. 110/22.
53. Ito, Y. u. K. Bessyo: Weldability formula of high strength steels. IIW doc. IX-576-68.
54. IIW doc. IX-115-55.
54a. Baumgardt, H., H. de Boer u. F. Heisterkamp: Thyssen Techn. Ber. (1983) S. 24/39.
55. Müsgen, B., u. J. Degenkolbe: Bänder Bleche Rohre 10 (1976) S. 401/05.
56. Sage, M. A., u. F. E. L. Copley: J. Iron Steel Inst. 195 (1960) S. 422/38.
57. Biener, E., u. W. Lauprecht: Stahl u. Eisen 96 (1976) S. 1044/50.
57a. Müsgen, B., u. U. Schriever: Thyssen Techn. Ber. (1984) S. 45/54.
58. Ito, Y., u. K. Bessyo: Sumitomo Search 1969, Nr. 1, S. 59/70.
59. Million, A.: In: DVS-Ber. Nr. 64. 1980. S. 19/23.
60. Nehl, F.: Stahl u. Eisen 72 (1952) S. 1261/67.
61. Wiester, H.-J., W. Bading, H. Riedel u. W. Scholz: Stahl u. Eisen 77 (1957) S. 773/84.
62. Vogels, H. A., P. König u. K.-H. Piehl: Arch. Eisenhüttenwes. 35 (1964) S. 339/51.
63. Lauprecht, W, u. P. Kremmers: Thyssenforsch. 1 (1969) S. 30/39.
64. Meyer, L., F. Schmidt u. C. Straßburger: Stahl u. Eisen 89 (1969) S. 313/30.
65. Degenkolbe, J., u. B. Müsgen: In: Herstellung und Verwendung von Grobblech. Informationstagung. [Hrsg.] Kommission der Europäischen Gemeinschaften. Düsseldorf 1979. (EUR 6189 d, e, f) S. 471/90.

66. Meyer, L., H.-E. Bühler u. F. Heisterkamp: Thyssenforsch. 3 (1971) S. 8/43.
67. DAS 2 116 357 vom 26.3.1971.
67a. Baumgardt, H.: Verbesserung der Zähigkeitseigenschaften in der Wärmeeinflußzone von Schweißverbindungen aus Feinkornbaustählen. Clausthal 1984 (Dr.-Ing.-Diss. Techn. Univ. Clausthal). – Baumgardt, H. u. C. Straßburger: Thyssen Techn. Ber (1985) S. 42/49.
68. Treppschuh, H., A. Randak, H. H. Domalski u. J. Kurzeja: Stahl u. Eisen 87 (1967) S. 1355/68.
69. Grange, R. A. u. J. B. Mitchell: Trans. Amer. Soc. Metals 53 (1961) S. 157/85.
70. Bungardt, K., u. G. Lennartz: Arch. Eisenhüttenwes. 34 (1963) S. 531/46.
71. Kalwa, G., R. Pöpperling, H. W. Rommerswinkel u. P. J. Winkler: Vorgetragen anläßlich der „Offshore North Sea Technology Conference". 21. bis 24.9.1976 in Stavanger. Norwegen.
72. Kalla, U., H. W. Kreutzer u. E. Reichenstein: Stahl u Eisen 97 (1977) S. 382/93.
73. Fröber, H.: In: Werkstoffkunde der gebräuchlichen Stähle. T. 1. Hrsg. vom Verein Deutscher Eisenhüttenleute. Düsseldorf 1977. S. 175/204.
74. Mahadevan, A., u. M. Thiele: Stahl u. Eisen 101 (1981) S. 111/15.
75. Technische Lieferbedingungen für Spundbohlen – Fassung 1985. Verkehrs- u. Wirtschafts-Verlag Dr. Borgmann. Dortmund.
75a. DIN 21 544. Stahl für Grubenausbau; Technische Lieferbedingungen. Ausg. Jan. 1985.
75b. Heller, W., u. J. Flügge: Stahl u. Eisen 104 (1984) S. 95/97.
76. Stahl-Eisen-Werkstoffblatt 087–81. Wetterfeste Baustähle. Ausg. Juni 1981.
77. DASt-Richtlinie 007. Lieferung, Verarbeitung und Anwendung wetterfester Baustähle. Ausg. Febr. 1979.
78. Petersen, J., u. M. Thiele: Hansa 14 (1977) S. 2030/35.
78a. Baumgardt, H., H. de Boer u. B. Müsgen: Metal Construction. Jan. 1984. S. 15/19.
79. DIN 17 102. Schweißgeeignete Feinkornbaustähle, normalgeglüht. Ausg. Okt. 1983.
80. Baumgardt, H., H. de Boer u. B. Müsgen: Thyssen Techn. Ber. 14 (1982) S. 136/45.
81. Werkstoffkunde der gebräuchlichen Stähle. T. I. Hrsg. vom Verein Deutscher Eisenhüttenleute. Düsseldorf 1977.
82. DAST Richtlinie 011, Hochfeste schweißgeeignete Feinkornbaustähle St E 460 und St E 690. Anwendung für Stahlbauten. Ausg. Febr. 1979.
83. DIN 1615. Geschweißte kreisförmige Rohre aus unlegiertem Stahl ohne besondere Anforderungen. T. L. Ausg. Okt. 1984. – DIN 1626. Geschweißte kreisförmige Rohre aus unlegierten Stählen für besondere Anforderungen. T. L. Ausg. Okt. 1984. – DIN 1628. Geschweißte kreisförmige Rohre aus unlegierten Stählen für besonders hohe Anforderungen. T. L. Ausg. Okt. 1984.
84. DIN 1629. Nahtlose kreisförmige Rohre aus unlegierten Stählen für besondere Anforderungen. T. L. Ausg. Okt. 1984. – DIN 1630. Nahtlose kreisförmige Rohre aus unlegierten Stählen für besonders hohe Anforderungen. T. L. Ausg. Okt. 1984.
85. DIN 17 119. Kaltgefertigte geschweißte quadratische und rechteckige Stahlrohre (Hohlprofile) für den Stahlbau. T. L. Ausg. Juni 1984. – DIN 17 120. Geschweißte kreisförmige Rohre aus allgemeinen Baustählen für den Stahlbau. T. L. Ausg. Juni 1984. – DIN 17 121. Nahtlose kreisförmige Rohre aus allgemeinen Baustählen für den Stahlbau. T. L. Ausg. Juni 1984.
86. DIN 17 123. Geschweißte kreisförmige Rohre aus Feinkornbaustählen für den Stahlbau. T. L. Ausg. Mai 1986. – DIN 17 124. Nahtlose kreisförmige Rohre aus Feinkornbaustählen für den Stahlbau. T. L. Ausg. Mai 1986. – DIN 17 125. Quadratische und rechteckige Rohre (Hohlprofile) aus Feinkornbaustählen für den Stahlbau. T. L. Ausg. Mai 1986. – DIN 17 178. Geschweißte kreisförmige Rohre aus Feinkornbaustählen für besondere Anforderungen. T. L. Ausg. Mai 1986. – DIN 17 179. Nahtlose kreisförmige Rohre aus Feinkornbaustählen für besondere Anforderungen. T. L. Ausg. Mai 1986.

C 3 Reinforcing Steel for Reinforced Concrete and Prestressed Concrete Structures

1. Hütte, Bautechnik I. Berlin, Heidelberg, New York 1974. S. 714 ff.
2. DIN 1045. Beton und Stahlbeton. Ausg. Dez. 1978.
3. DIN 4227. T. I. Spannbeton; Bauteile aus Normalbeton, mit beschränkter oder voller Vorspannung. Ausg. Dez. 1979; Änderung 1. Entwurf Sept. 1984. T. 2. Spannbeton; Bauteile mit teilweiser Vorspannung. Vornorm. Ausg. Mai 1984. Weiterhin: T. 3. Vornorm. Ausg. Dez. 1983. T. 4. Ausg. Febr. 1986. u. T. 6. Vornorm. Ausg. Mai 1982.
4. Lasbaarheid Betonstaal, Kruislasverbinding. [Hrsg.:] Stichting Nederlands Instituut vor Lastechniek. Den Haag 1973. (Rapport Nr. 60.)

5. Lasbaarheid Betonstaal, Overlapen, stompe Lasverbinding. [Hrsg.:] Stichting Nederlands Instituut vor Lastechniek. Den Haag 1975. (Rapport Nr. 72.)
6. Sage, A. M.: Metals Technol. 3 (1976) Nr. 2, S. 65/70.
6a. Hey, A., H. Weise u. W. G. Wilson: Production and Application of High-strength Concrete Reinforcing Bar. In: Stuart, H. (ed.): Niobium, Proceedings of the International Symposium. Warrendale: The Met. Soc. of AIME. 1984. p. 967/87.
7. Weise, H.: In: Microalloying '75. Proceedings. [Hrsg.:] Union Carbide Corporation. Metals Division. New York 1977. S. 676/83.
8. Economopoulos, M., Y. Respen et al.: Application of the Tempcore Process to the Fabrication of High Yield Strength Concrete-reinforcing Bars. C. R. M. No. 45, December 1975.
9. Funke, P., H. F. Meyer, H. Striepens u. H. Schulte: Stahl u. Eisen 96 (1976) S. 1015/20.
10. DIN 488. T. I. Betonstahl: Sorten, Eigenschaften, Kennzeichen. Ausg. Sept. 1984. Weiterhin: T.2 bis 7. Ausg. Juni 1986. – Euronorm 80. Betonstahl für nicht vorgespannte Bewehrung. Ausg. April 1986.
11. DIN 4099. Schweißen von Betonstahl: Ausführung und Prüfung (Neuausg. Herbst 1985).
12. Leonhardt, F.: Spannbeton für die Praxis. Berlin. 4. Aufl. in Vorbereitung.
13. Geithe, W.: Betonwerk u. Fertigteil-Techn. 40 (1974) S. 579/90.
14. Rehm, G., U. Nürnberger u. R. Frey: Werkst. u. Korrosion 32 (1981) S 211/21.
15. Martin, H.: Zusammenhang zwischen Oberflächenbeschaffenheit. Verbund und Sprengwirkung von Bewehrungsstählen unter Kurzzeitbelastung. Berlin, München, Düsseldorf 1973. (Deutscher Ausschuß für Stahlbeton. H. 228.)
16. Stress corrosion cracking resistance test for prestressing tendons. Report on Prestressing Steel. Nr. 5 [Hrsg:] Fédération Internationale de la Précontrainte. Wexham Springs 1980.
17. Spannungsrißkorrosion in Spannbetonbauwerken. Neue Forschungsergebnisse. Hrsg.: Verein Deutscher Eisenhüttenleute. Düsseldorf 1983.
18. Rieche, G. u. J. Delille: Erfahrungen bei der Prüfung von temporären Korrosionsschutzmitteln für Spannstähle, Berlin, München, Düsseldorf 1978 (Deutscher Ausschuß für Stahlbeton. H. 298) S. 5/19.
19. Euronorm 138. Spannstähle. Ausg. Sept. 1979.

C 4 Steels for Hot Rolled, Cold Rolled and Surface Coated Flat Rolled Products for Cold Forming

1. Euronorm EN 10079. Benennung und Einteilung von Stahlerzeugnissen nach Formen und Abmessungen.
2. DIN 1614 Teil 2. Warmgewalztes Band und Blech; T. L.; Weiche unlegierte Stähle zum unmittelbaren Kaltformgeben. Ausg. Febr. 1986.
2a. DIN 1614 Teil. 1. Warmgewalztes Band und Blech; T. L.; Weiche unlegierte Stähle zum Kaltwalzen. Ausg. Febr. 1986.
3. DIN 50 103. Härteprüfung nach Rockwell. Ausg. März 1942. Folgeausg. Teil 1. Verfahren C, A, B, F. Ausg. März 1984. – Teil 2. Verfahren N u. T. Ausg. März 1984. Teil 3. Modifiziertes Rockwell-Verfahren Bm, Fm und 30 Tm für Feinblech aus Stahl. Ausg. Febr. 1985. – Euronorm 4-79. Härteprüfung nach Rockwell für Stahl (Verfahren A-C-B-F). Ausg. März 1979. – Euronorm 109-80. Vereinbarte Härteprüfverfahren nach Rockwell HRN und HRT. Ausg. Juni 1980.
4. Euronorm 49-72. Rauheitsmessungen an kaltgewalztem Flachzeug aus Stahl ohne Überzug. Ausg. Dez. 1972.
5. Toda, K., H. Gondoh, H. Takechi u. M. Abe: Stahl u. Eisen 96 (1976) S. 1320/26.
6. DIN 1623 Teil 1. Kaltgewalztes Band und Blech. Technische Lieferbedingungen. Unlegierte Stähle zum Kaltumformen. Ausg. Febr. 1983.
7. Straßburger, C.: Entwicklungen zur Festikeitssteigerung der Stähle – unter besonderer Berücksichtigung der unlegierten und mikrolegierten Stähle. Düsseldorf 1976. (Habil-Schr. Techn. Univ. Clausthal.)
8. DIN/EN 10 025 Wärmojewalzte Erzengnisse aus unlegierten Baustählen, Jan. 1991
9. Stahl-Eisen-Werkstoffblatt 082. Begriffsbestimmungen Normalisierendes Umformen und Thermomechanisches Umformen 2. Ausg. Juni 1989
10. Stahl-Eisen-Werkstoffblatt 092. Warmgewalzte Feinkornstähle zum Kaltumformen. Gütevorschriften. Ausg. Jan. 1990
11. Euronorm 149-80. Flachzug aus Stählen mit hoher Streckgrenze für Kaltumformung – Breitflachstahl. Blech und Band. Ausg. Sept. 1980.

12. ISO-Empfehlung 5951 – 1980. Hot-rolled steel sheet of higher yield strength with improved formability.
13. Perlitarme Feinkornstähle für kaltgeformte und für geschweißte Bauteile. Hrsg. von der Beratungsstelle für Stahlverwendung. Düsseldorf 1978. (Merkblätter über sachgemäße Stahlverwendung. Nr. 498.)
14. Herr, D., u. B. Müsgen: Thyssen Techn. Ber. 11 (1979) S. 77/85.
15. Massip, A., u. L. Meyer; Stahl u. Eisen 98 (1978) S. 989/96.
16. Massip, A., L. Meyer u. G. Stich: Stahl u. Eisen 106 (1986) S. 115/21.
17. Rashid, M. S.: GM 980 X-A unique high strength sheet steel with superior formability. SAE Preprint 760 206, 1976.
18. Rigsbee, J. M., u. P. J. Vanderarend: Laboratory studies of microstructures and structure-property-relationships in DUAL-PHASE HSLA steels, Presented at the Fall meeting of the Metallurgical Society of the AIME, Chicago/111., Oct. 24–27, 1977.
19. Becker, J., X. Cheng u. E. Hornbogen: Z. Werkstoffkde 12 (1981) S. 301/08.
20. Drewes, E. J., u. D. Daub: In: Alloys for the eighties. Ed.: R. Q. Barr. Greenwich/Conn. 1981. S. 59/67.
20a. Kunishige, K., u. N. Nakao: The Sumitomo Search 31 (1985) S. 53/62.
21. Meyer, L., H.-P. Fickel u. D. Stender: Thyssen Techn. Ber. 11 (1969) S. 86/93.
22. DIN 17172. Stahlrohre für Fernleitungen für brennbare Flüssigkeiten und Gase. Technische Lieferbedingungen. Ausg. Mai 1978.
23. Müschenborn, W., L. Meyer u. C. Straßburger: Thyssen Techn. Ber. 6 (1974) S. 22/27.
24. Straßburger, C.: Stahl u. Eisen 95 (1974) S. 409/19.
25. Beck, G.: Ind. -Anz. 96 (1974) S. 2241/42.
26. Brockhaus, J. G., H. Singer u. W. Felber: Mitt. Dt. Forsch.-Ges. Blechverarb. u. Oberflächenbehandl. 24 (1973) S. 38/42.
27. Hiyami, S., u. T. Furakuwa: In: Microalloying 1975. [Hrsg.:] Union Carbide Corporation. Metals Division. New York 1977. S. 78/87.
28. Furakuwa, T.: Metal Progr. 116 (1979) Nr. 8, S. 36/39.
29. Magee, C. L., u. R. G. Davies: Same as [20], S. 25/35.
29a. Engl, B., u. E. J. Drewes: Stahl u. Eisen 103 (1983) S. 819/24.
29b. Prum, N., U. Meers u. G. Verhoeven: Bake hardenable continuous annealed drawing steel grades. SAE Paper 86 04 39.
29c. Okamoto, A., M. Takahashi u. T. Hino: Trans. ISIJ 21 (1981) S. 821/11.
30. DIN 1623. T. 2. Flacherzeugnisse aus Stahl; Kaltgewalztes Band und Blech; Technische Lieferbedingungen; Allgemeine Baustähle. Ausg. Febr. 1986.
31. Stahl-Eisen-Werkstoffblatt 093. Kaltgewalztes Band und Blech mit höherer Streckgrenze zum Kaltumformen aus mikrolegierten Stählen, Techn. Lieferbedingungen 2. Ausg. März 1987
32. DIN 1624. Kaltgewalztes Band in Walzbreiten bis 650 mm aus weichen unlegierten Stählen. Technische Lieferbedingungen. Ausg. Juni 1987.
33. DIN 17 200. Vergütungsstähle. Gütevorschriftenn. Ausg. Nov. 1984.
34. DIN 17 230. Wälzlagerstähle. Technische Lieferbedingungen. Ausg. März 1987.
34a. Streidl, M., u. P. M. Wollrab: Stahl u. Eisen 105 (1985) S. 65/70.
35. Albrecht, J., H.-E. Bühler u. L. Meyer: Mater-Prüf. 16 (1974) S. 59/64.
36. Heubner, U.: Erzmetall 32 (1979) S. 130/35.
37. Junkers, D., u. B. Meuthen: Z. wirtsch. Fertigung 68 (1978) S. 539/44.
38. Koenitzer, J., u. H. Schmitz: In: METEC 79. Internationaler Kongreß für Hüttentechnik – Verfahren und Anlagen –, 18, -20. Juni 1979 in Düsseldorf. Hrsg.: Düsseldorfer Messegesellschaft mbH – Nowea – in Zusammenarb. mit dem Verein Deutscher Eisenhüttenleute. Düsseldorf 1979. Ber. 8.4. S. 1/23.
39. Thiele, W.: Mitt. Arb.-Ber. Metallgesellschaft AG 1980. Ausg. 22, S. 67/73.
40. Meyer, L., u. C. Straßburger; Thyssen Techn. Ber. 11 (1979) S. 110/19.
41. Beachum, E. P.: Blast Furn. Steel Plant 51 (1963) S. 210/221 u. 368/72.
42. Steckenborn, B.: Fachber. Oberflächentechn. 8 (1970) S. 12/18.
43. Wiegand, H., u. K. H. Kloos: Mitt. Dt. Forsch.-Ges. Blechverarb. u. Oberflächenbehandl. 20 (1969) S. 65/77.
43a. Koenitzer, J., H. Schmitz, H. Klotzki u. C. Schneider: Stahl in Eisen 104 (1984) S. 381/56.
44. Fukuzuka, T., M. Urai u. K. Wakayama: Trans. Iron Steel Inst. Japan 20 (1980) S. B-526.
45. Emond, C., u. V. Leroy: Influence des conditions de galvanisation sur les propriétés de depot de zinc. Bericht Nr. EUR 6727 der EGKS. Brüssel 1980.
45a. Coutsouradis, D., F. E. Goodwin, J. Pelerin u. A. F. Skenazi: Stahl u. Eisen 104 (1984) S. 1073/80.
45b. Shibuya, A., T. Kurimoto u. M. Kimoto: The Sumitomo Search 31 (Nov. 1985) S. 75/90.

45c. Adaniya, T., M. Sagiyama u. T. Honma: Nippon Kokan Techn. Rep. Overseas Nr. 43 (1985) S. 35/39.
45d. Hada, T. and coworkers: Nippon Steel Techn. Rep. Nr. 25 (April 1985) S. 11/18.
46. Charakteristische Merkmale für elektrolytisch verzinktes Feinblech in Tafeln und in Rollen. (Hrsg.:] Deutscher Verzinkerei-Verband. Ausg. 1983. Düsseldorf 1983.
47. Charakteristische Merkmale für feuerverzinktes Band und Blech. [Hrsg.:] Deutscher Verzinkerei-Verband. Ausg. 1981. 2. Aufl. Düsseldorf 1981.
48. Meuthen, B.: Stahl u. Eisen 101 (1981) S. 1499/504.
49. Kontinuierlich feuerverzinktes Blech und Band aus weichen Stählen zum Kaltumformen, Techn. Lieferbedingungen DIN/EN 10 142, Ausg. März 1991
50. DIN 17 162 Teil 2. Feuerverzinktes Band und Blech. Techmsche Lieferbedingungen. Allgemeine Baustähle. Ausg. Sept. 1980. New ed. is being prepared; for the time being: Entwurf Jan. 1987.
51. Klotzki, H.: In: VDI-Ber. Nr. 372, 1980, S. 51/60.
52. Meyer, L., u. H.-E. Bühler: Aluminium 43 (1967), S. 733/38.
53. Denner, S.G., R.D. Jons u. R.J. Thomas: Iron & Steel internat. 48 (1975) S. 241/52.
54. Warnecke, W., u. S. Baumgartl: Metalloberfläche 31 (1977) S. 476/81.
55. Albrecht, J., W. Kaeseler u. H. Hilgenpahl: Blech Rohre Profile 19 (1972) S. 41/44.
55a. Allegre, L., R.J. Dotton u. A. Homayon: Metalloberfläche 40 (1986) S. 329/35.
56. Warnecke, W., u. H.-W. Birmes: Ind.-Anz. 98 (1976) S. 1849/50.
57. EURONORM 154-80. Feueraluminiertes Band und Blech aus weichen unlegierten Stählen für Kaltumformung. Technische Lieferbedingungen. Ausg. Dez. 1980.
58. Hoare, W. E., E. S. Hedges u. B. T. K. Barry: The technology of tinplate. London 1965. S. 102/12.
59. EURONORM 145-78. Weißblech und Feinstblech in Tafeln; Sorten, Maße und zulässige Abweichungen. Ausg. Okt. 1978.
60. Prospekt der Firma CCC 633, 3rd Avenue New York 17, NY, Ausg. Mai 1963.
61. Singer, H.: Blech Rohre Profile 9 (1979) S. 341/46.
62. Siewert, J.: Verpackungs-Rdsch. 3 (1970) S. 19/23.
63. Kunze, C. T., u. A. R. Willey: J. electrochem. Soc. 99 (1952) S. 354/59.
64. Kamm, G.G., A.R. Willey, R.E. Beese u. J.L. Krickl: Corrosion, Houston 17 (1961) S. 84t/92t.
65. Willey, A. R., J. L. Krickl u. R. R. Hartwell: Corrosion, Houston 12 (1956) S. 433/40.
66. Britton, S.C.: Fachber. Oberflächenetchn. 9 (1971) S. 77/80.
67. Hoare, W. E., E. S. Hedges u. B. T. K. Barry: Same as [58], S. 370/74.
68. Murray, T.R.: Rev. sci Instrum. 33 (1961) S. 172/76.
69. Rocquet, P., u. P. Aubrun: Corrosion, Traitements, protection, finition 16 (1968) S. 229/34.
70. Willey, A. R., u. D. F. Kelsey: Anal. Chem. 30 (1958) S. 1804/06.
71. Britton, S.C.: Brit. Corrosion J. 1 (1965) S. 91/97.
72. First international tinplate conference, London, Oct. 5–8, 1976. Proceedings. Publ. by the International Tin Research Institute. Greenford/Middx. (1977) S. 295/302.
73. Hoare, W. E., u. S. C. Britton: Tinplate testing. Hrsg.: Tin Research Institute. Greenford/Middx. (1960) S. 40/41.
74. Azzaroli, G.: Ind. Conserve 4 (1967) S. 319/21.
75. Nehring, P., u. H. Krause: Konserventechn. Handbuch 15 (1969) S. 717/19.
76. Hotchner, S.J., u. C.J. Poole: Vortrag beim 5. Weltkongreß der Konservenindustrie. Wien, 3.-6. Okt. 1967.
77. Siewert, J.: Stahl u. Eisen 89 (1969) S. 547/50.
78. Siewert, J.: Stahl u. Eisen 89 (1969) S 833/35.
79. Schade, L.: Blech 10 (1963) S. 518/24.
80. DIN 1616. Weißblech und Feinstblech in Tafeln. Sorten, Maße und zulässige Abweichungen. Ausg. Okt. 1984 (see [59]).
81. ASTM Standard. A626-68. Double-reduced electrolytic tin plate.
82. Habenicht, G.: Verpackungs-Mag. 1968, S. 18, 20 u. 22.
83. Kamm, G.G.: J. electrochem. Soc. 116 (1969) S. 1299/305.
84. Hoare, W. E., u. B. T. K. Barry: Blech 15 (1968) S. 586/94.
85. EURONORM 153-80. Kaltgewalztes feuerverbleites Flachzeug (Terneblech und -band) aus weichen unlegierten Stählen für Kaltumformung. Technische Lieferbedingungen. Ausg. Dez. 1980.
86. ISO 4999. Continuous hot-dip terne (lead alloy) coated cold-reduced carbon steel sheet of commercial and drawing qualities. First Ed. 1978-08-01, Annex B, S. 13/16.
87. DIN 50021. Korrosionsprüfungen. Sprühnebelprüfungen mit verschiedenen Natriumchloridlösungen. Ausg. Mai 1975. New ed. is being prepared; for the time being: Entwurf Jan. 1987.

88. Bersch, B.: Blech Rohre Profile 27 (1980) S. 215/21.
89. Pappert, W., E. J. Drewes u. K. Jürgig: Estel-Ber. 1974, Nr. 1, S. 24/32.
90. Falkenhagen, G.: In: Herstellung von kaltgewalztem Band. T. 2. Hrsg.: Verein Deutscher Eisenhüttenleute. Düsseldorf 1970. S. 192/238.
91. DRP 573 300 vom 12. Juli 1929.
92. Zeerleder, A. v.: Korrosion u. Metallschutz 12 (1936) S. 275/83.
93. VDM-Handbuch. 2. Aufl. Hrsg.: Vereinigte Deutsche Metallwerke. Frankfurt/M. 1964.
94. Jargon, F., u. H.-J. Langhammer: Blech Rohre Profile 27 (1980) S. 676/80.
95. Eisenmenger, F. C. L., u. F. A. Schmidt: Blech 13 (1966) S. 314/17.
96. Kutzelnigg, A.: Die Prüfung metallischer Überzüge. 2. Aufl. Saulgau/Württ. 1965.
97. Plog, H.: Schichtdicken-Messung, Verfahren und Geräte. 2. Aufl. Saulgau/Württ. 1968.
98. DIN 1623 T. 3. Flacherzeugnisse aus Stahl: Kaltgewalztes Band und Blech. Technische Lieferbedingungen. Weiche unlegierte Stähle zum Emaillieren. Ausg. Jan. 1987.
99. Warnecke, W., u. L. Meyet: Mitt. Ver. Dt. Emailfachl. 28 (1978) S. 131/38.
100. EURONORM 169-85. Organisch bandbeschichtetes Flachzeug aus Stahl. Ausg. Jan, 1986. See also: Meuthen, B., u. G. Vogtenrath: In: VDI-Ber. 450, 1982, S. 63/66, und Meuthen, B.: Blech Rohre Profile 29 (1982) S. 453/54.
101. Charakteristische Merkmale für bandbeschichtetes Feinblech in Tafeln und in Rollen. [Hrsg.:] Deutscher Verzinkerei Verband. Ausg. 1979. 5. Aufl. Düsseldorf 1979. New ed, 1986. See also: Meyer zu Bexten, J. H., u. U. Kirchner: In: VDI-Ber. 449, 1982, S. 109/15.
102. EURONORM 130-77. Kaltgewalztes Flachzeug ohne Überzug aus weichen unlegierten Stählen für Kaltumformung. Ausg. Mai 1977.
102a. ISO 3574. Kaltgewalztes Band und Blech aus weichen unlegierten Stählen. Ausg. 1986.
102b. ISO 4997, Kontinuierlich kaltgewalztes Flachzeug aus allgemeinen Baustählen. Ausg. 1978.
103. EURONORM 147-79. Kontinuierlich feuerverzinktes Blech und Band aus weichen unlegierten Stählen für Kaltumformung. Technische Lieferbedingungen. Ausg. April 1979.
103a. ISO 3575. Kontinuierlich feuerverzinktes Flachzeug aus weichen unlegierten Stählen. Ausg. 1976.
104. EURONORM 147-79. Kontinuierlich feuerverzinktes Blech und Band aus unlegierten Baustählen mit vorgeschriebener Mindeststreckgrenze. Gütenorm. Ausg. Nov. 1979.
104a. ISO 4998. Kontinuierlich feuerverzinktes Flachzeug aus allgemeinen Baustäblen. Ausg. 1977.
105. Klotzki, H., u. B. Meuthen: Stahl u. Eisen (1982) S. 21/25.
106. Vogtenrath, G., u. U. Feldmann: Stahl u. Eisen 103 (1983) S. 793/98.
107. Bandbeschichtetes Feinblech. Hrsg. von der Beratungsstelle für Stahlverwendung. Düsseldorf 1980 (Merkblätter über sachgemäße Stahlverwendung. Nr. 325).

C 5 Heat-Treatable and Surface Hardening Steels used for Vehicle and Machine Construction

1. DIN 17014. Wärmebehandlung von Eisenwerkstoffen. T. 1. Fachbegriffe und -ausdrücke. Ausg. März 1975 (New edition is being prepared, at the time being: Entwurf Okt. 1986).
2. Rapatz, F.: Die Edelstähle. 5. Aufl. Berlin, Göttingen, Heidelberg 1962. S. 348.
3. Tauscher, H.: Berechnung der Dauerfestigkeit: Einfluß von Werkstoff und Gestalt. Leipzig 1964.
4. Arnold G.: Masch.-Schaden 35 (1962) S. 151/56.
5. Gerber, W., u. U. Wyss: Roll-Mitt. 7 (1948) Nr. 2/3, S. 13/47.
6. Legat, A., u. A. Moser: Härterei-techn. Mitt. 23 (1968) S. 10/14.
7. Brisson, J., R. Blondeau, Ph. Mainier u. J. Dollet: Mém. sci. Rev. Métallurg. 72 (1975) S. 115/31.
8. Vetter, K.: Unveröff. Untersuchungen der Krupp Stahlwerke Südwestfalen AG, Siegen.
9. Frodl, D., A. Randak u. K. Vetter: Härterei-techn. Mitt. 29 (1974) S. 169/74.
10. Steinen, A. von den, S. Engineer, E. Horn u. G. Preis: Stahl u. Eisen 95 (1975) S. 209/14.
11. Hollomon, H., u. L. D. Jaffe: Trans. Amer. Inst. min. metallurg. Engrs., Iron Steel Div., 162 (1945) S. 223/49.
12. Grange, R. A., C. R. Hribal u. L. F. Porter: Metallurg. Trans. 8 A (1977) S. 1775/85.
13. SAE handbook. [Hrsg.:] Society of Automotive Engineers. New York (jährliche Ausgabe).
14. Kösters, R., D. Frodl, S. Engineer, G. Naumann u. E. Wetter: Stahl u. Eisen 101 (1981) S. 707/12.
15. French, H. J.: J. Metals, Trans., 8 (1956) S. 770/82.
16. Kroneis, M.: Schweiz. Arch. angew. Wiss. u. Techn. 28 (1962) S. 298/309.

17. Spies, H.-J., S. Wittig u. G. Münch: Neue Hütte 23 (1978) S. 421/23.
18. Rose, A., A. Krisch u. F. Pentzlin: Stahl u. Eisen 91 (1971) S. 1001/20.
19. Kochendörfer, A., K. E. Hagedorn, B. Schlatte u. H. Ibach: Arch. Eisenhüttenwes. 50 (1979) S. 123/28.
20. Berns, H.: Z. Werkst.-Techn. 9 (1978) S. 189/204.
21. Ineson, E.: Bibliography on the effects of microstructure on the mechanical properties of heat treated low alloy wrought steels. BISRA MG/A136/62 (1965/66).
22. Bernstein, M. L., u. K. E. Hensger: Thermomechanische Behandlung und Festigkeit von Stahl. Leipzig 1977.
23. Archer, R. S., J. Z. Briggs u. C. M. Loeb jr.: Molybdän. Stähle, Gußeisen, Legierungen. [Hrsg.:] Climax Molybdenum Company. Zürich 1951.
24. Bungardt, K., H. Kiessler u. E. Kunze: Stahl u. Eisen 74 (1954) S. 71/75.
25. Kern, R. F.: Metal Progr. 94 (1968) Nr. 5, S. 60/73.
26. Kiessler, H.: Stahl u. Eisen 61 (1941) S. 509/16.
27. Cornelius, H., u. H. Krainer: Stahl u. Eisen 61 (1941) S. 871/77.
28. Leslie, W. C., R. J. Sober, S. G. Babcock u. S. J. Green: Trans. Amer. Soc. Metals 62 (1969) S. 690/710.
29. Gerber, W.: Schweiz. Arch. angew. Wiss. Techn. 27 (1961) S. 493/502.
30. Kiessler, H.: Stahl u. Eisen 71 (1951) S. 433/40.
31. DIN 17200. Vergütungsstähle. Ausg. Nov. 1984. (Now superseded by a new ed.: Ausg. März 1987).
32. Borik, F., R. D. Chapman u. W. E. Jominy: Trans. Amer. Soc. Metals 50 (1958) S. 242/57.
33. Hänchen, R.: Dauerfestigkeitsbilder für Stahl und Gußeisen. München 1963. (Betriebsbücher 4.)
34. VDI-Richtlinie 2228. Festigkeit bei wiederholter Beanspruchung. Zeit- und Dauerfestigkeits-Schaubilder von Werkstoffen. Ausg. 1972.
35. DDR-Standards TGL 19340. 1975. Dauerfestigkeit. Werte, Ermittlung, Berechnung für glatte und gekerbte Bauteile aus Stahl.
35a. Kloos, K. H.: Z. Werkst.-Techn. 12 (1981) S. 134/42.
35b. Kloos, K. H., u. P. K. Braisch: Härterei-techn. Mitt. 37 (1982) S. 83/91.
35c. Bernstein, G., u. B. Fuchsbauer: Z. Werkst.-Techn. 13 (1982) S. 103/09.
36. Niemann, G., u. H. Rettig: VDI-Z. 102 (1960) S. 193/202.
37. Finnern, B.: Bad- und Gasnitrieren. München 1965. (Betriebsbücher 18.)
38. Roempler, D.: Härterei-techn. Mitt. 34 (1980) S. 220/29.
39. Kloos, K. H.: Verschließfeste Werkstoffe. VDI-Bericht 194 (1973) S. 5/21.
40. DIN 50191. Stirnabschreckversuch. Ausg. Sept. 1987.
41. DIN 17021. Stahlauswahl aufgrund der Härtbarkeit. Ausg. Febr. 1976.
42. EURONORM 83-70. Vergütungsstähle. Ausg. März 1970. (New edition is now being prepared: Entwurf 1987).
43. ISO/R 683. Für eine Wärmebehandlung bestimmte Stähle. T. 1: Unlegierte Vergütungsstähle. Ausg. 1968. – T. 2: Warmverformte Vergütungsstähle mit 1% Cr und 0.2% Mo. Ausg. 1968. – T. 5: Mangan-Vergütungsstähle, Ausg. 1970. – T. 6: Warmverformte Vergütungsstähle mit 3% Cr und 0.5% Mo. Ausg. 1970. – T. 7: Warmverformte Vergütungsstähle mit 1% Cr. Ausg. 1970. – T. 8: Warmverformte Chrom-Nickel-Molybdän-Vergütungsstähle. Ausg. 1970. (All these rec. are replaced by ISO 683/1 April 1987).
44. Tacke, G., u. W. Knorr: In: Werkstoffkunde der gebräuchlichen Stähle. T. 2. Hrsg. vom Verein Deutscher Eisenhüttenleute. Düsseldorf 1977. S. 1/10.
45. Wyss, U.: Härterei-techn. Mitt. 6. (1951) H. 2, S. 9/40.
46. Metals handbook. Vol. 1, 8th ed. [Hrsg.:] American Society for Metals. Metals Park/Novelty, Ohio 1961.
47. Vetter, K.: Härterei-techn. Mitt. 33 (1978) S. 84/89.
48. Engineer, S.: TEW Techn. Ber. 2 (1976) S. 125/29. – Cook, W. T., u. P. T. Arthur: In: Heat Treatment 79. [Hrsg.:] The Metals Society. Book Nr. 261. London 1980. S. 126/31. – Beneš, F., u. E. Pribil: Hutn. listy 32 (1982) S. 125/29.
49. Andre, K. H., L. Ettenreich u. F. Früngel: Härterei-techn. Mitt. 27 (1972) S. 45/51.
50. Stähli, G.: Mater. u. Techn. (1977) H. 2, S. 59/66.
51. DIN 50190. Härtetiefe wärmebehandelter Teile. T. 1. Ermittlung der Einsatzhärtungstiefe. Ausg. Nov. 1978. T. 2. Ermittlung der Einhärtungstiefe nach Randschichthärten. Ausg. März 1979. T. 3. Ermittlung der Nitrierhärtetiefe. Ausg. März 1979.
52. Induction hardening and tempering. [Hrsg.:] American Society for Metals. Metals Park/Ohio 1964. (ASM monograph on heat treating) S. 130.
53. Rettig, H.: Ind.-Bl. 63 (1963) S. 347/54.

54. Forrest, P. G.: Fatigue of metals. Oxford/London/New York 1962.
55. Tauscher, H., u. H. Buchholz: Inst. Leichtbau-Mitt. 8 (1969) S. 368/76.
56. Wilshaw, C. T.: Heat Treatm. Metals 5 (1978) Nr. 1, S. 13/16.
57. DIN 17212. Stähle für Flamm- und Induktionshärten. Ausg. Aug. 1972.
58. EURONORM 86-70. Stähle für Flamm- und Induktionshärten. Ausg. März 1970.
59. ISO/R 683 T. 12: Stähle für Flamm- und Induktionshärten. Ausg. 1972. (Withdrawn).
60. DIN 17211. Nitrierstähle. Ausg. Aug. 1970 (superseded by a new ed.: Ausg. April 1987).
61. EURONORM 85-70. Nitrierstähle. Ausg. März 1970.
62. ISO/R 683 T. 10: Nitrierstähle Ausg. 1972. (superseded by a new edition: ISO 683/10 Ausg. April 1987).
63. Jonck, R., u. G. Kunze: Z. wirtsch. Fertigung 74 (1979) S. 445/54.
64. Böhmer, S., W. Schröter, J. M. Lachtin, W. Lerche u. I. D. Kogan: Neue Hütte 24 (1979) S. 7/12.
65. Homerberg, V. O.: Iron Age 136 (1936) 15. Okt. S. 49/52, 54, 56, 61/62, 64, 66 U. 98. – Wiegand, H., u. M. Koch: Metalloberfläche 12 (1958) S. 69/74.
66. Vetter, K.: VDI-Z. 118 (1976) S. 997/1003. Derselbe: VDI-Z. 119 (1977) S. 163/65.
67. Liedtke, D.: Nitrieren. Hrsg. von der Beratungsstelle für Stahlverwendung. Düsseldorf 1974. (Merkblätter über sachgerechte Stahlverwendung Nr. 447).
68. Hodgson, C. D., u. H. O. Waring: J. Iron Steel Inst. 151 (1945) S. 55/70.
69. Kunze, E.: Härterei-techn. Mitt. 17 (1962) S. 233/36.
70. Domalski, H.-H., u. K. Vetter: In: Werkstoffkunde der gebräuchlichen Stähle. T. 2. Düsseldorf 1977. S. 55/74.
71. Spies, H. J.: Inst. Leichtbau-Mitt. 20 (1981) S. 200/05.
72. Wiegand, H.: Härterei-techn. Mitt. 21 (1969) S. 263/70.
73. Wiegand, H., u. M. Koch: Metalloberfläche 12 (1958) S. 97/101.
74. Schinn, R.: Die deutsche Entwicklung von Schmiedestücken für Wellen von Turboaggregaten. VGB Tech. wiss. Berichte „Wärmekraftwerke" VGB-TW 503. Hrsg. von VGB Technische Vereinigung der Großkraftwerksbetreiber e. V. 1983.
75. Schinn, R.: Zur Entwicklung der Homogenität und Homogenitätsprüfung sowie der Zähigkeit von Generator- und Turbinenwellen auf dem europäischen Kontinent. VGB Tech. wiss. Berichte „Wärmekraftwerke" VGB-TW 502. 1981.
76. Schieferstein, U., u. W. Wiemann: VGB-Kraftwerkstechn. 56 (1976) S. 268/73 u. 340/46.
77. Wolf, H., E. Stücker u. H. Nowack: Arch. Eisenhüttenwes. 48 (1977) S. 173/78.
78. Stahl-Eisen-Prüfblatt 1950–66. Warmrundlauf-Versuch an Turbinenläufern. 1. Ausg. Dez. 1966. (New ed. is being prepared).
79. Tix, A.: Stahl u. Eisen 76 (1956) S. 61/68.
80. Knorr, W.: In: 6. Internationale DH-BV-Vakuum-Konferenz, Stresa (Italien) 1972. S. 81/95.
81. Steffen, R.: Stahl u. Eisen 96 (1976) S. 840/46. – Jacobi, H.: Gießen und Erstarren von Stahl I. Zusammenfassender Bericht über die Ergebnisse des Gemeinschaftsprogrammes. Vertrag-Nr. 6210-50. Düsseldorf 1975.
82. Flemmings, M. C.: Scand. J. Metallurg. 5 (1976) S. 1/15.
83. Hochstein, F.: Stahl u. Eisen 95 (1975) S. 785/89.
84. Hölschermann, H., u. D. Reiber: Stahl u. Eisen 101 (1981) S. 573/76.
85. Kühnelt, G., u. P. Machner: Berg- u. hüttenm. Mh. 121 (1976) S. 179/86.
86. Kawaguchi, S., Y. Nakagawa, J. Watanabe, S. Shikano, K. Maeda u. N. Kanno: In: Congrès international de la grosse forge, Paris, 20–25 April 1975. Bd. 1. Paris 1976. S. 187/221.
87. Austel, W., H. Heymann, Ch. Maidorn u. W. Mogendorf: Wie unter [86], S. 143/63.
88. Jauch, R., A. Choudhury, A. Löwenkamp u. F. Regnitter: Stahl u. Eisen 95 (1975) S. 408/13.
89. Piehl, K. H.: Stahl u. Eisen 95 (1975) S. 837/46.
90. Haumer, H. Ch., u. W. Knorr: In: II. Kolloquium über Spurenelemente im Stahl, deren Einfluß auf die plastische Verformung und auf die Eigenschaften des Stahles im warmen und kalten Zustand. [Hrsg.:] Institut de Recherches de la Sidérurgie Française, Max-Planck-Institut für Eisenforschung u. Metalurški Inštitut Ljubljana. Portoroz 1967. S. 157/75.
91. Forch, K.: In: Grundlagen der Wärmebehandlung von Stahl. Berichte, gehalten im Kontaktstudium „Werkstoffkunde Eisen und Stahl II." Hrsg. von W. Pitsch. Düsseldorf 1976. S. 156/71.
92. Stahl-Eisen-Werkstoffblatt 550–76. Stähle für größere Schmiedestücke, Gütevorschriften. 3. Ausg. Aug. 1976.
93. Stahl-Eisen-Werkstoffblatt 555. Stähle für größere Schmiedestücke für Bauteile von Turbinen- und Generatorenanlagen. Ausg. Aug. 1984.
94. ASTM A 469 Specification for Vacuum-Treated Steel Forgings for Generator Rotors. [Hrsg.:] American Society for Testing and Materials. Philadelphia, Pa. 1984. Vol. 01. 05.

95. Schinn, R., u. U. Schieferstein: VGB-Kraftwerkstechn. 53 (1973) S. 182/95.
96. Ricken, H. G.: Unveröff. interner Forschungsbericht Krupp Stahl AG, 15.12.64: siehe auch Stahl u. Eisen 95 (1975) S. 842.
97. Stahl-Eisen-Werkstoffblatt 640–75. Stähle für Bauteile im Primärkreislauf von Kernenergie-Erzeugungsanlagen. I. Ausg. Mai 1975. (Withdrawn, partly replaced by KTA 3201.1 (Data sheet of the „Kerntechnischer Auschnuss")).
98. Tacke, G.: Stahl u. Eisen 94 (1974) S. 792/804.
99. Schaaber, O.: Härterei-techn. Mitt. 21 (1966) S. 55/63.
100. DIN 17 210. Einsatzstähle, Technische Lieferbedingungen. Ausg. Dez. 1969. (new ed.: Ausg. Sept. 1986).
101. DIN 50 601. Metallographische Prüfverfahren. Bestimmung der Austenit- und der Ferritkorngröße von Stahl und Eisenwerkstoffen. Ausg. Aug. 1985. – EURONORM 103. Mikroskopische Ermittlung der Ferrit- und Austenitkorngröße von Stählen. Ausg. Nov. 1971.
102. Burns, J. L., T. L. Moore u. R. S. Archer: Trans. Amer. Soc. Met. 26 (1938) S. 1/36.
103. Just, E.: Härterei-techn. Mitt. 23 (1968) S. 85/100.
104. Steinen, A. von den, u.W. Schmidt: In: Grundlagen des Festigkeits- und Bruchverhaltens. Berichte, gehalten im Kontaktstudium „Werkstoffkunde Eisen und Stahl". Hrsg. von W. Dahl. Düsseldorf 1974. S. 255/79.
105. Randak, A., u. R. Eberbach: Härterei-techn. Mitt. 24 (1969) S. 201/09.
106. Metha, K. K., u. L. Rademacher: TEW Techn. Ber. 2 (1976) S. 118/24.
107. Randak, A., u. E. Kiderle: Härterei-techn. Mitt. 21 (1966) S. 190/98.
108. Dressel, P. G., u. K. Vetter: Härterei-techn. Mitt. 27 (1972) S. 100/09.
109. Steinen, A. von den, u. H. P. Wisniowski: DEW Techn. Ber. 11 (1971) S. 222/29.
110. Kölzer, H.: Härterei-techn. Mitt. 16 (1961) S. 72/78.
111. Randak, A., R. Eberbach u. H. Langenhagen: Stahl u. Eisen 85 (1965) S. 1452/61.
112. Bierwirth, G.: Erörterungsbeitrag zu [105]; s. bes. S. 207/08.
113. Reimers, H.: Untersuchungen über den Einfluß des Restaustentitgehaltes auf die mechanischen Eigenschaften einsatzgehärteter Stähle. Aachen 1981 (Dr.-Ing.-Diss. Techn. Hochsch. Aachen.)
114. Razim, C.: Härterei-techn. Mitt. 22 (1967) S. 317/29.
115. Razim, C.: Härterei-techn. Mitt. 23 (1968) S. 1/9.
116. Brugger, H.: Schweiz. Arch. angew. Wiss. Techn. 36 (1970) S. 219/29.
117. Beumelburg, W.: Das Verhalten von einsatzgehärteten Proben mit verschiedenen Oberflächenzuständen und Randkohlenstoffgehalten im Umlaufbiege-, statischen Biege- und Schlagbiegeversuch. Karlsruhe 1973. (Dr.-Ing.-Diss. Univ. Karlsruhe.)
118. Beumelburg, W., H. Brugger, H. Gulden, A. Randak u. K. Vetter: Härterei-techn. Mitt. 29 (1974) S. 159/69.
119. Funatani, K.: Härterei-techn. Mitt. 25 (1970) S. 92/98.
120. Liedtke, D.: Einsatzhärten. Merkblatt Nr. 452 der Beratungsstelle für Stahlverwendung. Düsseldorf 1981.
121. Bungardt, K., H. Brandis u. P. Kroy: Härterei-techn. Mitt. 19 (1964) S. 146/53.
122. Brandis, H., u. P. Kroy: Draht-Welt 51 (1965) S. 501/09.
123. Meyer, H. U.: Stahl u. Eisen 76 (1956) S. 68/78.
124. Bungardt, K., H. Preisendanz u. Th. Mersmann: Arch. Eisenhüttenwes. 36 (1965) S. 709/24 u. 809/16.
125. Albrecht, C.: Härterei-techn. Mitt. 9 (1955) S. 9/26.
126. Chatterjee-Fischer, R.: Härterei-techn. Mitt. 28 (1973) S. 259/66.
127. Chatterjee-Fischer, R.: Erörterungsbeitrag zu [118]; s. bes. S. 166/67.
128. Jindal, P. C.: Metal Progr. 103 (1973) Nr. 4, S. 78/79.
129. Brandis, H.: DEW Techn. Ber. 5 (1965) S. 49/57.
130. Schmidtmann, E., M. Majdic u. H. Schenck: Arch. Eisenhüttenwes. 32 (1961) S. 851/56.
131. Bungardt, K., E. Kunze u. H. Brandis: DEW Techn. Ber. 5 (1965) S. 1/12.
132. Brugger, H.: Härterei-Techn. u. Wärmebehandl. 2 (1956) S. 9/12.
133. Rose, A.: Klepzig Fachber. 78 (1970) S. 424/29.
134. Finnern, B.: Arch. Eisenhüttenwes. 25 (1954) S. 345/50.
135. Berns, H.: Z. Werkst.-Techn. 8 (1977) S. 149/57.
136. Brugger, H.: Härterei-techn. Mitt. 19 (1964) S. 100/14.
137. Just, E.: Härterei-techn. Mitt. 17 (1962) S. 148/59.
138. Stenzel, W.: Härterei-techn. Mitt. 4 (1949) S. 27/51.
139. Staudinger, H.: Erörterungsbeitrag zu [136]; s. bes. S. 114.
140. Sigwart, H.: Härterei-techn. Mitt. 12 (1958) S. 9/22.

141. Klein, H.: Deutsche Kraftfahrforschung, Technischer Forschungsbericht, Zwischenbericht Nr. 128, 1944.
142. Wicke, D.: Das Festigkeitsverhalten von legierten Einsatzstählen bei Schlagbeanspruchung. Berlin 1976. (Dr.-Ing.-Diss. Techn. Univ. Berlin.)
143. Vetter, K.: VDI-Z, 118 (1976) S. 997/1003.
144. Feddern, G., u. E. Macherauch: Z. Metallkde. 65 (1974) S. 785/88.
145. Just, E.: Z. wirtsch. Fertigung 67 (1972) S. 311/19.
146. Scherer, R., K. Bungardt u. E. Kunze: Stahl u. Eisen 72 (1952) S. 1433/42.
147. Treppschuh, H., A. Randak, H. H. Domalski u. J. Kurzeja: Stahl u. Eisen 87 (1967) S. 1355/68.
148. Legat, A., u. A. Moser: Härterei-techn. Mitt. 19 (1964) S. 21/30.
149. Schreiber, R., H. Wohlfahrt u. E. Macherauch: Arch. Eisenhüttenwes. 48 (1977) S. 653/57. – Dieselben: Arch. Eisenhüttenwes. 49 (1978) S. 37/41. – Dieselben: Arch. Eisenhüttenwes. 49 (1978) S. 265/69.
150. Ulrich, M., u. H. Glaubitz: VDI-Z. 91 (1949) S. 577/83.
151. Beumelburg, W.: Erörterungsbeitrag zu [119]; s. bes. S. 98.
152. Rose, A., H. Sigwart u. E. Theis: Stahl u. Eisen 81 (1961) S. 800/08.
153. EURONORM 84–70. Einsatzstähle: Gütevorschriften. Ausg. März 1970.
154. ISO 683. Heat-treatable steels, alloy steels and free-cutting steels. P. 11. Wrought case hardening steels. Ed. 1987-04-01.
155. Büttinghaus, A., W. Peter u. A. Randak: Stahl u. Eisen 26 (1964) S. 1768/76.
156. DIN 1662. Nickel- und Chrom-Nickel-Stahl. Ausg. Juli 1928. (Withdrawn).
157. Kiessler, H.: Schweiz. Arch. angew. Wiss. Techn. 22 (1956) S. 93/96.

C 6 Steels for Cold Forming

1. Stahl-Eisen-Prüfblatt 1123 Zylinderstauchversuch-zur Ermittlung von Kaltfliess presskurven Ausg. Aug. 1986.
2. Straßburger, C., u. G. Robiller: Stahl u. Eisen 93 (1973), S. 1164/70.
3. VDI-Richtlinien 3200. Fließkurven metallischer Werstoffe. Ausg. Okt. 1978.
4. Heil, H.-P., u. A. Lienhart: Draht-Welt 56 (1970) S. 205/13.
5. Robiller, G., W. Schmidt u. C. Straßburger: Stahl u. Eisen 98 (1978) S. 157/63.
6. Bühler, H., u. H. Meyer-Nolkemper: Ind-Anz. 88 (1966) S. 269/73.
7. Domalski, H., u. H. Schmücker: Stahl u. Eisen 90 (1970) S. 1087/96.
8. Jonck, R., E. Just u. D. Wicke: Z. wirtsch. Fertigung 69 (1974) S. 419/24.
9. Leykamm, H.: Draht 29 (1978) S. 648/51.
10. Karl, A.: Neue Hütte 21 (1976) S. 501/02.
11. Billigmann, J.: Stauchen und Pressen. 2. Aufl. von H.-D. Feldmann. München 1973.
12. Stahl-Eisen-Prüfblatt 1520 Mikroskopische Prüfung der Carbidausbildung in Stählen mit Bildreihen. Ausg. März 1978.
13. Stanz, A., u. R. Krefting: Arch. Eisenhüttenwes. 49 (1978) S. 325/31
14. DIN 1654 Teil 1. Kaltstauch- und Kaltfließpreßstähle. Technische Lieferbedingungen. Allgemeines. Ausg. März 1980. (New ed. is being prepared: Ausg. 1988).
15. Esselborn, K.: Techn. Mitt. Stahlwerke Röchling-Burbach Nr. 41, 1975, S. 1/5.
16. Stahl-Eisen-Lieferbedingungen 055E. Warmgewalzter Stabstahl und Walzdraht. Oberflächen-Güteklassen. Ausg. März 1980.
16a. DIN 50191. Prüfung von Eisenwerkstoffen, Stirnabschreckversuch. Ausg. Sept. 1987.
17. DIN 1654 Teil 4. Kaltstauch- und Kaltfließpreßstähle. Technische Lieferbedingungen für Vergütungsstähle. Ausg. März 1980.
18. Schuster, M.: Werkstatt u. Betr. 107 (1974) S. 642/44.
19. Desalos, Y., R. Laurent, D. Thivellier u. D. Rousseau: Rev. Métallurg. 77 (1980) S. 1011/25.
20. Cooksey, R. J.: Metal Forming 35 (1968) S. 98/111.
21. Bäcker, L., u. X. Chevrant: Rev. Métallurg. 72 (1975) S. 163/76.
22. Irvine, K. J., F. B. Pickering, W. C. Heselwood u. M. Atkins: J. Iron Steel Inst 186 (1957) S. 54/67.
23. Tricot, R., B. Champin u. D. Thivellier: Rev. Métallurg. 69 (1972) S. 721/35.
24. Domalski, H. H.: In: VDI-Ber. Nr. 266, 1976, S. 95/100.

25. Engineer, S.: TEW Techn. Ber. 2 (1976) S. 125/29.
26. Koul, M. K., u. C. L. McVicker: Metal Progr. 110 (1976) Nr. 6, S. 40/44.
27. Jonck, R.: Z. wirtsch. Fertigung 66 (1971) S. 503/07.
28. Köstler, J., u. H. Sidan: Z. wirtsch. Fertigung 72 (1977) S. 207/10.
29. Jonck, R.: Wie unter [8], S. 525/32.
30. Köstler, H. J., u. M. Fröhlke: Arch. Eisenhüttenwes. 46 (1975) S. 655/59.
31. Soraya, S., S. Lukas u. E. Möbius: Draht 24 (1973) S. 470/77.
32. Möbius, E., u. S. Soraya: Techn. Mitt. Stahlwerke Röchling-Burbach Nr. 42, 1975, S. 1/7.
33. Domalski, H., u. H. Schücker: Wie unter [7], S. 1115/20.
34. Reissner, J., H. Mülders u. E. Plänker: Bänder Bleche Rohre 11 (1979) S. 487/92.
35. Bäcker, L., R. El Haik u. Y. Roger: Rev. Métallurg. 76 (1979) S. 305/22.
36. Kaiser, W., u. W. Weingarten: Stahl u. Eisen 101 (1981) S. 897/903.
37. DIN 17 111. Kohlenstoffarme unlegierte Stähle für Schrauben, Muttern und Niete. Technische Lieferbedingungen. Ausg. Sept. 1980.
38. DIN 1654 Teil 2. Kaltstauch- und Kaltfließpreßstähle. Technische Lieferbedingungen für nicht für eine Wärmebehandlung bestimmte beruhigte unlegierte Stähle. Ausg. März 1980.
39. DIN 1654 Teil 3. Kaltstauch- und Kaltfließpreßstähle. Technische Lieferbedingungen für Einsatzstähle. Ausg. März 1980.
40. DIN 1654 Teil 5. Kaltstauch- und Kaltfließpreßstähle. Technische Lieferbedingungen für nichtrostende Stähle. Ausg. März 1980.
41. Euronorm 119. Kaltstauch- und Kaltfließpreßstähle. Gütevorschriften. Ausg. Juni 1974.
42. ISO 4954. Kaltstauch- und Kaltfließpreßstähle. Gütevorschriften. Ausg. 1979.
43. DIN 267 Teil 3. Mechanische Verbindungselemente. Technische Lieferbedingungen. Festigkeitsklassen für Schrauben aus unlegierten oder legierten Stählen; Umstellung der Festigkeitsklassen. Aüsg. Aug. 1983.
44. DIN ISO 898 Teil 1. Mechanische Eigenschaften von Verbindungselementen: Schrauben. Ausg. April 1979.
45. DIN 17 240. Warmfeste und hochwarmfeste Werkstoffe für Schrauben und Muttern. Gütevorschriften. Ausg. Juli 1976.
46. DIN 17 211. Nitrierstähle. Gütevorschriften. Ausg. April 1987.
47. DIN 17 230. Wälzlagerstähle. Technische Lieferbedingungen. Ausg. Sept. 1980.

C 7 Unalloyed Wire Rod for Cold Drawing

1. Stahldraht-Erzeugnisse. Hrsg.: Ausschuß für Drahtverarbeitung im Verein Deutscher Eisenhüttenleute. 2. Bde. Düsseldorf 1956.
2. Herstellung von Stahldraht. T. 1 und 2. Hrsg. vom Verein Deutscher Eisenhüttenleute. Düsseldorf 1969.
3. Steel Wire Handbook. Hrsg.: The Wire Association. Branford. Con. Vol. 1. 1965. Vol. 2. 1969. Vol. 3. 1972. Vol. 4. 1980.
4. Stahldraht, Herstellung und Anwendung. Leipzig 1973.
4a. Steel Products Manual, Wire and Rods, Carbon steel, American Iron and Steel Inst. (ed). New York, 3rd, Ed. 1977 (or latest ed.).
5. Beck, H.: Stahl u. Eisen 86 (1966) S. 1005/14.
6. Flügge, J., W. Heller, E. Stolte u. W. Dahl: Arch. Eisenhüttenwesen 47 (1976) S. 635/40.-Flügge, J., u. W. Heller: Forschungsvorhaben der Kommission der Europäischen Gemeinschaft für Kohle und Stahl. Vetrags-Nr. 6210 KC/1/102. Hrsg: Verein Deutscher Eisenhüttenleute. Düsseldorf 1979.
7. DIN 17 140 Teil 1. Walzdraht zum Kaltziehen. Technische Lieferbedingungen für Grundstahl und unlegierte Qualitätsstähle. Ausg. März 1983. – DIN 17 140 Teil 2. Walzdraht zum Kaltziehen. Technische Lieferbedingungen für unlegierte Edelstähle sowie für legierte Edelstähle für Federn. In Vorbereitung.
7a. Stumm, W.: Drahtwelt 11 (1978) S. 433/37. – Thiery, D. u. J. Neuhauss: Stahl u. Eisen 98 (1978) S. 867/73.-Tassin, S. u. F. Tesche: Wire J. Int. 16 (1983) Nr. 9, S. 156/58. 161/62 u. 165/66.
7b. Kremer, K.-J.: Stahl u. Eisen, 103 (1983) S. 359/65.
8. McLean, D. W., A. B. Dove u. J. H. Hitchcock: Wire W. Prod. 39 (1964) S. 1606/15. 1622/23 u. 1668/70.
8a. Dietl, W.: Stahl u. Eisen 90 (1970) S. 1223/34

8b. Marcol, J. u. Z. Mikulec: Drahtwelt 72 (1986) Nr. 1/2, S. 3/7.
 9. Beck, H.: Stahl u. Eisen 101 (1981) S. 541/51.
10. Jauch. R., W. Courths, R. Hentrich, H.-P. Jung, H. Litterscheidt u. E. Sowka: Stahl u. Eisen 104 (1984) S. 429/34.
10a. Bersch, B. u. a.: Stahl u. Eisen 106 (1986) S. 323/31.
10b. Hoss, K. F.: Stahl u. Eisen 106 (1986) S. 313/16.
11. Bombeke, M., W. Storme u. C. Vandenbussche: Proceedings of the technical conference on wire rods. Harrowgate 1974. Hrsg: Brit. Ind. Steel Prod. Ass. Sheffield 1975. S. 67/82.
12. Nakamura, Y., E. Takahashi, N. Hatsuoka u. S. Ashida: Wire J. 11 (1978) Nr. 9. S. 110/13.
13. Berns, H.: Stahl u. Eisen 98 (1978) S. 662/64.
14. Ammerling, W. J., H. Muckli u. K. D. Richter: Draht 20 (1978) S. 51/61.
15. Walzdraht-Fehleratlas. Hrsg.: Verein Deutscher Eisenhüttenleute. Düsseldorf 1969.
16. Pawelski, O., J. Becker u. A. Punter: Stahl u. Eisen 98 (1978) S. 1082/88.
17. Lewis, D.: Wire W. Prod. 32 (1957) S. 1179/82 u. 1262/64.
18. Dietl, W.: Stahl u. Eisen 99 (1979) S. 1168/72.
19. a) Wie unter [3], Vol. 1. S. 75/79.
    b) Funke, P., u. M. Heinritz: Stahl u. Eisen 87 (1967) S. 293/300. – Böckenhoff, H., F. Schwier, G. Rockrohr u. E. Viebahn: Stahl u. Eisen 87 (1967) S. 300/311.
20. Prediger, P. J., u. I. M. Park; Wire W. Prod. 43 (1968) Nr. 5. S. 43/46, 48/50, 52/53 u. 111/12. – Beck, H., u. W. Dietl: Wire J. 2 (1969) Nr. 12. S. 54/59.–Beck. H.: Wire J. 5 (1972) Nr. 8, S. 48/55.
21. Beck, H.: Stahl u. Eisen 87 (1967) S. 316/17.-Beck, H., u. W.-D. Brand: Wire W. Prod. 43 (1968) Nr. 2, S. 78/84 u. 157/58. – Malmgren, N.-G., u. S. G. Tärnblom: Draht 26 (1975) S. 220/27.
22. Geitz, W.: Stahl u. Eisen 88 (1968) S. 14/21. – Kosmider, H., u. G. Geck: Draht-Welt 57 (1971) S. 548/58, – Ammerling, W. J.: Draht-Welt 57 (1971) S. 62/68.
23. Buch, E., H. D. Hirschfelder, R. A. Marzinkewitsch u. W. Krenn: Draht-Welt 57 (1971) S. 371/75. – Vgl. Buch, E.: Stahl u. Eisen 97 (1977) S. 125/27.
24. D. Pat. Schr. Nr. 15 08 404 vom 25.10.1966, 15 08 405 vom 5.10.1966, 12 62 324 vom 3.11.1966, 16 67 097 vom 26.1.1967, 12 81 469 vom 10.2.1967, 16 02 081 vom 4.3.1967, 17 58 303 vom 10.5.1968 u. 19 37 918 vom 25.7.1969. – Yamakoshi, N., T. Kaneda, A. Suzuki, E. Niina, Y. Yanagi u. N. Hatsuoka: Tetsu to Hagane 58 (1972) S. 1969/83. – Nagai, Ch., Y. Nakamura u. E. Takahashi: Wire J. 10 (1977) Nr. 6, S. 56/62. – DAS 15 83 987 vom 10.1.1968.
25. D. Pat. Schr. OS 21 52 514 vom 21.10.1971. Draht 24 (1973) S. 386. – Anon. Draht 24 (1973) S. 386: Economopoulos, M. u. N. Lambert: Wire (14) 1981, Nr. 3, S. 90/95. – Yamada, K. u. a. Trans. ISIJ 26 B (1986) S. 102/103.
25a. D. Pat. Schr. Nr. 24 35 830, 24 35 831 Vom 25.7.1974.
25b. Heinke, R.: Stahl u. Eisen 106 (1986) S. 129/36.
25c. Tominga, J., K. Matsuoka u. S. Inoue: Wire J. Int. 18 (1985) Nr. 2, S. 62/66 u. 69/72.
26. Vlad. C. u. H. Paulitsch Vlad, C. M., K. Klimpel u. U. Feldmann: Proc. Int. Conf. on Technology and Applications of HSLA. Steels, Philadelphia 1983, ASM Publ. 8306-047. S. 1003/15. – D, Pat Schr. Schr. Nr. 23 45738 vom 11.9.1973. – Frommann, K. u. C. M. Vlad: Proc. Int. Conf. on HSLA Steels, Beijing 1985, Processing and Mechanical Properties of Temprimar Rod and Bar Products, ASM Publ. in print. – Vlad, C. M.: ibid. Comparison between the Tempcore and Temprimar Processes, ASM Publ. in print.
26a. Rigaut, G. u., S. Tassin: Wire J. Ind. 18 (1985) Nr. 8, S. 34, 36, 38 u 42.
26b. Mocbius, H. E. u. S. P. Soraya: Härterei-Techn. Mitt. 27 (1972) S. 323/28 u. Z. wirtsch. Fertigung 68 (1973) S. 47/52.
27. Kiefer, M., u. R. L. Randall: Wire J. 11 (1978) Nr. 4, S. 58/61.
27a. Schifferl, H.-A.: Stahl u. Eisen 106 (1986) S. 317/21.
28. Burggaller, W.: Draht-Welt 27 (1934) Nr. 13, S. 195/97 u. Nr. 14, S. 211/13.
29. Siebel, E., u. W. Panknin: Stahl u. Eisen 72 (1952) S. 1193/95.
30. Werkstoff-Handbuch Stahl u. Eisen. Hrsg. vom Verein Deutscher Eisenhüttenleute. 4. Aufl. Düsseldorf 1965. Blatt Q 21.
31. DIN 2078. Stahldrähte für Drahtseile. Ausg. März 1978.
32. DIN 779. Formstahldrähte für verschlossene Spiralseile; Maße und Technische Lieferbedingungen. Ausg. Dez. 1980.
33. DIN 3051 Teil bis 4. Drahtseile aus Stahldrähten. Teil 1, 3 u. 4 Ausg. März 1972, Teil 2 Ausg. April 1972.
34. DIN 21 254 Teil 1. Förderseile, Bühnenseile; Technische Lieferbedingungen, Litzenseile und Flachseile. Ausg. April 1980.
35. DIN 15 020 Teil 1. Hebezeuge; Grundsätze für Seiltriebe, Berechnung und Ausführung. Ausg. Febr. 1974.

35a. Tassin, S.: Wire J.. Int. 17 (1984) Nr. 7, S. 72/74, 77 u. 79.
36. DIN 17 223 Teil 1. Runder Federstahldraht. Gütevorschriften. Patentiert gezogener Federdraht aus unlegierten Stählen. Ausg. Dez. 1984. – DIN 17 223 Teil 2. Runder Federstahldraht. Gütevorschriften. Vergüteter Federdraht und vergüteter Ventilfederdraht aus unlegierten Stählen. Ausg. März 1964 (Neuausgabe in Vorbereitung).
37. Krickau, O., u. J. Huhnen: Draht 23 (1972) S. 586/92 u. 653/59.
38. DIN 2088. Zylindrische Schraubenfedern aus runden Drähten und Stäben; Berechnung und Konstruktion von Drehfedern (Schenkelfedern). Ausg. Juli 1969.
39. DIN 2089 Teil 1. Zylindrische Schraubendruckfedern aus runden Drähten und Stäben; Berechnung und Konstruktion. Ausg. Dez. 1984. – DIN 2089 Teil 2. Zylindrische Schraubenfedern aus runden Drähten und Stäben; Berechnung und Konstruktion von Zugfedern (Vornorm). Ausg Febr. 1963. (In the meantime Teil 2 has been withdrawn).
40. Vgl. Walz, K.: Draht 19 (1968) S. 604/12 u. 783/90.
41. Beck, H., R. W. Simon u. R. A. Weber: Wire J. 9 (1976) Nr. 9, S. 160/68.
41a. Baroux, M. u. G. Mangel: Wire J. Int. 17 (1984) Nr. 4, S. 66/71.
41b. Yamada, Y., S. Shimazu, Y. Oki u. K. Mizutani; Wire J. Int. 19 (1986) Nr. 4, S. 53, 55/56, 59/60, 62 u. 64/65.

C 8 Ultra High-Strength Steels

1. Randak, A., A. von den Steinen u. E. Gondolf: In: Werkstoffkunde der gebräuchlichen Stähle. T. 2. Düsseldorf 1977. S. 75/95.
2. Randak, A., u. K. Vetter: Arch. Eisenhüttenwes. 40 (1969) S. 285/95.
3. Engell, H.-J., u. M. O. Speidel: Werkst. u. Korrosion 20 (1969) S. 281/300.
4. Beachem, C. D.: Metallurg. Trans. 3 (1972) S. 437/51.
5. Kennedy, J. W., u. J. A. Whittaker: Corrosion Sci. 8 (1968) S. 359/75.
6. Lieurade, H. P., A. Boucher u. P. Rabbe: Circ. Inform. techn. 32 (1975) S. 225/72.
7. Schmidt, W.: TEW Techn. Ber. 1 (1975) S. 39/55.
8. Schmidt, W.: Draht 29 (1978) S. 185/91 u. 245/50.
9. Ritchie, R. O., u. R. M. Horn: Met. Trans. 9 A (1978) S. 331/41.
10. Jones, F. W., u. W. I. Pumphrey: J. Iron Steel Inst. 163 (1949) S. 121/31.
11. Decker, R. R., J. T. Eash u. A. J. Goldman: Trans. Amer. Soc. Metals 55 (1962) S. 58/76.
12. Peters, D. T., u. C. R. Cupp: Trans. metallurg. Soc. AIME 236 (1966) S. 1420/29.
13. Detert, K.: Arch. Eisenhüttenwes. 37 (1966) S. 579/89.
14. Bungardt, K., u. W. Spyra: Luftfahrttechn. – Raumfahrttechn. 13 (1967) S. 63/67. – Dieselben: DEW Techn. Ber. 9 (1969) S. 316/69.
15. Bourgeot, J., Ph. Meitrepièrre, J. Manenc u. B. Thomas: Mém. sci. Rev. Métallurg. 70 (1973) S. 125/38.
16. Courrier, R., G. Le Caer: Mém. sci. Rev. Métallurg. 71 (1974) S. 692/709.
17. Vetter, K.: Unveröffentlichte Untersuchungen der Metallurgischen Zentrale der Krupp Stahlwerke Südwestfalen AG, Siegen.
18. Altstetter, C. J., M. Cohen u. B. L. Averbach: Trans. Amer. Soc. Metals 55 (1962) S. 287/300.
19. Horn, R. M., u. R. O. Ritchie: Metallurg. Trans 9 A (1978) S. 1039/53.
20. Materkowski, J. P. u. G. Krauss: Metallurg. Trans. 10 A (1979) S. 1643/51.
21. Steinen, A. v. d.: Härterei-techn. Mitt. 17 (1962) S. 210/19.
22. Murray, J. D.: In: High-strength steels. London 1962 (Spec. Rep. Iron Steel Inst. No. 76.) S. 41/50.
23. Sejnoha, R.: Freiberg. Forsch.-H., Reihe B, Nr. 121, 1966, S. 7/81.
24. Matas, S. J., M. Hill u. H. P. Munger: Metals Engng. Quart. 3 (1963) Nr. 3, S. 7/17.
25. Baus, A., J. C. Charbonnier, H.-P. Lieurade, B. Marandet, L. Roesch u. G. Sanz: Rev. Metallurg. 72 (1975) S. 891–935.
26. Gulden, H.: Unveröffentlichte Untersuchungen, Qualitätswesen und Forschung, Krupp Stahlwerke Südwestfalen AG, Siegen.
27. Bungardt, K., O. Mülders u. R. Meyer-Rhotert: Arch. Eisenhüttenwes. 37 (1966) S. 381/89.
28. Roberts, G. A., u. J. C. Hamaker jr.: Härterei-techn. Mitt. 16 (1961) S. 65/75.
29. Schempp, C. G., u. W. A. Morgan: Trans. metallurg. Soc. AIME 224 (1962) S. 420/29.
30. Vetter, K.: Härterei-techn. Mitt. 20 (1965) S. 113/25.
31. Hamaker jr., J. C., u. E. J. Vater: Proc. Amer. Soc. Test. Mater. 60 (1960) S. 691/720.

32. Shannon, J. L., G. B. Espey, A. J. Repko u. W. F. Brown: Proc. Amer. Soc. Test. Mater. 60 (1960) S. 761/77.
33. Carter C. S.: Metallurg. Trans. 1 (1970) S. 1551/59.
34. Weßling, W., u. K. Vetter: Klepzig Fachber. 76 (1968) S. 677/83 u. 744/52.
35. Bungardt, K., W. Spyra u. A. von den Steinen: Arch. Eisenhüttenwes. 39 (1968) S. 719/31.
36. Magnée, A., J. M. Drapier, J. Dumont, D. Coutsouradis u. L. Habraken: Cobalt-containing high-strength steels. Brüssel 1974.
37. Kunitake, T., u. Y. Okada: The Sumitomo Search No. 20, Nov. 1978, S. 55/64.
38. Soeno, K., T. Kuroda u. K. Taguchi: Trans. Iron Steel Inst. Japan 19 (1979) S. 484/89.
39. Grange, R. A.: Trans. Amer. Soc. Metals 59 (1966) Nr. 1, S. 26/48.
40. Peter, W., u. H. Finkler: Härterei-techn. Mitt. 24 (1969) S. 210/16.
41. Frodl, D., E. Plänker u. K. Vetter: Stahl u. Eisen 101 (1981) S. 75/80.
42. Rose, A., u. H. P. Hougardy: Z. Metallkde. 58 (1967) S. 747/52.
43. Lehnert, W.: Neue Hütte 13 (1968) S. 716/22.
44. Marshall, C. W., J. H. Gehrke, A. M. Sabroff u. F. W. Boulger: J. Metals 18 (1966) S. 328/36.
45. Zackay, V. F., u. E. R. Parker: In High-strength materials. Ed.: V. F. Zackay. New York, London, Sydney 1965. S. 130/54.
46. Bernstein, M. L., u. K. E. Hensger: Thermomechanische Behandlung und Festigkeit von Stahl. Leipzig 1977.
47. Mulherin, J. H.: Trans. AIME, Ser. D, J. basic Engng., 88 (1966) S. 777/82.
48. Carter, C. S.: Corrosion, Houston 25 (1969) S. 423/31.
49. Hanna, G. L., u. E. A. Steigerwald: Influence of environment on crack propagation and delayed failures in high-strength steels. Techn. Doc. Report RTD-TDR-63-4225 (1964) AD 433 286.
50. Kroupa, K. M., u. P. S. Venkatesan: Engng. Fracture Mech. 8 (1976) S. 547/53.
51. Steigerwald, E. A.: Proc. Amer. Soc. Test. Mater. 60 (1960) S. 750/60.
52. Brandis, H., S. Engineer, W. Spyra u. A. v. d. Steinen: TEW Techn. Ber. 4 (1978) S. 54/62.
53. Schütz, W.: Über eine Beziehung zwischen der Lebensdauer bei konstanter und veränderlicher Beanspruchungsamplitude und ihre Anwendbarkeit auf die Bemessung von Flugzeugbauteilen. München 1965. (Dr.-Ing.-Diss. Techn. Hochsch. München)
54. Laborde, L., u. D. Douillet: Rev. Métallurg. 61 (1964) S. 671/81.
55. Ladoux, G.: Härterei-techn. Mitt. 23 (1968) S. 14/21.
56. Pomey, G., u. P. Rabbe: Rev. Métallurg. 67 (1970) S. 87/97.
57. Tuffnell, G. W., D. L. Pasquine u. J. H. Olson: Trans. Amer. Soc. Metals 59 (1966) S. 769/83.
58. Hall, A. M., u. C. J. Slunder: The metallurgy, behavior and application of the 18-percent nickel maraging steels. NASA SP-5051 (1968).
59. Weigel, K.: Z. wirtsch. Fertigung 61 (1966) S. 107/11.
60. Bailey, N., u. C. Roberts: Weld. J. 57 (1978) S. 15/28.
61. Machining difficult alloys. [Hrsg.:] American Society for Metals. Ohio 1962. (ASD-TR 62-7-634.)
62. Olofson, C. T., J. A. Gurklis u. F. W. Boulger: Machining and grinding of ultrahigh-strength steels and stainless steel alloys. NASS-SP-5084, AD 639 654 (1970).
64. Munger, H. P.: Kobalt 44 (1969), S. 110/21.
65. Coutsouradis, D., N. Lambert, J. M. Drapier u. L. Habraken: Kobalt 37 (1967) S. 170/79.
66. Kunze, G., H.-D. Steffens, R. Deska u. H. Zeilinger: Radex-Rdsch. 1984, S. 428/36.
67. Kawabe, Y., S. Muneki u. K. Nakazawa: Trans. Iron Steel Inst. Japan 20 (1980) S. 682/89.
68. Muneki, S. Y. Kawabe, K. Nakazawa u. H. Yaji: Trans. Iron Steel Inst. Japan 20 (1980) S. 309/17.
69. Brandis, H., S. Engineer u. W. Schmidt: Härterei-techn. Mitt. 39 (1984) S. 224/32.

C 9 Elevated-temperature Steels (Creep Resistant and High Creep Resistant Steels) and Alloys

1. Atomwirtsch/Atomtechn. 23 (1978) Nr. 1, S. 1/2.
2. Wiegand, H., G. Granacher u. M. Sander: Arch. Eisenhüttenwes. 46 (1975) S. 533/49. – Kloos, K. H., J. Granacher, u. E. Abelt: Arch Eisenhüttenwes. 49 (1978) S. 259/63.
3. ASME boiler code NB 2300. Sect. III Ausg. 1974.

4. Wellinger, K., u. K. Lehr: Mitt. Verein. Großkesselbetr. 49 (1969) S. 190/201 – Lehr, K.: Technisch-wissenschaftliche Berichte der Staatlichen Materialprüfungsanstalt an der Technischen Hochschule Stuttgart 1969, H. 69–02.
5. Schoch, W.: In: VGB-Werkstofftagung 1969. [Hrsg.:] Vereinigung der Großkesselbetreiber. [Essen] 1970. S. 30/41. – Schoch, W., H. Spähn u. H. Kaes: In: VGB-Werkstofftagung 1971. [Hrsg.:] Technische Vereinigung der Großkraftwerksbetreiber. Essen [1972]. S. 93/104.
6. [Technische Regeln für Dampfkessel] TRD 301. Zylindrische Schalen. Berechnung. Ausg. Sept. 1972.
7. Pich, R.: VGB-Kraftwerkstechn. 61 (1981) S. 593/610.
8. Böhm, H.: Arch. Eisenhüttenwes. 45 (1974) S. 821/30.
9. Schmidt, W.: In: Festigkeits-und Bruchverhalten bei höheren Temperaturen. Berichte, gehalten im Kontaktstudium „Werkstoffkunde Eisen und Stahl IV". Hrsg. von W. Dahl u. W. Pitsch. Bd. 1. Düsseldorf 1980. S. 277/342.
10. DIN 50 118. Prüfung metallischer Werkstoffe. Zeitstandversuch unter Zugbeanspruchung. Ausg. Jan. 1982. – DIN 51 226. Werkstoffprüfmaschinen; Schwingprüfmaschinen. Begriffe. allgemeine Anforderungen. Ausg. Dez. 1977.
11. Granacher, J., u. H. Wiegand: Arch. Eisenhüttenwes. 43 (1972) S. 699/704.
12. Wickens, A., A. Strang u. G. Oakes: In: International conference on engineering aspects of creep. Sheffield, Sept. 1980. Bd. 1. S. 11/18.
13. ISO-Standard 6303. März 1981.
14. Arbeitsgemeinschaft für warmfeste Stähle, Arbeitsgemeinschaft für Hochtemperaturwerkstoffe. Vgl. Verein Deutscher Eisenhüttenleute, Mitgliedsverzeichnis. Düsseldorf 1980. S. 55/56. – Krause, M., J. Granacher, K.-H. Keienburg u. K.-H. Mayer: VGB-Kraftwerkstechn. 61 (1981) S. 19/25.
15. Arch. Eisenhüttenwes. 28 (1957) S. 245/323; 33 (1962) S. 27/60.
16. Ergebnisse deutscher Zeitstandversuche langer Dauer. Hrsg.: Verein Deutscher Eisenhüttenleute. Düsseldorf 1969.
17. Keienburg, K.-H., H. Granacher, H. Kaes, M. Krause, K.-H. Mayer u. H. Weber: In: VGB-Werkstofftagung 1980. [Veranst.] VGB Technische Vereinigung der Großkraftwerksbetreiber. Essen 1980. S. 61/128.
18. DIN 50 117. Prüfung von Stahl und Stahlguß. Bestimmung der DVM-Kriechgrenze. Ausg. Juni 1952. (In the meantime withdrawn).
19. Nechtelberger, E., F. Kreitner u. E. Krainer: Arch. Eisenhüttenwes. 44 (1973) S. 135/41.
20. Ruttmann, W., P. Bettzieche, E. Jahn, E.-O. Müller u. U. Schieferstein: Schweißen u. Schneiden 21 (1969) S. 8/17.
21. Jakobeit, W., J. P. Pfeiffer u. G. Ullrich: In: Energiepolitik in Nordrhein-Westfalen. Bd. 14. Düsseldorf 1982.
22. Kloos, K. H., J. Granacher u. H. Demus: Vortragsveranstaltung Arbeitsgemeinschaft warmfester Stähle, Düsseldorf, 4. Dez. 1981.
23. Grünling, H. W., B. Ilschner, S. Leistikow, A. Rahmel u. M. Schmidt: Werkst. u. Korrosion 29 (1978) S. 691/701.
24. Erker, A., Klotzbücher u. K.-H. Mayer: In: MAN Forsch.-H. Nr. 13, 1966/67. S. 62/76.
25. Kußmaul, K., J. Ewald u. G. Maier: Schweißen u. Schneiden 28 (1976) S. 250/55. – Kußmaul, K., J. Ewald u. V. Braun: Stahl u. Eisen 99 (1976) S. 244/50.
26. Dahl, W.: Arch. Eisenhüttenwes. 51 (1980) S. 7/14 u. 37/39.
27. Kloos, K. H., u. H. Diehl: VGB-Kraftwerkstechn. 59 (1979) S. 724/31.
28. Newhouse, D. L., u. D. R. Forest: In: 5. Internationale Schmiedetagung Terni. 6–9 Maggio 1970. [Hrsg.:] Camera di Commercio, Industria, Artigianato e Agricoltura di Terni. Terni. 1971. S. 739/55.
29. Granacher, J.: In: Festigkeits- und Bruchverhalten bei höheren Temperaturen. Berichte. gehalten im Kontakstudium „Werkstoffkunde Eisen und Stahl IV". Hrsg. von W. Dahl u. W. Pitsch. Bd. 2. Düsseldorf 1980. S. 324/777.
30. Vogels, H. A., P. König u. K.-H. Piehl: Arch. Eisenhüttenwes. 35 (1964) S. 339/51.
31. Straßburger, Chr., u. L. Meyer: Thyssenforsch. 3 (1971) S. 2/7.
32. Meyer, L., H.-E. Bühler u. F. Heisterkamp: Thyssenforsch. 3 (1971) S. 8/43.
33. Adrian, H.: In: Werkstoffkunde der gebräuchlichen Stähle, T. 1. Berichte, gehalten im Kontaktstudium „Werkstoffkunde Eisen und Stahl II". Hrsg. von W. Pitsch. Düsseldorf 1977. S. 205/21.
34. Meyer, L.: Stahl u. Eisen 97 (1977) S. 410/16.
35. Pense, A. W.: Diss. Lehigh Univ. 1962.
36. Forch, K.: In: Grundlagen der Wärmebehandlung von Stahl. Hrsg.: Verein Deutscher Eisenhüttenleute. Düsseldorf 1976. S. 155/71.
37. Sawada, S., u. T. Ohashi: Tetsu to Hagané 63 (1977) S. 1126/33.

38. Piehl, K.-H.: Mitt. Verein Großkesselbetreiber 50 (1970) S. 304/14.
38a. Vinckier, A. G., u. A. W. Pense: WCR Bull. 1975. Nr. 197, S. 1/35. – Schulze, G.: Schweißen u. Schneiden 27 (1975) S. 416/17.–van den Boom, J. E., u. J. P. Mulder: Weld. Res. internat. 2 (1972) S. 20/36. – Cerjak, H., u. W. Debray: In: VGB-Werkstofftagung 1971. [Hrsg.:] Technische Vereinigung der Großkraftwerksbetreiber. Essen [1972]. S. 23/31.
39. Forch, K.: Festigkeits- und Bruchverhalten bei höheren Temperaturen. Berichte. gehalten im Kontaktstudium „Werkstoffkunde Eisen und Stahl IV". Hrsg. von W. Dahl u. W. Pitsch. Düsseldorf 1980. S. 70/121.
40. Murray, J. D.: Brit. Weld. J. 14 (1967) 447/56.
41. Detert, K., R. Banga u. W. Bertram: Arch. Eisenhüttenwes. 45 (1974) S. 245/55.
42. Vougioukas, P.: Über die selektive und kumulative Wirkung von Legierungselementen auf das Relaxationsverhalten von Feinkornbaustählen. Clausthal 1973. (Dr.-Ing.-Diss. Techn. Univ. Clausthal.).
43. Vougioukas, P., K. Forch u. K.-H. Piehl: Stahl u. Eisen 94 (1974) S. 805/13.
44. Piehl, K.-H.: Stahl u. Eisen 93 (1973) S. 568/77.
45. Baird, J. D., u. A. Jamieson: J. Iron Steel Inst. 210 (1972) S. 847/56.
46. Glen, J., R. F. Johnson, J. M. May u. D. Sweetman: In: High-temperature properties of steels. London 1967. (Spec. Rep. Iron Steel Inst. No. 97.) S. 159/224.
47. Baker, T. N.: J. Iron Steel Inst. 205 (1967) S. 315/20.
48. Glen, J., J. Lessels, R. R. Barr u. G. G. Lightbody: In: Structural processes in creep. London 1961. (Spec. Rep. Iron Steel Inst. No. 70.) S. 222/45.
49. Glen, J., u. R. R. Barr: Wie unter [46]. S. 225/26.
50. Argent, B. B., M. N. Niekerk u. G. A. Redfern: J. Iron Steel Inst. 208 (1970) S. 830/43.
51. Fabritius, H., H. Imgrund u. H. Weber: Stahl u. Eisen 91 (1971) S. 1073/80.
52. Krisch, A., F. K. Naumann, H. Keller u. H. Kudielka: Arch. Eisenhüttenwes. 42 (1971) S. 353/57. – Naumann, F. K., H. Keller, H. Kudielka u. A. Krisch: Arch. Eisenhüttenwes. 42 (1971) S. 439/47.
53. Farrow, M.: In: Steels for reactor pressure circuits. London 1961. (Spec. Rep. Iron Steel Inst. No. 69) S. 89/100.
54. Stahl-Eisen-Werkstoffblatt 470–76. Hitzebeständige Walz- und Schmiedestähle. Ausg. Febr. 1976.
55. Petri, R., E. Schnabel u. P. Schwaab: Arch. Eisenhüttenwes. 51 (1980) S. 355/60.
56. Glen, J., u. J. D. Murray: Wie unter [53]. S. 40/53.
57. Yukitoshi, T., T. Abe, K. Nishida, H. Makiura, H. Yuzawa u. K. Fukushima: Sumitomo Search No. 13, 1975, S. 35/55.
58. Baerlecken, E., u. H. Fabritius: Arch. Eisenhüttenwes. 33 (1962) S. 261/67.
59. Kloos, K.-H., J. Granacher, H. Diehl u. Th. Polzin: Arch. Eisenhüttenwes. 48 (1977) S. 645/48.
60. Krause, M.: VDI-Z. 123 (1981) S. S98/S100 u. S105/S109.
61. Buchi, G. I., I. H. R. Page u. M. P. Sidey: J. Iron Steel Inst. 203 (1965) S. 291/98 u. 485.
62. Viswanathan, R., u. C. G. Beck: Metallurg. Trans. 6A (1975) S. 1997/2003.
63. Marrison, T., u. A. Hogg: In: Creep strength in steels and high temperature alloys. London 1974. (The Metals Society. Book No. 151.) S. 242/48.
64. Aronsson, B.: In: Die Verfestigung von Stahl. Symposium veranstaltet von der Climax Molybdenum Company in Zürich, 5. u. 6.5.1969. Greenwich/Conn. 1970. S. 37/87.
65. Fabritius, H.: In: VDI-Ber. 428. 1981. S. 79/91.
66. Houdremont, E., u. G. Bandel: Arch. Eisenhüttenwes. 16 (1942/43) S. 85/100.
67. Mellor, G. A., u. S. M. Barker: J. Iron Steel Inst. 194 (1960) S. 464/74.
68. Bungardt, K., H. Krainer u. H. Schrader: Stahl u. Eisen 84 (1964) S. 1796/811.
69. Siehe auch Steinen, A. von`den: Wie unter [9], Bd 2. S. 176/210.
70. Oppenheim, R., u. G. Lennartz: DEW Techn. Ber. 4. (1964) S. 1/8.
70a. Koch, W., G. Krisch, A. Schrader u. H. Rohde: Stahl u. Eisen 78 (1958) S. 1251/62.
70b. Schüller, J., P. Schwaab u. H. Ternes: Arch. Eisenhüttenwes. 35 (1964) S. 659/66.
70c. Bungardt, K., u. G. Lennartz: Arch. Eisenhüttenwes. 27 (1956) S. 127/33 u. 29 (1958) S. 359/64.
71. Kirkby, H. W., u. R. J. Truman: Iron and Steel 34 (1961) S. 625/29.
72. Richard, K., u. G. Petrich: Chem-Ing-Techn. 35 (1963) S. 29/36.
73. Goodell, P. D., T. M. Cullen u. J. W. Freeman: Trans. ASME, Ser. D, J. basic Engng., 89 (1967) S. 517/24.
74. Steinen, A. von den: DEW Techn. Ber. 9 (1969) S. 134/46.
75. Jesper, H., W. Wessling u. K. Achtelik: Stahl u. Eisen 86 (1966) S. 1408/18.
76. Lorenz, K., H. Fabritius u. E. Kranz: Stahl u. Eisen 92 (1972) S. 393/400.
77. Arbeitsgemeinschaft für warmfeste Stähle. Vgl. Arch. Eisenhüttenwes. 28 (1957) S. 245/323 u. 673/730 sowie 33 (1962) S. 27/60.
78. Lennartz, G.: DEW Techn. Ber. 3 (1963) S. 63/67.

79. Gerlach, H.: In: Werkstoffkunde der gebräuchlichen Stähle, T. 2, Hrsg. vom Verein Deutscher Eisenhüttenleute. Düsseldorf 1977. S. 106/20.
80. Rocha, H.J.: DEW Techn. Ber. 2 (1962) S. 16/24.
81. Ruttmann, W., u. N. Brunzel: Mitt. Verein, Großkesselbes. Nr. 80, 1962, S. 310/26.
82. Brennecke, C., u. R. Schinn: Z. VDI 99 (1957) S. 1165/71, 1233/44, 1275/83, 1335/42 u. 1611/19.
83. Murray, J. D.: Iron & Steel 34 (1961) Nr. 14, S. 634/40.
84. Class, I.: Chem.-Ing.-Techn. 29 (1957) S. 372/86.
85. Beattie, H.J., u. W.C. Hagel: Trans. metallurg. Soc. AIME 209 (1957) S. 911/17.
86. Castro, R., u. R. Tricot: Mem. Sci. Rev. Métallurg. 61 (1964) S. 573/91.
87. Weiss, B., u. R. Stickler: Metallurg. Trans. 3 (1972) S. 851/66.
88. Wiegand, H., u. M. Duruk; Arch. Eisenhüttenwes. 33 (1962) S. 559/66.
89. Jäger, W., R. Petri u. P. Schwaab: In: Fortschritte in der Metallographie. (Praktische Metallographie. Sonderbd. 10.) Stuttgart 1979. S. 384/400.
90. Fabritius, H.: Arch. Eisenhüttenwes. 48 (1977) S. 443/46.
91. Bungardt, K., A. von den Steinen u. G. Lennartz: DEW Techn. Ber. 9 (1969) S. 218/29.
92. Gerlach, H., u. E. Schmidtmann: Arch. Eisenhüttenwes. 39 (1968) S. 139/49.
93. Lennartz, G., u. A. von den Steinen: DEW Techn. Ber. 9 (1969) S. 163/76.
94. Levitin, V. V.: Fiz. met. i metalloved. 10 (1960) S. 294/97. u. 11 (1961) S. 564/67.
95. Bungardt, K., u. A. von den Steinen: DEW Techn. Ber. 1 (1961) S. 138/50. – Henry. G. A. Mercier, J. Plateau u. G. Hochmann: Rev. Métallurg. 60 (1963) S. 1221/32.
95a. Kautz, H. R., u. H. Gerlach: Arch. Eisenhüttenwes. 39 (1969) S. 151/58. – Engell, L.: Wie unter [46]. S. 460.
96. Hehemann, R. F., u. G. N. Ault: High temperature materials, New York 1959.
97. Decker, R. F.: In: Die Verfestigung von Stahl. Symposium. Zürich, 5. u. 6. Mai 1969. Veranst. von der Climax Molybdenum Company. Greenwich/Conn. 1970. S. 147/70.
98. Coutsouradis, D., P. Felix, H. Fischmeister, L. Habraken, Y. Lindblom u. M.O. Speidel: High temperature alloys for gas turbines. London 1978.
99. Sims, C. T., u. W. C. Hagel: The superalloys. New York 1972.
100. Bungardt, K., A. von den Steinen u. F. Schubert: Z. Werkst.-Techn. 3 (1972) S. 176/84.
101. Betteridge, W., u. H. Heslop: The Nimonic alloys 2. ed. London 1974.
102. Sahm, P. R., u. M. O. Speidel: high temperature materials in gas turbines. Amsterdam 1974.
103. David, K., E. Kohlhaas u. H. Müller: In: Warmformung und Warmfestigkeit. Berichte zum Symposium der Deutschen Gesellschaft für Metallkunde, Bad Nauheim 1975. [Hrsg.:] Deutsche Gesellschaft für Metallkunde. Oberursel [1976]. S. 181/202.
104. Schubert, F.: Wie unter [103]. S. 97/121.
105. Steinen, A. von den, u. E. Kohlhaas: In: Werkstoffkunde der gebräuchlichen Stähle. T. 2. Hrsg. vom VereinDeutscher Eisenhüttenleute. Düsseldorf 1967. S. 121/38.
106. Coutsouradis, D., J.-M. Drapier u. G. Davin: Z. Metallkde. 63 (1972) S. 306/14.
107. Piearcey, B. I., u. B. E. Terkelsen: Trans. metallurg. Soc. AIME 239 (1967) S. 1143/50.
108. DIN 17 102, Schweißgeeignete Feinkornbaustähle, normalgeglüht; Technische Lieferbedingungen für Blech, Band, Breitflach-, Form- und Stabstahl. Ausg. Okt. 1983.
109. Reumont, G. A. von: VGB-Kraftwerkstechn. 54 (1974) S. 418/30.
110. [Technische Regeln für Dampfkessel] TRD 301 – Berechnung – Ausg. April 1979.
111. KTA Regel 3201. 1 Abschn. 7. Anh. I. April 1978.
112. DIN 17 155. Blech und Band aus warmfesten Stählen; Technische Lieferbedingungen. Ausg. Okt. 1983. – DIN 17 175. Nahtlose Rohre aus warmfesten Stählen; Technische Lieferbedingungen. Ausg. Mai 1979. – DIN 17 177. Elektrisch preßgeschweißte Rohre aus warmfesten Stählen; Technische Lieferbedingungen. Ausg. Mai 1979. – ISO-International Standard 2604: Steel products for pressure purposes. Quality requirements. 1. Ausg. Mai 1975. Part II: Wrought seamless tubes. Part III: Electric resistance and induction-welded tubes. Part IV: Plates.
113. Fabritius, H.: In: VDI-Ber. Nr. 428, 1981, S. 79/91.
114. Schwaab, P., u. H. Weber: In: Eigenschaften warmfester Stähle. Internationale Tagung. Düsseldorf, 3 bis 5, Mai 1972. Zusammenstellung der Fachberichte. Bd. 2. [Hrsg.:] Arbeitsgemeinschaft für warmfeste Stähle, Verein Deutscher Eisenhüttenleute. Düsseldorf 1972. Ber. 6.2.
115. Baerlecken, E., u. H. Fabritius: Arch. Eisenhüttenwes. 33 (1962) S. 261/67.
116. Baerlecken, E., K. Lorenz, P. Bettzieche u. A. Raible: Über den Einfluß der Zusammensetzung und Wärmebehandlung auf die Eigenschaften des Stahles 14 MoV 63. Mitt. Verein. Großkesselbes. H. 88, 1964, S. 1/26.
117. Geiger, T.: In: VGB-Werkstofftagung 1971. [Hrsg.:] Technische Vereinigung der Großkraftwerksbetreiber. Essen [1972]. S. 79/84 u. 104/09.

118. Stahl-Eisen-Werkstoffblatt 555. Stähle für größere Schmiedestücke als Bauteile von Turbinen und Generatoranlagen. 1. Ausg. Aug. 1984.
119. DIN 17 240. Warmfeste und hochwarmfeste Werkstoffe für Schrauben und Muttern: Gütevorschriften. Ausg. Juli 1976.
120. Class, I.: Stahl u. Eisen 80. (1960) S. 1117/35; s. bes. S. 1122/25.
121. Weßling, W.: Sie & Wir [Werksz. d. Stahlwerke Südwestfalen AG] Nr. 17, 1976. S. 4/12.
122. Stahl-Eisen-Werkstoffblatt 670–69. Hochwarmfeste Stähle. Gütevorschriften. Ausg. 1969 (A DIN-Standard DIN 17 460, is being prepared).
123. DIN 17 243. Warmfeste schweißgeeignete Stähle in Form von Schmiedestücken oder gewalztem oder geschmiedetem Stabstahl. Technische Lieferbedingungen. Entwurf Aug. 1984. In the meantime: Schmiedestücke und gewalzter oder geschmiedeter Stabstahl aus warmfesten schweifs geeigneten Stählen Ausg. Jan. 1987.
124. DIN 17 245: Warmfester ferritischer Stahlguß. Ausg. Dez. 1987.
125. Jackson, W. J.: Gieß.-Praxis 1965, S. 312/19.
126. Zeuner, H.: Gießerei 44 (1957) S. 1/7.
127. Uhlitzsch, H., u. G. Radomski: Technik 16 (1961) S. 631/33.
128. Mayer, K. H., u. W. Rieß: VGB-Kraftwerkstechn. 56 (1976) S. 150/54.
128a. Christianus, D., K-H. Keienburg, H. König u. F. Staif: In: VGB-Werkstofftagung 1983. [Hrsg.:] Technische Vereinigung der Großkraftwerkstetreiber. Essen [1983]. S. 165/93.
129. Mayer, K. H., W. Gysel, A. Trautwein u. D. Tremmel: In: VGB-Werkstofftagung 1978 [Veranst.] VGB Technische Vereinigung der Großkraftwerksbetreiber. Essen 1978. S. 204/41.
130. Böhm, H.-J., W. Gysel, K.-H. Mayer u. A. Trautwein: In: VGB-Werkstofftagung 1983. [Hrsg.:] Technische Vereinigung der Großkesselbetreiber. Essen [1983]. S. 194/230.
130a. König, H., K. Niel, D. Christianus u. W. Gysel: 8. Vortragsveranstalhung der Arbeitsgemeinschaft für warmfeste Stähle und Hochtemperaturwerkstoffe. Düsseldorf, 1985. S. 45/53.
131. Errington, T., u. M. C. Murphy: Brit. Foundrym. 66 (1973) S. 294/304.
132. Batte, A. D., J. M. Brear, R. Holdsworth, J. Myers u. P. E. Reynolds: Philos. Trans. R. Soc., London, A 295 (1980) S. 253/64.
133. Miller, R. C., u. A. D. Batte: Metal Constr. 7 (1975) S. 559/58.
134. Ergebnisse Deutscher Zeistandversuche langer Dauer an Stahlgussorten nach DIN 17 245. Düsseldorf. 1986.
135. Schinn, R., E. O. Müller u. U. Schieferstein: In: VGB-Werkstofftagung 1969. [Hrsg.:] Vereinigung der Großkesselbetreiber. Essen [1970]. S. 54/89.
135a. Christianus, D., Fabritius, K. Forch, B. Huchtemann, K.-H. Keienburg, K. H. Mayer u. H. Weber: Stahl u. Eisen 107 (1987) S. 697/705.
136. Schinn, R., F. Staif u. W. Wiemann: VGB-Kraftwerkstechn. 54 (1974) S. 456/71.
137. Stahl-Eisen-Werkstoffblatt 410-81. Nichtrostender Stahlguß. Gütevorschriften. Ausg. März 1981. – Stahl-Eisen-Werkstoffblatt 470–76. Hitzebeständige Walz-und Schmiedestähle. Ausg. Febr. 1976. – Stahl-Eisen-Werkstoffblatt 590–61. Druckwasserstoffbeständige Stähle. Ausg. Dez. 1961. – Stahl-Eisen-Werkstoffblatt 640–75. Stähle für Bauteile im Primärkreislauf von Kernenergie Erzeugungsantagen. Ausg. Mai 1975. – Stahl-Eisen-Werkstoffblatt 670–69. Hochwarmfeste Stähle; Gütevorschriften. Ausg. Febr. 1969. – Stahl-Eisen-Lieferbedingungen 675–69. Nahtlose Rohre aus hochwarmfesten Stählen. Ausg. Febr. 1969. – Werkstoff-Handbuch der Deutschen Luftfahrt, 2 Bde. Hrsg. vom Bundesverband der Deutschen Luftfahrtindustrie. [Köln] 1956. – AD-Merkblatt W 2. Ausg. Dez. 1977. – ISO-International Standard 2604. Stähle für Druckbehälter. Ausg. Mai 1975.
138. Ruttmann, W., K. Baumann u. M. Möhling: Schweißtechn., Wien, 11 (1957) S. 6/10 u. 19/22.
139. Class, I., H. R. Kautz u. H. Gerlach: In: VGB-Werkstofftagung 1969. [Hrsg.:] Vereinigung der Großkesselbetreiber. Essen [1970]. S. 42/54.
140. Lorenz, K., H. Fabritius u. E. Kranz: Schweißen u. Schneiden 20 (1968) S. 459/64.
141. Gerlach, H.: In: Werkstoffkunde der gebräuchlichen Stähle. T. 2. Hrsg. vom Verein Deutscher Eisenhüttenleute. Düsseldorf 1977. S. 106/20.
142. Lorenz, K., H. Fabritius u. E. Kranz: Stahl u. Eisen 92 (1972) S. 393/400.
143. Kautz, H. R., H. F. Klärner u. E. Schmidtmann: Arch. Eisenhüttenwes. 36 (1965) S. 571/82.
144. Arbeitsgemeinschaft für warmfeste Stähle, Vgl. [77].
145. Bungardt, K., A. von den Steinen u. G. Lennartz: DEW Techn. Ber. 9 (1969) S. 218/29.
146. Engell, L.: Wie unter [46]. S. 460.
147. Schirra, M.: Kernforschungszentrum Karlsruhe, KfK-Bereicht 2296, Juni 1976.
148. Böhm, H.: Kernforschungszentrum Karlsruhe. KfK-Bericht 985, Juli 1969. – Derselbe: Arch. Eisenhüttenwes. 45 (1974) S. 821/30.

149. DIN 17 480. Ventilwerkstoffe. Technische Lieferbedingung. Ausg. Sept. 1984. – EURONORM 90–71. Stähle für Auslaßventile von Verbrennungskraftmaschinen Ausg. Aug. 1971. – ISO-Empfehlung ISO/R 683 T.XV. Stähle für Ventile von Verbrennungskraftmaschinen. 1976.
150. See [54]. In edition: Euronorm 95-79. Hitzebeständige Stähle. Ausg. Febr. 1979. – ISO 4955. Hitzebeständige Stähle und Legierungen Aug. Dez. 1983.
151. DIN 17 225. Warmfeste Stähle für Federn, Güteeigenschaften. Vornorm April 1955. (In the mean time withdrawn).
152. Bandel, G., u. K. Gebhard: Ber. Dt. Lilienthal-Ges. Luftfahrt-forsch Nr. 172. 1943. S. 138/52.
153. Materials and processing databook. Metal Progr. 122 (1982) Mid-June, Nr. 1, S. 46:124 (1983) Mid-June, Nr. 1, S. 60; 126 (1984) Nr. 1, S. 82.
154. Sims, T. S., u. W. C. Hagel: The superalloys. New York 1972.
155. Betteridge, W., u. H. Heslop: The Nimonic alloys. London 1974.
156. Metal Progr. 121 (1982) Nr. 1, S. 38, 40, 42, 46u. 49.
157. Sahm, P. R., u. M. O. Speidel: High temperature materials in gas turbines. Amsterdam 1974.

C 10 Low Temperature Steels (Steels with Good Toughness at Low Temperatures)

1. Haneke, M., u. W. Middeldorf: Sonderdruck der HOESCH Hüttenwerke AG. Dortmond. 31 S. aus: Die Kälte 1974, Nr. 12 u. 1975, Nr, 1, 2, 3 u. 4.
2. Degenkolbe, J., u. M. Haneke: Rohre Rohrleitungsbau Rohrleitungstransport 17 (1978) S. 514/20.
3. Bersch, B., J. Degenkolbe, M. Heneke u. W. Middeldorf: Stahl u. Eisen 98 (1978) S. 763/78.
4. Müsgen, B., u. J. Degenkolbe: Bänder Bleche Rohre 14 (1973) S. 245/52.
5. AD-Merkblatt W 10 Bl. 1. Werkstoffe für tiefe Temperaturen, Eisenwerkstoffe. Ausg. Nov. 1976.
6. Degenkolbe, J., u. B. Müsgen: Mater.-Prüf. 11 (1969) S. 365/72.
7. Degennkolbe, J.: Techn. Überwachung 10 (1969) S. 259/69.
8. Degenkolbe, J., u. B. Müsgen: Mater.-Prüf. 12 (1970) S. 413/20.
9. Haneke, M., B. Müsgen u. J. Petersen: Fachber. Hüttenprax. Metallweiterverarb. 19 (1981) S. 646/48, 654/60 u. 663/65.
10. DIN 17 102. Schweißgeeignete Feinkornbaustähle normalgeglüht. Technische Lieferbedingungen für Blech, Band, Breitflach-, Form- und Stabstahl. Ausg. Okt. 1983.
11. DIN 17 280. Kaltzähe Stähle. Technische Lieferbedingungen für Blech, Breitflachstahl, Formstahl, Stabstahl und Schmiedestücke. Ausg. April 1985. – Siehe auch DIN 17 173. Nahtlose Kreisförmige Rohre aus kaltzähen Stählen; Technische Lieferbedingungen. Ausg. Febr. 1985.
12. DIN 17 440 Nichtrostende Stähle. Technische Lieferbedingungen für Blech, Warmband, Walzdraht, gezogenen Draht, Stabstahl, Schmiedestrücke und Halbzeug. Ausg. Juli 1985. – DIN 17 441 Nichtrostende Stähle. Technische Lieferbedingungen für kaltgewalzte Bänder und Spaltbänder sowie daraus geschnittene Bleche. Ausg. Juli 1985.
13. Jesper, H., u. K. Achtelik: Techn. Ber. Nr. 35 der Stahlwerke Südwestfalen AG, Geisweid 0.0.1965. 44 S.
14. Behrenbeck, H., M. Haneke u. P. Sapp: Schweißen u. Schneiden 1 (1970) Bericht 21. S. 3/8.
15. Eichelmann, G. H. jr., u. F. C. Hull: Trans. Amer. Soc. Metals 45 (1953) S. 77/104.
16. Angel, T.: J. Iron Steel Inst. 177 (1954) S. 165/74 u. 1 Taf.
17. Schumann, H., u. H. J. v. Fircks: Arch. Eisenhüttenw. 40 (1969) S. 561/68.
18. Schumann, H.: Eisenhüttenw. 41 (1970) S. 1169/75.
19. Tamura, J., T. Maki u. H. Hator: Trans. Iron Steel Inst. Japan 10 (1970) S. 163/72.
20. Sanderson, G. P., u. D. T. Llewellyn: J. Iron Steel Inst. 207 (1969) S. 1129/40.
21. Fujikura, M., K. Takada u. K. Ishida: Trans. Iron Steel. Japan 15 (1975) S. 464/69.
22. Gueussier, A., u. R. Castro: Rev. Métallurg. 55 (1958) S. 107/22.
23. Haynes, A. G., K. Firth, G. E. Hollex u. J. Buchan: In: Properties of Material for liquified natural gas tankage. Philadelphia/Pa. 1975 (ASTM Spec. Techn. Publ. No. 579) S. 288/293.
24. Euronorm 129-76. Blech and Band aus nickellegierten Stählen für die Verwendung bei tiefen Temperaturen. Gütevorschriften. Ausg. März 1976.
25. Euronorm 141-79. Blech und Band aus austenitischen nichtrostenden Stählen zur Verwendung bei tiefen Temperaturen. Technische Lieferbedingungen. Ausg. Nov. 1979. – Euronorm 88-86. Nichtrostende Stähle Technische Lieferbedingungen Ausg. Dez. 1986.
26. Binder, W. O.: Metal Progr. 58 (1950) Nr. 2, S. 201/07.
27. Randak, A., W. Weßling, H. E. Bock, H. Steinmaurer u. L. Faust: Stahl u. Eisen 91 (1971) S. 1225/70.

28. Swales, G. L., u. A. G. Haynes: 2nd Liquified Natural Gas Transportation Conference Oct. 1972. The International Nickel Company of Canada, Ltd. (Inco Publ. 4376c) S. 1/11.
29. Degenkolbe, J., H. Höhne u. D. Uwer: In: DVS-Berichte Bd. 64. 1980. S. 69/79.
30. DIN 8556 Teil 1. Schweißzusätze für das Schweißen nichtrostender und hitzebeständiger Stähle; Bezeichnung, Technische Lieferbedingungen. Ausg. Mai. 1986.
31. DIN 1736 Teil 1. Schweißzustätze für Nickel und Nickellegierungen; Zusammensetzung. Verwendung und Technische Lieferbedingungen. Ausg. Aug. 1985. – DIN 1736 Teil 2. Schweißzusätze für Nickel und Nickellegierungen; Prüfstück, Proben, mechanisch-technologische Gütewerte. Ausg. Aug. 1985.
32. DIN 32 525 Teil 1. Prüfung von Schweißzusätzen mittels Schweißgutproben: Lichtbogengeschweißte Prüfstücke; Proben für mechanisch-technologische Prüfungen. Ausg. Dez. 1981 – DIN 32 525 Teil 2. Prüfung von Schweißzusätzen mittels Schweißgutproben; Prüfstücke für die Ermittlung der chemischen Zusammensetzung bei geringem Wärmeeinbringen. Ausg. Aug. 1979.

C 11 Tool Steels

1. DIN 17 350. Werkzeugstähle. Technische Lieferbedingungen. Ausg. Okt. 1980. See also EURONORM Eu 96 and ISO 4957.
2. EURONORM 20-74. Begriffsbestimmungen und Einteilung der Stahlsorten. Ausg. Sept. 1974. New edition EN 10 020 is being prepared.
3. DIN 8580. T.2 Fertigungsverfahren; Übersicht. Entw. Juni 1983.
4. Wilmes, S.: In: Werkstoffkunde der gebräuchlichen Stähle. T.2. Hrsg.: Verein Deutscher Eisenhüttenleute, Düsseldorf 1977. S. 200/04.
5. Hamaker jr., J. C., V. C. Stang u. G. A. Roberts: Trans. Amer. Soc. Metals 49 (1957) S. 550/75.
6. Wilmes, S.: Wie unter [4], S. 247/64.
7. Roberts, G. A.: Trans. Metallurg. Soc. AIME 236 (1966) S. 950/63.
8. DIN 50 351. Prüfung metallischer Werkstoffe; Härteprüfung nach Brinell. Ausg. Febr. 1985.
9. DIN 50 103. T. 1. Prüfung metallischer Werkstoffe; Härteprüfung nach Rockwell, Verfahren C, A, B, F. Ausg. März 1984.
10. DIN 50 133. T. 1. Prüfung metallischer Werkstoffe; Härteprüfung nach Vickers, Prüfkraftbereich: 49 bis 980 N (5 bis 100 kp). Ausg. Febr. 1985.
11. ASTM E 448-82. Scleroscope hardness. Testing of metallic materials.
12. Hengemühle, W.: Arch. Eisenhüttenwes. 42 (1971) S. 201/11.
13. Koester, R. D., u. D. P. Moak: J. Amer. Ceram. Soc. 50 (1967) S. 290/96.
14. DIN 50 150. Prüfung von Stahl und Stahlguß. Umwertungstabelle für Vickershärte. Binellhärte, Rockwellhärte und Zugfestigkeit. Ausg. Dez. 1976.
15. Lütjering, G., u. E. Hornbogen: Z. Metallkde. 59 (1968) S. 29/46.
16. Grosch, J.: In: Grundlagen der technischen Wärmebehandlung von Stahl. Hrsg. J. Grosch. Karlsruhe 1981, S. 25/49.
17. Bain, E. C., u. H. W. Paxton: Alloying elements in steel. 2nd ed. (Hrsg.) American Society of Metals. Metals Park/Ohio. 1966.
18. Hodge, J. M., u. M. A. Orehoski: Trans. Amer. Inst. min. met. Engrs. 167 (1946) S. 627/42.
19. Mitsche, R., u. K. L. Maurer: Arch. Eisenhüttenwes. 26 (1955) S. 563/65.
20. Just. E.: Härterei-techn. Mitt. 23 (1968) S. 85/100.
21. Just. E.: In: [16] S. 167/89.
22. Ilschner, B.: Werkstoffwissenschaften. Berlin, Heidelberg, New York 1982. S. 81.
23. DIN 17 014, T. 1. Wärmebehandlung von Eisenwerkstoffen; Fachbegriffe und-ausdrücke. Ausg. März 1975 (New edition 1988).
24. Peter, W., A. Klein u. H. Finkler: Arch. Eisenhüttenwes. 38 (1967) S. 561/69.
25. Rose, A.: Härterei-techn. Mitt. 21 (1966) S. 1/6.
26. Bühler, H., u. A. Rose: Arch. Eisenhüttenwes. 40 (1969) S. 411/23.
27. Stahl-Eisen-Prüfblatt 1665-71. Prüfung der Härtbarkeit von Edelstählen mit Härtebruchproben. Ausg. Dez. 1971.
28. DIN 50 191. Stirnabschreckversuch 25 mm. Ausg. Sept. 1987.
29. Rose, A., u. L. Rademacher: Stahl u. Eisen 23 (1956) S. 1570/73.
30. Hollomon, J. H., u. L D. Jaffe: Trans. Amer. Inst. min. metallurg. Engrs., Iron Steel Div., 162 (1945) S. 223/49.

31. Stahl-Eisen-Werkstoffblatt 250-63. Legierte Warmarbeitsstähle. Ausg. März 1963 (Withdrawn, see [1]) – S. a. Briefs, H., u. M. Wolf: Warmarbeitsstähle. Düsseldorf 1975. S. 8/11.
32. Bungardt. K., O Mülders u. G. Lennartz: Arch. Eisenhüttenwes. 32 (1961) S. 823/41.
33. Bungardt, K., u. O. Mülders: Stahl u. Eisen 86 (1966) S. 150/60. – Bungardt. K., O. Mülders u. G. Lennartz: Arch. Eisenhüttenwes. 32 (1961) S. 823/41.
34. Simcoe, C. R., u. A. E. Nehrenberg: Trans. Amer. Soc. Metals 58 (1965) S. 378/90.
35. Simcoe, C. R., A. E. Nehrenberg, V. Biss u. A. Coldren: Trans. Amer. Soc. Metals 61 (1968) S. 834/42.
36. Haberling, E.: See [4], S. 233/46.
37. Weigand, H. H., u. E. Haberling: TEW Techn. Ber. 1 (1975) S. 110/21.
38. Wilmes, S.: See [4], S. 247/64.
39. Brandis, H., P. Gümpel, E. Haberling u. H. H. Weigand: TEW Techn. Ber. 7 (1981) S. 221/30.
40. DIN 50 106. Prüfung metallischer Werkstoffe; Druckversuch. Ausg. Dez. 1978.
41. Stahl-Eisen-Prüfblatt 1320. Statischer Biegeversuch an Stählen geringen Verformungsvermögens. Entwurf 1962 (In the mean-time withdrawn).
42. Wilmes, S.: Arch Eisenhüttenwes. 35 (1964) S. 649/57.
43. Randak, A., A. Stanz u. W. Verderber: Stahl u. Eisen 92 (1972) S. 981/93.
44. Wilmes, S.: Stahl u. Eisen 81 (1961) S. 676/84.
45. Arch. Eisenhüttenwes. 33 (1962) S. 461/83.
46. Bungardt, K., O. Mülders u. W. Schmidt: Stahl u. Eisen 81 (1961) S. 670/75.
47. Mülders, O., u. R. Meyer-Rhotert: DEW Techn. Ber. 1 (1961) S. 96/106.
48. Brandis, H., u. K. Wiebking: DEW Techn. Ber 11 (1971) S. 158/65.
49. Wilmes, S.: Research results (not published).
50. Kroneis, M., E. Krainer u. F. Kreitner: Berg-u. hüttenm. Mh. 113 (1968) S. 416/25.
51. Bayer, E., u. H. Seilsdorfer: Arch. Eisenhüttenwes. 53 (1982) S. 494/500.
52. Bungardt, K., O. Mülders u. R. Mayer-Rhotert: Arch. Eisenhüttenwes. 37 (1966) S. 381/89.
53. Kulmburg, A., G. Schöberl u. K. Koch. Österr. Ing.-Z. 24 (1981) S. 404/08.
54. Randak, A., u. W. Verderber: Berg-u. hüttenm. Mh. 121 (1976) S. 39/91.
55. Habig, K.-H.: Verschleiß und Härte von Werkstoffen. München u. Wien 1980. S. 35 u. 216.
56. Habig, H.-K.: Z. Werkstoffkd. 4 (1973) S. 33/40.
57. Danzer, R., u. F. Sturm. Arch. Eisenhüttenwes. 53 (1982) S. 245/50.
58. Voss, H., E. Wetter u. F. Netthöfel: Arch. Eisenhüttenwes. 38 (1967) S. 379/86.
59. Bühler, H., F. Pollmar u. A. Rose: Arch. Eisenhüttenwes. 41 (1970) S. 989/96.
60. Elsen, E., G. Elsen u. M. Markworth: Metall 19 (1965) S. 334/45.
61. Berns, H.: Wie unter [4], S. 205/13.
62. Berns, H., P. Gümpel, W. Trojan u. H. H. Weigand: Arch. Eisenhüttenwes. 55 (1984) S. 267/70.
63. Trojan, W.: Gefüge und Eigenschaften ledeburitischer Chromstähle mit Niob und Titan. (Fortschrittsberichte VDI. Reihe 5. Nr. 90.) Düsseldorf 1985. S. 64.
64. Berns, H., u. W. Trojan. Radex Rdsch. 1985, H. 1/2, S. 560/67.
65. Trojan, W.: Diss. Ruhr-Universität Bochum 1985.
66. Liedtke, D.: Z., wirtsch. Fertigung 75 (1980) S. 33/48.
67. Wahl, G.: VDI-Z. 117 (1975) S. 785/89.
68. Rapatz, F.: Die Edelstähle 5. Aufl. Berlin, Göttingen, Heidelberg 1962.
69. Liedtke, D.: Z. wirtsch. Fertigung 65 (1970) S. 234/37..
70. Finnern, B., u. H. Kunst: Härterei-techn. Mitt. 30 (1975) S. 26/33.
71. Atens, H. von, u. H. Kunst: Härterei-techn. Mitt. 28 (1973) S. 266/70.
72. Fichtl, W.: Härterei-techn. Mitt. 29 (1974) S. 113/19.
73. Hutterer, K., u. E. Krainer: Berg-u. hüttenm. Mh. 121 (1976) S. 187/92.
74. Inzenhofer, A.: Fachber. Hüttenpraxis Metallweiterverarb. 22 (1984) S. 318/30 u. 819/33.
75. Ruppert, W.: Metalloberfläche 14 (1960) S. 193/98.
76. Hintermann, H. E., u. H. Gass: Schweiz. Arch. angew. Wiss. u. Techn. 33 (1967) S. 157/66.
77. Peterson, D.: Z. Werkstoffe u. ihre Veredlung 2 (1980) S. 173/82.
78. Schintlmeister, W., O. Pacher, W. Wallgram u. J. Kanz: Metall 34 (1980) S. 905/09.
79. Straten, P. J. M. van der, u. G. Verspui: VDI-Z. 124 (1982) S. 693/98.
80. Liedtke, D.: Harterei-Techn. Mitt. 37 (1982) S. 160/65.
81. Demny, J., u. G. Wahl: Härterei-Techn. Mitt. 37 (1982) S. 166/73.
82. Bosch, M., u. E. Boecker: Ind.-Anz. 105 (1983) Nr. 35, S. 30/33.
83. König, W., u. J. Fabry: Metall 37 (1983) S. 709/17.
84. Münz, W. D., u. G. Hessberger: Werkstoffe u. ihre Veredlung 3 (1981) S. 108/13.
85. Frey, H.: VDI-Z. 123 (1981) S. 519/25.
86. Kübert, M., u. R. Woska: Werkst. u. Betr. 116 (1983) S. 91/96, 117/26 u. 241/50.

87. Woska, R.: Stahl u. Eisen 102 (1982) S. 1013/17.
88. Bungardt, K., u. W. Spira: Arch. Eisenhüttenwes. 36 (1965) S. 257/67.
89. Krainer, H., K. Swoboda u. F. Rapatz: Arch. Eisenhüttenwes. 20 (1949) S. 111/15.
90. Keil, E.: Stahl u. Eisen 80 (1960) S. 1805/11.
91. Volmer, H.: Unpublished lecture hold on Unterausschuß für Werkzeugstähle des Werkstoffausschusses des Vereins Deutscher Eisenhüttenleute. Febr. 1961.
92. Williams, D. N., M. L. Kohn, R. M. Evans u. R. I. Jaffee: Mod. Castings 37 (1969) S. 19/25.
93. Rädeker, W.: Stahl u. Eisen 74 (1954) S. 929/43.
94. Bungardt, K., H. Preisendanz u. O. Mülders: Arch. Eisenhüttenwes. 32 (1961) S. 561/72.
95. Kindbom, L.: Arch. Eisenhüttenwes. 35 (1964) S. 773/80.
96. Kasak, A., u. G. Steven: Vortrag Nr. 112 beim 6. SDCE-Congress, Cleveland/Ohio, 1970.
97. Briefs, H., u. M. Wolf: Warmarbeitsstähle. Düsseldorf 1975. S. 8/11.
98. Schindler, A.: Leoben 1981 (Diss. Montan. Hochsch. Leoben). S. a. Gießerei-Rdsch. H. 12. 1981, S. 1/6.
99. Krebs, W.: Gießerei 65 (1978) S. 645/52 u. 733/40.
100. Bungardt, K., K. Kunze u. E. Horn: Arch. Eisenhüttenwes. 29 (1958) S. 193/203.
101. Bungardt, K., K. Kunze u. E. Horn: Arch. Eisenhüttenwes. 38 (1967) S. 309/312; 39 (1968) S. 863/67 u. 949/51.
102. Bäumel, A., u. C. Carius: Arch. Eisenhüttenwes. 32 (1961) S. 237/49.
103. Koenig, R. F.: Iron Age 172 (1953) Nr. 8, S. 129/33.
104. Schönert, K.: Gießerei 48 (1961) S. 257/60.
105. Horstmann, D.: Stahl u. Eisen 73 (1953) S. 659/65.
106. Berns, H.: Z. wirtsch. Fertigung 66 (1971) S. 289/95.
107. Dechema-Werkstoff-Tabellen A 49. 235.420.1 u. A 49. 235.420.2.
108. Mott, N. S.: Chem. Engng. Progr. 50 (1954) S. 45 u. S. 532.
109. Heumann, Th., u. S. Dittrich: Z. Metallkde. 50 (1959) S. 617/25.
110. EURONORM 52-83. Begriffe der Wärmebehandlung von Eisenwerkstoffen. Ausg. Mai 1983.
111. Hribernik, B., u. F. Russ: Arch. Eisenhüttenwes. 53 (1982) S. 373/77.
112. Schuhmacher, B. M.: Techn. Zbl. prakt. Metallberab. 78 (1984) S. 23/28.
113. Frehser, J., u. O. Lowitzer: Stahl u. Eisen 77 (1957) S. 1221/33.
114. Frehser, J., u. O. Lowitzer: Werkstattstechn. u. Masch. Bau 47 (1957) S. 558/63.
115. Bühler, H., u. E. Herrmann: VDI-Z. 103 (1961) S. 436/42. – VDI-Z. 103 (1961) S. 1229/35. – Arch. Eisenhüttenwes. 35 (1964) S. 1089/95.
116. Berns, H., A. Kulmburg u. E. Staska: VDI-Z. 114 (1972) S. 1229/33.
117. Berns, H.: Z. Werkst. Techn. 8 (1977) S. 149/57.
118. Haberling, E., u. H. H. Weigand: Thyssen Edelstahl Techn. Ber. 9 (1983) S. 89/95.
119. Lemont, B. S.: Distortion in tool steels. Publ. by the American Society for Metals. Metals Park, Novelty, Ohio 1959.
120. Wilmes, S., u. A. Kulmburg: not yet published
121. Krainer, E., A. Schindler, A. Kulmburg u. K. Hutterer: Metall 29 (1979) S. 487/92.
122. Kroneis, M., E. Krainer u. F. Kreitner: Berg- u. hüttenm. Mh. 113 (1968) S. 416/25.
123. Swoboda, K., A. Kulmburg, E. Staska u. R. Blöck: Berg- u. hüttenm. Mh. 116 (1971) S. 94/98.
124. Haberling, E., R. Bennecke u. K. Köster: TEW Techn. Ber. 5 (1979) S. 121/28.
125. Pawelski, O.: Z. Metallkde. 68 (1977) S. 79/89.
126. TEW Techn. Ber. 5 (1979) S. 17/21.
127. Künelt, G., u. H. Straube: Berg.-u. hüttenm. Mn. 111 (1966) S. 398/405.
128. Staska, E., u. A. Kulmburg: Z. Werkstofftechn. 4 (1973) S. 41/49.
129. Thelning, K.-E.: Werkstattstechn. u. Masch.-bau 48 (1958) S. 209/15.
130. Thelning, K.-E.: Schweiz. Arch. angew. Wiss. u. Techn. 27 (1961) S. 503/10.
131. Hoischen, C.: Belastbarkeit und Abformgenauigkeit der Stempel beim Kalteinsenken. Hannover 1966. (Dr.-Ing.-Diss. Techn. Hochsch. Hannover.) Also in: Forschungsberichte des Landes Nordrhein-Westfalen. Nr. 1625.
132. Brandis, H., P. Gümpel u. E. Haberling: Thyssen Edelstahl Techn. Ber. 7 (1981) S. 123/33.
133. Ortmann, R., u. E. Haberling: TEW Techn. Ber. 1 (1975) S. 142/46.
134. Hellmann, P.: Werkst. u. Betr. 108 (1975) S. 277/79.
135. Fachber. Hüttenprax. Metallweiterverarb. 15 (1977) S. 169/73.
136. Stahleinsatzliste 171. Stähle für Werkzeuge der Kunststoffverarbeitung. 4. Aufl. Ed.: Verein Deutscher Eisenhüttenleute Düsseldorf 1972 (In the mean-time withdrawn).
137. Verderber. W., u. B. Leidel: Kunststoffe 69 (1979) S. 719/26.
138. Sidan, H.: Berg-u. hüttenm. Mh. 122 (1977) S. 93/98.
139. Jänichen, H.: Neue Hütte 21 (1976) S. 97/101.

140. Becker, H.-J.: VDI-Z. 113 (1971) S. 385/90.
141. Becker, H.-J.: VDI-Z. 114 (1972) S. 527/32.
142. Kortmann, W.: TEW Techn. Ber. 9 (1983) S. 71/80.
143. Krumpholz, R., u. R. Meilgen: Kunststoffe 63 (1973) S. 286/91.
144. Dittrich, A., u. W. Kortmann: Thyssen Edelstahl Techn. Ber. 7 (1981) S. 190/99.
145. Steel products manual. Tool steels. Editor: American Iron and Steel Institute. New York 1978.
146. Breitler, R.: Gießerei Prax. Nr. 1981, S. 1/8.
147. Stein, H. K.: VDI-Z. 115 (1973) S. 275/85.
148. Stäbli, G., H. Schlicht u. E. Schreiber: Z. Werkst.-Techn. 7 (1976) S. 198/208.
149. Horstmann, D.: Arch.Eisenhüttenwes. 46 (1975) S. 137/41.
150. Haberling, E., u.K. Rasche: Thyssen Edelstahl Techn. Ber. 9 (1983) S. 111/20.
151. Schindler, A., K. Hutterer, A. Kulmburg u. G. Preininger: Berg.-u.hüttenm. Mh. 120 (1975) S. 285/93.
152. Barten, G.:Z. Wirtschaftl. Fertigung 72 (1977) S. 307/13.
153. Stahl-Einsatzliste 198. Düsseldorf 1975.
154. Schindler, A., u. A. Kulmburg: Gießerei Rdsch. 21 (1974) S. 65/71.
155. Hiller, H.: Gießerei 60 (1973) S. 206/14.
156. Becker, H.-J.: Wirtschaftl. Fertigung 64 (1969) S. 353/57 u. 451/55.
157. Verderber, W.: Maschine 10 (1969) S. 41/42.
158. Köster, R.: Fachber. Hüttenprax. u. Metallweiterverarb. 13 (1975) S. 399/407.
159. Höpken, H.: Sie & Wir (Werksz. der Fried. Krupp Hüttenwerk AG) Nr. 3, 1978. S. 37/39.
160. Köster, R.: Fachber. Hüttenprax. Metallverarb. 13 (1975) S. 399/407.
161. Lange, K.: Bleche Rohre Profile 26 (1979) S. 511/13, 576/80 u. 649/52. – The same: Draht 30 (1979) S. 612/14, 664/68 u. 763/66.
162. Müller, C. A., u. R. Fizia: In: Werkstoff-Handbuch Stahl u. Eisen. 4. Aufl. Editor: Verein Deutscher Eisenhüttenleute. Düsseldorf 1965. S. H 31-1/31-2.
163. Feldmann, H.-D.: VDI-Z. 121 (1979) S. 72/78.
164. Brandis, H., E. Haberling, W. Hückelmann u. H. Kempkens: Thyssen Edelstahl Techn. Ber. 9 (1983) S. 153/65.
165. Heinrich, E.: Fachber. Hüttenprax. Metallverarb. 16 (1978) S. 874/80.
166. Haberling, E.: Thyssen Edelstahl Techn. Ber. 3 (1977) S. 135/39.
167. Jonck, R.: Z. wirtsch. Fertigung 76 (1981) S. 496/502.
168. VDI-Richtlinie 3186. Werkzeuge für das Kaltpressen von Stahl. Ausg. Juni 1974.
169. Zapf, G., G. Hoffmann u. K. Dalal: Z. Werkst.-Techn. 6 (1975) S. 384/90 u. 424/32.
170. Becker, H.-J.: Vortrag beim Seminar „Neue Entwicklung in der Massivumformtechnik" der Forschungsgesellschaft Umformtechnik, 1981.
171. Kiefer, J., u. A. Schindler: Berg- u. hüttenm. Mh. 121 (1976) S. 449/53.
172. Köster, R., u. F. Schubert: TEW Techn. Ber. 1 (1975) S. 154/61.
173. Berns, H., u. F. Pschenitzka: Z. Werkstofftechn. 11 (1980) S. 258/66.
174. Berns, H.: In: Festigkeits-und Bruchverhalten bei höheren Temperaturen. Bd. 2. Berichte, gehalten im Kontaktstudium „Werkstoffkunde Eisen und Stahl IV". Editor: W. Dahl u. W. Pitsch. Düsseldorf 1980. S. 281/301.
175. Berns, H.:Z. wirtsch. Fertigung 71 (1976) S. 64/69 u. 401/06.
176. Karagöz, S., R. Riedl, M. R. Gregg u. H. Fischmeister: Prakt. Metallogr. 14 (1983) S. 376.
177. Karagöz, S., K. Schur u. H. Fischmeister: Beitr. elektronenmikroskop. Direktabb. Oberfl. 15 (1982) S. 235.
178. Horn, E.: DEW Techn. Ber. 12 (1972) S. 217/24.
179. Fredriksson, H., M. Hillert u. M. Nica: Scand. J. Metall 8 (1979) S. 115/22.
180. Wilmes, S., u. E. Weber: Industriebl. 64 (1964) S. 8/12.
181. Bennecke, R., u. H. H. Weigand: Thyssen Edelstahl Techn. Ber. 7 (1981) S. 107/14.
182. Weigand, H.: DEW Techn. Ber. 7 (1967) S. 209/15.
183. Brandis, H., E. Haberling u. H. H. Weigand: Thyssen Edelstahl Techn. Ber. 4 (1978) S. 79/84.
184. Brandis, H., E. Haberling u. R. Ortmann: TEW Techn. Ber. 1 (1975) S. 106/09.
185. Brandis, H., E. Haberling, R. Ortmann u. H. H. Weigand: Thyssen Edelstahl Techn. Ber. 3 (1977) S. 81/99.
186. Brandis, H., P. Gümpel u. E. Haberling: Thyssen Edelstahl Techn. Ber. 7 (1981) S. 123/33.
187. Haberling, E., u. H. Martens: Thyssen Edelstahl Techn. Ber. 3 (1977) S. 100/04.
188. Gümpel, P., u. E. Haberling: TEW Techn. Ber. 5 (1979) S. 129/35.
189. Pacyna, J.: Arch, Eisenhüttenwes. 55 (1984) S. 291/99.
190. Brandis, H., u. E. Haberling: Thyssen Edelstahl Techn. Ber. 4 (1978) S. 85/90.
191. Roberts, G. A.: Trans. metallurg. Soc. AIME 236 (1966) S. 950/63.

192. Heisterkamp, F., u. S. R. Keown: Proc. Symposium 109. AIME Annual Meeting Metallurg. Soc. AIME Las Vegas Febr. 1980. S. 103/23.
193. Berkenkamp, E.: Thyssen Edelstahl Techn. Ber. 7 (1981) S. 134/38.
194. Kulmburg, A., W. Wilmes u. F. Korntheuer: Arch. Eisenhüttenwes. 47 (1976) S. 319/24.
195. Geiser, W.: Fertigungstechnik. Hamburg 1973.
196. Ferstl, G., F. Russ u. A. Schindler: Berg.-u. hüttenm. Mh. 12 (1976) S. 464/69.
197. Kottsieper, E.-E., u. D. Jäger: Fachber. Metallbearb. 60 (1983) S. 17/23.

C 12 Wear resistant steels

1. Tribologie BMFT-FBT 76-38, 1976, S. 50.
2. Habig, K.-H.: Verschleiß und Härte von Werkstoffen, München 1980.
3. Uetz, H. (ed): Abrasion und Erosion. München u. Wien 1986.
4. Zum Gahr, K.-H.: Microstructure and Wear of Materials. Amsterdam 1987.
5. Hornbogen, E., K. Rittner u. J. Becker: Prakt. Metallographie 15 (1984) S. 342/52.
6. Henke, F.: Gießerei-Praxis 1975, S. 378/407.
7. Berns, H.: Gießerei 64 (1977) S. 323/28.
8. Berns, H., u. A. Fischer: New Abrasion-resistant Alloys. Proc. Int. Conf. on Tribology, Inst. Mech. Eng. London 1987.
9. Jost, N., u. I. Schmidt: Friction-induced Martensite in Austenitic Fe-Mn-C-Steels. In: Ludema K. C. (ed): Wear of Materials, 1985, ASME, New York, 185, pp. 205/11.
10. Fairhurst, W., u. K. Röhrig: Foundry Trade J. (1974) 2999, pp. 685/98.
11. Uetz, H. W.a.: Sprechsaal ceramics, glass, cement 2 (1978).
12. Berns, H., u. W. Trojahn: Wear of Ledeburitic Chromium Steels. In: Ludema K. C. (ed): Wear of Materials, 1985, ASME. New York, 1985, pp. 186/93.
13. Berns, H., u. A. Fischer: Abrasive Wear Resistance and Microstructure of Fe-Cr-C-B Hard Surfacing Weld Deposits. In: Ludema, K. C. (ed): Wear of Materials, 1985, ASME, New York. 1985. pp. 625/33.

C 13 Stainless Steels

1. Class, J.: Chem.-Ing.-Techn. 36 (1964) S. 131/41.
2. Allsop, H., u. C. Frith: Iron Steel 23 (1950) S. 309/11.
3. Bäumel, A.: Werkst. u. Korrosion 18 (1967) S. 289/302.
4. Lennartz, G.: In: VDI-Ber. Nr. 235, 1975, S. 169/82.
5. Strauß, B., H. Schottky u. J. Hinnüber: Z. anorg. allg. Chem. 188 (1930) S. 309/24. – DIN 50 914. Prüfung nichtrostender Stähle auf Beständigkeit gegen interkristalline Korrosion: Kupfersulfat-Schwefelsäure-Verfahren; Strauß-Test. Ausg. Juni 1984.
6. Rocha, H.-J.: DEW Techn. Ber. 2 (1962) S. 16/24.
7. Stahl-Eisen-Prüfblatt 1877. Prüfung der Beständigkeit hochlegierter korrosionsbeständiger Werkstoffe gegen interkristalline Korrosion. Ausg. Juni 1979.
8. DIN 50921. Prüfung nichtrostender austenitischer Stähle auf Beständigkeit gegen örtliche Korrosion in stark oxidierenden Säuren. Ausg. Okt. 1984.
9. Versuchsbericht der Thyssen Edelstahlwerke A. G., unveröffentlicht.
10. Bock, H. E., u. W. Weßling: Sie & Wir [Werksz. d. Stahlwerke Südwestfalen A. G.] Nr. 10, 1973, S. 1/7. – S. auch Z. Werkstofftechn. 4 (1973) S. 186/95.
11. Wendler-Kalsch, E.: Werkst. u. Korrosion 29 (1978) S. 703/20. – Herbsleb, G., u. R. Pöpperling: Werkst. u. Korrosion 29 (1978) S. 732/39.
12. Brauns, E., u. H. Ternes: Werkst. u. Korrosion 19 (1968) S. 1/19.
13. Kohl, H., G. Rabensteiner u. G. Hochörtler: In: Alloys for the eighties. Ann Arbor, Michigan 17./18. Juni 1980. Ed.: R. Q. Barr. Greenwich, Conn. 1981. S. 343/51.
14. Schwenk, W.: Stahl u. Eisen 89 (1969) S. 535/47.
15. Oppenheim, R.: Kontaktstudium Umformtechnik. „Warmwalzen auf freier Bahn" vom 9. bis 14 Nov. 1980 in Winterscheid über Hennef/Sieg. T1. S. 127/80.

16. Michel, K. H., H. M. Mozek u. H. Mülders: Kaltbreitband aus nichtrostenden Stählen. Eigenschaften ferritischer, martensitischer und austenitischer Werkstoffe. Düsseldorf 1984 (Stahleisen-Sonderberichte H. 13). See also: DIN 17441. Technische Lieferbedingungen für Blech, Warm band, Walzdrahl, gezogenen Draht, Stabstahl; Schmiedestücke und Halbzeug. Ausg. Juli 1985.
17. Küppers, W.: Bliech Rohre Profile 25 (1978) S. 453/61.
18. Küppers, W., u. W. Schmidt: DEW Techn. Ber. 14 (1974) S. 49/55.
19. Weßling, W., u. H. E. Bock: In: Stainless steel '77. Ed. R. Q. Barr. New York 1977. S. 217/25.
20. Oppenheim, R.: DEW Techn. Ber. 14 (1974) S. 5/13.
21. Zingg, E., u. T. Geiger: Schweiz. Arch. angew. Wiss. Techn. 23 (1957) S. 71/78 u. 121/27.
22. Oppenheim, R.: DEW Techn. Ber. 2 (1962) S. 87/92.
23. Spähn, H.: Metalloberfläche 16 (1962) S. 369/73.
24. Lorenz, K., u. G. Medawar: Thyssenforsch. 1 (1969) S. 97/108; s. a. Bock, H.-E., A. Kügler, G. Lennartz u. E. Michel: Stahl u. Eisen 104 (1984) S. 557/63.
25. Herbsleb, G.: Werkst. u. Korrosion 33 (1982) S. 334/40.
26. Kowaka, M., H. Nagano, T. Kudo u. K. Yamanaka: Boshoku Gijutsu 30 (1981) Nr. 4, S. 218.
27. Gräfen, H.: Chem.-Ing.-Techn. 54 (1982) S. 108/19.
28. Kubaschewski, O.: Iron-binary phase diagrams. Berlin, Heidelberg, New York u. Düsseldorf. 1982. S. 32 u. 75.
29. Baerlecken, E., W. A. Fischer u. K. Lorenz: Stahl u. Eisen 81 (1961) S. 768/78.
30. Bungardt, K., E. Kunze u. E. Horn: Arch. Eisenhüttenwes. 29 (1958) S. 193/203.
31. Strauss, B., u. E. Maurer: Kruppsch Mh. 1 (1920) Aug., S. 129/46.
32. Schaeffler, A. L.: Weld. Res. 1947, S. 601-s/20-s.
33. Long, C. J., u. W. T. Delong: Weld. Res. 1973, S. 281-s/97-s.
34. Delong, W. T.: Weld. Res. 1974, S. 273-s/80-s.
35. Niederau, H. J.: Stahl u. Eisen 98 (1978) S. 385/92.
36. Randak, A., A. von den Steinen u. E. Gondolf: In: Werkstoffkunde der gebräuchlichen Stähle. T. 2. Hrsg. vom Verein Deutscher Eisenhüttenleute. Düsseldorf 1977. S. 75/95.
37. Barker, R.: Metallurgia, Manch., 76 (1967) S. 49/54.
38. Irvine, K. V., D. T. Llewellyn u. F. B. Pickering: J. Iron Steel Inst. 192 (1959) S. 218/38.
39. Mende, A. B.: Über den Einfluß von Kupferzusätzen auf die Werkstoffeigenschaften und das Kaltumformverhalten austenitischer Chrom-Nickel-Stähle. Aachen 1973. (Dr.-Ing.-Diss. Techn. Hochsch. Aachen.)
40. Küppers, W.: Thyssen Edelstahl Techn. Ber. 8 (1982) S. 153/61.
41. Schafmeister, P., u. R. Ergang: Arch Eisenhüttenwes. 12 (1938/39) S. 459/64.
42. Hoffmeister, H., u. R. Mundt: Arch. Eisenhüttenwes. 52 (1981) S. 159/64.
43. Krainer, H.: Stahl u. Eisen 82 (1962) S. 1527/43.
43a. Stahl-Eisen-Werkstoffblatt 400. Nichtrostende Walz- und Schmiedestähle. Entwurf Jan. 1985
44. DIN 17440. Nichtrostende Stähle.
45. Kiesheyer, H.: Thyssen Edelstahl Techn. Ber. 8 (1982) S. 111/14.
46. Kiesheyer, H.: Über das Ausscheidungsverhalten hochreiner ferritischer Stähle mit 20 bis 28% Cr und bis zu 5% Mo. Aachen 1974. (Dr.-Ing.-Diss. Techn. Hochsch. Aachen.)
47. Oppenheim, R.: Thyssen Edelstahl Techn. Ber. 8 (1982) S. 97/110.
48. Weßling, W.: Unveröff. Bericht der Krupp Südwestfalen A. G.
49. Brezina, P.: Escher Wyss Mitt. 53 (1980) S. 218/35.
50. Knorr, W., u. H. J. Köhler: Werkst u. Korrosion 32 (1981) S. 371/76.
51. Heimann, W., u. F. H. Strom: Thyssen Edelstahl Techn. Ber. 8 (1982) S. 115/25.
52. Heimann, W., u. M. Hoock: Thyssen Edelstahl Techn. Ber. 8 (1982) S. 126/34.
53. Weßling, W.: Blech Rohre Profile (1977) S. 142/45.
54. Dietrich, H., W. Heimann u. F. H. Strom: Thyssen Edelstahl Techn. Ber. 2 (1976) S. 61/69.
55. Oppenheim, R.: Werkst. u. Korrosion 16 (1965) S. 1/11.
56. McNeely, V. J., u. D. T. Llewellyn: Sheet Metal Ind. 49 (1972) S. 17/18 u. 21/25.
57. Heimann, W., W. Schmidt, G. Lennartz u. E. Michel: DEW Techn. Ber. 13 (1973) S. 94/107.
58. Lorenz, K., H. Fabritius u. G. Medawar: Thyssenforsch. 1 (1969) S. 10/20.
59. Bungardt, K., G. Lennartz u. R. Oppenheim: DEW Techn. Ber. 7 (1967) S. 71/90.
60. Lennartz, G.: Mikrochim. Acta. Wien, (1965) S. 405/28.
61. Lennartz, G.: DEW Techn. Ber. 5 (1965) S. 93/105.
62. Bäumel, A., E. M. Horn u. G. Siebert: Werkst. u. Korrosion 23 (1972) S. 973/83.
63. Thier, H., A. Bäumel u. E. Schmidtmann: Arch. Eisenhüttenwes. 40 (1969) S. 333/39.
64. Bungardt, K., H. Laddach u. G. Lennartz: DEW Techn. Ber. 12 (1972) S. 134/54.
65. Laddach, H., G. Lennartz u. G. Preis: DEW Techn. Ber. 13 (1973) S. 75/84.
66. Brandis, H., W. Heimann u. E. Schmidtmann: Thyssen Edelstahl Techn. Ber. 2 (1976) S. 150/66.

67. Bock, H.-E.: Techn. Mitt. Krupp. Forsch.-Ber. 36 (1978) H. 2, S. 49/60.
68. Wehner, H., u. H. Speckhardt Z. Werkst.-Techn. 10 (1979) S. 317/32.

C 14 Steels for Hydrogen Service at Elevated Temperatures and Pressures (High-pressure Hydrogen Resistant Steels)

1. Stahl-Eisen-Werkstoffblatt 590. Druckwasserstoffbeständige Stähle. Ausg. Dez. 1961. Neuausgabe in Vorbereitung; derzeit Entwurf Aug. 1984. – Stahl-Eisen-Werkstoffblatt 595. Stahguß für Erdöl- und Erdgasanlagen. Ausg. Aug. 1976. DIN 17 176 Nahtlose Kreisförmige Rohre aus druckwasserbeständigen Stählen (Nov. 1990)
2. Reuter, M.: Techn. Überwachung 15 (1974) S. 10/18, 65/70 u. 101/10.
3. Bosch, C.: Z. Ver. Dt. Ing. 77 (1933) S. 305/17; vgl. Stahl u. Eisen 53 (1933) S. 1187/89.
4. Huijbregts, W. M. M., J. H. N. Jelgersma u. A. Snel: VGB-Kraftwerkstechn. 55 (1975) S. 26/39.
5. Zapffe, C. A.: Trans. Amer. Soc. Mech. Engrs. 66 (1944) S. 81/126.
6. Class, I.: Stahl u. Eisen 80 (1960) S. 1117/35.
7. Class, I.: Stahl u. Eisen 85 (1965) S. 149/55 u. 204/11.
8. Smialowsky, M.: Hydrogen in steel. Oxford, London, New York, Paris 1962.
9. Jäkel, U.: In: Werkstoffkunde der gebräuchlichen Stähle. T. 2. Hrsg. vom Verein Deutschr Eisenhüttenleute. Düsseldorf 1977. S. 151/58.
10. Rahmel, A., u. W. Schwenk: Korrosion und Korrosionsschutz von Stählen. Weinheim, New York 1977.
11. Naumann, F. K.: Stahl u. Eisen 58 (1938) S. 1239/50.
12. DIN 17 175. Nahtlose Rohre und Sammler aus warmfesten Stählen. Technische Lieferbedingungen. Ausg. Mai 1979.
13. DIN 17 243. Schmiedestrücke aus warmfesten Stählen. Gütevorschriften. In Vorbereitung, derzeit Entwurf Aug. 1984. In the meantime: Schmiedestucke und gewalzter oder geschmederer Stabstahl aus warm festen schweifsgeeigneten Stählen. Ausg. Jan 1987.
14. Stahl-Eisen-Werkstoffblatt 670. Hochwarmfeste Stähle. Gütevorschriften. Ausg. Febr. 1969. (A DIN-standard, DIN 17 460 is being prepared).
15. Stahl-Eisen-Werkstoffblatt 675. Nahtlose Rohre aus hochwarmfesten Stählen. Technische Lieferbedingungen. Ausg. Febr. 1969 (Will be replaced by DIN 17 459).
16. Vgl. z. B. ASTM A 200 Specifications for Seamless Intermediate Alloy-Steel Still Tubes for Refinery Service. – ASTM A 213. Seamless Ferritic and Austenitic Alloy-Steel Boiler, Superheater and Heat Exchanger Tubes – ASTM A 335. Seamless Ferritic Steel Tubes for High Temperature Service. – ASTM A 336 Alloy Steel Forgings for Pressure Vessels with High Service Tempratures.
17. DIN 17 155. Blech und Band aus warmfesten Stählen. Technische Lieferbedingungen. Ausg. Okt. 1983.
18. Nelson, G. A.: Trans. Amer. Soc. Mech. Engrs. 73 (1959) S. 205/19; Werkst. u. Korrosion 14 (1963) S. 65/69. – American Petroleum Institut (API), Refining Department Publication 941. Washington 1983 and 1990.
19. Janzon, W.: Thyssen Techn. Ber. 6 (1974) S. 14/21. – Bauman, Th. C.: J. Metals 29 (1977) Nr. 8, S. 8/11. – Shih, H.-M., u. H. H. Johnson: Acta metallurg. 30 (1982) S. 537/45.
20. Class, I.: Mitt. Verein. Großkesselbes. H. 58, 1959, S. 38/59.
21. Jäkel, U., u. W. Schwenk: Werkst. u. Korrosion 22 (1971) S. 1/7.

C 15 Heat Resisting Steels

1. Oppenheim, R.: Stahl u. Eisen 94 (1974) S. 426/34. – Derselbe: DEW Techn. Ber. 13 (1973) S. 11/18.
2. Siehe Stahl-Eisen-Werkstoffblatt 470–76. Hitzebeständige Walz- und Schmiedestähle. Ausg. Febr. 1976.
3. Rahmel, A., H.-J. Schüller, P. Schwaab u. W. Schwenk: Bänder Bleche Rohre 1964. S. 245/52.
4. Siehe z. B. Ledjeff, K., A. Rahmel u. M. Schorr: Werkst. u. Korrosion 30 (1979) S. 767/84: 31 (1980) S. 83/97.
5. Bandel, G., u. K. E. Volk: Arch. Eisenhüttenwes. 15 (1942) S. 369/78.
6. Kohl, H., u. H. Zitter: Arch. Eisenhüttenwes. 39 (1968) S. 855/62.
7. Houdremont, E.: Handbuch der Sonderstahlkunde. 3 Aufl. Bd. I. Berlin, Göttingen, Heidelberg u. Düsseldorf 1956. S 684.

8. Oppenheim, R.: In: VDI-Ber. Nr. 318, 1978, S. 65/77.
9. Richard, K., u. G. Petrich: Chem.-Ing.-Techn. 35 (1963) S. 29/36.
10. Brandis, H., E. Haberling, H. Hellmonds u. S. Engineer: Thyssen Edelstahl Techn. Ber. 7 (1981) S. 176/89.
11. Kohl, H.: Z. Werkstofftechn. 8 (1977) S. 125/30.
12. Pfeiffer, H., u. H. Thomas: Zunderfeste Legierungen. Berlin, Göttingen, Heidelberg 1963.
13. Aydin, I., H.-E. Bühler u. A. Rahmel: Werkst. u. Korrosion 31 (1980) S. 675/82.
14. Baerlecken, E., u. H. Fabritius: Stahl u. Eisen 78 (1958) S. 1389/95.
15. Bungardt, K., H. Borchers u. D. Kölsch: Arch. Eisenhüttenwes. (1963) S. 465/76.
16. Schüller, H.-J.: Arch. Eisenhüttenwes. 36 (1965) S. 513/16. – Derselbe: J. Iron Steel Inst. 171 (1952) S. 345/53. – Derselbe: Trans. metalllurg. Soc. AIME 212 (1958) S. 497/502.
17. Oppenheim, R.: Unveröff. Untersuchung 1964.
17a. DIN 50 601. Ermittlung der Ferrit – oder Austenitkorngröße von Stahl und Eisenwerkstoffen. Ausg. Aug. 1985.
18. Mülders, H., I. Stellfeld u. H. J. Köhler: Techn. Mitt. Krupp 33 (1975) S. 45/50.
19. Bandel, G.: Arch. Eisenhüttenwes. 11 (1937) S. 139/44.
20. DIN 1736 Teil 1. Schweißzusätze für Nickel und Nickellegierungen; Zusammensetzung. Verwendung und Technische Lieferbedingungen. Ausg. Juni 1979. Entwurf Febr. 1984.
21. Schnaas, A., u. J. J. Grabke: Werkst. und Korrosion 29 (1978) S. 635/44.
22. Moran, J. J., J. R. Mihalisin u. E. N. Skinner: Corrosion, Houston 17 (1961) S. 191t/95t.
23. Herda, W., u. G. L. Swales: Werkst. u. Korrosion 19 (1968) S. 679/90.
24. Kobalt Nr. 56, 1972, S. 99/113.

C 16 Heating Conductor Alloys

1. Pfeiffer, H., u. H. Thomas: Zunderfeste Legierungen. 2. Aufl. Berlin/Göttingen/Heidelberg 1963. S. 90 ff.
2. See [1], S. 116 ff.
3. See [1], S. 185.
4. Rohn, W.: Elektrotechn. Z. 48 (1927) S. 227/30 u. 317/20.
5. Franz, H., H. Pfeiffer u. I. Pfeiffer: Z. Metallkde. 58 (1967) S. 87/92.
6. See [1], S. 236 f. u. 245.
7. Rhines, F. N., u. P. J. Wray: Trans. Amer. Soc. Metals 54 (1961) S. 117/28. – Gibbons. T. B., u. B. E. Hopkins: Metal Sci. 8 (1974) S. 203/08.
8. Pfeiffer, I.: Z. Metallkde. 51 (1960) S. 322/26.
9. See [1], S. 265 ff.
10. Elliott, R. P.: Constitution of binary alloys. I. Suppl. New York. St. Louis, San Francisco. Toronto, London, Sydney 1965. S. 345 f. – Kubaschewsky, O.: Iron-Binary Phase Diagrams. Berlin. Heidelberg, New York und Düsseldorf 1982, S. 31 ff.
11. See [1], S. 131 ff.
12. Fisher, R. M., E. J. Dulis u. K. G. Carroll: J. Metals. Trans. AIME 5 (1953) S. 690/95. – Marcinowski, M. J., R. M. Fisher u. A. Szirmae: Trans. metallurg. Soc. AIME 230 (1964) S. 676/89.
13. Shortsleeve, F. J. u. M. E. Nicholson: Trans. Amer. Soc. Metals 43 (1951) S. 142/60.
14. Tagaya, M., u. S. Nenno: Techn. Rep. Osaka Univ. 5 (1955) S. 149/52.
15. Tagaya, M., S. Nenno u. M. Kawamoto: Nippon Kinzoku Gakkai-Si 22 (1958) S. 387/89.
16. See [1], S. 163.
17. Houdremont, E.: Handbuch der Sonderstahlkunde. 3. Aufl Bd. 2 Berlin, Göttingen, Heidelberg u. Düsseldorf 1956. S. 815.
18. Chubb, W., S. Alfant, A. A. Bauer, E. J. Jablonowski, F. R. Shober u. R. F. Dickerson: Battelle Mem. Inst., Rep. No. BMI-1298 (1958).
19. See [1], S. 246 ff.
20. Scheil, E., u. E. H. Schulz: Arch. Eisenhüttenwes. 6 (1932/33) S. 155/60 – Bandel. G.: Arch. Eisenhüttenwes. 15 (1941/42) S. 271/84. – Pfeiffer, I.: Z. Metallkde. 53 (1962) S. 309/12. – Gulbransen, E. A., u. K. F. Andrew: J. electrochem. Soc. 106 (1959) S. 294/302.
21. See [1], S. 263 ff.
22. Wenderott, B.: Z. Metallkde 56 (1965) S. 63/74. – Pfeiffer, H., u. G. Sommer; Z. Metallkde. 57 (1966)

S. 326/31. – Pfeiffer, H.: Werkst. u. Korrosion 21 (1970) S. 977/82. – Hillinger. H.: Werkst. u. Korrosion 22 (1971) S. 504/09.
23. Nickel und Nickellegierungen. Hrsg. von K. E. Volk. Berlin. Heidelberg, New York 1970. S. 151 ff.
24. Amano, T., S. Yajima u. Y. Saito: Trans. Japan Inst. Metals 20 (1979) S. 431/41. – Whittle, D. P., u. J. Stringer: In: Residuals, additives and materials properties [Proc. Conf.], London, May 1978, 1980, S. 309/29. Vgl. Whitle, D. P., u. J. Stringer: Phil. Trans. R. Soc., Lond., A 295 (1980) S. 309/29.
25. DIN 17 470. Heizleiterlegierungen. Technische Lieferbedingungen für Rund- und Flachdrähte. Ausg. Okt. 1984.
26. See [1], S. 166 u. 231.
27. Arnold, A. H. M.: Proc. Instn. electr. Engrs., P. B., 103 (1956) S. 439 ff. – Starr, C. D.: Proc. Instn. electr. Engrs. P.B. 104 (1957) S. 515 ff. – Herman, F.: Elektrotechn. Z., Ausg. B, 16 (1964) S. 670/72.
28. Thomas, H.: Z. Metallkde. 41 (1950) S. 185/90.
29. Thomas, H.: Z. Phys. 129 (1951) S. 219/32. – See [1]. S. 174 ff. u. 227 ff. – See [23], S. 147 ff. – Warlimont, H. u. G. Thomas: Metal Sci. J. 4 (1970) S. 47/52. – Heidsiek, H., K. Lücke u. R. Scheffel: J. Phys. Chem. Solids 43 (1982) S. 825/36.
30. Thomas, H.: Z. Metallkde, 52 (1961) S. 813/16.
31. See [1], S. 294 ff. – Pfeiffer, H., u. G. Sommer: Werkst. u. Korrosion 13 (1962) S. 667/77; Elektrotechn. Z., Ausg. B, 15 (1963) S. 568/72; Metall 19, (1965) S. 108/12. – Pfeiffer, H.: Werkst. u. Korrosion 21 (1970) S. 977/82 – Wilke-Dörfurt, U.: Prakt. Metallogr. 9 (1972) S. 119/28.

C 17 Steels for Valves in Internal Combustion Engines

1. Handbuch, Teves-Thompson GmbH. 4. Aufl. Barsinghausen, Hannover 1977, S. 9.
2. Kocis, J. F., u. W. M. Matlock: Metal Progress, 108 (1975) Nr. 3, S. 58/60 u. 62.
3. Umland, F.: Ventilwerkstoffkorrosion. Frankfurt 1978 (Forschungsberichte Verbrennungskraftmaschinen. H. R. 332.)
4. Kocis, J. F., u. W. M. Matlock: Z. Werkst. Techn. 9 (1978) S. 132/40.
5. Cowley, W. E., P. J. Robinson u. J. Flack: Proc. Instn. Mech. Engrs. 179 (1964–65) Pt. 2A. No 5 S. 145/80.
6. Schönlau, H.: In: Technische Tagung 1968. [Hrsg.:] Teves Thompson GmbH. Barsinghausen/Hannover 1969. S. 9/18.
7. Johnson, V. A., u. R. A. Wilde: J. Mater. 4 (1969) S. 556/65.
8. Chaudhuri, A.: In: SAE-Meeting, Juni 1973, Chicago. Ber.-Nr. 730679.
9. Wegner, K. W.: Ventilwerkstoffkorrosion. Frankfurt 1979. (Forschungsberichte Verbrennungskraftmaschinen. H. R. 361.)
9a. Held, G.: Masch.-Schad. 53 (1980) S. 15/19.
10. Tauschek, M. J.: In: Symposium 72. [Hrsg.:] Teves-Thompson GmbH. Barsinghausen/Hannover 1974. S. 17/21.
11. Tauschek, M. J.: In: Symposium 72. [Hrsg.:] Teves-Thompson GmbH. Barsinghausen/Hannover 1974. S. 17/21.
11. Tauschek, M. J.: Wie unter [6], S. 19/22.
12. Jahr, O.: Hochwarmfeste Werkstoffe. Frankfurt 1975. (Forschungsberichte Verbrennungskraftmaschinen. H. 177.)
13. Milbach, R.: Wie unter [10], S. 26/28.
14. Wakuri, Y., M. Tsuge u. T. Hamatake: Bull. JSME 17 (1974) S. 1313/20.
15. Heumann, Th.: Ventilwerkstoffkorrosion. Frankfurt 1976. (Forschungsberichte Verbrennungskraftmaschinen. H. R. 294.)
16. Thompson, R. F., D. K. Hauiuk, E. B. Etchell u. K. B. Valentine: SAE J. 1955. Aug. S. 54/56.
16a. DIN 17 200. Vergütungstähle. Ausg. März 1987.
17. DIN 17 480. Ventilwerkstoffe. Technische Lieferbedingungen. Ausg. Sept 1984.
18. Hodgson, C. C., u. H. G. Baron: J. Iron Steel Inst. 161 (1949) S. 81/85. – Allsop, H., u. P. W. Bygate: J. Iron Steel Inst. 161 (1949) S. 318/25 u. 14 Taf.
19. Hsiao, C. M., u. E. J. Dulis: Trans. Amer. Soc. Metals 52 (1960) S. 855/77.
20. Nacken, M., u. K. Müller, Arch. Eisenhüttenwes. 33 (1962) S. 863/72.

C 18 Spring Steels

1. Ammareller, S.: Stahl u. Eisen 72 (1952) S. 475/89.
2. Rapatz, F.: Die Edelstähle. 5. Aufl. Berlin/Göttingen/Heidelberg 1962.
3. Ammareller, S.: Stahl u. Eisen 84 (1964) S. 926/31.
4. Schreiber, D., u. H. Ziehm: HOESCH-Ber. 1 (1966) S. 34/41.
5. Hempel, M.: Draht 11 (1960) S. 429/37.
6. Graßhoff, H., u. D. Schreiber: HOESCH-Ber. 5 (1970) S. 163/72.
7. Hodge, J. U., u. M. A. Orehoski: Trans. Amer. Inst. min. metallurg. Eng. 167 (1946) S. 627/42.
7a. Symposium für Federnwerkstoffe 1983, Nürnberg. [Veranst.: Verband der Deutschen Federnindustrie.]
8. Just, E.: Z. wirtsch. Fertigung 73 (1978) S. 95/102.
9. Krautmacher, H.: In: Herstellung von Stahldraht. T1. Hrsg. vom Verein Deutscher-Eisenhüttenleute. Düsseldorf 1969. S. 219/64.
10. Walz, K.: Blech 10 (1964) S. 512/14.
11. Schreiber, D., u. H. Weise: Draht 28 (1977) S. 199/202.
12. Linhart, V.: IFL-Mitt. 7 (1968) Seite 268/76.
13. Sikora, E., P. Funke, W. Heye, u. A. Randak: Stahl u. Eisen 96 (1976) S. 28/32.
14. Einfluß der Abkohlung auf die Dauerschwingfestigkeit von Federstahl 55 Cr 3. Darmstadt 1965. (Forschungsarbeit der Arbeitgemeinschaft Industrieller Forschungsvereinigungen (AIF Nr. 3765) der Technischen Hochschule Darmstadt).
15. Noll, G. C., u. C. Lipson: Proc. Soc. Exp. Stress Anal. 3 (1946) S. 89/109.
16. Siebel, E., u. M. Gaier: VDI-Z. 98 (1956) S. 1715/23.
17. Kloos, K. H., u. B. Kaiser: Draht 28 (1977) S. 415/21 u. 539/45.
17a. Kloos, K. H., u. B. Kaiser: Härtereitechn. Mitt. 37 (1982) S. 7/16.
18. Muhr, K.-H.: Stahl u. Eisen 88 (1968) S. 1449/55.
19. Lepand, H.: Änderung des Dauerschwingverhaltens von Federstahl durch Oberflächenverfestigungen mit Strahlmittel verschiedener Härte bei unterschiedlichen Flächenbedeckungen. Clausthal 1965. (Dr.-Ing.-Diss. Techn. Univ. Clausthal.)
20. Kreinberg, W., H. Ziehm u. J. Ulbricht: Automobilind. 12 (1967) S. 3/11.
21. Keding, H.: Kraftfahrzeugtechn. (1970) S. 104/06.
22. Gillet, H. W., u. E. L. Mack: Proc. Amer. Soc. Test. Mater. 24 (1924) S. 476/577.
23. Ransom, J. T.: Trans. Amer. Soc. Metals 46 (1954) S. 1254/69.
24. Ineson, E., J. Clayton-Cave u. R. J. Tayler: J. Iron Steel Inst. 184 (1956) S. 178/85; 190 (1959) S. 277/83.
25. Atkinson, M.: J. Iron Inst. 195 (1960) S. 64/75.
26. Cummings, H. N., F. B. Stulen u. W. C. Schulte: Trans. Amer. Soc. Metals 49 (1957) S. 482/516.
27. Cummings, H. N., F. B. Stulen u. W. C. Schulte: Trans. Amer. Soc. Test. Mater. 58 (1958) S. 504/14.
28. Buch, A.: Mater.-Prüf. 7 (1965) S. 1/5.
29. Kiessling, R.: Non metallic inclusions in Steel. III. London 1968 (ISI-Publ. No. 115) S. 87/118.
30. Schreiber, D., u. H. Ziehm: HOESCH-Ber. 11 (1976) S. 182/90.
31. Kloos, K. H., B. Kaiser u. D. Schreiber: Z. Werkstofftechn. 12 (1981) S. 206/18.
32. Pomp. A., u. M. Hempel: Arch. Eisenhüttenwes. 21 (1950) S. 67/76.
33. Pomp. A., u. M. Hempel: Arch. Eisenhüttenwes. 21 (1950) S. 53/66.
34. Castagné, J. L., J. H. Davidson, F. Duffaut u. J. Morlet: In: Production and applications of clean steels. London 1972. (ISI-Publ. 134) S. 221/26.
35. Baustähle der Welt. Bd. 2. Von einem Autorenkollektiv. Leipzig 1968. S. 349/406.
36. DIN 17221. Warmgewalzte Stähle für vergütbare Federn; E, Technische Lieferbedingungen. Ausg. Sept. 1987. – DIN 17222, Kaltgewalzte Stahlbänder für Federn; Technische Lieferbedingungen. Ausg. Aug. 1979. – DIN 17223 Teil I. Runder Federstahldraht; Gütevorschriften; Patentiertgezogener Federdraht aus unlegierten Stählen. Ausg. Dez. 1984. – DIN 17223 Blatt 2. Runder Federstahldraht; Gütevorschriften; Vergüteter Federdraht und vergüteter Ventilfederdraht aus unlegierten Stählen, Ausg. März 1964. – DIN 17224. Federdraht und Federbond aus nichtrostenden Stählen. Technische Lieferbedingungen. Ausg. Febr. 1982. – V DIN 17225. Warmfeste Stähle für Federn; Güteeigenschaften. Ausg. April 1955. (V DIN 17225 has been with drawn).
37. Wiesenecker-Kreig, I.: In: Werkstoffkunde der gebräuchlichen Stähle T.2. Hrsg. vom Verein Deutscher Eisenhüttenleute. Düsseldorf 1977. S. 29/40.
37a. Metals Handbook. Amer. Soc. for Metals (ed) 1961 (or latest edtn.) Vol. 1: Properties and selection of metals, steel springs, pp. 160/75. Metals Park Novelly, Ohio.
37b. DIN 17140 Teil 1. Walzdraht zum Kaltziehen. Technische Lieferbedingungen für Grundstahl und unlegierte Qualitätsstähle. Ausg. März 1983. – DIN 17140 Teil 2. Walzdraht zum Kaltziehen.

Technische Lieferbedingungen für unlegierte Edelstähle sowie für legierte Edelstähle für Federn. In Vorbereitung.
37c. DIN 1624. Kaltgewalztes Band in Walzbreiten bis 650 mm aus weichen unlegierten Stählen. Gütenorm. Ausg. Juli 1977.
38. Walz, K.: Draht 19 (1968) S. 604/12 u. S. 783/90.
39. Wahl, A.: Draht-Welt 57 (1971) S. 376/79.
39a. See [37a] pp. 635/36.
40. Walz, K.: Draht 27 (1976) S. 91/98.
41. Hempel, M., u. H. Luce: Mitt. Kais.-Wilhelm-Inst. Eisenforsch. 23 (1941) S. 53/79.
42. Kenneford A. S., u. R. W. Nichols: J Iron Steel Inst. 195 (1960) S. 13/18.
43. Mehta, K. K., H. Kemmer u. L. Rademacher: Thyssen Edelstahl Techn. Ber. 5 (1979) S. 175/80.
44. Kayser, K. H.: Draht 19 (1968) S. 827/38 u. S. 906/15.
45. Sjöberg, J.: Draht 23 (1972) S. 772/75.
46. Weßling, W.: Blech Rohre Profile 24 (1977) . 142/45.

C 19 Free-cutting Steels

1. Bartholome, W., u. H. Sutter: In: Werkstoffkunde der gebräuchlichen Stähle. T. 1 Hrsg. vom Verein Deutscher Eisenhüttenleute. Düsseldorf 1977. S. 259/77.
2. Vlack, L. H. van: Trans. Amer. Soc. Metals 45 (1953) S. 741/57.
3. Gaydos, R.: J. Metals 16 (1964) S. 972/77.
4. Radtke, D., u. D. Schreiber: Stahl u. Eisen 86 (1966) S. 89/99.
5. Dahl, W., H. Hengstenberg u. C. Düren: Stahl u. Eisen 86 (1966) S. 782/95. – The same: Stahl u. Eisen 86 (1966) S. 796/817.
6. Bäcker, L., M. Rolin u. C. Messager: Mém. sci. Rev. Métallurg. 63 (1966) S. 319/28.
7. Poyet, P., u. R. Lévêque: Rev. Métallurg. 64 (1967) S. 653/73.
8. Kiessling, R.: Nonmetallic inclusions in steel. P. 3. London 1968 (Spec. Rep. Iron Steel Inst. No. 115.) S. 51/115.
9 Marston, G. J., u. J. D. Murray: J. Iron Steel Inst. 208 (1970) S. 568/75.
10. Baker, T. J., u. A. Charles: J. Iron Steel Inst. 210 (1972) S. 680/90, 211 (1973) S. 187/92.
11. Brunet, J.-C., J. Frey, J. Bellot u. M. Gantois: C. R. hebd. Séances Acad. Sci., Ser. C., Sci. chim., 273 (1971) S. 620/22.
12. Stahl-Eisen-Prüfblatt 1572–71. Mikroskopische Prüfung von Automatenstähle auf sulfidische nichtmetallische Einschlüsse mit Bildreihen. Ausg. Aug. 1971.
13. Fröhlke, M.: Microscope 19 (1971), S. 403/14.
14. McClymonds, N. L.: Metal Progr. 92 (1967) Nr. 2, S. 183/84.
15. Aborn, R. H.: The role of metallurgy, particularly bismuth, selenium and tellurium in the machinability of steels. Hrsg.: American smelting and refining company. New York 1979.
16. Reh, B., U. Finger, W. Voigt u. W. Schultz: Neue Hütte 27 (1982) S. 121/24.
17. Schroer, H.: Techn. Zbl. prakt. Metallbearb. 62 (1968) S. 514/23.
18. Kämmer, K.: Z. wirtsch. Fertigung 64 (1969) S. 286/95.
19. Müller, Ch. A., A. Stetter u. E. Zimmermann: Arch. Eisenhüttenwes. 37 (1966) S. 27/41 u. 145/58.
19a. Becker, G., E. von Blumenstein, W.-D. Brand, R. Jauch, D. Prem u. H. Weise: Stahl u. Eisen 105 (1985) S. 411/16.
20. DIN 1651. Automatenstähle. Technische Lieferbedingungen. Ausg. 1988.
21. ISO 683/Part 9. Wrought Free-Cutting Steels. Dec. 1988
22. DIN 17111. Kohlenstoffarme unlegierte Stähle für Schrauben, Muttern und Niete. Technische Lieferbedingungen. Ausg. Sept. 1980.
23. Heinritz, M.: Werkstattstechn. 63 (1973) S. 623.
24. Bersch, B., H. Fröber u. H. Weise: Z. Werst. Techn. 7 (1966) S. 181/89.
25. Becker, G., Ch. Kowollik u. H. Schroer: Thyssen Techn. Ber. 2 (1977) S. 59/66.
26. Warke, W. R., u. N. N. Breyer: J. Iron steel Inst. 209 (1971) S. 779/84.
27. Klaus, F., W. König, W. Lückerath u. H. Siebel: Stahl u. Eisen 85 (1965) S. 1669/86.
28. Bellot, J., M. Hugo, E. Schirrecker u. E. Herzog: Rev. Métallurgie 63 (1966) S. 959/75.
29. Kaluza, E.: Ind.-Anz. 86 (1964) S. 1347/52.
30. Weigl, K.: Z. wirtsch. Fertigung 63 (1968) S. 502/10, 556/65 u. 607/14. – Wuich, W.: Werkstatt u. Betr. 104 (1971) S. 550/52.
31. Vöge, H.: Fachber. Metallbearb. 60 (1983) S. 375/81.

C 20 Soft magnetic materials

1. Pawlek, F.: Magnetische Werkstoffe. Berlin. Göttingen, Heidelberg, 1952.
2. Boll, R.: Elektrotechn. Z., Ausg. B, 26 (1974) S. 696/98.
3. Boll, R.: Soft Magnetic Materials. London, 1978.
4. Reinboth, H.: Neue Hütte 19 (1974) S. 156/61.
5. Heck, C.: Magnetische Werkstoffe und ihre technische Anwendung. 2. Aufl. Heidelberg 1974.
   – Tebble, R. S. and D. J. Craik: Magnetic Materials. London, New York, Sydney, Toronto 1969. p. 726.
6. Reinboth, H.: Technologie und Anwendung magnetischer Werkstoffe. 2. Aufl. Berlin 1963. 3 Aufl. Berlin 1970.
7. Reichel, K.: Praktikum der Magnettechnik. München 1980.
8. Radeloff, C.: In: Magnettechnik, (Technik und Anwendung der weichmagnetischen Werkstoffe). Grafenau 1980 (Kontakt + Studium Bd. 56) S. 13/32.
9. Kneller, E.: Ferromagnetismus. Berlin, Göttingen, Heidelberg 1962. – Berkowitz, A. E. and E. Kneller (ed.): Magnetism and Metallurgy. New York, London 1969. p. 838.
10. Bozorth, R. M.: Ferromagnetismus. New York 1951.
11. DIN 1325. Magnetisches Feld; Begriffe, Ausg. Jan. 1972.
12. Publ. CEI 50 (901): Advanced edition of international electrotechnical vocabulary. Chapter 901; Magnetism. Ausg. 1973.
13. Jellinghaus, W.: Magnetische Messungen an ferromagnetischen Stoffen. Berlin 1952.
14. ASTM Standards. In: Annual Book of ASTM Standards. Vol. 03.04. Magnetic properties. Philadelphia/Pa.
15. Direct-current magnetic measurements for soft magnetic materials. Philadelphia/Pa. 1970. (ASTM Spec. Techn. Publ. No. 371–S1.)
16. DIN 50 462. Prüfung von Stahl; Verfahren zur Ermittlung der magnetischen Eigenschaften von Elektroblech und -band im 25-cm-Epstein-Rahmen. T.1 bis 6. – T.1 bis 3 u. T.6 Ausg. Aug. 1986 u. T.5 Ausg. Juni 1979.
17. Dijkstra, L. J. u. C. Wert: Phys. Rev. 79 (1950) S. 979/85.
18. Hoffmann, A.: Arch. Eisenhüttenwes. 40 (1969) S. 999/1003.
19. DIN 17 405. Weichmagnetische Werkstoffe für Gleichstromrelais; Technische Lieferbedingungen. Ausg. Sept. 1979.
20. DIN 46 400. T. 1. Flacherzeugnisse aus Stahl mit besonderen magnetischen Eigenschaften: Elektroblech und -band, kaltgewalzt, nicht kornorientiert, schlußgeglüht; Technische Liefebedingungen. Ausg. April 1983. -T.2. Flacherzeugenisse aus Stahl mit besonderen magnetischen Eigenschaften; Elektroblech und- band aus unlegierten Stählen; kaltgewalzt, nicht schlußgeglüht; Technische Liefebedingungen. Ausg. Jan. 1987. – T.3. Flacherzeuguisse aus Stahl mit besonderen magnetischen Eigenschaften; Elektroblech und -band, kornorientiert; Technische Liefebedingungen. Ausg. Nov. 1975 (New edition is being prepared; for the time being: Entwarf Aug. 1986). – T.4. Flachertengnisse aus Stahl mit besonderen magnetischen Eigenschaften; Elektroblech und-band aus legierten Stählen, Kallgewalzt, nicht kornorientiert, nicht schlupgeglüht; Technische Lieferbedingungen. Ausg. Jan. 1987.).
21. Knorr, W.: Vorgetragen auf der Tagung der Arbeitsgemeinschaft Ferromagnetismus am 13. u. 14.4. 1965 in Marburg/Lahn. S. a. Z. angew. Phys. 21 (1966) S. 438/41.
22. Ricken, H. G., u. H. M. Thimmel: In: The Second International conference on Magnet-Technology. Hrsg.: The Rutherford Laboratory. Oxford 1967. S. 256/61.
23. Weichmagnetische Stähle für schwere Schmiedestücke. Produktinformation 1.11.1980, ARBED Saarstahl GmbH.
24. Pry, R. H., u. C. B. Bean.: J. appl. Phys. 29 (1958) S. 532/33. – Haller, T. R., u. J. J. Kramer: J. appl. Phys. 41 (1970) S. 1034/36. – Sun, J. N., T. R. Haller u. J. J. Kramer: J. appl. Phys. 42 (1971) S. 1789/91.
25. Mayer, A., u. F. Bölling: J. Magn. Magnetic Mater. 2 (1976) S. 151/61.
26. Die Hauptforschungsarbeit zum Thema „anomaler Verlust" wird zur Zeit von A. Ferro-Milone und Mitarbeitern am Instituto elettrotecnico Nazionale Galileo Ferraris in Turin geleistet.
27. Amer. Pat. 1 965 559 vom 7. Aug. 1933. – Yensen, T. D.: Stahl u. Eisen 56 (1936) S. 1545/50; 57 (1937) S. 123.
28. May, J. E., u. D. Turnbull: Trans. metallurg. Soc. AIME 212 (1958) S. 769/81.
29. Taguchi, S.: Trans. Iron Steel Inst. Japan 17 (1977) S. 604/15.
30. Taguchi, S., A. Sakakura, F. Matsumoto, K. Takashima u. K. Kuroki: J. Magn. Magnetic Mater. 2 (1976) S. 121/31.

31. DOS 23 51 141 (1974).
32. DOS 25 31 515 (1975).
33. Littmann, M. F.: J. appl. Phys. 38 (1967) S. 1104/08.
34. Bölling, F., u. A. Mayer: ETG-Fachber. 8 (1981) S. 60/64.
35. Bär, N., A. Hubert u. W. Jillek: J. Magn Magnetic Mater. 6 (1977) S. 242/48.
36. Nozawa, T., T. Yamamoto, Y. Matsuo u. Y. Ohya: IEEE Trans. Magnetics 15 (1979) S. 972/81.
37. Phillips, R., u. K. J. Overshott: IEEE Trans. Magnetics 10 (1974) S. 168/69.
38. Bölling, F.: In: Magnettechnik – Technik und Anwendung der weichmagnetischen Werkstoffe. Grafenau 1980. (Kontakt + Studium Bd. 56) S. 57/76.
39. Mayer, A.: Stahl u. Eisen 83 (1963) S. 1169/76.
40. IEC Standard: Publication 404-8-3, Part 8: Specification for individual materials, Section 3 – Specification for cold-rolled magnetic non-alloyed steel strip delivered in the semi processed state. First edition 1985. IEC-Standard: Publication 404-8-2, Part 8: Specification for individual materials, Section 2 – Specification for cold-rolled magnetic alloyed steel and strip delivered in the semi processed state. First edition 1985. IEC Standard: Publication 404-8-4, Part 8: Specification for individual materials, Section 4 – Specification for cold-rolled non-oriented magnetic sheet and strip. Under consideration. IEC Standard: Publication 404-8-5, Part 8: Specification for individual materials, Section 5 – Specification for cold rolled grain-oriented magnetic sheet and strip. Under consideration.
41. DIN 50466. Prüfung der magnetischen Eigenschaften von Elektroblechen und Blechkernen; Bestimmung der komplexen Permeabilität und ihres Kehrwerts im magnetischen Wechselfeld. Ausg. Sept. 1975.
42. Bölling, F., H. Pottgießer u. K.-H. Schmidt: Stahl u. Eisen 102 (1982) S. 833/37.
43. Naumann, F.: Unveröff. Mitt. der Thyssen Grillo Funke GmbH, Gelsenkirchen.
44. Pepperhoff, W., u. W. Pitsch: Arch. Eisenhüttenwes. 47 (1976) S. 685/90.
45. Stahl-Eisen-Werkstoffblatt 550. Stähle für größere Schmiedestücke; Gütevorschriften. Ausg. Aug. 1976. – Stahl-Eisen-Werkstoffblatt 555. Stähle für größere Schmiedestücke als Bauteile von Turbinen- und Generatorenanlagen. Ausg. Aug. 1984.
46. DIN 17100. Allgemeine Baustähle; Gütenorm. Ausg. Jan. 1980.
47. Stahl-Eisen-Werkstoffblatt 092. – Warmgewalzte Feinkornstähle zum Kaltumformen: Gütevorschriften. Ausg. Juli 1982.
48. Opel, P., C. Florin, F. Hochstein u. K. Fischer: Stahl u. Eisen 90 (1970) S. 465/75.
49. Forch, K.: Technica 1972, Nr. 1, S. 39/45.
50. Neidhoefer, G., u. A. Schwengeler: J. Magn. Magnetic Mater. 9 (1978) S. 112/22.
51. Ricken, H. G.: Untersuchung über die magnetischen Eigenschaften von Induktorwellenstählen. Unveröff. Bricht der Krupp Hüttenwerke AG vom 15.12.1964.
52. Blower, R., C. A. Clark u. G. Mayer: Metallurgia 70 (1964) S. 207/12.
53. Toitot, M., C. Roques u. P. Bastien: Rev. Métallurg. 59 (1962) S. 631/37.
54. Downing, G. S., W. E. Jones u. L. E. Osman: Metal Progr. 53 (1948) Nr. 1, S. 87/90: Nr. 2, S. 235/40.
55. Blower, R., u. M. J. Fleetwood: In: Proceedings of the 1st italian meeting on heavy forgings, 26–29 Sept. 1961, Treni. Hrsg.: Camera di Commercio, Industria e Agricoltura di Terni. Terni 1962. S. 253/76.
56. Lüling, H., u. K. Gut: Gießerei-Prax. 1963, S. 39/45.
57. Dietrich, H.: Gießerei, techn.-wiss. Beih., 14 (1962) S. 79/91.
58. Jackson, W. J.: J. Iron Steel Inst. 194 (1960) S. 29/36.
59. DIN 1681. Stahlguß für allgemeine Verwendungszwecke; Technische lieferbedingüngen. Ausg. Juni 1985.
60. Haneke, M., u. H. Auktun: Handelsblatt, Techn. Linie, 31 (1978) Nr. 23, S. 89.
61. Eberly, W. S.: Iron Age 183 (1959) Nr. 17, S. 106/08.
62. Kiesheyer, H., u. H. Brandis: Z. Werkst.-Techn. 9 (1978) S. 14/18.
63. Werksprospekt DEW, Physikalische Stähle, Nr. 12, 1964.
64. Chikazumi, S.: Physics in magnetism. New York, London, Sydney 1964.
65. Pfeifer, F., u. C. Radeloff: J. Magn. Magnetic Mater. 19 (1980) S. 190/207.
66. Assmus, F.: In: Proceedings Third International Conference on Soft Magnetic Materials. Bratislava 1977. S. 83/91.
67. Kunz, W., u. F. Pfeifer: In: Amer. Inst. Phys. Conf. Proc. Ser. 1976, Proc. Nr. 34.
68. Chin, G. Y.: IEEE Trans. Magnetics, MAG-7, 1971, S. 102/13.
69. Mager, R., u. F. Pfeifer: In: Nickel und Nickellegierungen. Hrsg. von K. E. Volk. Berlin, Heidelberg, New York 1970.
70. DIN 41301. Elektrobleche. Magnetische Werkstoffe für Übertrager. Ausg. Juli 1967. – DIN 41302 T 2. Kleintransformatoren, Übertrager und Drosseln; Kernbleche; Technische Lieferbedingungen.

Ausg. Mai 1986. – T. 100. Kleintransformatoren, Übertrager und Drosseln; Kernbleche nach IEC. Ausg. Mai 1981. (T.100 withdrawn, is contained in T.1. Ausg. Mai 1986) – ASTM A 753–85. Nickel-Iron Soft Magnetic Alloys. Philadelphia 1985. 1986 Annual Book of ASTM Standards, p. 10. – IEC Publication 404-8-6: Magnetic materials Part 8: Soft Magnetic Metallic Materials. International Electrotechnical Commission, Geneva, 1986. p. 25.
71. Miyazaki, T., Sawada, R. u. Y. Ishijima: IEEE Trans. Magnetics, MAG-8, 1972. S. 501/02.
72. Pfeifer, F., u. R. Deller: Elektrotechn. Z., Ausg. A, 89 (1968) S. 601/04.
73. Rassmann, G., u. H. Wich: In: Berichte der Arbeitsgemeinschaft Ferromagnetismus 1959. Hrsg. vom Gemeinschaftsausschuß der Deutschen Gesellschaft für Metallkunde, Werkstoffausschuß des Vereins Deutscher Eisenhüttenleute, Verband Deutscher Physikalischer Gesellschaften. Düsseldorf 1960. S. 181/89.
74. Pfeifer, F.: Z. Metallkde. 57 (1966) S. 240/44.
75. Hall, R.C.: J. appl. Phys., Suppl. 31 (1960) S. 1575/85.
76. Narita, K., Teshima, N. u. M. Mori: IEEE Trans. Magnetics, MAG-17, 1981, S. 2857/62.
77. Assmus, F.: Siemens Forsch. u. Entwickl. Ber. 7 (1978) S. 118/23. – Hasegawa, R. (Edit.): Glassy Metals; Magnetic, Chemical and Structural Properties. Bora Raton (Florida) 1983. p. 263.
78. Hilzinger, H.R.: NTG-Fachber. 76 (1980) S. 283/306. – Steeb, S. and W. Warlimond (Edit.): Rapidly Quenched Metals. Proceedings of the Fifth International Conference on Rapidly Quenched Metals, Würzburg (Germany) Sept. 1984. Amsterdam, Oxford, New York, Tokyo 1985. p. 1799.
79. Boll, R., u. H. Warlimont: Trans. Magnetics, MAG-17, 1981, S. 3053/58.
80. Hilzinger, R.: Transact. on Magn. MAG 21 (1985) pp. 20 20/.25.

C 21 Permanent Magnet Materials

1. DIN 17410. Dauermagnetwerkstoffe. Ausg. Mai 1977.
1a. IEC Publication 404-8-1: Magnetic materials, Specification for individual materials, Standard specifications for magnetically hard materials. International Electrotechnical Commission, Geneva, 1986.
2. Schüler, K., u. K. Brinkmann: Dauermagnete, Werkstoffe und Anwendungen. Berlin. Heidelberg, New York 1970.
3. Parker, R.J., u. R.J. Studders: Permanent magnets and their application. New York 1962.
4. Permanent magnets and magnetism. [Hrsg.:] D. Hadfield. London, New York 1962.
5. Reinboth, H.: Technologie und Anwendung magnetischer Werkstoffe, 3. Aufl.: Berlin 1969.
6. Heck, C.: Magnetische Werkstoffe und ihre technische Anwendung. Heidelberg 1967.
7. Heimke, G.: Keramische Magnete. Berlin, Heidelberg, New York 1976.
8. Smit, J., u. H.P.J. Wijn: Ferrite. Eindhoven 1962. (Philips technische Bibliothek.)
9. Kneller, E.: Ferromagnetismus. Berlin, Göttingen, Heidelberg 1962.
10. Magnetism and metallurgy. Vol. 1.2. Ed. by A.E. Berkowitz and E. Kneller, New York, London 1969.
11. McCaig, M.: Permanent magnets in theory and practice. Plymouth 1977.
12. Brown, W.F.jr.: Micromagnetics, New York 1963.
13. Fischer, J.: Abriß der Dauermagnetkunde. Berlin, Göttingen, Heidelberg 1949.
14. DIN 50470. Bestimmung der Entmagnetisierungskurve und der permanenten Permeabilität in einem Joch. Ausg. Sept. 1980. – DIN 50471. Bestimmung der Entmagnetisierungskurve und der permanenten Permeabilität im Doppeljoch. Ausg. Sept. 1980.
14a. IEC Publication 404-5: Magnetic materials, Part 5: Methods of measurements of the magnetic properties of magnetically hard (permanent magnet) materials.
15. DIN 50472. Prüfung von Dauermagneten; Bestimmung der magnetischen Flußwerte im Arbeitsbereich. Ausg. März 1981.
16. Dietrich, H.: Kobalt Nr. 35, 1967, S. 71/87.
17. Luborsky, F.E.: J. appl. Phys. 33 (1962) S. 2385/90.
18. Stäblein, H.: Techn. Mitt. Krupp, Forsch.-Ber., 29 (1971) S. 101/09.
19. Koch, A.J.J., M.G. van der Steeg u. K.J. de Vos: In: Conference on magnetism and magnetic materials. Boston 1956, S. 173/83.
20. Planchard, E., C. Bronner u. J. Sauze: Z. angew. Phys. 21 (1966) S. 63/65.
21. Cahn. J.W.: Trans. metallurg. Soc. AIME 242 (1968) S. 166/80.
22. Vos, K.J. de: Z. angew. Phys. 17 (1964) S. 168/74.
23. Kronenberg, K.J.: J. appl. Phys. 31 (1960) S. 80S/82S.

24. Fahlenbrach, H.: Techn. Mitt. Krupp 12 (1954) S. 177/84.
25. Mason, J.J., D.W. Ashall u. A.N. Dean: Kobalt Nr. 46, 1970, S. 20/24.
26. Baran, W.: Techn. Mitt. Krupp 17 (1959) S. 150/52.
27. Pant, P.: Techn. Mitt. Krupp, Forsch.-Ber., 35 (1977) S. 59/64.
28. IEC-Publication 404-1: Magnetic materials, Part 1, Classification. 1979.
29. McCaig, M.: Kobalt Nr. 5, 1959, S. 26/28.
30. Fahlenbrach, H.: Kobalt Nr. 49, 1970, S. 174/82.
31. Pfeiffer, I.: Siemens Forsch.- u. Entw.-Ber. 1 (1971) S. 71/79.
32. Joffe, I.: J. Mater. Sci. 9 (1974) S. 315/22.
33. Josso, E.: IEEE Trans. Magnetics 10 (1974) S. 161/65.
34. Shur, Ya. S., M.G. Luzhinskaya u. L.A. Shibina: Phys. Metals Metallogr. 4 (1957) S. 40/44 u. 45/52.
35. Baran, W., W. Breuer, H. Fahlenbrach u. K. Janssen: Techn. Mitt. Krupp 18 (1960) S. 81/90.
36. Fahlenbrach, H.: electronic ind. 1/2 (1974) S. 11/13.
37. Kaneko, H., M. Homma, K. Nakamura u. M. Miura: IEEE Trans. Magnetics 8 (1972) S. 347/48.
38. Kaneko, H., M. Homma, T. Fukunaga u. M. Okada: IEEE Trans. Magnetics 11 (1975) S. 1440/42.
39. Chin, G.Y., S. Jin, M.L. Green, R.C. Sherwood u. J.H. Wernick: J. appl. Phys. 52 (1981) S. 2536/41.
40. Ervens, W.: Techn. Mitt. Krupp, Forsch.-Ber., 40 (1982) S. 109/16.
41. Cremer, R., u. I. Pfeiffer: Physica 80 B (1975) S. 164/176.
42. Okada, M., G. Thomas, M. Homma u. H. Kaneko: IEEE Trans. Magnetics 14 (1978) S 245/52.
43. Jin, S.: IEEE Trans. Magnetics 15 (1979) S 1748/50.
44. Nicholson, R.B., u. P.J. Tufton: Z. angew. Phys. 21 (1966) S. 59/62.
45. Adelsköld, V.: Ark. Kemi Mineralogie och Geologi 12A (1938) S. 1/9.
46. Kneller, E., u. F.E. Luborsky: J. appl. Phys. 34 (1961) S. 2318/28.
47. Wullkopf, H.: Ber. Dt. keram. Ges. 55 (1978) S. 292/93.
48. Haberey, F.: Ber. Dt. keram. Ges. 55 (1978) S. 297/301.
49. Stäblein, H., u. W. May: Ber. Dt. keram. Ges. 46 (1969) S. 69/74 u. 126/28.
50. Stuijts, A.L., G.W. Rathenau u. G.H. Weber: Philips techn. Rdsch. 16 (1954/55) S. 221/28.
51. Stäblein, H., u. J. Willbrand: Z. angew. Phys. 21 (1966) S. 47/51.
52. Denes, P.A.: Amer. ceram. Soc. Bull. 41 (1962) S. 509/12.
53. Richter, H., u. H. Völler: DEW Techn. Ber. 8 (1968) S. 214/21.
54. Bäder, E.: Feinwerktechn. 64 (1960) S. 79/84.
55. Hubler, E.: Elektrotechn. Z., Ausg. B, 17 (1965) S. 817/19.
56. Dietrich, H.: Kobalt Nr. 35, 1967, S. 71/87.
57. Stäblein, H.: Hard Ferrites and Plastoferrites. In: Ferromagnetic Materials. Vol. 3 Ed. E.P. Wohlfarth. Amsterdam/London 1982. S. 441/602.
57a. Sagawa, M., S. Fujimura, N. Togawa, H. Yamamoto u. Y. Matsuura: J. appl. phys. 55 (1984) S. 2083/87.
57b. Croat, J.J., J.F. Herbst, R.W. Lee u. F.E. Pinkerton: J. appl. phys. 55 (1984) S. 2078/82.
57c. Lee, R.W.: Appl. Phys. Lett. 46 (8), (1985) pp. 790/91.
57d. Nd-Fe permanent magnets, their present and future applications. Report and proceedings of a workshop meeting, held in Brussels on 25 Oct. 1984. [Hrsg.:] I. V. Mitchell. Commission of the European Communities, Directorate General for Science, Research and Development.
58. Wallace, W.E.: Rare Earth Intermetallics. New York, 1973.
59. Lemaire, R.: Kobalt Nr. 32, 1966, S. 117/24; Nr. 33, 1966, S. 175/84.
60. Strnat, K.: Kobalt Nr. 36, 1967, S. 119/28.
61. Martin, D.L., J.T. Geertsen, R.P. Laforce u. A.C. Rockwood: In: Proceedings II. rare earth conference, Michigan, 1974, S. 342/52.
62. Nagel, H., u. A. Menth: Goldschmidt informiert Nr. 35, 1975, S. 42/46.
63. Schuchert, H.: Internat. J. Magnetism 5 (1973) No. 1/2/3, S. 215/22.
64. Menth, A., H. Nagel u. R.S. Perkins: Ann. Rev. Mater. Sci. 8 (1978) S. 21/47.
65. Ojima, T., S. Tomizawa, T. Yoneyama u. T. Hori, IEEE Trans. Magnetics 13 (1977) S. 1317/19.
66. Ervens, W., W. Baran u. H. Schuchert: Metall 33 (1979) S. 727/32.
67. Herget, C., u. H.G. Domazer: Goldschmidt informiert Nr. 35, 1975, S. 3/33.
68. Schäfer, G.: Karlsruhe 1974. (Diss. Techn. Hochsch. Karlsruhe).
69. Den Broeder, F.J.A., u. K.H.J. Buschow: J. Less-Common Metals 29 (1972) S. 65/71.
70. Schäfer, G., u. W.W. Spyra: TEW Techn. Ber. 7 (1975) S. 40/48.
71. Ervens, W.: Techn. Mitt. Krupp, Forsch.-Ber., 40 (1982) S. 99/107.
72. Cech, R.E.: J. appl. Phys. 41 (1970) S. 5247/49.
73. Das, D.: AFML – TR 71-151 (1971).

74. Paladino, A. E., M. J. Dionne, P. F. Weihrauch u. E. C. Wettstein: Goldschmidt informiert Nr. 35, 1975, S. 63/74.
75. Ojima, T., S. Tomizawa, T. Yoneyama u. T. Hori: IEEE Trans. Magnetics 13 (1977) S. 1317/19.
76. Ervens, W.: In: Proceedings of the VIth international workshop on rare earth-cobalt permanent magnets and their applications. August 31-Sept. 2, 1982, Baden/Vienna, Austria. RCO-6, S. 319/27.
77. Newkirk, J. B., A. H. Geisler, D. L. Martin u. R. Smoluchowski: Trans. metallurg. Soc. AIME 188 (1950) S. 1249/60.
78. Martin, D. L.: Wie unter [19], S. 188/202.
79. Köster, W., u. E. Wachtel: Z. Metallkde. 51 (1960) S. 271/80.
80. Ravdjel, M. P., In: Arbeiten des zentralen Forschungsinstitutes für Metallurgie. Bd. 71. (Moskau 1969) S. 93/108.
81. Gödecke, T., u. W. Köster: Z. Metallkde. 62 (1971) S. 727/32.
82. Koch, A. J. J., P. Hokkeling, M. G. v. d. Steeg u. K. J. de Vos: J. appl. Phys. 31 (1960) S. 75s/77s.
83. Kojima, S.: AIP conference proceedings magnetism and magnetic materials Nr. 24, 1974, S. 768/69.
84. Ohtani, T.: IEEE Trans. Magnetics 13 (1977) S. 1328/30.
85. Ervens, W.: Techn. Mitt. Krupp. Forsch.-Ber. 40 (1982) S. 117/22.
86. Amer. Patent 3 661 567 vom 6. Dez. 1967.
87. DAS 1458 411 vom 30. Okt. 1963.
88. Dietrich, H., u. W. Schmidt: DEW Techn. Ber. 13 (1973) S. 189/92.

C 22 Nonmagnetizable Steels

1. Bluhm, P.: Der nichtmagnetisierbare Stahl und seine Anwendung, Hamburg, Berlin 1967.
2. Heimann, W., I. Bischoff u. J. Buckstegge, Thyssen Edelstahl Techn. Ber. 5 (1979) S. 194/200.
3. Eichelmann, G. H., u. F. C. Hull: Trans. Amer. Soc. Metals 45 (1953) S. 77/104.
4. Dietrich, H., W. Heimann u. F. H. Strom: TEW Techn. Ber. 2 (1976) S. 61/69.
5. Lorenz, K., H. Fabritius u. G. Médawar: Thyssenforsch. 1 (1969) S. 97/108.
6. Hochmann, J.: Matér. & Techn. 65 (1977) Dez. (Sondernr. Manganese), S. 69/87.
7. Jesper, H., W. Weßling u. K. Achtelik: Stahl u. Eisen 86 (1966) S. 1408/18.
8. Weßling, W., u. H. E. Bock: Stahl u. Eisen 91 (1971) S. 1442/45.
9. Lorenz, K., u. G. Médawar: Thyssenforsch. 1 (1969) S. 97/108.
10. Heinrich, E., G. Kröncke u. G. Tacke: Stahl u. Eisen 102 (1982) S. 1183/88.
11. Riedl, J., u. H. Kohl: Berg- u. hüttenm. Mh. 122 (1977) S. 62/66.
12. Kohl, H.: Werkst. u. Korrosion 14 (1963) S. 831/837. – Bäumel, A., u. F. Bachmann: Arch. Eisenhüttenwes. 43 (1972) S. 631/37.
13. Speidel, M. O.: VGB-Kraftwerkstechn. 61 (1981) S. 417/27. – Derselbe: VGB-Kraftwerkstechn. 61 (1981) S. 1048/53. – Derselbe: VGB-Kraftwerkstechn. 62 (1982) S. 424/28.
14. DP 1957 375 vom 14. Nov. 1969.
15. Dietrich, H.: DEW Techn. Ber. 4 (1964) S. 111/32.
16. Heimann, W., W. Schmidt, G. Lennartz u. E. Michel: DEW Techn. Ber. 13 (1973) S. 94/107.

C 23 Steels with Defined Thermal Expansion and Special Classic Properties

1. Ebert, H.: Die Wärmeausdehnung fester und flüssiger Stoffe. Braunschweig 1940. S. 53.
2. Otto, J., u. W. Thomas: Z. Phys. 175 (1963) S. 337/44.
3. Partridge, J. H.: Glass-to-metal seals. Sheffield 1949.
4. Hermann, H.: Glas- u. Hochvakuumtechn. 2 (1953) S. 189/200. Z. angew. Phys. 7 (1955) S. 174/76. – Metall 9 (1955) S. 407/10. Vakuumtechn. 4 (1955) S. 115/17. Glastechn. Ber. 33 (1960) S. 252/57.
5. Geyer, F.: Siemens-Z. 46 (1972) S. 709/10.
6. Masiyama, Y.: Sci. Rep. Tôhoku Univ. 20 (1931) S. 574/93. – Becker, R., u. W. Döring: Ferromagnetismus. Berlin 1939. S. 305 ff. – Kneller, E.: Ferromagnetismus. Berlin, Göttingen, Heidelberg 1962. S. 217 ff.
7. Chevenard, P.: Trav. Mém. Bur. Int. Poids et Mesures 17 (1927) S. 1 ff.
8. Guillaume, C. E.: Rev. Métallurg. Mém. 25 (1928) S. 35/43.
9. Lement, B. S., B. L. Averbach u. M. Cohen: Trans. Amer. Soc. Metals 43 (1951) S. 1072/97.

10. Hoffrogge, C.: Z. Phys. 126 (1949) S. 671/88.
11. Kußmann, A., u. K. Jessen: Arch. Eisenhüttenwes. 29 (1958) S. 585/94.
12. Hausch, G., u. H. Warlimont: Phys. Letters 36 A (1971) S. 415/16.
13. Masumoto, H.: Sci. Rep. Tôhoku Univ. 20 (1931) S. 101/23.
14. Kase, T.: Sci. Rep. Tôhoku Univ. 16 (1927) S. 491/513.
15. Köster, W., u. W.-D. Haehl: Arch. Eisenhüttenwes. 40 (1969) S. 569/74; Arch. Eisenhüttenwes. 40 (1969) S. 575/83.
16. Herrmann, H., u. H. Thomas: Z. Metallkde. 48 (1957) S. 582/87. – Nickel und Nickellegierungen. Edited by K. E. Volk. Berlin, Heidelberg, New York 1970. S. 35 ff.
17. Scott, H.: Trans. Amer. Inst. min. metallurg. Eng. 89 (1930) 506/37; J. Franklin Inst. 220 (1935) S. 733/54.
18. Hansen, M.: Constitution of binary alloys. New York, Toronto, London 1958. S. 677 ff. – Heumann, T., u. G. Karsten: Arch. Eisenhüttenwes. 34 (1963) S. 781/85.
19. Tino, Y., u. H. Kagawa: J. Phys. Soc. Japan 28 (1970) S. 1445/51.
20. Bendick, W., H. H. Ettwig, F. Richter u. W. Pepperhoff: Z. Metallkde. 68 (1977) S. 103/07.
21. Masumoto, H.: Sci. Rep. Tôhoku Univ. 23 (1934) S. 265/80.
22. Kußmann, A.: Phys. Z. 38 (1937) S. 41/42.
23. Masumoto, H. u. T. Kobayashi: Trans. Japan Inst. Metals 6 (1965) S. 113/15.
24. Masumoto, H., H. Saito u. T. Kobayashi: Trans. Japan Inst. Metals 4 (1963) S. 114/17. – Kußmann, A., u. K. Jessen: Z. Metallkde. 54 (1963) S. 504/10.
25. Colling, D. A., u. M. P. Mathur: J. appl. Phys. 42 (1971) S. 5699/5703.
26. Fujimori, H.: J. Phys. Soc. Japan 21 (1966) S. 1860/65. – Fukamichi, K., u. H. Saito: Phys. status sol. (a) 10 (1972) S. K 129/K 131. – Saito, H., u. K. Fukamichi: IEEE Trans. Magnetics 8 (1972) S. 687/88.
27. Richter, F., u. W. Pepperhoff: Arch. Eisenhüttenwes. 47 (1976) S. 45/50.
28. Richter, F.: Arch. Eisenhüttenwes. 48 (1977), S. 239/41.
29. Aßmus, F.: In: 40 Jahre Vacuumschmelze AG. 1923–1963. Hanau 1963 S. 47 ff.
30. Hausch, G., u. H. Warlimont: Z. Metallkde. 64 (1973) S. 152/60.
31. Nakamura, Y.: IEEE Trans. Magnetics 12 (1976) S. 278/91.
32. Köster, W.: Z. Metallkde. 35 (1943) S. 194/99.
33. Chevenard, P.: Trav. Mém. Bur. Int. Poids et Mesures 17 (1927) S. 142 ff. – Guillaume, C. E.: Rev. Métallurg. Mém. 25 (1928) S. 35/43.
34. Fine, M. E., u. W. C. Ellis: Trans. AIME 191 (in J. Metals 3) (1951) S. 761/64.
35. Krüger, G.: Metallurg. Rev. 8 (1963) S. 427/59.
36. Albert, H.: IEEE Trans. Magnetics 9 (1973) S. 346/48.
37. Ochsenfeld, R.: Z. Phys. 143 (1955) S. 357/73. – Derselbe: Z. Phys., 143 (1955) S. 375/91.
38. Schneider, W., u. H. Thomas: Metallurg. Trans. 10 A (1979) S. 433/38.
39. Krächter, H., u. W. Pepperhoff: Arch. Eisenhüttenwes. 39 (1968) S. 541/43.
40. Steinemann, S. G.: J. Magnetism Magnetic Mater. 7 (1978) S. 84/100.
41. Saito, H., K. Wakaoka u. K. Fukamichi: Trans. Japan Inst. Metals 17 (1976) S. 844/48.
42. Stahl-Eisen-Liste. 7. Aufl. Hrsg.: Verein Deutscher Eisenhüttenleute. Düsseldorf 1981. – Bungardt, K.: In: Werkstoff-Handbuch Stahl und Eisen. 4. Aufl. Hrsg.: Verein Deutscher Eisenhüttenleute. Düsseldorf 1965. Blatt O 11-1/-11-6. – Werkstoff-Handbuch Nichteisenmetalle. 2. Aufl. Hrsg.: Deutsche Gesellschaft für Metallkunde u. Verein Deutscher Ingenieure. Düsseldorf 1960. Teil III. Ni 2.2.
43. Espe, W.: Werkstoffkunde der Hochvakuumtechnik. Bd. 1. Berlin 1959. S. 340 ff u. 378 ff.
44. Landolt-Börnstein: Zahlenwerte und Funktionen aus Physik, Chemie, Astronomie, Geophysik und Technik. 6. Aufl. 4. Bd. 2. Teil, Bandteil b. Berlin, Göttingen, Heidelberg, New York 1964. S. 512 ff. u. 520 ff.
45. Marsh, J. S.: The alloys of iron and nickel. Vol. I. New York und London 1938. S. 107 ff. u. 135 ff.
46. Stahl-Eisen-Werkstoffblatt 385–57. Eisenlegierungen mit besonderer Wärmeausdehnung. Ausg. Aug. 1957. – DIN 17 745. Knetlegierungen aus Nickel und Eisen; Zusammensetzung. Ausg. Jan. 1973.
47. DIN 1715. Thermobimetalle. Technische Lieferbedingungen. Teil 1. Ausg. Nov. 1983.
48. Guillaume, C. E.: C. R. hebd. Séances Acad. Sci. 171 (1920) S. 1039/41. – Siehe auch [45]. S. 159.
49. Wie [44]. S. 517.
50. Wie [44]. S. 516; [45]. S. 131 ff.
51. Dietrich, H.: Nickel-Ber. 25 (1967) S. 37/45.
52. Kašpar, F.: Thermobimetalle in der Elektrotechnik. Berlin 1960. – Engstler, D.: Arch. techn. Messen. Blatt Z 972-3 u. 972-4. Ausg. Okt. u. Nov. 1972.
53. Dean, R. S.: Electrolytic manganese and its alloys. New York 1952. S. 134 ff.
54. Nickel-Ber. 25 (1967) S. 117. – Albert, H.: Nickel-Ber. 25 (1967) S. 240/41.

C 24 Steels with Good Electrical Conductivity

1. DIN 43137. Elektrische Bahnen; Drähte für Erdung und Stromrückleitung. Ausg. Juni 1978.
2. VDE 0203/XII. 44. Vorschriften für Stahlkupfer- (Staku-) Leiter in der Elektrotechnik. Neuere Ausg. Jan. 1947. Nachdruck 1951.
2a. Shirai, H. u. M. Hirose: Wire Ind. 50 (1983) Nr. 1, S. 43, 46.
3. Wanser, G.: Draht 31 (1980) S. 525/32.
4. Steel wire handbook. Hrsg.: The Wire Association. Inc. Bd. 2. Branford/Conn. 1969. S. 169/209.
5. DIN 48300. Drähte für Fernmeldefreileitungen. Ausg. April 1981.
6. ASTM A 111-66. Zinc coated (galvanized) "iron" telephone and telegraph line wire. Ausg. 1966 (Reapproved 1980).
7. GOST 4231-48. Gewalzter Telegraphendraht. Ausg. 1948. Neuausg. 1952. Latest ed.: Runder gewalzter Draht für Fernmeldeleitungen. Ausg. 1970.
8. ASTM A 326-67. Zinc coated (galvanized) high tensile steel telephone and telegraph line wire. Ausg. 1967 (Reapproved 1985).
9. DIN 48200 Teil 7. Drähte für Leitungsseile; Drähte aus Stahlkupfer (Staku). Ausg. April 1981.
10. DIN 48200 Teil 8. Drähte für Leitungsseile; Drähte aus aluminium-unmanteltem Stahl. Ausg. April 1977.
11. DIN 17122. Stromschienen aus Stahl für elektrische Bahnen; Technische Lieferbedingungen. Ausg. März 1978.
12. Werkstoff-Handbuch Stahl und Eisen. 4. Aufl. Hrsg.: Verein Deutscher Eisenhüttenleute. Düsseldorf 1965. Abschn. B 11. Vgl. auch DIN 1324. Elektrisches Feld, Begriffe. Ausg. Jan. 1972.
12a. ASTM B 193-78. Standard Test Method for Resistivity of Electrical Conductor Materials. 1978. Reapproved 1983.
13. Landolt/Börnstein: Zahlenwerte und Funktionen aus Physik, Chemie, Astronomie, Geophysik und Technik. 6. Aufl. Bd. 4. T. 2a. Grundlagen, Prüfverfahren, Eisenwerkstoffe. Berlin, Göttingen, Heidelberg 1963. S. 229.
14. Richter, F.: Thyssenforsch. 1 (1969) S. 70/76.
15. Yensen, T. D.: Trans. Amer. Soc. Metals 27 (1939) S. 797/820. S. bes. S. 801.
16. Hütte. Taschenbuch für Eisenhüttenleute. 5. Aufl. Hrsg. vom Akademischen Verein Hütte. Berlin/Düsseldorf 1961. S. 31/32 u. 56.
17. Dahl, W., u. K. Lücke: Arch. Eisenhüttenwes. 25 (1954) S. 241/50.
18. Maurer, E., u. F. Stäblein: Z. allg. u. anorg. Chem. 137 (1924) S. 115/24.
19. Bardenheuer, P., u. H. Schmidt: Mitt. Kais.-Wilh.-Inst. Eisenforsch. 10 (1928) S. 193/212.
20. Köster, W., u. H. Tiemann: Arch. Eisenhüttenwes. 5 (1932) S. 579/86.
21. Radcliffe, S. V., u. E. C. Rollason: J. Iron Steel Inst. 189 (1958) S. 45/48.
22. Richter, F.: Die wichtigsten physikalischen Eigenschaften von 52 Eisenwerkstoffen. Düsseldorf 1973. (Stahleisen-Sonderberichte H. 8)
23. Nach: Nemkina, E. D., D. V. Vostrikova, D. I. Zaletov, A. A. Zborovskij u. E. M. Furman: Stal 27 (1967) S. 1038/40.
24. Tammann, G., u. G. Moritz: Ann. Phys., Ser. 5, 16 (1933) S. 667/79.
25. Ueda, T.: Sci. Rep. Tôhoku Univ. 19 (1930) S. 473/98.
26. Messkin, W. S.: Arch. Eisenhüttenwes. 3 (1929) S. 417/25.
27. Andrew, J. H., H. Lee, P. L. Chang, B. Fang u. R. Guenot: J. Iron Steel Inst. 165 (1950) S. 145/65 u. 2 Taf. – Andrew, J. H., H. Lee, P. L. Chang u. R. Guenot: Ebenda 165 (1950) S. 166/84.
28. Frommeyer, G.: Z. Werkst.-Techn. 10 (1979) S. 166/71.
29. Köster, W.: Arch. Eisenhüttenwes. 2. (1929/30) S. 503/22.
30. Cottrell, A. H., u. A. T. Churchman: J. Iron Steel Inst. 162 (1949) S. 271/76.
31. Balicki, M.: J. Iron Steel Inst. 151 (1945) S. 181/224.
32. Stahl-Eisen-Liste. 7. Aufl. Hrsg.: Verein Deutscher Eisenhüttenleute. Düsseldorf 1981.

C 25 Steels for Line Pipe

1. Maxey, W. A.: In: 5th symposium on line pipe research. [Hrsg.:] American Gas Association, AGA Houston 1974. S. J1/J30.
2. Wiedenhoff, W. W., u. G. H. Vogt: Rohre Rohrleitungsbau Rohrleitungstransport 22 (1983) S. 492/96.
3. Herbsleb, G., R. Pöpperling u. W. Schwenk: Corrosion, Houston, 37 (1981) S. 247/56.

3a. DIN 50115. Kerbschlagbiegeversuch. Ausg. Febr. 1975.
4. Stahl-Eisen-Prüfblast 1326-83. Fallgewichtsversuch nach Battelle. Ausg. Jan. 1983.
5. ASTM E 436-74. Drop weight tear test of ferritic steels. American Petroleum Institute RP 5 L3.
6. Junker, G., F. O. Koch, J. Kügler, W. A. Maxey, A. Peeck, P. A. Peters, K. Seifert u. G. H. Vogt: Rohre Rohrleitungsbau Rohrtransport 23 (1984) S. 512/17.
7. Wiedenhoff, W. W., u. G. H. Vogt: Rohre Rohrleitungsbau Rohrleitungstransport 26 (1987) S. 522/527.
8. Buzzichelli, G., G. Fearnehough, F. Nicolazzo, G. Re., S. Venzi u. G. H. Vogt: Rohre, Rohrleitungsbau Rohrtransport 26 (1987) S. 177/83.
9. Coors, P. Ph. C., G. D. Fearnehough, F. O. Koch, J. Kügler, S. Venzi u. G. H. Vogt: Rohre Rohrleitungsbau Rohrleitungstransport 18 (1979) S. 380/86.
10. Vogt, G.: In: 6th symposium on line pipe research. [Ed.] American Gas Association, AGA. Houston 1979 S. MI/M20.
11. Bersch, B. u. F. O. Koch: Rohre Rohrleitungsbau Rohrleitungstransport 17 (1978) S. 772/79.
11a. Düren, C: Equations for the prediction of cold cracking resistance in field-welding large-diameter pipes. iiW. document IX-1356–85.
12. Düren, C. in: Schweißen von Baustählen. Hrsg. Verein Deutscher Eisenhüttenleute. Düssldorf 1983. S. 114/51. Siehe auch Stahl-Eisen-Werkstoffblatt 063. Empfehlungen für das Verarbeiten, besonders für das Schweißen von Stahlrohren für den Bau von Fernleitungen. In Vorbereitung, derzeit Entwurf Februar 1985.
13. Granjon, H.: Metal Constr. & Brit. Weld. J. (1969) S. 509/15.
14. Moore, E. M., u. J. J. Warga: Mater. Performance 15 (1976) No. 6, S. 17/23.
15. Kalwa, G., R. Pöpperling, H. W. Rommerswinkel u. P. J. Winkler: Offshore North Sea Technology Conference (ONS) Stavanger 1976, Paper T-1/21.
15a. Haumann, W., W. Heller, H. A. Jungblith, H. Pircher, R. Pöpperling u. W. Schwenk: Stahl u. Eisen 107 (1987) S. 585/94.
16. Pöpperling, R., u. W. Schwenk: Werkst. u. Korrosion 31 (1980) S. 15/20.
17. Schwenk, W., u. R. Pöpperling: Rohre Rohrleitungsbau Rohrleitungstransport 19 (1980) S. 571/77.
18. Meyer, L.: Stahl u. Eisen 101 (1981) S. 483/91.
19. Lorenz, K., W. M. Hof, K. Hulka, K. Kaup, H. Litzke u. U. Schrape: Stahl u. Eisen 101 (1981) S. 593/600.
20. Wiedenhoff, W. W.: In: Symposium on the interrelation between the iron and steel industry and the steel consuming sectors. Steel/SEM 3 R.18. New grades of steel for the production of large-dimension steel pipe. [Ed.] Economic Commission for Europe. Genf 1977. S. 1/27.
21. Gray, J. M.: Metallurgy of high strength low-alloy pipeline steels. Present and future possibilites. [Ed.] Molybdenum Corporation of America. Application Report 7201, Jan. 1972.
22. Haumann, W., u. K. Kaup: In: DVS-Ber. Nr. 62, 1980, S. 88/95.
23. DIN 17172. Stahlrohre für Fernleitungen für brennbare Flüssigkeiten und Gase. Technische Lieferbedingungen. Ausg. Mai 1978.
24. American Petroleum Institute (API), Standards 5 L, 5 LX u. 5 LS, also RP 513.

C 26 Ball and Roller Bearing Steels

1. Zwirlein, O., u. H. Schlicht: Z. Werkst.-Techn. 11 (1980) S. 1/14. – Eberhard, R., H. Schlicht u. O. Zwirlein: Härterei-techn. Mitt. 30 (1975) S. 338/45. – Broßeit, E., F. Schmidt u. H. J. Schröder: Z. Werkst.-Techn. 9 (1978) S. 210/14.
2. Mannot, J., R. Tricot u. A. Gueussier: Rev. Métallurg. 67 (1970) S. 619/37.
3. Randak, A., A. Stanz u. W. Verderber: Stahl u. Eisen 92 (1972) S. 981/93.
4. Lundquist, M. R. C.: In: Weiss, V.: Proc. Amer. Soc. Test. Mater. 59 (1959) S. 655/61, bes. S. 657.
5. DIN 50602. Metallographische Prüfverfahren; Mikroskopische Prüfung von Edelstählen auf nichtmetallische Einschlüsse mit Bildreihen. Ausg. Sept. 1985.
6. Stahl-Eisen-Prüfblatt 1520–78. Mikroskopische Prüfung der Carbidausbildung in Stählen mit Bildreihen. Ausg. März 1978. – Siehe auch ISO 5949. Tool steels and bearing steels-Micrographic method for assessing the distribution of carbides using reference photomicrographs. Ed 1984.
7. DIN 50192. Ermittlung der Entkohlungstiefe. Ausg. Mai 1977.
8. Diergarten, H.: Gefüge-Richtreihen im Dienste der Werkstoffprüfung in der stahlverarbeitenden Industrie. 4. Aufl. Düsseldrof 1960.
9. Barteld, K., u. A. Stanz: Arch. Eisenhüttenwes. 42 (1971) S. 581/97.
10. DIN 17 230. Wälzlagerstähle; Technische Lieferbedingungen. Ausg. Sept. 1980.
11. Krefting, R., u. A. Stanz: Arch. Eisenhüttenwes. 49 (1978) S. 325/31.

12. DIN 50 191. Stirnabschreckversuch. Ausg. Sept. 1987.
13. Brandis, H., u. P. Kroy: Klepzig Fachber. 72 (1964) S. 434/40.
14. Barteld, K.: Fachber. Hüttenparx. Metallweiteverarb. 13 (1975) S. 792/94, 796/98 u. 800/01.
15. Randak, A., A. Stanz u. H. Vöge: Influence of melting practice and hot forming on type and amount of non metallic inclusions in bearing steels. In: Bearing steels. Philadelphia/Pa. 1975. (ASTM Spec. Techn. Publ. No. 575.) S. 150/62.
16. Orlich, J.: Härterei-techn. Mitt. 29 (1974) S. 231/36.
17. EURONORM 103-71. Mikroskopische Ermittlung der Ferrit-oder Austenitkorngöße von Stählen. Ausg. Nov. 1971. – DIN 50 601. Metallographische Prüfverfahren; Bestimmung der Austenit-und der Ferritkorngröße von Stahl und Eisenwerkstoffen. Ausg. 1985.
18. Franz, M., u. E. Hornbogen: Arch. Eisenhüttenwes. 49 (1978) S. 449/53.
19. EURONORM 94-73. Wälzlagerstähle; Gütevorschriften. Ausg. Nov. 1973.
20. ISO 683. Part 17: Bell and Roller bearing steels. Ed. 1976

C 27 Steels for permanent-way Materials

1. Fastenrath, F.: Die Eisenbahnschiene. Berlin, München, Düsseldorf 1977.
2. Internationale Schienentagung 1979, Heidelberg, 18. u. 19. Okt. 1979. Berichte. Veranst. Schienenausschß beim Verein Deutscher Eisenhüttenleute. Düsseldorf 1979.
3. Jäniche, W.: Stahl u. Eisen 72 (1952) S. 758/66.
4. Oettel, R.: Braunkohle Wärme u. Energie 13 (1961) S. 7/18.
5. Schmedders, H., R. Hammer u. U. Schrape: Arch. Eisenhüttenwes. 37 (1966) S. 551/60.
6. Eisenmann, J., G. Oberweiler, R. Schweitzer u. W. Heller: Eisenbahntechn. Rdsch. 23 (1974) S.122/26.
7. Technishce Lieferbedingungen des Internationalen Eisenbahnverbandes. Code UIC 8608. Ausg. 8. Juli 1987.
8. Schweitzer, R., J. Flügge u. W. Heller Stahl u. Eisen 105 (1985) S. 1451/1456.
9. Gladman, T., I. D. Mclvor, u. F. B. Pickering: J. Iron Steel Inst. 210 (1972) S. 916/30.
10. Pickering, F. B.: Physical metallurgy and the design of steels. London 1978.
11. Flügge, J., W. Heller, E. Stolte u. W. Dahl: Arch. Eisenhüttenwes. 47 (1976) S. 635/40.
12. Hyzak, J. M., u. I. M. Bernstein: Metallurg. Trans. 7A (1976) S. 1217/24.
13. Heller, W.: Techn. Mitt. Krupp, Werksber., 33 (1975) S. 73/77.
14. Heller, W., R. Schweitzer u. L. Weber: In: 19th annual conference of metallurgists. [Hrsg.:] Canadian Institute of Mining and Metallurgy. – Canadian Metallurgical Quarterly 21, Nr. 1, S. 3/15.
15. Schmedders, H., K. Wick u. H.-J. Quell: Thyssen Techn. Ber. 1 (1979) S. 60/69.
16. Thermitschweißen von Schienen. Hrsg. von der Beratungsstelle für Stahlverwendung. Düsseldorf 1964. (Merkblätter über sachgemäße Stahlverwendung. Nr. 241).
17. Abbrennstumpfschweißen von Schienen. Hrsg. von der Beratungsstelle für Stahlverwendung. Düsseldorf 1981. (Merkblätter über sachgemäße Stahlverwendung. Nr. 258).
18. Stahl u. Eisen 90 (1970) S. 922/28.
19. Schweitzer, R., u. O. Huber: Techn. Mitt. Krupp, Werksber., 37 (1979) S. 105/108.
20. Technische Lieferbedingungen de Internationalen Eisenbahnverbandes. Code UIC 865-1. Ausg. Jan. 1967.
21. Entwicklung und Erprobung einer Stahlschwelle für Strecken der Deutschen Bundesbahn mit schwerem and schnellem Verkehr. Bericht der Studiengesellschaft für Anwendungstechnik von Eisen und Stahl e. V. zum Projekt 47. Düsseldorf 1982. Pietzko, G., u. H. Schmedders: Thyssen Techn. Ber. 2 (1980) S. 126/36.
22. Technische Lieferbedingungen des Internationalen Eisenbahnverbandes Code UIC 864-4. Ausg. April 1963.
23. Technische Lieferbedingungen des Internationalen Eisenbahnverbandes Code UIC 864-6. Ausg. April 1963.

C 28 Steels for Rolling Stock

1. Wirner, R.: In: 2. Internationaler Radsatzkongreß, München 1966. Radsätze. Der Radsatz in Gegenwart und Zukunft. Hrsg. von der Gruppe Rollendes Eisenbahnzeug der Wirtschaf-

tsvereinigung Eisen- und Stahlindustrie und vom Verein Deutscher Eisenbahnleute. Essen 1967. S. 273/81.
2. Swaay, J. L. van: In: 3. Internationaler Radsatzkongreß, Sheffield 1969. Vortrag Nr. 8, 8 S.
3. Egelkraut, K., H. Lange u. V. Mussnig: Eisenbahntechn. Rdsch. 15 (1966) S. 346/60.
4. Rudolph, W.: Glas.-Ann. 88 (1964) S. 98/109.
5. Müller, C. T.: Österr. Ing.-Z. 7 (1964) S. 215/24.
6. Nishioka, K., u. Y. Morita: The strength of railroad wheels. Bull. JSME 12 (1969) S. 738/46; 13 (1970) S. 1165/71; 14 (1971) S. 11/19.
7. Eck, B. J., u. M. G. Novak: In: 4. Internationaler Radsatzkongreß, Paris 1972. Bd. 3. S. 67/73.
8. Raquet, E.: Glas.-Ann. 99 (1975) S. 249/55.
9. Forch, K.: Forschungsvorhaben Erforschung der Grenzen des Rad-Schiene-Systems: Einflußgrößen auf das Verschleißverhalten verschiedener legierter Stähle für Eisenbahnräder. Statusseminar München 1974, Vortag 10. Vergl. Forch, K.: Thyssen Techn. Ber. II (1979) Š. 70/76.
10. Stolte, E.: Techn. Mitt. Krupp, Forsch.-Ber., 20 (1962) S. 143/51. – Stolte, E.: Stahl u. Eisen 83 (1963) S. 1363/69.
11. Forch, K.: In: 8. Internationaler Radsatzkongreß, Madrid 1985.
12. Hegenbarth, F.: Wie unter [1], S. 179/94.
13. Lange, H., F. Hildebrandt u. F. Hogenkamp: Glas.-Ann. 98 (1974) S. 93/100.
14. Fox, M. P., u. M. G. Hewitt: Wie unter [7], Bd 2. S. 67/91.
15. Bröhl, W., u. G. Oedinghofen: Werkstofffragen bei Radreifen. Veröff. Klöckner-Werke AG. Georgsmarienwerke, Q 1.4316, 1976/77.
16. DIN 17 200. Vergütungsstähle. Gütevorschriften. Ausg. Nov. 1984.
17. ISO 1005. Railway rolling stock material Part 1: Rough-rolled tyres for tractive and trailing stock, quality requirements. Ed. 1982. – Part 2: Tyres, wheel centres and tyred wheels for tractive and trailing stock – Dimensional, balancing and assembly requirements. Ed. 1986. – Part 3: Axles for tractive and trailing stock-quality requirments Ed. 1982. – Part 4: Rolled and forged wheel centres for tyred wheels for traction and trailing stock. Ed. 1986. – Part 6: Solid wheels for tractive and trailing stock, quality requirements. Ed. 1982. – Part 7: wheels for tractive and trailing stock, quality requirements Ed. 1982.
18. UIC 810-1 V. Technische Lieferbedingungen für Rohrradreifen aus gewalztem. unlegiertem Stahl für Triebfahrzeuge und Wagen. 4. Ausg. 1981. – UIC 810-2 V. Technische Lieferbedingungen: Rohrradreifen für Wagen, Abmessungen und Toleranzen. 3. Ausg. 1963. 4. Ausg. in Vorberetiung.-UIC 811 V. Technische Lieferbedingungen für Wagenachswellen. 3 Ausg. 1968. – UIC 812-1 V. Technische Lieferbedingungen für gewalzte oder geschmiedete Radkörper für bereifte Wagenradsätze 3. Ausg. 1968. – UIC 813-3 V. Technische Lieferbedingungen für Vollräder aus gewalztem, unlegiertem Stahl für Triebfahrzeuge und Wagen. 5. Ausg. 1984. – UIC 813-1 V. Technische Lieferbedingungen für Wagenradsätze. 3. Ausg. 1968. – UIC 813-2 V. Technische Lieferbedingungen für unlegierten Flach- und Formstahl für Radreifensprengringe. 1. Ausg. 1969.

C 29 Steels for Screws, Bolts, Nuts and Rivets

1. Domalski, H., u. H. Schücker: Stahl u. Eisen 90 (1970) S. 1087/96.
2. Karl, A., u. K. H. Kiesel: Technik 27 (1972) S. 681/84.
3. DIN ISO 898 Teil 1. Mechanische Eigenschaften von Verbindungselementen; Schrauben. Ausg. April 1979. (New ed. is being prepared, for the time being: Draft Aug. 1985).
3a. Kloos, K.-H., u. W. Schneider: VDI-Z. 128 (1986) S. 101/09.
4. Illgner, K. H.: In: VDI-Ber. 220, 1974, S. 135/44.
5. Blume, D.: Masch.-Markt 82 (1976) S. 350/52.
6. Schuster, M.: Masch.-Markt 79 (1973) S. 1318/21.
7. DIN 50 601. Metallografische Prüfverfahren. Ermittlung der Austenit- und der Ferritkorgröße von Stahl und anderen Eisenwerkstoffen. Ausg. Juni 1985. – Euronorm 103. Mikroskopische Ermittlung der Ferrit- oder Austenitkorngröße von Stählen. Ausg. Nov. 1971.
8. Blume, D.: Masch.-Markt 76 (1970) S. 1693/98.
9. Enke, Chr. G.: Maschine 31 (1972) S. 623/25.
10. Richter, E.: Draht 28 (1977) S. 142/47.
11. DIN 1013 Teil 1. Stabstahl. Warmgewalzter Rundstahl für allgemeine Verwendung: Maße, zulässige Maß- und Formabweichungen. Ausg. Nov. 1976. – DIN 1013 Teil 2: Stabstahl. Warmgewalzter Rundstahl für besondere Verwendung; Maße, zulässige Maß- und Formabweichungen. Ausg. Nov. 1976.

12. DIN 59 115. Walzdraht aus Stahl für Schrauben, Muttern and Niete; Maße, zulässige Abweichungen, Gewichte. Ausg. Nov. 1972.
13. DIN 59 130. Stabstahl. Warmgewalzter Rundstahl für Schrauben and Niete; Maße, Gewichte, zulässige Abweichungen. Ausg. Sept. 1978.
14. Bauer, C. O.: Draht-Welt 55 (1969) S. 365/75.
15. Stahl-Eisen-Prüfblatt 1520. Mikroskopische Prüfung der Carbidausbildung in Stählen mit Bildreihen. Ausg. März 1978.
16. Illgner, K. H: Draht-Welt 56 (1970) S. 706/11.
17. DIN 51 212. Prüfung metallischer Werkstoffe. Verwindeversuch an Drähten. Ausg. Sept. 1978.
18. DIN 1654 Teil 1. Kaltstauch- und Kaltfließpreßstahle. Technische Lieferbedingungen. Allgemeines. Ausg. März 1980. (New ed. is being prepared Ausg. 1988).
19. Engineer, S.: TEW Techn. Ber. 2 (1976) S. 125/29.
20. Schuster, M.: Draht-Welt 58 (1972) S. 649/51.
21. Esselborn, K.: Kaltstauchstähle für Schrauben und Muttern unter besonderer Berücksichtigung der borlegierten Güten. Röchling-Burbach Techn. Mitt. Nr. 41, Ausg. 1975.
21a. Niobium Information Nr. 1/85.
21b. Rofés-Vernis, I., B. Héritier, P. Maitrepierre u. A. Wyckaert: Rev. Metallurg. 80 (1983). S. 879/85.
22. DIN 17 111. Kohlenstoffarme unlegierte Stähle für Schrauben, Muttern und Niete. Technische Lieferbedingungen. Ausg. Sept. 1980.
23. DIN 1654 Teil 3. Kaltstauch- und Kaltifließpreßstähle. Technische Lieferbedingungen für Einsatzstähle. Ausg. März 1980.
24. DIN 1654 Teil 4, Kaltstauch- und Kaltifließpreßstähle. Technische Lieferbedingungen für Vergütungsstähle. Ausg. März 1980.
25. DIN 267 Teil 3. Mechanische Verbindungselemente. Technische Lieferbedingungen. Festifkeitsklassen für Schrauben aus unlegierten oder legierten Stählen; Umstellung der Festigkeitsklassen. Ausg. Aug. 1983.
26. DIN 1654 Teil 5. Kaltstauch- und Kaltfließpreßstähle. Technische Lieferbedingungen für nichtrostende Stähle. Ausg. März 1980.
27. DIN 267 Teil 11. Mechanische Verbindungselemente. Technische Lieferbedingugen mit Ergänzungen zu ISO 3506, Teile aus rost- and säurebeständigen Stählen. Ausg. Jan. 1980.
28. DIN 17 240. Warmfeste und hochwarmfeste Werkstoffe für Schrauben and Muttern. Ausg. Juli 1976.
29. DIN 267 Teil 13. Mechanische Verbindungselemente. Technische Lieferbedingungen, Teile für Schraubenverbindungen, vorwiegend aus kaltzähen oder warmfesten Werkstoffen. Ausg. März 1980.
30. Bauer, C. O.: Z. Werkst.-Techn. 7 (1976) S. 279/92.
31. Fleischer, N., u. E. E. Fastenrath: Mutternfiebel. Plettenberg 1969.
32. Zwahr, A.: Blech 17 (1970) Nr. 10, S. 69/74.
33. Arnim, H. von: Draht 24 (1973) S. 58/62.
34. Andrews, D. S.: Sheet Metal Ind. 53 (1976) S. 223/25.
35. Beelisch, K. H.: Masch.-Markt 77 (1971) S. 2261/63.
36. Großberndt, H.: Dachdecker-Handwerk 1977, Nr. 12, S. 857/71; Nr. 13, S. 1001/03; Nr. 14, S. 1070/71.
37. Großberndt, H., u. H. Kniess: Stahlbau 44 (1975) S. 289/300 u. 344/51.
38. Großberndt, H., u. K. Kayser: Blechschraubenhandbuch. Essen 1968.
39. Wiegand, H., u. W. Thomala: Draht-Welt 59 (1973) S. 542/51.
40. Peters, W.: Draht 26 (1975) S. 629/33.

C 30 Steels for Welded Round Link Chains

1. Stahldraht-Erzeugnisse. Hrsg. vom Ausschuß für Drahtverarbeitung im Verein Deutscher Eisenhüttenleute. Bd. 2. Düsseldorf 1956. S. 142/91.
2. Werkstoff-Handbuch Stahl und Eisen. Hrsg.: Verein Deutscher Eisenhüttenleute. 4. Aufl. Düsseldorf 1965. Blatt Q 25.
3. Smetz, R., u. K. Niederberger: Ing. Digest 16 (1977) Nr. 3, Taf. 1.
4. Wellinger, K., u. A. Stanger: VD1-Z. 101 (1959) S. 1425/31.
5. Schaefer, W.: Glückauf 96 (1969) S. 550/62.

6. Rieger, W., u. W. Rieß: Fördern u. Heben 11 (1961) S. 599/604.
7. Schaefer, W.: Glückauf 98 (1962) S. 915/24.
8. DIN 685 Teil 1 bis 4. Geprüfte Rundstahlketten Begriffe, Anforderungen, Prüfung, Prüfbescheimigüngen. – DIN 685 Teil 5. Geprüfte Rundstahlketten. Anwendung.
9. DIN 766. Rundstahlketten. Güteklasse 3; lehrenhaltig, geprüft. Aüsg. Jan. 1986.
10. DIN 5684 Teil 1. Rundstahlketten für Hebezeuge; Güteklasse 5, lehrenhaltig, geprüft. Ausg. Mai 1984. – DIN 5684 Teil 2. Rundstahlketten für Hebezeuge; Güteklasse 6, lehrenhaltig, geprüft. Ausg. Mai 1984. – DIN 5684 Teil 3. Rundstahlketten für Hebezeuge; Güteklasse 8, lehrenhaltig, geprüft. Ausg. Mai 1984.
11. DIN 22252. Rundstahlketten für Förderer und Gewinnungsanlagen im Bergbau, lehrenhaltig, geprüft. Ausg. Sept. 1983..
12. Germanischer Lloyd. Vorschriften für Klassifikation und Bau von stählernen Seeschiffen. Bd. III,Kap. 6 – Werkstoffe. Ausg. 1981.
13. Schneeweiß, G.: Mater.-Prüf. 8 (1966) S. 217/22.
14. Rieger, W.: Dt. Hebe-u. Fördertechn. 13 (1967) S. 646/48.
15. Oechsle, D.: Konstruktion 28 (1976) S. 483/88.
16. Seidemann, A.: Ind.-Anz. 85 (1963) S. 2054/56.
17. Schaefer, W.: Glückauf 106 (1970) S. 17/26.
18. Minuth E., u. E. Hornbogen: Prakt. Metallogr. 13 (1976) S. 584/98.
19. Schaefer, W.: Glückauf 100 (1964) S. 897/904.
20. Seidemann, A.: Techn. Überwachung 5 (1964) s. 341/43.
21. Kurrein, M.: Stahl u. Eisen 85 (1965) S. 148/49.
22. Naumann, F. K., u. F. Spies: Prakt. Metallogr. 9 (1972) S. 706/07. – Dieselben: Prakt. Metallogr. 15 (1978) S. 309/13.
23. Schmidtmann, E., u. R. Schumann: Schweißen u. Schneiden 19 (1967) S. 352/61.
24. Krüger, A., u. J. Müller: Draht-Welt 48 (1962) S. 362/68.
25. Püngel, W., u. P. Koch: Draht-Welt 48 (1962) S. 368/71. – Dieselben: Draht-Welt 52 (1966) S. 453/54.
26. Atlas zur Wärmebehandlung der Stähle. Hrsg. vom Max-Planck-Institute für Eisenforschung in Zusammenarbeit mit dem Werkstoffausschuß des Vereins Deutscher Eisenhüttenleute. Bd. 1. T. 2. Von A. Rose, W. Peter, W. Straßburg u. L. Rademacher. Düsseldorf 1954–1958. S. 24/27:
27. Mehta. K. K.: Thyssen Edelstahl Techn. Ber. 4 (1978) S. 38/46.
28. DIN 17 115. Stähle für geschweißte Rundstahlketten. Technische Lieferbedingungen. Ausg. Febr. 1987.
28a. DIN 17 210. Einsatzstähle; Technische Lieferbedingungen. Ausg. Sept. 1986
29. a) ASTM-A 466-80. Weldless C-Steel Chain
    b) ASTM-A 413-80. Carbon Stell Chain
    c) ASTM-A 391-85. Alloy Steel Chain
    d) ASTM-A 467-85. Machine and Coil Chain
30a. British Standard Specification BS 6405: 1984 Non-calibrated Short Link Steel Chan (grade 30) for General Engineering Purposes: Class 1 and 2.
30b. British Standard Specification BS 4942. Part 2, 1981: Short link chain for lifting purposes (Grade M (4) non-calibrated chain). – Part 3, 1981: Short link chain for lifting purposes (Grade M (4) calibrated chain). – Part 4, 1981: Short link chain for lifting purposes (Grade S (6) non-calibrated chain). – Part 5, 1981: Short link chain for lifting purposes (Grade T (8) non-calibrated chain). - Part 6, 1981: Short link chain for lifting purposes (Grade T (8) calibrated chain.
30c. British Standard Specification BS 1633: 1950. Electrically Welded Higher Tensile Steel Chain (Short Link and Pitched or Calibrated for Lifting purposes).
30d. British Standard Specification BS 3113: 1959. Alloy Steel Chain Grade 60 (Short Link for Lifting Purposes).
30e. British Standard Specification BS 3114: 1959. Alloy Steel Chain Grade 60 (Polished Short Link Calibrated Chain for Pulley Blocks).
31. Japanese Institute of Standards Handbook 1986: JIS-G 3105 Steel Bars for Chains
32. ISO 1835. 1st ed. 1980-09-15 $\equiv$ BS 4942 Part 2
    ISO 1836. 1st ed. 1980-09-15 $\equiv$ BS 4942 Part 3
    ISO 3075. 1st ed. 1980-08-01 $\equiv$ BS 4942 Part 4
    ISO 3076. 2nd ed. 1980-08-01 $\equiv$ BS 4942 Part 5
    ISO 3077. 2nd ed. 1980-08-01 $\equiv$ BS 4942 Part 6
33. Just, E.: Härterei Techn. Mitt. 23 (1968) S. 85 100
34. Mehta, K. K.: Thyssen Edelstahl Techn. Ber. 6 (1980) S. 104/10 u. S. 111/16.

D 2 Crude Steel Production

1. Kalla, U., H. W. Kreutzer u. E. Reichenstein: Stahl u. Eisen 97 (1977) S. 382/93.
2. Haastert, H. P., E. Köhler u. E. Schürmann: Stahl u. Eisen 83 (1963) S. 204/12.
3. Haastert, H. P., E. Köhler u. E. Schürmann: Stahl u. Eisen 85 (1965) S. 1588/95.
4. Mahn, G., P. Ottmar u. H. Voigt: Stahl u. Eisen 89 (1969) S. 262/73.
5. Haastert, H. P., W. Meichsner, H. Rellermeyer u. K. H. Peters: Thyssen Techn. Ber. 7 (1975) S. 1/7; Iron and Steel Eng. Oct. 1975, S. 71/77.
6. Schulz, H. P.: Stahl u. Eisen 89 (1969) S. 249/62.
7. Köhler, E., K. Nürnberg, W. Ullrich, R. A. Weber u. F. Winterfeld: Thyssenforsch. 3. (1971) S. 118/24.
8. Haastert, H. P.: Thyssen Techn. Ber. 15 (1983) S. 1/14.
9. Emi, T.: Stahl u. Eisen 100 (1980) S. 998/1011.
10. Ottmar, H., H. Schenck u. W. Dahl: Stahl u. Eisen 97 (1977) S. 731/41.
11. Haastert, H. P., F. Winterfeld, E. Höffken, G. Bauer u. R. A. Weber: Stahl u. Eisen 97 (1977) S. 723/31.
12. Richardson, F. D., u. J. H. E. Jeffes: J. Iron Steel Inst. 160 (1948) S. 261/70; 163 (1949) S. 397/420.
13. Vacher, H. C., u. E. A. Hamilton: Trans. Amer. Inst. min. metallurg. Engrs. Iron Steel Div., 95 (1931) S. 124/40.
14. Willems, J.: Radex-Rdsch 1965, S. 425/31.
15. Schäfer, K.: Stahl u. Eisen 99 (1979) S. 412/20.
16. Haastert, H. P., E. Köhler, F. Regneri u. E. Schürmann: Stahl u. Eisen 89 (1969) S. 24/20. – Haastert, H. P., E. Köhler u. K. Nürnberg: Thyssenforsch. 2 (1970) S. 127/37.
17. Die physikalische Chemie der Eisen- und Stahlerzeugung. Hrsg. vom Verein Deutscher Eisenhüttenleute Düsseldorf 1964.
18. Chino, H., u. K. Wada; Yawata techn. Rep. Nr. 251, 1965, S. 5817/42.
19. Knüppel, H.: Desoxydation und Vakuumbehandlung von Stahlschmelzen. Bd. 2. Textteil u. Bildteil. Düsseldorf 1983.
20. Grabner, B., u. H. Höffgen: Radex-Rdsch. 1983, S. 179/209 u. 1 Falttaf.
21. Pluschkell, W.: Stahl u. Eisen 99 (1979) S. 398/404 u. 404/11.
22. Kataura, Y., u. D. Oelschlägel: Stahl u. Eisen 100 (1980) S. 20/29.
23. Förster, E., W. Klapdar, H. Richter, H. W. Rommerswinkel, E. Spetzler u. J. Wendorff: Stahl u. Eisen 94 (1974) S. 474/85.
24. Spetzler, E., u. J. Wendorff: Thyssen Techn. Ber. 7 (1975) S. 8/13.
25. Haastert, H. P.: Thyssen Techn. Ber. 1 (1980) S. 8/14; Metallurg. Plant & Techn. 3 (1980) Nr. 5, S. 26, 28, 30, 32, 34 u. 36.
26. Kreutzer, H. W.: Stahl u. Eisen 92 (1972) S. 716/24.
27. Hochstein, F.: Stahl u. Eisen 95 (1975) S. 785/89.
28. Schmidt, M., O. Etterich, H. Bauer u. H. J. Fleischer: Stahl u. Eisen 88 (1968) S. 153/68.
29. Bauer, H., O. Etterich, H. J. Fleischer u. J. Otto: Stahl u. Eisen 90 (1970) S. 725/35.
30. Bauer, H., K. Behrens u. M. Walter: Stahl u. Eisen 97 (1977) S. 938/44.
31. Behrens, K., E. Köhler u. K.-D. Unger: Stahl u. Eisen 99 (1979) S. 1302/10.

D 3 Casting and Solidification

1. Nilles, P.: In: Gießen und Erstarren von Stahl. Informationstagung, Luxembourg, 29.11–1.12. 1977. Bd. 2. [Hrsg.:] Kommission der Europäischen Gemeinschaften. Düsseldorf 1977. (EUR 5903 d, e, f.) S. 20/60.
1a. Schwerdtfeger, K.: Stahl u. Eisen 106 (1986) S. 5/12.
1b. Stadler, P.: Stahl u. Eisen 106 (1986) S. 1187/95.
2. Ebneth, G., Haumann, K. Rüttiger u. F. Oeters: Arch. Eisenhüttenwes. 45 (1974) S. 353/59.
3. Diener, A., G. Ebneth u. A. Drastik: Estel-Ber. 10 (1975) S. 149/61.
4. Diener, A.: Stahl u. Eisen 96 (1976) S. 1337/40.
5. Oeters, F., H. J. Selenz u. K. Rüttiger: Wie unter [1], Bd. 1. S. 144/95.
6. Tiller, W. A.: J. Iron Steel Inst. 192 (1959) S. 338/50.
7. Jacobi, H.: Stahl u. Eisen 96 (1976) S. 964/68.
8. Jacobi, H., u. K. Wünnenberg: Stahl u. Eisen 97 (1977) S. 1075/81.

9. Rellermeyer, H.: Der Erstarrungsverlauf bei Stahlblöcken. In: Gießen und Erstarren von Stahl. Hrsg.: Verein Deutscher Eisenhüttenleute. Düsseldorf 1967. S. 111/41.
10. Ebneth, G., W. Haumann u. K. Rüttiger: ESTEL-Ber. 9 (1974) S. 13/24.
11. Waudby, P. E., P. C. Morgan u. P. Waterworth: Segregation in wide-end-up ingots. Final report, P. 1. Commission of the European Communities EU 7723 (1982).
11a. Kitagawa, T., Nakada, M., Komatsu, M., Yamada, M., Asano, S. u. Yano, K.: Transactions ISIJ 25 (1983) S. 1227/36.
12. Pesch, R. u. A. Etienne: Wie unter [1], Bd. 1. S. 198/245.
13. Plöckinger, E., u. A. Randak: Stahl u. Eisen 78 (1958) S. 1041/58.
14. Steinmetz, E., H.-U. Lindenberg, W. Mörsdorf u. P. Hammerschmid: Stahl u. Eisen 97 (1977) S. 1154/59.
15. Delmore, J., M. Laubin u. H. Maas: Wie unter [1], Bd. 1. S. 248/318.
16. Langhammer, H.-J., u. H. G. Geck: In: Vorgänge beim Gießen und Erstarren von unberuhigtem Stahl. Gießen u. Erstarren von Stahl. Hrsg.: Verein Deutscher Eisenhüttenleute, Düsseldorf 1967. S. 33/75.
17. Steinmetz, E., u. H. U. Lindenberg: Arch. Eisenhüttenwes. 47 (1976) S. 521/24.
18. Sims, C. E.: Trans. metallurg. Soc. AIME 215 (1959) S. 367/93.
19. Jacobi, H.: Wie unter [1], Bd. 1. S. 108/42.
20. Straßburger, Chr. u. L. Meyer: Thyssenforsch. 3 (1971). S. 2/7.
21. Pantke, H.-D., u. H. Neumann: Vorgänge beim Gießen und Erstarren von halbberuhigtem Stahl. Wie unter [16], S. 76/90.
22. Volker, W.: Stahl u. Eisen 88 (1968) S. 1455/63.
22a. Schwerdtfeger, K.: Archiv Eisenhüttenwes. 54 (1983) S. 87/98.
22b. Nemoto, H., Kawawa, T., Sato, H., Sakamoto, E.: Tetsu to Haganè: 58 (1972) Nr. 3 S. 387/94.
23. Ushijima K, A. Yoshida, M. Mizutani u. H. Okajima: South East Asia Iron & Steel Inst. Quart. 11 (1982) S. 40/47.
23a. Ohno, T., Ohashi, T., Matsunaga, H., Hiromoto, T. u. Kumai, K.: Trans. Iron Steel Inst. Japan 15 (1975) S. 407/16.
24. Steinmetz, E., U. Lindenberg, P. Hammerschmid u. W. Glitscher: Stahl u. Eisen 103 (1983) S. 539/45.
24a. Tacke, K.-H. u. Ludwig, J.: Steel Research 58 (1987) Nr. b. S. 262/70.
24b. Hammerschmid, P., Tacke, K.-H., Popper, H., Weber, L., Dubke, M. u. Schwerdtfeger, K.: Ironmaking and Steelmaking 11 (1984) Nr. 6, S. 332/39.
25. Nashiwa, H., K. Yoshida, A. Mori, H. Tomono u. K. Kimura: Iron & Steelmaker 7 (1980) Nr. 10, S. 17/22.
26. Hagen, K., u. H. Litzke: Techn. Mitt. Krupp, Werksber. 33 (1975) S. 95/100.
27. Stadler, P., K. Hagen, P. Hammerschmid u. K. Schwerdtfeger: Stahl u. Eisen 102 (1982) S. 451/59.
27a. Haastert, H. P.: Thyssen Techn. Ber. 1983 Nr. 1, S. 1/13.
28. A study of the continuous casting of steel. [Hrsg.:] International Iron and Steel Institute, Committee on Technology. Brussels 1977. S. 2–27.
29. Kohno, T., M. Wake, T. Yamamoto, T. Kuwabara, T. Shima u. A. Tsuneoka: Iron & Steelmaker 9 (1982) Nr. 10, S. 37/40.
29a. Jeschar, R., Reiners, U. u. Scholz, R.: 5 Intern. I. Steel. Congr., 6.-9. April 1986, Vol. 1, I Steel Soc. AIME Proc. 69. Steel M. Conf. S. 511/21.
30. Ohashi, T., u. K. Asano: In: Japan – USA joint seminar on solidification of metals and alloys, Tokyo, Jan. 17–19, 1977.
31. Schwerdtfeger, K.: Stahl u. Eisen 98 (1978) S. 225/35.
32. Miyazawa, K., u. K. Schwerdtfeger: Arch. Eisenhüttenwes. 52 (1981) S. 415/22.
33. Wünnenberg, K., u. H. Jacobi: Stahl u. Eisen 101 (1981) S. 875/82.
34. Alberny, R., u. J. P. Birat: Continuous casting of steel. London 1977. (The Metals Society. Book No. 184.) S. 116/24.
34a. Böcher, G., Jacobi, H., Litterscheidt, H., Rüttiger, K., Steffen, R., Weber, L. u. Wünnenberg, K.: In: Elektromagnetisches Rühren beim Brammenstranggießen von Stahl. Verl. Stahleisen, Düsseldorf 1985, S, 3/106.
35. Shah, N. A., u. J. J. Moore: Iron & Steelmaker 9 (1982) Nr. 10, S. 31/36; Nr. 11, S. 42/47.
36. Sorimachi, K., A. Kawaharda, K. Hamagami, K. Kinoshita, Y. Yoshii, M. Shiraishi: Tetsu to Haganè 67 (1981) S. 1345/53.
37. Rellermeyer, H.: Stahl u. Eisen 103 (1983) S. 415/20.
38. Ende, H. vom, u. G. Vogt: J. Iron Steel Inst. 210 (1972) S. 889/94.
39. Flender, R., u. K. Wünnenberg: Stahl u. Eisen 102 (1982) S. 1169/76.
39a. Steinmetz, E., Lindenberg, H.-U. u. Loh, J.: Stahl u. Eisen 105 (1985) S. 1049/54.

40. Gallucci, F., u. E. S. Szekeres: Iron & Steelmaker 7 (1980) Nr. 10, S. 23/28.
41. Fastner, T., u. L. Pochmarski: Berg- u. Hüttenm. Mh. 127 (1982) S. 227/33.
42. Kawakami, K., T. Kitagawa, H. Mitzukami, H. Uchibari, S. Miyahara, M. Suzuki u. Y. Shiratani: Tetsu to Hagané 67 (1981) S. 1190/99.
42a. Brimacombe, J. K., Samarasekera, I. V. u. Bommaraju, R.: Wie unter [29a] S. 409/23.
42b. Yasunaka, H., Mori, T., Nakata, H., Kamei, F. u. Harada, S.: Wie unter [29a] S. 497/502.
43. Mills, N. T., u. B. N. Bhat: Iron & Steelmaker 5 (1978) Nr. 10, S. 18/24.
44. Ribaud, P. V., Y. Roux, L.-D. Lucas u. H. Gaye: Fachber. Hüttenprax. u. Metallverarb. 19 (1981) S. 859/60, 863/66 u. 868/69.
45. Nakato, H., Sakuraya, T., Nozaki, T., Emi, T. u. Nishikawa, H.: Wie unter [29a] S. 137/43.
46. Lindenberg, H.-U. u. Loh, J.: Wie unter [29a] S. 161/67.
47. Koyana, T., Terada, O., Uchida, S. u. Ishikawa, M.: Wie unter [29a] S. 449/59.
48. Iso, H., Narita, S., Honda, M. u. Isogami, K.: Wie unter [29a] S. 457/66.
49. Carlsson, L.-E., Pettersson, H. u. Söderhäll, L.: Wie unter [29a] S. 487/94.
50. Robert, S., G. Klages u. R. W. Simon: Thyssen Techn. Ber. 14 (1982) S. 19/28.
51. Schwerdtfeger, K.: Stahl u. Eisen 106 (1986) S. 65/70.
52. Thielmann, R. u. Steffen, R.: Stahl u. Eisen 100 (1980) S. 401/07.
53. Buch, E., Figge, D., Heinke, R. u. Nonn, H.: Stahl u. Eisen 107 (1987) S. 125/28.
54. Hornich, H. u. Maschlanka, W.: Wie unter [29a] S. 425/34.
55. Höffken, E., Kappes, P. u. Lax, H.: Stahl u. Eisen 106 (1986) S. 1253/59.
56. Steffen R, u. J. P. Birat: La Revue de Metallurgie – CIT (1991) S. 543/56.

D 4 Special Methods for Melting and Casting

1. Plöckinger, E.: Die Umschmelzverfahren. In: Internationaler Eisenhüttenischer Kongreß, Düsseldorf 1974, [27–30 Mai]. Bd. 3 [Düsseldorf] 1974. Ber. 4.2.1. 14 S.
2. Wahlster, M., A. Choudhury u. K. Forch: Stahl u. Eisen 88 (1968) S. 1193/202.
2a. United States Pat. Nr. 4.261.412, April 14, 1981.
3. Wahlster, M., u. H. Spitzer: Stahl u. Eisen 92 (1972) S. 961/72.
4. Jauch, R., A. Choudhury, H. Löwenkamp u. F. Regnitter: Wie unter [1], Ber. 4.2.2.1. 13 S.
4a. Stahl-Eisen-Prüfblatt 1570–71 Mikroskopische Prüfung von Edelstählen auf nichtmetallische Einschlüsse mit Bildreihen. Ausg. 1971. Verein Deutscher Eisenhüttenleute (Ed.) Düsseldorf.
5. Randak, A., A. Stanz u. W. Verderber: Stahl u. Eisen 92 (1972) S. 981/93.
6. Plöckinger, E.: Stahl u. Eisen 92 (1972) S. 972/82.
7. Wessling, W.: Sie & Wir [Werksz. d. Stahlwerke Südwestfalen AG] Nr. 6, 1970.
8. Hochstein, F.: Stahl u. Eisen 95 (1975) S. 777/89.
9. Kühnelt, G., u. P. Machner: Stahl u. Eisen 101 (1981) S. 1311/16.
10. Jauch, R., A. Choudhury u. F. Tince: In: 9. Internationale Schmiedetagung [Düsseldorf 4.–9. Mai 1981]. Veranst. vom Verein Deutscher Eisenhüttenleute in Zsarb. mit Vereinigung Deutscher Freiformschmieden mit Unterstützung der Kommission der Europäischen Gemeinschaften. Düsseldorf 1981. Ber. 1.4.
11. Tarmann, R., P. Machner u. G. Kühnelt: Berg- u. hüttenm. Mh. 124 (1979) S. 212/21.
12. Ramaciotii, A., und Mitarbeiter: In: 8. International forgemasters meeting, Kyoto, Japan 1977.
13. Austel, W., u. Ch. Maidorn: In: Proceedings of the 5th international conference on vacuum metallurgy and electroslag remelting processes, Munich, Oct. 11–15, 1976 [Hrsg.:] Leybold-Heraeus GmbH & Co KG Hanau 1977. S. 241/42 u. 4 Taf.
14. Kawaguchi, S., u. S. Sawada: In: EPRI workshop on rotor forgings for turbines and generators, Palo Alto, Calif. 13–17, Sept. 1980.
15. Olette, M.: Mém sci. Rev. Métallurg. 57 (1960) S. 467/80.
16. Turillon P. P.: Transactions of the vacuum metallurgy conference meeting. New York 1963.
16a. Katcher, P.: IAMI 1/1982, S. 16MP2, 16MP4.

D 5 Hot Rolling

1. Riemann, W., u. K.-H. Bald: Stahl u. Eisen 94 (1974) S. 15/22.
2. Koenigsmann, F., u. F. Oeters: Werkst. u. Korrosion 29 (1978) S. 10/16.
3. Paulitsch, H., G. Schönbauer, E. Sikora u. H. Voigt: Stahl u. Eisen 94 (1974) S. 8/15.

4. Schrader, H.: Techn. Mitt. Krupp (1934) S. 136/42.
5. Birks, N.: In: Decarburization. London 1970. (ISI-Publ. No. 133) S. 1/12.
5a. DIN 17 221, Hot rolled steels for quenched and tempered springs; technical delivery conditions. Ed. Dec. 1972 and draft, Ed. Sep. 1987.
6. Melford, D. A.: J. Iron Steel Inst. 200 (1962) S. 290/99.
7. Amelung, E., u. H. Schütt: Stahl u. Eisen 93 (1973) S. 740/41.
8. Einfluß dispersoider Phasen auf das Austenitkornwachstum von Baustählen. In: Forschungshefte „Stahl". Hrsg. von der Kommission der Europäischen Gemeinschaften. Luxemburg 1975. Vertrags-Nr. 6210/62/1/011.
9. Täffner, K., u. L. Meyer: Grundlagen des Festigkeits- und Bruchverhaltens. Düsseldorf 1974. S. 240/53.
9a. DIN 17 172. Steel pipes for pipelines for the transport of combustible fluids and gases. May 1978.
10. Ishihara, S.: In: IISI/Techco/14, June 14 (1982) Helsinki 9.
11. Anker, F., u. M. Vater: Einführung in die technische Verformungskunde. Düsseldorf 1974.
12. Spittel, M.: Neue Hütte 27 (1982) S. 55/60.
13. Spittel, M., u. Th. Spittel: Freiberg. Forsch. – H., Reihe B, Nr. 231, 1982, S. 7/20.
14. Schmidt, W., u. H. Hüskes: Blech Rohre Profile 25 (1978) S. 5/11.
15. Pursian, G., F. Zeise u. K.-H. Weber: Neue Hütte 21 (1976) S. 231/35.
16. Fuchs, A.: ESTEL-Ber. 8 (1975) S. 127/35.
17. Fuchs, A.: Hoesch Hüttenwerke. Interner Bericht QE 61953 (1973).
18. Zidek, M.: Hutn. Listy 25 (1970) S. 342/50.
19. Kösters, F.: Klepzig Fachber. 77 (1969) S. 630/39.
20. Nerger, D., u. H. Reinhold: Neue Hütte 23 (1978) S. 400/03.
21. Weise, H., u. W. Haumann: In: Preprints Tempcore-Lizenznehmertagung, 1982, Vortragsnummer B4D.
22. Lorenz, K., W. Hof, K. Hulka, K. Kaup, H. Litzke u. U. Schrape: Stahl u. Eisen 101 (1981) S. 593/600.
23. Neuhauß, J., u. D. Thiery: Stahl u. Eisen 92 (1972) S. 1106/13.
24. Funke, P., R. Kulbrok u. H. Wladika: Stahl u. Eisen 92 (1972) S. 1113/22.
25. Bading, W., P. Funke u. Th. Kootz: Stahl u. Eisen 97 (1977) S. 1307/14.
26. Kopineck, H.-J., u. H. Wladika: Stahl u. Eisen 102 S. 1053/60.
27. Stahl-Eisen-Lieferbedingungen 071–77. Oberflächenbeschaffenheit von warmgewalztem Grob- und Mittelblech sowie Breitflachstahl. Ausg. Dez. 1977.
28. Stahl-Eisen-Lieferbedingungen 055 E-80. Warmgewalzter Stabstahl und Walzdraht mit rundem Querschnitt und nicht profilierter Oberfläche, Oberflächen-Güteklassen. Technische Lieferbedingungen. Ausg. März 1980.
29. Stahl-Eisen-Lieferbedingungen 025. Halbzeug mit quadratischem Querschnitt aus unlegierten und legierten Baustählen zum Gesenkschmieden (in Vorbereitung).
30. Kremer, K.-J.: Stahl u. Eisen 103 (1983) S. 359/65.
31. Lorenz, J., E. Raeder u. W. Schierloh: Stahl u. Eisen 103 (1983) S. 367/73.
32. Schneider, H.: Stahl u. Eisen 103 (1983) S. 615/18.

D 6 Hot Forming by Forging

1. Kawaguchi, S., Y. Nakagawa, J. Watanabe, S. Shikano, K. Maeda u. N. Kanno: Aufgabe der Herstellung von schweren Schmiedestücken aus 500-t-Blöcken. In: Congrès international de la grosse Forge. Paris, 20–25. avril 1975. T.1. [Hrsg.:] Chambre Syndicale de la Grosse Forge Française. Paris 1976. S. 187/220.
2. Knorr, W.: Beeinflussung von Blockseigerungen durch die Vakuum-Kohlenstoff-Desoxydation. Sixth International DH/BV Vacuum Conference, Stresa, Italy May 9–12, 1972. S. 81/95.
3. Hochstein, F., A. Choudhury, K. Fischer, H. Heymann, W. Knorr, E. Ogiewa u. R. Rischka: Stahl u. Eisen 95 (1975) S. 777/89.
4. Ambaum, E.: Untersuchungen über das Verhalten innerer Hohlstellen beim Freiformschmieden. Aachen 1979. (Dr.-Ing.-Diss. Techn. Hochsch. Aachen).
5. Kopp, R., E. Ambaum u. T. Schultes: Optimierung von Umformprozessen durch Verknüpfung empirischer und theoretischer Erkenntnisse am Beispiel des Freiformschmiedens. Stahl u. Eisen 99 (1979) S. 495/503.

6. Kopp, R., u. F. Stenzhorn: Optimierung des Freiformschmiedens hinsichtlich Qualität, Energie- und Rohstoffeinsparung. Abschlußbericht zum Forschungsvorhaben Förderungskennzeichen 01ZG 067 - ZA/NT/NTS 1011, Aachen, Febr. 1982.
7. Vater, M., u. H.-P. Heil: Stahl u. Eisen 91 (1971) S. 864/76.
8. Tateno, M., u. S. Shikano: Tetsu to Hagané 48 (1962) S. 495/98. – Tetsu to Hagané Overseas 3 (1963) S. 117/29. – Jap. Soc. Techn. Plasticity J. 65 (1966) S. 299/308.
9. Vater, M., G. Nebe u. H.-P. Heil: Stahl u. Eisen 86 (1966) S. 892/905.
10. Haller, W.: Handbuch des Schmiedens. München 1971.
11. Schinn, R.: Stahl u. Eisen 72 (1952) S. 676/83.
12. Ammareller, S., u. P. Grün: Stahl u. Eisen 72 (1952) S. 653/62.
13. Maurer, E., u. H. Korschan: Stahl u. Eisen 53 (1933) S. 209/15.
14. Burton, H. H.: In: Comptes rendus des journées de la grosse forge. Organisées par le Centre Technique de la Grosse Forge (Chambre Syndicale de la Grosse Forge Française) á Paris 27.–29. Mai 1948. Paris 1948. S. 157/75.
15. Nitschke, K.: In: Internationale Schmiedetagung 1965, Berlin [9.–11. Juni]. Hrsg. vom Verein Deutscher Eisenhüttenleute. Düsseldorf 1966. S. 11/14.
16. Feeg, F.E.: Masch.-Markt 66 (1960) Nr. 15, S. 15/20.
17. Feeg, F. E.: Masch.-Markt 69 (1963) Nr. 66, S. 11/19.
18. Oehler H. H., H. Robra u. H. J. Bargel: VDI-Z. 111 (1969) S. 773/78 u. 1043/46.
19. Piehl, K. H., O. W. Buchholtz, H. Finkler, K. Forch, O. Jacks, H. G. Ricken, R. Rischka u. J. Venkateswarlu: Stahl u. Eisen 95 (1975) S. 837/46.
20. Krautkrämer, J., u. H.: Werkstoffprüfung mit Ultraschall. Berlin/Heidelberg/ New York (1980).

D 7 Cold Forming by Rolling

1. Pawelski, O.: Grundlagen des Kaltwalzens von Band. In: Herstellung von kaltgewalztem Band. Teil 1. Hrsg. vom Verein Deutscher Eisenhüttenleute. Düsseldorf 1970. S. 236/64.
1a. EURONORM 79. Definition and classification of steel products by shape and dimensions. March 1982 and draft European Standard 10079: 1988.
1b. DIN 46400 Part 1 to 4: Steel flat products with special magnetic properties: Part 1: Non oriented April 1983, Part 2: Nonoriented, semi-processed 1987, Part 3: grain oriented, 1988, Part 4: Alloyed steels non oriented, Jan. 1987.
2. Pawelski, O., u. V. Schuler: Versuche zum Regeln der Planheit beim Kaltwalzen von Band. Stahl u. Eisen 90 (1970) S. 1214/22.
2a. DIN 1541. Flat products of steel; cold rolled wide mill strip and sheet of unalloyed steels; dimensions, tolerances on dimensions and form 1975.
2b. DIN 59 382. Flat products of steel; cold rolled wide mill strip and sheet of stainless steels; dimensions, tolerances on dimensions and form Aug. 1975.
3. Grassl, D., u. J. Wünning: Entwicklung der Inertgasverwendung bei der Wärmebehandlung. Z. wirtsch. Fertigung. 65 (1970) S. 187/98.
4. Funke jr., P.: Einfluß des Walzens und Nachwalzens auf die Bandeigenschaften. Wie unter 1. S. 280/82.
5. Müschenborn, W., H.-M. Sonne u. L. Meyer: Die erzeugungsbedingten Gütemerkmale von kaltgewalztem Feinblech unter dem Blickwinkel der Kaltumformbarkeit. Thyssenforsch. 4 (1972) S. 43/55.
6. Kranenberg, H.: Untersuchungen über die Walzenrauheit und deren Übertragung auf das Blech beim Walzen im Bereich kleiner Formänderungen. Berlin 1967. (Dr.-Ing.-Diss. Techn.-Univ. Berlin.)
7. Oberflächenfehler an kaltgewalztem Band und Blech. Hrsg.: Verein Deutscher Eisenhüttenleute. Düsseldorf 1967.

D 8 Heat Treatment

1a. DIN 1654 Cold heding and cold extruding steels. March 1980.
1b. DIN 17 100 Steels for general structural purposes. Jan 1980 and draft DIN 17 108. Dec 1987 as draft European stand. EN 10025.

1c. DIN 17 200 Steels for quenching and tempering. March 1987 and draft DIN 17 200 quenched and tempered steels. March 1988 as European Standard EN 10083.
1d. DIN 17210. Case hardening steels. Sept. 1986.
1e. DIN 17230. Ball and rolled bearing steels. Sept. 1980.
1f. DIN 17440. Stainless steels; technical delivery in conditions for plates, hot rolled strip wire rods, drawn wire, bars, forgings and semi-finished products. June 1985.
2. Ruhfus, H.: Wärmebehandlung der Eisenwerkstoffe. Düsseldorf 1958. (Stahleisen-Bücher. Bd. 15.)
3. Technologie der Wärmebehandlung von Stahl. Hrsg. von H.-J. Eckstein. Leipzig 1977.
4. Connert, W.: Stahl u. Eisen 80 (1960) S. 1049/60.
5. Plöckinger, E., u. A. Randak: Stahl u. Eisen 78 (1958) S. 1041/58.
6. Horn, W., u. H.-J. Horn: Wärmebehandlung von Stahl. Düsseldorf 1968.
7. Spies, H. J., G. Münch u. A. Prewitz: Neue Hütte 22 (1977) S. 443/45.
8. Thelning, K. E.: Z. wirtsch. Fertigung 66 (1971) Nr. 3, S. 13/22.
9. Nichtrostende Stähle. Bearb. von P. Schierhold. Düsseldorf 1977. S. 40 u. 120.

D 9 Quality Management in the Production of Steel Works Products

1. Treppschuh, H.: Stahl u. Eisen 84 (1964) S. 1714/23.
1a. DIN 55 350 Part 11 Concepts in quality management and statistics; basic concepts relating to quality management. Draft March 1986.
1b. ISO Standard 8402 Quality – Vocabulary. 1st Ed. 1986-06-15. – ISO Draft Proposal 3534/1 Statistics – Vocabulary and symbols, Part 1, Probability and general statistical terms. 1986-05-23. – ISO Draft international standard ISO/DIS 3534/2 Statistics – Vocabulary and symbols Part 2, Statistical quality control. 1986-05-01
2. Jäniche, W., u. W. Heller: In: Werkstoffkunde der gebräuchlichen Stähle. T.2. Hrsg.: Verein Deutscher Eisenhüttenleute. Düsseldorf 1977. S. 357/78.
3. Feldmann, U.: Stahl und Eisen 103 (1983) S. 351/57.
4. Kremer, K.-J.: Stahl u. Eisen 103 (1983) S. 359/65.
5. Lorenz, J., E. Raeder u. W. Schierloh: Stahl u. Eisen 103 (1983) S. 367/73.
6. Qualitätssicherungs-Handbuch. Rahmenrichtlinie für die Werke der Stahlindustrie. Vom Werkstoffausschuß des Vereins Deutscher Eisenhüttenleute in einer Gemeinschaftsarbeit erstellte Richtlinie. Düsseldorf 1978.
7. Jäniche, W., P. Hammerschmid u. M. Kühlmeyer: Z. Metallkde. 64 (1973) S. 1/7.
8. Heller, W., u. P. Hammerschmid: Techn. Mitt. Krupp, Werksber. 33 (1975) S. 89/93.
9. Handbuch der Qualitätssicherung. Hrsg. von W. Masing. München, Wien 1980.
10. Crosby, Ph. B.: Qualität kostet weniger. 1971.
11. Zink, K. J., u. G. Schick: Quality Circles-Problemlösungsgruppen. München, Wien 1984.
12. Daeves, K.: Stahl u. Eisen 43 (1923) S. 462/66.
13. Daeves, K., u. A. Beckel: Großzahlforschung und Häufigkeitsanalyse. Weinheim/Bergstr., Berlin 1948.
14. Knüppel, H., A. Stumpf u. A. Fricke: Arch. Eisenhüttenwes. 32 (1961) S. 883/91; 33 (1962) S. 67/76.
15. Jäniche, W.: Mater.-Prüfung 6 (1964) S. 418/25.
16. Kühlmeyer, M.: In: Werkstoffkunde der gebräuchlichen Stähle. T.2. Hrsg.: Verein Deutscher Eisenhüttenleute. Düsseldorf 1977. S. 379/93.
17. Baumann, H.-D., u. P. Greis: Statistische Prüfung. Düsseldorf 1978.
18. Meyer, D., H.-J. Kläring, J. Wernstedt u. W. Winkler: Neue Hütte 26 (1981) S. 179/84.
19. Helck, F., u. W. Rohde: Statistische Methoden bei der Prüfung von Stahlerzeugnissen – Eine kritische Analyse. Bericht zur Tagung „Werkstoffprüfung 1984", Bad Nauheim. [Hrsg.:] Deutscher Verband für Materialprüfung (DVM), Berlin.
20. Kühlmeyer, M.: Die Rolle von Stichprobenplänen bei der Prüfung von Stahlerzeugnissen. Wie unter [19].
20a. DIN 17 102 Weldable Fine grain steels, normalized; technical delivery conditions for strip, plate, universal plate, sections and merchant bars. Oct. 1983.
21. Kremer, K.-J., u. H. Spitzer: Einsatz der Datenverarbeitung bei der Steuerung der Edelstahlherstellung hinsichtlich qualitativer Zielvorgaben. Bericht einer Fachtagung des KfK Karlsruhe über „Prozeßlenkung mit Datenverarbeitungsanlagen" in Karlsruhe vom 9. bis 19. Mai 1979.
22. Kühlmeyer, M.: Frontiers in statistal quality control. Würzburg (1981) S. 148/64.
23. Kühlmeyer, M.: Wie [20] (1984), S. 136/45.

24. Spitzer, H., u. K.-J. Kremer: In: VDI-Ber. Nr. 428, 1981, S. 1/25.
25. Handbuch für das Eisenhüttenlaboratorium. 5 Bde. Hrsg.: Chemikerausschuß des Vereins Deutscher Eisenhüttenleute. Bd. 2. Düsseldorf 1966.
26. Koch, K.-H. Stahl u, Eisen 103 (1983) S. 449/52.
27. Born, A., J. Bewerunge, J. Brauner, M. Heinen u. K.-J. Kremer: Arch. Eisenhüttenwes. 52 (1981) S. 289/94.
28. Pluschkell, W.: Stahl u. Eisen 99 (1979) S. 404/11.
29. Hougardy, H. P., K. Barteld, P. G. Dressel, M. Heyder, H.-J. Nierhoff u. D. Schreiber: Stahl u. Eisen 103 (1983) S. 509/12.
30. Kügler, J., G. Geimer, G. Naumann, W. Rohde u. W. Schmidt: Stahl u. Eisen 103 (1983) S. 559/64.
31. Schneider, H.: Stahl u. Eisen 103 (1983) S. 615/18.
32. Ahrens, H., B. Hesse, F. Meuters u. H. Schmedders: Thyssen Techn. Ber. 12 (1980) S. 111/16.
33. DIN 1599. Kennzeichnungsarten für Stahl. Ausg. Aug. 1980.
33a. DIN 17 155 Plate and strip steels for elevated temperatures; Technical delivery conditions. October 1983.
34. Tschorn, G.: Schleiffunkenatlas für Stähle, Gußeisen, Roheisen, Ferrolegierungen und Metalle. Leipzing 1961.
35. Brauner, J., K.-D. Glaubitz u. K.-J. Kremer: Stahl u. Eisen 100 (1980) S. 1323/28.
35a. DIN 50 049 Documents on material testing Aug. 1986.
36. Smit, H., L. Schulz, H. Paaßen, D. Küpper u. J. Mahn: Thyssen Techn. Ber. 12 (1980) S. 143/51.

Subject Index

Preliminary remark

Symbols and abbreviations see glossary in volume 1. Comparison of German steel grades mentioned in this subject index with corresponding steel grades according to international standards: see table 1 in the annex of volume 1

300 °C brittleness 127
475 °C embrittlement 410, 438
500 °C brittleness 127
α-martensite in low-temperature steels 283
acceptance inspections 724
adhesion of a zinc coating 102
"after pouring"-process 676
aftertreatment of steel melts in the ladle 646
-, process and operating elements 647
aging
-, of flat products for cold forming 94
aging susceptibility 46, 623, 625
aim of volume 2
-, presentation of the application of the fundamentals on the development of steels 3
AlNiCo magnets 539, 540
alloy calculation in connection with quality management 718
alloying of steel melts 645
aluminium for
-, desoxydation 645
-, normal and high strength structural steels 46
-, precipitation hardening of alloys with defined modulus of elasticity 561
-, soft magnetic materials 504
-, steels for chains 627
aluminum-coated flat products **105**
-, cold formability 107
-, corrosion resistance 107
aluminum-nickel and aluminum-nickel-cobalt alloys as permanent magnet materials 538
aluminum nitride, importance for
-, aging non - susceptibility 46
-, controlling secondary grain growth in soft magnetic materials 505
-, grain refinement 48, 79, 85, 201
-, production of textures 79, 83, 505
amorphous metals, see metallic glasses
angular tensile test (for screws and bolts) 611
anisotropy energies, importance for magnetic materials 496, 535
anisotropy of properties, dependence on forging conditions 694, 696
anisotropy of the mechanical properties of
-, free cutting steels 487
-, normal and high strength structural steels 50
-, steels for line pipe 579
-, their dependence on the ratio of stretching during forging 696
anisotropy of the toughness, importance of sulfur 50

anisotropy (vertical) 81
annealing of cold rolled flat products for cold forming 83
antimony, as steering phase for the formation of Goss-texture 505
AOD = Argon Oxygen Decarburization 649
ausforming of UHS steels 218
austenitic grain growth controlling 708
austenitic stainless steels 416
-, details see also stainless steels
A_V = notch impact energy (see below)

B_{10}-value: see Weibull diagram
bainitic steels for flat products for cold forming 86, 88, 93
bake-hardening steels 91, 94
ball and roller bearing steels **584**
-, required properties of use and their characterization 584
-, -, mechanical properties (above all hardness and fatigue strength) 584, 585
-, several measures of physical metallurgy to attain the required properties 587
-, -, chemical composition 587
-, -, importance of carbide distribution 588
-, -, importance of cleanness 587
-, -, -, influence of the melting process 587
-, -, microstructure 586, 587
-, typical steel grades proven in service 589
banded structures 708
barium ferrites for permanent magnets 542, 545
bars (reinforcing steels) 61, 62
Battelle drop-weight tear test 575
Bauschinger-effect 10, 24
beam test (reinforcing steels) 63
bending yield stress of various tool steels 318, 319
berylium, for the precipitation hardening of alloys with a high modulus of elasticity and defined temperature coefficient 561
Bessemer converter 640
BEST = Böhler Electroslag Topping 675
"BG-annealing" 182
bismuth, in free-cutting steels 482
blackplate 77
blast furnace process 638
Bloch-wall movements, importance for coercivity 495, 534
-, influence of inclusions or precipitations and of grain size 495

Subject Index

Bloch-walls, importance for magnetic materials 495, 503, 524, 534
blowholes in steel, formation during solidification 659
bonding properties (reinforcing steels and prestressing steels) 62, 65, 70, 74
boron, influence on the properties of
-, case hardening steels 172
-, -, influence on hardenability, impact strength and impact toughness 172
-, elevated-temperature steels 249
-, free-cutting steels 482
-, heat-treatable and surface hardening steels 137
-, -, influence on the cold formability 133
-, normal and high strength structural steels 50
-, steels for cold forming 182
-, -, influence on the flow curve 183
-, -, influence on the hardenability 183
-, steels for flat products for cold forming 88
-, steels for screws, nuts and rivets 613
bottom air refining 640
bottom blowing converter 640
bottom pouring 650, 651
box annealing of cold rolled flat products 704
bright annealing of cold rolled flat products 703
"burn-through" (damage of valves) 453
"burned" condition of hot rolled steel 684

calcium
-, in free-cutting steels 480
-, in heating conductor alloys 448
capability for carburization of case hardening steels 159
carbides in
-, elevated - temperature steels 235, 236, 239, 241, 243, 245, 247, 251
-, high - pressure hydrogen resistant steels 426
-, stainless steels 393, 394
-, tool steels 231
-, wear - resistant steels 380
carbon equivalent (new) to characterize weldability 577
carbon, influence on the properties of
-, ball and roller bearing steels 587
-, case hardening steels 161, 162
-, elevated-temperature steels 234, 237, 243
-, free cutting steels 479
-, heat treatable and surface hardening steels 122, 127, 133, 170
-, low-temperature steels
-, -, austenitic 283, 287
-, -, ferritic 278, 280
-, non magnetizable steels 550
-, normal and high strength structural steels 29, 40, 42
-, reinforcing steels 63
-, soft magnetic materials 495, 500, 520
-, spring steels 469
-, stainless steels 387, 394
-, steels for cold forming 181

-, steels for flat products for cold forming 85
-, steels for permanent-way (track)material (rails) 596
-, steels for rolling stock 602
-, steels for screws, nuts and rivets 612
-, steels for valves 456
-, steels for welded round link chains 627
-, steels with good electrical conductivity 570
-, tools steels 308, 341
-, ultra high-strength steels 214
-, wear resistant steels 374
-, wire rod for cold drawing 200, 203
"carbon stabilization" in austenitic steels 245, 386, 396
carburizing depth (case hardening steels) 167
carburizing of case hardening steels 163
case hardening 161
case-hardening of free-cutting steels 485, 486
case hardening steels **158**
-, required properties of use and their characterization 158 (core), 161 (case hardened layer)
-, several measures to attain the required properties 159 (core), 161, 163 (case hardened layer)
-, -, carburizing 163
-, -, hardening 165
-, -, -, direct 165
-, -, -, double
-, -, -, single 165
-, -, stress relieving 168
-, -, typical steel grades proven in service 174
cast steel treatment 660
casting and solidification **650**
see also melting and casting, special methods
-, common methods of casting, characteristics 650
-, -, bottom pouring 650, 651
-, -, top pouring of ingots 650, 651
-, -, -, fully killed 651
-, -, -, rimmed 651
-, -, -, semi killed
-, continuous casting 661
-, -, reoxidation flow, and superheat 661
-, -, heat transfer during solidification 662
-, -, formation of
-, -, -, internal cracks 665
-, -, -, -, influence of alloying elements 666
-, -, -, segregations 663
-, -, -, -, influence of electromagnetic stirring 664
-, -, -, surface defects 666
-, -, processing (hot charging and hot direct rolling) 668
-, -, significance of casting fluxes 667
-, steel casting and solidification in ingot molds
-, -, crystallization process 653
-, -, formation of
-, -, -, blowholes 659
-, -, -, oxide inclusions 657
-, -, -, pipes 660
-, -, -, segregations 655

Subject Index 815

-, -, -, sulfide inclusions 658
-, -, heat transfer and solidification 654
-, -, reoxidation, flow and superheat 651
casting fluxes, behaviour in continuous casting 667
castings: see steel castings
CAT = crack arrest temperature of Robertson test 21, 33, 276
center segregation in continuously cast slabs 663
cerium in
-, free-cutting steels 480
-, heating conductor alloys 448
-, heat resisting steels 437
chains 630
-, distribution of stress in a link 623
-, final heat treatment 629, 631
-, their working stress limit in comparison to the chain steel tensile strength 622
chain steels: see steels for welded round link chains
chemical composition, influence on the steel properties: see the various elements and the various steel types under: measures to attain the required properties. In addition: more general comments on the influence of the chemical composition on the properties of
-, elevated - temperature steels 237, 243
-, heat - treatable steels 127, 151
-, low - temperature steels 278, 283
-, normal strength and high strength structural steels 39
-, stainless steels 393, 395
-, steels for cold forming 181
-, tool steels 308
chemical composition of various steel grades, data in tables
-, alloys with defined thermal expansions 563
-, ball and roller bearing steels 591
-, case hardening steels 175
-, constant - modulus alloys 565
-, elevated - temperature materials 255, 257, 260, 262, 265, 271
-, flat products for cold forming 88
-, free cutting steels 486
-, heat resisting materials 435
-, heat treatable steels 136, 141, 145, 151, 152, 175
-, heating conductor alloys 449
-, high - pressure hydrogen resistant steels 430, 431
-, low - temperature steels 288
-, nitriding steels 145
-, non - magnetizable steels 554
-, normal and high strength structural steels 52, 56, 57, 58, 59
-, prestressing steels 76
-, reinforcing steels 68
-, soft magnetic steels for generator shafts 517
-, spring steels 474
-, stainless steels 409
-, steels with good electrical conductivity 571

-, steels for cold forming 188, 189, 191, 193
-, steels for heavy forgings 152
-, steels for quenching and tempering 136
-, steels for screws, bolts, nuts and rivets 616, 617, 618
-, steels for surface hardening 141
-, steels for welded round link chains 629
-, tool steels 352, 356, 358, 360, 362, 364, 366, 367, 371, 372
-, ultra high - strength steels 223
-, valve materials 458
-, wear - resistant steels 380
chi-phase ($Fe_{36}Cr_{12}Mo_{10}$) in
-, elevated-temperature austenitic steels 247
-, stainless steels 414
chromated flat products 114
chromium coated flat products 111
chromium equivalent (stainless steels) 398
chromium, influence on the properties of
-, ball and roller-bearing steels 587
-, case hardening steels 160
-, elevated-temperature steels 236, 239, 243, 254, 260, 261
-, heat resisting steels 437
-, heat treatable steels 127, 137, 151, 153, 160, 164, 167
-, heating conductor alloys 446
-, low-temperature austenitic steels 283
-, nitriding steels 143
-, nonmagnetizable steels 550, 551, 553
-, normal and high strength structural steels 45
-, spring steels 475
-, stainless steels 393, 398, 399, 401
-, steels for cold forming 182
-, steels for flat products for cold forming 85
-, steels for line pipe 579
-, steels for permanent-way (track) material (rails) 595, 598
-, steels for use with hydrogen at elevated temperatures and pressures 427
-, steels for valves 456, 457
-, steels with defined thermal expansion 558, 560, 561
-, tool steels 308, 311, 316, 324, 329, 330, 338
-, ultra high-strength steels 214, 220
-, wear resistant steels 379
chromium-iron-cobalt alloys as permanent magnet materials 541
chromizing strip 114
circulating processes in steelmaking 648
cladding 113
cleanness, influence on the properties of
-, bearing steels 588
-, heat treatable steels 125
-, steels for cold forming 185
-, wire rod for cold drawing 200
cleanness of steels, influence of remelting 673
CLU = Creusot Loire Uddeholm process 649
clusters, formation during steel crystallization 653

816   Subject Index

coarse grain annealing of case hardening steels 161
coil-coated steel flat products 116
cold cracking behaviour in
-, welding of low-temperature steels 295
-, welding of structural steels 40
cold cracking susceptibility 47
cold formability (see also formability) of
-, flat products for cold forming 78, 80, 81, 82, 85, 98, 101, 107
-, heat treatable steels 133
-, steels for cold forming 177, 178, 181
-, steels for screws, bolts, nuts and rivets 608, 612
-, tool steels 343
-, wire rod for cold drawing 195, 196, 199
cold forming by rolling **699**
-, aim of cold rolling and definition of cold rolled products 699
-, cold rolling mill, sequences for the production of various flat products 699, 700
-, -, equipment 699, 700
-, -, several auxiliary materials 699, 702
-, effect of process variables on material properties
-, -, anisotropy (r - value) 702
-, -, flatness 702
-, -, waviness 702
-, heat treatment of cold strip 703
-, -, aim of annealing 703
-, -, -, recrystallization 703
-, -, -, solution 703
-, -, type of annealing 703
-, -, -, continuous 703
-, -, -, box annealing 704
-, temper rolling 704
-, -, aim 704
-, -, results: influence on the properties of
-, -, -, electrical steel sheet 705
-, -, -, deep drawing steels 704, 705
-, -, -, stainless steels 705
-, the finishing for shipment 706
cold forming, effect on the mechanical properties of
-, reinforcing steel 64, 66
-, StE 355 and StE 690 23, 24, 25
cold forming steels: see steels for hot rolled, cold rolled and surface treated flat products for cold forming
cold forming tools, steels for them 357
cold rolling: see cold forming by rolling
cold-rolled flat products for cold forming **80**, 84
-, mild unalloyed steels **80**, 84, 88
-, -, required properties of use and their characterization 80, 81, 82
-, -, -, cold formability in particular 80, 81, 82
-, -, several measures to attain the required properties 82
-, -, typical steel grades 83, 84
-, normal and high strength steels 88, **90**, 95
-, -, required properties of use and their characterization 90

-, -, -, cold formability in particular 91
-, -, typical steel grades 94, 95
cold-upsetting test (compression test) 178
columnar crystals, formation during steel crystallization 653, 664
combined top-bottom blowing converter 640
composite material of case hardened steel 169
-, behaviour of the case and the tough core under service conditions 169
compression casting 665
compression strength (tool steels) 317
constant-modulus alloys 561, 565
continuous annealing of cold rolled flat products 703
-, electrical steel sheet 704
-, soft steels 703
-, stainless steels 703
continuous casting
-, approaching the final shape of the rolled product 668
-, comparison with ingot casting 668
continuous casting machines 652
continuously cast steel processing
-, hot charging 668
-, hot direct rolling 668
cooling from forging heat 697
-, influence on steel properties 697
cooling time t 8/5 in welding 30
-, influence on the mechanical properties of the welded joints 30, 31
copper, influence on the properties of
-, normal and high strength steels 46
-, stainless steels 395
copper-nickel-iron alloys as permanent magnet materials 542
core loss 492, 501
core (power) loss of electrical steel sheets 502, 507, 509
Core-Zone-Remelting process 676
corrosion equivalent 552
corrosion fatigue cracking 389
corrosion resistance of stainless steels 393
-, fundamentals 393
-, -, chemical composition 393, **395**, 396
-, -, microstructure **396**
-, testing 384
corrosion resistance of surface treated flat products
-, aluminum coated 107
-, lead coated 111
-, tin coated 110
-, zinc coated 103
corrosion resistance of the various types of stainless steels
-, austenitic steels 416
-, ferritic-austenitic steels 421
-, ferritic steels 408
-, martensitic steels 411
corrosion types 382
-, schematic representation 385
corrugations (roaring rails) 594

Subject Index 817

crack initiation 19,20
see also fracture behaviour
crack parameter after Ito 41
crack propagation 20
see also fracture behaviour
creep 226, 229
cracks in weldeds joints 30
creep limit 230
-, values for heat resisting steels 436
creep resistance 226, 229, 237
creep resistant steels: see elevated-temperature steels
creep rupture strength 227, 229
-, procedure of testing 229
-, test results for
-, -, austenitic steels (general) 224, 245, 250
-, -, cobalt- and nickel alloys
-, -, -, hot forged
    CoCr 20 Ni 20 W 272
    CoCr 20 W 15 Ni 272
    NiCo 18 Cr 15 MoAlTi 272
    NiCo 19 Cr 18 MoAlTi 272
    NiCr 18 CoMo 272
    NiCr 19 NbMo 272
    NiCr 20 Mo 272
    NiCr 20 TiAl 272, 466
-, -, -, precision cast
    G-CoCr 22 Ni 10 WTa 272
    G-NiCo 10 W 10 CrAlTa 272
    G-NiCo 15 Cr 10 MoAlTi 272
    G-NiCr 13 Al 6 MoNb 272
    G-NiCr 16 Co 8 AlTiN 272
-, -, steel grades
    21 CrMoW 5 7 232
    X 5 NiCrTi 26 15 266
    X 6 CrNi 18 11 266, 267
    X 6 CrNiMo 17 13 266, 267
    X 6 CrNiWNb 16 16 266
    X 8 CrNiMoBNb 16 16 249, 266, 267
    X 8 CrNiMoNb 16 16 266, 267
    X 8 CrNiMoVNb 16 13 266
    X 8 CrNiNb 16 13 249, 266
    X 10 NiCrAlTi 32 20 266
    X 12 CrCoNi 21 20 266, 465
    X 20 CrMoWV 12 1 230
    X 22 CrMoV 12 1 230
    X 40 CrNiCoNb 17 13 266
    X 45 CrSi 9 3 465
    X 45 CrNiW 18 9 465
    X 53 CrMnNiN 21 9 465
    X 60 CrMnMoVNbN 21 10 465
-, -, unalloyed steels 238
-, -, test results scattering 230
creep-stress rupture test 229
creep test 229
crevice corrosion 387
critical limiting concentrations (of special carbide forming elements in elevated-temperature steels) 236
crude steel production **638**
-, raw materials for steelmaking, concerning

-, -, direct reduction of iron ores 638
-, -, production of pig iron in a blast furnace 638
-, -, -, possibilities to improve pig iron 638
-, charged materials, importance for steelmaking, regarding
-, -, desired elements 641
-, -, undesired elements 641
-, steelmaking, process steps **642**
-, -, melting and refining 643
-, -, -, decarburization and correlation with the oxygen concentration 643, 644
-, -, -, desulfurization 644
-, -, -, removal of hydrogen, nitrogen and phosphorus 643
-, -, -, deoxidizing and alloying deoxidation 645
-, -, -, in the ladle 646
-, -, -, in the steel making vessel 645
-, -, ladle metallurgy processes for steel aftertreatment 646
-, -, -, gas rinsing/stirring treatment 640, 646
-, -, -, heating 640, 648
-, -, -, injection treatment 640, 646
-, -, -, vacuum treatment 640, 648
crude steel production, process alternatives 640
crystal anisotropy 535
crystallization of steel during casting 653, 662
cube texture in electrical steel sheet 505
cumulative frequency curves, application in quality control
-, general 713
-, curves of
-, -, carbon content in 45 heats of steel grade 21 Mn 4 714
-, -, notch impact energy at - 60 °C of a fine-grained structural steel 714
-, -, sulphur content in fine-grained structural steels 716
-, -, yield strength values of steel St 37-3 715
Curie temperature
-, importance for
-, -, permanent magnet materials 534
-, -, steels with defined thermal expansion 558
-, -, steels with special elastic properties 558, 561
cutting tool life 330
cutting tools, steels for them 368
CVD = chemical vapour deposition (coating of tools) 332

decarburization in steelmaking 643
deep-drawing steels 79, 83, 92
defect depths in rolling flat and round steel bars 688
defects in forgings 697
-, avoidance 698
-, testing 697
deformation degree, influence on the properties of patented wire 205
deformation resistance in hot rolling 686
degenerated pearlite steels 87
degree of grain orientation (texture) 505
"delayed brittle fracture" 209

delivery conditions, possibilities for the various steels 707
"Demag-Yawata"-process (wire) 203
dendritic crystals, formation during steel crystallization 653, 654
deoxidation by
-, diffusion 645
-, oxide precipitation 645
deoxidation of steel melts 645, 659
-, influence on the steel properties 645, 659
deoxidation products removal 645
deoxidizing potential of different elements at 1600 °C 645
depth of case (case hardening steels) 167
desulfurization in steelmaking 644
DH = Dortmund-Hoerde process 648
diffusion annealing 327, 336, 342
direct hardening of case hardening steels 165
direct reduction of iron ores 638
direct rolling of continuously cast steel 668, 685
directional solidification 253
documentation (quality management) 712
domain structure in soft magnetic materials 506, 507
double hardening of case hardening steels 165
double tension test 20
drawability of
-, wire rod for cold drawing 195, 196, 199
dual-phase steels 86, 87, 88, 93, 95, 97
DVM-creep limit 230

ε-martensite in low-temperature steels 283
EB = Electron Beam process 671
EBCHR = Electron Beam Cold Hearth Refining 677
"ED"-process (wire) 203
"EDC"-process (Easy Drawing Conveyor) (wire) 203
edge breaks in hot rolling 687
edge cracks in hot rolling 687
ELA (extra-low-additions) ferritic steel 619
elastic modulus: see modulus of elasticity 561
ELC (extra-low-carbon) steels 619
electric arc furnace 640
electrical resistivity of the heating conductor alloys 449
electrical steel sheet **501**
see also soft magnetic materials
-, production process of electrical steel sheets **508**, 510
-, -, grain-oriented electrical steel sheet (GO) 508
-, -, non-grain-oriented electrical steel sheet (NO) 508
-, required properties of use 501
-, several considerations and measures of physics and physical metallurgy to influence
-, -, eddy currents 502
-, -, magnetization processes 502
 and corresponding losses 502, 503 by

-, -, -, chemical composition, importance of silicon 504, 508, 509
-, -, -, grain orientation **504**, 506
-, -, -, microstructure 508
-, -, -, texture 504
-, typical electrical steel sheet grades 512
-, -, currently available 512
-, -, recent developments 514
-, -, -, amorphous metals 514
-, -, -, steel with 6 % Si 514
 and other silicon-alloyed materials 515
elements, undesired
 influence on steel properties 641
elevated-temperature alloys: see elevated-temperature steels
elevated-temperature steels **226**
-, required properties of use and their characterization 226, 229
-, -, general 226
-, -, in particular: behaviour under mechanical loads at high temperature (creep) and relaxation resistance 226, 229
-, several measures of physical metallurgy to attain the required properties **233**
-, -, austenitic steels 242
-, -, -, chemical composition and microstructure 243
-, -, -, importance of the lattice 242
-, -, ferritic steels for slightly elevated temperatures 233
 creep resistance 237
-, -, -, chemical composition and microstructure 234, 237, 241
-, -, nickel and cobald alloys 249
-, -, -, chemical composition and microstructure 251
-, -, -, importance of the lattice 250
-, typical material types proven in service **253**
-, -, austenitic steels 264, 265
-, -, ferritic steels for
-, -, -, castings 261, 262
-, -, -, forgings and bars 259, 260
-, -, -, plates and tubes 256, 257
-, -, -, slightly elevated temperature 253, 255
-, -, nickel and cobalt alloys 269, 271
-, -, -, forging alloys 269
-, -, -, precision cast alloys 269
embrittlement by sigma-phase 410, 432, 438, 440
enamelled flat products 115
end - quench test curves of
-, 13 MnCrB 5 183
-, 16 MnCr 5 183
-, 35 B 2 614
-, 42 CrMo 4 (also effect of annealing) 123
-, 50 CrV 4 475
-, 50 Si 7 475
-, 51 CrMoV 4 475
-, 55 Cr 3 475
-, 55 Si 7 475
-, 65 Si 7 475
-, Cq 35 614

Subject Index 819

end quench test to characterize hardenability 132
equiaxed crystals, formation during steel crystallization 653, 654
ESD (= elongated single domain)-magnets 539
ESR = Electro-Slag-Remelting 671

"false" nitrogen pearlite in heat resisting steels 442
fatigue fractures of machine components 121
fatigue resistance 14
fatigue strength 15
-, dependence on surface condition 15
fatigue strength for finite life 14
fatigue strength (general data) of
-, heat treatable steels 129
-, normal and high strength structural steels 14
-, rail steels 598, 599
-, spring steels 470
-, tool steels 320
-, UHS steels (sensitivity to mean stress) 221
fatigue strength of various steel grades
-, Wöhler - curves of
-, -, 15 CrNi 6, influence of surface oxydation 163
-, -, 16 MnCr 5, influence of the case hardening conditions 173
-, -, 17 NiCrMo 14, influence of the melting process 585
-, -, 50 CrV 4, influence of surface decarburization 471
-, -, 100 Cr 6, influence of the melting process 585
-, -, S 6-5-2, influence of the melting process 320
-, -, StE 460, influence of welding 17
-, -, StE 690, influence of welding and weldseam remelting 17, 18
-, -, X 10 CrNiNb 18 9, influence of corrosion conditions 390
-, fatigue strength under fluctuating (tension) stresses
-, -, St 37-3 16
-, -, St 52-3 16
-, -, StE 690 16
-, rotating bar bending fatigue strength of
-, -, X 41 CrMoV 5 1, influence of salt - bath nitriding 221
-, fatigue strength diagram acc. to Smith of prestressing steels
-, -, St 1080/1230 72
-, -, St 1420/1570 72
fatigue testing 14
FATT = fracture appearance transition temperature 233
-, of elevated-temperature steels 233
ferritic-austenitic stainless steels 421
-, details see also stainless steels
ferritic stainless steels 408
-, details see also stainless steels
filler metals for steels for low-temperature service 295, 296

final release of steel products (quality management) 723, 725
fine - grained microstructure, importance for
-, heat treatable steels 125
-, low-temperature steels 280, 282
-, -, measures to attain it 280
-, normal and high strength structural steels 35, 48, 49, 55
-, -, for cold forming 85, 92
-, ultra high - strength steels 218
-, wire rod for cold drawing 200, 201
fine-grained steels 42, 45, 49, **55**, 85, 89, 92, 233
fine grain structural steels 45, 55
-, normalized 55
-, quenched and tempered 59
-, thermomechanically treated 59
fine-grain structure, importance for toughness 125
flakes 642
-, importance of hydrogen 642, 688, 697
-, in forgings 697
-, in hot rolled products 688
-, prevention during cooling 688, 697
flat products for cold forming
-, cold rolled 80, 90
-, hot rolled 77, 84
-, surface treated 97
flat products for cold forming: see also steels for hot rolled, cold rolled and surface treated flat products for cold forming
flatness of cold rolled flat products 702
-, influence of temper rolling 705
flow - curve: see true stress - true strain curve
flow of steel during pouring
-, casting into ingot molds 651
-, continuous casting 662
flow stress (true stress for a given true strain) of steels for cold forming 182, 183, 185
-, influence of alloying elements 182
-, influence of the microstructure 179, 183, 185
"Fluidized-bed"-process (wire) 203
forging and pressing dies, tool steels for them 361
forging ingots 691
-, importance of their geometric shape 692
forging methods: see hot forming by forging
formability (general) 686
-, dependence on temperature 687
-, importance for hot rolling 687
formability (see also cold formability) of
-, normal and high strengt structural steels 10, 22, 25
-, reinforcing steels 62
-, steels for welded round link chains 623, 628
-, wire rod for cold drawing 195
forming analysis test 78
fracture behaviour of
-, low-temperature steels 275, 276
-, normal and high strength structural steels 20, 21
-, pipelines 576
-, rail steels 594

-, welded joints of StE 690 33
fracture mechanics 21, 156
fracture toughness (K_{Ic} - values) of
-, elevated - temperature steels 233
-, rail steels 595
-, steels for heavy forgings 157
-, ultra high-strength steels 215, 219, 224
fracture toughness under corrosive conditions K_{ISCc} 210
free-cutting steels **478**
-, required properties 478
-, -, machinability 485
-, -, mechanical properties 487
-, -, -, influence of anisotropy 488
-, -, surface quality 489
-, several measures of physical metallurgy to attain the required properties 479
-, -, chemical composition 479, 480
-, -, -, importance of sulfur and sulfide shape and of lead 479, 480, 481
-, -, -, influence of tellurium and other elements 480, 482
-, -, microstructure 479
-, -, smelting process 482
-, -, -, subsequent processing: cold drawing; hot forming; quenching and tempering 489
-, typical steel grades proven in service 485
-, -, case hardening free-cutting steels 485
-, -, low carbon free-cutting steels 485
-, -, quenched and tempered free-cutting steels 485
"full-hard" condition of flat products for cold forming 92
"full-hard" material 102

Galfan 103
Galvalume 103, 107
galvannealed sheet 103
gas rinsing/stirring aftertreatment of steel in the ladle 640, 646
general structural steels **51**
"GKZ-annealing": see spheroidizing annealing
Goss texture in electrical steel sheet 504, 505
grain orientation in electrical steel sheet 504
-, degree of orientation 505
-, -, improvements 506
grain-oriented electrical steel sheet **504**, 508, 511, 513
grain refinement 48, 85, 201
-, by controlled precipitation of aluminum nitride 85
-, of heavy forgings 698
grain size, dependence on the number of austenitic transformations 150
-, of electrical steel sheet, influence on magnetic properties 509, 511
graphite forming in elevated temperature steels 239
green rot 441, 450
grindability of tool steels 347

"Gütegruppen" of general structural steels 53

hard ferrites: see oxide materials for permanent magnets 542
hand tools, steels for them 373
hardenability 132, 135, 159, 165, 174, in particular of
-, heat - treatable steels 132, 135, 159, 165, 174
-, spring steels 475
-, steel for screws, bolts, nuts and rivets 613, 614
-, steels for cold forming 180
-, tool steels 311
hardening 37
-, of case hardening steels 165
hardening degree 125, 135
-, influence on the ratio of
-, -, tensile strength to proof stress 123
-, -, toughness to strength 125
hardening grade 122
hardness, importance for tool steels 305
-, dependence on temperature 312, 314, 315, 316
-, influence of chemical composition 316
-, influence on toughness 331
-, relationship with bending yield stress 306, 318, 319, 324
-, values of various tool steel grades 352, 356, 358, 360, 362, 364, 366, 367, 371, 372
hardness of martensite 122
hardness penetration 137, 140, 311
hardness retention of
-, ball and roller bearing steels 592
-, tool steels 314, 316
-, UHS steels 216, 225
HAZ = heat affected zone, see also welding and weldability
hearth and electric processes for crude steel making 640
heat resistance
-, definition 431
heat resisting steels **433**
-, required properties of use and their characterization 431, 437
-, -, mechanical properties 433, 435
-, -, resistance to scaling 433, 434
-, -, -, its definition 431, 434
-, several measures of physical metallurgy to attain resistance to scaling 437
-, -, chemical composition
-, -, microstructure
-, -, -, importance of chromium 437
-, typical steel grades proven in service
-, -, austenitic steels and alloys 440
-, -, ferritic steels 439
-, -, selection of heat resisting materials 444
heat transfer and solidification of steel during
-, casting into ingot molds 654, 655
-, continuous casting 662
heat treatability of steels for quenching and tempering 132
heat treatable steels **118**

-, definition and subdivision 118
-, -, case hardening steels 158
-, -, nitriding steels 119, 140
-, -, steels for flame, induction and immersion surface hardening 119, 137
-, -, steels for heavy forgings 147
-, -, steels for quenching and tempering 119, 134
-, their classification according to hardenability 135
heat treatment 707,
in addition see details for the various steel types and grades
heat treatment, short principles of
-, hardening (quenching) and tempering 709
-, heat treatment for a ferritic-pearlitic structure 708
-, heat treatment for definite tensile strength 709
-, normalizing 708
-, normalizing forming 708
-, precipitation hardening 707
-, recrystallization annealing 710
-, soft annealing 708
-, solution annealing 710
-, spheroidize annealing 708
-, stress relieving annealing 709
-, treatment for cuttability 709
heat tratement to obtain specific microstructures and properties of
-, ball and roller bearing steels 587
-, flat products for cold forming 91
-, free cutting steels 489
-, heat - treatable steels 125, 127, 137, 140, 150, 165
-, low temperature steels 281
-, normal and high strength structural steels 35
-, soft magnetic materials 497, 511
-, spring steels 473
-, stainless steels 399
-, steels for cold forming 183
-, steels for line pipe 579
-, steels for screws, bolts, nuts and rivets 613
-, steels for welded round link chains 630
-, tool steels 311, 388
-, ultra high - strength steels 211, 216
-, valve materials 458
heating conductor alloys **445**
-, required properties and their testing
-, -, electrical resistance 445, 449
-, -, other physical and mechanical properties 450
-, -, scaling resistance 445, 448
-, several measures of physical metallurgy to attain the required properties 446
-, -, chemical composition 446
-, -, microstructure 446
-, -, -, austenitic alloys 446
-, -, -, ferritic alloys 447
-, technically established heating conductor alloys 449
-, -, austenitic materials 449
-, -, ferritic steels 449

heating during aftertreatment of steel in the ladle 640, 648
heating for
-, hot forging 692
-, hot rolling 679
heavily temper-rolled steels for flat products for cold forming 99, 91
heavy forgings, special steel melting methods for them 675
-, influence on the steel properties 676
high creep resistant steels: see elevated-temperature steels
highly permeable electrical steel sheet 513
high-performance free-cutting steels 485
high-pressure hydrogen attack 425
high-pressure hydrogen resistant steels **423**
-, definition 423
-, importance of high-pressure hydrogen diffusion 424
-, measures of physical metallurgy to attain high-pressure hydrogen resistance of steel 424
-, -, chemical composition 425
-, -, -, importance of carbide forming elements 425, 426
-, -, heat treatment 427
-, processing properties 427
-, typical steel grades and their selection 427, 430
-, -, operating limit curves (Nelson-diagram) 429, 432
high-speed steels 365, 367
high strength structural steels **6**, 58, 59
-, definition 6
-, details see normal and high strength structural steels
"Hoko-acid" 396
Hollomon parameter 26
homogeneity degree, influence on toughness of heat treatable steels 125
hot charging of continuously cast steel 669, 685
hot cracking in welding of
-, heat resisting steels 441
-, low-temperature steels 293, 301
hot extrusion press tools, steels for them 363
hot formability of
-, normal and high strength structural steels 10, 25
-, tool steels 342
hot forming by forging **691**
-, aims 691
-, forging methods 691
-, -, closed-die forging 691
-, -, open-die forging 691
-, material for forging 691
-, -, forging ingots 692
-, -, their yield 692
-, working conditions in forging and defects 692, 697
-, -, conditions of heating 692
-, -, forging conditions 693
-, -, -, importance of the die form for
-, -, -, -, closing cavities 693

-, -, -, -, through - forging 694
-, -, -, -, open-die or closed-die forging, effect on the mechanical properties (above all on their anisotropy) 694
-, -, the cooling from forging heat 697
-, -, -, complete transformation 697
-, -, -, non-degassed steel 697
-, -, -, vacuum-degassed steel 697
hot forming by rolling **678**
-, fundamentals 678
-, -, aims 678
-, -, various rolling processes acc. to the product 678
-, -, -, processing sequencies 680
-, heating 679
-, -, heating conditions 679
-, -, hot feeding and direct rolling 685
-, -, influences on the microstructure 684
-, -, scaling 682
-, -, solder brittleness 684
-, -, -, influence of Cu and Sn 684
-, -, surface decarburization 682
-, -, -, influencing factors 683
-, -, -, influence on the finished products properties 684
-, rolling processes 685
-, -, deformation resistance 685
-, -, -, importance of work - hardening 686
-, -, -, influence on the mill design 686
-, -, formability 686
-, -, improved surface characteristics 688
-, -, microstructure and metal properties controll 687
-, -, yield 689
-, treatment after rolling 690
-, -, importance of surface, internal and shape defects 690
hot-rolled flat products for cold forming **77, 79**
-, mild unalloyed steels 77, 88
-, -, required properties of use and their characterization 78
-, -, -, cold formability in particular 86
-, -, several measures to attain the required properties 79
-, -, typical steel grades 79
-, normal and high strength steels 84, 88, 89
-, -, required properties of use and their characterization 84
-, -, -, above all cold formability 78, 79
-, -, several measures to attain the required properties 85
-, -, typical steel grades 89
hot rolling of steel, processing sequences 680
hot scarfing 689
"hot shortness" of hot rolled steel 684
hydrogen embrittlement of UHS steels 219
hydrogen-induced cracking 423, 578, 581
-, influence of carbon content 581
-, influence of sulfur content 581
hydrogen in
-, high - pressure hydrogen resistant steels 423

-, steels for heavy forgings 150
-, UHS - steels 219
hysteresis loops 503
-, specific shapes of soft magnetic materials 524, 528, 533

IF (interstitial free)-steel 82, 84
IGC = intergranular corrosion 383, 385
IIW-test 41
impact bending test 144
-, results on different steels after case hardening or nitriding 144
inclusion value (bearing steels) 588
induction furnace 640
ingot casting methods 651
ingot casting, comparison with continuous casting 668
ingot segregations 655, 673
-, A - segregations 656
-, V - segregations 656
inhibitors in electrical steel sheet to control
-, secondary grain growth 504, 510
-, texture 505, 510
injection processes for steel aftertreatment in the ladle 640, 647
inner oxidation in heat resisting steels 442
intergranular corrosion (IGC) 385, 386
internal cracks, formation in continuously cast strands 665
-, influence of alloying elements 666
internal defects, formation in cast steel 660
"Invar alloys" 558
"Invar Effect" 560
iron-cobalt-vanadium-chromium alloys as permanent magnet materials 541
"irreversible alloys" 560
Ito - crack parameter 41
ITRP = Ingot Top Remelting Process 675

killed steel 650, 651, 655, 660
K_{Ic} - values: see fracture toughness
knife-line corrosion 389
"Kohärazie" test 275
"Krupp"-process (wire) 203

ladle furnace (LF) 648
ladle metallurgy 646
lamellar tearing 32
Laves-phase (Fe_2Mo) in
-, elevated-temperature steels 247
lead coated flat products 111
lead, influence on the properties of
-, free-cutting steels 481, 485
lead oxide test for testing valve steels 455
Ledeburitic tool steels 343
lifting processes in steelmaking 648
"light treatment" (of steel melts) 648

liquid gas technology, application of low-temperature steels 274
local corrosion phenomena 383
long-term embrittlement of elevated-temperature steels 235
low-carbon free-cutting steels 485, 486
low energy tear 38
low pearlite steel 86, 88
low temperature steels **271**
-, required properties of use and their characterization 271, 272
-, -, general 271
-, -, in particular: toughness at low temperature and weldability 271, 272
-, several measures of physical metallurgy to attain the required properties 277
-, -, austenitic steels 282
-, -, -, chemical composition 283
-, -, -, microstructure 283, 284
-, -, ferritic steels
-, -, -, chemical composition 277
-, -, -, heat treatment 281
-, -, -, microstructure 278, 280
-, -, -, rolling 282
-, typical steel grades proven in service 287
-, -, austenitic steels 288, 292
-, -, ferritic steels 275, 287, 288
-, -, processing of the steels (above all :welding) 294
-, -, their application in liquid gas technology 274

machinability of free-cutting steels **485**, 487, 488
machining tools, steels for them 365
macrosegregations 655, 663
magnetic and other characteristics of some permanent magnet grades 536
magnetic properties, characteristics for their evaluation **492**
magnetic properties of
-, electrical steel sheet 507
-, permanent magnetic materials 536
-, soft iron grades 499, 501
-, -, importance of purity 500
-, soft magnetic nonferrous alloys 527
-, stainless chromium steels 522
-, various steel grades 515, 517, 521
magnetizability of iron-chromium-nickel alloys 550
magnetization of low-temperature austenitic steels during welding 296
magnetostriction of grain-oriented electrical steel sheet 514
manganese-aluminum materials for permanent magnets 545
manganese, influence on the properties of
-, ball and roller bearing steels 587
-, free-cutting steels 479, 482
-, -, importance of the ratio manganese to sulfur content for the

-, -, -, behaviour-during hot rolling (red shortness) 489
-, -, -, machinability 482, 483
-, -, -, solidification 485
-, elevated temperature steels 234
-, heat treatable steels 127
-, low- temperature steels
-, -, austenitic 285, 286, 293,
-, -, ferritic 278
-, non magnetizable steels 550
-, normal and high strength structural steels 42, 43, 44
-, stainless steels 396
-, steels for cold forming 181
-, steels for flat products for cold forming 85, 87, 92
-, steels for line pipe 579
-, steels for permanent-way (track) material (rails) 595, 598
-, steels for screws, bolts, nuts and rivets 612
-, steels for valves 457
-, wear resistant steels 379
-, wire rod for cold drawing 200
maraging 211
maraging steels **211**, 216, 225
-, UHS steels 224
-, stainless steels 416
martensite embrittlement in UHS steels 211
martensitic stainless steels 411
martensitic temperature M_d30, formula to calculate it 283
martensitic temperature M_S, formula to calculate it 283, 550
materials for thermobimetals 565
maximum achievable hardness 311
mechanical alloying 272
mechanical pre-strain hardening of elevated-temperature austenitic steels 249
mechanical properties of StE 355 and StE 690 in dependence on cold forming 23, 24, 25
mechanical properties of various steel grades, data in tables
-, flat products for cold forming 79, 84, 89, 95, 114, 115
-, heat resisting materials 435
-, high - pressure hydrogen resistant steels 430, 431
-, nitriding steels 146
-, non-magnetizable steels 555
-, normal and high strength structural steels 52, 56, 57, 58
-, prestressing steels 75
-, reinforcing steels 67
-, stainless steels 412
-, steels for cold forming 188, 189, 191, 193
-, steels for heavy forgings 152
-, steels for quenching and tempering 138
-, steels for screws, nuts and rivets 616, 617, 618
-, steels for surface hardening 141
-, ultra high - strength steels 224
-, valve materials 466

mechanical properties: see the various steel types and see
-, fatigue strength, general data
-, fatigue strength of various steel grades
-, mechanical properties of various steel grades, data in tables
-, notch impact energy of some steel grades
-, tensile test results for some steel grades
mechanical properties (their anisotropy) related to the ratio of stretching 696
melt-spinning technique 515
melting and casting, special methods **671**
see also casting and solidification (common methods)
-, remelting methods 670
-, -, influence on the steel properties (above all cleanness and segregations) 671, 672, 673, 674
-, special melting methods for heavy forgings 675
-, -, influence on the steel properties 676
-, vacuum melting methods 676
-, -, influence on the steel properties 677
metallic glasses as soft magnetic materials 532
microalloyed steels 87, 88, 91
microalloying 86, 92
microsegregation during solidification of steel 656, 673
microstructure, dependence on hot rolling conditions 687
microstructure diagram for stainless steels acc. to
-, Schaeffler 401
-, Strauss and Maurer 400
microstructure, importance for the properties: see the various steel types under: measures to attain the required properties. In addition: more general comments on the influence of microstructure on the properties of
-, flat products for cold forming 82, 85, 92
-, heat - treatable steels 124, 125, 129, **151**, 160, 218
-, normal and high strength structural steels 35, 40
-, prestressing steels 72
-, reinforcing steels 63
-, stainless steels 396
-, steels for cold forming 181, 182
-, steels for screws, bolts, nuts and rivets 612, 613
-, steels for wire rod for cold drawing 200
-, tool steels 380, 324, 343
"mini-ingots", formation during continuous casting 664
modulus of elasticity 557
molybdenum, influence on the properties of
-, ball and roller bearing steels 587
-, elevated-temperature steels 233, 236, 238, 246
-, heat treatable and case hardening steels 128, 154, 160
-, low-temperature steels 283, 293
-, non magnetizable steels 550, 551
-, normal and high strength structural steels 43
-, stainless steels 395, 404, 411

-, steels for cold forming 181
-, steels for flat products for cold forming 87
-, steels for permanent-way (track) material (rails) 595
-, steels for use with hydrogen at elevated temperatures and pressures 427, 429
-, steels for valves 460
-, steels with defined thermal expansion 561, 562
-, tools steels 308, 316, 326, 333, 341, 365
-, ultra high-strength (maraging)steels 217
-, wear resistant steels 374, 379
monoblock-wheels: see solid wheels
"multi-Pouring"-process 676

n-value of flat products for cold forming 83, 85, 97, 98
-, dependence on temper rolling 704, 705
NDT-temperature = nil-ductility-transition temperature (Pellini drop weight test) 21
-, values of high-strength fine-grained structural steels 21
Nelson-diagram 429, 432
neodymium-iron-materials for permanent magnets 543
nickel equivalent (stainless steels) 398
nickel, influence on the properties of
-, elevated temperature steels 234, 243, 246
-, heat resisting steels 437
-, heat treatable and case hardening steels 127, 143, 160
-, heating conductor alloys 446
-, low-temperature steels 279, 283, 285
-, non magnetizable steels 547, 550
-, normal and high strength structural steels 44
-, soft magnetic materials 523, 524, 525
-, stainless steels 395
-, steels for cold forming 181
-, steels for valves 457
-, steels fur use with hydrogen at elevated temperatures and pressures 426
-, steels with defined thermal expansion 558
-, tool steels 313
-, ultra high-strength steels 211, 214
nickel martensite, importance for UHS steels 208, 211, 216
nickel martensitic stainless steels 415
-, details see also stainless steels
niobium, influence on the properties of
-, elevated-temperature steels 234, 245
-, heat treatable steels 120
-, normal and high strength structural steels 49
-, reinforcing steels 64
-, stainless steels 386, 396, 404
-, steels for flat products for cold forming 86
-, steels for use with hydrogen at elevated temperatures and pressures 426
-, ultra high-strength steels 216
-, wear resistant steels 379
nitriding 140
nitriding steels **140**, 145

Subject Index 825

-, required properties, their characterization and measures to attain them: see steels for quenching and tempering
-, -, special aspects of nitriding steels (a.o. wear resistance and fatigue strength) 140
-, typical steel grades 146
nitrogen, influence on the properties of
-, elevated-temperature steels 233, 243, 245
-, free cutting steels 484
-, heat resisting steels 442
-, heat treatable steels 120, 122
-, low-temperature steels, austenitic 285
-, non magnetizable steels 550
-, normal and high strength structural steels 46
-, stainless steels 394, 418, 420
-, steels for valves 457
nitrogen pickup during pouring steel into ingot molds 651
nonferrous alloys as soft magnetic materials **523**
-, production of the alloys 526
-, required properties of use 523
-, several aspects of physics and physical metallurgy to attain the required properties 523
-, -, magnetic constants and ordering processes 524
-, -, phases and transformations 523
-, -, specific shapes of the hysteresis loop 524
-, some commercial nonferrous alloys
-, -, iron-cobalt alloys with
-, -, -, 27 % to 50 % Co 526
-, -, iron-nickel alloys with
-, -, -, 35 % to 40 % Ni 530
-, -, -, 45 % to 65 % Ni 529
-, -, -, 70 % to 80 % Ni 528
-, -, -, nickel contents near 30 % 530
non-grain-oriented electrical steel sheet 503, 508, 511, 512
nonmagnetizable steels **547**
-, required properties of use and their characterization 547
-, -, corrosion resistance 552
-, -, magnetic permeability 548
-, -, mechanical properties 551
-, -, -, influence of cold deformation 554, 552
-, -, polarization 548, 549
-, several measures of physical metallurgy to attain the required properties 549
-, -, chemical composition 550, 553
-, -, microstructure 550
-, typical steel grades proven in service 552
non metallic inclusions, formation in steel making 642, 657
-, endogenous inclusions 642, 657
-, exogenous inclusions 642, 657
non-metallic inclusions, influence on the properties of
-, bearing steels 587
-, -, values of steel 100 Cr 6, produced by various methods 672
-, heat treatable steels 125, 131, 150
-, low-temperature steels 281

-, spring steels 472
-, steels for cold forming 185
-, wire rod for cold - drawing 197
normal and high strength structural steels **6**, **7**
-, definition 6
-, required properties of use and their characterization 7
-, -, application properties:
-, -, -, mechanical properties under cyclic loads 8, 14
-, -, -, mechanical properties under static loads 7, 12
-, -, -, resistance against brittle fracture 9, 18
-, -, -, resistance to weathering 9, 22
-, -, -, wear resistance 9, 22
-, -, processing properties
-, -, -, formability 10, 22
-, -, -, suitability for shearing, machining and flame cutting 10, 27
-, -, -, suitability for zinc coating (galvanizing) 11, 33
-, -, -, uniformity 12
-, -, -, weldability 11, 28
-, several measures of physical metallurgy to attain the required properties 34
-, -, chemical composition and its effect on microstructure 39
-, -, heat treatment and the resulting microstructure 35
-, -, -, normalizing 35
-, -, -, quenching and tempering 36, 41
-, typical steel grades proven in service 51
-, -, fine grain structural steels 55
-, -, -, normalized 55, 58
-, -, -, quenched and tempered 59
-, -, -, thermomechanically treated 59
-, -, general structural steels 51
-, -, ship building steels 55, 56, 57
-, -, steels for colliery arches 54
-, -, steels for sheet piling 54
-, -, steels resistant to weathering 54
normalizing of normal and high strength structural steels 35
normalizing rolling (forming) 25, 36, 85
normal strength structural steels **6**, **52**
-, definition 6
-, details see normal and high strength structural steels
notched-specimen tensile strength
-, of StE 690 33
-, of UHS steels 209
notch effect in UHS steels 220
notch impact energy
-, at room temperature
-, -, 20 MnCr 5 160
-, -, 23 MnNiCrMo 5 2 626
-, -, 23 MnNiCrMo 6 4 632
-, -, 23 MnNiMoCr 6 4 626
-, -, 38 NiCrMoV 7 3 213
-, -, 41 SiNiCrMoV 7 6 213
-, -, 42 CrMo 4 124, 126

826   Subject Index

-, -, 50 CrV 4 126
-, -, Ck 45 124
-, -, St 52-3
-, -, -, influence of the sulfur content 50
-, -, StE 690
-, -, -, influence of annealing time 27
-, -, -, influence of cooling time 38
-, -, X 3 CrNiMoN 18 14 293
-, -, X 3 CrNiN 18 10 293
-, -, X 20 CrNiMo 17 2 415
-, -, X 32 NiCoCrMo 8 4 215
-, -, X 40 CrMoV 5 1
-, -, -, influence of the smelting process 326
-, -, X 41 CrMoV 5 1 215
-, -, X 55 MnCrN 18 5
-, -, -, influence of the stretch - forming degree 552
-, at elevated temperatures
-, -, X 10 CrAl 24 439
-, -, X 20 CrNiN 21 12 461, 462
-, -, X 45 CrNiW 18 9 462
-, -, X 45 CrSi 9 3 464
-, -, X 53 CrMnNiN 21 9 461, 462
-, -, X 85 CrMoV 18 2 464
-, at low temperatures
-, -, 20 MnCr 5 at - 100 °C 160
-, -, St 52-3 at - 20 °C
-, -, -, influence of the annealing temperature 27
-, -, StE 500 at - 40 °C 39
-, -, StE 690 at - 40 °C 39
-, -, StE 890 at - 40 °C 39
-, -, -, influence of the tempering temperature 39
notch impact energy of structural steels, dependence on the ratio of longitudinal rolling to transverse rolling 19
notch impact energy - temperature curves of
-, heat resisting steels
-, -, X 10 CrAl 24 439
-, -, -, influence of a 1000h - annealing 439
-, heat treatable steels
-, -, 50 CrV 4 and SAE 1340
-, -, -, influence of the microstructure 126
-, -, various SAE steels
-, -, -, influence of the alloying 130
-, low temperature steels
-, -, 12 Ni 19 290
-, -, 13 MnNi 6 3 280, 297
-, -, -, in the welded condition (HAZ) 297
-, -, TStE 355 280
-, -, X 3 CrNiMoN 18 14 293
-, -, X 3 CrNiN 18 10 293
-, -, X 8 Ni 9 280, 291
-, -, -, in the welded condition (HAZ) 298
-, normal and high strength structural steels
-, -, RSt 37-2 19, 53
-, -, St 37-3 53
-, -, St 52-3 19
-, -, StE 355 19
-, -, StE 460 19
-, -, StE 690 19
-, -, USt 37-1 53

-, -, USt 37-2 53
-, -, laboratory steels 48
-, stainless steels
-, -, X 4 CrNi 13 4 416
-, -, X 4 CrNiMo 16 5 416
-, -, various grades, scatter bands 391
-, steels for chains
-, -, 21 Mn 4 Al 625
-, -, -, influence of the annealing temperature 625
notch impact energy - transition temperature
-, general comments in connection with
-, -, heat treatable steels 127
-, -, rail steels 596
-, -, structural steels 18
-, test results on
-, -, 11 MnNi 5 3 282
-, -, 26 NiCrMoV 14 5 186
-, -, Cr-Mo-V-steels
-, -, -, influence of molybdenum 154
-, -, Cr-Ni-Mo-V-steels
-, -, -, influence of chromium and nickel 153
-, -, StE 355
-, -, -, influence of cold forming 25
-, -, StE 500,
-, -, StE 690 and
-, -, StE 890 in the welded condition 31, 32
-, -, StE 690
-, -, -, influence of the annealing temperature 27

open coil annealing 83, 703
open hearth furnace 640
Ostwald ripening 310
over-aging of UHS steels
overheating (during heat treatment) 709
overtiming (during heat treatment) 709
oxide inclusions in steel, formation during solidification 657
-, endogenous inclusions 657
-, exogenous inclusions 657
oxide materials (hard ferrites) as permanent magnet materials 542
oxygen blowing process for crude steel making 640

partly recrystallized steels for flat products for cold forming 88, 91, 92, 95
passive layer (importance for chemical resistance of stainless steels) 384
patenting treatment of wire rod for cold drawing 201
PCHR = Plasma Cold Hearth Refining process 677
pearlite content, effect on the notch impact toughness transition temperature 40
pearlite formation (shape) in rail steels 596
pearlite-free fine-grained steel 86
Pellini drop weight tear test 21
permanent magnet materials **534**
-, applications 545

Subject Index   827

-, required properties of use 534
-, -, magnetic properties, above all high coercivity 534, 535
-, -, other characteristics 535, **536**
-, several measures of physical metallurgy and production to attain the properties of
-, -, high-iron materials
-, -, -, pure iron or iron with 30 % Co (ESD magnets) 539
-, -, -, steels 538
-, -, medium-iron materials
-, -, -, Al-Ni and Al-Ni-Co alloys 537
-, -, -, Cr-Fe-Co alloys 541
-, -, -, Cu-Ni-Fe alloys 542
-, -, -, Fe-Co-V-Cr alloys 541
-, -, -, Nd-Fe-B materials 543
-, -, -, oxyde materials (hard ferrites) 542
-, -, several materials containing little or no iron
-, -, -, Mn-Al materials 545
-, -, -, Pt-Co materials 544
-, -, -, rare earth - Co materials 544
-, some grades of permanent magnetic materials 536
Perrin effect 646
phase diagram for
-, iron - chromium 397, 399, 438
-, iron - nickel 398
-, iron - chromium-nickel 406, 555
-, -, at 550 °C and 650 °C 447
-, -, at 800 °C 441
-, -, at 1200 °C 407
-, iron - chromium - aluminium at 700 °C 447
-, iron - chromium - carbon 400
-, nickel - aluminium 252
phase diagram for stainless steels 396
phosphated flat products 114
phosphorus, influence on the properties of
-, free-cutting steels 484
-, low-temperature steels, ferritic 281
-, normal and high strength structural steels 50
-, steels for flat products for cold forming 91, 92, 94, 95, 96
-, steels for welded round link chains 628
phosphorus steels (high strength steels with increased phosphorus content) for cold forming 88, 91, 92, 95, 97
pig iron 638
-, refining steps 639
pipelines fracture behaviour 576
pipes in steel, formation during solidification 660
pitting corrosion 385, 387
pitting potential of stainless steels 389
pitting resistance equivalent 395
pitting tendency at the surface of case hardened steels 162
plastics molds, tool steels for them 350
plate production
-, production steps and testing for quality management 724, 725
platinum-cobalt materials for permanent magnets 544

polishability of tool steels 349
pouring stream, its shielding during
-, ingot casting 652
-, continuous casting 661
powder-metallurgy application to produce
-, high-temperature alloys 270
-, permanent magnet materials 539, 542, 543
-, tool steels 325, 327, 368, 370
precipitation cracks 235
precipitation hardening of
-, elevated - temperature steels 240, 248
-, flat products for cold forming 86, 92
-, high-strength structural steels 39, 46
-, spring steels 477
-, stainless steels 402, 405
-, steels for quenching and tempering 120
-, ultra high-strength steels 216
precision cast alloys for elevated-temperature service 269, 272
preferencial (selective) corrosion 385
pressure casting molds, tool steels for them 354
prestressing steels **67**, 76
-, required properties of use and their characterization 67, 70
-, -, bonding properties 70
-, -, mechanical properties 67, 73
-, -, shape 73, 74
-, several measures of physical metallurgy to attain the required properties 72
-, -, chemical composition 73
-, -, microstructure 73
-, typical steel grades proven in service 76
precision cast alloys 269
process steps in steelmaking 642
processes for crude steel making 640
production processes, influence on the steel properties, overview **635**
-, casting methods 636
-, heat treatment 636
-, hot and cold working 636
-, steelmaking technologies to produce the desired chemical composition 635
production steps for
-, fabrication of cold strip 700, 701
-, fully processed electrical steel sheet 510
-, hot rolling of various steel products 680
purity degree: see cleanness
purity of soft iron as soft magnetic material 500
PVD = physical vapour deposition (coating of tools) 332

quality control: see quality management
quality groups: see "Gütegruppen"
quality inspection: .712
quality management **711**
-, definition and general 711
-, measures 711
-, -, planning 711
-, -, -, aim 711
-, -, -, documentation 712
-, -, -, quality characteristics 712

828   Subject Index

-, -, quality inspection 712
-, -, -, testing during production 712
-, -, -, testing of the final product 713
-, -, the quality control 713
-, -, -, application of statistical methods 713
quality management in
-, steel making 718
-, -, production steps and critical points for measurements 720
-, deforming of crude steel 719
-, -, testing quality characteristics 720
-, -, maintaining the material identity 721
-, steel bar production 721
-, -, production steps and testing sequence 722
-, plate production 724
-, -, production steps and testing sequence 725
quality planning: see quality management
quenched and tempered steels 88, 91, 93, **119**, 125, 134, 208, 341, 458, 473, 485, 489, 579, 615
quenching and tempering of high-strength structural steels 36, 41

r-value, dependence on
-, cold rolling 702
-, temper rolling 705
r-value of flat products for cold forming 83, 85, 92, 97, 98
rail steels: see steels for permanent-way (track) material
rail strength, influence of rail shape 599
Rajacovicz evaluation (creep) 230
rare earth metal-cobalt materials for permanent magnets 544
raw materials for steelmaking 638
rebend test (reinforcing steels) 63
RECo magnets 544
red shortness of free-cutting steels 484
reduction steps for the production of metallic iron materials from iron ores 639
refining of steel melts 643
refining steps (pig iron) 639
reheat-cracking 235
reinforcing steels **62**, 68
-, required properties of use and their characterization 62
-, -, bonding properties 63
-, -, formability 62, 63
-, -, mechanical properties 62, 64
-, -, shape 65
-, -, weldability 62, 63
-, several measures of physical metallurgy to attain the required properties 63
-, -, chemical composition 63
-, -, microstructure 63
-, typical steel grades proven in service 66
relaxation behaviour of
-, elevated - temperature steels 226, 231, 236
-, prestressing steels 71
relaxation test 229, 231
remelting methods 670

-, schematic representation 671
-, influence on the steel properties (above all cleanness and segregations) 671, 672, 673, 674
reoxidation of steel during pouring 651
-, casting into ingot molds 651
-, continuous casting 661
-, -, measures to prevent it during continuous casting 661
residual (internal) stresses of different types of steel after heat treatment 313
resistance to
-, alternating temperatures of heat resisting steels 437
-, brittle fracture 18
-, its testing 18
-, high-pressure hydrogen 426
-, -, influence of chemical composition 426
-, scaling
-, -, definition 431, 434, 435
-, thermal fatigue of tool steels 334
-, thermal shock of heat resisting steels 437
RH = Ruhrstahl-Heraeus process 648
RH-OB = Ruhrstahl Heraeus-Oxygen Blowing 649
RHO = Ruhrstahl Heraeus Oxygen process 649
rimmed steel 650, 651, 656, 660
Robertson test (see also CAT = crack arrest temperature) 19, 20, 275
rolling: see hot forming by rolling
roughness of temper rolled steel sheet 705, 706

"Salt-Bath-Patenting" (wire) 203
Sandelin-effect 102
"sandwich materials" 117
scale loss of heat resisting steels 434, 442
scaling during heating for hot rolling 682
scaling resistance 242, 438
-, limiting temperatures 435
scattering of creep-rupture test results 229
SCC = stress corrosion cracking 383
Schaeffler-diagram (microstructure of stainless steels) 401
"Schloemann"-process (wire) 203
secondary hardening 127 of
-, structural steels 25
-, tool steels 319
-, UHS steels 215, 216
secondary hardness 319
segregations in
-, continuously cast slabs 663
-, -, influence of electromagnetic stirring 665
-, ingots 655
-, -, influence of the ingot shape 655, 656
-, steels for heavy forgings 149
-, tool steels 324
-, wire rod for cold drawing 200
selenium
-, in stainless steels 396
"semi-free-cutting steels" 485
semikilled steel 650, 651, 660

Subject Index   829

"sendust" (soft magnetic material) 532
Sendzimir process 699, 703
sensitivity to pick up of sulphur (austenitic alloys for heating conductors) 450
shape
-, anisotropy 535
-, stability (tool) 306, 317
shipbuilding steels 55, 56, 57
side blowing converter 640
sigma-phase 437, 438
-, embrittlement by it 410, 432, 438, 440
sigma-phase in
-, elevated-temperature austenitic steels 243
-, heat resisting steels 437, 440
-, stainless steels 410, 432
single hardening of case hardening steels 165
silicon, influence on the properties of
-, low-temperature steels, ferritic 278
-, normal and high strength structural steels 42
-, soft magnetic materials 504, 514, 515
-, steels for cold forming 100, 181
-, ultra high-strength steels 213
skin pass rolling (temper rolling) 83
slenderness degree, influence on stability behavior 14, 15
soft iron
-, for heavy forgings 499
-, for relays 499
-, softmagnetic material **494**
soft magnetic materials **491**
-, definition and classification 491
-, -, characteristics for the evaluation of soft magnetic materials **492**
-, materials
-, -, electrical steel sheet 501
-, -, metallic glasses 532
-, -, non ferrous alloys 523
-, -, soft iron 494
-, -, special iron alloys 532
-, -, various steels 515
-, required properties of use 494
-, -, in particular: magnetic properties 499
-, several findings and measures of physics and physical metallurgy to influence
-, -, anisotropy energies **496**
-, -, coercivity **495**
-, -, saturation polarization 494
-, several measures of production technique to attain the required properties 497
-, soft iron grades proven in service 499
-, -, relay materials 499
-, -, soft iron for heavy forgings 499
soft magnetic steels, various types and grades **515** see also soft magnetic materials
-, required properties of use 515
-, -, influence of chemical composition 516, 519
-, -, influence of microstructure 516, 518
-, steel grades for
-, -, castings 520
-, -, corrosive conditions (stainless ferritic chromium steels) 521

-, -, generator shafts 516
-, -, pole sheets 520
solder brittleness 642
-, of hot rolled steel 684
-, -, influence of the copper and tin content 684
solidification of steel **650**
-, cast into ingot molds 650, 653
-, continuously cast 650, 662
solidification structure of steel 653
-, influence of oxygen concentration 660
solid wheels 604, 605
special iron alloys as soft magnetic materials 532
spheroidizing annealing **183**, 613, 615
-, influence on the
-, -, flow curve 183, 613
-, -, formability 183, 613, 614, 708
-, -, properties of
-, -, -, steels for cold forming 183, 194
-, -, -, steels for screws, nuts and rivets 613, 615, 708
spheroidizing degree 184
sponge iron 638
spring steels **468**
-, required properties of use 468
-, -, mechanical properties, above all fatigue strength 468, 470, 471
-, several measures of physical metallurgy to attain the required properties 469
-, -, chemical composition 469
-, -, microstructure 469
-, typical steel grades proven in service 472
-, -, elevated-temperature strength steels for springs 475
-, -, spring steels with good thoughness at low temperatures 476
-, -, stainless steels for springs 476
-, -, steels for cold formed springs 473
-, -, steels for quenched and tempered springs 473
"Stainless Invar" 560
stainless steels **382**
-, classification 397
-, required properties of use and their characterization 382, 383
-, -, corrosion resistance 383
-, -, -, its testing 384
-, -, mechanical and technological properties 383
-, -, -, their testing
-, several measures of physical metallurgy to attain the properties, above all corrosion resistance 393
-, -, chemical composition 393, **395**
-, -, -, decisive importance of chromium 393
-, -, -, phase diagrams for stainless steels: iron-chromium 397, 399, iron-nickel 398, iron-chromium-carbon 400, iron-nickel-chromium 406, 407
-, -, microstructure 393
-, -, -, heat treatment to produce the characterizing microstructure groups 399
-, -, -, structure diagrams acc. to Strauss and Maurer and acc. to Schaeffler

-, typical steel grades proven in service 408
-, -, austenitic steels 416
-, -, ferritic steels 408
-, -, ferritic-austenitic steels 421
-, -, martensitic steels (incl. maraging steels) 411
stainless steels classification acc. to microstructure **397**
-, austenitic 402, 416
-, ferritic 399, 408
-, ferritic-austenitic 404, 421
-, martensitic 399, 411
stainless steels for flat products for cold forming **96**
statistical evaluation methods, application in quality control 713
steel aftertreatment, processes and operating elements 647
steel bar production
-, production steps and testing for quality management 721, 722
steel castings for
-, elevated - temperature materials 261, 266, 269
-, high - pressure hydrogen resistant steels 429
-, magnetic applications 520
steel cord wire 207
steel grade (see also chemical composition of various steel grades, data in tables)
-, 10 Ni 14 288
-, -, TTT-diagram for continuous cooling 290
-, -, yield stress at 20 °C and low temperatures 277
-, 11 MnNi 5 3 288
-, -, notch impact energy - transition temperature 282
-, -, tensile test results 277, 282
-, -, -, at room temperature 282
-, -, -, at low temperatures 277
-, 12 Ni 19
-, -, A_v-T- curve 291
-, -, TTT-diagram for continuous cooling 291
-, -, yield stress at 20 °C and low temperatures 277
-, 13 MnCrB 5 183, 189
-, -, end-quench hardness curve 183
-, -, flow curve 183
-, 13 MnNi 6 3 288
-, -, A_v-T- curve 280
-, -, -, HAZ 297
-, -, TTT-diagram for continuous cooling 289
-, -, yield stress at 20 °C and low temperatures 277
-, 15 CrNi 6 175, 627
-, -, Wöhler-curve 163
-, 16 Mn Cr 5 175, 183, 616
-, -, end-quench hardness curve 183
-, -, flow curve 179, 183, 614
-, -, TTT-diagram for continuous cooling 166
-, -, Wöhler-curve 173
-, 17 NiCrMo 14 591
-, -, Wöhler curve 585
-, 20 MnCr 5

-, -, notch impact energy at 20 °C and at 100 °C 160
-, -, tempering diagram 160
-, 20 MoCr 4 175
-, -, flow curve 614
-, 21 CrMoV 5 7
-, -, creep behaviour 232
-, 21 Mn 4 Al 627
-, -, A_v-T- curve 625
-, 23 MnNiCrMo 5 2 627
-, -, notch impact energy at 20 °C 626
-, 23 MnNiCrMo 6 4 628
-, -, notch impact energy at 20 °C 628, 632
-, -, tempering diagram 632
-, 23 MnNiMoCr 6 4 628
-, -, notch impact energy at 20 °C 626
-, 26 NiCrMoV 8 5 152
-, -, fracture toughness (K_{Ic} values) 157
-, 26 NiCrMoV 14 5 152
-, -, fracture toughness (K_{Ic} values) 157
-, -, magnetic flux density 157
-, -, notch impact energy-transition temperatures 156
-, -, -, influence of microstructure 157
-, 26 NiMoV 14 5 157
-, -, fracture toughness (K_{Ic} values) 157
-, 28 NiCrMoV 8 5
-, -, magnetic flux density 518
-, -, -, influence of microstructure 518
-, 35 B 2 191, 617
-, -, end-quench hardness curve 614
-, -, flow curve 614
-, 38 Cr 2 191
-, -, flow curve 614
-, 38 NiCrMoV 73 223
-, -, notch impact energy at 20 °C 213
-, -, tempering diagram 213
-, 41 Cr 4 136, 138, 191, 617
-, -, flow curve 179
-, 41 SiNiCrMoV 7 6 223
-, -, notch impact energy at 20 °C 213
-, -, tempering diagram 213
-, 42 CrMo 4 136, 138, 145, 152, 380, 617
-, -, end-quench hardness curve 123
-, -, flow curve 179
-, -, notch impact energy at 20 °C 124, 126
-, -, tensile test results 124
-, 50 CrV 4 474
-, -, A_v-T- curve 126
-, -, end-quench hardness curve 475
-, -, notch impact energy at 20 °C
-, -, tensile test results 126
-, -, -, influence of the microstructure 126
-, -, Wöhler curve 471
-, 50 Si 7 474
-, -, end-quench hardness curve 475
-, 51 CrMoV 4 474
-, -, end-quench-hardness curve 475
-, 55 Cr 3 474
-, -, end-quench hardness curve 475
-, 55 Si 7 474

Subject Index   831

-, -, end-quench hardness curve 475
-, 65 Si 7 474
-, -, end-quench hardness curve 475
-, 100 Cr 6 366, 380, 591
-, -, time-temperature - dissolution diagram for continuous heating 590
-, -, Wöhler curve 585
-, BSt 420 S 67
-, -, tensile test results 64
-, BSt 500 S 67
-, -, tensile test results 64
-, chromium steels with 13% Cr
-, -, tensile test results, influence of the carbon content 403
-, Cq 15 617
-, -, flow curve 179, 614
-, Cq 35 191
-, -, end-quench hardness curve 614
-, -, flow curve 179, 614
-, Ck 45 136, 145
-, -, notch impact energy at 20 °C 124
-, -, tensile test results 124
-, -, -, influence of the microstructure 124
-, Cq 45 179
-, -, flow curve 179
-, dual-phase steel 88
-, -, tensile test results 97
-, phosphorns steel (steel with increased phosphorns content)
-, -, tensile test results 97
-, RSt 37-2 52, 88
-, -, A_V-T- curve 53
-, S 6-5-2 360, 366, 369
-, -, Wöhler curve 320
-, -, -, influence of the smelting process 320
-, ST 37-3 52, 88
-, -, A_V-T - curve 19, 53
-, -, fatigue strength 16
-, -, stress - strain curve 13
-, -, tensile test results (frequency curves) 715
-, St 52-3 52, 88, 380
-, -, A_V-T- curve 19
-, -, fatigue strength 16
-, -, notch impact energy at -20 °C 27
-, -, -, effect of sulfur content 50
-, -, stress - strain curve 13
-, -, tensile test results 27
-, St 835/1030 75
-, -, stress - strain curve 69
-, St 1080/1230 75
-, -, fatigue strength diagram (acc. to Smith) 72
-, -, stress - strain curve 69
-, St 1420/1570 75
-, -, fatigue strength diagram (acc. to Smith) 72
-, -, stress - strain curve 69
-, -, tempering diagram 73
-, St 1470/1670 75
-, -, stress - strain curve 69
-, St 1570/1770 75
-, -, stress - strain curve 69
-, StE 355 58

-, -, A_V-T- curve 19
-, -, notch impact energy-transition temperature 25
-, -, -, influence of cold forming 25
-, -, tensile test results 24
-, -, -, influence of cold forming 25
-, -, -, influence of sulfur and sulfide shape (only reduction of area) 51
-, StE 460 58
-, -, A_V-T- curve 19
-, -, stress - strain curve 13
-, StE 500 58
-, -, notch impact energy at -40 °C 39
-, -, notch impact energy - transition temperature (HAZ) 31
-, -, tempering diagram (yield stress) 39
-, StE 690 59, 380
-, -, A_V -T- curve 19
-, -, fatigue strength 16
-, -, notch impact energy at -40 °C 39
-, -, notch impact energy - transition temperature 27
-, -, -, HAZ 31
-, -, stress - strain curve 13
-, -, tempering diagram (yield stress) 39
-, -, tensile test results 24, 39
-, -, -, influence of cold forming 24
-, -, TTT - diagram for continuous cooling 37, 38
-, -, Wöhler curve (welded joints) 18
-, StE 890 59
-, -, notch impact energy at -40 °C 39
-, -, notch impact energy transition temperature (HAZ) 31
-, -, tempering diagram (yield stress) 39
-, TStE 355 288
-, -, A_V-T - curve 280
-, -, tensile test results (only yield stress) at low temperatures 277
-, USt 37-1 88
-, -, A_V-T- curve 53
-, USt 37-2 52, 88
-, -, A_V-T- curve 53
-, X I CrMo 18 2 619
-, -, true stress - true strain curve 619
-, X 2 CrNi 18 9
-, -, tensile test results 420
-, -, -, influence of cold forming 420
-, X 2 CrNiCu 18 9 3
-, -, tensile test results 420
-, -, -, influence of cold forming 420
-, X 2 CrNiMoN 18 13 193
-, -, magnetic properties 419
-, -, -, influence of cold forming 419
-, -, tensile test results 419
-, -, -, influence of cold forming 419
-, X 2 CrNiN 25 7 407
-, -, TTT - diagram for continuous cooling 407
-, X 2 NiCr 18 16 619
-, -, tensile test results 420
-, -, -, influence of cold forming 420
-, -, true stress - true strain curve 420

-, X 3 CrNiMoN 18 14 288
-, -, A$_V$-T- curve 293
-, -, notch impact energy at 20 °C and low temperatures 29
-, -, tensile test results at room and low temperatures 293
-, X 3 CrNiN 1810 288
-, -, A$_V$-T- curve 293
-, -, notch impact energy at 20 °C and low temperatures 293
-, -, tensile test results at room and low temperatures 277, 293
-, X 4 CrNiMo 16 5 409, 413
-, -, A$_V$-T - curve 416
-, -, tempering diagram 417
-, X 4 CrNi 13 4 409, 412
-, -, A$_V$-T - curve 416
-, X 5 CrNi 18 9
-, -, magnetic properties 419
-, -, -, influence of cold forming 419
-, -, tensile test results 419, 420
-, -, -, influence of cold forming 419, 420
-, X 5 CrNiMo18 10 193, 618
-, -, true stress - true strain curve 619
-, X 5 CrNiTi 26 15 265
-, -, creep rupture strength 266
-, X 6 CrNi 18 11 265
-, -, creep rupture strength 266, 267
-, X 6 CrNiMo 17 13 265
-, -, creep rupture strength 266, 267
-, X 6 CrNiWNb 16 16 265
-, -, creep rupture strength 266
-, X 7 Cr 13 409
-, -, tensile test results 403
-, X 7 CrAl 13
-, -, tensile test results 403
-, X 8CrNiMoBNb 16 16 265
-, -, creep rupture strength 250, 266, 267
-, X 8 CrNiMoNb 16 16 265
-, -, creep rupture strength 260, 267
-, X 8 CrNiMoVNb 16 13 265
-, -, creep rupture strength 266
-, X 8 CrNiNb 16 13 265
-, -, creep rupture strength 250, 266
-, X 8 Ni 9 288, 292
-, -, A$_V$-T - curve 280, 291
-, -, -, HAZ 299
-, -, tensile test results at low temperatures (only yield stress) 277
-, -, TTT- diagram for continuous cooling 292
-, X 10 Cr 13 618
-, -, tensile test results 403
-, X 10 CrAl 24 435
-, -, A$_V$-T - curve 439
-, -, notch impact energy at elevated temperatures 439
-, X 10 CrNiNb 18 9 390
-, -, Wöhler curve 390
-, -, -, influence of corrosive attack 390
-, 10 CrNiTi 18 9 619
-, -, true stress - true strain curve 619

-, X 10 CrNiTi 18 10
-, -, tensile test results at low temperatures (only yield stress) 277
-, X 10 NiCrAlTi 32 20 265
-, -, creep rupture strength 266
-, X 12 CrCoNi 21 20 458
-, -, creep rupture strength 266, 466
-, X 15 Cr 13
-, -, tensile test results 403
-, X 20 Cr 13 409, 412
-, -, tempering diagram 394
-, -, tensile test results 403
-, X 20 CrMoWV 12 1
-, -, creep rupture strength 230
-, X 22 CrMoV 12 1
-, -, creep rupture strength 230
-, X 20 CrNiMo 17 2 409
-, -, notch impact energy at 20°C 415
-, -, tempering diagram 415
-, X 20 CrNiN 21 12 458
-, -, notch impact energy at 20°C 461, 462
-, -, tensile test results at 20°C 461,462
-, -, -, at elevated temperatures 465
-, X 30 Cr 13
-, -, tensile test results 403
-, X 30 WCrV 5 3 356, 364
-, -, TTT-diagram for continuous cooling 328
-, X 32 CrMoV 3 3 356, 363, 364
-, -, TTT-diagram for continuous cooling 328
-, X 32 NiCoCrMo 8 4 223
-, -, notch impact energy at 20°C 215
-, -, tempering diagram 215
-, X 38 CrMoV 5 1 353, 362, 364
-, -, TTT-diagram for continuous cooling 328
-, X 40 Cr 13 409, 412
-, -, corrosion weight loss curves 386
-, -, tensile test results 403
-, X 40 CrMoV 5 1 356, 360
-, -, notch impact energy at 20°C 326
-, -, -, influence of smelting process 326
-, X 40 CrMoV 5 3
-, -, TTT-diagram for continuous cooling 328
-, X 40 CrNiCoNb 17 13
-, -, creep rupture strength 266
-, X 41 CrMoV 5 1 223
-, -, fatigue strength 221
-, -, fracture toughness (K$_{Ic}$-values) 219
-, -, notch impact energy at 20°C 215
-, -, tempering diagram 215
-, X 45 CrNiW 18 9 458
-, -, creep rupture strength 466
-, -, notch impact energy at 20 °C 462
-, -, tensile test results at 20 °C 462
-, -, -, at elevated temperatures 465
-, X 45 CrSi 9 3 458
-, -, creep rupture strength 466
-, -, notch impact energy at 20 °C and at elevated temperatures 464
-, -, tempering diagram 460
-, -, tensile test results at 20°C and at elevated temperatures 464

Subject Index 833

-, -, TTT-diagram for continuous cooling 459
-, X 53 CrMnNiN 21 9 458
-, -, creep rupture strength 466
-, -, notch impact energy at 20°C 461, 462
-, -, tensile test results at 20 °C 461, 462
-, -, -, at elevated temperatures 466
-, X 55 MnCrN 18 5 554
-, -, notch impact energy at 20°C 552
-, -, -, influence of the stretch-forming degree 552
-, -, tensile test results 552
-, -, -, influence of the stretch-forming degree 552
-, X 60 Cr 13
-, -, tensile test results 403
-, X 60 CrMnMoVNbN 21 10 458
-, -, creep rupture strength 466
-, X 85 CrMoV 18 2 458
-, -, notch impact energy at 20°C and at elevated temperatures 464
-, -, tempering diagram 460
-, -, tensile test results at 20°C and at elevated temperatures 464
-, -, TTT-diagram for continuous cooling 459
steel making: see crude steel production
steel making
-, production steps and testing for quality management 718, 720
steels for cold extrusion: see steels for cold forming
steels for cold forming **177**, 178
-, required properties and their characterization 177
-, -, cold formability 177, 178
-, -, mechanical properties 178
-, -, surface quality 178, 179, 186
-, several measures to attain the required properties 180
-, -, chemical composition 181
-, -, cleanness 185
-, -, heat treatment 183
-, -, microstructure 182
-, typical steel grades proven in service 187, 188, 189, 191, 193
steels for cold heading: see steels for cold forming
steels for colliery arches 54
steels for flame, induction and immersion surface hardening **137**, 141
-, required properties, their characterization and measures to attain them: see steels for quenching and tempering
-, -, special aspects of steels for flame, induction and immersion surface hardening (a.o. wear resistance and fatigue strength) 137, 139
-, typical steel grades 140, 141
steels for heavy forgings **147**, 152
-, required properties of use and their characterization 147
-, several measures to attain the required properties 148
-, -, chemical composition 151
-, -, heat treatment 150
-, -, melting and casting 149

-, -, microstructure 151
-, typical steel grades proven in service 152, 155
steels for hot rolled, cold rolled and surface treated flat products for cold forming 77, 80, 97
steels for line pipe **573**
-, development 582
-, required properties of use and their characterization 573, 574
-, -, resistance to (corrosion) cracking 574, 578
-, -, strength 573, 574
-, -, toughness 573, 575
-, -, weldability 574, 576
-, several measures of physical metallurgy to attain the required properties 578
-, -, chemical composition 579
-, -, -, importance of desulfurization 579
-, -, microstructure 578
-, -, -, influencing it by
    quenching and tempering 579
    thermomechanical rolling 578
-, typical steel grades proven in service 581
steels for permanent-way (track) material **593**
-, definition 593
-, required properties of use and their characterization 593, 594
-, -, importance of mechanical properties 593
-, -, -, dependence on shape 594, 599
-, -, wear resistance 594, 595
-, several measures of physical metallurgy to attain the required properties 595
-, -, chemical composition 595
-, -, microstructure 595
-, typical steel grades and their application 599
-, -, rail steels (also for switches) 599, 560
-, -, steels for fishplates 601
-, -, steels for soleplates 601
steels for quenching and tempering **119**, **134**, 136
-, required properties of use and their characterization 119
-, -, cold formability 133
-, -, heat treatability (hardenability) 132
-, -, machinability 132
-, -, strength properties under cyclic loads (fatigue) 129
-, -, strength properties under static stress 120
-, -, toughness properties 124
-, -, weldability 133
-, several measures to attain the required properties 120
-, -, chemical composition 127, 133
-, -, heat treatment 120, 127
-, -, microstructure 120, 124, 129
-, typical steel grades **134**, 136
steels for rolling stock **602**
-, definition 602
-, required properties of use and their characterization 602, 603
-, -, mechanical properties 602
-, -, resistance to heat cracking 602
-, -, resistance to wear 602

-, several measures to attain the required properties by
-, -, chemical composition 604 and
-, -, microstructure 604, 605 of
-, -, -, solid wheels 604, 605
-, -, -, wheel discs 604
-, -, -, wheelset shafts 606
-, -, -, wheel tires 604
-, -, overcoming of the contradictory demands for resistance to wear and resistance to heat cracking 605
-, typical steel grades proven in service for
-, -, solid wheels 606
-, -, wheel discs 606
-, -, wheelset shafts 606
-, -, wheel tires 606
steels for screws, bolts, nuts and rivets **608**
-, required properties of use and their characterization 608, 610
-, -, cold formability 609, 610
-, -, corrosion resistance 609
-, -, hot formability 609, 610
-, -, machinability 610
-, -, mechanical properties 611
-, -, suitability for resistance welding 609
-, several measures of physical metallurgy to attain the required properties 612
-, -, importance of the microstructure 612
-, typical steel grades proven in service 615
-, -, case hardening steels 615, 617
-, -, non-heat-treatable steels 615, 616
-, -, stainless steels 615, 618
-, -, steels for quenching and tempering 615, 617
steels for sheet piling 54
steels for use with hydrogen at elevated temperatures and pressures: see high-pressure hydrogen resistant steels
steels for valves in internal combustion engines: see valve materials
steels for welded round link chains **621**
-, required properties of use and their characterization 621, 624
-, -, formability 623, 626
-, -, strength properties 621, 624
-, -, -, tensile strength in comparison to the working stress limit of chains 622
-, -, suitability for cold shearing 624, 626
-, -, susceptibility to strain aging 623, 626
-, -, toughness 623, 625
-, -, wear resistance 624, 625
-, -, weldability 624, 626
-, several measures to attain the required properties 626
-, -, importance of heat treatment to produce adequate microstructures 627, 630
-, -, importance of the chemical composition, above all of the carbon content 627, 628, 630
-, typical steel grades proven in service 629
-, -, applications 629, 631
-, -, classification 629
steels resistant to weathering 46, 54, 85

steels (with Co) as permanent magnetic materials 538
steels with defined thermal expansion and special elastic properties **556**
-, applications 556
-, correlation between the two properties 561
-, properties and their characterization by test results 556
-, -, coefficient of expansion 556
-, -, modulus of elasticity (Young's modulus) 557
-, some considerations of physical metallurgy 558
-, -, importance of lattice and microstructure 558
-, -, -, dependence on chemical composition 558, 560
-, technically proven materials 562, 563
-, -, constant-modulus alloy 565
-, -, materials for thermobimetals 565
-, -, materials with special thermal expansion 562
steels with good electrical conductivity **566**
-, applications 566
-, properties of use and their characterization 566
-, -, electrical conductivity 567, 568
-, -, tensile strength 567
-, several measures of metallurgy and physical metallurgy to attain a good conductivity 569
-, -, chemical composition 569
-, -, heat treatment 572
-, -, microstructure 569
-, typical steel grades 572
steels with special elastic properties 556
"Stelmor"-process (wire) 203
sticking scale 682
stirring, electromagnetic, during continuous casting 664
-, influence on the steel properties 665
strain-aging treatment 218
-, susceptibility of steels for welded round link chains 627
strain hardening 93
-, exponent n 81
-, exponent n_m 80, 97
Strauss and Maurer-diagram (microstructure of stainless steels) 400
strengthening mechanisms 34
strength properties: see mechanical properties
stress corrosion cracking (SCC) 385, **387**, 395, (intergranular 385, transgranular 385) of
-, high-pressure hydrogen resistant steels 423
-, non magnetizable steels 553
-, prestressing steels 71, 74
-, stainless steels 387
-, steels for line pipe 574, 578
-, UHS steels 210, **219**
stress relief annealing, influence on the properties of
-, case hardening steels
-, elevated - temperature steels 235
-, low - temperature steels 298
-, normal and high strength structural steels 23, 24, 26
stress-relief cracking 235

stress - strain - curves of
-, deep drawing steel, Al-killed 94
-, -, influence of annealing 94
-, dual - phase steel 94
-, precipitation hardened steel 94
-, prestressing steels 69
-, solid solution strengthened steel 94
-, steels with different strength 13
-, St 37-3 13
-, St 52-3 13
-, St 835/1030 69
-, St 1080/1230 69
-, St 1420/1570 69
-, St 1470/1670 69
-, St 1570/1770 69
-, StE 460 13
-, StE 690 13
strontium ferrites for permanent magnets 542, 545
structural steels 6
-, definition 6
sub-zero cooling of UHS steels 218
sulfidation 432
sulfide inclusions in steel, formation during solidification 658
-, possibilities to influence the kind and shape 658, 659
sulfides, influence on the properties of
-, flat products for cold forming 85
-, free-cutting steels 480, 484, 487
-, heat treatable steels 150
-, normal and high strength structural steels 50
-, steels for cold forming 186
sulfide shape, influence on the
-, machinability of free-cutting steels 479, 480
-, toughness properties 51, 85, 150
sulfur, influence on the properties of
-, free-cutting steels 479, 485
-, low-temperature steels 281
-, normal and high strength structural steels 50
-, stainless steels 396
-, steels for line pipe 580
-, steels for welded round link chains 628
-, tool steels 347
superferritic stainless steels 399, 410
"superfine grain" of UHS steels 218
superheat
-, casting into ingot molds 652
-, continuous casting 662
"Superinvar" 560
surface characteristics, dependence on hot rolling conditions 685
surface condition of flat products for cold forming 81
surface decarburization during heating for hot rolling 682
-, influence of alloying elements 683
-, influence of the conditioning of the billet surface 683
-, influence on the properties of the finished products 684

surface defects, formation on
-, continuously cast steel 666
-, on ingot castings 652
surface factor m in fatigue tests of spring steel
surface hardenability of case hardening steels 159
surface hardening 137, 139
surface oxidation during carburizing of case hardening steels 164
surface protection of valves 457, 464
surface quality, importance for
-, flat products for cold forming 80, 81
-, steels for cold forming 179, 186
-, steels for screws, bolts, nuts and rivets 609
-, wire rod for cold drawing 197
surface treated flat products for cold forming **97**
-, treatment with metallic coatings of
-, -, aluminum 105
-, -, chromium 109
-, -, lead 111
-, -, tin 109
-, -, various metals 112
 (electrolytically - deposited 113,
 deposited by cladding 113)
-, -, zinc 99
-, treatment with inorganic coatings of
-, -, chromate 115
-, -, enamel 115
-, -, phosphate 114
-, treatment with organic coatings (coil-coated steel flat products) 116
switches, steels for them 599

TEC = linear coefficient of thermal expansion 556
-, values 563
Tekken test 47
tellurium, influence on the properties of
-, free cutting steels 480
Tempcore process 688
temper brittleness 127, 151, 241, 710
-, influence of various (alloying) elements 127, 151, 235, 460, 710
temper embrittlement 235, 241
-, see also temper brittleness
temper rolling, influence on the properties of
-, cold rolled flat products 95, 704
-, electrical steel sheet 705
-, soft steels (deep drawing steels), importance of roughness 705, 706
-, stainless steels 705
tempering 120
-, influence on the properties of
-, -, case hardening steels 160
-, -, steels for quenching and tempering 125, 127
-, -, UHS-steels 213, 215
tempering diagrams of
-, 20 MnCr 5 160
-, 23 MnNiCrMo 6 4 632
-, 38 NiCrMoV 7 3 213
-, 41 SiNiCrMoV 7 6 213

836 Subject Index

-, St 1420/1570 73
-, StE 500 39
-, StE 690 39
-, StE 890 39
-, X 4 CrNiMo 16 5 417
-, X 20 Cr 13 394
-, X 20 CrNiMo 17 2 415
-, X 32 NiCoCrMo 8 4 215
-, X 41 CrMoV 5 1 215
-, X 45 CrSi 9 3 460
-, X 85 CrMoV 18 2 460
tempering parameter P 315
"Temprimar"-process (wire) 203
tensile strength, formula to calculate it on the base of chemical composition 198
tensile strength: see mechanical properties
tensile testing 12
tensile test results (values)
-, at room temperature
-, -, 11 MnNi 5 3 282
-, -, 20 MnCr 5 160
-, -, 23 MnNiCrMo 6 4 632
-, -, 38 NiCrMoV7 3 213
-, -, 41 SiNiCrMoV 7 6 213
-, -, 42 CrMo 4
-, -, -, influence of the microstructure 124, 126
-, -, 50 CrV 4
-, -, -, influence of the microstructure 126
-, -, BSt 420 S 64
-, -, BSt 500 S 64
-, -, Ck 45
-, -, -, influence of the microstructure 124
-, -, wire with 0,15 % C
-, -, -, influence of cold forming 66
-, -, dual - phase steel 97
-, -, P 260 (steel with increased phosphorus content) 97
-, -, St 37-3 (frequency curve) 715
-, -, St 52-3 (frequency curve) 715
-, -, -, influence of the annealing time 27
-, -, St 1420/1570 73
-, -, StE 355 24, 25
-, -, -, influence of cold forming 24, 25
-, -, -, influence of sulfur content and sulfide shape (only reduction of area) 51
-, -, StE 500 39
-, -, StE 690 23, 24, 39
-, -, -, influence of annealing temperature 23, 27
-, -, -, influence of cold forming 23
-, -, StE 890 39
-, -, electrolytically zinc coated steel 105
-, -, wire rod for cold drawing 197
-, -, -, influence of cold forming 205
-, -, X 2 CrNi 18 9,
-, -, X 2 CrNiCu 18 9 3,
-, -, X 2 CrNiMoN 18 13 and
-, -, X 2 NiCr 18 16
-, -, -, influence of cold forming 419, 420
-, -, X 3 CrNiMoN 18 14 and
-, -, X 3 CrNiN 18 10
-, -, -, also at low temperatures 213

-, -, X 4 CrNiMo 16 5 417
-, -, X 5 CrNi 18 9
-, -, -, influence of cold forming 419, 420
-, -, X 7 Cr 13,
-, -, X 7 CrAl 13,
-, -, X 10 Cr 13,
-, -, X 15 Cr 13 and
-, -, X 20 Cr 13
-, -, -, influence of the carbon - content 403
-, -, X 20 CrNiMo 17 2 415
-, -, X 20 CrNiN 21 12 462
-, -, -, influence of the solution - annealing temperature 461
-, -, X 30 Cr 13 and
-, -, X 32 NiCoCrMo 8 4 218
-, -, X 40 Cr 13
-, -, -, influence of the carbon - content 403
-, -, X 41 CrMoV 5 1 218
-, -, X 45 CrNiW 18 9 462
-, -, X 45 CrSi 9 3 460
-, -, X 53 CrMnNiN 21 9 462
-, -, -, influence of the solution - annealing temperature 461
-, -, X 55 MnCrN 18 5
-, -, -, influence of the stretch - forming degree 552
-, -, X 60 Cr 13
-, -, -, influence of the carbon content 403
-, -, X 85 CrMoV 18 2 460, 464
-, at elevated temperatures
-, -, hot - dip aluminium coated sheet 109
-, -, scatter bands of heat resisting steels 436
-, -, X 20 CrNiN 21 12 465
-, -, X 45 CrNiW 18 9 465
-, -, X 45 CrSi 9 3 464, 465
-, -, X 53 CrMnNiN 21 9 465
-, -, X 85 CrMoV 18 2 464
-, -, -, influence of the precipitation hardening 462
-, -, -, influence of solution - annealing temperature 461
-, at low temperatures (only yield strength)
-, -, 10 Ni 14,
-, -, 11 MnNi 5 3,
-, -, 12 Ni 19,
-, -, 13 MnNi 6 3,
-, -, TStE 355,
-, -, X 3 CrNiN 18 10,
-, -, X 8 Ni 9 and
-, -, X 10 CrNiTi 18 10 277
tension cracks in forgings 697
test certificates 724
texture 505
thermal conductivity of tool steels 334
thermal expansion, correlation between it and
-, Curie temperature 561
-, elastic properties 561
thermal expansion of pure iron-nickel alloys and other alloys 557, 558, **559**, 560
thermomechanical forming, (e.g. rolling),
-, influence on the physical metallurgical processes 579, 688

Subject Index 837

thermomechanical treatment 36
-, flat products for cold forming 86
-, heat treatable steels 125
-, high - strength structural steels 36, 59
-, low - temperature steels 287
-, steels for pipelines 578
-, ultra high-strength steels 218
"thermoshock" test 603
through-hardenability of bearing steels 587
-, influence of chromium content 587
tight coil annealing 83
time - temperature - austenitization diagrams
-, for isothermal transformation of
-, -, wire rod with 0,75 % C 198
-, for continuous heating of
-, -, 100 Cr 6 590
-, -, wire rod with 0,75 % C 198
time - temperature - properties diagram of
-, StE 690 38
time - temperature - transformation diagrams
-, for isothermal transformation of
-, -, wire rod with 0,75 % C 198
-, for continuous cooling of
-, -, 10 Ni 14 290
-, -, 12 Ni 19 291
-, -, 13 MnNi 6 3 289
-, -, 16 MnCr 5
-, -, -, influence of the carbon content 166
-, -, elevated - temperature Cr - Mo - V - steel with about 1 % Cr 241
-, -, rail steels with 0,7 % C 600
-, -, reinforcing steel with 0,13 % C and 1,21 % Mn 65
-, -, StE 690 37, 38
-, -, wire rod, unalloyed 198
-, -, X 2 CrNiN 25 7 407
-, -, X 8 Ni 9 241
-, -, X 30 WCrV 5 3 328
-, -, X 32 CrMoV 3 3 328
-, -, X 38 CrMoV 5 1 328
-, -, X 40 CrMoV 5 3 328
-, -, X 45 CrSi 9 3 459
-, -, X 85 CrMoV 18 2 459
tin-coated flat products 109
tinplate 77, 110
titanium, influence on the properties of
-, alloys with special elastic properties 561
-, elevated-temperature steels 242, 245, 246
-, maraging ultra high-strength steels 216
-, normal and high strength structural steels 49
-, stainless steels 386, 396, 404
-, steels for use with hydrogen at elevated temperatures and pressures 426
-, wear resisting steels 379
TN = Thyssen Niederrhein process 647
tool steels **302**
-, definition and classification 302
-, required properties of use and their characterization 303
-, -, application properties 303
-, -, -, general 304

-, -, -, hardenability 311
-, -, -, hardness and its retention 305, 314
-, -, -, -, some other mechanical properties (a.o. compression strength, fatigue) 317, 320, 335
-, -, -, resistance to corrosion 337
-, -, -, thermal conductivity 334
-, -, -, toughness 321
-, -, -, wear resistance 329
-, -, processing properties 339
-, -, -, formability (incl. suitability for hobbing) 342, 343
-, -, -, grindability 347
-, -, -, machinability 346
-, -, -, others (e.g. dimensional accuracy during heat treatment of tools) 339
-, several measures of physical metallurgy to attain the required properties
-, -, chemical composition 308, 313, 316, 319, 330, 333, 338
-, -, heat treatment 341, 342
-, -, microstructure 308, 324, 331, 343, 347
-, typical tool steel grades proven in service, steels for
-, -, cold-forming tools 357
-, -, cutting tools 368
-, -, forging and pressing dies 361
-, -, glass processing 355
-, -, hand tools 373
-, -, hot extrusion press tools 363
-, -, machining tools 365
-, -, plastics molds 350
-, -, pressure casting molds 354
top blowing converter 640
top pouring 650, 651
toughness properties, dependence on microstructure 125
toughness properties (in general notch impact energy) of various steel grades, data in tables
-, high pressure hydrogen resistant steels 430, 431
-, non magnetizable steels 555
-, normal and high strength structural steels 52, 56, 57, 58
-, stainless steels 412, 413
-, steels for cold forming 188, 191, 193
-, steels for screws, bolts, nuts and rivets 616
-, ultra high - strength steels 224
-, see also notch impact energy, test results for some steel grades
toughness properties: see the various steel types. In addition: universally valid comments regarding toughness of
-, elevated - temperature steels 227, 231, **234**
-, heat resisting steels 439
-, heat - treatable steels **124**, 126, 129, 130
-, low - temperature steels 274, **279**
-, -, welded joints 295, 297
-, normal and high strength structural steels 9, 18, 38
-, -, welded joints 31

-, rail steels 596, 597
-, spring steels 470
-, stainless steels 402, 416
-, steels for line pipe 596, 597
-, steels for rolling stock 605
-, tool steels **321**, 326
-, ultra high - strength steels 210, **212**, 217
transformation behaviour: see time - temperature - transformation diagrams
transition temperature T_{tr} 18
transverse cracks in pass rolling 687
TREST = Terni Refractory Electroslag Topping 675
true stress - true strain curves of
-, austenitic steels for cold forming 186
-, low alloy steels for cold forming 179
-, -, influencing factors
-, -, -, boron 183
-, -, -, heat treatment 183
-, steels for screws, bolts, nuts and rivets 610, 613, 614, 619
-, steel grade
-, -, 13 MnCrB 5 183
-, -, 16 MnCr 5 179, 183
-, -, -, influence of annealing 183, 613
-, -, 20 MoCr 4
-, -, -, influence of annealing 613
-, -, 35 B 2 614
-, -, 38 Cr 2 614
-, -, 41 Cr 4 179
-, -, 42 CrMo 4 179
-, -, Cq 15 179,
-, -, Cq 35 179,
-, -, -, influence of annealing 610
-, -, Cq 45 179
-, -, X 1 CrMo 18 2 619
-, -, X 2 NiCr 18 16 619
-, -, X 5 CrNiMo 18 10 619
-, -, X 10 CrNiTi 18 9 619
-, unalloyed steels for cold forming 179
TTA - diagram: see time - temperature - austenitization diagram
TTP - diagram: see time - temperature - properties diagram
T_{tr}: see notch impact energy - transition temperature
TTT - diagram: see time - temperature - transformation diagram
tungsten in
-, elevated-temperature steels 242, 247
twin belt Hazelett machine 669
type tests 275

UHS steels: see ultra high - strength steels
ULC (ultra-low carbon) steels for cold forming 194
ultra high-strength steels (UHS-steels) **208**, 223
-, definition 208

-, required properties of use and their characterization: see steels for quenching and tempering. In addition: 208, 212, 219, 220 and
-, -, resistance to hydrogen embrittlement 219
-, -, resistance to stress corrosion cracking 219
-, several measures to attain the required properties: see steels for quenching and tempering. In addition 221 and
-, -, maraging 211
-, typical steel grades **223**
-, -, conventionally heat-treatable UHS steels 223
-, -, maraging steels 224
under-cladding cracks 236
uniform corrosion 383, 385, 386
uniformity of steel and steel productrs 12

Vacher-Hamilton relationship 643
vacuum deoxidation 645
vacuum melting methods 676
-, influence on the steel properties 677
vacuum refining processes 649
vacuum treatment of steel in the ladle 640, 648
vacuum-carbon-deoxidation 648
VAD = Vacuum Arc Degassing process 648
VADER = Vacuum Arc Double Electrode Remelting 671
valve materials **452**
-, required properties of use and their characterization 453, 455
-, -, hot-corrosion resistance 453, 455
-, -, mechanical properties (above all fatigue strength) 453, 455
-, -, wear resistance 453, 455
-, several measures to attain the required properties
-, -, measures of physical metallurgy 456
-, -, -, chemical composition 457
-, -, -, microstructure 456, 459, 463
-, -, measures of shaping valve parts 457
-, -, measures of surface protecting 457
-, typical valve materials 458
-, -, austenitic steels 460
-, -, nickel alloys 461
-, -, steels for quenching and tempering 458
-, -, -, comparison of the high-temperature behavior of the three material groups 462
vanadium, influence on the properties of
-, case hardening steels 120, 123, 142
-, elevated-temperature steels 234, 240, 246
-, heat treatable steels 120, 123, 142
-, normal and high strength structural steels 48
-, prestressing steels 74
-, reinforcing steels 64
-, stainless steels 396
-, steels for flat products for cold forming 86, 92
-, steels for use with hydrogen at elevated temperatures and pressures 426
-, steels for valves 457
-, tool steels 308, 313, 316, 330
-, ultra high-strength steels 216

-, wear-resistant steels 381
VAR = Vacuum-Arc-Remelting 671
Vicalloy alloys 541
VOD = Vacuum Oxygen Decarburization 649
VODC = Vacuum Oxygen Decarburization Converter 649

wear mechanisms 374
wear resistance, schematic representation 376
wear-resistant steels **374**
-, chemical composition 374, 379
-, hardness 380
-, microstructure 374
-, typical steel grades 380
Weibull diagram (testing of bearing steels) 586
weldability, methods to test and characterize it **28**
-, importance of the carbon content 29
weldability of
-, elevated - temperature steels 234
-, flat products for cold forming 81
-, heat - treatable steels 133
-, low - temperature steels 271, 275, 294
-, normal and high strength structural steels 11, 28, 40
-, rail steels 598
-, reinforcing steels 62, 63
-, stainless steels 392, 401
-, steels for line pipe 573, 576, 577
-, steels for welded round link chains 624, 626, 627
-, wire rod for cold drawing 195
welding of
-, low - temperature steels 294
-, normal and high strength structural steels 40
wheel caster 669

wheel rim heat treatment 604
"white bands" 665
wide plate test results on StE 355 and StE 690 20
wire-fabric (reinforcing steels) 61, 62, 65
wire rod (unalloyed) for cold drawing **195**
-, required properties of use and their characterization 195, 196
-, -, drawability in particular 196
-, several measures to attain the required properties 199
-, -, importance of the microstructure 199, 200
-, -, metallurgical measures 200
-, -, mill practices 201, 203
-, typical steel grades proven in service 204
Wöhler-curve 17
Wöhler's method of fatigue testing 14

yield of
-, hot rolled products 689
-, -, possibilities to improve it 689
-, ingots for hot forging 692
yield strength: see mechanical properties
-, in addition
-, -, yield strength in relation to
-, -, -, carbon content (schematically) 716
-, -, -, product thickness (schematically) 717
-, dependence on temper rolling 705
Young's modulus: see modulus of elasticity

zinc coated flat products 99
zirconium in
-, free cutting steels 480
-, steels for use with hydrogen at elevated temperatures and pressures 42